Advances in Systems, Computing Sciences and Software Engineering

# Advances in Systems, Computing Sciences and Software Engineering

Proceedings of SCSS 2005

Edited by

Tarek Sobh
*School of Engineering, University of Bridgeport, USA*

Khaled Elleithy
*School of Engineering, University of Bridgeport, USA*

Springer

A C.I.P. Catalogue record for this book is available from the Library of Congress.

ISBN-10 1-4020-5262-6 (HB)
ISBN-13 978-1-4020-5262-0 (HB)
ISBN-10 1-4020-5263-4 (e-book)
ISBN-13 978-1-4020-5263-7 (e-book)

---

Published by Springer,
P.O. Box 17, 3300 AA Dordrecht, The Netherlands.

*www.springer.com*

*Printed on acid-free paper*

# Contents

# Acknowledgements

The International Conference on Systems, Computing Sciences and Software Engineering (SCSS 2005) would not have been possible to conduct without the work, efforts and dedication of many individuals and organizations.

The editors would like to acknowledge the technical co-sponsorship provided by the University of Bridgeport and the Institute of Electrical and Electronics Engineers (IEEE). We would like to express our gratitude to Prof. Toshio Fukuda, the Chair of the International Advisory Committee and Prof. Srinivas Ramaswamy, the SCSS 2005 Conference Chair. The efforts of the CISSE Webmaster, Mr. Andrew Rosca, have been instrumental in the success of the conference. The work of Mr. Tudor Rosca in managing and administering the conference on-line presentation system has been crucial in conducting the world's first real-time on-line high caliber research conference. We also wish to recognize the roles played by Ms. Susan Kristie and Mr. Sarosh Patel, our administrative and technical support team.

Finally, and most importantly, we would like to express our thanks to our colleagues, the reviewers and technical committee members who did an exceptional job in reviewing the submitted manuscripts. In particular, we would like to acknowledge the contributions of Abdelaziz Almulhem, Ehab Elmallah, Julius Dichter, Michael Lemmon, Mohammed Younis, Natalia Romalis and Rodney Roberts.

# Preface
# Advances in Systems, Computing Sciences and Software Engineering

This book includes the proceedings of the International Conference on Systems, Computing Sciences and Software Engineering (SCSS'05). The proceedings are a set of rigorously reviewed world-class manuscripts addressing and detailing state-of-the-art research projects in the areas of computer science, software engineering, computer engineering, systems sciences and engineering, information technology, parallel and distributed computing and web-based programming.

SCSS'05 was part of the International Joint Conferences on Computer, Information, and Systems Sciences, and Engineering (CISSE'05) (www.cisse2005.org), the World's first Engineering/Computing and Systems Research E-Conference.

CISSE'05 was the first high-caliber Research Conference in the world to be completely conducted online in real-time via the internet. CISSE'05 received 255 research paper submissions and the final program included 140 accepted papers, from more than 45 countries. The concept and format of CISSE'05 were very exciting and ground-breaking. The PowerPoint presentations, final paper manuscripts and time schedule for live presentations over the web had been available for 3 weeks prior to the start of the conference for all registrants, so they could choose the presentations they want to attend and think about questions that they might want to ask. The live audio presentations were also recorded and were part of the permanent CISSE archive, which also included all power point presentations and papers.

SCSS'05 provided a virtual forum for presentation and discussion of the state-of the-art research on Systems, Computing Sciences and Software Engineering. The virtual conference was conducted through the Internet using web-conferencing tools, made available by the conference. Authors presented their PowerPoint, audio or video presentations using web-conferencing tools without the need for travel. The Conference sessions were broadcasted to all the conference participants, where session participants were able to interact with the presenter during the presentation and (or) during the Q&A slot that followed the presentation. This international conference was held entirely on-line. The accepted and presented papers were made available after the conference both on a CD and as a book publication by Springer.

The SCSS conference audio room provided superb audio even over low speed internet connections, the ability to display PowerPoint presentations, and cross-platform compatibility (the conferencing software runs on Windows, Mac, and any other operating system that supports Java). In addition, the conferencing system allowed for an unlimited number of participants, which in turn granted us the opportunity to allow all SCS2 participants to attend all presentations, as opposed to limiting the number of available seats for each session.

This volume of the conference proceedings includes 66 papers that were presented in the conference. The papers cover an interesting range of topics such as fuzzy algorithms, parallel computing, multimedia applications, grid computing, distributed software programming, semantic web, web mining, semantic web knowledge management, pervasive grids, non visual interfaces, character recognition, and self evolving software.

We hope that you will find the selected papers interesting and covering the state-of-the-art advances in the area of Systems, Computing Sciences and Software Engineering. We are looking forward to your participation in CISSE'06 (www.cisse2006.org).

**Editors**

**Prof. Tarek Sobh**
**Vice Provost for Graduate Studies & Research**
**Dean, School of Engineering**
**University of Bridgeport**

**Prof. Khaled Elleithy**
**Associate Dean, School of Engineering**
**Dept. of Computer Science and Engineering**
**University of Bridgeport**

# An Elastic Display Method for Visualizing and Navigating a Large Quantity of Alarms in a Control Room of a Nuclear Power Plant

Sang Moon Suh, Gui Sook Jang, Geun Ok Park, Hee Yun Park, In Soo Koo
Korea Atomic Energy Research Institute
P.O. Box 105
Yusong, Taejon, Korea

***Abstract*-In a conventional control room of a Nuclear Power Plant, a great number of tiled alarms are generated especially under a plant upset condition. As its conventional control room evolves into an advanced one, an annunciator-based tile display for an alarm status is required to be removed and replaced by a computer-based tile display. Where this happens, it becomes a bothering task for plant operators to navigate and acknowledge tiled alarm information, because it places an additional burden on them. In this paper, a display method, Elastic Tile Display, was proposed, which can be used to visualize and navigate effectively a large quantity of tiled alarms. We expect the method to help operators navigate alarms with a little cost of their attention resources and acknowledge them in a timely manner.**

## I. Introduction

In the control room of an NPP(nuclear power plant), an alarm system is one of the primary means which provides the operating personnel with status information of the process abnormalities and failures. According to [1], it should be designed (a)to alert the operators to off-normal conditions which require them to take actions, (b)to guide the operators, to the extent possible, to the appropriate response, (c)to assist the operators in determining and maintaining an awareness of the state of the plant and its systems or functions, and (d)to minimize distraction and unnecessary workload placed on the operators by the alarm systems.

The alarm information is presented to the plant personnel in a various ways of a display format. While the annunciator-based tile display is a main format for the alarm information in the conventional control room, a computer-based tile display, message list display, and integration display are additional or dominant formats in a hybrid control room or the advanced one.

### *Annunciator-based Tile Display*

Fig. 1 shows a typical type of the annunciator-based tile display for the alarm information installed in the upper section of the control panel. The annunciator-based alarm system is usually found in the conventional or hybrid control room of an NPP. This type of display presents alarms in a spatially dedicated position. The spatially dedicated alarms are continuously visible whether in an alarmed or cleared state.

The annunciator-based tile display has been generally found to be superior to the other two types of displays during plant upset conditions of which a great number of alarms are generated [2]. It allows the plant operating personnel a rapid detection and pattern recognition, because the operator can access alarm information in parallel via the display. Also it does not put additional burden on the plant personnel caused by secondary tasks like navigating numerous pages.

The disadvantage of using the spatially dedicated alarm is that the operators usually have difficulties finding the alarm contents because of its limited space for the alarm message.

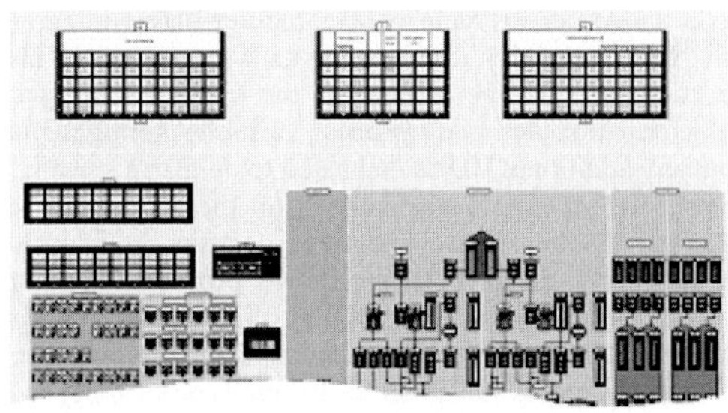

Fig. 1. Typical control panel with the annunciator-based alarm display in the conventional control room.

### *Computer-based Message List Display*

Fig. 2 shows an example of a computer-based message list display for alarm information. In this type of display, alarms are not presented in a spatially dedicated position as they are usually presented according to some logic such as time or priority. It often presents many different alarms via the same display device in contrast with an annunciator-based tile display where a given location presents only one alarm. A scrolling display is required to acknowledge the hidden alarms.

The major advantage of the computer-based message list display is the flexibility to present the many attributes of alarm information[2]. It can provide the plant operators with more descriptive alarm messages in contrast with a tile display. Also it can provide them with the higher priority alarms prior to the lower priority ones.

The disadvantages of the message list display are generally from the limited viewing area on the VDUs. This type of



*T. Sobh and K. Elleithy (eds.), Advances in Systems, Computing Sciences and Software Engineering*, 1–3.

display format is usually considered as creating difficulties of acknowledging the hidden alarms for the plant operators especially in high density alarm conditions. And operators typically prefer to use SDCV (spatially dedicated and continuously visible) displays in these conditions.

Fig. 2. An example of the computer-based message list display.

*Computer-based Integration Display*

Fig. 3 shows an example of a computer-based integration display with a process mimic diagram for presenting alarm information. In this type of display, the operators can access alarm information and plant process variables simultaneously without any additional VDUs dedicated to an alarm system.

Despite the operators preference for the computer-based integration display that integrates alarm and process information in a display, there are two problems associated with the computer-based integration display for an alarm system [3]. There are usually several hundreds of display pages for the process information. In this type of display, the operators have to navigate many pages to acknowledge newly generated alarms. When he/she moves to the next page to acknowledge new alarm, he/she should remember the alarms he has already acknowledged.

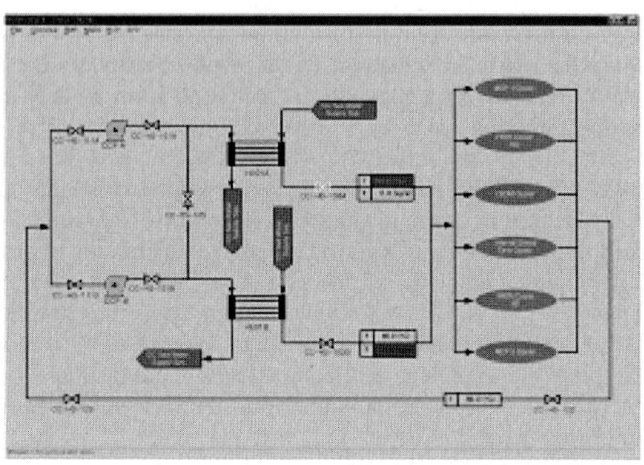

Fig. 3. An example of a computer-based integration display.

*Purpose of the Study*

In contrast with the conventional type of a control room shown in Fig. 1, the advanced control room is expected to be some form of the one shown in Fig. 4. As shown in Fig. 4, it is expected that the advanced control room consists of VDU-based interfaces mainly, and the conventional annunciator-based tile display will be removed and replaced by the computer-based alarm display system.

But as discussed in the previous section, the conventional annunciator-based tile display has several advantages of a good human performance such as a rapid detection and pattern recognition. So its predominant features should be incorporated into the computer-based tile display in the advanced control room.

Fig. 4. Typical type of advanced control room of a complex process plant.

This paper presents a new method to be used when the annunciator-based tile display for alarm information is implemented into the computer-based tile display. It was motivated from the work that E. Kandogan and B. Shneiderman had done [4].

The next section provides some requirements which are the basis for the ETD(elastic tile display) we proposed in this paper, followed by a brief description of the ETD. Next, the further studies to verify its impact on a human performance by an experiment will be described.

## II. Requirements

When we incorporate the alarm information of the annunciator-based tile display into the computer-based tile display, we have to consider the key advantages of the annunciator-based tile display on a human performance such as a rapid detection and pattern recognition. To maximize these advantages, the following requirements should be established.

*The method of a computer-based tile display method should be assured so that only one display page accommodates all the alarms to take advantage of a pattern recognition.*

It is a very troublesome task for operators to navigate alarm information on a VDU-based user interface to find out what's really going on the plant. Management and navigation of the displays can impose a significant additional burden on the operator which are not related to the primary task of monitoring and controlling the plant [2].

*The navigating task of the display method should be done with a small amount of attention to help the operators do the primary task of monitoring and controlling the plant.*

When we implement or design the computer-based tiled alarm system, it should satisfy the guidelines or standards on the density of alarm information. The following requirement described in [5] should be satisfied to make a tile matrix.

*The alarm tile display of the method should contain a maximum of 50 alarms.*

## III. Elastic Tile Display

In the previous section, we established three key requirements that the method of the computer-based tile display should meet. Based on these requirements, the computer-based alarm display method, ETD, was designed, which can be used to visualize and navigate effectively lots of tiled alarm information (see Fig. 5). It was motivated from the work that E. Kandogan and B. Shneiderman had done [4]. They proposed a rapid window management method called Elastic Windows.

The Elastic Windows is based on three principles: *hierarchical window organization, space-filling tiled layout,* and *multi-window operations*. The animating characteristics of ETD are similar to Elastic Windows. But the purpose of an application is quite different from each other. The way of navigating the tiled alarm information is illustrated in Fig. 5.

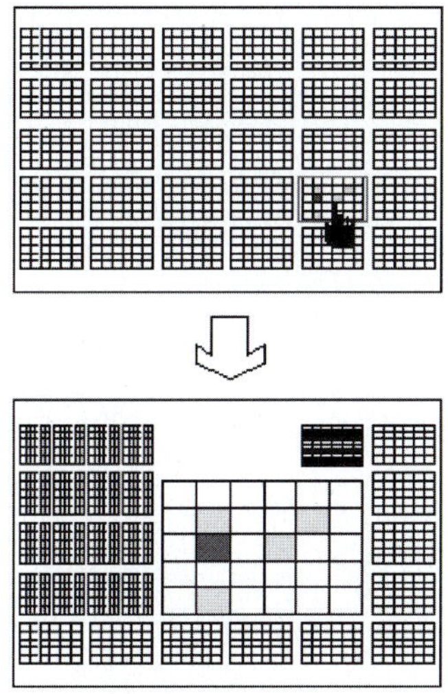

Fig. 5. Elastic Tile Display: an effective technique of visualizing and navigating many tiled alarm information.

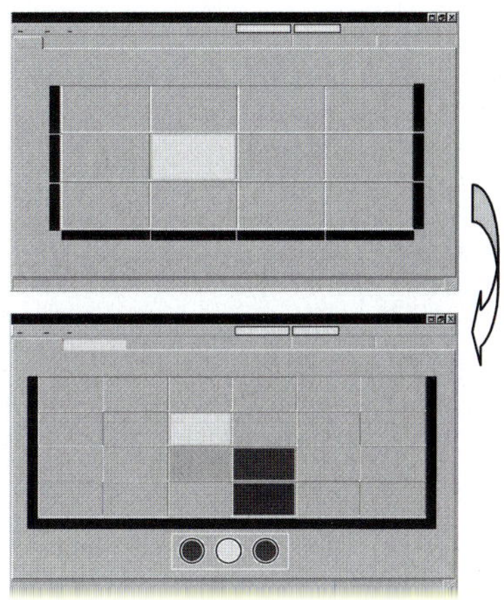

Fig. 6. An independent overlapped tile display.

## IV. Further Works

In this paper, we established the key requirements to implement tiled alarm information onto the VDU-based user interface. Based on these requirements, a display method, ETD, was proposed. We expect that the ETD is better than the independent overlapped tile display shown in Fig. 6. It is constructed in a hierarchical manner which has several display levels from a system to a component. It will be the base line for a comparison experiment.

The experiment will be set to validate the following hypotheses.

*Elastic Tile Display a yields faster performance than an independent overlapped tile display for an alarm acknowledgement task.*

*Elastic Tile Display yields a faster performance than an independent overlapped tile display for a situation awareness task.*

## References


[1] EPRI-ALWR URD, "Chapter 10: Man-Machine Interface System," *Advanced Light Water Reactor Utility Requirements Document*, vol. III, pp. 10.4-15, December 1995. (Electric Power Research Institute).

[2] J. M. O'Hara, W. F. Stubler, and J. C. Higgins, "Hybrid Human- System Interfaces: Human Factors Considerations," Brookhaven National Laboratory, December 1996.

[3] Shaw, J., "Distributed control systems: Cause of cure of operator errors," *Reliability Engineering & System Safety*, 39(3), pp. 263-271.

[4] E. Kandogan, B. Shneiderman, "Elastic Windows: Improved Spatial Layout and Rapid Multiple Window Operations," *ACM AVI'96 Advanced Visual Interfaces, Gubbio, Italy*, pp. 29-38, May 1996.

[5] NUREG-0700, Rev. 2, "U.S. NRC, Human-System Interface Design Review Guide," U.S. Nuclear Regulatory Commission, May 2002.

# An EAI Technology Framework

Jing Liu, Lifu Wang, Wen Zhao, Shikun Zhang, Ruobai Sun
School of Electronics Engineering and Computer Science, Peking University
Beijing 100871 China
liujing@cs.pku.edu.cn

***Abstract*—EAI issue is constantly a significant aspect of enterprise computing area. Furthermore, recent advances in Web Service and integration sever provide a promising pushing to EAI issues. However, few investigations have focused on a general view of EAI technologies and solutions. To provide a clear perspective about EAI, a technology framework-ABCMP is presented in this paper. ABCMP attempts to describe the structure of diverse EAI technologies, which allows a general comprehension of the EAI issue. Moreover, process topic about ABCMP is also discussed to provide a wider vision of the technology framework.**

***Key words*—software engineering, EAI, technology framework, software process, software architecture.**

## I. Introduction

With the emergence of a growing number of enterprise applications, there has been an increasingly focus on EAI (enterprise application integration) concepts and technologies. By reducing time consumed in transferring data between existing information systems, EAI provides real time business processing ability to a company and greatly promotes its efficiency. As EAI develops, the vast emerging technologies for EAI create a lot of complexity to the EAI knowledge system as well as successful solutions to several cases. Also the combination of a series of problems and individual requirements from different organizations and trading communities ask for a number of different solutions. However, although approaches to application integration vary considerably, it is possible to create some general categories, which include information-oriented, business process integration-oriented, service-oriented and Portal-oriented integration. [1]

Information-oriented integration refers to the integrations accomplished through databases, which means that databases or information-producing APIs are the main channels for integration. Information-oriented integration includes three categories: data replication, data federation and information and interface processing. [1] Information-oriented integration is a relatively straightforward and explicable EAI approach which widely adopted in EAI cases, especially the earlier ones. However, its deficiencies in maintainability and business logic describing ability make it less popular recent years. Technologies supporting this kind of integration include ADO, JDBC, EJB, and ORM, which often provides a series of drivers for accessing different kinds of databases. Information-oriented integration related business entities are mainly business data and their interrelationships.

Business process integration-oriented products lay a set of easily defined and centrally managed processes on top of existing sets of processes contained within a set of enterprise applications. [1] Earlier business process integration products are message queue servers which link a number of applications through message queues. By increasingly combined to the business domain and supports from the workflow technology, business process integration products are becoming more related to the business processes in organizations, and consequently allow it to be more directly designed and maintained [11]. Business process integration is increasingly adopted through a growing number of supporting products, such as WBI of IBM and Biztalk of Microsoft.

Service oriented application integration allows applications to share common business logic methods. This type of integration is supported by remote method and remote object technologies, such as DCOM and web service. Among these technologies, Web service is the most prevailing for its widely accepted standards, which lead to an excellent ability of crossing platforms. Unlike information oriented integration, service oriented integration requires some changes to the existing applications, which demands a greater investment and process.

Portal oriented integration is a specific integration approach aimed at the integration of user interface, which closely related to the B/S structure and Portal server. Even it is very different from the other three kinds of integration which directly connect applications, it is really suit for some situations such as enterprise portal site and B2B exchange. [13] Many companies are providing their Portal server products, which generate a series of open standards such as portlet.

Moreover, the widely spreading trend of web service in recent years has greatly influenced the EAI approaches. Firstly, the commonly supported SOAP and WSDL protocol provide basic connectivity, which are open standards and have the ability of platform independent. [5] Besides, the UDDI protocol supports the publishing service for web services. SOA and Service Bus are presented for an entire architecture for integration. [3][4] Also, WS-T, WS-C, WSPR and many other open standards are provided for Web Service, which launches a strong impact to service oriented integration. Web Service-based solutions tend to be broke down into a series of small and low-risk steps towards a more legible process. With a growing amount of software supporting these standards above, service-oriented integration is being more and more widely adopted.

At the same time, the application integration server emerged as another trend in EAI technology domain. Many companies have developed their own integration server, such as WBI from IBM and Weblogic Integration from BEA. These integration



*T. Sobh and K. Elleithy (eds.), Advances in Systems, Computing Sciences and Software Engineering*, 5–10.

servers provide a series of solutions for EAI in addition to several common integration functions and services. For example, most of them provide the Business Process Integration-oriented solution, and offer a message server and a workflow server supporting it. Portal-Oriented integration solution and Portal server are also supported by most of the integration severs. Combined with developing environment, EAI applications based on integration server could be developed with great ease.

In this paper an EAI technology framework-ABCMP is presented to provide a global view of these various technologies in diverse areas of EAI. Besides, some relevant processes and features are discussed in detail.

The paper is organized as follows. In Section 2, related literature is presented Section 3 introduces the ABCMP technology framework, including a detail description. Section 4 presents a common feature analysis of the ABCMP. Section 5 discusses important processes of the framework. In Section 6, we present a sample application based on ABCMP and an analysis. In Section 7, the conclusion of this paper is presented.

## II. Related Literature

Many frameworks and architectures have been proposed to support the development of various types of EAI applications. In this paper, we argue that although existing frameworks are useful for certain cases of EAI applications, they have often ignored establishing a global view of EAI issues and technologies, which is especially valuable for designing complex solutions needing the combination of different integration technologies.

Many previous researches are centered on special categories or even specific products. In [15], an architecture for dynamic data source integration is proposed, which is useful for information oriented integration. Nevertheless, it is incompatible with combined integrations requiring both information and service connectivity. In [14], the integration framework provided by BEA, which is closely related to the Weblogic Integration, provides solutions for data integration and application integration. However, it does not describe relationship between these solutions and how to combine them together when necessary. In [7], many more models, which are useful for solving certain problems of EAI, are described and classified.

The UML profile for Enterprise Distributed Object Computing (EDOC) and the UML profile for EAI try to provide a comprehensive view of the EAI issues. [8] They successfully describe most concepts in the EAI domain through the object oriented way. On the other hand, a clear and distinctive description of the relationships of these concepts and technologies is not presented, which is very critical for a universal view of EAI.

Denver Robert presents the adaptive EAI framework, which has a fine structure and can adapt most existing technologies well. [6] The framework organizes many EAI elements in a layered way and generates a comprehensive view which includes many technologies. However, its software architecture is not flexible enough and other EAI services are not considered.

All these EAI models, architecture and frameworks provide meaningful description of many aspects of EAI, which makes the foundation of our research. In this paper, we try to presents an EAI technology framework which organize elements in previous researches and provides a global view of EAI technologies.

## III. ABCMP – An Technology Framework of EAI

### *A. Basic elements in EAI*

EAI is a strategy approach that connects internal information systems of an enterprise. EAI refers to the existing applications that being integrated and an EAI application connecting these systems. Basic elements in EAI are showed in the figure below.

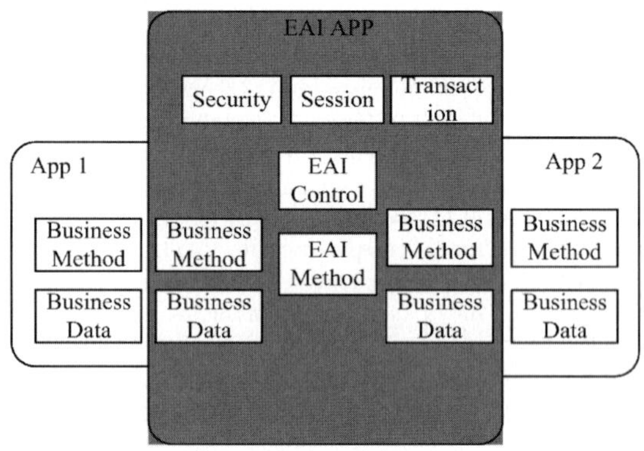


Fig. 1. Basic elements in EAI

As the figure shows above, there are four parts of an EAI application: business data and method (the overlapping area of EAI application and integrated applications), EAI method, EAI control and EAI basic support. Firstly, Business data and method refers to several business data and method in existing applications which need to interact with other applications. Secondly, EAI method refers to the new developed methods used to connect these business data and method which comes from different applications of different platforms. Thirdly, EAI control refers to the part which controls the invocation of EAI methods and business methods. The EAI control part drives the static methods to an EAI application. Furthermore, EAI basic support part includes several basic supporting components in EAI runtime environment, such as security and transaction, which are critical for a strong and effect integration.

Figure 1 depicts a number of necessary basic elements in EAI. An EAI technology framework ABCMP is presented below, which describes the relationships between these elements and technologies related to these elements.

*B. ABCMP- An EAI Technology Framework*

Various technologies are adopted for basic elements in EAI, which add complexity to the EAI solutions. We present the ABCMP, an EAI technology framework, to depict the structure of these technologies and bring a comprehensive view to the EAI solutions. The structure of ABCMP is showed in the following figure.

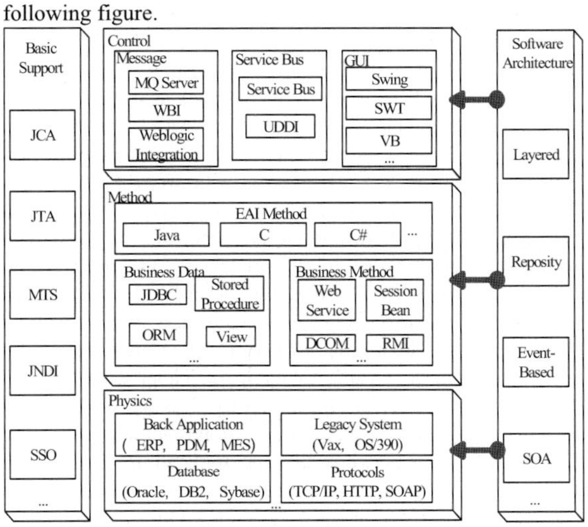


Fig. 2. ABCMP

ABCMP depicts EAI in five aspects: physics, method, control, software architecture and basic support. The major code part of the application appear in three layers: the Physic layer depicts the physic platforms and connections for existing application; the Method layer depicts the business method and data in existing applications and the integration application; the Control layer is responsible for organizing elements in model layer to an application. The Software Architecture part describes how to select different elements in the three layers to form an integration application. There is also the Basic Support part offering basic supports to the EAI environment. Technologies used for each part are given in detail as follows.

**Physics:** The Physics layer depicts the platforms and protocols used by applications which need to be integrated. Back Applications, such as ERP (Enterprise resource plan), PDM (Product data relationship management), MES(Manufacture execution system), are among the major applications which need to be integrated. Usually, these applications are strictly tied to important departments of a company and are now required to be integrated with each other. Legacy Applications is another kind of applications needing integration. These applications usually run on mainframes such as IBM OS/390. Therefore, they can not be easily replaced, and need to be integrated with new applications to form more automated and wider area processes. Database is also a key source for integrated applications, which are used in almost every enterprise applications in recent years, among them there are Oracle, DB2, Sybase and SQL Sever. In addition, there are lots of protocols maintaining the connections between these existing applications and databases, most of them are open standards and are supported by most companies.

**Method:** The Method part consists of business data, business method and EAI method. Business data are mainly stored in databases, which can be accessed via several technologies, such as JDBC, ADO, Stored Procedure, ORM tools. However, most of these technologies are closely related to specific platforms. Business method wraps the behavior of business. Technologies supporting business method are developing rapidly, from RMI to EJB, from DCOM to Web Service. Besides existing business data and method, EAI method is one important part of Model layer. EAI method includes accessing to existing business data and method, and its own business logic. EAI methods are often local methods written in languages such as Java, C and C#.

**Control:** The control layer is mainly used for the driving and organization of Model layer. There are three major styles of control: the message style control, the service bus style control and the GUI style control. The message style control is usually driven by a Message Queue Sever or Workflow Engine. In this style, the invocation of EAI method is controlled by the Workflow process and the method is invoked when message arrives. MQ, WBI, Web Logic Integration are some famous software supporting this kind of control. Service bus style control is based on the SOA. EAI method and business method are wrapped to web service and exposed on service bus, while other applications control these services through the service bus. The major standards used in this style are UDDI, WSDL and WS-I. Moreover, there are also some implementations for Service Bus. GUI style control is a more common and simple style of control. By developing GUI, user can control the integration application by interacting with GUI, in which the invocation of EAI method is caused by operation in user interface. Also, a lot of technologies can be used for GUI developing, such as JSP, Swing, and SWT.

**Software Architecture:** The software architecture part depicts the structure of the application, or in another word, the structure of the Physic, Model and Control layer. Although it is the only part in the technology framework which refers to little technology, it is still essential for the framework for it depicts the structure of the units from other parts. Since there are several software architecture styles for common applications, software architecture styles for EAI applications vary accordingly. [2] Event based style is frequently used in message style control system, in which message producer and message subscriber are largely separated. Repository style is combined with database centered applications, in which distributed application is integrated by interacting with the same database. There are also many other software architecture style in EAI applications, such as Layered and Client-Server. Also the combinations of several software architecture styles are common in complex EAI applications.

**Basic Support:** The Basic Support part consists of common functions and services needed for EAI. Transaction is one important basic support, which is usually associated with several distributed systems in EAI environment. MTS and JTS are major technologies used for transaction support in EAI. Security is also an important basic support, especially in some applications for key departments. SSO is a major technology supporting EAI security, which connects the security systems in distributed applications. Event log and error handler are some other basic supports, and log4j is one developing tool supporting it.

As described above, most technologies related to EAI can be classified according to the framework. Consequently, a clear structure of the EAI application is formed, which is helpful for designing the EAI application. How to adopt EAI based on ABCMP is discussed in detail in Section 5.

## IV. Common Features of ABCMP

Each part of ABCMP refers to a number of similar technologies, which have some common features behind the variety of technologies. Some of these common features are more closely related to business domain, which helps providing a profound view of solutions. Major common features are displayed in the figure below.

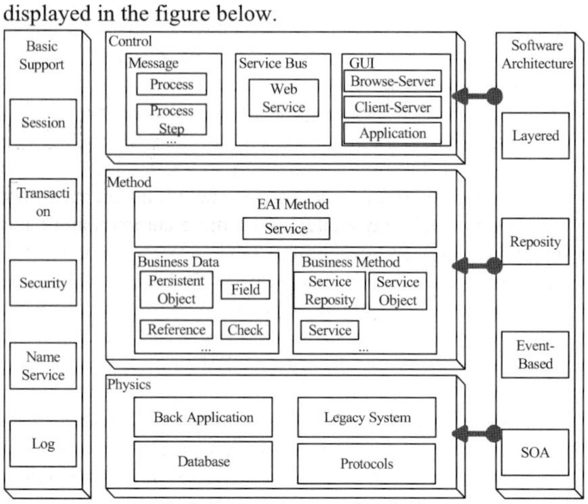


Fig. 3. Common features of ABCMP

As the figure shows, business data part is mainly responsible for describing, storing and accessing business data in EAI. In systems using database for storing, table and other concept are used for describing data, and JDBC and other technologies are used for storing and accessing data. The commonness of business data can be described by Persistent Object, Field, Reference and other elements. As to database, based on the ORM technologies, Persistent Object can be compatible with RDBMS and OODBMS. As to the file system and other storing mode, business data can also be easily described as Persistent Object. By using object-oriented concepts as Persistent Object, one can manipulate business data more concisely, and make semantic features more consistent with business domain.

Business method and EAI method are mainly in charge of offering of business functions. RMI, EJB and other technologies expose business functions through remote call or objects, making them available for other systems. This common feature can be depicted by a number of relevant concepts, such as Service Repository, Service and Service Object. These concepts can well describe different kinds of business method and EAI method, despite that they are implemented by different technologies such as RMI or EJB.

Message style control part adopts message and workflow technologies. Among these technologies, every transaction step is activated by some specific message. These specific messages are raised by application when using message server, and are also generated by workflow engine according to transaction results and predefined rules with workflow engine. The common features in this part can be described by concepts as Process and Process Step. Every transaction in workflow process is a Process Step, which is the same with every invocation caused by a specific message in a message queue.

Basic supports can be described by Session, Transaction, Security Name Service and Log. Session maintains status and data in one session between different applications. Transaction maintains transactions related to EAI method, business method and business data. Security depicts security mechanism in EAI applications. All these basic supports have different implementation technologies on different platforms.

There are also some common features in other parts of the technology framework. In SOA style control part, Web Service can be used to depict composition parts. Meanwhile, in GUI style control part, technologies can be sorted to three kinds: B/S, C/S and common application. In addition, applications in physics part can be categorized into back application, legacy application and database. Moreover, in software architecture part, software architecture for different EAI application can be classified through different styles.

Most of these common features can be used to simplify the design task. After mapping the existing technologies to the ABCMP, business entities besides the technologies can be described through these common features and differences among the technologies are diminished. The processes guide in Section 4 will describe how to develop EAI applications based on ABCMP and its common features.

## V. ABCMP Related Processes

To efficiently develop applications based on ABCMP, some important processes should be followed. These processes could be executed one by one or by iteration, according to specific process model.

**Technology analysis and software architecture definition.** The first important step is to define the software architecture based on technologies analysis according to ABCMP. By mapping the technologies used by applications which need to be integrated to ABCMP, a clear structure of the application could be formed. Based on this structure, some software architecture styles can be utilized to design the software architecture of the EAI application. For example, it is recommended to utilize repository style when most applications are based on database. Moreover, the software architecture part of ABCMP should be basically defined in this process. Also, the control style should be decided as part of the software architecture. Besides, modules and connections between modules should be preliminarily defined in this process.

**Business analysis and business data and service definition.** In this process, business data and method in method part should be defined. By analyzing business logic in existing applications, business data and method which are related to the integration should be recognized and defined. The recognition and definition of these business entities can be done according to common features in the Method part. Clear understanding of existing business logic and integration requirement is demanded in this process.

**EAI method definition and development.** After defining the existing business data and service, definition of EAI methods used for manipulating them is also consequential. These EAI methods are closely related to the integration requirements. A clear definition of EAI methods is productivity for further development and test. After this process, the method part of ABCMP will be fully developed.

**Control part definition and development.** After defining and developing the method layer, the control part should be developed accordingly. While the control style has been defined in software architecture, only specific design and development are required in this process. For example of using message style control, the work of this process is to define processes in GUI develop environment.

These processes are well suited for modern iterated software process models such as Agile Modeling, XP programming, and RUP. However, it is recommended that the first process be executed earlier than the other three processes. For instance of executing these processes in RUP, it is recommended that the first process be completed in the Inception phase; the second process be mainly executed during the inception and elaboration phases; and the last two processes be mainly executed during the elaboration and construction phases.

## VI. Sample Application and Analysis

### *A. Sample Application.*

In this section, a sample application based on ABCMP is described. The requirement is an EAI application which needs integrate an existing ERP system based on Oracle database, an OA system based on Domino and an Enterprise Portal running on WBI. Processes for developing the application based on ABCMP are as follow:

**Technology analysis and software architecture definition:** By analyzing and mapping the existing technologies, event-based style based software architecture is selected. The reason is that most existing technologies are based on process. Based on the event-based style, the software architecture is basically defined, which is presented in Figure 4.

**Business analysis and business data and service definition:** By analyzing existing business logic, several business entities are defined. There are business data entities such as Product, Order Form Contract, Customer, and so on. Also, existing service are recognized: Product Subscribe, Product Deliver, Contract and many other services that have relationship with integration.

**EAI method definition and development:** After investigating integration requirements, EAI methods are defined, such as Product Add method to add product produce plan when a contract is signed, Product Deliver Notification method to notify that products are ready and customer should be contacted for receiving products.

**Control part definition and development:** For message style control is selected and WBI platform is used for EAI application, the development of control part is to define processes in WSAD platform. The GUI for deploying applications and monitoring running process are already provided by WBI platform, so the control part development is very simple here.

The security and transaction support is also provided by WBI platform. There is SSO support that can be used to access Domino and Oracle. There is also Transaction support that provided by WebSphere, which can be interacted through JTA interface.

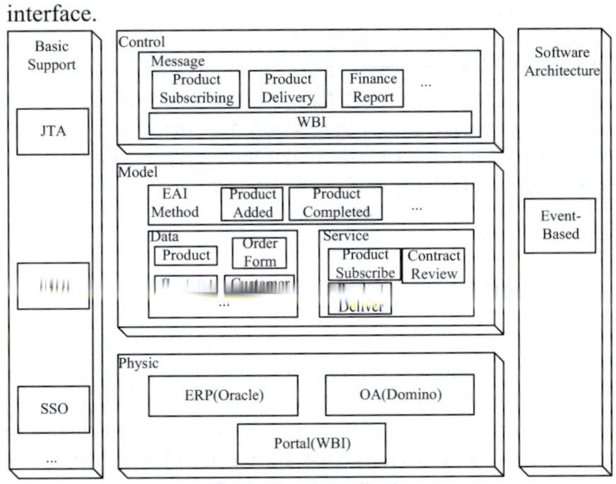


Fig. 4. Sample application

From the development of the sample application, we can see that developing EAI application based on the ABCMP is a well-organized process. By separating the tasks and processes according to the model, the development is free from complex existing technologies, so more energy could be concentrated on the business logic analysis and definition.

### *B. Analysis*

An analysis of the solutions based on the ABCMP is presented in this section. The analysis is based on the integration requirements factors presented by Themistocleous [10], such as maturity and security. Major properties enhanced by adopting ABCMP are as follow:

**Maintainability:** Maintainability is a critical property of software system, referring to the capability of software to allow changes without infecting the other parts of the application. Regarding the EAI, major changes come from two aspects: changes in business domain and changes of integrating new systems. As to the changes in the business domain, mainly referring to the changes in business data and method, ABCMP can satisfy it well. In regard to the adding of systems, new systems are compatible with ABCMP in most situations, which makes the integration simple and cause little change to the existing parts.

**Complexity:** Complexity refers to the complexity of the EAI solution. A complex solution will bring too many difficulties in implementation. ABCMP provide a reasonable partition for the solution, which brings a clear structure to the application and greatly reduces the complexity.

**Non-Invasive:** Non-Invasive shows the changes to the existing systems. An excellent EAI solution should try to avoid changes to existing systems. Applications based on ABCMP access existing business data and method through mature technologies, not referring to the alteration of existing applications.

**Security:** Security is one of the basic supports in ABCMP, including mainly SSO and log mechanisms. Security systems in applications are connected through these security supports, and provide functions as access control, integrity, usability and non-repudiation. Many security supports can be added into applications based on ABCMP.

**Transaction:** Transaction is another basic support in ABCMP. By connecting distributed transaction through a distributed manager, a sound support for transactions is provided. Through ABCMP, transaction support can be easily considered and adopted.

## VII. Conclusion

This paper presents an EAI technology framework- ABCMP, trying to describe the structure of EAI technologies. ABCMP is useful for clarifying and analyzing EAI, which contributes a lot to design decisions. Furthermore, this paper discusses some other relevant issues about ABCMP, including common features and process guiding of ABCMP. The analyzing of these issues helps solving possible problems of ABCMP, which makes the basic of ABCMP more stable.

In the further research of EAI, apart from advances on individual technologies, we will work on an improved knowledge structure. As EAI matures, better technology and technology framework will be provided, and more automated design tools and platforms based on them will be produced. Moreover, process and management related knowledge will also increase. All these improvements will finally drive EAI to produce significant benefits to enterprise by a small investment.

## References


[1] David S.Linthicum. "Next generation application integration: from simple information to Web service". Boston: Addison-Wesley, 2003, 6~21.

[2] Mary Shaw, David Garlan. "Software Architecture: Perspectives on an Emerging Discipline". Beijing: Tsinghua University Press, 1998, 19~32.

[3] Mark Endrei, Jenny Ang, Ali Arsanjani, etal. "Patterns: Service-Oriented Architecture and Web Services". http://www.redbooks.ibm.com/.

[4] Martin Keen, Amit Acharya, Susan Bishop, etal. "Patterns: Implementingan SOA Using anEnterprise Service Bus". http://www.redbooks.ibm.com/.

[5] IBM Inc. Web Service. http://alphaworks.ibm.com/webservices.

[6] Denver Robert Edward Williams. "An adaptive integration architecture for software reuse". PhD Thesis. Florida: University of Central Florida, 2001.

[7] Francisca Losavio, Dinarle Ortega and María Pérez. "Modeling EAI". In Proceedings of the XXII International Conference of the Chilean Computer Science Society, 2002.

[8] OMG. UML, http://www.uml.org/.

[9] Kostas Kontogiannis, Dennis Smith, Liam O'Brien. "On the role of services in enterprise application integration". In Proceedings of the 10th International Workshop on Software Technology and Engineering Practice, 2002.

[10] M. Themistocleous. "Evaluating and Adoption of Enterprise Application Integration", PhD Thesis. London: Brunel University, 2002.

[11] Ying Huang, Anthosh Kumaran and Kumar Bhaskaran. "Platform-Independent Model Templates for Business Process Integration and Management Solutions". Information Reuse and Integration, 2003.

[12] Sudip Bhattacharjee, R. Ramesh, and Stanley Zionts. "A Design Framework for e-Business Infrastructure Integration and Resource Management". IEEE TRANSACTIONS ON SYSTEMS, MAN, AND CYBERNETICS—PART C: APPLICATIONS AND REVIEWS, 2001.

[13] Clark III, I and Flaherty, T.B., (2003), "Web-based B2B Portals", Industrial Marketing Management, 32, pp.15-23.

[14] "Introduction to Using Application Integration". http://e-docs.bea.com/.

[15] Ian Gorton2, Justin Almquist, Kevin Dorow, etal. "An Architecture for Dynamic Data Source Integration". Proceedings of the 38th Hawaii International Conference on System Sciences – 2005.

# A Fuzzy Algorithm for Scheduling Soft Periodic Tasks in Preemptive Real-Time Systems

Mojtaba Sabeghi, Mahmoud Naghibzadeh,
Toktam Taghavi

*Ferdowsi University of Mashhad*
*Mashhad, Iran*

{sabeghi, naghib, taghavi}@um.ac.ir

***Abstract*** **- Most researches concerning real-time system scheduling assumes scheduling constraint to be precise. However, in real world scheduling is a decision making process which involves vague constraints and uncertain data. Fuzzy constraints are particularly well suited for dealing with imprecise data. This paper proposes a fuzzy scheduling approach to real-time system scheduling in which the scheduling parameters are treated as fuzzy variables. A simulation is also performed and the results are compared with both EDF and LLF scheduling algorithms. The latter two algorithms are the most commonly used algorithms for scheduling real-time processes. It is concluded that the proposed fuzzy approach is very promising and it has the potential to be considered for future research.**

**Keywords**: Fuzzy scheduling, real-time systems, EDF, LLF, MFDF, MFLF

## I. Introduction

Real-time systems are vital to industrialized infrastructure such as command and control, process control, flight control, space shuttle avionics, air traffic control systems and also mission critical computations [1, 3]. In all cases, time has an essential role and having the right answer too late is as bad as not having it at all.

In the literature, these systems have been defined as: "systems in which the correctness of the system depends not only on the logical results of computation, but also on the time at which the results are produced" [1]. Such a system must react to the requests within a fixed amount of time which is called deadline.

In general, real-time systems can be categorized into two important groups: hard real-time systems and soft real-time systems. In hard real-time systems, meeting all deadlines is obligatory, while in soft real-time systems missing some deadlines is tolerable.

In both cases, when a new task arrives, the scheduler is to schedule it in such a way that guaranties the deadline to be met. As stated in [1] scheduling involves allocation of resources and time to tasks in such a way that certain performance requirements are met.

These tasks can be classified as periodic or aperiodic. A periodic task is a kind of task that occurs at regular intervals, and aperiodic task occurs unpredictably. The length of the time interval between the arrivals of two consecutive requests in a periodic task is called period.

Another aspect of scheduling theory is to decide whether the currently executing task should be allowed to continue or it has had enough CPU time for the moment and should be suspended. A preemptive scheduler can suspend the execution of current executing request in favor of a higher priority request. However, a nonpreemptive scheduler executes the currently running task to completion before selecting another request to be executed. A major problem that arises in preemptive systems is the context switching overhead. The higher number of preemptions a system has, the more context switching needed [5].

There are a plenty of real-time scheduling algorithms that are proposed in the literature. Each of these algorithms bases its decision on certain parameter while attempting to schedule tasks to satisfy their time requirements. Some algorithms use parameters that are determined statically such as the Rate Monotonic algorithm that uses the request interval of each task as its priority [7, 15]. Others use parameters that are calculated at run time. Laxity and deadline are among those parameters that are the most considered. Laxity says the task execution must begin within a certain amount of time while deadline implies the time instant at which its execution must be completed [2].

In the following, there are descriptions of two famous algorithms which are commonly used in real-time systems and are proved to be optimal for uniprocessor systems when the system load factor is less than one. System load factor is defined as follow:

$$L=\sum_{i=1}^{n}\frac{e_i}{r_i}$$

Earliest Deadline First (EDF) is a dynamic algorithm that does not require processes to be periodic. Whenever a process needs the CPU time, it announces its presence and its deadline. This algorithm keeps a list of running



*T. Sobh and K. Elleithy (eds.), Advances in Systems, Computing Sciences and Software Engineering*, 11–16.

processes that is sorted on deadlines. It always runs the first process on the list that is, the one with the closest deadline. When a new process becomes ready, the algorithm first checks its deadline. If this deadline occurs before the currently running process, then the algorithm preempts the current one and starts the new process.

The Least-Laxity-First (LLF) scheduling algorithm assigns higher priority to a task with the least laxity. The algorithm, however, is impractical to implement because laxity tie results in the frequent context switches among the tasks [4].

Static scheduling works perfect when there is enough information in advance about what has to be done, but dynamic scheduling does not have this restriction. Although, the dynamic algorithms focus on timing constraints but there are other implicit constraints in the environment, such as uncertainty and lack of complete knowledge about the environment, dynamicity in the world, bounded validity time of information and other resource constraints. In real world situations, it would often be more realistic to find viable compromises between these objectives. For many problems, it makes sense to partially satisfy objectives. The satisfaction degree can then be used as a parameter for making a decision. One especially straightforward method to achieve this is the modeling of these constraints through fuzzy constraints.

The scope of the paper is confined to scheduling of preemptive periodic tasks in soft real-time systems with fuzzy constraints. The rest of the paper is organized as follow. In section II the fuzzy inference system is discussed. Section III covers the proposed model and section IV contains the experimental results. Conclusion and future works are debated in Sections V.

## II. Fuzzy Inference System

Fuzzy logic is an extension of Boolean logic dealing with the concept of partial truth which denotes the extent to which a proposition is true. Whereas classical logic holds that everything can be expressed in binary terms (0 or 1, black or white, yes or no), fuzzy logic replaces Boolean truth values with a degree of truth. Degree of truth is often employed to capture the imprecise modes of reasoning that play an essential role in the human ability to make decisions in an environment of uncertainty and imprecision.

Fuzzy Inference Systems (FIS) are conceptually very simple. They consist of an input, a processing, and an output stage. The input stage maps the inputs, such as deadline, execution time, and so on, to the appropriate membership functions and truth values. The processing stage invokes each appropriate rule and generates a corresponding result. It then combines the results. Finally, the output stage converts the combined result back into a specific output value [6].

The membership function of a fuzzy set corresponds to the indicator function of the classical sets. It is a curve that defines how each point in the input space is mapped to a membership value or a degree of truth between 0 and 1. The most common shape of a membership function is triangular, although trapezoidal and bell curves are also used. The input space is sometimes referred to as the universe of discourse [6].

As discussed earlier, the processing stage which is called inference engine is based on a collection of logic rules in the form of IF-THEN statements where the IF part is called the "antecedent" and the THEN part is called the "consequent". Typical fuzzy inference systems have dozens of rules. These rules are stored in a knowledgebase. An example of a fuzzy IF-THEN rule is: IF *laxity* is *critical* then *priority* is *very high*, which laxity and priority are linguistics variables and critical and very high are linguistics terms. Each linguistic term corresponds to membership function.

An inference engine tries to process the given inputs and produce an output by consulting an existing knowledgebase. The five steps toward a fuzzy inference are as follows:

- Fuzzifying Inputs
- Applying Fuzzy Operators
- Applying Implication Methods
- Aggregating All Outputs
- Defuzzifying outputs

Bellow is a quick review of these steps but a detailed study is not in the scope of this paper.

Fuzzifying the inputs is the act of determining the degree to which they belong to each of the appropriate fuzzy sets via membership functions. Once the inputs have been fuzzified, the degree to which each part of the antecedent has been satisfied for each rule is known. If the antecedent of a given rule has more than one part, the fuzzy operator is applied to obtain one value that represents the result of the antecedent for that rule. The implication function then modifies that output fuzzy set to the degree specified by the antecedent. Since decisions are based on the testing of all of the rules in an FIS, the results from each rule must be combined in order to make a decision. Aggregation is the process by which the fuzzy sets that represent the outputs of each rule are combined into a single fuzzy set. The input for the defuzzification process is the aggregated output fuzzy set and the output is a single value. This can be summarized as follows: mapping input characteristics to input membership functions, input membership function to rules, rules to a set of output characteristics, output characteristics to output membership functions, and the output membership function to a single-valued output.

There are two common inference processes [6]. First is called Mamdani's fuzzy inference method proposed in 1975 by Ebrahim Mamdani [8] and the other is Takagi-Sugeno-Kang, or simply Sugeno, method of fuzzy inference Introduced in 1985 [9]. These two methods are the same in many respects, such as the procedure of fuzzifying the inputs and fuzzy operators.

The main difference between Mamdani and Sugeno is that the Sugeno output membership functions are either linear or constant but Mamdani's inference expects the output membership functions to be fuzzy sets.

Sugeno's method has three advantages. First it is computationally efficient, which is an essential benefit to real-time systems. Second, it works well with optimization and adaptive techniques. These adaptive techniques provide a method for the fuzzy modeling procedure to extract proper knowledge about a data set, in order to compute the membership function parameters that best allow the associated fuzzy inference system to track the given input/output data. However, in this paper we will not consider these techniques. The third, advantage of Sugeno type inference is that it is well-suited to mathematical analysis.

## III. The Proposed Model

The block diagram of our inference system is presented in Figure 1.

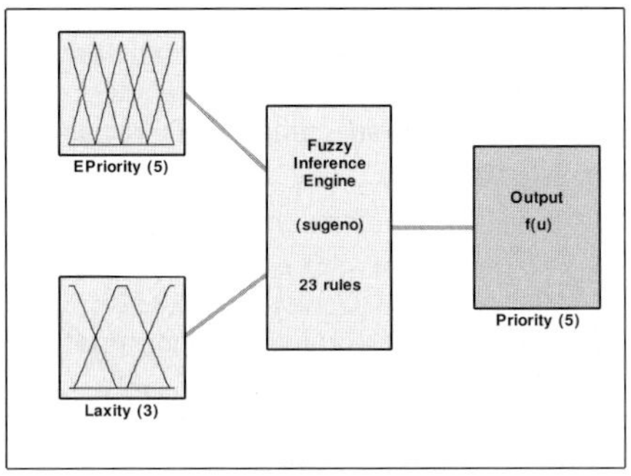


Fig.1. Inference system block diagram.

In the proposed model, the input stage consists of two linguistic variables. The first one is an external priority which is the priority assigned to the task from the outside world. This priority is static. One possible value can be the tasks interval, as rate monotonic algorithm does. For Figure 1, the other input variable is the laxity. This input can easily be replaced by deadline, wait time, or so on, for other scheduling algorithms. Each parameter may cause the system to react in a different way. The only thing that should be considered is that by changing the input variables the corresponding membership functions may be changed accordingly.

For the simulation purposes, as it is discussed later, two situations are recognized: First, by using laxity as a secondary parameter and, second, by replacing the laxity parameter with deadline. In fact, two algorithms are suggested: one with laxity as the second parameter. This algorithm is called MFLF[1]. The other algorithm is with deadline as the second parameter. This one is called MFDF[2].

The input variables mapped into the fuzzy sets as illustrated in Figures 2 and 3.

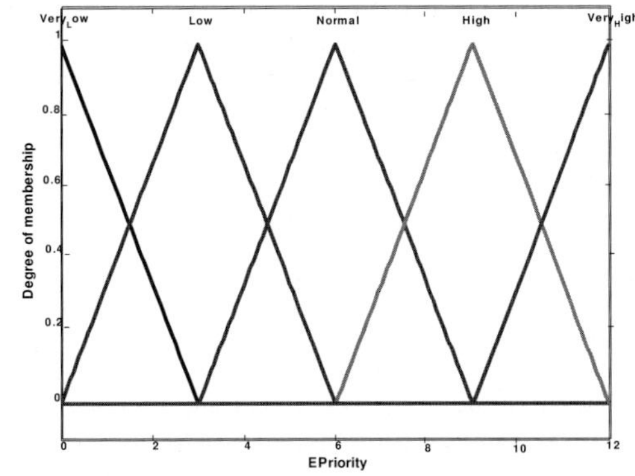


Fig.2. Fuzzy sets corresponding to external priority

The shape of the membership function for each linguistic term is determined by the expert. It is very difficult for the expert to adjust these membership functions in an optimal way. However, there are some techniques for adjusting membership functions [10, 13]. In this paper, we will not consider these techniques. They can be further studied in a separate paper.

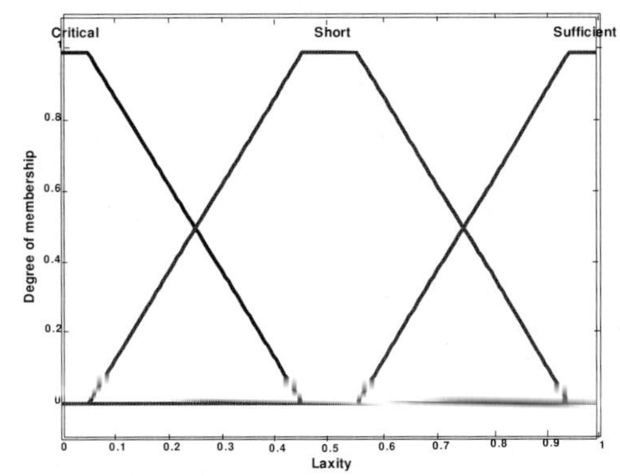


Fig.3. Fuzzy sets corresponding to laxity

We have produced 23 rules for our proposed system. Some of these rules are mentioned here:

- If (EPriority is high) and (laxity is critical) then (Priority is very high)
- If (EPriority is normal) and (laxity is critical) then (Priority is high)
- If (EPriority is very low) and (laxity is critical) then (Priority is normal)

---

[1] Minimum fuzzy laxity first

[2] Minimum fuzzy deadline first

- If (EPriority is high) and (laxity is sufficient) then (Priority is normal)
- If (EPriority is very low) and (laxity is sufficient) then (Priority is very low)

In fuzzy inference systems, the number of rules has a direct effect on its time complexity. So, having fewer rules may result in a better system performance.

### The Proposed Algorithms

The MFLF algorithm is as follows:

```
Loop
    1. For each task T, feed its external priority and
    laxity into the inference engine. Consider the
    output of inference module as priority of task T.
    2. Execute the task with highest priority until an
    scheduling event occurs (a running task finishes,
    a new task arrives)
   3. Update the system states (laxity, deadline, etc)
End loop
```

The algorithm for the MFDF is similar to the MFLF with laxity replaced by deadline.

## IV. Experimental Results

The simulation consists of two parts. First, the system was examined for the case where the system load factor is less than one. Second, the system was observed in overloaded conditions. These divisions are suggested because, first, both EDF and LLF algorithms has been proved to be optimal in situations where the system load factor is less than one. The results of this phase shows whether or not the simulation is performed correctly. A correct simulation will reveal that there is no task misses for either of EDF and LLF algorithms. At the same time, it will show whether or not our algorithms perform as well as the EDF and LLF. Second, recall that soft real-time systems, as their definition implies, can tolerate some deadline misses. In real situations, there is no guarantee for soft real-time systems not to be overloaded. Evaluating systems in overloaded conditions is important in comparing the behavior of our scheduling algorithms with the existing EDF and LLF algorithms. As it was discussed earlier, LLF is impractical to implement so we decided to use a modified version of it that solves the problem of frequent context switches. This modified algorithm is fully discussed in reference [4] and is proved to be optimal.

To compare these algorithms, we need to automatically generate some sample systems. The system generation methods will be discussed later.

Performance metrics, which are used to compare different algorithms, must be carefully chosen to reflect the real characteristics of a system. These metrics are as follows.

Response time, which is defined as the amount of time a system takes to react to a given input, is one of the most important factors in most scheduling algorithms.

Number of missed deadlines is an influential metric in scheduling algorithms for soft real-time systems.

When task preemption is allowed, another prominent metric comes into existence and that is the number of preemptions. Each of preemptions requires the system to perform a context switching which is a time consuming action.

CPU utilization is also an important metric because the main goal of a scheduling algorithm is to assign and manage system resources so that a good utilization is achieved.

Yet another metric, which is considered in our study, is the number of missed deadlines from the class of highest priority tasks. This corresponds to the external priority being *very high*.

### A. Comparison in Non-overloaded Conditions

This comparison was mainly performed to show the correctness of the simulations. To do the evaluation, 2500 test cases with load factors less than one were generated. In each test case, the number of tasks and the corresponding execution time and request interval randomly generated.

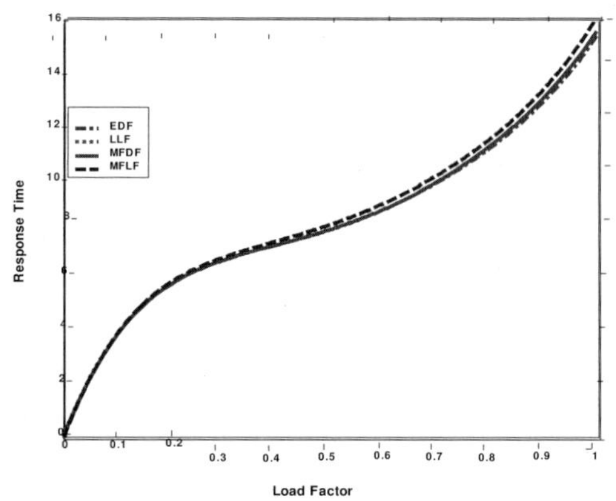


Fig.4. Response time in non-overloaded conditions

For this simulation phase, the goal is to compare average response time. As Figure 4 states all four algorithms show approximately the same performance with respect to the response time. The results are exactly what we have expected. The average response time of the test cases is summarizes in Table 1.

Table 1. Average Response time

| EDF | LLF | MFDF | MFLF |
|---|---|---|---|
| 8.728 | 8.966 | 8.762 | 8.958 |

## B. Comparison in Overloaded Conditions

Comparison parameters which are used here are average response time, number of tasks missing their deadlines, number of preemptions, and CPU utilization.

The simulation was done on 2500 test cases. These test cases were randomly generated. In each test case, the number of tasks and the corresponding execution time and request interval randomly generated. Also, each task has been assigned a priority according to the rate monotonic principle (tasks with shorter request interval are given higher priorities) [7].

As Figure 5 states, when the load factor is less than one, all the algorithms have the similar performance. However, when the system becomes overloaded, the response time of both EDF and LLF is much tardier than MFLF and MFDF.

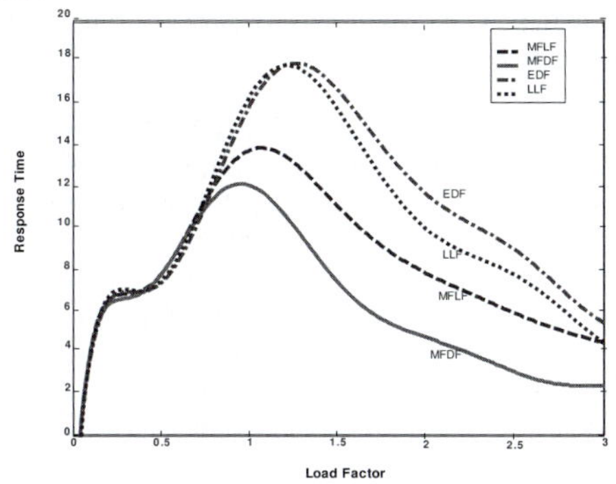


Fig.5. Response time in overloaded conditions

Figure 6 states that for load factors less than one the number of misses is zero. This is because it has already been proved that any system with a load factor less than or equal to one runs safe under either of EDF and LLF.

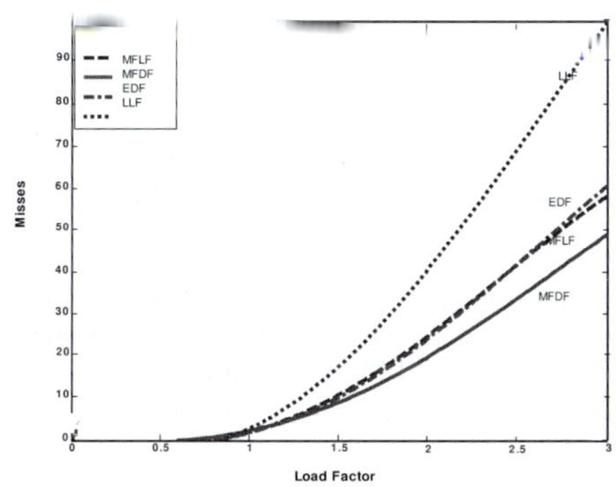


Fig.6. Number of Misses

Fortunately, MFLF and MFDF perform as well as either of EDF and LLF. In this case, the number of misses is exactly zero for all four algorithms. Because in drawing diagrams some curve fitting techniques is used, it seems that number of misses for algorithms when the load factor is a little bit less than one is a positive number. However, we have examined the numerical results and confirm that the number of misses is exactly zero.

When the load factor is more than one the MFDF has the best performance and MFLF has a performance similar to EDF. The LLF has the worse performance among all four algorithms.

As the Figure 7 shows, there is an opposite relation between the numbers of preemptions on the one hand and response time on the other hand. As the response time gets better number of preemptions comes to worse value.

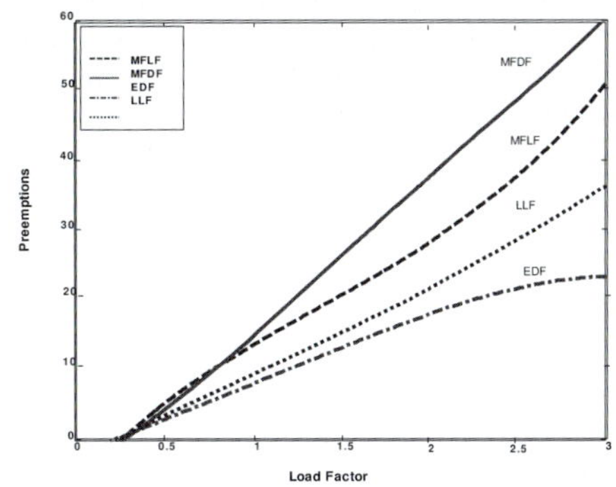


Fig.7. Number of Preemptions

MFDF that has the best performance with respect to response time has a larger number of preemptions. But there is something good about it, and that is, its behavior is predictable as it acts in a linear way. Having higher number of preemptions is reasonable because it eventually leads to having better response time and also better CPU utilization. There should be a balance between the number of preemptions and other factors. Reference [11] argues why such a balance is needed.

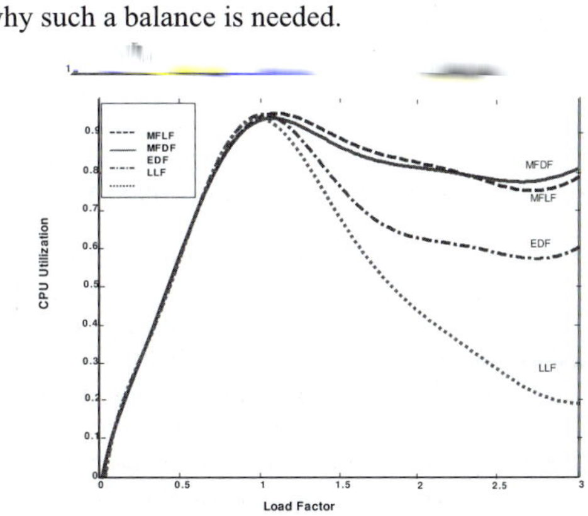


Fig.8. CPU Utilization

Figure 8 demonstrates that with the fuzzy methods CPU utilization is much higher than non-fuzzy methods. When

the load factor is about 3, the MFDF and MFLF use about 80 percent of CPU time while EDF uses 60 percent of CPU time and the LLF just uses about 20 percent of CPU time.

Considering the number of missed deadlines from the class of highest priority tasks, Figure 9 shows that both MFDF and MFLF perform much better than EDF and LLF. Comparing Figure 9 with Figure 6 shows that in load factor 3 about 80 percent of missed deadlines in both EDF and LLF are from the class of highest priority tasks while in MFDF and MFLF just about 30 percent of misses are among highest priority tasks. This is because external priority is considered as a decision parameter in the latter two algorithms. It should be mentioned that highest priority tasks in this simulation as discussed earlier, are those with shorter request intervals. These kinds of tasks since their deadline is too short may miss their deadline easier than the others. This is why in EDF and LLF about 80 percent of misses are among these tasks.

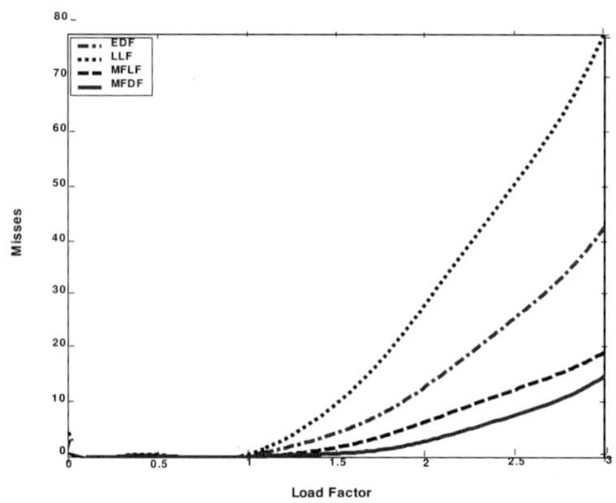


Fig.9. Number of missed deadlines from the class of highest priority tasks

## V. Conclusion and Future Works

Using the fuzzy concept in real-time scheduling, as it was shown, has the following advantages: (1) it better utilizes system resources such as CPU, (2) it decreases the number of missing deadlines, (3) it improves the system response time, and (4) it serves more important tasks better.

In the future, for improving the time complexity of the system, rule reduction techniques are going to be applied to the system. Also, to improve performance, adjusting membership functions with adaptive methods of inference is required [10, 13]. Fuzzy scheduling is well suited for parallel and distributed systems as some parallel and distributed fuzzy inference systems have been introduced [14]. A detailed analysis of fuzzy scheduling for parallel systems is in progress. Also a non-preemptive version of these algorithms is under publication.

## References

[1] Ramamritham K., Stankovic J. A., Scheduling algorithms and operating systems support for real-time systems, Proceedings of the IEEE, vol. 82, no. 1, pp. 55--67, January 1994.

[2] Hong J., Tan X., Towsley D., A Performance Analysis of Minimum Laxity and Earliest Deadline Scheduling in a Real-Time System, *IEEE Trans. on Comp.*, vol. 38, no. 12, Dec. 1989

[3] Sha, L. and Goodenough, J. B., Real-Time Scheduling Theory and Ada, IEEE Computer, Vol. 23, No. 4, pp. 53-62 (April 1990).

[4] Oh S.-H., Yang S.-M., A Modified Least-Laxity-First Scheduling Algorithm for Real-Time Tasks, *rtcsa*, p. 31, Fifth International Conference on Real-Time Computing Systems and Applications (RTCSA'98), 1998.

[5] Tanenbaum A. S., Modern Operating Systems, Second Edition, Prentice-Hall, 2001.

[6] Wang Lie-Xin, A course in fuzzy systems and control, Prentice Hall, Paperback, Published August 1996.

[7] Liu C. L., Layland J. W., Scheduling Algorithms for Multiprogramming in a Hard Real-Time Environment. Journal of the ACM, 20(1):46-61, 1973.

[8] Mamdani E.H., Assilian S., An experiment in linguistic synthesis with a fuzzy logic controller, International Journal of Man-Machine Studies, Vol. 7, No. 1, pp. 1-13, 1975.

[9] Sugeno, M., Industrial applications of fuzzy control, Elsevier Science Inc., New York, NY, 1985.

[10] Jang, J.-S. R., ANFIS: Adaptive-Network-based Fuzzy Inference Systems, IEEE Transactions on Systems, Man, and Cybernetics, Vol. 23, No. 3, pp. 665-685, May 1993.

[11] Hamidzadeh B., Shekhar S., Specification and Analysis of Real-time Problem Solvers, IEEE Transactions on Software Engineering, Volume 19, pages 788-803, 1993

[13] Simon D, Training fuzzy systems with the extended Kalman filter, Fuzzy Sets and Systems, Volume 132, Number 2, 1, pp. 189-199, December 2002.

[14] Lee S.G., Lee H.H., Miyazaki M., Parallel Fuzzy Inference on Hypercube Computer, IEEE International Fuzzy Systems Conference Proceedings August 22-25, 1999, Seoul, Korea

[15] Naghibzadeh M, Kim K. H. , A Modified Version of Rate-Monotonic Scheduling Algorithm and its Efficiency Assessment, Seventh IEEE International Workshop on Object-Oriented Real-Time Dependable Systems (WORDS'02), 2002.

# Parallel Construction of Huffman Codes

S. Arash Ostadzadeh[1], M. Amir Moulavi[2], Zeinab Zeinalpour[3], B. Maryam Elahi[4]

[1]ostadzadeh@mshdiau.ac.ir,
Islamic Azad University of Mashhad, Faculty of Engineering, Computer Engineering Department
Ostad Yousefi St. Ghasem Abad, Mashhad, Iran

[2]amir.moulavi@computer.org, [3]zeinabzt@gmail.com, [4]melahi@acm.org,
Young Researchers Club, Islamic Azad University of Mashhad,
Emamiyeh 59, Emamiyeh Blvd., Ghasem Abad, Mashhad, Iran

*Abstract* - **For decades, different algorithms were proposed addressing the issue of constructing Huffman Codes. In this paper we propose a detailed time-efficient three-phase parallel algorithm for generating Huffman codes on CREW PRAM model exploiting *n* processors, where *n* is equal to the number of symbols in the input alphabet. First the codeword length for each symbol is computed concurrently with a direct parallelization of the Huffman tree construction algorithm eliminating the complexity of dealing with the original tree-like data structure. Then the Huffman codes corresponding to symbols are generated in parallel based on a recursive formula. The performance of the proposed algorithm depends directly on the height of the corresponding Huffman tree. It achieves an O(*n*) time complexity in the worst case which is rarely encountered in practice.**

*Keywords*: Data Structures, Parallel Algorithms, Huffman Codes, Optimal Prefix Codes, PRAM

## 1. INTRODUCTION

Since the introduction of the Huffman encoding scheme by D. A. Huffman [8] in 1952, which elegantly addresses the problem of constructing optimal prefix codes for a given alphabet, Huffman [illegible] in information processing systems particularly in text, image and video compression. Huffman encoding uses variable-length codes to compress the input data, where frequently used symbols are represented with a short code while less often used symbols have longer representations.

Consider $S = \{s_0, s_1, ..., s_{n-1}\}$ a set of $n$ source symbols with frequencies $F = \{f_0, f_1, ..., f_{n-1}\}$ such that $f_0 \geq f_1 \geq ... \geq f_{n-1}$, where symbol $s_i$ has the frequency $f_i$. Using the Huffman algorithm to construct the Huffman tree $T$, the codeword $c_i$, $0 \leq i \leq n-1$, for symbol $s_i$ can then be determined by traversing the path from the root to the leaf node associated with the symbol $s_i$, where a left branch corresponds to '0' and a right one corresponds to '1'. Let the level of the root be zero, and the level of any other node in the tree be equal to summing up its parent's level and one. Codeword length $l_i$ for $s_i$ can be considered as the level of $s_i$. The weighted external path length $\sum_{i=0}^{n-1} fi * li$ in Huffman codes is the least possible.

Huffman proposed an algorithm to generate optimal prefix codes in O($n$log$n$) time. Later it was proved that it needs only linear time provided that the frequencies of appearance for symbols are sorted in advance [17, 22]. It should be noted that prefix codes have the nice property that a message can be decoded in only one way.

Huffman coding theory has attracted many researchers regarding its analyses, improvements and applications [2, 3, 4, 5, 6, 7, 9, 10, 13, 15, 17, 19, 22]. There is also a considerable amount of work addressing the relevant issues, particularly the Huffman codes construction and decoding methods in parallel and distributed environments [1, 11, 14, 16, 18, 21]. We briefly investigate several attempts to construct the Huffman codes in parallel.

Teng [21] proposed the first *NC* algorithm to generate Huffman codes using $n^6$ processors in O(log$n$) time which seems rather impractical due to the huge number of processors. Atallah et al. [1] showed how to reduce the number of processors to $n^3$ while maintaining the same time complexity. They also presented an O($\log^2 n$) time, $n^2$/log$n$ processors as well as O(log$n$) time, $n^3$/log$n$ processors CREW deterministic parallel algorithms for the construction of Huffman codes. [illegible] reduced to O(log$n$(loglog$n$)$^2$) on a CRCW model using only $n^2$/(loglog$n$)$^2$ processors.

Kirkpatrick and Przytycka [11] presented several approximated parallel algorithms for the construction of Huffman codes. Later Larmore and Przytycka [16] proposed an O($\sqrt{n}$ log$n$) time algorithm that uses $n$ processors. The original algorithm is developed as a solution for the *concave least weight subsequence* problem but it is extended for the Huffman coding problem.

Milidiú et al. [18] proposed a work efficient parallel algorithm on CREW to address the problem. Their algorithm runs in O($H$ loglog($n/H$)) time with $n$ processors where $H$ is the length of the longest Huffman code. Since $H$ is in the interval $[\lceil \log n \rceil, n-1]$, the algorithms requires O($n$) time in the worst case.



*T. Sobh and K. Elleithy (eds.), Advances in Systems, Computing Sciences and Software Engineering*, 17–23.

The major problem with Milidiú et al.'s algorithm and some other parallel Huffman construction solutions is that instead of actually generating the Huffman codes, they rather construct the Huffman tree in parallel and it is not clearly stated how the codes should be built from the Huffman tree in parallel.

In this paper we particularly address this problem by proposing a practical three-phase CREW algorithm based on a sophisticated parallelization of the direct Huffman tree constructing simulation with the elimination of the need to store the nodes in a tree-like structure. We first compute the path length for each symbol in the Huffman tree then focus on generating the Huffman codes in parallel by exploiting Hashemian's [5] recursive formula based on the single-side growing Huffman tree. Our algorithm can be implemented in O($n$) time on the CREW PRAM model incorporating $n$ processors.

The rest of this paper is organized as follows. In Section 2, we describe the fundamental data structures used in our pseudocode and their definitions. In section 3, we first draw an outline for our algorithm and then its description is given in details. We discuss the performance of the algorithm in section 4. The performance analysis of each part is stated in details. We conclude in section 5.

## 2. Data Structure

We assume that the input to the first phase of our algorithm is a symbol table including $S = \{s_1, s_2, \ldots, s_n\}$ an array of symbols and $F = \{f_1, f_2, \ldots, f_n\}$ an array of the corresponding frequencies. This symbol table is sorted based on frequencies in non-decreasing order. Each symbol corresponds to a leaf in the Huffman tree. We define a structure for these leaves, including *freq*, a field for frequency value, and *leader*, a pointer to the root of the subtree that this leaf belongs to. *lNodes* is an array of the mentioned structure, in which $lNodes_i$ corresponds to $s_i$. It shows the leader of the leaves that have already participated in the construction of the tree levels. We define a similar structure for internal nodes. *iNodes* behaves as a queue of the mentioned data structure which contains the internal nodes.

As new tree levels are generated, new internal nodes are added to the queue. Array *Temp* is a data structure for temporary storage of a merged list of internal nodes from *iNodes* and leaf nodes from *lNodes* that are the nodes participating in the construction of each new tree level. Each element of *Temp* contains three fields: *Freq*, *isLeaf* and *index*. The chosen leaf nodes who participate in the construction of the current level are first copied to *Copy*, which is an array with the same structure as *Temp*.

The first phase generates an array of codeword lengths $CL=\{cl_1, cl_2, \ldots, cl_n\}$ in which $cl_i$ is the codeword length for $s_i$, such that $cl_i \leq cl_{i+1}$. To know how many codewords have the same length, *PackedArray* is filled in the second phase with the index of those elements in *CL*, which have a different value compared to the next element.

The third phase of our algorithm receives the array of codeword lengths *CL*, and the *PackedArray* as inputs and generates the final codewords in *CW*, in which $cw_i$ is the Huffman codeword for $s_i$.

## 3. Algorithm

In this section, the outline of the proposed algorithm is described and the details of implementation issues concerning each phase are presented afterwards.

### 3.1. Outline

We propose an algorithm for time-efficient construction of Huffman codes in CREW PRAM model. Our algorithm requires $n$ processors, which is equal to the number of input symbols. This task could be divided into three phases: Codeword Length Generation (CLGeneration), Length Count (LC) and Codeword Generation (CWGeneration). The outline is illustrated in Fig. 1.

In the first phase, the CLGeneration algorithm computes the codeword length for each symbol $s_i$ that is equal to the path length of $s_i$ in the Huffman tree. This is done without maintaining a tree structure in practice and by generating the tree levels, one at a time, in a bottom-up fashion. With the generation of each new level in the tree, the codeword length of each symbol whose leader has participated in the construction of the new level is incremented. At each level, the two nodes with smallest frequencies are found among the internal nodes constructed in the previous iterations and leaf nodes who have not participated in the construction of the tree yet. These two are combined to form a new internal node in the next higher level. The combined value indicates the minimum frequency of the next higher level, which plays a crucial role in the selection of leaf nodes for participation in the current level. At this point, all the internal nodes and leaf nodes that are meant to participate in the current level are selected and merged, and combined pair-wise (melded) to form the new internal nodes of the next higher level. This process is repeated until only one internal node remains which is the root of the tree.

To be able to generate the codewords in parallel, we need an intermediate step to reverse *CL* and to decide for each symbol the number of symbols that have shorter codeword lengths. In order to know how many symbols have the same codeword length, those processors whose corresponding elements in *CL* have a different value from their next element set their indexes in *PackedArray*. Next, a parallel array packing algorithm is performed on the *PackedArray*. The details of the algorithm are presented in [20]. The resulting array has the indexes of those elements in *CL*, which have a different value compared to the next element.

In the third step, the final Huffman code for each symbol is generated from its codeword length with the help of a parallel version of a recursive formula introduced in [5]. To make $CW_i$ correspond to its symbol in *S*, *CW* is reversed in the end.

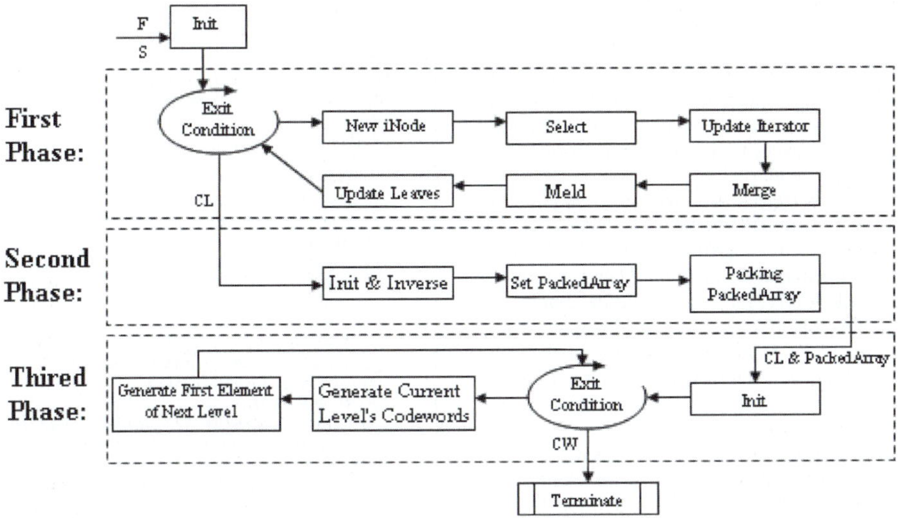


Fig. 1. Algorithm Outline

*3.2. Description*

In the following three sections, each phase of the algorithm is discussed in details. To clarify the process each phase is accompanied by an example showing the result of performing each step of the algorithm on the sample input illustrated in Fig. 2.

*3.2.1. CLGeneration*

CLGeneration is depicted in Fig. 3. First the following arrays are initialized in parallel. *lNodes* is an array that keeps the statuses of the leaf nodes. Each leaf is initialized with its corresponding frequency and its leader is set to -1. *CL*, the array that shows the codeword length for each symbol is initialized with 0. Next, processor $p_1$ sets the following variables. *lNodesCur* points to the last leaf that has participated in the construction of a level so far and it is initialized to 0. *iNodesFront* and *iNodesRear* are the front and rear indicators for *iNodes*, which behaves like a queue and shows the newly generated internal nodes who have not participated in the construction of a level yet. They are initialized to zero, indicating that the *iNodes* queue is empty.

After initialization, the following operations are performed within a loop until all leaves are processed and no internal node is left in the *iNodes* queue, except for the root. The sum of the two minimal frequency values within the current list of leaf nodes and internal nodes, *MinFreq*, determines the frequency value of the next internal node. This internal node could be constructed from the combination of two new leaves, an internal node and a new leaf or two internal nodes. In case of ties, leaves are preferred to participate in the construction of the new internal node, and the new internal node is added to the *iNodes* queue. The leaders of the two internal nodes or leaves who have been combined are set to the index of the newly generated internal node in the *iNodes* queue.

Selection of the participating leaves in the current level is performed in parallel. Those leaves whose index is more than *lNodesCur* and their *Freq* is less than or equal to *MinFreq* are copied to the *Copy* array. *Copy* array has three fields: *Freq*, *isLeaf* and *index*. *Freq* is the value of the selected leaf; *isLeaf* in this step has the value true for all elements, because they are all leaves and *index* is the index of the selected leaves in *lNodes*. *CurLeavesNum* is the number of selected leaves that have been copied to the *Copy* array. *CurLeavesNum* and *Copy* are next passed to the Merge function.

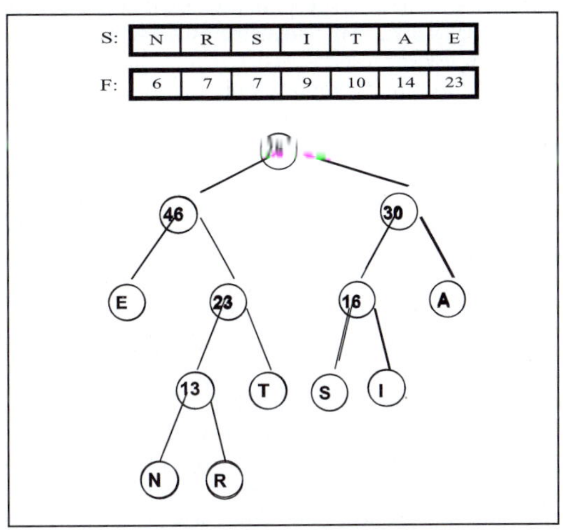


Fig. 2. Sample input and its corresponding Huffman tree

Before merging the selected leaves and internal nodes, we need to make sure the number of nodes that are going to be merged and then combined pair-wise is even. Thus, in case an odd total of internal nodes and leaf nodes are selected for this level, queue indicators are updated in a way that a leaf or internal node with the maximum frequency is left out for the next level, regarding leaves to have higher priorities to participate in the current level.

The Merge function performs the task of combining two sorted lists in O(loglog*n*) on CREW PRAM model [12]. The function accepts *Copy*, *CurLeavesNum*, *iNodes*, *iNodesFront*, *iNodesRear* as input parameters and *Temp* and *TmpLn* as outputs. The main task of this function is to fill the *Temp* array which is the combination of *Copy* array and the section in *iNodes* that is indicated by *MergeFront* and *MergeRear*. The value of *isLeaf* for elements taken from *iNodes* is set to 0. The result in *Temp* is sorted based on the *Freq* field in non-decreasing order.

In the next step, selected nodes are combined pair-wise. Each processor is assigned to two consecutive elements of *Temp* according to its index. These pairs of elements are melded to form new internal nodes whose *Freq* is the sum of the combined pairs' frequencies. Then the responsible processor updates the *leader* fields of the two participating nodes. The location of each node is shown by the *isLeaf* and *index* fields indicating whether the element resides in *iNodes* or *lNodes* and what its index is. In the end, $p_1$ increments the *iNodeRear* based on the number of the newly added internal nodes which is equal to the half of *TmpLn*. *lNodesCur* is incremented by the value of *CurLeavesNum* which is equal to the number of leaves who were used in the current level.

If the leader of an internal node is changed, all its children need to update their leader and set it to the index of the new internal node and the codeword length corresponding to these leaf nodes is also incremented. In this step all leaves check their leaders in parallel and they figure out whether or not their leaders have been changed. In the case of a change in the index of their leaders, they update their leaders by getting the value of their leader's leader.

The mentioned process repeats until only one internal node remains, which is the root. At this point, *CL* has the codeword lengths for all symbols.

To clarify the process, the result of the first cycle of this phase performed on the sample input in Fig.2 is illustrated in Fig. 4a-f. Fig. 4g shows the final *CL* that is passed to the next phase.

### *3.2.2. Length Count*

Length Count is depicted in Fig. 5. We generate the final codewords in a top-down fashion, so we need to reverse *CL*. Reversing *CL* is performed in parallel. To generate the final codes in parallel, we need to figure out the exact location of each symbol in the Huffman tree. The level that each symbol resides in is equal to the number of symbols with shorter lengths. In order to accomplish this task, *n* processors are assigned to *CL*. Every processor $p_i$, except the first one, checks

```
Forall Pi (1 ≤ i ≤ n) do in parallel
  lNodes[i].freq ← F[i], lNodes[i].leader ← -1, CL[i] ← 0
P1 sets
  iNodesFront, iNodesRear, lNodesCur ← 0
While (lNodesCur < n ||  iNodesRear – iNodesFront > 1)
  P1 sets
    mid ← {∞, ∞, ∞, ∞}
    if (lNodesCur ≤ n – 1)
        mid [1] ← lNodes[lNodesCur+1].Freq
    if (lNodesCur ≤ n – 2)
        mid [2] ← lNodes[lNodesCur+2].Freq
    if (iNodesRear > iNodesFront)
        mid [3] ← iNodes[iNodesFront+1].Freq
    if (iNodesRear > iNodesFront + 1)
        mid [4] ← iNodes[iNodesFront+2].Freq
    sort (mid)
    MinFreq ← mid[1] + mid[2]
    iNodes [iNodesRear + 1].freq ← MinFreq
    iNodes [iNodesRear + 1].leader ← -1
    if (isLeaf (mid[1]))
        lNodes[lNodesCur+1].leader ← iNodesRear + 1
        CL[lNodesCur+1]++, lNodesCur++
    else
        iNodes[iNodesFront + 1].leader ← iNodesRear + 1
        iNodesFront++
    if (isLeaf(mid[2]))
        lNodes[lNodesCur+1].leader ← iNodesRear + 1
        CL[lNodesCur+1]++,  lNodesCur++
    else
        iNodes[iNodesFront + 1].leader ← iNodesRear + 1
        iNodesFront++
  Forall Pi (lNodesCur < i ≤ n)
    if (lNodes[i].freq ≤ MinFreq)
        Copy[i – lNodesCur].freq ← lNodes[i].freq
        Copy[i – lNodesCur].index ← i
        Copy[i – lNodesCur].isLeaf ← true
        if (i = n || lNodes[i+1].freq > MinFreq)
                CurLeavesNum ← i – lNodesCur
  P1 Sets
    mergeRear ← iNodesRear, mergeFront ← iNodesFront
    if((CurLeavesNum+ iNodesRear - iNodesFront)%2=0)
        iNodesFront ← iNodesRear
    else if ((iNodesRear - iNodesFront != 0) &&
    (F[lNodesCur+CurLeavesNum]≤ iNodes[iNodesRear].freq))
        mergeRear--, iNodesFront ← iNodesRear - 1
    else
        iNodesFront ← iNodesRear, CurLeavesNum--
    lNodesCur ← lNodesCur + CurLeavesNum, iNodesRear++
  tmpln← merge(Temp,Copy, CurLeavesNum,mergFront, mergRear)
  Forall Pi (1 ≤ i ≤ tmpln/2) do in parallel
    iNodes[iNodesRear + i].freq←temp[2*i-1].freq+ temp[2*i].freq
    iNodes[iNodesRear + i].leader ← -1
    if (temp [2*i-1].isLeaf)
        lNodes [temp [2*i – 1].index].leader ← iNodesRear +  i
        CL[temp [2*i – 1].index]++
    else
        iNodes [temp [2*i – 1].index].leader ← iNodesRear +  i
    if (temp [2*i].isLeaf)
        lNodes [temp [2*i].index].leader ← iNodesRear +  i
        CL[temp [2*i].index]++
    else
        iNodes [temp [2*i].index].leader ← iNodesRear +  i
  P1 sets
    iNodesRear ← iNodesRear + (tmpln/2)
  Forall Pi (1 ≤ i ≤ n) do in parallel
    if(lNodes[i].leader != -1)
    if(iNodes[lNodes[i].leader].leader != -1)
        lNodes[i].leader← iNodes[lNodes[i].leader].leader
        CL[i]++
```

Fig. 3. CLGeneration

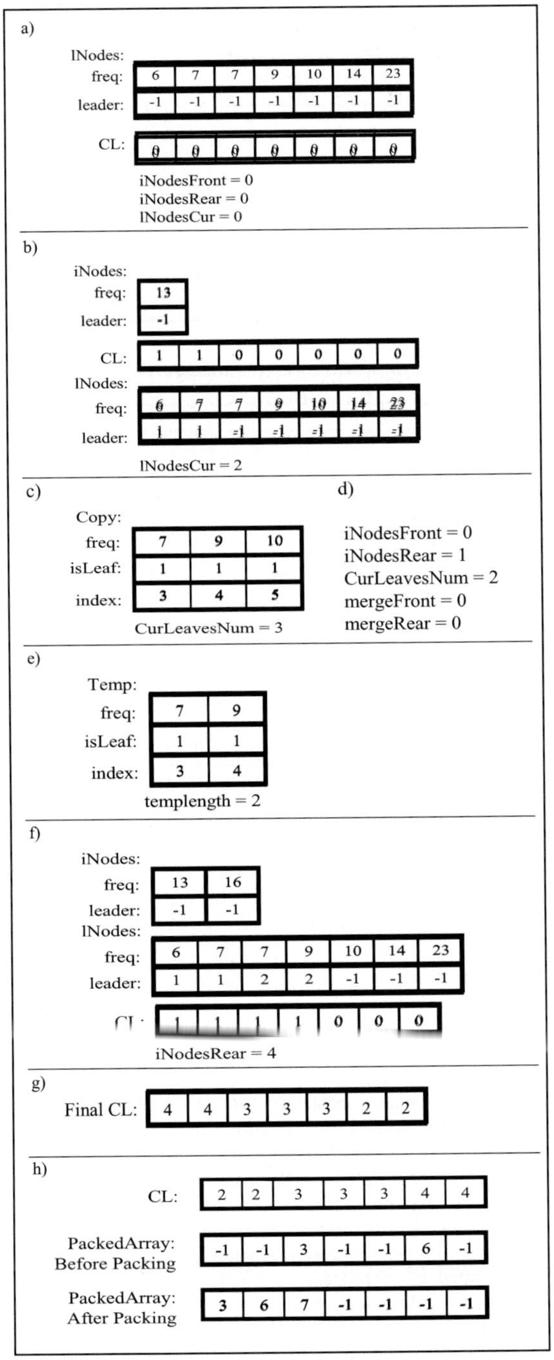


Fig. 4. a) CLGeneration Initialization, [b – f are the result of the first cycle of CLGeneration on the sample input] b) New iNode c) Select d) Update Iterators e) Merge f) Meld g) Final CL h) Length Count

the two $cl_i$ and $cl_{i-1}$; if these two elements have different values, then $p_i$ will copy its index to *PackedArray* in the corresponding location. The result is a sparse array. Next, we apply a parallel array packing algorithm on *PackedArray* as presented in [20]. The resulting array has the indexes of those elements in *CL*, which have a different value compared to the next element, sorted in non-decreasing order. The result of length count on our sample input is present in Fig. 4h.

### *3.2.3. CWGeneration*

In the CWGeneration phase, we generate the final codewords by introducing a parallel version of the algorithm proposed by Hashemian [5]. *CL* is the input array for this phase with the length of *n* and *CW* is the output array with the same number of elements. In the pseudocode depicted in Fig. 6, the following variables are used. *CCL* and *CDPI* show the current codeword length and current done-processor index respectively. In each cycle, *CCL* (Current Code Length) has the length of the codewords in the current level and *CDPI* (Current Done-Processor Index) indicates the index of the last processor who has finished generating its codeword. This phase is accomplished in two steps, assignment and codeword generation.

If we have the codeword lengths for all symbols, we can compute the final Huffman codewords by incorporating the following recursive formula [5]:

$$C_{i+1} = (C_i + 1) * 2^{cl_{i+1} - cl_i} \tag{1}$$

Initialization is performed by the first processor. The value of *CCL* is set to the first element of *CL*. A string of zeros with length of *CCL* is put in the beginning of the *CW* array. For the processors whose corresponding symbol's codeword length is equal to *CCL*, the construction of the codewords can be performed in parallel because they only have to add a number to the codeword of the first symbol in their series. $p_1$ generates the first codeword of the next series to set *CCL* for the next level, hence the processors assigned to next level are able to generate their codewords in the next cycle. These iterations continue until the codewords for all symbols are constructed.

```
For all processors P_i 1 ≤ i ≤ n/2
        LocalVariable ← CL[i]
        CL[n – i + 1] ← LocalVariable
For all processors P_i 1 ≤ i ≤ n
        PackedArray[i] ← -1

For all processors P_i 2 ≤ i ≤ n
        if (CL[i-1] != CL[i])
                PackedArray[i] ← i
P_1 Sets
        PackedArray[n+1] ← n+1
PackedArray ← ArrayPacking (PackedArray)
```

Fig. 5. Length Count

```
P1 sets
   CCL ← CL [1]
   CW [1] ← bit string of CCL zeros
   CDPI ← 1
   Level ← 1
While (CDPI < n)
   For all processor Pi 1≤i≤n do in parallel
       if (i > CDPI && CL[i] = CCL)
             CW[i] ← CW [CDPI] + (i - CDPI)
       else be idle
   P1 sets
       CDPI ← PackedArray [Level] - 1
       Level ++
       CW[CDPI+1]←(CW[CDPI]+1)* 2^(CL[CDPI + 1]–CL[CDPI])
       CDPI ← CDPI + 1
       CCL ← CL[CDPI]

For all processors Pi (1 ≤ i ≤ n/2)
   LocalVariable ← CW[i]
   CW [n – i + 1] ← LocalVariable
```

Fig. 6. Codeword Generation

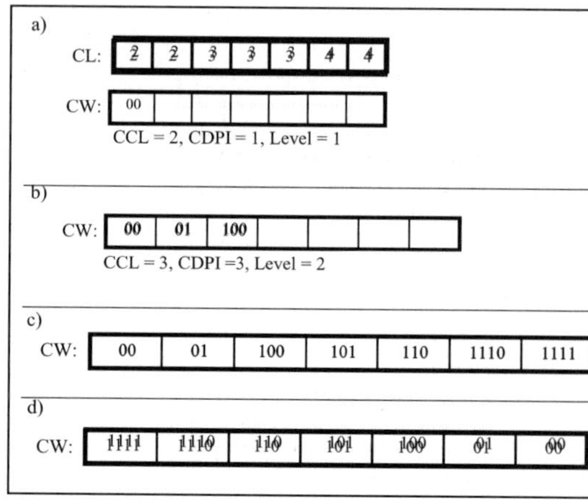


Fig. 7. a) CWGeneration Initialization b) Result of the first cycle c) Resulting CW d) The final reversed CW that corresponds to the symbol table of the sample input

Fig. 7b depicts the result of the first cycle of CWGeneration on our sample. For this example, after three iterations, the final codewords are constructed in *CW*. Because we reverse *CL* in the second phase, the resulting *CW* is reversed at the end to correspond to the input symbol table. The resulting *CW* does not directly match with the expected Huffman tree of the sample given in Fig. 2, however, the lengths of the codewords match. This is due to the fact that Huffman codes are not uniquely generated. The only important issue for the resulting codewords is that their codeword lengths should comply with the expected Huffman tree.

## 4. Performance Analysis

First, we analyze each phase of the algorithm separately and then we discuss the performance of the algorithm as a whole.

**Theorem 1**. *The CLGeneration runs in* O(*L* loglog(*n*/*L*)) *time, where L is the number of levels in the corresponding Huffman tree, i.e. the height of the Huffman tree.*

**Proof**. CLGeneration is comprised of a set of operations that are performed within a number of cycles, one cycle per each level of the tree. It can be proved that the number of these cycles is L, which is equal to the height of the Huffman tree [18].

Before the main loop, initialization is performed which has a parallel section and a sequential section, both of O(1) time. Next, at each cycle, a number of operations are performed. New *iNodes* contains a few sequential comparisons on four nodes in O(1). Updating Iterators is also of constant order. Updating Leaders, Selecting leaves and Melding, which all execute in parallel, are of O(1) parallel time. This is because a constant number of operations are performed on *i* variables by *i* processors in parallel; such that $1 \le i \le n$.

The only part that is not of constant order is the Merge operation that can be performed in O(loglog*M(i)*) parallel time [12], in which *M(i)* is the number of internal nodes and leaf nodes that are melded at cycle *i*. As a result the time required by CLGeneration is

$$T = O\left(\sum_{i=1}^{L} \log\log M(i)\right)$$

where *M(i)* is constrained to $\sum_{i=1}^{L} M(i) = (n-1)$. It is seen that the upper bound directly depends on *L*, the height of the tree that is in the interval $[\lceil \log n \rceil, n-1]$. If *L* equals *n-1*, which means we have a one sided tree, *M(i)* is of O(1), hence the algorithm runs in O(*n*). If the tree is balanced, the time complexity is computed as follows:

$$\begin{aligned} T &= \log\log 2 + \log\log 2^2 + \log\log 2^3 + \ldots + \log\log 2^{(\log n)-1} \\ &= (0 + \log 2 + \log 3 + \ldots + \log(\log n - 1)) \\ &= \log(1 * 2 * 3 * \ldots * (\log n - 1)) \\ &= \log(\log n - 1)! \\ &\equiv O(\log(\log n - 1)!) \end{aligned}$$

In general, we can find an upper bound, by maximizing *T*. It can be proved that *T* is maximized when *M(i)* = (*n*-1)/*L* [18]; therefore the time complexity of the CLGeneration is O(*L*loglog(*n*/*L*)) parallel time, in which O(*n*) is the worst case and O(log(log*n*-1)!) is the best case. ■

Time complexity of the second phase consists of Initializing *PackedArray* and inverting *CL*, filling *PackedArray*, and packing *PackedArray*. *PackedArray* initialization and *CL* inversion are of O(1), because a constant number of operations are performed by *n* processors. Filling *PackedArray* is the same as the previous section and for the same reason, it is of constant order. Array Packing takes O(log*n*) time to complete [20]. Hence the time complexity of this phase is O(log*n*).

The CWGeneration phase contains an initialization that is performed in constant time. Reversing *CW* at the end of this phase is accomplished in parallel and is of O(1). The major time consuming operation of this phase is the loop. The operations within the loop are performed in O(1), thus the worst case for this phase is equal to the worst case for the main loop cycles. Similar to theorem 1, and since CWGeneration completes the generation of codewords for one level at each cycle, CWGeneration performs in O($L$) where $\log n \leq L \leq n-1$, hence the worst case is O($n$).

Having the time complexity of each of the three phases, it is seen that the algorithm runs in $O(L \log\log(n/L) + \log n + L)$, which again depends directly on $L$, the height of the tree. Since $L$ is in the interval $[\lceil \log n \rceil, n-1]$, the algorithm as a whole runs in O($n$) time in the worst case and runs in $O(\log(\log n - 1)!)$ in the best case.

## 5. CONCLUSION

The significant and influential role of the Huffman coding is due to its broad applications in information processing and especially in image and data compression algorithms. Our contribution in this paper is presenting a time-efficient parallel algorithm on CREW PRAM model, incorporating $n$ processors, for constructing the Huffman codes without dealing with the complexity of using a tree structure. This algorithm is devised in three separate phases and its output is the final Huffman codeword for each given input symbol. The algorithm runs in O($n$) time in the worst case where the corresponding Huffman tree is a one-sided tree that occurs rarely. When the corresponding tree is nearly balanced, which is the case that happens more often in practice, the algorithm runs in $O(\log(\log n - 1)!)$. The proposed algorithm is not cost optimal, so our future research would focus on optimizing the time complexity and also the number of required processors towards devising a cost optimal parallel algorithm.

## 6. REFERENCES


[1] M. J. Atallah, S. R. Kosaraju, L. L. Larmore, G. L. Miller and S-H. Teng, "Constructing trees in parallel", *ACM SIGACT, Proc. 1st Annual ACM Symposium on Parallel Algorithms and Architectures*, June 1989, pp. 421-[illegible]

[2] M. Buro, "On the maximum length of Huffman codes", *Information Processing Letters*, 45(5), April 1993, pp. 219-223.

[3] H. C. Chen, Y. L. Wang and Y. F. Lan, "A memory efficient and fast Huffman decoding algorithm", *Information Processing Letters*, 69(1999), pp. 119-122.

[4] T. J. Fexguson and J. H. Rabinowitz, "Self-synchronizing Huffman codes", *IEEE Trans. Inform. Theory*, vol. IT-30, July 1984, pp. 687-693.

[5] R. Hashemian, "Memory Efficient and High Speed search Huffman codes", *IEEE Trans. On Comm.*, Vol. 43, No. 10, Oct. 1995.

[6] R. Hashemian, "Direct Huffman coding and decoding using the table of code-lengths", *Proc. International Conf. on Inform. Technology, Computers and Communications* (ITCC'03), 2003.

[7] S. Ho and P. Law, "Efficient hardware decoding method for modified Huffman code", *Electronics Letters*, vol. 27, no. 10, May 1991, pp. 855-856.

[8] D. A. Huffman, "A method for the construction of minimum redundancy codes", *Proc. IRE*, vol. 40, pp. 1098-1101, Sept. 1952.

[9] S. T. Klein, "Skeleton trees for the efficient decoding of Huffman coded text", *Kluwer Journal of Inform. Retrieval*, 3 (2000), pp. 7-23.

[10] D. E. Knuth, "Dynamic Huffman coding", *Journal of Algorithms*, 6(1985), pp. 163-180.

[11] D. G. Kirkpatrick and T. M. Przytycka, "Parallel construction of near optimal binary search trees", *ACM SIGACT, Proc. 2nd Annual ACM Symp. on Parallel Algorithms and Architectures*, July 1990, pp. 234-243.

[12] C. Kruskal. "Searching, merging and sorting in parallel computations", *IEEE Trans. Comput.*, pp. 942-946, 1983.

[13] L. L. Larmore, "Height restricted optimal binary trees", *SIAM Journal on Computing*, 16(1987), pp. 1115-1123.

[14] Y. Lin, K. Chung, "A space-efficient Huffman decoding algorithm and its parallelism", *Journal of Theoretical Computer Science*, Vol. 246 (2000), pp. 227-238.

[15] L. L. Larmore and D. S. Hirschberg, "A fast algorithm for optimal length-limited Huffman codes", *Journal of ACM*, 37(1990), pp. 464-473.

[16] L. L. Larmore and T. M. Przytycka, "Constructing Huffman trees in parallel", *SIAM Journal on Computing*, 24(6), December 1995, pp. 1163-1169.

[17] A. Moffat and J. Katajainen, "In-place calculation of minimum-redundancy codes", *4th Intl. Workshop, Algorithms and Data Structures*, vol. 955, Aug. 1995, pp. 393-402.

[18] R. L. Milidiú, E. S. Laber and A. A. Pessoa, "A work efficient parallel algorithm for constructing Huffman codes", DCC '99, Proc. Data Compression Conference, Mar. 1999.

[19] A. Moffat and A. Turpin, "On the implementation of the minimum redundancy prefix codes", *IEEE Trans. Commun.* 45, 1997, pp. 1200-1207.

[20] S. A. Ostadzadeh, M. A. Moulavi, Z. Zeinalpour, "Massive Concurrent Deletion of Keys in B*-tree", Technical Report, Azad University of Mashhad, Mashhad, Iran, Sep. 2005.

[21] S-H. Teng, "The construction of Huffman-equivalent prefix code in NC", *ACM SIGACT*, 18(4), 1987, pp. 54-61.

[22] J. Van Leeuwen, "On the construction of Huffman trees", *3rd International Colloquium on Automata, Languages and Programming*, Jul. 1976, pp. 382-410.

# Semantic Description of Multimedia Content Adaptation Web services

Surafel Lemma
Department of Computer Science
Faculty of Informatics
Addis Ababa University
lsurafel@yahoo.com

Dawit Bekele
Department of Computer Science
Faculty of Informatics
Addis Ababa University
dawit@math.aau.edu.et

Girma Berhe
Lab. d'InfoRmatique en Images et Systèmes d'information (LIRIS)
INSA de LYON
girma.berhe@insa-lyon.fr

*Abstract-* **Multimedia content adaptation can be performed by third party services that can be implemented using web services. In order to use these web services, one has to be able to find them on the Web. However, today, finding the right web service for a specific task (multimedia content adaptation in our case) is difficult. This is mainly because of the type of description used to advertise the web services.**

**Current standards that are used to describe web services are syntactic. Syntactic description of web services contains some keywords and signature of the web service, which is later, used for searching the services. However, this information does not allow high rate of success during service discovery. These limitations of the current standards can be improved by using semantic description methods. Semantic description gives the meanings and relationships of the terms and concepts used in describing a web service, which improves its discovery. In order to achieve this, we have developed an ontology for multimedia content adaptation web services.**

## I. Introduction

Pervasive computing, also called ubiquitous computing, is a technological evolution heading to allow computation from anywhere and anytime on various computing devices. This technology is also sometimes referred as the third wave of computing (many computers per person). The idea of this evolution is first conceived in Xerox PARK laboratory by Mark Weiser in 1988 [4].

In pervasive computing, the electronic devices, which are used to access information, vary from small devices like PDAs to much more powerful computational devices such as personal computers. Due to this diversity of display devices, a major challenge of pervasive computing is to be able to display any content that a user may want to access in a form that can be displayed by his/her device. For example, the content may have an image that is too big to display on the small PDA of the user. In order to display the content on the user's device, its presentation can be modified to fit on the device without loosing its meaning. For Example, the size of the image can be reduced. This process is called content adaptation.

Current researches propose the following three alternatives with regards to where the content adaptation should be performed: by the client machine at arrival, by the server before it is sent, or by the proxy on the way. A newly proposed approach to adapt the content of the information is using third party adaptation services. In this approach, the adaptation services can be implemented using web services [1].

In frameworks that use the new approach, if a client receives a content that needs to be adapted, it searches for a web service that can perform the specified content adaptation in a web service registry. To search for web services the client uses the advertised (published) description of the web services.

Currently, the most common web service description is the syntactic description. The service is described by giving its signature (name, parameter, data types, etc.) and keywords that identify it. To use the web service, one has to discover it by providing some of this information. This way of describing services has some major shortcomings. In particular, it requires knowledge of the terminology that is used to describe the services in order to search and find it. For example, the service may not be found just because the right keywords are not used for the search. The above limitation of syntactic description can be alleviated by using semantic descriptions. This is because in this case, we describe what the web services do using standard semantics description system, which gives a better success rate for the searches. Hence, semantic descriptions are increasingly considered as better alternatives [6].

The objective of this research is to develop an adequate service description for multimedia content adaptation web services that enhances lookup of these services. We have preferred to use semantic description method that we consider more efficient in this context.

## II. Motivation

In a pervasive environment, the devices which are used to view multimedia information have different display sizes and different capabilities such as processing power, storage size, etc. The bandwidth of the network that connects the devices also varies greatly. Therefore, in order to access multimedia information using these devices, the content has to be modified in order to meet the device's requirement(s) and the limitations of the network. For example, if a user wants to see an image compressed using JPEG (Joint Picture Experts Group) with a device that is capable of displaying only images compressed using GIF (Graphics Interchange Format), the image has to be modified so as to meet the device's requirement. Similarly, if a user wants to read in French a document written in English, the content has to be converted according to the user's requirement. If a user also wants to hear a radio broadcast over the Internet and is in an area where the bandwidth of the network is very small, then the audio has



*T. Sobh and K. Elleithy (eds.), Advances in Systems, Computing Sciences and Software Engineering*, 25–31.

to be adapted so as to meet the networks requirement. These can be achieved using adaptation techniques which can be implemented using web services.

Web services are well-defined, reusable software components that perform specific, encapsulated tasks via standardized web-oriented mechanisms [8]. The provider of the web service has to make it available on the web and publish (advertise) it so that users can find it. To advertise a web service, different frameworks such as the service-based distributed content adaptation framework described in [1, 2] use UDDI (Universal Description, Discovery and Integration) or UDDI like registries.

These registries however have limitations in service discovery. They allow lookup of services only using keywords such as provider name, service name, or service category. In addition, they use direct match to identify web services [12, 14]. Hence, these make searching a service difficult as the user has to remember names or keywords of all web services advertised and choose the right one.

Web service interfaces are described using WSDL (Web Service Description Language) which is a W3C[1] recommended language [17]. The description contains information about the operation supported by the service, the kind of data or message being communicated, binding, etc.

Using elements of WSDL, it is possible to identify what type of messages or data types are exchanged by the web service, how it is technically implemented i.e. how it is bind to a specific communication protocols like SOAP (Simple Object Access Protocol), and where it is located. However, this syntactic information is not adequate for automatic web service discovery and composition [2]. Web service discovery and composition require semantic information like type of the web service, the meaning of the input/output, how much it costs, or its execution time etc. Due to these, web service discovery and composition require human intervention.

For example, the description shown in Fig. 1 does not tell anything to software agents or other services that are looking for such web services, about the parameters except that they are of type string which can be used to contain information. For software agents or other services the names of the parameters do not give a meaning as they may do to humans. Consequently, it is difficult to identify the web service they need by using the information in the WSDL description.

```
<message name="TextToSpeechFromURLHttpPostIn">
  <part name="fileURL" type="s:string" />
  <part name="outputFormat" type="s:string" />
  <part name="language" type="s:string" />
</message>
```

Fig. 1: Part of a WSDL description of text to speech translation web service

[1] W3C (World Wide Web Consortium) is a group that develops interoperable technologies (specifications, guidelines, software, and tools). http://www.w3.org/

In this paper, we try to address the above problems by developing an ontology for semantic description of multimedia content adaptation web services. This ontology will serve as a means to annotate and share the semantic information, such as the service performed by the web service, its inputs/outputs, etc.

## III. Semantic Description of Web Services

Different approaches are developed to semantically describe web services [10, 11]. The major once are OWL-S (Semantic Markup for Web Services), IRS (Internet Reasoning Service) and WSMF (Web Service Modeling Framework).

IRS has strong user and application integration support but it does not have semantic description of composed services. WSMF is an effort towards developing a comprehensive conceptual architecture, which covers requirements of e-commerce. Compared to these two, OWL-S has a richer service ontology which builds on the semantic web stack of standards. In addition, the service ontology of OWL-S is public.

As OWL-S has richer service ontology that provides a good basis to our work, we will discuss it in greater depth in the next sub-section.

### A. *Semantic markup for web services*

OWL-S is an ontology consisting of a set of basic classes and properties for declaring and describing services. This ontology tells “What a service provides”, “How the service is used” and “How one interacts with it” using its three sub-ontologies (see Fig. 2) “ServiceProfile”, “ServiceModel”, and “ServiceGrounding” respectively [9].

Of the three sub-ontologies, *Service Profile* sub-ontology helps to present the information required by the service provider to publish a service. It is also used by a service requester to discover a service. Since we are interested in semantic description of web services that is used for web service advertisement and discovery, we will focus on this sub-ontology.

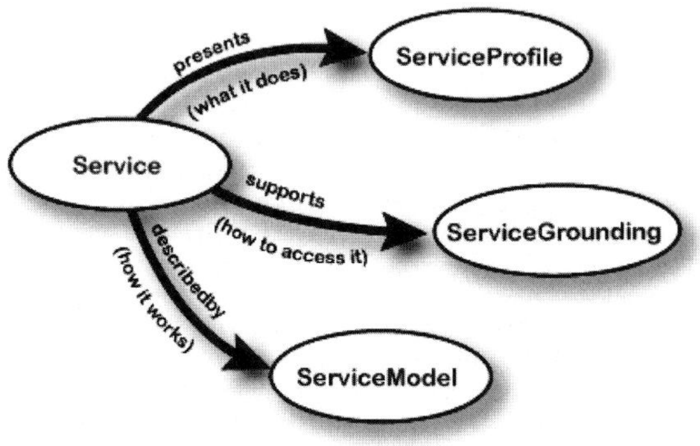


Fig. 2[2]: Sub-ontologies of OWL-S

[2] ***Source:*** OWL-S: Semantic markup for web services [9]

The service profile does not mandate any representation of services; but using the OWL sub-classing, it is possible to create specialized representations of services that can be used as service profiles. OWL-S provides one possible representation through the class *Profile* [9]. The service profile ontology specifies web service descriptions based on their provider, functional and non functional descriptions. The provider information includes the name of the service being offered, a summarized text description on what the service offers, what the service requires to work and any additional information that the compiler of the profile wants to share with the receiver. It also includes contact information of the owner of the service or the customer representative that may provide additional information about the service.

The non-functional description presents information on the characteristics of a service. It incorporates information on the category of a given service, and an unbounded list of service parameters that can contain any type of information about the service like quality guarantee provided by the service.

The functional information has two aspects: the information transformation and the state change. The information transformation is represented by the inputs and outputs. It tells what the web service takes and what it returns to the caller. The state change is represented by preconditions and the effects. The precondition tells what has to be satisfied before invoking the web service in order to get the desired output. The effect tells the change that is going to happen as a result of a service execution. This functional information about the web service (i.e. inputs, outputs, preconditions and effects) is presented using the service profile sub-ontology. However, this sub-ontology does not provide a schema to describe the instances of this information. Instead it defines how the information can be linked to the service model [9].

The service profile sub-ontology is designed to describe and advertise web services' properties and capabilities. However, this representation does not provide a detailed description of the functionalities of web services which can be used as basic criteria for the multimedia content adaptation web services lookup.

## IV. Proposed Semantic Description of Multimedia Content Adaptation Web Services

The discovery of the "right" web service for a specific purpose basically involves the matching of the properties of the required service with the properties of advertised services and/or the matching of functionality of the required service with the advertised service. Hence, advertising web services by using both their functionality and properties is an advantage. In this section, we present the proposed ontology that can be used to advertise and lookup audio content adaptation web services by using their functionality and properties.

### A. *Functionalities of audio content adaptation web services*

In the proposed ontology, functionalities of web services are represented as classes. To identify functionalities of audio content adaptation web services, we have used the techniques that are used to adapt the content. These techniques are listed in various papers [1, 2, 3, 5, 7, 13, 15] based on which, we have identified possible audio content adaptation web services as exhaustively as possible, as shown in table I.

TABLE I
Types of Audio Content Adaptation Web Services

| Name | Description |
|---|---|
| *AudioReductionAdaptationServices* | Services that reduce the bit rate of an audio file |
| *AudioSamplingRateReductionAdaptationServices* | Services that reduce the rate at which an audio file is sampled. |
| *StereoToMonoAdaptaionServices* | Services that convert an audio which has two channels to a single channel which in effect reduces the bandwidth required to transmit the audio. |
| [illegible] | Services that change the format of an audio file. For example, an audio which is in WAV format can be converted to an mp3 format using a [illegible] format transcoding techniques |
| *AudioLanguageTranslationAdaptaionServices* | These services translate an audio in one language to another language. For example, an audio file which contains a speech made about pervasive computing in English language can be translated to a speech in French language using a service which implements language translation technique. |
| *AudioMediaTranslationAdaptationServices* | Using a service that implements translation adaptation technique an audio file is translated to another media element like text or animation. |
| *AudioToTextAdaptatinServices* | These services are specific types of *AudioMediaTranslationAdaptationServices*. They convert an audio file to a text document. |

TABLE II
CONCEPTS OF MULTIMEDIA DATA TYPES

| Name | Description |
|---|---|
| *MultimediaDatatype* | This concept is used to represent different multimedia data types: audio, video, image, text, graphics or animation. |
| *ImageDatatypeFormat* | It is used to represent image data type formats. |
| *AudioDatatypeFormat* | It is used to represent audio data type formats. |
| *VideoDatatypeFormat* | It represents the format of a video data type. |
| *TextDatatypeFormat* | It represents the format of a text data type. |
| *GraphicsDatatypeFormat* | It is used to represent the format of graphics data type. |
| *AnimationDatatypeFormat* | It is used to represent the format of animation data type. |

In addition to the above concepts that are used to represent the functionalities of audio content adaptation web services, we need to identify concepts of multimedia data types. The multimedia data type concepts help to represent the input and/or output of the adaptation web services in a standard way. These concepts are listed in table II.

The functionalities listed above can be classified based on different perspectives of the adaptation techniques used to list the services. These classifications are presented in [3, 7].

The taxonomy of audio adaptation services will be developed using some of these broad classifications as its base. One of the classification categorize the adaptation techniques as static or dynamic based on when they are used to generate the adapted content.

*Static adaptation techniques*: are techniques used to generate versions that differ in quality and processing requirement. The techniques which are categorized here always produce the same result and use the same amount of resource for the same input. The method which they use to adapt the content is always fixed, irrespective of the resource available at the given time. For example, if they have to sample an audio file in order to adapt it, the interval with which they take the samples is always the same, i.e. it does not change with time.

*Dynamic adaptation techniques*: are techniques that are used to generate a content that meets the user and/or device constraints on-the-fly. Depending on the resource available, adaptation techniques under this category use various techniques to adapt the content. For example, if the multimedia file has to be sampled in order to be adapted, then the service which implements this technique may sample too many or very few samples depending on the resource available to process the content of the multimedia file.

Another classification categorizes adaptation techniques as unimodal or multimodal based on the number of media types involved during the adaptation process.

*Unimodal adaptation techniques*: This category incorporates techniques used to create different versions of a media element with different qualities, formats and resource requirements. For example, techniques that are used to convert an audio from one format, WMA (Windows Media Audio) file to another, mp3 are considered as unimodal adaptation techniques.

*Multimodal adaptation techniques*: This category incorporates techniques that are used to convert one media type to another so that the converted content meets the particular device specification. For example, audio to text conversion.

Using the concepts identified above, we have developed our ontology which is represented using OWL (Web Ontology Language). These concepts are represented as classes in the ontology, which can be arranged in a hierarchy that is useful for service discovery (see Fig. 3). This hierarchy will be used with the service properties to give a complete ontology of services.

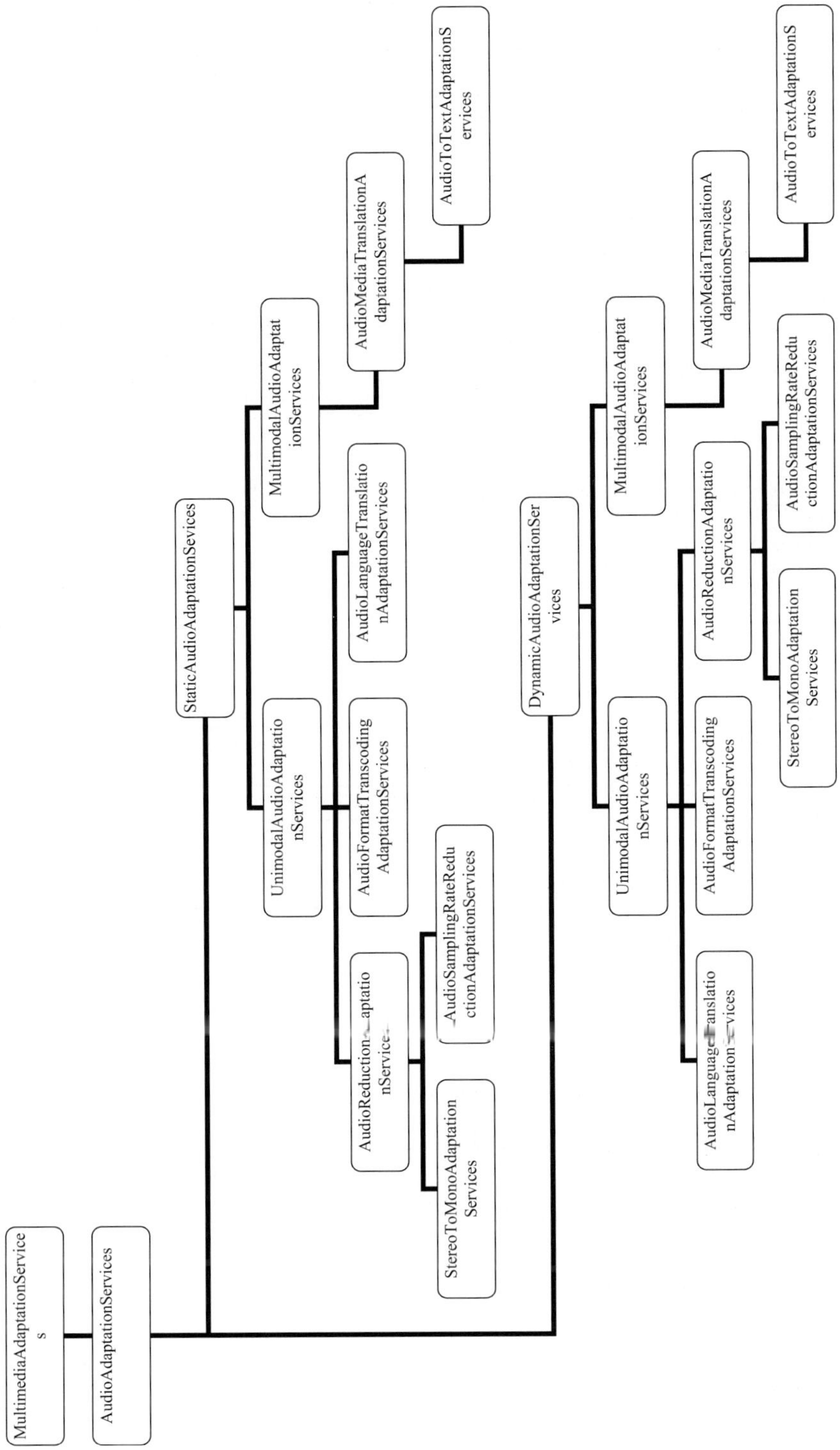


Fig. 3: Hierarchy of concepts and terms

B. *Properties of audio content adaptation web services*

For audio adaptation web services we have identified the following properties, most of which are similar to those of OWL-S profile class. These properties, shown in table III, are general to all multimedia adaptation web services. In order to increase the extensibility of this ontology to other media elements, we have put these common properties to be properties of the general class *MultimediaAdaptationSevices*.

TABLE III
PROPERTIES OF MULTIMEDIA ADAPTATION WEB SERVICES

| **MultimediaAdaptationServices** | | |
|---|---|---|
| **Property name** | **Value** | **Description** |
| adaptationServiceName | String | This property holds the actual name of the service |
| adaptationTime | Duration | This property holds the time or duration (in seconds) that the service takes in order to do the adaptation. |
| adaptedMediaQuality | String | This property takes one of the five subjective quality rating scales: Excellent, Good, Fair, Poor, or Bad (ITU-T recommendations stated in [16]). |
| hasEffect | Instance of a class | This property holds a value which is an instance of a class. The class varies depending on the web service. For example, if the effect of the service is to change the language of an audio file, the class can be from an external language ontology which describes the language may be based on ISO639 (i.e. two or three letter codes)[3] or if the effect of the service is to change the format of the service then the class can be an instance of the *MultimediaDataTypeFormat*. |
| hasInputType | Instance of MultimediaDatatype class | The value of this property is an instance of the *MultimediaDatatype* class. In all of the sub-classes described in this ontology the instance value for this property is audio. |
| hasOutputType | Instance of MultimediaDatatype class | The value of this property is also an instance of the *MultimediaDatatype* class. |
| hasServicePrice | Instance of a class | This property is a property which can be used to link the service with an external price ontology's class. |
| hasPrecondition | Instance of a class | This property holds an instance of a class that can be used to describe the prerequisite of the service like format or language of the audio. |
| hasServiceProvider | Instance of a class | This property is a property that can be used to give information about the service provider. It can be used as a link to external ontologies like the Actor ontology[4] that is used to record address and other information of the provider. |
| serviceDescription | String | This parameter holds any description about the service. |

TABLE IV
PROPERTIES OF MULTIMEDIA DATA TYPE FORMAT CLASS

| **MultimediaDatatypeFormat** | | |
|---|---|---|
| **Property name** | **Value** | **Description** |
| hasDatatype | Instance of the *MultimediaDatatype* class | This property is an instance of the class *MultimediaDatatype*. It is used to link the data type format with one of the instances of the multimedia data types. |
| hasFormatName | String | This is a property which is used to hold the name of the format type such as mp3 or wav for audio data type. |
| additionalInformation | String | This is a property which is used to hold additional information that the service describer wants to incorporate with the definition of the format for the specified multimedia data type. |

[3] http://www.loc.gov/standards/iso639-2/

[4] http://www.daml.org/services/owl-s/1.1/ActorDefault.owl

All types of multimedia data type formats have similar properties. Therefore, we have put all these properties (Table IV) to be properties of the general class *MultimediaDatatypeFormat*.

## V. Discussion

We have presented an ontology for audio content adaptation web services. This ontology allows audio content adaptation web services to advertise themselves by specifying the type of functionality they give and providing additional information, such as an input they take, through their properties. This allows a better lookup success since the service seeker can look for services by using the properties of the service and/or the functionality of this ontology. By specifying the functionality, the service seeker, narrows down the domain in which the service is going to be searched. The service seeker can also be more specific by providing information about the required service's arguments.

We have also presented a multimedia data type ontology that can be used to standardize the parameters of the web services. In addition to the names of the data types, this ontology provides a means to describe the formats of the data types. This helps to precisely advertise and look for a specific web service and hence improves the lookup success.

## VI. Conclusion and Future Works

In this paper, we have explored current standards and ontologies that can be used to describe web services, and shown their limitations. These limitations arise mainly due to the fact that web services are described using syntactic descriptions. Syntactic description does not tell the meanings of the words used in describing a web service and their relationship which are important information for web service discovery. In addition, they allow lookup of web services only using keywords which make web service discovery a very difficult task for the user.

To improve these limitations that are especially related to multimedia content adaptation web services, we have developed an ontology that can be used to describe audio content adaptation web services semantically. This ontology allows to describe both functional and non functional properties of audio content adaptation web services. This helps to easily describe these adaptation web services. It has also incorporated a taxonomy that can be used to facilitate searching of the web services. Moreover, it proposes a multimedia data type ontology that can be used to standardize parameters of the web services.

To semantically describe web service, we have proposed an ontology for audio content adaptation web services. This ontology can be extended to other multimedia content adaptation web services.

The ontology is developed mainly for description of audio content adaptation web services, more specifically for web service lookup. It does not provide information on how one uses the service or interacts with it. Therefore, it needs to be extended or linked to OWL-S's service model and/or service grounding to incorporate this information.

## References


[ 1 ]. Girma, B., Brunie, L. and Pierson, J-M., "Modeling Service-Based Multimedia Content Adaptation in Pervasive Computing", ACM Proceedings of the first conference on computing frontiers (CF'04), pp. 60-64, Ischia, Italy, April 14-16, 2004.

[ 2 ]. Girma, B., Brunie, L. and Pierson, J-M., "Realization of Distributed Content Adaptation with Service-Based Approach for Pervasive Systems", Unpublished, INSA de LYON.

[ 3 ]. Lei, Z. and Georganas, N. D., "Context-Based Media Adaptation in Pervasive Computing", Proc. ACM Multimedia 2001 Doctoral Symposium, Ottawa, September 2001.

[ 4 ]. Meaning of Ubiquitous Computing. (URL: http://www.hyperdictionary.com).

[ 5 ]. Mohan, R., Smith, J. R. and Li, C. S., "Adapting Multimedia Internet Content for Universal Access", IEEE Transactions on Multimedia, Vol. 1, No. 1, pp. 104-114, 1999.

[ 6 ]. Moreau, L., Miles, S., Papay, J., Decker, K. and Payne, T., "Publishing Semantic Description of Services", Semantic Grid Workshop at GGF9, 2005.

[ 7 ]. Mulugeta, L., "Metadata Supported Content Adaptation in Distributed Systems", Ph. D. Thesis, University of Klagenfurt, Austria, June 2004.

[ 8 ]. Ollermann, W. L., "Architecting Web Services", Springer-Verlag New York, Inc., 175 Fifth Avenue, New York, NY, 2001.

[ 9 ]. Paolucci, M., et al., "OWL-S: Semantic Markup for Web Services", W3C Member Submission, November 2004. (URL: http://www.w3.org/Submission/2004/subm-owl-s-20041122/).

[ 10 ]. Payne, T., Bandara, A., Roure, D. and Clemo, G., "An Ontological Framework for Semantic Description of Devices", 3rd International Semantic Web Services (ISWC 2004), Hiroshima, Japan, November 7-11, 2004.

[ 11 ]. Payne, T., Cabral, L., Domingue, J., Motta, E. and Hakimpour, F., "Approaches to Semantic Web Services: An Overview and Comparisons", In proceedings of the First European Semantic Web Symposium (ESWS 2004), Herakilon, Crete, Greece, May 10-12, 2004.

[ 12 ]. Roeder, S., Goodwin, R., Doshi, P. and, Akkiraju, R., "A method for semantically Enhancing the Service Discovery capabilities of UDDI", In proceedings of the Workshop on Information Integration on the Web (IJCAI 2003), Mexico, August 9-10, 2003.

[ 13 ]. Shaha, N., Desai, A. and Parashar, M., "Multimedia Content Adaptation for QoS Management over Heterogeneous Networks", Proc. of International Conference on Internet Computing 2001, pp. 642-648, Computer Science Research Education and Applications (CSREA) press, June 2001.

[ 14 ]. Siberski, W., Thaden, U., and Nejdl, W., "A Semantic Web Base Peer-to-Peer Service Registry Network", Technical Report, Learning Lab Lower Saxony, University of Hannover, Germany, February 17, 2003.

[ 15 ]. Smith, J. R., Mohan R. and Li, C. -S., "Transcoding Internet Content for Heterogeneous Client Devices", Proc. IEEE Int. Conf. on Circuits and Syst. (ISCAS), May 1998.

[ 16 ]. Voelcker, R. M., Hollier, M. P., Rimell, A. N. and Hands, D. S., "Multi-modal Perception", BT Technol J, Vol 17, No 1, January 1999.

[ 17 ]. Weerawarana, S., Christensen, E., Curbera, F. and Meredith, G., "Web Services Description Language (WSDL) 1.1", W3C Note, March 15, 2001. (URL: http://www.w3.org/TR/2001/NOTE-wsdl-20010315).

# Grid Computing Communication Strategies for Cross Cluster Job Execution

Jens Mache, Chris Allick, André Pinter, Damon Tyman
*Lewis & Clark College*
*Portland, OR 97219 USA*
*{jmache, callick, apinter, dtyman} @lclark.edu*

## Abstract

*With the advent of grid computing, some jobs are executed across several clusters. In this paper, we show that (1) cross cluster job execution can suffer from network contention on the inter-cluster links and (2) how to reduce traffic on the inter-cluster links. With our new all-to-all communication strategies and host file mappings, it is possible to reduce the volume of data between two clusters from (n/2)^2 to n/2, for a job with size n.*

*Our measurements confirm this reduction in data volume on the inter-cluster link and show a reduction in runtime. With these improvements it is possible to increase the size of grids without significantly compromising performance.*

**Keywords:** Grid Computing, Cluster Integration, All-to-All Communication, Network Contention, Optimization, Performance Evaluation.

## 1. Introduction

There is considerable excitement about cluster integration and grid computing. *Grid computing* uses the Internet to allow sharing of computational and data resources among geographically dispersed users within and across institutional boundaries [2].

Several geographically distinct clusters can be linked together to share resources and act as one cohesive, more powerful unit. It is not uncommon to run jobs across several clusters, examples include [1, 3]. However, a job run across several clusters does not have as much bisection bandwidth available as a job run within a single cluster (even if the inter-cluster links have the same bandwidth as the intra-cluster links). Using vanilla all-to-all communication strategies (that do not take the grid topology into consideration), there is considerable data volume on the inter-cluster links. This can cause network contention and performance bottlenecks. In this paper, we show how to decrease the data volume that travels over the inter-cluster links.

This paper is organized as follows: In Section 2, we analyze network contention on the inter-cluster links. In Section 3, we describe and analyze communication strategies. In Section 4, we consider host file mappings. In Section 5, we report on performance measurements. After a discussion in Section 6, we conclude in Section 7.

## 2. Performance Degradation due to Contention on Inter-Cluster Links

In Figure 1, the typical topology of two integrated clusters is shown.

**Figure 1: Grid topology.**

In our earlier work [5] we ran NAS Parallel Benchmarks and saw significant performance differences when jobs were run within one cluster versus run across two clusters (same total number of nodes). Communication performance was limited by the bandwidth available on the link between the clusters. How can this be explained?

We assume All-to-All (non-personalized) communication (called All_Gather in MPI [6]) and each node passing its data directly to the other n-1 nodes. Intuitively, the communication graph has to be embedded into the network topology. Figure 2 shows the demand for passing messages across the link between the clusters. This demand is known as link-contention [4], which is defined as two different edges of the communication graph are mapped to the same edge of the physical topology.



*T. Sobh and K. Elleithy (eds.), Advances in Systems, Computing Sciences and Software Engineering*, 33–38.

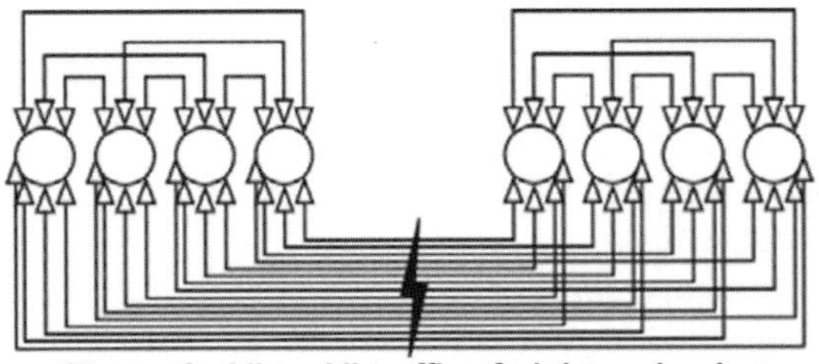

Figure 2: All-to-All traffic of eight nodes in two clusters: 16 out of 28 message pairs contend for the inter-cluster link.

The total number of messages passed across the inter-cluster link is *(n/2)^2* per direction (half duplex). Observing traffic that travels from the left cluster to the right cluster, n/2 nodes on the left side are sending to n/2 nodes on the right side. Reducing traffic on the inter-cluster link was the impetus for this paper.

## 3. Communication Strategies

We first assume two clusters of equal size. In section 6, we undo those assumptions.

The algorithms presented here (aside from simple All-to-All) are aimed at improving performance for parallel applications run across multiple clusters. We tested these algorithms on our own mini-grid and have presented the data in Section 5. Whenever '*n*' is referenced, it refers to the job size (the number of nodes used).

### 3.1 Simple All-to-All

Simple All-to-All ($ATA_{simple}$) is what we call above mentioned strategy of each node passing its data directly to the other *n-1* nodes.

Figure 3 shows how the data of the first node is sent. All the other nodes send their data in the same manner.

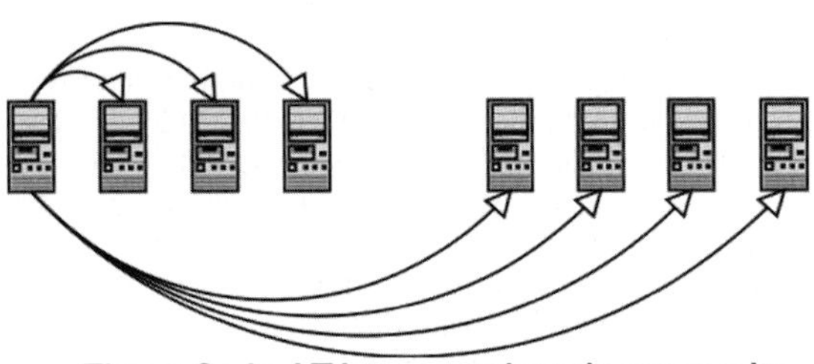

Figure 3: In $ATA_{simple}$, each node passes its data directly to all the other nodes.

Since n/2 nodes on one side of the inter-cluster link are sending messages to the n/2 nodes on the other cluster, (n/2)^2 communications must cross the inter-cluster link per direction. The link contention of $ATA_{simple}$ is shown in Figure 2, above. $ATA_{simple}$ does not take topology into consideration, and thus was used as a basis of comparison for the other algorithms.

### 3.2 Optimized All-to-All

The Optimized All-to-All ($ATA_{optimized}$) strategy is designed to minimize link contention (overlap) of traffic across the inter-cluster link. This algorithm operates by each node having a "buddy" in the other cluster.

The cluster on the left is side A and the cluster on the right is side B. Node0A will pass its message to Node0B. Once Node0B receives the message from Node0A, Node0B will pass the message to the other nodes in its cluster, see Figure 4. Similarly, Node0A receives Node0B's message and forwards it to all the other nodes in cluster A. This same process occurs for the remaining *n/2-1* "buddy" pairs. Since each node sends only one message across the inter-cluster link, only n/2 messages are sent in any one direction, see Figure 5.

The key point to take from $ATA_{optimized}$ is that data is sent across the inter-cluster link as few times as possible, and then distributed locally within the cluster. Once the buddy gets the data, it helps distributing the data (in its cluster). This "helping" behavior has similarities to (1) peer-to-peer computing and (2) broadcast trees.

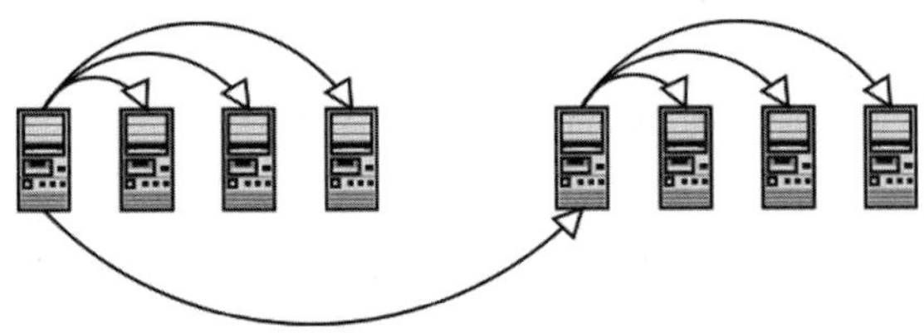

Figure 4: In $ATA_{optimized}$, data is sent across the inter-cluster link as few times as possible, and then distributed locally within the cluster.

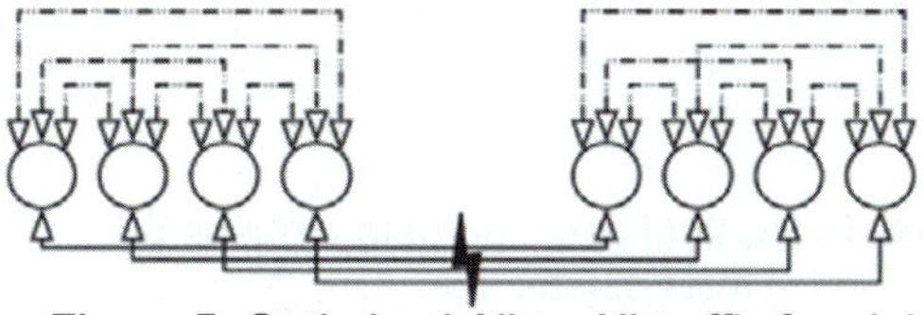

**Figure 5: Optimized All-to-All traffic for eight nodes and two clusters: only four message pairs contend for the inter-cluster link.**

### 3.3 N-Body Communication

N-Body communication is an alternative to $ATA_{simple}$. The N-Body strategy was originally designed by physicists looking for increased performance when all nodes have to exchange their data with all the other nodes. To understand the N-Body communication, envision that all nodes are arranged in a circle. Each node first sends its original data to the node on the right, while receiving the original data from the node on the left. In the next iteration, the just received data is passed on. After n/2-1 iterations, each node has data from half the total nodes.

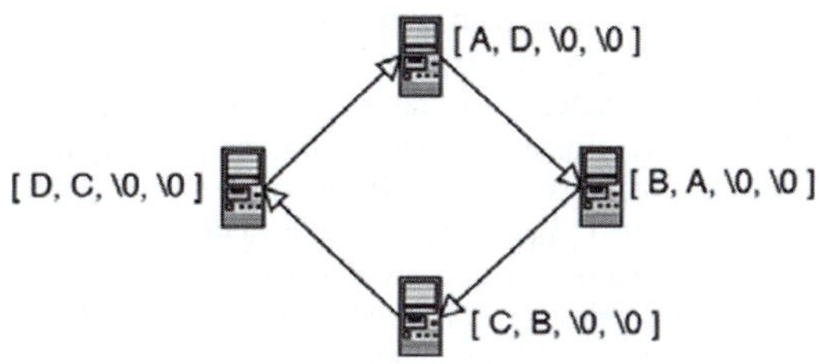


**After the first pass, each node contains the data from one neighbor in addition to their own.**

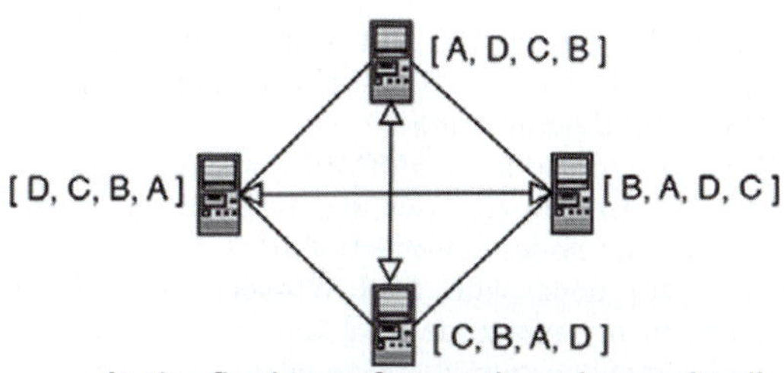


**In the final transfer, each node sends all data so far received to the node directly across the circle and from that node receives all data from the other side of the circle.**

**Figure 6: Step-by-step illustration of how the passing works with N-body.**

A final data exchange then takes place where each node sends n/2 pieces of information (all of the data it has so far collected) to the node directly across the circle, see Figure 6. Prior to the final data pass each node already had *n/2* pieces of information. The node opposite any particular node in the circle had the missing *n/2* pieces of information. Thus, all nodes have the necessary data.

The big difference is that whereas $ATA_{simple}$ sends 1 piece of information to all other n-1 nodes, N-Body only sends to two other nodes (n/2-1 and n/2 pieces of information, respectively).

Since the way in which the nodes from different clusters are assigned positions in the circle is important to how N-Body performs, the various formulas for message overlap under different mapping conditions are discussed next.

## 4. Host File Mapping Considerations

The performance of the communication algorithms used was drastically affected by the way in which the available compute nodes were assigned positions in the host file.

**Table 1: Three possible host file configurations**

| Config 1 | Config 2 | Config 3 |
|---|---|---|
| Red0 | Red0 | Red0 |
| Red1 | Green0 | Red1 |
| Red2 | Red1 | Green0 |
| Red3 | Green1 | Green1 |
| Green0 | Red2 | Red2 |
| Green1 | Green2 | Red3 |
| Green2 | Red3 | Green2 |
| Green3 | Green3 | Green3 |

We first examine how different node assignments for N-Body affect the amount of data that has to travel between clusters. Since in the final pass, each node has to send *n/2* pieces of information half way across the circle (all the data that node has so far received), preventing this traffic from crossing the inter-cluster link is the most important consideration. With Configuration 1 (see Table 1), where all the nodes in one cluster are listed first followed by all the nodes of the second cluster, this is not achieved. Instead, it limits the traffic that has to cross the inter-cluster link in the first *n/2-1* passes. There are only two places in the circle where a node's neighbor is not a member of the same cluster. Since the traffic in these two cases occurs in opposite directions, they do not conflict.

Thus, before the final pass, only *n/2-1* messages have to cross the bottleneck link in one direction. In the final pass though, all pieces of information must be sent from one cluster to another. With *n/2* nodes each sending *n/2* pieces of information, this results in an additional *(n/2)^2* messages crossing the inter-cluster link in the same direction. The total link contention on the inter-cluster link is <u>*(n/2)^2 + n/2-1*</u>.

Using a configuration where the nodes are listed by alternating cluster (such as Configuration 2) with N-Body does alleviate the strain on the cross-cluster link in the final pass, but increases the strain in the initial passing phase. This time, each node's neighbor is on a different cluster, but their partner in the final pass is not. This means the only messages that can tax the bottleneck come before the last data exchange. With half of the total nodes sending each message of the first phase in the same direction, this results in *(n/2-1)*n/2 = (n^2-2n)/4* messages that have to be carried over to the other cluster.

While this is an improvement over using Configuration 1, alternating cluster halves (Configuration 3) is better. Any node at position j in a host file of size n=4i will have a partner in the final pass at position j+(n/2) mod n, which is always a member of the same cluster. This means no data in the final sharing will have to cross the hot spot, similar to Configuration 2. Moreover, the only time information has to travel between clusters is during the initial n/2-1 passes, but with Configuration 3 only four nodes have neighbors that are not on the same cluster. Since only two go in the same direction, the total link contention on the inter-cluster link is 2(n/2-1) = <u>n – 2</u>.

The following figures illustrate N-Body using a host file that interleaves cluster halves.

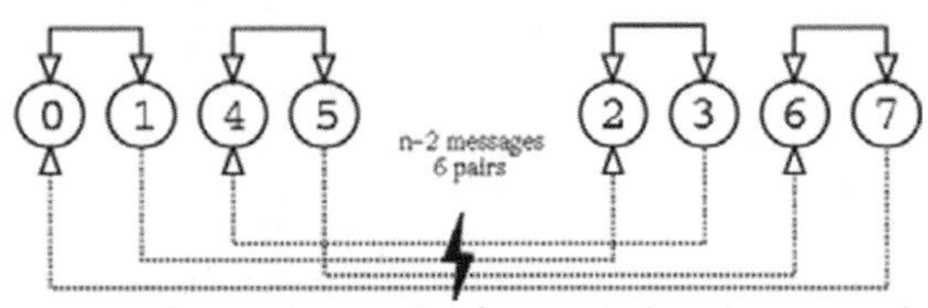


**Figure 7: Illustrates the first *n/2-1* exchanges of the N-Body algorithm with an optimal configuration. When n=8, there are three such passes, each with two overlaps resulting in a total of 6 messages crossing the hot spot.**

**Figure 8: In the final pass, no data crosses the inter-cluster link.**

In $ATA_{simple}$, the communication is symmetric. Thus, host file mapping does not matter.

That is different in $ATA_{optimized}$. The $ATA_{optimized}$ code was written assuming that the buddy is n/2 nodes away. This assumes that first half of the host file belongs in one cluster and the second half belongs in the other cluster (Configuration 1). If the order that the nodes are listed in is changed, link contention increases.

## 5. Experiments

### 5.1 Experimental Setup

Having designed $ATA_{optimized}$ and implemented all algorithms as MPI programs, we sought to test the actual performance of each with the hope of verifying our analysis. We ran our MPI programs with two small homogeneous clusters. All tests utilized four nodes, two from each cluster. Nodes communicated within their cluster over a switch at Gigabit Ethernet speeds. Each cluster also had a head node with a second network connection to the inter-cluster switch. For communication from node1 on cluster A to node2 on cluster B to occur, node1's message will first go to the head node of cluster A through A's switch, out to the inter-cluster switch, and back through cluster B's head node before finally arriving at node2 via B's switch, see Figure 1. Neither of the head nodes were ever used as compute nodes when performing tests. All computers on both clusters had Athlon XP processors clocked at 1.4 GHz, 1 GB of RAM, and PCI Intel e1000 Gigabit Ethernet cards.

We chose to use the LAM/MPI implementation of the MPI standard. In all cases, the message each node sent to all other nodes consisted of 10^7 integers. The timings were done with MPI_Wtime() calls. They began before any inter-node communication, and ended when all communications were complete. We measured traffic volume reflecting the amount of data that was sent through the head node of the cluster from which the job was not originated. The numbers are those reported by the switch on that particular port.

### 5.2 Results

**Table 2: Results: Traffic volume and runtimes**

| Algorithm | Data volume (MB) | Time (sec) |
|---|---|---|
| $ATA_{simple}$ | 343 | 16.2 |
| $ATA_{optimized}$ | 171 | 11.53 |
| NBody | 428 | 18.79 |
| $NBody_{optimized}$ | 171 | 11.28 |

Recall from section 3 that $ATA_{optimized}$ should reduce the amount of data crossing the hot spot from (n/2)^2 for $ATA_{simple}$ to n/2. With four nodes, this translates to a 1/2 reduction. This is consistent with what we measured. Running $ATA_{optimized}$ resulted in only 171 MB of data having to travel between clusters, whereas when running $ATA_{simple}$, we saw 343 MB going over the bottleneck link. As predicted, the inter-cluster communication required for $ATA_{optimized}$ was very close to half of what $ATA_{simple}$ required. Table 2 also shows a significant runtime improvement from using $ATA_{optimized}$. N-Body with optimal configuration was expected to produce hot spot traffic volume of n-2. For the special case of n=4, n-2 = n/2 = 2. Again, this is exactly what we saw.

Considering each message contains 10 million integers, a single message without extra baggage such as communication headers should be 40MB of data (a total of 80 MB for each upload/download message pair). For $ATA_{optimized}$, then, we predicted 160 MB and saw 171 MB. Similarly, we expected $ATA_{simple}$ to have to send 320 MB instead of 343 MB. Our measured traffic volumes are all 7% higher than what we predicted. We attribute this to communication overhead such as TCP/IP headers on each packet that is sent. As the amount of data crossing the bottleneck was reduced, so was the runtime.

Using the wrong host file mapping (Configuration 2) decreased the performance of $ATA_{optimized}$. It increased the amount of data that has to travel across the bottleneck link from 171 MB to 343 MB.

Table 3 summarizes the formulas from our analysis plus our traffic volume measurements side by side.

**Table 3: Summary**

| Algorithm | Formula | n=4 | Data volume (MB) | |
|---|---|---|---|---|
| | | | predicted | observed |
| $ATA_{simple}$ | (n/2)^2 | 4 | 320 | 343 |
| $ATA_{optimized}$ | n/2 | 2 | 160 | 171 |
| NBody | (n/2)^2 +n/2-1 | 5 | 400 | 428 |
| $NBody_{optimized}$ | n-2 | 2 | 160 | 171 |

## 6. Discussion

Performance improvements depend on job characteristics like communication intensity and communication pattern. The importance of the All-to-All pattern is largely dependent upon the communication requirements of the program being run across clusters. The communication patterns for such things as the NAS Parallel benchmarks are a key factor in their speed on being run across the clusters.

Sometimes the inter-cluster link is further reduced by services like a VPN. A VPN is necessary for security, and routing of network services such as NFS, NIS, and rsh between clusters. The bandwidth use of a VPN makes it all the more necessary to minimize inter-cluster traffic as described in this paper.

We implemented our communication strategies with standard MPI point-to-point calls (no lower level optimizations). We hope that our techniques will find their way into the MPI standard libraries for collective operations. That way, programmers can use normal MPI calls and receive the benefit of our algorithm. The host file mapping should be done by job scheduler.

### 6.1 More than two Clusters

In our analysis above, we use our new and modified algorithms for two clusters of computers. While this may be a common configuration, having more than two clusters is a likely situation.

Running N-Body on more than two clusters, the best host file mapping is still interleaving cluster halves. There will still be 2*(n/2-1)=n-2 messages between the clusters and there will be no pieces of information passed between the clusters in the final stage.

Running $ATA_{optimized}$ on *c* clusters, each of the n/c nodes of one cluster sends to a buddy in each of the *(c-1)* other clusters. Thus, the formula for the number of messages being passed across the inter-cluster links is *(c-1) n/c*. In the worst case, *c* = *n* and there will be *n-1* messages taxing the bottleneck. Consequently, whenever multiple clusters of the same size are integrated, our optimized All-to-All communication strategy will never exceed an overlap on order of *n*. This is a huge improvement over $ATA_{simple}$ that always yields an overlap on the order of *n^2*.

### 6.2 Integrating clusters of different sizes

Another scenario is that the clusters being integrated are not of the same size, or due to availability of nodes, a job of size *n* runs on *x* nodes on one cluster plus *n-x* nodes on a second cluster. For $ATA_{optimized}$, nodes on the smaller cluster will have to serve as a buddy to more than one node on the other cluster. The bottleneck will now be in one direction going from the bigger cluster to the smaller cluster. The link contention will be max(*x*, *n-x*), again on the order of *n*.

For the N-Body strategy, the host file should still interleave cluster halves. That way, the overlap is n-2.

## 7. Conclusions

With the advent of grid computing, some jobs are executed across several clusters. We have shown in this paper that (1) cross cluster job execution can suffer from network contention on the inter-cluster links and (2) how to reduce traffic on the inter-cluster links for non-personalized all-to-all communication. Our analysis and experimental results lead us to conclude the following:

- Our new Optimized All-to-All ($ATA_{optimized}$) communication strategy sends data across inter-cluster links as few times as possible, and distributes data locally within each cluster.
- If an *n*-node job runs on two clusters, $ATA_{optimized}$ reduces the traffic volume on the inter-cluster link from *(n/2)^2* to *n/2*. That is an improvement of a factor of 32 for a 64 node job.
- Our measurements show that $ATA_{optimized}$ reduces the traffic volume on the inter-cluster links and thus reduces runtime of cross cluster jobs.
- An alternative to $ATA_{optimized}$ is the n-body communication strategy. However, without the optimal host file mapping, n-body can perform worse than even the simple strategy $ATA_{simple}$. The proper host file mapping for n-body is interleaving cluster halves.
- Mismatched host file mappings can worsen the hotspot for most communication strategies, including $ATA_{optimized}$.

This study presents strategies for improving grid computation communication. The results of improving such communication are a noticeable increase in performance. With these improvements it is possible to increase the size of grids without significantly compromising performance.

## Acknowledgements

This work is supported in part by the W.M. Keck Foundation, by the John S. Rogers Program, and by the National Science Foundation under grant DUE 0411237.

## References

[1] L. Amar, A. Barak, A. Shiloh, "An Organizational Grid of Federated MOSIX Clusters," *Proceedings of CCGrid*, 2005.

[2] I. Foster and C. Kesselman, "The Grid 2: Blueprint for a New Computing Infrastructure", Morgan Kaufmann, 2004.

[3] W.M. Jones, L.W. Pang, D Stanzione, and W.B. Ligon III, "Bandwidth-Aware Co-Allocating Meta-Schedulers for Mini-Grid Architectures" *Proceedings of Cluster*, 2004.

[4] J. Mache, V. Lo, and S. Garg, "The Impact of Spatial Layout of Jobs on I/O Hotspots in Mesh Networks", *Journal of Parallel and Distributed Computing*, Volume 65, Issue 10, pages 1190-1203, Elsevier, October 2005.

[5] J. Mache, D. Tyman, A. Pinter, and C. Allick, "Parallelizing OpenVPN for High-Bandwidth Cluster Integration", *Proceedings of SC|05 - 18th ACM/ IEEE Conference for High-Performance Computing, Networking and Storage*, 2005.

[6] Message Passing Interface Forum, "MPI: A Message-Passing Interface standard", *International Journal of Supercomputer Applications and High Performance Computing*, 1994

# Glue Code Synthesis for Distributed Software Programming

Jian Liu, Farokh B. Bastani, and I-Ling Yen
*Department of Computer Science*
*University of Texas at Dallas*
*Richardson, TX, 75083*
*jian.liu@student.utdallas.edu, {bastani, ilyen}@utdallas.edu*

## Abstract

In this paper, we propose a method for synthesizing the glue code for distributed programming. The goal of this method is to completely automate the synthesis of code for handling distributed computing issues, such as remote method calls and message passing. By using this method, the software for migrating the objects, synchronizing the communication, and supporting remote method calls can be automatically generated. From the programmer's point of view, remote accesses to objects are syntactically and semantically indistinguishable from local accesses. The design of the system is discussed and the whole system is based on the Linda notation. A prototype has been developed using JavaSpaces for code synthesis in Java. Experiments show that this method can help developers generate the code for handling distributed message passing and remote procedure calls.

## 1. Introduction

With advances in computer hardware technologies, substantial changes have occurred in computing practice as powerful computers and networks have become increasing popular. This also raises the complexity and difficulty of software development and, at the same time, the software must have a shorter development time and satisfy more stringent non-functional requirements [1]. Software systems are increasing being developed based on distributed computing environments. With distributed computing, the programmer has to take care of many extra issues besides the basic business logic, such as distributed message passing, synchronization, process or object migration, load balancing, and so on.

Many techniques have been proposed to help the programmer address these issues and improve the quality of the system. Among these, two methods have gained the most significant success, namely, distributed component models [2] and distributed programming languages [3].

Modern distributed component models, such as CCM [4], COM [5], and EJB [6], provide the basic communication structure of the system by defining the interfaces of the components and the services that are needed in distributed computing. These models have been successfully used in industry and are being enhanced with more and more features for supporting specific application domains, such as real-time and mobile computing. This technique is mainly for large enterprise applications and not very suitable for medium scale system development. Also, to learn how to use and program using these models is time consuming.

Distributed programming languages, such as Emerald [7] and Linda [8], support distributed programming by providing language primitives and control statements to handle distributed environments. Methods, such as remote procedure calls (RPC), are often used in these languages to provide basic communication mechanism. But as with the case of distributed component models, there is a significant learning curve for using these languages and they are relatively too large for systems that have limited distributed computing requirements.

To overcome these limitations, we propose a more lightweight and helpful method for medium scale distributed software development, namely, automated glue code synthesis for distributed programming, called DCS (Distributed Code Synthesis).

DCS is a method that can be used by the programmer to automatically generate the code for distributed applications. With DCS, the programmer can start coding the program as if all the components are running on the same machine. DCS uses two steps to synthesize the code for handling distributed programming: First, the components are automatically migrated to the remote location together with their attributes; second, the original code is automatically transformed into a new version that has distributed support. At the same time, the logic of the code remains the same. DCS statically migrates the component to the remote location and generates the code to perform the computation with proper synchronization control.

DCS is a dual purpose platform. It serves as a code generator for developing small to medium scale distributed systems and software prototypes by generating the code for handling network communication and synchronization. Also, it can serve as the basic platform for future enhancements that deal with other distributed computing issues, such as concurrency and asynchronizing communication.



*T. Sobh and K. Elleithy (eds.), Advances in Systems, Computing Sciences and Software Engineering*, 39–46.

The rest of the paper is organized as follows. Section 2 gives the background and presents an overview of the DCS method. Section 3 presents the design of the system. Section 4 discusses the implementation of the method with a case study. Section 5 evaluates DCS in terms of glue code synthesis. Section 6 reviews the related works. Finally, Section 7 concludes this paper and identifies some future research directions.

## 2. Motivation and overview

To automatically generate the code for distributed computing, we need to use either a distributed model (J2EE), a low level communication mechanism (Sockets), or a distributed programming language (Linda) as the underlying communication engine. We choose a distributed language over the other methods based on the following reasons:

1. A distributed programming language (DPL) [3] is more flexible than a component model. Typically, a DPL is well defined with mechanisms for dealing with communication issues at a relatively lower level than a component model. This typically gives the programmer better performance and flexibility in writing their programs.

2. Modern distributed component models [2] are usually based on object-oriented techniques, e.g., EJB. Even though non-object-oriented programs can be wrapped for reuse in CORBA, an interface must be provided for this purpose. Also, DCS is not specifically for object-oriented programming. It can be used with either object-oriented or procedural languages.

3. Low level communication methods, such as TCP Socket, have good performance over other communication mechanisms. But programming using Sockets is time-consuming and many low level communication issues, such as data representation, need to be taken care of by the programmer.

Based on these observations, a distributed programming language, specifically the Linda programming language [8], is chosen as the underlying computation platform for DCS. Linda is a distributed programming language based on generative communication, developed originally for the SBN network computer [8]. It is a coordination language consisting of a small number of primitive operations that are added into existing sequential languages.

DCS is intended for loosely coupled distributed systems similar to the ones discussed in [9]. We view a distributed system as a collection of computers linked by communication channels. A node is defined as a virtual processor with a single memory space. A node resides on a single physical machine. A node can reside on at most one machine. Also, nodes do not share any memory. Nodes communicate with each other by sending messages.

DCS migrates the component (a procedure or an object) to a remote location and synthesizes the glue code for coordinating the communication. We show how DCS works using an example. Assume that a program written in *C* uses a procedure that reads an integer from some input source, for example, a file (*readInt(x,y,z)*). The procedure has no free variables. Instead of making a local procedure call, the actual input resides on a remote node. This is where an RPC could be used. As discussed earlier, we want to use DCS to relieve the developer from having to write these code segments. So, instead of rewriting the program to use RPC or other methods, we can apply DCS to procedure *readInt(x,y,z)* to automatically obtain a new version of the program where distributed communication is enabled and the procedure is migrated to the target remote node.

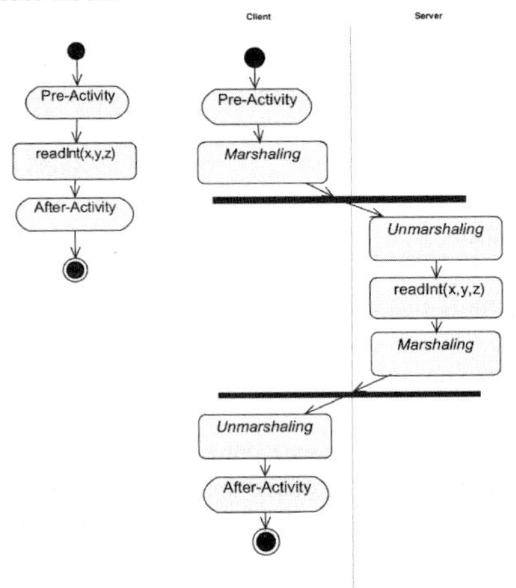


Figure 1. An example of applying DCS

Figure 1 shows the process underlying this example using UML activity diagrams. At the beginning (left hand side of Figure 1), the program is sequential and runs on one node. After using DCS (right hand side of Figure 1), the program is distributed over two nodes, namely, the client and the server nodes where the procedure runs. DCS generates two major parts of the code for the client, namely, marshaling and unmarshaling. *Marshaling* evaluates the procedure, marshals the arguments and the name of the procedure, and sends the data to the server. *Unmarshaling* receives the result from the server and unmarshals it. Given the target node, DCS migrates the procedure code to the server and generates the corresponding unmarshaling and marshaling code for the server.

In the generated program, the client evaluates the procedure and creates a message containing the arguments by using the code generated by DCS. It then sends the message to the server. The server receives the message, unmarshals the arguments, makes the method call, and returns the result. Finally, the client receives the result and proceeds.

Please note that this process is static, which means that the program transformation is done prior to compilation. The programmer needs to recompile the generated code (both the client and the server programs).

## 3. System architecture

Having discussed DCS in general terms, we now present the architecture and design for DCS. As discussed in Section 2, the system is designed and implemented based on Linda. DCS is designed as an agent-oriented system where the developer program is preprocessed, the generated server code is migrated to the remote location, and the resulting code is compiled by a regular language compiler.

As stated in Section 1, the system consists of a set of nodes. A DCS agent runs on top of the JVM at each node. The central node, which is the node where the client is running and the program is being developed, has knowledge of all the agents and their locations.

Since the system is based on Linda, we first introduce basic Linda communication structures and then show how it is used as the underlying mechanism for code generation.

The basic communication structures of Linda are as follows:

```
Remote procedure call:
    out(P, me, out-actuals);
    in(me, in-formals);

Remote procedure:
    in(P, who:name, in-formals) =>
    [ body of procedure P;
      [illegible]
    ]
```

In Linda, *in()* and *out()* have semaphore-like nature, so no executable code block needs to be written to implement the *P()* and *V()* operations. Semaphores are primitives in Linda.

The above communication structure is for asynchronous message-exchange. For synchronizing message-exchange procedure call, [8] provides a macro implementation:

```
def SYNCH_SEND(s:tuple ) [in(get); out(s); in(got)]
def SYNCH_RECV(s:tuple) [out(get); in(s); out(got)].
```

Since we are dealing with sequential imperative languages, such as Java and C, the synchronous method will be used in DCS for preserving the program logic.

The Linda code for the caller (client) is shown as follows.

To create a remote object:

```
out(ID, type:Type, name:String [, parameters:List]);
in(ID, type:Type, name:String, o:type)
```

To call a method on a remote object:

```
out(ID, type:Type, name:String, method:String [,parameters]);
in(ID, type:Type, name:String, method:String
    [, result:return_type])
```

*ID* is the identification dynamically generated by DCS to identify the current communication between the client and the server. Each *ID* denotes a specific connection between the client and the server. It also helps to guarantee that they can receive the correct messages.

The Linda code (Linda agent) for the remote node is as follows:

```
OBJECTS&*[CREATION / METHOD]
def OBJECTS
   List of objects in the host.

def CREATION
   [ in(ID, type:Type, name:String
        [, parameters:List]);
   type obj_name = new type([parameters]);
   OBJECTS.add(obj_name);
   out(ID, type, name, obj_name)
   ]

def METHOD
   [ in(ID, _type:Type, _name:String,
         _method:String[,parameters:List]);
   if(OBJECTS has obj_name whose name is _name,
              type is _type and has
    _method([parameters]) ){
      [result = ] name.method([parameters]);
      out(ID, _type, _name, _method[, result]);
   } else {
      out(ID, _type, _name, _method, ERROR);
   }
   ]
```

where & is the and-statement, / is the or-statement, and * is the star-statement in Linda [8].

In order to correctly distribute the objects, DCS and the programmer need to follow four steps:

1. For each component that needs to be distributed, obtain the address of the remote DCS agent.

2. If the remote agent is alive (i.e., is part of the DCS framework), DCS migrates the component to the specified node.

3. For the selected component, DCS automatically generates the code for using the component at the remote node.

4. For every statement in the original program that uses this component, DCS generates the code for synchronizing distributed message passing.

Before using DCS, the program must be successfully compiled, and the migration process treats each component as a stand-alone piece of code that can be used independently. A DCS agent should be running on each machine in the framework.

Several constraints are considered during the design of the DCS system:

1. Ease of use: DCS should provide the developer a simple way to generate code.

2. Language adaptation: The design should be easy to apply to any programming languages, procedure-oriented or object-oriented.

3. Correctness: If the original program is syntactically sound, the generated program should also be syntactically sound.

4. Consistency: The program logic remains the same after the code transformation.

Other concerns of the system include the security of the generated program and the performance of the synthesized system.

### 3.1. Usability

The system provides a graphical interface for the user to select the procedure that needs to be migrated to a remote node. Only two steps are needed for the developer to generate a distributed program from a sequential one:

1. Compile the original program and ensure that it is free of errors.

2. Select the component that the developer wants to migrate and provide the target location.

Based on the user input, the selected component is migrated to the target location together with its resources. Also, the communication code is generated for both the client and the server by DCS (Figure 2).

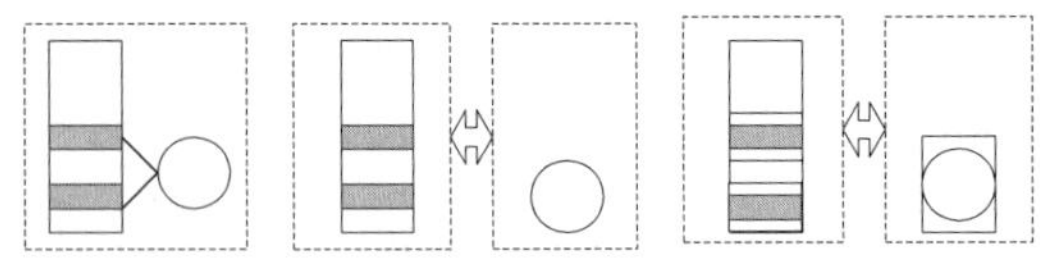


Figure 2. Distribute a unit using DCS

To use DCS, the developer only needs to know which component should be distributed and the address of the remote node. The other aspects are automatically handled by DCS.

### 3.2. Language independence

The target language of this method can either be procedure-oriented or object-oriented. Three issues should be addressed regarding using this method for different languages:

1. Different Linda languages need to be used for handling different programming languages. For example, to use C in DCS, C-Linda can be used, while in the case of Java, JavaSpaces can be used.

2. For object-oriented languages, DCS supports object migration, which is the process of migrating an object to the remote location and generating the code for accessing all the methods of the object remotely. For procedural languages, the migration unit is a program procedure instead of an object. For example, in the case of Java, the source and class files of the object and its related resources are migrated to the target node. While in the case of C, the code of the C procedure is transferred to the remote node.

3. In case of procedural languages, there are two ways that the developer can use to distribute a procedure, namely, by selecting the procedure from its definition or by selecting the procedure from its application. If a procedure is selected for distribution from its definition, all the calls of the procedure will become remote calls, otherwise, only the specific procedure call will become a remote call.

For objects, since an object needs to be instantiated before it is used and it may be stateful, the object has to be migrated as a whole. As a result of this, all the uses of the object within the current scope should be transformed.

### 3.3. Correctness

The correctness of the generated program depends on the migration process and the synthesized code.

The migration process should ensure that all the needed resources are migrated to the remote node. DCS uses the dependence relation between the components in the program and the program call graph to make sure that the components are migrated using the correct order.

Since the generated code is mainly for parameter marshaling and communication synchronization, the correctness of the generated program depends on how DCS handles these issues using Linda.

For the parameters to be correctly marshaled, they should be defined as a marshalable type in the specific language. For example, if the language is Java, DCS requires that all the parameters are serializable. This ensures that the parameters can be sent through the network and can be retrieved without any changes.

If the original program is correct, all the related resources are migrated, and all the parameters that need to be sent are marshalable, then the generated program is also correct.

### 3.4. Consistency

The preservation of the program logic is guaranteed by DCS because the code generated by DCS does not change the system logic. The UML sequence diagrams of the original code and the synthesized program are exactly the same.

For example, suppose we have a procedure (proc) with the following specification:

*{Pre} proc {Post}*.

The program is

*{p1} proc {p2}*, and we have

$p1 \Rightarrow Pre$, $Post \Rightarrow p2$.

Now we want to use DCS to distribute *proc* to a remote node and transform the original program. Since the code generated by DCS performs marshaling and synchronization, the sequence of the statements remains the same and the pre-condition *Pre* and post-condition *Post* will still hold after code generation:

*{p1} DCS_A {p1⇒Pre} proc {Post} DCS_B {Post⇒p2}*, where *DCS_A* and *DCS_B* are the code fragments generated by DCS.

## 4. Implementation and case study

A prototype of DCS has been implemented using Java and JavaSpaces. With Java, DCS focuses on the distribution of the OO objects; however, procedural languages can follow the same way to get a similar DCS system.

Since the current DCS uses JavaSpaces, the generated code is a Java code. After the user specifies which object will be distributed and identifies the target location, the object is migrated and the Java program is pre-processed by DCS. It is then compiled by a regular Java compiler to generate the bytecode that can run on a Java virtual machine (JVM) (Figure 3).

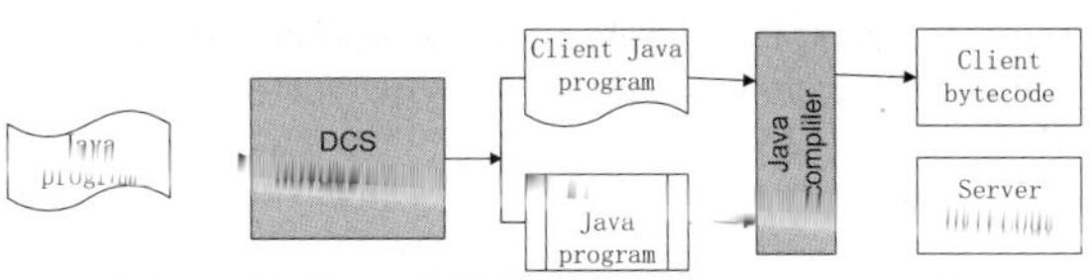


Figure 3. Using DCS and regular Java compiler

Next, we use a case study to present the central ideas of DCS. Also, we discuss how the migrated component is represented on the remote node and how DCS generates the code for coordinating the communication.

Suppose we have the following Java class *MC* defined in our program:

```
public class MC extends Base implements Interf {
  MC();
  MC(T1 p);
  T1 v1 = v1_I;
  static void m3(T1 p1, T2, p2);
  T1 m4(T1 p);
  … …
}
```

and the following code for using the class *MC*:

```
… …
T1 mp1 = I1;
T2 mp2 = I2;
MC x = new MC(mp1);
…
mp1 = x.m4(mp1);
…
MC.m3(mp1, mp2);
… …
```

Both the class definition and the code reside on one single machine and the program is compiled without errors. Now, suppose we want to run object *x*, which is of type *MC*, at a different location where the actual resource resides (for example, suppose *MC* is defined for handling a complex file structure).

Based on the discussion in Section 3, four steps need to be followed to distribute *x*, namely, object selection, object migration, server code synthesis, and client code transformation. These steps are presented briefly in the following subsections.

### Step 1. Object selection

During this step, the user selects the object that needs to be distributed. In our case, the object *x* is selected by the user for migration. DCS will ask the user to specify the remote node where the object will run.

### Step 2. Object migration

After the user chooses the object and specifies the target node, DCS checks whether the node is alive. If it is alive, it will migrate the object to the remote node. This includes all the classes and interfaces related to the object. In our example, the Java class files, which include *Base.java*, *Interf.java*, *Base.class*, and *Interf.class*, will be migrated to the remote node.

### Step 3. Server code synthesis

Once the object is migrated to the remote node, DCS generates the code for the server to use the object. The code invokes the same methods as in the client and follows the same order. The following shows a portion of the code generated by DCS for using *MC*:

```
… …
try {
  MC x;
  parameter_list = null;
  WriteEntry cweT = new WriteEntry(ServerID, ClientID,
     ProcessID, parameter_list);
  WriteEntry cwe = (WriteEntry) space.take(cweT, null,
     10000);
  if (cwe != null) {
    T1 p1 = (T1)getParameter(cwe);
    x = new MC(p1);
    ReadEntry cre = new ReadEntry(ServerID, ClientID,
       ProcessID, r);
    space.write(cre, null, 10000);
  }
  cweT.parameter_list = null;
  cwe = (WriteEntry) space.take(cweT, null, 10000);
```

```
if (cwe != null) {
  T1 p1 = (T1)getParameter(cwe);
  T1 r = x.m4(p1);
  ReadEntry cre = new ReadEntry(ServerID, ClientID,
     ProcessID, null);
  cre.setResult(r);
  space.write(cre, null, 10000);
}
... ...
```

Here, the code for initializing and freeing JavaSpaces is omitted. Since DCS knows what object should be created and what methods will be invoked for the objects and their corresponding calling sequences, it can automatically generate the code for the server to manipulate the object in the same way as the original program.

*WriteEntry* defines the template that can be used by both the client and the server. A connection between a client and a server can be uniquely identified by a tuple (*serverId*, *clientId*, *processId*). *serverId* is the identification of the server node, *clientId* is the id of the client node, and *processId* identifies the client program. *parameter_list* defines all the parameters of the method.

As we can see, the expressions that are related to the object are embedded in the generated code with the same sequence as that in the original program. The difference is that each statement is surrounded with code for coordinating the communication between the server and the client.

**Step 4. Client code transformation**

To complete the transformation, the client code needs to be synthesized to work with the server. The expressions that use the object are transformed to the corresponding code for communicating with the server. As in the case of the server side, a connection between a client and a sever is represented by a tuple. The communication between them is handled by reading and writing tuples using Linda space.

The following is the code generated for the client. Each expression related to the object is transformed into a *try* code block that writes a request to the space and reads the result from it. Combined with the server, this yields a program that is equivalent to the original one.

```
... ...
T1 mp1 = I1;
T2 mp2 = I2;
try {
  parameter_list.add(mp1);
  WriteEntry cweT = new WriteEntry(ServerID, ClientID,
     ProcessID, parameter_list);
  space.write(cweT, null, 10000);
  ReadEntry creT = new ReadEntry(ServerID, ClientID,
     ProcessID, null);
  ReadEntry cre = (ReadEntry) space.take(creT, null,10000);
} catch (Exception e){e.printStackTrace(); return;}
try {
  parameter_list.add(mp1);
  WriteEntry cweT = new WriteEntry(ServerID, ClientID,
     ProcessID, parameter_list);
  space.write(cweT, null, 10000);
  ReadEntry creT = new ReadEntry(ServerID, ClientID,
     ProcessID, null);
  ReadEntry cre = (ReadEntry) space.take(creT, null, 10000);
  mp1 = (T1)cre.getResult();
} catch (Exception e){e.printStackTrace(); return;}
... ...
```

## 5. System evaluation

In order to demonstrate that DCS can help a programmer to generate code for software development, the system has been evaluated with an application that has several distributed nodes. The result shows that DCS provides good code synthesis support for distributed software development.

Real-time online testing is becoming more and more popular in education. The online test system considered here can dynamically generate the test questions based on the students' previous performance during the test. A real-time testing system is a distributed application that has three major nodes, namely, the test client, the test generator, and the question database.

This experimental study illustrates how to apply DCS for code synthesis in typical distributed applications. In particular, we focus the analysis on how DCS can be used to synthesize the glue code for distributed software development.

The system has three major components (implemented as Java classes). One component is to present the questions to the students and obtain the answers from them. Another one is to generate the next question based on the test generation rules. The last component is to query the database and find the appropriate test question. Each of these components will run on a different machine. We wish to do the programming as if all of these components are running on a single machine and then distribute the components using DCS.

Table 1 compares the code before and after using DCS. The portion of the new code generated by DCS coordinates the communication between the distributed components. The generated system runs on three different nodes and is based on JavaSpaces.

Currently, DCS has some limitations. First, the generated code could have some code redundancy and the possibility of reducing the readability of the program. This is due to the possible complex relationships between the remote objects. This problem can be solved by constraining the distributed objects to have well defined interfaces and require limited or no additional services.

Another concern with the current DCS is the performance of the generated program. Sometimes Linda tuple space does not provide good performance for the communication. We are investigating other methods that

Table 1. Code synthesis result

| | Lines of code | Distribute support | Need run-time support | Nodes |
|---|---|---|---|---|
| Original program | 1532 | No | No | 1 |
| Generated program | 2104 | Yes | Yes (DCS run-time) | 3 |

Table 2. Distribute code generation systems

| | Code generation | Underlying mechanism | Code migration | Load balance | Language extension | Language support | Restrict to OO |
|---|---|---|---|---|---|---|---|
| JavaParty | Yes | RMI | Yes | No | Yes | Java | Yes |
| AdJava | Yes | TCP/UDP | Yes | Yes | Yes | Java | Yes |
| JR | Yes | | No | No | No | Java | Yes |
| DCT | Yes | | No | Yes | No | Java/C/C++/etc. | No |
| Java/DSM | Yes | DSM | No | No | Yes | Java | Yes |
| Charlotte | Yes | DSM | No | Yes | Yes | Java | Yes |
| Remote Evaluation | No | | Yes | No | Yes | LISP/etc. | No |
| TCDOJ | Yes | RPC | No | No | No | Java | Yes |
| DCS | Yes | Linda | Yes | No | No | Java/C/etc. | No |

can be more efficient in communication than a tuple space, while at the same time, having a similar flexibility.

## 6. Related works

Considerable research has been done in the field of code generation for distributed systems and Java based distributed programming.

JavaParty [10] is a system for using a class modifier, *remote*, to denote a class that should be used in a distributed fashion. JavaParty is a centralized system and each host in the system runs a JP agent. JavaParty is solely for distributed Java classes and it is possible for it to generate ten Java files for one distributed class.

AdJava [11] can be used to convert a multi-threaded Java program into a distributed Java application. Java Serialization is used to transfer objects and UDP is used for message passing. A special keyword, *distribute*, is introduced to define remote objects. AdJava supports dynamic object migration.

[9] presents a native-code implementation of Java that supports distributed objects. Remote reference is used to provide the transparency of the remote access of the objects. RPC is used as the underlying communication mechanism. A conceptually global shared object-space is introduced to specify the distributed programs. The system modifies the RPC pass-by-value method to use implicitly generated remote references. The programming model consists of a set of *bases* and supporting APIs.

JR [12], a programming language that extends Java to provide a concurrency model, is based on SR [13]. It provides dynamic remote virtual machine creation, remote object creation, and remote method invocation. A JR program is written in an extended Java notation and then translated to a regular Java program. JR provides asynchronous message passing between the objects and an SR-like structure in an object-oriented fashion.

Distributed component technologies, such as CCM [4] and EJB [6], provide frameworks for using components in a distributed fashion. Many distributed programming issues, such as synchronization and load balancing, are handled by these frameworks. Component deployment activities are needed to transfer the components into the run-time environment and to do the component and framework configuration.

In Java/DSM [14], Yu and Cox extend Java to include mechanisms for distributed shared memory. Its goal is to integrate heterogeneous machines into a coherent distributed system. The message-passing code can be automatically generated by the system. A customized JVM runs on each host in the system.

Charlotte [15] is another Java based distributed shared memory implementation. It is developed as a metacomputing resource for parallel computations. Automatic load balancing and fault tolerance is introduced in the system as distributed computing services.

Remote evaluation [16], introduced by J. W. Stamos and D. K. Gifford at MIT, is another way of dynamically migrating code to a remote host and evaluating it for the result. The authors argue why remote evaluation is needed in addition to remote procedure calls and discuss the related issues for this new technique. A LISP based prototype is discussed with some experimental results.

Table 2 compares all the systems described above. As we can see, except for remote evaluation, all the other systems can generate code for distributed programming. Some systems use RPC or RMI as the underlying mechanism, some use TCP/UDP as the communication basis. Several systems select Java as the programming language and make extensions to support distribution. Also, most of these systems provide either a set of APIs to use or base classes to extend from. In most cases, DCS can be viewed as a tool for automated component deployment and distributed code generation. DCS does not need to modify the original programming language and is conceptually language-independent. It uses a static code migration method and is not restricted to object-oriented techniques. Also, at this time, no distributed services are provided in DCS.

## 7. Conclusion

DCS is a system that enables a program unit to be distributed to a remote location and synthesizes the necessary glue code for this purpose. It focuses on code generation for distributed software development. If the original program is correct, then DCS can generate a program that is correct and consistent with the original one. DCS is language independent and it can be used for either procedure-oriented or object-oriented programming languages. A Java/JavaSpaces based prototype has been implemented and used to help generate code for distributed programming. The evaluation of the system is discussed using a case study together with its limitations.

In order to improve the performance of the generated code, we are exploring other methods besides Linda. The primary purpose of DCS is to synthesize code for distributed process, and we are investigating extensions of DCS for asynchronous communication and for use in large concurrent systems in the future.

## 8. References

[1] U. Aβmann, *Invasive Software Composition*. Springer-Verlag, Feb. 2003.

[2] W. Emmerich, "Distributed component technologies and their software engineering implications", in *ICSE '02: Proceedings of the 24th International Conference on Software Engineering*. ACM Press, 2002, pp. 537-546.

[3] G. R. Andrews, "Distributed programming languages", in *ACM 82: Proceedings of the ACM '82 conference*. New York, NY, USA: ACM Press, 1982, pp. 113-117.

[4] O. M. Group, *Corba Components*, vol. I, December 1999.

[5] D. Box, *Essential COM*. Addison Wesley, 1998.

[6] R. Monson-Haefel, *Enterprise JavaBeans*. O'Reilly UK, 1999.

[7] E. Jul, H. Levy, N. Hutchinson, and A. Black, "Fine-grained mobility in the Emerald system", *ACM Trans. Comput. Syst.*, vol. 6, no. 1, pp. 109-133, 1988.

[8] D. Gelernter, "Generative communication in Linda", *ACM Trans. Program. Lang. Syst.*, vol. 7, no. 1, pp. 80-112, 1985.

[9] M. Hicks, S. Jagannathan, R. Kelsey, J. T. Moore, and C. Ungureanu, "Transparent communication for distributed objects in Java", in *JAVA '99: Proceedings of the ACM 1999 conference on Java Grande*. ACM Press, 1999, pp. 160-170.

[10] M. Philippsen and M. Zenger, "JavaParty: Transparent remote objects in Java", *Concurrency: Practice and Experience*, vol. 9, no. 7, 1997.

[11] M. M. Fuad and M. J. Oudshoorn, "AdJava: Automatic distribution of Java applications," in *CRPITS '02: Proceedings of the twenty-fifth Australasian conference on Computer science*. Australian Computer Society, Inc., 2002, pp. 65-75.

[12] A. W. Keen, T. Ge, J. T. Maris, and R. A. Olsson, "JR: Flexible distributed programming in an extended Java", *ACM Trans. Program. Lang. Syst.*, vol. 26, no. 3, pp. 578-608, 2004.

[13] G. R. Andrews, M. Coffin, I. Elshoff, K. Nilson, G. Townsend, R. A. Olsson, and T. Purdin, "An overview of the SR language and implementation", *ACM Trans. Program. Lang. Syst.*, vol. 10, no. 1, pp. 51-86, 1988.

[14] W. Yu and A. Cox., "Java/DSM: A platform for heterogeneous computing", in *ACM 1997 Workshop on Java for Science and Engineering Computation*, vol. 43.2, Jun. 1997, pp. 65-78.

[15] A. Baratloo, M. Karaul, Z. M. Kedem, and P. Wijckoff, "Charlotte: Metacomputing on the web", *Future Gener. Comput. Syst.*, vol. 15, no. 5-6, pp. 559-570, 1999.

[16] J. W. Stamos and D. K. Gifford, "Remote Evaluation", *ACM Trans. Program. Lang. Syst.*, vol. 12, no. 4, pp. 537-564, 1990.

# Interactive Elicitation of Relation Semantics for the Semantic Web

Cartik R. Kothari, David J. Russomanno and Phillip N. Tran
Department of Electrical and Computer Engineering
The University of Memphis, Memphis TN 38152 USA

**Abstract - This paper presents the workflow and architecture of the Relation Semantics Elicitation Prototype (RSEP). RSEP has been developed to address the limitations of the Description Logic based constructs of the Web Ontology Language (OWL) with respect to capturing the intrinsic nature of binary relations. After extracting relation definitions from an input OWL ontology, RSEP interactively elicits the intrinsic semantics of these relations from knowledge providers and appends this elicited knowledge in OWL Full syntax to the input ontology. RSEP has been tested on the IEEE Suggested Upper Merged Ontology (SUMO) and the results are presented in this paper. Preliminary results from using the elicited relation semantics to cluster relations from SUMO and to arrange them taxonomically are also presented to highlight their potential contribution to knowledge interoperability and reuse on the Semantic Web.**

## I. Introduction

The Web Ontology Language (OWL) [1, 2] has been adopted by the World Wide Web Consortium (W3C) as the knowledge representation standard for ontologies on the Semantic Web [3]. In the interests of interoperability, OWL syntactic constructs have standard semantics grounded in Description Logic (DL) formalisms [4]. Specifically, the semantics of OWL constructs are mapped to the constructs from the *SHOIN* (**D**) DL [5]. OWL extends the semantic expressiveness of the Resource Description Framework (RDF) [6] and the RDF Schema (RDFS) [7], which were among the very first knowledge representation formalisms that were developed for the Semantic Web. While RDF provides constructs to specify metadata about Web pages using Object-Attribute-Value triples, RDFS constructs can be used to define simple domain-specific classes and relations in an ontology.

OWL comes in three flavors viz. OWL Lite, OWL DL and OWL Full. The semantics of all of the constructs provided by OWL DL and OWL Lite are grounded in DL formalisms. However, these flavors are less expressive than OWL Full, which allows the free intermixing of constructs from RDF/RDFS with OWL DL. OWL Full is subject to inference complications since RDF and RDFS constructs lack model-theoretic semantics [8].

Certain limitations of the semantic expressiveness of OWL DL constructs with respect to capturing the semantics of relations, specifically pertaining to their intrinsic nature, have been pointed out by Kothari and Russomanno [9]. Specifying the intrinsic nature of relations in Semantic Web ontologies may enable increased expressiveness of knowledge representations, the creation of relation taxonomies and the development of algorithms that can analyze the similarity of relations. In addition, novel inference procedures with capabilities beyond deductive closure may be enhanced by specifying the intrinsic nature of relations.

The Relation Semantics Elicitation Prototype (RSEP) has been developed to elicit additional relation semantics from knowledge providers, and then append these semantics to existing OWL DL ontologies resulting in an OWL Full ontology. RSEP has been tested on the IEEE Suggested Upper Merged Ontology (SUMO) [10, 11] and the results are presented. Based upon these results, the effectiveness of elements on improving the understanding of the intrinsic nature of relations is discussed.

## II. Relation Elements

Relation Element Theory was proposed by Chaffin and Hermann [12] as a means to explicitly specify the intrinsic nature of relations. To describe the intrinsic nature of relations, RSEP uses a basis set of elements (called the RSEP set) drawn in-part from the proposals of Chaffin and Hermann [12], Storey [13], DiMarco [14] and Huhns and Stephens [15] that has been described in [16].

Each element in the RSEP set can take a value from {*yes*, *no*, *N/A*} to describe a relation. An element takes the value *yes* when the relation captures the specific aspect of the relationship between the domain and range entities; or *no* when this aspect is not captured by the relation. Lastly, an element takes the value *N/A* when the specific aspect is not applicable to the relationship between the domain and range entities. Subjectivity is involved in the assignment of element values; however, this subjectivity is made explicit in the OWL Full ontology created by RSEP. Therefore, elements provide a framework for explicit declaration of the intrinsic semantics of relations; enabling better analysis and understanding of their intrinsic nature.

OWL DL constructs alone cannot be used to specify elements to describe relations because this requires the treatment of relations as concepts themselves, which is explicitly disallowed by the DL formalisms that OWL DL constructs are based upon. Therefore in RSEP, elements are defined in RDF syntax [17]. Figure 1 displays the elements from the RSEP set as a hierarchy.



*T. Sobh and K. Elleithy (eds.), Advances in Systems, Computing Sciences and Software Engineering*, 47–52.

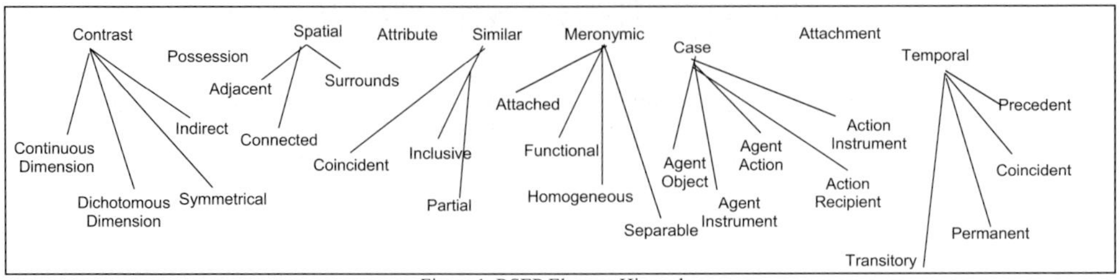


Figure 1. RSEP Element Hierarchy

## III. Prototype Architecture

The architecture of RSEP is shown in Figure 2. RSEP first converts the element definitions and relation definitions in an input ontology into sets of RDF "triples." This is done with the SWI-Prolog RDF Parser [18]. Next, relation definitions with associated domain and range entities are extracted from the RDF triples, using Prolog predicates. If RSEP element values have been previously asserted then they are also extracted in this step. The extracted relations are then displayed to the user for edit.

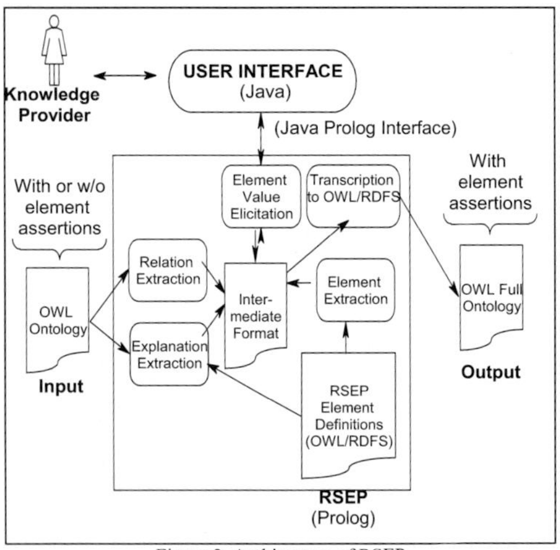


Figure 2. Architecture of RSEP

RSEP makes use of a GUI implemented in Java, which subsequently uses the Java Prolog Interface (JPL) [19] to invoke the Prolog predicates that extract the relations from the ontology. The Prolog predicates output the extracted relation definitions in an intermediate format that is accessed by the JPL and is displayed to the knowledge provider via a user interface as shown in Figure 3.

If available, RSEP also extracts relevant comments about the relations from the input ontology in addition to values asserted for the RSEP elements. This information is displayed to the knowledge provider as a description about the relations. The descriptions about each RSEP element can also be accessed by the knowledge provider to help her in the process of value assertion. The knowledge provider can choose to assert element values for any relation from the displayed set of relations. The root elements from the hierarchy (in Figure 1) are displayed first to the knowledge provider. If the provider asserts the value *yes* for the root element then child elements are displayed to the knowledge provider to elicit additional semantics. If the knowledge provider asserts *no* or *N/A* as the value for the root element then the child elements take the same value as the root element and are not displayed. The child elements further refine the semantics of a given relation.

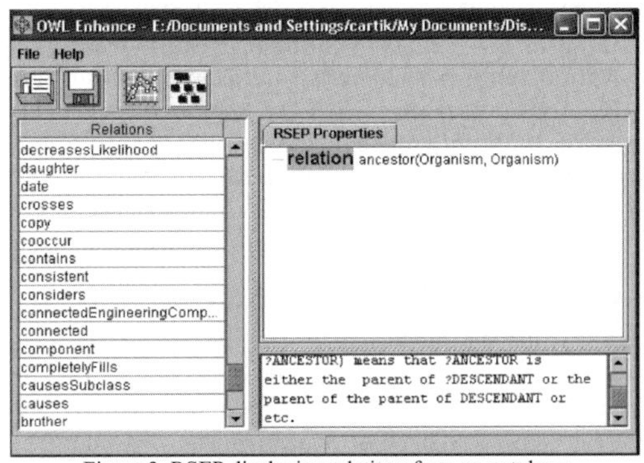


Figure 3. RSEP displaying relations from an ontology

The knowledge provider's assertions are saved in an intermediate format, converted by RSEP into OWL Full syntax and then appended to the original ontology file. The namespace that holds the definitions of the basis set of elements is also appended to the header section of the input ontology. Note that RSEP will display previously asserted element values to the provider if the elements in question belong to the RSEP set. The knowledge provider can choose to edit these values after which they can be saved and appended to the original ontology file.

## IV. RSEP Test

As a baseline test, RSEP was used to elicit element values to describe relations from the IEEE SUMO ontology. SUMO consists of approximately 145 relation definitions. Of these, RSEP does not extract relations whose domain and range entities have not been defined. Only the remaining relations are extracted and displayed to the knowledge provider. Figure 4a shows some relations from SUMO being displayed by

RSEP. In this figure, the *ancestor* relation has been chosen and its details are displayed in the bottom right frame.

The knowledge provider can choose to assert element values for any displayed relation by right clicking on the relation name in the top right frame. Figure 4b shows the root elements being displayed by RSEP to the knowledge provider for element value elicitation. If the provider asserts the value *yes* for a root element then the child elements are presented as shown in Figure 4c. If a value has already been asserted for an element then RSEP displays that value as shown in Figure 4d. Once the knowledge provider is finished with the element value assignments, she can choose to save them.

RSEP converts the knowledge provider's assertions into valid RDF and OWL syntax and appends these assertions to the input ontology resulting in an OWL Full ontology. The OWL and RDF/RDFS syntax of these assertions have been validated using the online OWL and RDF Validator [20, 21].

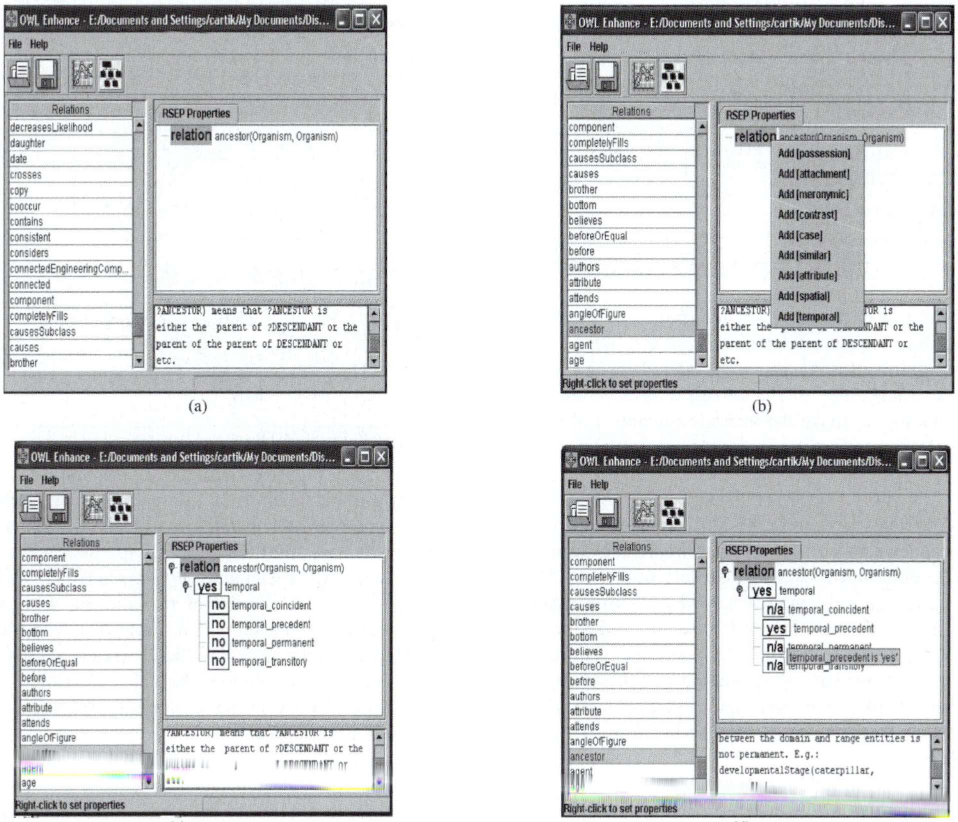


(a) (b) (c) (d)

Figure 4. Screenshots of RSEP

## V. ANALYSIS

Initial tests of RSEP on SUMO yielded new insights into the effectiveness of element values in disambiguating between relations. For example, sets of relations with identical element values after the elicitation process are shown in Figure 5. Note that the relations *sister* and *sibling* appear in the same set; as do the relations *spouse*, *husband* and *wife* and so do the relations *father*, *mother* and *ancestor*. The relations between time-based entities appear in their own set.

However, note that relations such as *wants* (*Cognitive-Agent*, *Physical*) and *needs* (*Cognitive-Agent*, *Physical*) also appear in the same set and are quite hard to disambiguate. The description of the *needs* relation from SUMO states that the *Cognitive-Agent* entity needs the *Physical* (i.e., physical object) entity for its continued existence. This is not the case with the same entities in the *wants* relation. Further disambiguation is possible between these relations if more elements are used. In this manner, the addition of more elements to the set will eventually result in orthogonal sets of element values for every relation in the ontology.

However, the addition of more elements may make the element sets unwieldy and add to the complexity of inference procedures that reason with these representations of relations. The generic, domain independent nature of the representation capabilities of the elements may also be compromised by this approach. Moreover, most Semantic Web ontologies are very domain-specific comprising relatively few relation definitions, such as the Beer ontology [22], in which all the elements in the RSEP set were not used by knowledge providers in some of our other tests. A compromise needs to be reached between complexity bounds and disambiguation requirements in such cases.

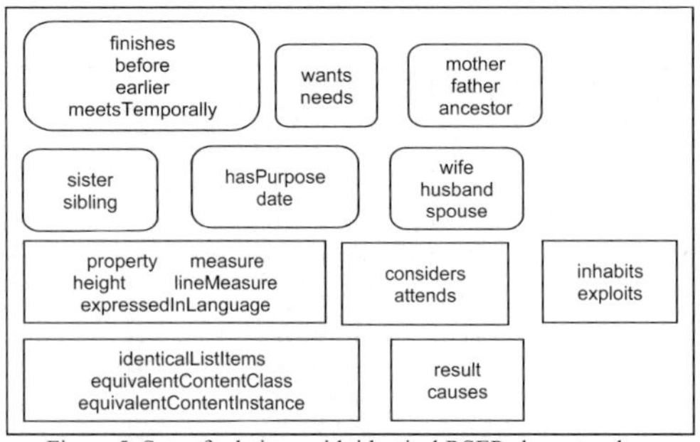

Figure 5. Sets of relations with identical RSEP element values

The RSEP set of elements is not intended to be a definitive and comprehensive set of elements capable of describing every known relation. But it can be considered to be a step towards the identification of such a universal basis set of elements and an attempt to make the intrinsic semantics of relations in an ontology more precise and explicit.

## VI. Applications

The use of elements to capture the intrinsic nature of relations has been shown to be useful in the implementation of novel inference procedures such as plausible inference [23]. Plausible inference derives implicit knowledge from assertions that are not entailed by formal logical inference procedures. These inferences are beyond the deductive closure of formal inference mechanisms. Russomanno and Kothari [24] have implemented plausible inference using elements in RDF/RDFS syntax to demonstrate its viability from the perspective of the Semantic Web.

Elements that capture the intrinsic nature of relations can be used to construct domain specific relation taxonomies using an attribute exploration methodology similar to the technique described by Ganter and Wille [25]. It should be noted that relation hierarchies can be specified using the RDFS construct `<rdfs:subPropertyOf>`. However, this specification is subjective to the interpretation of human knowledge providers and domain experts. The attribute exploration technique is a formal mathematical procedure that reduces the subjectivity of human assertion in the creation of the hierarchy. Relation hierarchies will contribute to the interoperability and reuse of asserted knowledge, furthering the cause of one of the primary objectives of the Semantic Web.

As a preliminary test, the element values elicited by RSEP to describe the intrinsic nature of 96 relations from SUMO were used with an attribute exploration algorithm to group these relations together as relation sets. The relations in every set are described by the identical sets of element values. Subsumption relations between these sets of SUMO relations were used to arrange them taxonomically as a lattice as shown in Figure 6. In this figure, the top of the lattice consists of all the 96 relations from SUMO for which element values were assigned by knowledge providers using RSEP. The bottom of the lattice is the empty set. The set of relations at each node of the lattice is subsumed by the set at the node below to which it is connected by an edge. Again, the set of relations at any node subsumes the set at the node above to which it is connected by an edge.

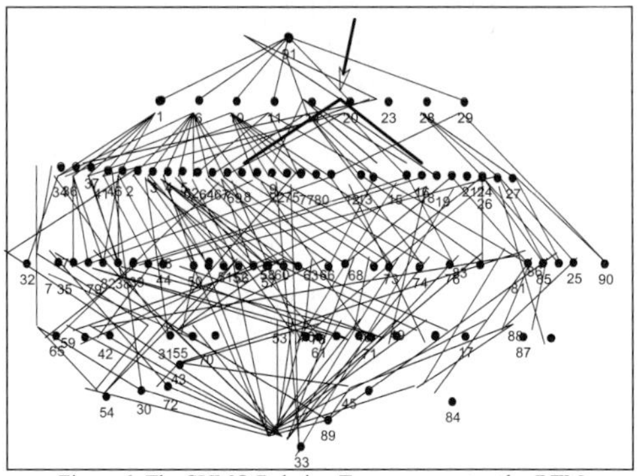
Figure 6. The SUMO Relation Taxonomy output by RTM

At node 23 in Figure 6 (with the arrow pointing to it), the relations *larger* and *smaller* are grouped together. Both these relations capture the comparative nature of the relationship between the domain and the range entities. Both relations are described by the RSEP elements *contrast* and *spatial*. In this manner, the implicit similarity of the intrinsic semantics of these two relations is made explicit. From this, it can be seen that elements not only promote a better understanding of the intrinsic semantics of relations but also contribute to the grouping together (and possible subsequent integration) of relations with similar semantics using methodologies such as attribute exploration.

Element value assertions can also be used to compare the semantic similarities and dissimilarities between relations on the basis of their intrinsic nature. Kothari and Russomanno [26] have presented a naïve algorithm that can be used to compute a discrimination coefficient between pairs of relations by comparing corresponding pairs of the asserted element values and asserting numerical scores based upon identical and different values. These scores are summed to compute a discriminator coefficient. A set of these coefficients coupled with a suitable threshold parameter can be used to create clusters of relations within a knowledge domain and also in the integration of ontologies with synonymously or polysemously named relations, again contributing to knowledge sharing and reuse on the Semantic Web.

As a preliminary test, this naïve algorithm was implemented with the element values assigned to 96 relations from

SUMO and then, the K-Means Hierarchical Clustering Algorithm [27] was used to cluster these relations together using a suitable threshold value. A sample output of the relation clusters is displayed in Figure 7.

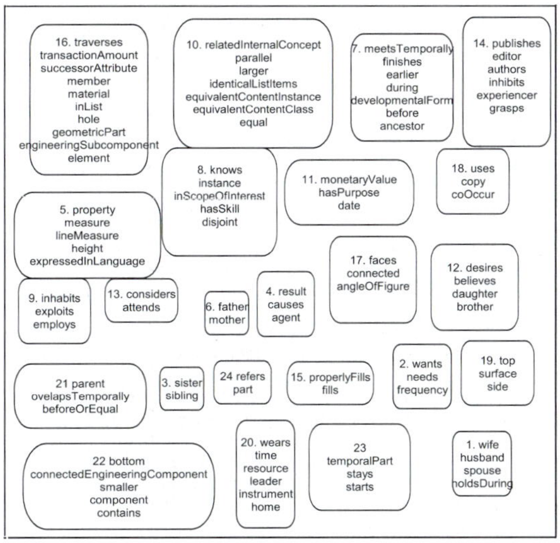


Figure 7. SUMO relation clusters obtained by using assigned RSEP element values

Grouping relations with similar semantics in clusters reduces the dimensionality of relations in equivalent ontologies during the ontology integration process. RCM clusters 96 relations from the SUMO ontology having assigned RSEP element values into 24 clusters when a threshold value of 12 is used as shown in Figure 7. If another high-level ontology is to be integrated with SUMO, it will be easier to identify a cluster to which a relation from the new ontology may belong than to look through all the 96 relations for a possible match.

These approaches to integrate relations across ontologies may complement the methods implemented by Maedche et al. [28] as part of the Mapping Framework (MAFRA) for Distributed Ontologies and the methods discussed in the technical report by De Bruijn and Polleres [29]

## VII. Related Work

The purpose of RSEP is to add semantics to relations in terms of their intrinsic nature. These relation semantics are captured using elements from the RSEP element set proposed in [16]. In a similar way, basic relations such as part-of and dependence proposed as part of the Descriptive Ontology for Linguistic and Cognitive Engineering (DOLCE) [30] may be used to describe the semantics of other relations in the same manner as RSEP elements. Comparing the capabilities of DOLCE with RSEP, RSEP elements are broader in scope than DOLCE with respect to specifying relation semantics. There appear to be no equivalents of the RSEP elements in DOLCE that can describe "similar," "contrast," "attachment" or "possession" relations. The strongest emphasis of DOLCE relations is in their ability to distinguish between short-term and long-term relations. For example, DOLCE relations can be used to distinguish between short-term and long-term meronymic relations However, this capability can be emulated by combining the *temporal* and *meronymic* root elements from RSEP and appropriate combinations of the child elements of *temporal* viz. *temporal-permanent* and *temporal-transitory*.

The DOLCE basic relations can be used to classify relations at a superficial level but may not provide the same extents of relation disambiguation that have been discussed in Section 5. The axiomatization of DOLCE relations is very desirable in the interests of interoperability and reuse. However, these axioms of relations (and their properties) can be represented only in First Order Logic (FOL) and not in OWL. DOLCE has been implemented using the Knowledge Interchange Format (KIF) [31], which is capable of representing axioms in FOL. Only part of DOLCE (called DOLCE-Lite) has been implemented in OWL DL. The axioms may be appended as KIF comments in OWL ontology files. Similar axiomatization of the RSEP elements may be taken up in the future.

The RSEP set of elements are analogous to the properties of concept attributes proposed in the OntoClean methodology [32]. The OntoClean methodology proposed properties such as "essence," "rigid," "identity" and "unity" that could be used to describe concept attributes and subsequently validate subsumptions between concepts in a concept hierarchy. RSEP elements promote deeper understanding of the semantics of relations in the same way that the OntoClean methodology enables a better understanding of concept attributes and concept subsumptions. However, RSEP elements promote a deeper understanding of all relations, not just subsumptions.

## VIII. Conclusions

Analyses of the semantic expressiveness of OWL have revealed limitations with respect to specifying the semantics of relations pertaining to their intrinsic nature. In addition to enabling more effective knowledge representation techniques and additional inference procedures on the Semantic Web, capturing the intrinsic nature of relations using element values has potential benefits in terms of novel inference procedures and will contribute significantly to the cause of knowledge sharing and reuse on the Semantic Web.

The Relation Semantics Elicitation Prototype (RSEP), which elicits intrinsic relation semantics from knowledge providers and appends them to input OWL ontologies is introduced in this paper. Initial results of applying this prototype to the SUMO ontology have been presented. Preliminary results from using assigned element values that describe relations from SUMO to cluster and taxonomically arrange these relations have also been presented to showcase their utility to knowledge interoperability and reuse. Future results from the implementation of the attribute exploration and the relation clustering algorithms that use element value descriptors of relations from other ontologies may validate this hypothesis.

The RSEP element set is capable of describing many binary relations, but is neither definitive nor comprehensive in its scope. The RSEP set can be considered to be a step toward the identification of a universally acceptable, suitably abstract set of elements that can describe the intrinsic nature of relations.

Using element value assertions for clustering relations may be dependent upon appropriate disambiguation among relations, which in turn may be improved by using more elements to describe relations. However, this may lead to more computational complexity of the inference procedures that reason upon the set of element value assertions. A compromise will need to be reached between complexity constraints and disambiguation requirements in such cases.

In summary, elements promote better understanding of relations by providing a framework within which intrinsic relation semantics can be captured and better understood. This framework can be used by quantitative methods to order and partition relations in an ontology and to disambiguate among relations within and between ontologies.

Future directions include testing RSEP on more domain-specific and/or medium–level ontologies. This is expected to provide more insights into its effectiveness at enhancing the understanding of the intrinsic nature of relations beyond the scope of simple definitions of relations that exist in most ontologies available today; in addition to evaluating its usefulness and efficiency as a knowledge elicitation tool.

## REFERENCES


[1] P. Patel-Schneider, P. Hayes and I. Horrocks, "OWL Web Ontology Language Semantics and Abstract Syntax," 2004. Available at: http://www.w3.org/TR/owl-semantics/.

[2] D. McGuinness and F. van Harmelen, "OWL Web Ontology Language Overview," 2004. Available at: http://www.w3.org/TR/owl-features/

[3] T. Berners-Lee, O. Lassila and J. Hendler, "The Semantic Web: A New Form of Web Content that is Meaningful to Computers will unleash a Revolution of New Possibilities," *Scientific American*, May 2001.

[4] D. Nardi and R. Brachman, "An Introduction to Description Logic," In F. Baader et al. (Eds.), *The Description Logic Handbook: Theory, Implementation and Applications*, Cambridge Press, Cambridge, UK, 2004, pp. 1 – 40.

[5] I. Horrocks, P. Patel-Schneider and F. van Harmelen, "From SHIQ and RDF to OWL: The Making of a Web Ontology Language," *Journal of Web Semantics*, 1(1): 7 – 26, 2003.

[6] G. Klyne and J. Carroll, "Resource Description Framework (RDF): Concepts and Abstract Syntax," W3C Recommendation, February 10, 2004. Available at: http://www.w3.org/TR/rdf-concepts/.

[7] D. Brickley and R. Guha, "RDF Vocabulary Description Language 1.0: RDF Schema," W3C Recommendation, February 10, 2004. Available at: http://www.w3.org/TR/rdf-schema/.

[8] J. Pan and I. Horrocks, "RDFS (FA) and RDF MT: Two Semantics for RDFS," In *Proc. of the 2nd Int'l Semantic Web Conf. (ISWC 2003)*, Sanibel Island, FL, 2003, pp. 30 – 46.

[9] D. Russomanno and C. Kothari, "Expressing Inter-Link Constraints in OWL Knowledge Bases," *Expert Systems*, 21(4): 217 – 228, 2004.

[10] I. Niles and A. Pease, "Towards a Standard Upper Ontology," In *Proc. of the 2nd Int'l Conf. on Formal Ontology in Inference Systems (FOIS 2001)*, Ogunquit, ME, 2001, pp. 2 – 9.

[11] A. Pease. I. Niles and J. Li, "The Suggested Upper Merged Ontology: A Large Ontology for the Semantic Web and its Applications," *Working Notes of the AAAI-2002 Workshop on Ontologies and the Semantic Web*, Edmonton, Canada, 2002.

[12] R. Chaffin and D. Hermann, "Relation Element Theory: A New Account of the Representation and Process of Semantic Relations," In D. Gorfein and R. Hoffman (Eds.), *Memory and Learning: The Ebbinghaus Centennial Conference*, Lawrence Erlbaum, Hillsdale, NJ, 1987, pp. 221 – 245.

[13] V. Storey, "Understanding Semantic Relationships," *Very Large Databases Journal*, 2(4): 455 – 488, 1993.

[14] C. DiMarco, "The Nature of Near-Synonymic Relations," In *Proc. of the 15th Int'l Conf. on Computational Linguistics (COLING 1994)*, Kyoto, Japan, 1994, pp. 691 – 695.

[15] M. Huhns and L. Stephens, "Plausible Inference using Extended Composition," In *Proc. of the 11th Int'l Joint Conf. on Artificial Intelligence (IJCAI 1989)*, Detroit, MI, 1989, pp. 1420 – 1425.

[16] C. Kothari and D. Russomanno, "A Relation Semantics Elicitation Prototype for the Semantic Web," In *Proc. of the 2005 Int'l Conf. on Artificial Intelligence (ICAI 2005)*, Las Vegas, NV, June 2005, pp. 682 – 688.

[17] C. Kothari, "Relation Element Definitions for RSEP," Available at: http://www.ee.memphis.edu/ksl/RSEP.owl, 2005.

[18] J. Wielemaker, G. Schreiber and B. Wielinga, "Prolog-based Infrastructure for RDF: Scalability and Performance," In *Proc. of the 2nd Int'l Semantic Web Conf. (ISWC 2003)*, Sanibel Island, FL, 2003, pp. 644 – 658.

[19] F. Dushin, "JPL: A Java Interface to Prolog," 2003. Available at: http://www.swi-prolog.org/packages/jpl/java_api/.

[20] D. Rager, "The OWL Validator," 2003. Available at: http://owl.bbn.com/validator.

[21] E. Prud'hommeaux and R. Lee, "W3C RDF Validator Service," 2003. Available at: http://www.w3.org/RDF/Validator.

[22] D. Aumueller, "An Ontology that Models Brewers and Types of Beers," 2005. Available at: http://www.purl.org/net/ontology/beer.owl

[23] P. Cohen and C. Loiselle, "Beyond ISA: Structures for Plausible Inference in Semantic Networks," In *Proc. of the 7th Nat'l Conf. on Artificial Intelligence*, St. Paul, MN, 1988, pp. 415 – 420.

[24] D. J. Russomanno and C. R. Kothari, "An Implementation of Plausible Inference for the Semantic Web," In *Proc. of the 2003 Int'l Conf. on Information and Knowledge Engineering (IKE 2003)*, Las Vegas, NV, 2003, pp. 246 – 254.

[25] B. Ganter and R. Wille, *Formal Concept Analysis: Mathematical Foundations*, Springer Verlag, Berlin, Germany, 1999.

[26] C. Kothari and D. Russomanno, "Relation Elements for the Semantic Web," In *Proc. of the 12th Int'l Conf. on Conceptual Structures (ICCS 2004)*, Huntsville, AL, 2004, pp. 275 – 286.

[27] A. Jain, M. Murthy and P. Flynn, "Data Clustering: A Survey," *ACM Computing Surveys*, 31(3): 264 – 323, 1999.

[28] A. Maedche, B. Motik, N. Silva and R. Volz, "MAFRA – A MApping FRAmework for Distributed Ontologies," In *Proc. of the 13th Int'l Conf. on Knowledge Engineering and Knowledge Management (EKAW 2002)*, Siguenza, Spain, 2002, pp. 235 – 250.

[29] J. De Bruijn and A. Polleres, "Towards an Ontology Mapping Specification Language for the Semantic Web," Technical Report DERI-TR-2004-06-30, 2004. Available at: http://www.deri.ie/publications/techpapers/documents/DERI-TR-2004-06-30.pdf

[30] A. Gangemi, N. Guarino, C. Masolo, A. Oltramari and L. Schneider, "Sweetening Ontologies with DOLCE," In *Proc. of the 13th Int'l Conf. on Knowledge Engineering and Knowledge Management (EKAW 2002)*, Siguenza, Spain, 2002, pp. 166 – 181.

[31] M. Genesereth, "Knowledge Interchange Format," In *Proc. of the 2nd Int'l Conf. on the Principles of Knowledge Representation and Reasoning*, 1991, pp. 238 – 249.

[32] N. Guarino and C. Welty, "An Overview of OntoClean," In S. Staab and R. Studer (Eds.), *Handbook on Ontologies*, Springer Verlag, Berlin. Germany, 2004, pp. 151 – 159.

# An Improved Configuration Similarity Retrieval Model

Lau Bee Theng
School of Information Technology and Multimedia
Swinburne University of Technology Sarawak

Wang Yin Chai
Faculty of Computer Science and Information Technology
Universiti Malaysia Sarawak

*Abstract* – **Modern information technology has equipped the world with more powerful telecommunication and mobility. We interact with various information systems daily especially those dealing with global positioning systems. Geographical information system is one of important element in those positioning systems. Hence spatial retrieval becomes more and more important in telecommunication systems, fleet management, vehicle navigation, robot automation, and satellite signal processing. Nowadays, spatial query is not longer done with lengthy and complicated syntax. However it is easily done with sketching and speech. The spatial queries are more dealing with daily routines like searching for buildings and routes, analyzing customer preferences and so on. This kind of query is called configuration or structural spatial query that needs powerful retrieval technique as it involves high volume of database accesses to locate the suitable objects. Configuration query retrieval is in content based retrieval family that is also an active research area in spatial databases. This research developed an enhanced configuration similarity retrieval model using single measure in contrast with the multimeasure models in the domain of spatial retrieval.**

**Keywords:** Content based retrieval, structural spatial query, configuration similarity, and spatial retrieval.

## I. Introduction

Presently multimeasure configuration similarity is widely used in object retrieval from spatial databases. This paper presents design of the enhanced configuration query retrieval model namely, Single Measure Structural Similarity (SMSS) model for spatial retrieval systems. The model is proven to be more effective than existing multimeasure models in several areas that are reduced similarity measures, reduced object associations, and eliminated object approximation. This research developed a unique structure, Spiral Web that caters the object arrangement need of a spatial query. It produces single measure for spatial object retrieval by configuration similarity. The single measure reduces the processing time, efforts and complexity in structural similarity assessment.

The existing researches [1, 4, 5 and 6] used multimeasure that is multi spatial relations like direction, distance and topology for structural similarity assessment. Hence there is no single measure that can substitute these multi relations as to date of this research. The invention of a single similarity measure is essential as it reduces processing time, effort and complexity. Furthermore the issues of integration and calibration of multi measures can be resolved. However the single measure structural similarity must be able to substitute the existing multi measures without sacrificing or trading off the efficiency of spatial retrieval.

In view of the loop holes found in the generic multimeasure models, the enhanced model is taken for comparisons in Table 1. The generic framework for configuration similarity includes all major types of spatial relations and also handles the fuzziness of the spatial relations in a query. It consists of multi relations defined in binary strings for topology, direction and distance, encoding of binary relations using conceptual neighborhood and algorithm to compute similarity for the binary relations. The claimed advantages of the generic framework are the expressiveness of the binary string encoding when given a binary string, a spatial configuration can be easily inferred, and vice versa; the efficient automatic calculation of neighborhoods and relation distance, and the uniform representation of all three types of relations (topological, directional, distance) in various resolution levels

Though the generic structural similarity models use multi spatial similarity measures like topology, cardinal direction and distance for spatial query retrieval by structural similarity retrieval in spatial databases [1][4], but the multi measures lack of proper integration mechanisms causing high inexactness in retrieved spatial configurations, increased computation time and processing efforts as the number of measures used increases. The number of objects to be searched in a spatial database, N for n number of objects in a query is [N!/(N-n)!]. If N>n, the candidate is $N^n$ where retrieval of structural queries is exponential to the query size. Structural query processing becomes more expensive if inexact matches are to be retrieved in common practical applications.



*T. Sobh and K. Elleithy (eds.), Advances in Systems, Computing Sciences and Software Engineering*, 53–60.

TABLE 1 GENERIC VERSES ENHANCED MODEL

| Components of Generic Model | Components of Enhanced Model |
|---|---|
| Definition of multi measures/relations like cardinal direction, distance and topology. | Definition of single measure |
| Determine conceptual neighborhoods and encoding of query object into binary string of multi measures | Definition of Spiral Web structure with encoding of query objects into object values (OV) |
| Algorithm to assess structural similarity with multi relations | Algorithm to assess structural similarity with single measure |
| **Problems** | **Solutions** |
| Multi measures lack of integration and increase similarity processing complexity | Introduces single measure that does not require integration and reduces similarity processing complexity |
| Conceptual neighborhood of objects in query increase processing time, effort and complexity | Introduces Spiral Web that has improved object association computation |
| Object approximation with bounding box cannot support concave objects | Introduces Spiral Web structure that is object approximation free |

In the encoding of binary relations and similarity assessment, the generic models associate the query objects for forming object pairs with either complete or reduced association relations. In fact, both the complete and reduced object association relation computation still have rooms for improvement on how to determine and define the most meaningful association computation that can improve the exactness of retrieved spatial configurations [1]. Hence this can reduce the number of associations for encoding and similarity assessment complexity of spatial query retrieval by structural similarity. On top of that, the objects in a spatial query are approximated into bounding boxes where those that are unlikely to satisfy the query are eliminated and a set of potential candidates are selected in the filtering step of spatial query retrieval. Since bounding rectangles are only the approximations, they cause some potential objects being eliminated at the early stage of spatial query retrieval. This is crude approximation that often leads to incorrect matching when concave region objects are involved [3].

The SMSS model covers spatial query representation in spiral web, encoding of configuration similarity into single measure and assessment of configuration similarity.

## II. How Does the Spiral Web Representation Work

The Spiral Web structure is the foundation component of the proposed model that can overcome the shortcomings in the existing spatial query retrieval by configuration similarity.

The existing structural similarity models utilize the bounding rectangles to approximate a query. Approximation filters and speeds up the searching of similar configurations from spatial databases. However there are situations where approximation causes incorrect filtering. Bounding rectangles are only the approximations; they cause some potential objects being eliminated at the early stage of spatial query retrieval. The approximated queries lose their original positioning when the topology, direction, distance and other metrical refinements between bounding boxes do not necessarily coincide with the topological relation between the actual objects. This crude approximation often leads to incorrect matching when concave region objects are involved.

Spiral Web structure does not have this problem, as it is object approximation free. It eliminates the use of approximation and bounding boxes and uses actual geometric structure in representing a spatial query. It is a unique way of spatial query representation. It provides single similarity measures to represent all targeted objects. Hence it is more sensitive to the relative positions of query objects in a query. Fig. 1 shows a query and its Spiral Web. In Fig. 2, the relative distances and geometry among the objects in the query change slightly, therefore the Spiral Web created for representing that query also changes. These two samples show that the Spiral Web is sensitive to the changes of relative settings in a query with the distinguishable values in Table 2.

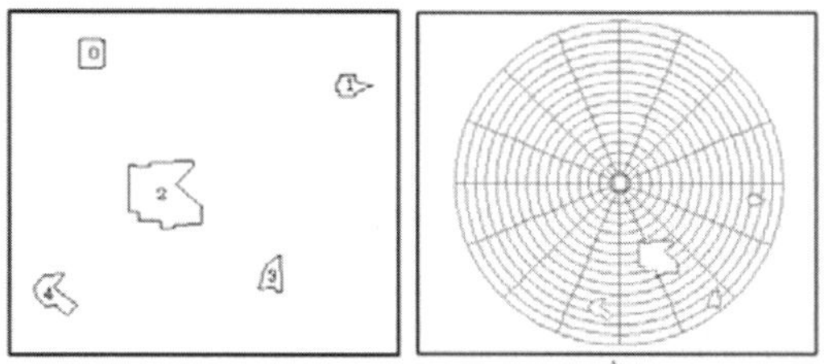


a b

Fig. 1. Sample 1: (a) Spatial Query (b) Spiral Web

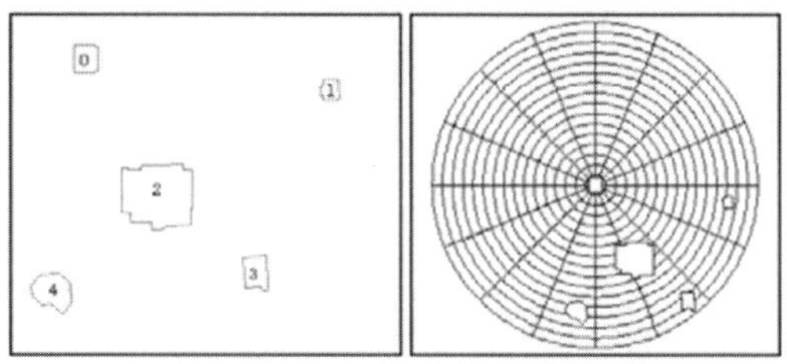


a b

Fig. 2. Sample 2: (a) Spatial Query (b) Spiral Web

TABLE 2 DISTINCTIVE MEASURED VALUE FOR QUERY 1 AND QUERY 2

| Object | OV for Query 1 | OV for Query 2 |
|---|---|---|
| 0 | 8.556370, 15.99990 | 8.556370, 15.99990 |
| 1 | 5.000010, 3.239480 | 5.000010, 3.116270 |
| 2 | 7.264850, 8.230100 | 7.311050, 8.290780 |
| 3 | 7.000010, 1.915820 | 7.000000, 1.693650 |
| 4 | 9.000000, 3.952420 | 8.999990, 3.842360 |

Every query is represented in a Spiral Web with multi cells, $SW_i = \{C_1, C_2, \ldots, C_{n-1}\}$and a single similarity value is computed for each object in the query, $SOV_i = \{OV_1, OV_2, \ldots, OV_{n-1}\}$ where every Spiral Web built would have a set of single similarity measures associated to it. A cell is identified by an index consists of zone and ring identity numbers, $C_i(Z_i, R_i)$. The height and partitions depend solely on the number of objects in a spatial query. An OV for each object in the spatial query has a zone value and ring value, OV (ZN, RN) encoded from the Spiral Web. The encoded OV is compared with the objects in the spatial database through similarity assessment. The Spiral Web simplified the encoding and similarity assessment of query as it eliminates the multi measures that are used in the generic models. A Spiral Web has a few components that play the important roles, {Q, RO, Ctr, D, Di, C, H, P}. Firstly, a spatial query, Q is taken in for processing, reference object, RO, center, Ctr, diameter, D and maximum distance, Di are extracted from the query. Then a Spiral Web consists of core, C that has height, H and partitions, P is created for each structural query.

### *A. Core*

A spatial query, $S_i$ has multiple, n targeted objects, $T_n$, therefore $S_i = \{T_1, T_2, \ldots, T_n\}$. Once a query input is accepted, identifying the reference object from the query is the initial task. The reference object (RO) is identified from the first drawn object in a query. This is because the first object is the initial or starting location when a user sketches or draws a spatial query for structural similarity assessment. It has no reference to base on when it is drawn while the rest of the objects are drawn relatively base on its location. The rationale of assigning a reference object in each query is to allow comparison of center and off-center objects in the query to preserve their relative direction, distance and topology details. The main idea of creating the Spiral Web is to represent a query with single measure that is meaningful for structural similarity assessment. Hence the core is important for determining that single measure. The extracted reference object is required for determining the center, diameter and maximum distance in constructing a Spiral Web.

A center (Ctr) is determined from the centroid of reference object. It consists of a pair of coordinates, Ctr(x, y). Ctr is important for constructing a Spiral Web for a query where it is the center of a Spiral Web. Once a reference object and its centroid and diameter are determined, the core of the Spiral Web can be constructed. The core determines the width and height of the whole Spiral Web; hence it determines the object values of a query. From the identified RO and Ctr, the diameter, D of a reference object is measured. The diameter is one of the components used to determine the height of a Spiral Web. The obtained diameter is used to determine the core of the Spiral Web.

The second component to determine the height of the Spiral Web is the maximum distance from the reference object to the furthest object. In a query, there is a set of distances from reference object to targeted objects, $D_i = \{d_1, d_2, \ldots, d_{n-1}\}$. $D_i$ is the highest measured distance value from the reference object to all the targeted objects in a query. The distance is measured with Euclidean Distance. The object pair is determined with the improved reduced association relation that is (N-1) where N is the number of object in a query.

### *B. Height*

Height is the number of rings of a Spiral Web. It is determined with the diameter and maximum distance obtained from the query. H=Di/D where H is the height, Di is the maximum distance and D is the diameter. From the H value obtained, a Spiral Web with rings is built. The height of each Spiral Web differs as the diameter and maximum distance vary in each spatial query. A ring is a circular structure derived specifically for a Spiral Web. The rationale for creation of ring for a Spiral Web is to complement the zone to describe a query. With the partitions with zone alone is insufficient to describe a query that has objects falling on more than a zone.

### *C. Partition*

People manipulate concrete relations rather than continuous quantitative direction that is angles to express and reason about directions. Most previous work defines qualitative directions using either object projections or centroids. Each approach has its advantages and shortcomings. In this research, a centroid-based method is applied where the direction between two objects is determined by the angle between their centroids since projections are more complicated. This is because the Spiral Web aims to encode the query by its relative relations between query objects that do not use any projections. The set of relative direction relations defined as north, north northeast, northeast, east northeast, east, east southeast, southeast, south southeast, south, south southwest, southwest, west southwest, west, west northwest, northwest, and north northwest. The directions are used to formulate the zone in the Spiral Web.

The zone is divided according to the direction model that is 4 zones consist of north, east, south and west. The 8 zones model consists of additional directions like northeast, southeast, northwest, and southwest. The 16 zones consist of north, northeast, north northeast, east northeast, south, southeast, south southeast, east southeast, east, northwest, north northwest, west northwest, south southwest, southwest, west southwest and west. The criterion used to derive zones is the number of objects exists in a structural query. The rationale for various numbers of zones is: The lesser number of partitions, lesser details can be provided; but the more number of partitions, the higher the processing effort is required. It is desirable to have more detailed partitions, but it is mandatory to minimize the processing effort required. Hence only an optimized number of partitions is used. Initially a Spiral Web with 4 zones is built for every query. Then computation of single measure value follows. The conditional checking continues for "IF two objects fall into same zone" and "IF two objects have same object values". These are the deterministic criteria for the number of zones to be created in a Spiral Web. The rationale for adopting these criteria is to ensure all objects in a query are represented with distinguishable values. If a Spiral Web fail to meet the criteria, then it will be recreated with more zones at an exponential factor of $4^n$ as the directional categories are also subdivided with $4^n$ where n is the number of times the Spiral Web is recreated and it starts with the value of 1. If there are only five objects exist in a structural query, the number of zones is computed accordingly from 4, 8 to 16 zones. The created Spiral Web consists only 8 zones as it is sufficient to determine a single measure value for the structural query where Object B falls in Zone 8, Object C falls in Zone 1 and Object D falls in Zone 2 and 3 and Object E falls in Zone 4 where no object fall into similar zone, and no similar single measure value for any object in the query.

### D. *Association*

An association relation is a link between two or more objects. It is formed with complete or reduced association methods. A standard relation includes a pair of objects. Multi relations are involved when a group of objects is brought into relation with single object like a house is related to a park in the housing estate. Association can be made with the most commonly used spatial relations in structural similarity assessment like topology, relative direction and relative distance [1][3][4][5].

With the complete association technique, the number of possible association relations in a spatial scene is R = (N*(N-1)) for N number of object exists in every query. A query consists of four objects; the complete association approach has 12 association relations, the reduced association with R = ([N*(N-1)]/2) reduces the association relations to 6 whereby the proposed improved approach assesses only 3 associations. The reduced association computation still allows room of improvement to further remove redundancies though it has reduced by half of the number as compared to complete association computation.

With the proposed computation, (N-1) is sufficient to assess the structural similarity of a query hence the reduced association, [(N-2)*(N-1)]/2 is further reduced. The number of redundant relations increases by the number of similarity measures, P used that is (((N-2)*(N-1))/2*P). For a query consists of 8 objects processed with multi measures consist of topology, direction and distance, there are 63 associations. However the number of association reduces from ((N*(N-1))/2) to (N-1) for a single similarity measure with the improved reduced association relation

The proposed structural similarity assessment uses the improved reduced object association technique. It has been improved in term of the number of association relations that needs to be assessed to further reduce redundancy. Instead of evaluating all the relations, the improved object association only considers the relation from reference object to all targeted objects in a query with $\alpha = \beta - 1$ where $\alpha$ is the number of relations to be created or assessed, $\beta$ is the number of objects in a query. This research aims to prove that this neighborhood relation arrangement is effective to represent a structural query and reduces complexity in processing as it reduces number of associations. The cyclometric complexity determination used in Software Engineering is applied to show the complexities in each object association relation to prove that the proposed improved object association is less complicated as compared to the complete and reduced approaches where C is Cyclometric Complexity, E is number of edges or links and N is number of nodes. The proposed approach has the cyclometric complexity of "1" in all queries despite the number of objects exist in them.

### E. *Area*

Area is one of the components in Spiral Web structure [2]. It is required to compute the single measure for a query. Area is one important descriptor of 2-D geometric structure [2]. Area of Occupied Zone (refer as $A_Z$) means the area of a zone being touched by an object and (refer as $A_T$) is the size of the object. Area of Occupied Ring (refer as $A_R$) means the area of a ring being touched by an object. Fig. 2 shows five objects obtained from a spatial query where Object 2 touches on ten rings and two zones, the area of occupied zone is the area of the object that falls on the two zones and the area of occupied ring is the area of the object that falls on the ten rings.

## III. How Does the Single Measure Work

The single similarity measure is developed to enhance the multi measures used in existing structural similarity assessment as multi measures grows exponentially complicated when number of query objects grows. Furthermore, the multi measures require higher processing effort and lack of proper integration. The improved reduced association relation technique improves on the number of association relations that needs to be assessed to further reduce redundancy in encoding of query similarity. The single measure structural similarity is made possible with a structure for mapping a spatial query, Spiral Web. It provides a significant single object value, OV to represent each spatial query and objects.

### A. *Multi Zones Single Ring (MZSR) Configurations*

The OV for a query object that falls into multi zones single ring is modeled as in Equation 1. As there are more than a zone involve, $\Sigma$(Zone) is computed by summing up all the zones being touched by the object. Since it involves only single ring, $\Sigma$(Ring) is computed by the only ring being touched by the object.

$$
\begin{aligned}
OV &= (ZV_n, RV_n) \\
ZV &= \sum_{n=1} \left[ [ZN_1 + ZN_2 + \ldots + ZN_n] \right] \pm tol_z \\
RV &= \sum_{n=1} \left[ [RN_1] \right] \pm tol_R
\end{aligned} \tag{1}
$$

### B. *Multi Zones Multi Rings (MZMR) Configurations*

The OV for an object that falls into multi zones single ring is modeled in Equation 2. Since the computation concerns on computing the OV for an object that falls into more than one zone and more than one ring, the number of zones and rings touched by the target object is crucial to determine the OV. For more than one zones involve, Σ(Zone) is the sum of all the zones being touched by the object. On the other hand, it also involves more than one ring, Σ(Ring) is computed by summing all the rings being touched by the object.

$$
\begin{aligned}
OV &= (ZV_n, RV_n) \\
ZV &= \sum_{n=1} \left[ [ZN_1 + ZN_2 + \ldots + ZN_n] \right] \pm tol_z \\
RV &= \sum_{n=1} \left[ [RN_1 + RN_2 + \ldots + RN_n] \right] \pm tol_R
\end{aligned} \tag{2}
$$

### C. *Single Zone Single Ring (SZSR) Configurations*

Equation 3 shows the designed computation of OV for an object that falls into single zone single ring. Consequently both the number of zone and the number of ring touched by the target object are used to determine the OV. Since there is one zone involve, therefore Σ(Zone) is modeled by the only one zone being touched by the object and Σ(Ring) is computed by the only ring being touched by the object.

$$
\begin{aligned}
OV &= (ZV_n, RV_n) \\
ZV &= \sum_{n=1} \left[ ZN_1 \right] \pm tol_z \\
RV &= \sum_{n=1} \left[ RN_1 \right] \pm tol_R
\end{aligned} \tag{3}
$$

### D. *Single Zone Multi Rings (SZMR) Configurations*

Equation 4 is designed to compute OV for an object that falls into multi zones single ring. The equation mainly concerns on computing the OV for an object that falls into single zone and multi rings. Due to only one zone involve, therefore Σ(Zone) is computed by the only one zone being touched by the object. It involves more than one ring, so Σ(Ring) is computed by summing up all the rings being touched by the object.

$$
\begin{aligned}
OV &= (ZV_n, RV_n) \\
ZV &= \sum_{n=1} \left[ ZN_1 \right] \pm tol_z \\
RV &= \sum_{n=1} \left[ [RN_1 + RN_2 + \ldots + RN_n] \right] \pm tol
\end{aligned} \tag{4}
$$

## IV. HOW DOES STRUCTURAL SIMILARITY ASSESSMENT WORK

The structural similarity assessment is based on the proposed similarity measure in Section III. The similarity assessment is straightforward and simple as compared to the multi measures similarity assessment, this help to reduce the complexity, time, and effort. Furthermore it eliminates the multi measure integration in similarity assessment. The most important one is it still preserves the essential direction, distance and topology contents of a query in the structural similarity assessment.

Generic structural similarity assessment employs a multi-step strategy that makes use of the dependencies among the different types of spatial relations like coarse topology, detailed topology, metrical refinements, cardinal directions and relative distances in the interpretation of a query. With the single measure structural similarity made available with the Spiral Web, the structural similarity assessment procedures are simplified. Instead of assessing the similarities from multi measures, the similarity is assessed with single measure only. A query description is obtained from the Spiral Web representation. Searching can start when spatial database queries are formulated. If a result is found, then the single similarity measure is compared. The false hits are eliminated while the remaining hits are prioritized and being presented to user.

Furthermore it eliminates the integration problem of multi measures while preserving the importance of topology, direction and distance details obtained from a spatial query. This proposed model emphasizes on preserving the importance of topology, direction, and distance details in a spatial query with a single measure structural similarity. Consequently, this section discusses on how topology, direction and distance affects the single measure structural similarity by looking into how the OVs change as these parameters change as in Equation 5. The change of OV is determined by the change in ZV, the Zone Value and RV, the Ring Value. ΔZ is the change of ZN, the Zone ID; ΔR is the change of Ring ID, RN, $A_Z$ means the area of a zone being touched by an object, $A_R$ is the area of a ring being touched by an object and TA is the size of the object. TZ is the total number of zone and TR is the total number of ring in a Spiral Web. $\Delta t_Z$ is the change in tolerance value for a zone value and $\Delta t_R$ is the change in tolerance value for a ring value.

$$OV_{new} = (ZV_{new}, RV_{new})$$

$$\begin{aligned} &where: \\ &ZV_{new} = ZV_{org} + \Delta Z \pm \Delta t_Z \\ &RV_{new} = RV_{org} + \Delta R \pm \Delta t_R \\ &\Delta t_Z = \left[\frac{1}{TZ_{new}} - t_{Z_{old}}\right] \\ &\Delta t_R = \left[\frac{1}{TR_{new}} - t_{R_{old}}\right] \end{aligned} \quad (5)$$

Equation 6 is the change of OV for an object that is affected by a change in topology relation with the reference object in a query. From the various changes of topological relation in the above queries, the changes of OV are obtained. The topological relation changes the OV because of the change in relative distance from the reference object to each object in a query. The change in relative distance also causes the change in maximum distance in a query. The maximum distance is the distance of the reference to the furthest object in a query. When maximum distance and relative distance change, the number of ring and zone in a Spiral Web also changes. Then the OV changes as a result of these changes. For object that falls into single zone single ring, multi zones multi rings, single zone multi rings or multi zone single ring, the change in OVs is different.

For MZSR and MZMR object, ΔZ is computed by adding the changes in all the zones being touched by an object. ΔZ is computed according to the proportion of each zone being touched. However as MZSR only involves single ring, ΔR is the change computed by the single ring being touched by the object. For MZMR that involves more than a ring, ΔR is the change computed by all the rings being touched by an object according to the proportion of each ring being touched. For SZMR and SZSR object, ΔZ is computed by adding the change in only single zone being touched by an object. For SZMR, the ΔR is computed according to the proportion of each ring being touched since it involves more than a ring. On the other hand, ΔR for SZSR is the value change computed by the single ring being touched by the object.

$$\begin{aligned} &\Delta Z = \sum_{n=1}[[ZN_1 + ZN_2 + \ldots + ZN_n]] \\ &\Delta R = \sum_{n=1}[[RN_1 + RN_2 + \ldots + RN_n]] \\ &ZN_n = (ZN_{new} - ZN_{old}) * [A_Z / TA] \\ &RN_n = (RN_{new} - RN_{old}) * [A_R / TA] \end{aligned} \quad (6)$$

Equation 7 shows the change of OV as a consequence of changes in relative directions between object pair. This computation is applicable for all MZMR, MZSR, SZSR and SZMR object.

$$\begin{aligned} &\Delta Z = \sum_{n=1}[ZN_1 + ZN_2 + \ldots + ZN_n] \\ &\Delta R = \sum_{n=1}[RN_1 + RN_2 + \ldots + RN_n] \\ &ZN_n = [ZN_{old} + ((RD_{old} - RD_{new})/(360/TZ_{old}))] * [A_Z/TA] \\ &RN_n = (RN_{new} - RN_{old}) * [A_R/TA] \end{aligned} \quad (7)$$

Equation 8 shows the change of OV when relative distance changes. The OV changes because the relative distance changes the maximum distance and the number of ring and zone in a Spiral Web. If the relative distance increases the maximum distance, it also increases the number of zone and ring. However if the relative distance reduces the maximum distance, it will reduce the number of zone and ring as well. Therefore the OV changes because the ZV, the Zone Value and RV, the Ring Value changes when relative distance changes.

$$\begin{aligned} &\Delta Z = \sum_{n=1}[[ZN_1 + ZN_2 + \ldots + ZN_n]] \\ &\Delta R = \sum_{n=1}[[RN_1 + RN_2 + \ldots + RN_n]] \\ &ZN_n = (ZN_{new} - ZN_{old}) * [A_Z / TA] \\ &RN_n = (RN_{new} - RN_{old}) * [A_R / TA] \end{aligned} \quad (8)$$

Equation 9 shows that the OV is influenced by the relative size of the object in a Spiral Web applicable to all MZMR, SZSR, MZSR and SZMR object. As the relative size changes, the $A_Z$ that is the area of a zone being touched by an object, $A_R$, the area of a ring being touched by an object and TA, the size of the object also change. Hence it affects ZV, the Zone Value and RV, the Ring Value.

$$\begin{aligned} &\Delta Z = \sum_{n=1}[[ZN_1 + ZN_2 + \ldots + ZN_n]] \\ &\Delta R = \sum_{n=1}[[RN_1 + RN_2 + \ldots + RN_n]] \\ &ZN_n = (ZN_{new} - ZN_{old}) * [(A_{Z_{new}} / TA_{new}) - (A_{Z_{old}} / TA_{old})] \\ &RN_n = (RN_{new} - RN_{old}) * [(A_{R_{new}} / TA_{new}) - (A_{R_{old}} / TA_{old})] \end{aligned} \quad (9)$$

On top of these, a Spiral Web representation is object approximation free; it makes the configuration retrievals more intuitive to the query. A Spiral Web representation computes less object association relations with the proposed improved reduced object association also reduces the number of assessment on objects in a spatial query.

For each query consists of a set of query objects with OV, $S=(S_1, S_2, \ldots, S_n)$, there are zero to many sets of retrieved configurations that is $R=\{R_1, R_2, \ldots, R_n\}$. For each retrieved configuration, $R_n$ there is at least more than one retrieved object, $R_n=\{r_1, r_2, \ldots, r_n\}$. The structural similarity of a retrieved configuration to a query, $S_Q$ is made up of a list of assessed structural similarity for individual object pair, $S_{OB}$ where $S_n$ is the OV of the query object, $R_n$ is the OV of the retrieved object, N is total number of associated object pairs, $SX_n$ is the zone value for the query object, $SY_n$ is the ring value for the query object, $RX_n$ is the zone value for the retrieved object, $RY_n$ is the ring value for the retrieved object, $T_z$ is the total number of zone exists and $T_r$ is the total number of ring exists in a Spiral Web. $S_{OBJ}$ determines the similarity of each object in a query to a matched object from a database; hence the similarity of a query is determined by averaging the similarity values of the matched objects from database.

$$S_{OB}(S_i, R_j) = 1 - [\frac{(SX_i - RX_j)^2}{T_Z} + \frac{(SY_i - RY_j)^2}{T_r}] \quad (10)$$
$$where \ S_i \leftarrow OV_i, R_j \leftarrow OV_j$$

Equation 10 computes the structural similarity of an object in query to a retrieved object in a database. It computes the differences with $[\frac{(SX_i - RX_j)^2}{T_Z} + \frac{(SY_i - RY_j)^2}{T_r}]$ where $(SX_i - RX_j)^2$ is the difference of zone value and $(SY_i - RY_j)^2$ is the difference of ring value. There are four types of single measure, therefore the structural similarity of query object also differed depending on whether it falls on multi zones multi rings, single zone single ring, multi zones single ring or single zone multi rings. The details are discussed in the following sections. Equation 11 is the structural similarity for a query. It is derived from the structural similarity of query object. Since there are four types of structural similarity of query object, the structural similarity of query is a summation of all of them that is $S_{OB_{MZSR}}(S_i, R_j)$, $S_{OB_{MZMR}}(S_i, R_j)$, $S_{OB_{SZSR}}(S_i, R_j)$ and $S_{OB_{SZMR}}(S_i, R_j)$.

$$S_Q = \sum_{\forall ij, i \supseteq 2, j \supseteq 0} \frac{S_{OB}(S_i, R_j)}{n - 1} \quad (11)$$
$$where \ S_i \leftarrow OV_i, R_j \leftarrow OV_j$$

The structural similarity assessment of query object falls on multi zones single ring is shown in Equation 12. The computation obtains the multi zone values of query by $\sum_{n=1}^{n}[ZN_n \pm tol_z]$ and the single ring value by $(RN_1 \pm tol_r)$. RX is the zone value and RY is the ring value of the retrieved spatial object. For each object from the database that is taken into comparison with the query has a pair of zone and ring value.

$$S_{OB_{MZSR}}(S_i, R_j) = 1 - [\frac{((\sum_{n=1}^{n}[ZN_n \pm tol_z])_i - RX_j)^2}{T_Z} + \frac{((RN_1 \pm tol_r)_i - RY_j)^2}{T_r}] \quad (12)$$

Equation 13 shows the assessment of structural similarity of query object that falls into multi zone multi rings in a Spiral Web. Since there are multi zones and multi rings involved, the computation of zone value is $\sum_{n=1}^{n}[ZN_n \pm tol_z]$ and ring value is $\sum_{n=1}^{n}(RN_1 \pm tol_r)$. The zone and ring values from spatial database remain as RX and RY.

$$S_{OB_{MZMR}}(S_i, R_j) = 1 - [\frac{((\sum_{n=1}^{n}[ZN_n \pm tol_z])_i - RX_j)^2}{T_Z} + \frac{((\sum_{n=1}^{n}[RN_n \pm tol_r])_i - RY_j)^2}{T_r}] \quad (13)$$

Equation 14 is the computation for similarity assessment for query object in single zone single ring. Since there is one zone and one ring involved, the zone value, $(ZN_1 \pm tol_z)$ is compared with RX and the ring value, $(RN_1 \pm tol_r)$ is compared with RY.

$$S_{OB_{SZSR}}(S_i, R_j) = 1 - [\frac{((ZN_1 \pm tol_z)_i - RX_j)^2}{T_Z} + \frac{((RN_1 \pm tol_r)_i - RY_j)^2}{T_r}] \quad (14)$$

Equation 15 shows the similarity assessment for query object that falls into single zone multi rings. There are more than one rings involved; hence the ring value is assessed by $\sum_{n=1}^{n}[RN_n \pm tol_r]$ and the zone value is assessed with $(ZN_n \pm tol_z)$. The spatial object from the spatial database determines RX and RY.

$$S_{OB_{SZMR}}(S_i, R_j) = 1 - [\frac{((ZN_1 \pm tol_z)_i - RX_j)^2}{T_Z} + \frac{((\sum_{n=1}^{n}[RN_n \pm tol_r])_i - RY_j)^2}{T_r}] \quad (15)$$

## V. EVALUATION

The outcome of a similarity assessment between a query and a spatial database is a set of retrieved results and a list of similarity values. Different models used affect the similarity values and rankings of retrieved objects. For instance, the reduced association relation model claimed to be better that complete object association technique as it reduces the number of binary relations assessed in a query. Consequently the performance evaluation of the proposed model is conducted by evaluating the ranking of the retrieved results using the proposed model compared with the complete association model and reduced association model [1].

There are three commonly used statistical measurements for comparing the retrieved results. The statistical analysis of the correlated similarity assessment between the proposed model with Blaser [1], and between the proposed model with conventional model are compared separately by using the well known Spearman Rank Correlation Test, Wilcoxon Signed Rank Test and, Mean and Standard Deviation Test.

## VI. Conclusion

This paper only managed to discuss the modeling of the concept. However the research has successfully proved the enhanced model is feasible and practical for configuration similarity retrieval of spatial objects from spatial databases. Furthermore the representation and similarity assessment proposed in the model have been tested and compared with the two main streams in configuration query retrieval that are Conventional and Blaser [1] models. The results proved the applicability and practicability of the model. The model has produced better results in overall situations.

## Acknowledgment

I would like to thank my supervisor, Assoc. Prof. Dr. Wang Yin Chai for his guidance throughout the research and Swinburne University of Technology Sarawak for the funding of the publication of this paper.

## References

[1] Blaser, A.D. (2000). *Sketching Spatial Queries*. Dissertation for the Degree of Doctor of Philosophy in Spatial Information Science and Engineering. University of Maine: Department of Spatial Information Science and Engineering.

[2] Ballard, D.H. and Brown, C.M. (1984). *Computer Vision*. Eaglewood Cliffs: Prentice Hall.

[3] Goyal, R. and Egenhofer, M.J. (2001). Similarity of Direction Relations. *Seventh International Symposium on Spatial and Temporal Databases (*Jensen, C., Schneider, M., Seeger, B., Tsotras, V.; eds.). LNCS 2121:36-55.

[4] Papadias, D. and Delis, V. (1997). Relation-based Similarity. In *Proceedings of the 5^th^ ACM-GIS*, pp.1-4. ACM Press.

[5] Papadias, D., Karacapilidis, N. and Arkoumanis, D. (1998a). Processing Fuzzy Spatial Queries: A Configuration Similarity Approach. *International Journal of Geographic Information Science (IJGIS)*, 13(2): 93-128.

[6] Papadias, D., Mamoulis, N. and Delis, V. (1998b). Algorithms for Querying by Spatial Structure. In *Proceedings of the 24th VLDB Conference*.

[7] Papadias, D. and Sellis, T. (1993). The Semantics of Relations in 2D Space Using Representative Points: Spatial Indexes. In *Proceedings of the European Conference on Spatial Information Theory* (Frank, A. and Campri, I.; eds.). Springer Verlag.

[8] Papadias, D. and Sellis, T. (1994). On the Qualitative Representation of Spatial Knowledge in 2D Space. *Very Large Data Bases Journal Special Issue on SpatialDatabases*, 3(4): 479-516.

[9] Egenhofer, M.J. (1991). Reasoning about Binary Topological Relations. In *Proceedings of Advances in Spatial Databases* (Gunther, O. an Schek, H.J.; eds.).

[10] Egenhofer, M.J. (1994a). Pre-Processing Queries with Spatial Constraints. *Photogrammetric Engineering and Remote Sensing*, 60(6): 783-790.

[11] Egenhofer, M.J. (1994b). Spatial SQL: A Query and Presentation Language. *IEEE Transactions on Knowledge and Data Engineering,* 6 (1): 86-95.

[12] Egenhofer, M.J. (1995). Modeling Conceptual Neighborhoods of Topological Line-Region Relations. *International Journal of Geographical Information Systems,* 9 (5): 555-565.

[13] Egenhofer, M.J. (1996). Spatial-Query-by-Sketch. In *Proceedings of VLDB'96*, pp. 60–67.

[14] Egenhofer, M.J. (1997). Query Processing In Spatial Query By Sketch. *Journal of Visual Languages and Computing*, 8(4):403-424.

# Automatic Generation of Software Component Wizards based on the Wizard Pattern

Hironori Washizaki [1], Shinichi Honiden [1], Rieko Yamamoto [2], Takao Adachi [3] and Yoshiaki Fukazawa [3]
[1] National Institute of Informatics, 2-1-2 Hitotsubashi, Chiyoda-ku, Tokyo 101-8430, Japan
{ washizaki, honiden }@nii.ac.jp
[2] Fujitsu Laboratories Limited, 4-1-1 Kamikodanaka, Nakahara-ku, Kawasaki-shi, Kanagawa 211-8588, Japan
r.yamamoto@jp.fujitsu.com
[3] Waseda University, 3-4-1 Ohkubo, Shinjuku-ku, Tokyo 169-8555, Japan
{ adachi, fukazawa }@fuka.info.waseda.ac.jp

***Abstract*-When a software component is used, it is often necessary to set initial values in many of its attributes. To set these initial values appropriately, the user of the component must ascertain which attributes are needed to be initialized, and set them programmatically to suitable initial values. The work involved in this sort of initialization can be alleviated by attaching a wizard interface to the target component itself and setting the initial values visually from the wizard. However, there are large development costs associated with devising suitable initial value candidates and producing a new wizard to use these initial values for each individual component. In this paper, we propose a system whereby application programs that use a target component are subjected to dynamic analysis to discover which attributes and initial values are set most often during the running of the component. The proposed system generates and attaches a wizard, which supports application programmers to initialize the component visually by using these initial values, to the component. The proposed system can be recognized as a system for applying the Wizard pattern to each component automatically. Experiments have shown that the attributes and their initial values chosen for initialization by generated wizards closely resemble the expectations of the component's original developers. We have thus confirmed that the proposed system can bring about a substantial reduction in wizard development costs.**

## I. INTRODUCTION

Component-based software development (CBD) has become accepted as a cost-effective approach to software development. In CBD, software development is considered to involve the composition of various software components. A software component is a self-contained unit of composition with contractually specified interfaces [1]. CBD is capable of reducing developmental costs and improving the reliability of an entire system. In this paper, we use object-oriented (OO) programming language for the implementation of components. CBD does not always have to be object-oriented; however, it has been indicated that using OO paradigm/language is a natural way to model and implement components [2]. In fact, some of practical component architectures, such as JavaBeans [3] and Enterprise JavaBeans (EJB) [4], are based on OO technologies.

In CBD, components are not only reused within organizations to which the components' developers belong, but are also distributed in the form of an object/byte code via the Internet and reused in other environments [5]. Therefore, to allow components to be reused in a variety of contexts and environments, the component developers provide these components with a set of attributes whose values can be externally modified. The provision of these attributes can improve the variability of components. On the other hand, component developers sometimes want to present a set of several suitable initial values for attributes that they have prepared. By presenting initial values in this way, the target components can be made to exhibit both variability and specificity.

A component wizard is a user interface mechanism for presenting a suitable set of initial values for attributes, which is added to the target component itself based on the Wizard pattern. A component developer can use the component wizard added to a component to indicate to the component's users (i.e., application programmers) a suitable set of initial values for component attributes that can be set at runtime so that the component is used as intended by the developers. However, there are large development costs associated with devising suitable candidate values for each individual component and creating a new wizard that uses these initial values. Consequently, only a few of reusable components in circulation on the Internet have wizards added to them.

In this paper, we propose a system for automatically adding wizards to components, which are based on JavaBeans framework (a framework for modularization and reuse in the Java language). The proposed system runs multiple Java application programs that use a JavaBeans component, analyzes the runtime data to discover which attributes and initial values are set most frequently during the execution of these programs, automatically generates a wizard that uses the discovered initial values to initialize the component visually, and attaches this wizard to the component. Our system can be recognized as a system for applying the Wizard pattern [6] to each component automatically. Users of the component can then suitably initialize it by visually selecting an initialization set from a connected group of dialogs presented by the attached wizard.



*T. Sobh and K. Elleithy (eds.), Advances in Systems, Computing Sciences and Software Engineering*, 61–68.

## II. COMPONENT WIZARDS

We will briefly introduce the concept of component wizards based on a specific component architecture, JavaBeans.

### *A. Fine-grained components and JavaBeans*

A software component is a self-contained unit of composition with contractually specified interfaces. Compared with an ordinary object-oriented class, a component is designed more with distribution in mind. In CBD, components are not only reused within organizations to which the components' developers belong, but are also distributed in the form of an object/byte code via the Internet and reused in other environments [5]. Therefore, users who want to reuse components often cannot obtain the source codes of the components.

The granularity of the component can be defined as the conceptual size of the component's functions [7]. Components are classified into the following according to granularity: coarse-grained, medium-grained and fine-grained.

- The enterprise component that encapsulates business logic concerning the remote processing is a coarse-grained component.
- The application component, which is composed of fine-grained components and specific logic, is a medium-grained component.
- GUI widgets and non-GUI components with minimum business logic are fine-grained components.

Currently, fine-grained components are the most widespread due to the success of the RAD tool. ActiveX [8] and JavaBeans [3] are component architectures suitable for treating fine-grained components. In this paper, we deal with JavaBeans as the target component architecture. JavaBeans is a practical component architecture, which provides a fine-grained component development model for Java. One JavaBeans component is a single Java class having the following features:

- Properties: Properties are the named attributes associated with a JavaBeans component, whose values can be read or written by invoking the appropriate getter/setter methods. Usually, properties correspond to the class's fields one-to-one [7]. Properties whose values can be get (set) are called "readable (writable) properties".
- Getter method: Getter methods are methods implemented within the class to get the properties' values from outside of the component.
- Setter methods: Setter methods are methods implemented within the class to set the properties' values from outside of the component.
- Business methods: Business methods are simply normal Java methods that can be invoked from outside of the component, except for the getter/setter methods, and are implemented within the class.

For example, Figure 1 shows the UML class diagram [9] of a typical JavaBeans component "`WEbean`" [10], which visually represent a flashing light. In Figure 1, the component is composed of one class, and has a field `shineColor` in the class. In Figure 1, we omitted other several fields and methods from the component due to the space limitation. Since there are a getter method `getShineColor()` and a setter method `setShineColor()` corresponding to the field, the introspection mechanism recognizes that the component has one readable and writable property, named ``shineColor."

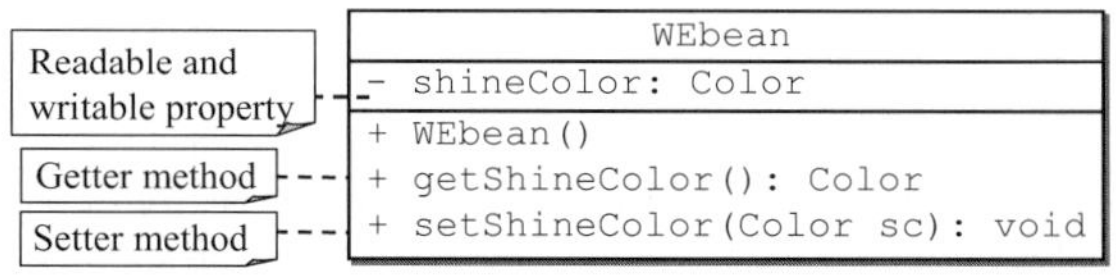


Fig. 1. Class diagram of `WEbean`

### *B. Software wizards*

A wizard is a mechanism that facilitates the definition of complex software settings [11]. There are two types of wizards: those that support the use of software by end users (such as installing software [12]), and those that support the development of software by developers. The latter reduces the costs associated with using software under development or from a development environment, ultimately reducing the overall development costs. In this paper, we will deal with the latter case of a wizard used during development.

A wizard promotes a suitable initialization process by presenting the user with a sequence of multiple dialogs for initializing the target software, and presenting a small amount of tasks in each dialog [6][13]. A task refers to a unit of work needed to initialize the target software. By presenting a sequence of multiple dialogs, the wizard breaks complex tasks down into small tasks, allowing the user to perform the initial setting process visually in a step-by-step fashion.

### *C. Building component wizards*

A component wizard is a wizard that is added to the component itself. This type of wizard is activated in an application development environment that can handle component architectures that target components conform to as standard. An activated component wizard provides application programmers with functions for visually setting the initial values of multiple properties provided by the component. In JavaBeans, a framework (named "`java.beans`" package) is prepared for developing wizards that can visually edit the properties of JavaBeans components. Using this framework makes it possible to generate wizards for JavaBeans components.

In this framework, the component wizards are produced as Java classes that implement a `java.beans.Customizer` interface. The operations needed to produce a wizard using this framework are the preparation of a wizard class incorporating the `Customizer` interface, and the association of this wizard class with a `java.beans.BeanInfo` object that stores data on the component's properties, methods and events. This association is performed by registering the wizard class in the `java.beans.BeanDescriptor` stored in the `BeanInfo` object. The above operations make it possible to use the wizard after an instance of the component has been created in an application development environment (RAD tool) compatible with JavaBeans.

For example, Figure 2 shows an example of executing the component wizard attached to the component `WEbean` [10] in BeanBox [14]. BeanBox is a visual Java application development environment in which multiple JavaBeans components can be arranged and associated with each other by means of events and the like. In the central part of the wizard, three specific initial values are presented as choices ("Arrangement") for the combination of foreground and background colors for `WEBean`. When a set of initial values displayed in the wizard is selected, these initial values are set in the properties of the component in the BeanBox environment. In this way, the user of the wizard can easily set complex component properties, and the component whose properties have been initialized can be used in RAD tools.

However, for JavaBeans components, large development costs are involved in devising suitable initial value candidates that the component properties should be set to, and in producing new wizards to use these initial setting values. Consequently, only a few of the JavaBeans components that are distributed over the Internet have wizards.

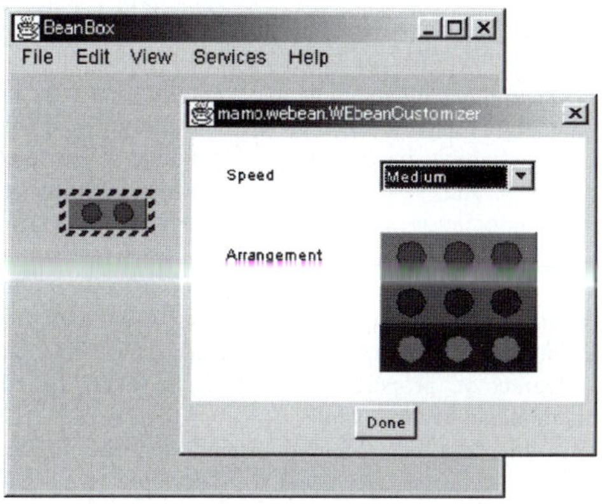


Fig. 2. Example of a component wizard

## III. ATTACHING WIZARDS AUTOMATICALLY

We propose a system (called "our system") that automatically generates wizards and attaches them to JavaBeans components.

Our system needs multiple application programs that actually use the component in the generation of a wizard. After these applications have been acquired, our system automatically generates and adds wizards according to the procedure shown below:

(1) Our system automatically generates data acquisition components to obtain setting information on the target component properties used within the applications.
(2) Our system runs the applications after substituting their target components with data acquisition components, and acquires the setting information and execution histories of these properties.
(3) The usage information of the components is analyzed based on the acquired setting information and execution histories. As a result of this analysis, our system obtains component setting information to be presented to the developer.
(4) Our system uses this setting information to generate a wizard automatically and add it to the component.

In below, we will show details of each step in the above-mentioned procedure.

### *A. Automatic generation of data acquisition components*

To record the change history of property values, data acquisition components are generated which are capable of being substituted for the target components. Specifically, a subclass (e.g., `ExtendFooBean`, which extends `FooBean`) that overrides the setter method of an target component class (e.g., `FooBean`) is automatically generated as a data acquisition component. Inside the overridden setter method, the property values before and after calling setter methods in the target component class or its superclass are written to an external file in CSV (comma-separated value) format. The information written to this file consists of the name and type of the setter method, the name and type of the property whose value is changed by this setter method, and the values of the property before and after change. When the property type is a primitive type such as `int` or `boolean`, its value is stored directly. When the property type is an ordinary object, then its value is stored by serializing the object state.

### *B. Substitution and execution of components*

For all the accumulated applications, the program statements where instances of target components are generated (e.g., `new FooBean()`) are manually substituted so as to generate instances of components for acquiring generated data (e.g., `new ExtendedFooBean()`). Since the data acquisition components are subclasses of the original components, there is no need to modify other statements apart from the instance generation statements. Accordingly, the above-mentioned work involved in substituting components in the application can be performed by the developer with low labor costs.

Also, for all the constituent classes of the application, a bytecode modification tool Javassist [11] is used to automatically add a trace function that writes out a history of the entire method calls.

After the components have been substituted and the trace functions have been added, each application is run so as to automatically record information in the change of property values in each component (property setting information) and the overall execution history of the application.

*C. Analysis of usage information*

The recorded property setting information and execution histories are analyzed to produce method call graphs (MCGs) representing the relationships among method calls, and a set of initial values to be presented in the wizard for the property group are determined from the trends in the frequency of appearance of values and the proximity of nodes in this graph. The procedure for determining these settings is shown below.

*C.1. Method call graph and degree of initial setting*

A method call graph is produced from the execution history of the application including the target components. MCG *G* comprises a set of nodes *M* representing methods and constructors that generate class instances, and a set of edges *E* representing calls from one method to another. *M* includes the application's entry point (i.e., `main()` method) which is called when the application is started up. The MCG is defined as shown below:

MCG: $G = (M, E)$
Set of methods/constructors: $M = \{ m_1, ..., m_n \}$
Set of calls: $E = \{ e_1, ..., e_m \}$
Call: $e_i = (m_j, m_k)$ (where $0<i<m$ and $0<j,k<n$)

Figure 3 shows an example of a MCG. The labels on each node in the MCG represent the name/type of method (or constructor) that each node corresponds to, and the property values of the target component that are changed by executing this method (or constructor). In Figure 3, the `App.main()` node corresponds to the application `App`'s method `main()`.

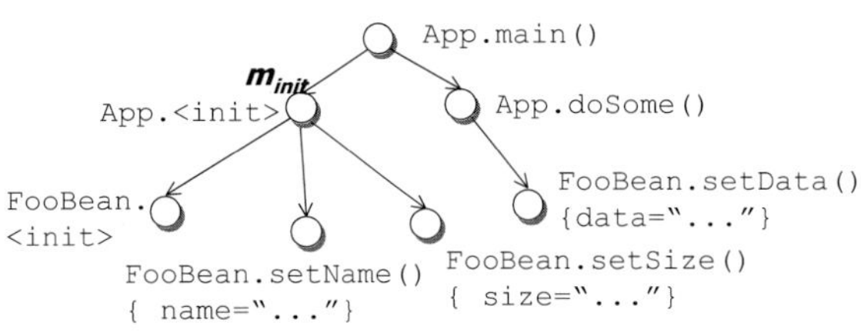


Fig.3. Example of a method call graph

Next, in the resulting MCG, the degree to which a method `m` of the component is used for the initial setting of the component is determined. The method that invokes an instantiation method (i.e., `Target.<init>`) of the target component is denoted as $m_{init}$. The degree of initial setting *init(m)* of method *m* represents its proximity in the MCG to the position where the component's instance was generated. The definition of *init(m)* is shown below.

$init(m) = 1/d(m_{init}, m)$
where $d(m_i, m_j)$ = ( number of nodes in the shortest path between $m_i$ and $m_j$ )

For example, in Figure 3, the degree of initial setting of the setter method `setSize()` in the `FooBean` class is *1*, and the degree of initial setting of the setter method `setData()` is *1/3*. In cases where the same setter methods are called multiple times and registered at different locations in the MCG, the arithmetic mean of each degree of initial setting is taken.

*C.2. Initial set*

Next, our system determines a set of properties (referred to as "initial set") to be presented in the wizard for the property group. The initial set constitutes a set of properties that are commonly and almost continuously initialized by most applications in the initial setting of the component.

The initial sets are derived using a measure called the "degree of property association", which expresses the degree to which two properties are set simultaneously. The degree of property association $pa(p_i, p_j)$ between two properties $p_i$ and $p_j$ is defined by the following formula using the setter methods $m_i$ and $m_j$ that set the values of the two properties respectively.

$pa(p_i, p_j) = 1 / ( d(m_i, m_j) - 1 )$

For example, in Figure 3, the degree of property association between the property `size` and the property `name` of `FooBean` (which have the corresponding setter methods `setSize()` and `setName()`) is `1`, whereas the degree of property association between the property `size` and the property `data` is *1/3*. In cases where the setter methods of the two properties are called multiple times in the same application, the degree of property association is calculated for all combinations of setter methods registered in the MCG, and the final degree of property association is taken to be the arithmetic mean of these values.

Next, the degrees of initial setting and the degrees of property association are used to determine the order of presenting each initial set from all initial sets. Here, "all initial sets" means the combination of all the properties whose values are set in all the collected applications. For example, when there are three properties $p_1$, $p_2$, $p_3$, the set *S* of the initial set *s* for all the properties consists of: $S = \{ \{p_1\}, \{p_2\}, \{p_3\}, \{p_1, p_2\}, \{p_1, p_3\}, \{p_2, p_3\}, \{p_1, p_2, p_3\} \}$. Among all initial sets, the degree of initial set recommendation *sr(s)* to which a certain initial set *s* is recommended is defined as follows.

$$sr(s)=\begin{cases} \dfrac{2\sum_{i=1}^{|s|-1}\sum_{j=i+1}^{|s|} pa(p_i,p_j)}{|s|(|s|-1)} \dfrac{\sum_{k=1}^{|s|} init(m_k)}{|s|} & (|s|>1) \\ init(m_1) & (|s|=1) \end{cases}$$

where s = { p1, ... , p|s| }, $|s|$ is the number of elements of *s*, and $m_x$ is the setter method that sets the value of the property $p_x$.

When there are multiple collected applications, the arithmetic mean of the degree of initial set recommendation calculated for each application is taken as the final degree of initial set recommendation.

The wizard presents the initial sets in order starting with the one that has the largest degree of initial set recommendation. When there are multiple initial sets that have the same degree of initial set recommendation, priority is given to the presentation of the initial set *s* where $|s|$ is greater.

In our system, the frequency with which values are set in the properties are also calculated from the results of running all the applications. When the initial sets are presented in the wizard, candidate sets of multiple specific values for a single initial set are presented in order starting with the property values that were set the greatest number of times.

*D. Automatic generation of wizards*

To generate wizards automatically in our system, we prepared a template in which specific values such as initial sets are turned into parameters and can be modified. Figure 4 shows the appearance of the wizard's template running by itself. In the template, the wizard's common factors, such as next/back buttons, are predefined. The final wizard is generated by externally applying items that make changes to the values of the component's properties and the like. Figure 5 shows a specific example of a generated wizard in action for the component `WEbean`.

When a component to which a wizard has been added by our system is used in a RAD tool, the automatically generated wizard provides the following functions:

- The component information (GUI image) set with the property values indicated by the initial sets is presented in descending order of the degree of initial set recommendation. When setting properties whose values are not reflected in the GUI image provided by the component, the property values are shown substituted with text strings in the wizard, and the reflected status is output after selecting and setting their values.
- When a wizard has finished running, an instance of the component set with the specific property values is generated and arranged in the RAD tool.
- The wizard is provided with facilities to allow users to undo/redo the setting of property values.

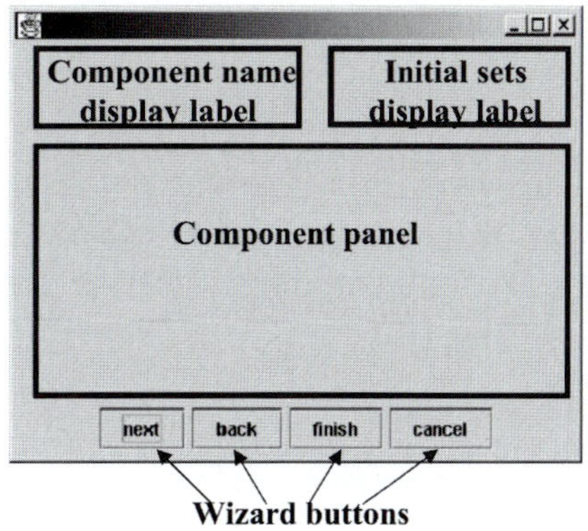


Fig. 4. Wizard template

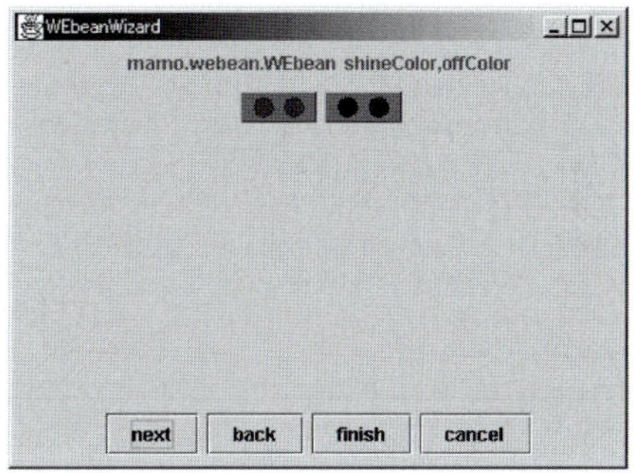


Fig.5. A generated wizard in action

## IV. EXPERIMENTAL EVALUATION

To evaluate the effectiveness of our system, we used it to add a wizard to `WEbean`. The experimental procedure and the resulting wizard are described below, together with the results of the evaluation.

*A. Experimental procedure*

As an application as a target of analysis for estimating the initial values in the wizard, we used the Java applications `App1`, `App2` and `App3` developed by three graduate students using `WEBean`. `App1` is an application in which a button is used to start a light flashing, `App2` is an application that can change the rate of flashing, and `App3` is an application that is identical to `App2` except that it sets the value of a background property. Figure 6 shows excerpts of the program statements where the `WEBean`'s methods are called in the program code of `App1`, `App2` and `App3`.

```
public class App1 { WEbean bean;
 public static void main(String[] args){
  App1 app = new App1();  }
 public App1() {
  bean = new WEbean();
  bean.setShineColor(Color.white);
  bean.setOffColor(Color.black);
  bean.setSpeed(100); } ...
}
public class App2 {  WEbean bean;
 public static void main(String[] args){
  App2 app = new App2();  }
 public App2() {
  bean = new WEbean();
  chooseColor();
  addAcitonListener(this);  }
 public void chooseColor(){
  bean.setShineColor(Color.yellow);
  bean.setOffColor(Color.blue);  }
 public void actionPerformed(
  ActionEvent ae){
   bean.setSpeed(50);  } ...
}
public class App3 {
 public App3() {
  bean = new WEbean();
  chooseColor();
  bean.setBackground(Color.gray);
  addAcitonListener(this); }
 // Other parts are the same as App1
}
```

Fig. 6. Program code of App1, App2 and App3

*B. Wizard generation experiments*

Figure 7 shows excerpts of the parts associated with property setter methods in the MCGs obtained by our system by running App1 and App3 respectively. The MCG obtained by running App2 was the same as the one obtained from App3 except that it had no node corresponding to the method WEbean#setBackground().

Tables I and Table II show the results of calculating the degree of initial setting and degree of property association for all the applications. Figure 8 shows the degree of initial set recommendation calculated from the degree of initial setting and degree of property association. In the following discussion, the WEbean's properties shineColor, offColor, speed and background are abbreviated to sc, oc, sp and bg respectively.

Our system determines the order in which the multiple initial sets are presented according to the results shown in Figure 8. Specifically, since { bg } has the largest degree of initial set recommendation (=*1*), this initial set is presented first. Next, the initial sets { sc }, { oc } and { sc, oc } have the same degree of initial set recommendation (=*0.67*), but { sc, oc } is presented second because priority is given to the initial set with more properties. Finally, the remaining set { sp } is presented third.

Table III shows the initial sets presented by our system and the order of their presentation (denoted as *O*), arrived at as a result of analyzing the three applications. It also shows specific values of the properties in each initial set *s*, and the frequencies with which these values are used in applications (denoted as *N*). According to Table III, the resulting wizard presents the set { bg=Color.gray } as the initial set for { bg }. Similarly, the wizard presents the sets { sc=Color.yellow, oc=Color.blue } and {pp sc=Color.white, oc=Color.black } in that order for { sc, oc }, and presents the sets { sp=50 } and { sp=100 } in that order for { sp }.

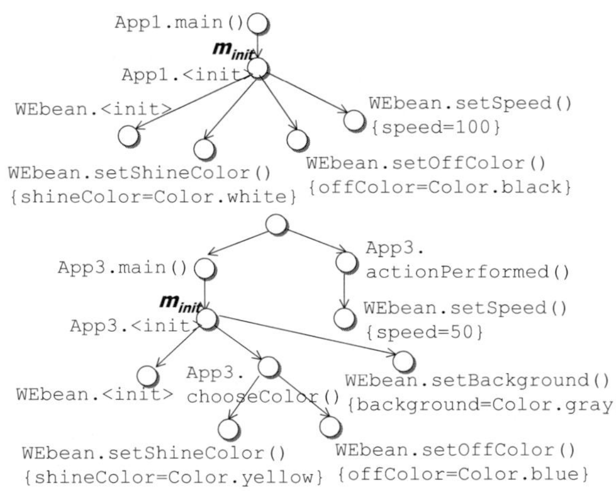


Fig. 7. Obtained method call graphs (Top: App1; Bottom: App3)

TABLE I
CALCULATED DEGREE OF INITIAL SETTING

| Method *m* | *init(m)* in the application: | | |
|---|---|---|---|
| | App1 | App2 | App3 |
| setShineColor() | 1 | 0.5 | 0.5 |
| setOffColor() | 1 | 0.5 | 0.5 |
| setSpeed() | 1 | 0.25 | 0.25 |
| setBackground() | | | 1 |

TABLE II
CALCULATED DEGREE OF INITIAL SETTING

| Properties $p_i$, $p_j$ | $pa(p_i, p_j)$ in the application: | | |
|---|---|---|---|
| | App1 | App2 | App3 |
| sc, oc | 1 | 1 | 1 |
| sc, sp | 1 | 0.25 | 0.25 |
| oc, sp | 1 | 0.25 | 0.25 |
| sc, bg | | | 0.5 |
| oc, bg | | | 0.5 |
| sp, bg | | | 0.33 |

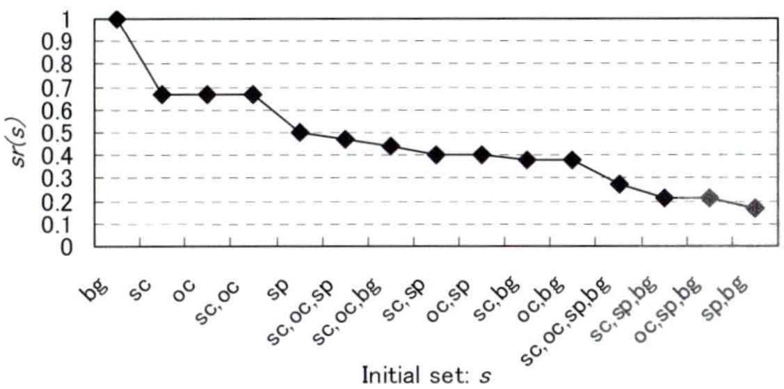


Fig. 8. Calculated degree of initial set recommendation

TABLE III
RECOMMENDED INITIAL SET WITH PROPERTY VALUES AND FREQUENCY

| O | s | sr(s) | Property value | N |
|---|---|---|---|---|
| 1 | { bg } | 1 | Color.gray | 1 |
| 2 | { sc, oc } | 0.67 | Color.yellow, Color.blue | 2 |
| | | | Color.white, Color.black | 1 |
| 3 | { sp } | 0.5 | 50 | 2 |
| | | | 100 | 1 |

*B. Comparative evaluation of wizards*

We performed a comparative evaluation of a WEbean's wizard (called "original wizard") provided by an original component developer [10] shown in Figure 2 and a wizard (called "generated wizard") generated automatically in our experiments. The original wizard consists of three classes and uses four WEBean's properties (sc, oc, sp and bg) in the presentation of the initial values. The original wizard presents these properties in two sets: { sc, oc, bg } and { sp }. On the other hand, the generated wizard presents three initial sets { bg }, { sc, oc } and { sp } in that order.

A common point in both wizards is that they present the speed property separately. As for the other three properties shineColor, offColor, and background, the original wizard presents them as a single set, whereas the generated wizard presents them as two initial sets. Although these wizards differ in terms of how the sets are combined, it can be seen that the initial settings are configured from the same properties in both cases. We have thus shown that our system was successful in automatically recognizing a combination of properties requiring initial setting and presenting them in a wizard in a similar form to the intentions of the component developers.

Furthermore, we have shown that by automating all the work except for the gathering of applications that use a target component and converting this component into a data acquisition component in the applications, our system is far more effective in terms of development costs for the development of wizards than the conventional manual approach. In the above-mentioned experiments, the component WEbean's developer has consumed the cost about *29* [ELOC] (Effective Lines of Code) for its wizard class; however, this cost is unnecessary for the user of our system.

## V. RELATED WORK

Our system is the first one to automatically generate wizard interfaces for JavaBeans components by dynamic analysis of application programs. Nonetheless, our system bares resemblance to several existing techniques [13][15].

Sheong proposed a wizard framework as a template having variable parts for the development of wizards for Java applications, and a wizard builder tool to support development using this framework [13]. However, Sheong's wizard framework and builder are intended to provide shared templates and a support environment for developing wizards for entire completed Java applications, and do not provide functions for generating wizards for JavaBeans components. Also, no mechanisms are provided for deriving/presenting the specific values to be set for the component properties.

Birngruber and Hof proposed a BeanPlan system that uses combinations of previously set JavaBeans component groups to support the rapid design of Java applications [15]. BeanPlan guides the application programmer semi automatically (similar to a wizard) through the component assembly process; however, BeanPlan does not provide a mechanism for deriving/presenting the specific values to be set for the component properties.

## VI. CONCLUSION

In this paper, we have proposed a system that analyzes the dynamic usage information of JavaBeans components in Java applications to acquire information needed for the initial setting of components, and automatically attaches wizards to these components to allow their initial values to be set according to the component developer's intentions. Our experimental results have shown that the properties presented for initialization by the wizards generated by the proposed system are close to the properties imagined by the original component developer. Accordingly, we have confirmed that the proposed system can bring about substantial reductions in the cost of generating wizards.

In the future, we intend to apply the proposed system to groups of complex applications that use large numbers of JavaBeans components, and to evaluate the convenience of the wizards automatically added to these components, thereby obtaining practical confirmation of the proposed system's effectiveness with a wide variety of components.

## REFERENCES

[1] C. Szyperski, Component Software: Beyond Object-Oriented Programming, Addison-Wesley, 1999.

[2] J. Hopkins: Component Primer, Communications of the ACM, Vol.43, No.10, 2000.

[3] G. Hamilton: JavaBeans 1.01 Specification, Sun Microsystems, 1997, `http://java.sun.com/products/javabeans/`

[4] L.G. DeMichiel: Enterprise JavaBeans 2.1 Specification, Sun Microsystems, 2003, `http://java.sun.com/products/ejb/`

[5] M. Aoyama and T. Yamashita: Software Commerce Broker over the Internet, Proc. 22nd IEEE Annual International Computer Software and Applications Conference, 1998.

[6] M. Welie: The Wizard Pattern, Proc. CHI 2000 Workshop on Pattern Languages for Interaction Design: Building Momentum, 2000.

[7] H. Washizaki, H. Yamamoto and Y.Fukazawa: A Metrics Suite for Measuring Reusability for Software Components, Proc. 9th IEEE International Symposium on Software Metrics, 2003.

[8] A. Denning: ActiveX Controls Inside Out, Microsoft Press, 1997.

[9] Object Management Group, OMG Unified Modeling Language Guide Specification, 1999, `http://www.uml.org/`

[10] M. Sakamoto: JavaBeans Programming Primer, Ohmsha, 1997. (in Japanese)

[11] S. Chiba: Javassist: A Reflection-based Programming Wizard for Java, Proc. Workshop on Reflective Programming in C++ and Java, 1998.

[12] I.H. Witten, D.Bainbridge and S.J. Boddie: Power to the people: End-user building of digital library collections, Proc. ACM/IEEE Joint Conference on Digital Libraries, 2001.

[13] C.S. Sheong: Build Wizards Quickly Using a Swing-Based Wizard Framework, JavaReport, May, 2001.

[14] Sun Microsystems, BeanBox, `http://java.sun.com/products/javabeans/`

[15] D. Birngruber and M. Hof: Using Plans for Specifying Preconfigured Bean Sets, Proc. 34th International Conference on Technology of Object-Oriented Languages and Systems, 2000.

# Content Based Image Retrieval Using Quadrant Motif Scan

Tsong-Wuu Lin and Chung-Shen Hung

Department of Computer and Information Science, Soochow University, Taipei, Taiwan R.O.C.

***Abstract*-Image retrieval based on Quadrant Motif Scan (QMS) is proposed in this paper. Motif traces from image pixels are the core idea to extract feature vectors and used for distinguishing images by region-based comparisons. We exploit recursive quadrant segmentation in images and derive representative motif for stratified regions. In this sense, a parent region is segmented into sub-regions until a predefined stratum threshold. Matching data for each region contains its motif code plus the result from uniformity detection. By the credit setting, the similarity mechanism proceeds in corresponding regions from two images in a top-down manner. Dynamic parameter adjustments to relevance feedback can help pursue best retrieval results. Besides, a peak inspection technique is also added in the QMS matching metric to enhance performance. Experimental results reveal effectiveness and efficiency comparable to the Motif Cooccurrence Matrix (MCM) method with invariance to image scaling.**

***Index Terms*—Content-based, image retrieval, motif scan**

## 1. Introduction

VISUAL information has proliferated for multiple purposes in recent years. With the Internet burst, a number of multimedia applications are evolving pervasively for intuitive information expression. Over the decades, many brilliant researches have produced a variety of outstanding techniques in image-related fields, and some of those became the de facto standards, such as JPEG in [1]. However, those mature studies largely reside in image encoding and storage format. By contrast, a wide range of approaches in [2, 3, 5, 6] using color, shape, or other factors for *Content-based image retrieval (CBIR)* is still under sprightly development. They focus on different attributes of images for retrieval on particular occasions while some existent CBIR systems afford versatile querying ability for users as in [4].

There are a plenty of works in the image retrieval area. Some of the renowned approaches are recognized as introductory examples for later works. For instance, color histogram in [5] is a well-known one among those precedents. It utilizes the statistics of global color distribution to calculate the similarity between two images. Generally speaking, this is quite an efficient way to retrieval, but in fact some flaws exist. Although color histogram has the advantage insusceptible to local parts of color differences between images, it otherwise provides the likelihood to misjudge the similarity. That is, dissimilar images with similar color distributions, as a coincidence, may occur even though different contents, objects or spatial arrangements are presented.

Other akin to but advanced techniques were devised to remedy color histogram afterwards. For instance, Color Coherence Vectors (CCV) in [6] adds the information of pixels coherence of one particular color and generates coherent regions. This enhances distinction of pixels with same colors but not distributed in the same regions. Moreover, another attractive technique, Color Correlogram (CC) in [7], highlights the spatial correlations of colors. It takes on the probability of joint occurrence from any two pixels in separate colors. Computation and the size of extracted features in CC are respectively easy and small. Both of the two methods appear to perform much better than traditional color histogram. In a word, spatial information is exactly a significant cue for image retrieval implementation. Not only the mentioned methods but more related studies have shown spatial property feasible and useful to retrieval refinement. From this point of view, we adopt the spatial factor into our work to reduce fidelity loss. Like Blobworld and SIMPLIcity in [8, 9], region-based techniques for image retrieval flourish in an alternate way.

Jhanwar et al. [10] retrieve images based on motif notion and capture the low level semantics of space filling curves. The Motif Cooccurrence Matrix (MCM) is the descriptive structure of motif features scanned from image pixels. At first, this method divides a whole image into 2 × 2 pixel grids. Each of these grids is then replaced by one motif from a set of six Peano scan (Z-scan) motifs [11, 12]. The particular version of applied motifs is shown in Fig. 3. For simplified computation consideration, only six diverse motifs traversed in a 2 × 2 grid are adopted, all starting from the top left corner. Then, the MCM is constructed from the motif transformed image, which is composed of a sequence of motifs. According to the definition, this method calculates the probability of finding a motif $i$ at a distance $k$ from a motif $j$. With the designated formula, the distance between the MCMs of two images is used as the similarity measure. Besides, this study also declares that the proposed method is invariant to any monotonic mapping of the image gray levels such as contrast stretching or histogram equalization.

Note that, there are some noticeable requirements about this



*T. Sobh and K. Elleithy (eds.), Advances in Systems, Computing Sciences and Software Engineering*, 69–76.

retrieval technique. Basically, it merely includes individual 6 × 6 motif matrices for every color plane irrespective of the image size hence makes MCMs very efficient for retrieval. Secondly, the distance *k* is fixed to capture the MCMs of two images in this study. For avoiding translation effects, they conceive the idea of adding 3 extra images, which are shifted by one pixel horizontally, vertically and diagonally from the original image. Therefore, multiple MCMs from the corresponding motif transformed image of these shifted images are consequently generated. On comparison stage, all four sets of MCMs of the query image should be compared with the non-shifted MCMs of target images in database. As a result, the minimum distance is pick out and regarded as the actual distance between the query and target images.

Through detailed observations, the MCM scheme is especially suitable for images in equal size for retrieval. Variation in image size can inevitably induce inconsistent motif statistics; the total amount of motifs is varied from image to image. Without the common gauge of image size, it does not make sense to matching process. Meanwhile, although the MCM contains spatial relationships between grids to a certain extent, it is possibly proper to apply in special retrieval scenarios. Those scenarios are likely a series of repetitive patterns, such as a block of buildings, windows and something else in a regular and neat arrangement. Because the granularity of motif san, in a 2 × 2 grid, is minute, a sequence of repetitive motif traces can be captured and accumulated regardless of their exact locations. For example, textured images are the best demonstration. Therefore, the more percentage the texture is stuffed in images, the more effective the method can present. Once the scenario is in outdoor scene, not textured, the performance might be down with unexpected retrieval results.

In this paper, we propose another image retrieval method based on motif symbols, called *Quadrant Motif Scan* (QMS). Consecutive quadrant segmentation on images is the main strategy in our scheme. With a motif extraction for each region, we collect all matching data, representative motifs from regions, for retrieval. Through the hierarchy of motif information inside an image, we devise a matching algorithm for similarity comparison with ranking result. Our experiments show that QMS has the merits comparable or even superior to the MCM method

This paper is organized as follows. In section 2, the proposed QMS is explored. Section 3 describes the matching metric used in the QMS. In section 4, the experimental results are presented with a related discussion. The final section gives the conclusion of our approach and prompts the relevant errands for future work

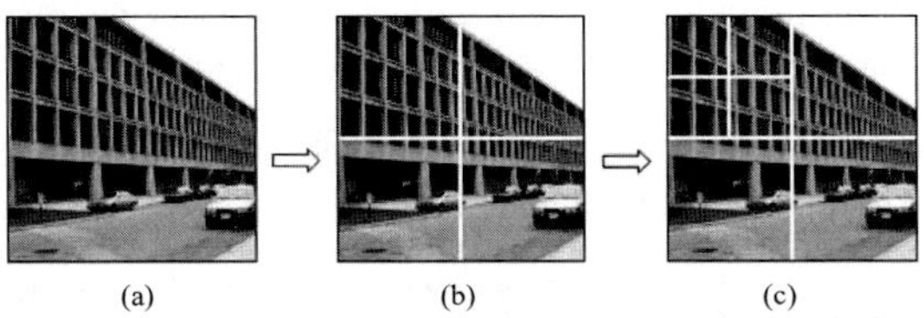


Fig. 1. The recursive segmentation process in Quadrant Motif Scan. (a) The original image. (b) 4 quads for the 1st stratum. (c) 4 quads for each successive stratum. Note that only the upper-left region is illustrated here, the same process will proceed on other three regions as well.

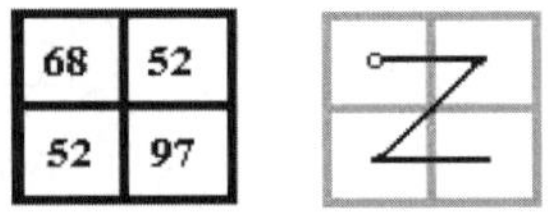


Fig. 2. An example of motif traverse.

## 2. QUADRANT MOTIF SCAN

Motif is an important feature to express the chromatic trend lying in a bounded region. In the MCM, the region is devised a 2 × 2 pixel grid. Instead of deriving a motif on such a tiny region, our QMS uses an entire image as the initial region, as shown in Fig. 1a. An image is first subdivided into four non-overlapping parts in Fig. 1b. The four segmented quads form the source generating a motif for the first region which belongs to the 1st stratum. The QMS then calculates the mean value of all pixels for each quad. These mean values range from 0 to 255, and quadrant motif scan proceeds to yield the order among them.

To avoid ambiguity of motif scan, we introduce the suggestion used in the MCM study in [10]. The QMS follows the breadth first strategy to regulate the motif recognition. For instance, Z-type motif is recognized rather than N-type in Fig. 2. Note that any kind of motif traverses starts from the top-left corner regardless of its actual value. After a motif is extracted from a region, a simple code is given to represent the specific motif (see Fig. 3). This arrangement implies the data storage consideration and facilitates further computation. Besides, during a motif extraction, the highest and lowest mean values from the four quads are recorded. A simple subtraction operation is then performed to retain the information whether the region is uniform or not. Finally, both of the motif and uniformity information for a region is collected and assembled together into a single data structure. The uniformity threshold in our scheme is set 35 in default; difference less than the threshold will lead a region to be uniform.

### *2.1 Stratified Motif Scan*

Basically, section 2 only describes a portion of operations in the QMS, which is concentrated on the detailed process of a single motif scan. However, there are following tasks to complete the whole procedure. As mentioned above, the

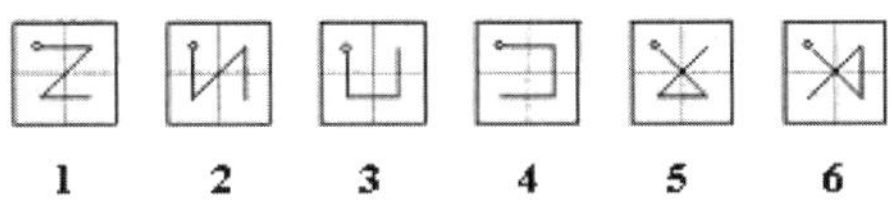


Fig. 3. Codes for each type of motifs

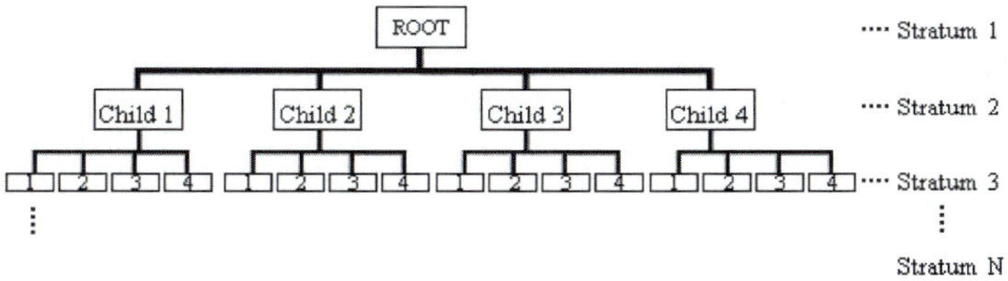


Fig. 4. Hierarchy diagram for segmentation of child regions.

motif scan process will continue to subdivide the image into more and more sub-regions by an increment of 4 each time. When the first region is subdivided into 4 child regions, a new stratum is created simultaneously and labeled stratum *i* (see Fig. 4). Assume that there is an 8 × 11 image, and its pixel values are shown in Fig. 5a. The image is divided into four quads (regions) to evaluate separate means, and then a motif is received. This motif will be saved in code (see Fig. 3) and if the region is uniform, the code is added 0x08. For instance, the motif in Fig. 6c is originally encoded 0x04. Uniformity for its corresponding region is calculated from these mean values in Fig. 5b. In the case, this region appears to be uniform, and the code is modified to 0x0C. As a result, this motif block, which contains a motif record, is associated with the stratum *i* for score calculation.

Likewise, successive subdivision operations from the current region continue until a pre-defined stratum threshold is reached. As shown in Fig. 6a, the current region is subdivided into four child regions, and the second stratum is then added. The same manipulation to evaluate mean values is carried out for every child region, and separate motifs are eventually derived in Fig. 6c. In particular, the four motif blocks belong to the same stratum (2nd) so they share a common credit for the matching metric. One thing deserved to be mentioned is that subdivided child regions from diverse parent regions may belong to the same stratum. For example, there are 16 motif blocks separately derived from 4 different parent regions in the 2nd stratum, but those 16 motif blocks exactly belong to the 3rd stratum. Thus, the motif data extracted from the child regions in the same stratum is all arranged on the same stratum shelf. In short, a parent's quad, used as an element for its motif scan source, is regarded as a child's intrinsic region.

This architecture for stratified motif scan is to find out local motif information throughout an image. Due to different significance of separate strata, varied credits, or so-called weights, are granted to reflect their power. In the QMS scheme, we designed an algorithm using these credits to match images. As a general rule, a broader region hence its four

| 16 | 23 | 58 | 77 | 32 | 81 | 124 | 49 |
|---|---|---|---|---|---|---|---|
| 63 | 21 | 29 | 66 | 34 | 198 | 75 | 74 |
| 82 | 84 | 65 | 18 | 5 | 19 | 78 | 96 |
| 53 | 55 | 107 | 123 | 93 | 52 | 27 | 44 |
| 68 | 61 | 40 | 37 | 41 | 50 | 22 | 28 |
| 75 | 87 | 64 | 33 | 20 | 11 | 10 | 96 |
| 81 | 79 | 34 | 25 | 88 | 57 | 43 | 30 |
| 99 | 83 | 15 | 23 | 59 | 180 | 71 | 156 |
| 47 | 34 | 51 | 158 | 221 | 72 | 6 | 48 |
| 168 | 54 | 93 | 67 | 48 | 24 | 27 | 14 |
| [illegible] | [illegible] | [illegible] | [illegible] | [illegible] | [illegible] | [illegible] | [illegible] |

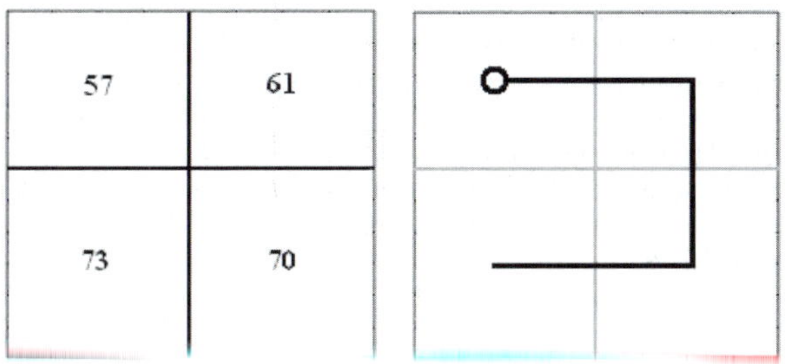


(a) (b) (c)

Fig. 5. Formation of a motif for a region. From left to right are (a) image pixel values, (b) means of four quads, and (c) a resultant motif

| 16 | 23 | 58 | 77 | 32 | 81 | 124 | 49 |
|---|---|---|---|---|---|---|---|
| 63 | 21 | 29 | 66 | 34 | 198 | 75 | 74 |
| 82 | 84 | 65 | 18 | 5 | 19 | 78 | 96 |
| 53 | 55 | 107 | 123 | 93 | 52 | 27 | 44 |
| 68 | 61 | 40 | 37 | 41 | 50 | 22 | 28 |
| 75 | 87 | 64 | 33 | 20 | 11 | 10 | 96 |
| 81 | 79 | 34 | 25 | 88 | 57 | 43 | 30 |
| 99 | 83 | 15 | 23 | 59 | 180 | 71 | 156 |
| 47 | 34 | 51 | 158 | 221 | 72 | 6 | 48 |
| 168 | 54 | 93 | 67 | 48 | 24 | 27 | 14 |
| 60 | 74 | 66 | 177 | 205 | 98 | 17 | 85 |

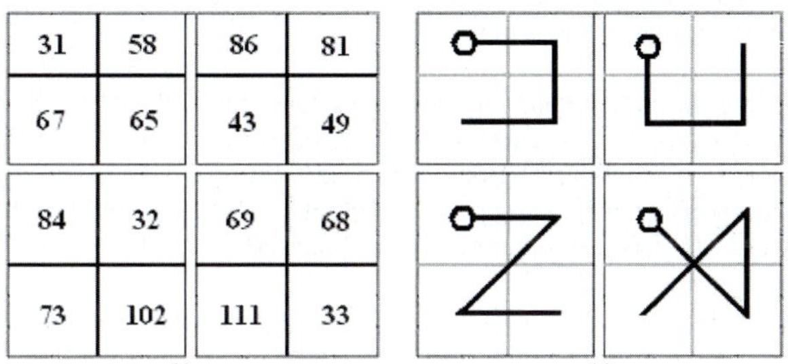


(a) (b) (c)

Fig. 6. Successive subdivisions into four child regions. There are 4 sets of (a) pixel values, (b) quad means, and (c) individual motifs.

TABLE I
VARIATIONS IN NUMBERS OF STRATA, REGIONS AND MOTIF BLOKS

| Stratum No. | 1 | 2 | 3 | 4 | 5 | 6 | 7 | 8 |
|---|---|---|---|---|---|---|---|---|
| Regions (Motif blocks) | 1 | 4 | 16 | 64 | 256 | 1024 | 4096 | 16384 |
| Accumulated motif blocks(A) | 1 | 5 | 21 | 85 | 341 | 1365 | 5461 | 21845 |
| Granularity | Coarse | ◄ | | | | | ► | Delicate |

quads may cause an inaccurate averaging effect, except a uniform region. The credit, set on a stratum basis, to that region should be lowered. However, this concern is sometimes unnecessary depending on the content of an image. On the other hand, what the exact stratum threshold is supposed to be becomes another issue. If a higher stratum threshold is designated, more blocks of motif are then received. The extra motifs obtained from narrowed sub-regions, especially in the bottom strata, have the ability of depicting an image more precisely. From this viewpoint, the retrieval performance can explicitly rise via adding more strata at the expense of computational costs. In practice, an adjustable stratum threshold might be needed according to the user or application's demand. Performance choice is usually recommended to come to a compromise between speed and effect. Moreover, the number of stratum to set is to be in connection with actual image size. Excessive setting can cause meaningless or even negative effect. Table 1 shows the cross-references among strata, regions, and motif blocks.

## 3. MATCHING AND RANKING METRICS

Before matching two images, the prerequisite preparations are as follows:

(i) A pre-defined stratum threshold for the motif scan process. All images are manipulated by this rule so that they have the consistent number of motif blocks stored in the database for the matching functionality.

(ii) A criterion for uniformity detection. This is used as a value range to recognize whether a region/motif block is uniform or not. The same as stratum threshold, this must be decided at the very beginning.

(iii) Varied credits (weights) for each stratum. These variables are in relation to retrieval results.

Item (i) and (ii) are the most important parameters for QMS implementation. Both of them can also be critical to retrieval performance. In turn, the third setting is about scoring schemes and is independent of the two previous settings. This can be dynamically adjusted at run-time to evaluate and rectify the retrieval performance.

Assumed that the required parameters are verified and the QMS database is then constructed, the matching process can launch on the final comparison stage. Motif blocks of two images, the query and target, are matched on the block-by-block basis. If a pair of corresponding blocks/regions is exactly the same, which have equal motif type and uniformity property, the target scores the specified credit. Otherwise, the target fails in this block and goes on next until all motif blocks are through. When the process is finished, a total score will be attached to that target. The higher the score is, the more similar the query and target images are. In effect, it will receive a perfect score when the query and target are exactly the identical image. After all images in the database are thoroughly matched with the query, a score list is made in a descending manner.

Given a mapping function, it can find out the *j*th stratum to which the *i*th block belongs and hence the corresponding credit $C_j$. Each pair of motif blocks, $M_i^q$ and $M_i^t$ for the query and target respectively, is compared to get a similarity status $\boldsymbol{S_i}$. If two blocks are equal, it leads to $\boldsymbol{S_i} = 1$, or 0 otherwise. The overall matching expression is as follows:

$$D(I_q, I_t) = \sum_{i=1}^{K} S_i C_j \quad \forall j \in [1 \ldots N], \ j = map(i)$$

$$with\ S_i = \begin{cases} 1, if\ M_i^q = M_i^t \\ 0, if\ M_i^q \neq M_i^t \end{cases}, \mathrm{A}[j-1] < i \le \mathrm{A}[j], \quad (1)$$

where $\boldsymbol{I_q}$ and $\boldsymbol{I_t}$ are the query and target images and $\boldsymbol{K}$ means the maximum of motif blocks assigned by the number of strata $\boldsymbol{N}$ specified for motif scan process. Array $\boldsymbol{A}$ gives the information the *i*th block is subsidiary to the *j*th stratum. This distance measure will result in a score sum for ranking.

## 4. EXPERIMENTAL RESULTS AND DISCUSSION

Query-By-Example is a common technique used in many image retrieval systems. For instance, IBM QBIC in [13] is one of the best-known systems. Our *Quadrant Motif Scan* also applies such a skill in the system design. On the side, we chose the MCM method to serve as the opposite to show the improvement of motif-based approaches.

### *4.1 Experimental setup*

We collected a huge amount of images to test the retrieval performances in different query scenarios. To demonstrate competitiveness, we only show the experiments with

TABLE III
IMAGE CLASSIFICATION IN THE 128 × 128 SET FROM THE MIT VISTEX COLLECTION

| Class | Bark | Brick | Buildings | Fabric | Flower | Food | Grass | Leaves | Metal |
|---|---|---|---|---|---|---|---|---|---|
| Pieces | 13 | 9 | 11 | 20 | 8 | 12 | 3 | 17 | 6 |

| Class | Misc. | Paintings | Sand | Stone | Terrain | Tile | Water | Wood | Total |
|---|---|---|---|---|---|---|---|---|---|
| Pieces | 4 | 13 | 7 | 6 | 11 | 11 | 8 | 3 | 162 |

25 images from the Aridi art collection and the MIT Media Lab's collection, which is also introduced in the MCM study. Especially, we made a combination of images approximate the specification for the images used in the MCM method. Textured and non-textured images are existent in our experiments for specific comparison purposes. Scaling manipulation for some samples is also done to examine the retrieval ability of the QMS. About the QMS system, the stratum threshold is adequately set to 5 in default with 341 motif blocks. We found this setting competent to satisfy general retrieval situations and a good balance between computational cost and speed. This setting is, however, still subject to variations of image size. Table 2 shows the related parameters used in our experiments. Again, these numeric values are changeable to pursue the best performance. If the stratum threshold is changed, it requires a reconstruction process of motif extractions for all images in the database. By contrast, the credit setting is flexible and could be dynamically adjusted to evaluate retrieval results. Since there might be many diverse images in the database, a universal credit setting for all kinds of retrieval is really hard to define. We therefore design this flexible facility to meet needs in our experiments. This feature is very helpful when we want to clarify the empirical credit setting toward a specific sort of content in a query image.

With different setups, two typical schemes (see Table 2a) are presented for performance analysis. The only difference is at the number of stratum (and its motif blocks). Scheme B is used to seek if there is a better solution after scheme A is tried out. Normally, a larger stratum threshold can exhibit more accurate motif information about images. A more satisfactory result is then prospective. Even though it has been proven useful raising the stratum threshold for enhancing retrieval, this advantage does not always remain true. Through our observations, some factors can still weaken the effect. For example, the content variation or textual semantics of the query image is. Heterogeneity of images in the same database is another key point. These factors more or less give rise to direct or indirect influences on our multi-stratum comparison. Sometimes it may result in a worse case when a larger stratum threshold is given. Therefore, even though this credit adjustment is available, it does not always bring out substantial effects.

TABLE II
SYSTEM PARAMETERS IN QUADRANT MOTIF SCAN

| Scheme | Strata | Motif Blocks | Uniformity | Peaks |
|---|---|---|---|---|
| A | 5 | 341 | 35 | 10 |
| B | 6 | 1365 | | |

(a)

CREDIT SETTING FOR SCORING

| Stratum No. | 1 | 2 | 3 | 4 | 5 | 6[a] |
|---|---|---|---|---|---|---|
| Credit | 4 | 4 | 4 | 4 | 4 | 4 |

(b)

[a]For scheme B only.

Besides the parameters in motif construction, we add a peak inspection function for retrieval supplement. This function uses a peak number to set up how many global peaks of same pixel values accumulated in an image are selected. These peaks are traversed from the highest with proximal peaks omission. To enhance the proposed QMS method, the peak inspection can refine retrieval results by promoting the matching score according to the peak similarity between images. Hence, the resulting similar images with peak difference will be ranked down. The computational cost is considerably trivial to this additional enhancement in exchange for the advancement of precision.

### 4.2 Results

We have performed a mass of queries using multiple combinations of images in order to identify which type of images is more favorable for the QMS scheme. Hypothetically, different types of images may induce different effects to retrieval.

The 128 × 128 set of images in the MIT Vistex[1] collection was set as the first database used to compare with the MCM method. Information for the classification and volume in the database is shown in Table 3. In advance, we would like to note that images of the same class are not necessarily similar even though they are previously classified into groups. In addition, we solely implemented the MCM method without adding color histogram so as to do pure comparison. Fig. 7a shows the query image same as the example used in the MCM study in [10]. The retrieval results in Fig. 7b expose the effectiveness of the MCM method, retrieving 10 out of 11 in the series of buildings. Although the results may cause

[1]http://vismod.media.mit.edu/vismod/imagery/VisionTexture/vistex.html

(a) (b)

Fig. 7. (a) Example of the query image. (b) Retrieval results using the MCM method. The retrieved images are ordered from left to right and top to bottom by their similarity to the query image.

different evaluations from individual perceptions, some characteristics are apparent to realize. For example, the intricate grids, which form the appearance of the buildings, are the most salient features in the query image. Among these grids, two colors, black and yellowish gray, are alternate regularly throughout the surface of the buildings. The extent of buildings also occupies most of the image. Hence, we can intuitively infer that a great deal of some repetitive motifs will be extracted during the motif scan process. Plus the uniform parts like the black grids, they all contribute to the similarity metric applied in the MCM. Because the granularity to form a motif is quite small, a 2 × 2 grid, any massive prominent patterns are crucial to retrieval; the MCM's similarity metric is based on their statistics. Despite the fact that there are a few deficiencies in the MCM retrieval results shown in Fig. 7b, the performance is brilliant. Referring to the definition for retrieval performance in [14], we made a slight modification and only the first 24 retrieved images in the first page in our system are shown for discussion. Other images are still ranked and remained below the retrieval manifest. We followed this suggestion to display the meaningful matching results.

Taking a look at the QMS examples, Fig. 8a shows the results using 5 strata as the motif scan parameter. The first retrieved, upper-left image is the query image itself, and the second one is unanimously the most similar image among all others. In this scheme, similarity is decided on an overall view, such as the 4th and 9th ones. Then, we set up another scheme using 6 strata. As we can see in Fig. 8b, the ranking deficiencies in the previous scheme have been evidently improved. To be more persuasive, the 6th image in Fig. 8b should be put ahead to the third or fourth position. In fact, we can achieve this through other credit settings, but here we only use the default setting in Table 2b for general-purpose queries. In comparison with the MCM, however, our QMS appears to be comparable. As mentioned earlier, though some of the retrieved images by the MCM belong to the same class, i.e. buildings, their similarity to the query image may be differentiated from individual to individual. At least speaking of quality, our QMS reveals its competitiveness in this query example.

The MCM method keeps robust when the query image is highly textured or regularly organized. As shown in Fig. 9, any one in the same pair as the query image can readily retrieve the other as the most prior result. Instead, the power to retrieving similar images by the MCM obviously diminishes when the other image is perceptually similar to the query one (see Fig. 10). Unlike the MCM, the proposed QMS is less sensitive to this constraint. The following examples show the retrieval performance when the first (left) one in each pair in

(a) (b)

Fig. 8. Retrieved images by the QMS with different stratum settings. The stratum threshold is set to (a) 5 and increased to (b) 6.

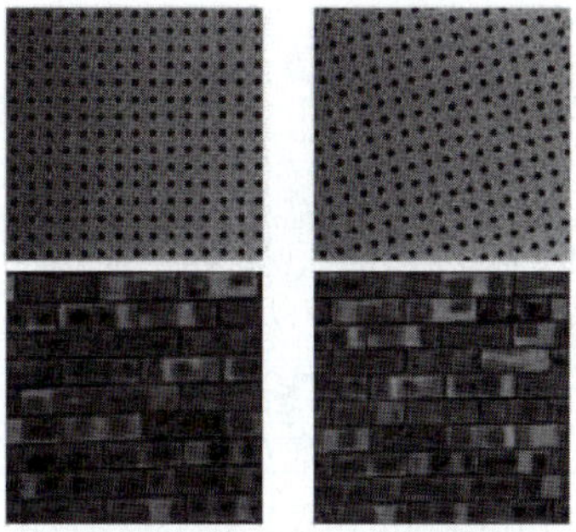

Fig. 9. Two pairs of closely similar images.

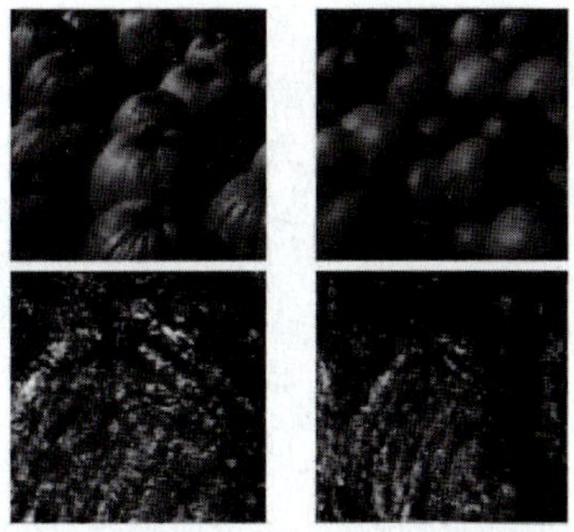

Fig. 10. Two pairs of perceptually similar images.

Fig. 10 is set as the query image, and the second (right) one is the only similar image in the database. Clearly, the ranking results are getting better from a to c in Fig. 11. After the peak inspection technique is applied to our method, the ranking result is further improved. Turn to another query example (see Fig. 12), this advantage of peak inspection sheerly emerges. In this case, both of the MCM and QMS did not retrieve the indicated image (the 4th in Fig. 12) within the first page. Actually, it was ranked 29 and 35 by the QMS and MCM respectively. Therefore, without peak inspection our QMS may be considered failed in this query example. Besides, we have performed multiple other queries to test its efficacy and we found it useful to refine retrieval results. More importantly, this technique merely causes a slight overhead in computation.

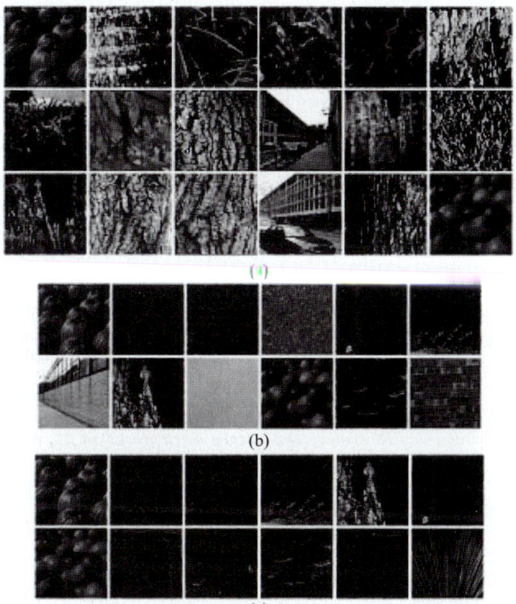

Fig. 11. Examples of retrieval performances by different methods when images are perceptually similar to each other. (a) MCM, (b) QMS (5 strata), and (c) QMS with peak inspection.

Above experiments are all processed on target images of equal size. In turn, we will show that the QMS has another feature aimed at databases containing images of unequal sizes. To account for the retrieval capability invariant to image scaling, a set of 25 color images was collected from the Aridi art collection.[2] These images differ in size ranging from 239 × 363 to 724 × 192 in disproportional ratios. Moreover, a sole entity, such as a crest or ribbon, located in the center is the common property of the images. From the query examples shown in Fig. 13, the performance of MCM obviously drops down. Due to unfixed image size, the MCM method can not match the query and other images on a consistent criterion. That is, the amount of motifs in the matrix varies image by image. The variation makes matches distorted and hence the invalid results. As shown in Fig. 13a, the query sample is expected to match those images with a circle-shaped object. However, the matching results almost lie in a disordered state. Likewise, the next example shown in Fig. 13b, with a cross-shaped object instead, presents the same problem. Thus, the MCM method seems incompetent to such a scenario.

Along with the previous scheme setting (5 strata without peak inspection), our proposed QMS apparently gave much better results than the MCM (see Fig. 14). Rather than that unorganized matching order by the MCM, the QMS make it more reasonable and acceptable. Ideally, images with a circular or cross object should be retrieved first before others in this case, and the QMS almost fulfills this requirement. Though both retrieval performances are not perfect, refinement could still be reached via changing credit setting. Thus, our QMS is capable of querying in such a scenario in which target images vary in size and outperforms the MCM. Furthermore, we also examined the sensitivity to image scaling to the query

Fig. 12. Example of retrieval results using the QMS with peak inspection.

[2] http://www.aridi.com/vol14.htm

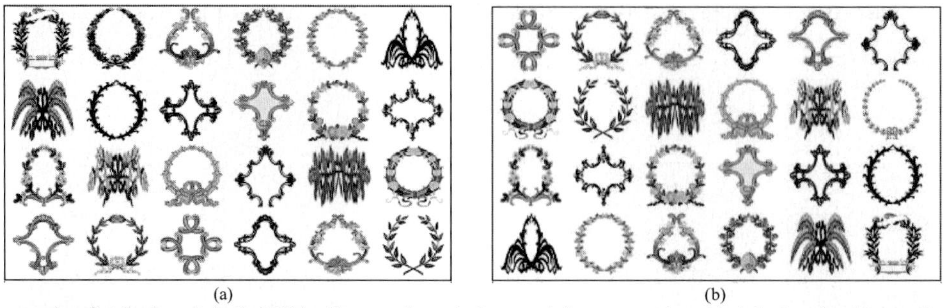

Fig. 13. Two examples of retrieval results by the MCM. The query image in the upper-left corner contains a (a) circle-shaped or (b) cross-shaped object. All images are displayed in thumbnail.

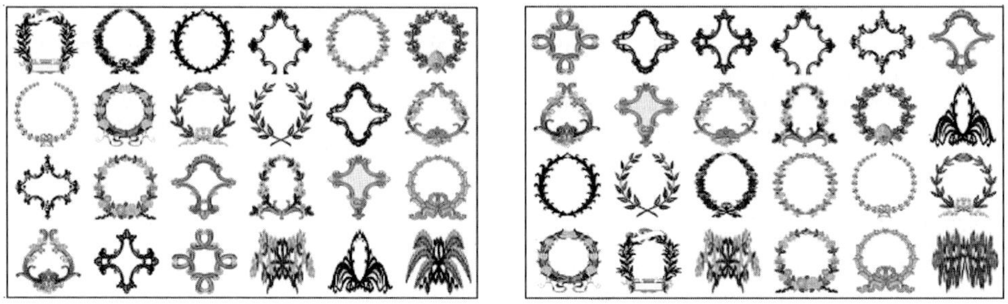

Fig. 14. Two examples of retrieval results by the QMS using the upper-left image as the query image.

image itself and the results showed that the QMS is tolerant of that. In another trial, however, our method is sensitive to image rotation.

## 5. CONCLUSIONS

In this paper, we have introduced another method, Quadrant Motif Scan, based on motif for image retrieval. Images are segmented recursively into sub-regions by an increment of four each time and motif scan proceeds on every region in a limited stratum threshold. Combining with uniformity detection, representative data for a single region is then made. With an arbitrary credit setting, scores for regions on a stratum basis, the matching metric is conducted in a simple, fast calculation process. Other enhanced techniques, such as the mentioned peak inspection in our experiments, can be incorporated in our method to promote precision. Nevertheless, there are existent drawbacks like rotation problem to overcome in future work.

## ACKNOWLEDGEMENT

We are grateful to the reviewers' appreciation and comments for revision. This work was funded by the Taiwan NSC under grant no. 94-2213-E-031-006.

## REFERENCES

[1] W.B. Pennebaker and J.L. Mitchell, "JPEG Still Image Data Compression Standard," Van Nostrand Reinhold, New York, 1993.

[2] F. Mahmoudi et al, "Image Retrieval Based on Shape Similarity by Edge Orientation Autocorrelogram," *Pattern Recognition*, vol. 36, no. 8, pp. 1725-1736(12), Aug. 2003.

[3] H. Nezamabadi-pour and E. Kabir, "Image Retrieval Using Histograms of Uni-color and Bi-color Blocks and Directional Changes in Intensity Gradient," *Pattern Recognition Lett.*, vol. 25, pp. 1547-1557, 2004.

[4] W. Niblack et al., "Querying Images by Content Using Color, Texture and Shape," *Proc. SPIE*, vol. 1908, pp. 173-187, 1993

[5] M. Swain and D. Ballard, "Color indexing," *Int'l J. Computer Vision*, vol. 7, pp. 11-32, 1991.

[6] G. Pass, R. Zabih and J. Miller, "Comparing Images Using Color Coherence Vectors," *Proc. ACM Conf. on Multimedia*, pp. 65-73, 1996.

[7] J. Huang et al., "Image Indexing Using Color Correlogram," in *Proc. CVPR97*, pp. 762-768, 1997.

[8] C. Carson, M. Thomas, S. Belongie, J.M. Hellerstein, and J. Malik, "Blobworld: A System for Region-Based Image Indexing and Retrieval," *IEEE Trans. Pattern Analysis and Machine Intelligence*, vol. 24, no. 8, pp. 1026-1038, Aug. 2002.

[9] J.Z. Wang, J. Li, and G. Wiederhold, "SIMPLIcity: Semantics-Sensitive Integrated Matching for Picture LIbraries," *IEEE Trans. Pattern Analysis and Machine Intelligence*, vol. 23, no. 9, pp. 947-963, Sep. 2001.

[10] N. Jhanwar et al., "Content Based Image Retrieval Using Motif Cooccurrence Matrix," *Image and Vision Computing*, vol. 22, pp. 1211-1220, 2004.

[11] G. Peano, Su rune courbe qui remplit toute une aire plaine, Mathematicshe Annalen 36 (1890) 157-160.

[12] G. Seetharaman, B. Zavidovique, Image processing in a tree of Peano coded images, in: Proceedings of the IEEE Workshop on Computer Architecture for Machine Perception, Cambridge, CA, 1997.

[13] M. Flickner et al., "Query by Image and Video Content: The QBIC System," *IEEE Computer*, vol. 28, no. 9, pp. 23-32, Sept. 1995.

[14] G. Qiu, "Color Image Indexing Using BTC", *IEEE Trans. Image Processing*, vol. 12, 2003.

# Parallel Implementation of MPEG-4 Encoding over a Cluster of Workstations

**Karthik Sankar. R, Shivsubramani. K. Moorthy and Soman. K. P**

*Amrita Vishwa Vidyapeetham (Deemed University)*

*Ettimadai (PO), Coimbatore–641105 Tamilnadu, India*

*{r_karthiksankar,ramanand}@ettimadai.amrita.edu, kp_soman@amrita.edu*

**Abstract**

*"Parallel Implementation of MPEG-4 Encoding over a Cluster of Workstations" is a research project and a project proposal to reduce the delay involved in the MPEG-4 encoding process.*

*As a case study we take Amrita Vishwa Vidyapeetham which is a multi-campus Deemed University, one of its own kinds in India. All the campuses are connected through satellite provided by ISRO (Indian Space Research Organization). A number of e-learning classes, guest lectures and meetings are conducted across the satellite network with the video being compressed in MPEG-4 standard.*

*The project optimizes the system by parallelizing the encoding process. For this a cluster of machines is created and the video frames are distributed among these nodes. This follows the SPMD (Single Program Multiple Data) model. Dedicated nodes within the cluster perform the encoding process, while there is a node that distributes the video to be encoded over these nodes. Another node collects the encoded video and places them back in the original sequence. The two nodes can be utilized for the encoding process also.*

## 1. Introduction

The encoding sequence contributes a significant delay in a video transmission/distribution system. The Amrita e-learning network is one such system. There are a number of solutions to the delay problem like exploring functional parallelism in the MPEG-4 algorithm and spatio-temporal parallelism. A more interesting solution is to decompose the video sequence into GOPs (Groups of Pictures), and then a dedicated processor independently processes every GOP. The basic idea for data distribution is to arrange the uncompressed video sequence in GOPS. Then, we decide (a) how processors get the GOPS, and (b) which GOPS correspond to each processor as reported by [1] A.Rodriguez, A. González and M.P. Malumbres. The performance of an



*T. Sobh and K. Elleithy (eds.), Advances in Systems, Computing Sciences and Software Engineering*, 77–82.

encoding system can be improved by employing a cluster of workstations. To take advantage of the potential processing power of clusters of workstations, we can use parallel programming techniques like MPI (Message passing Interfaces). The number of nodes required for optimum performance has been found to be 32 (page 2, Figure 1)[1]. Based on this we find that using 4 processing nodes we can reduce the delay of about 3 seconds in the existing system to about 1.5 seconds. Further, a distribution scheme consisting of 4 processing nodes utilizing a node for distribution, collection and saving or streaming the video can be employed.

## 2. The E-Learning system

The Amrita-ISRO e-Learning system is a dedicated satellite based system used for educational purposes. Currently it connects the four campuses of Amrita at Coimbatore, Cochin, Kollam and Bangalore in India. Seminars, classes, video lectures etc. are streamed over this network.

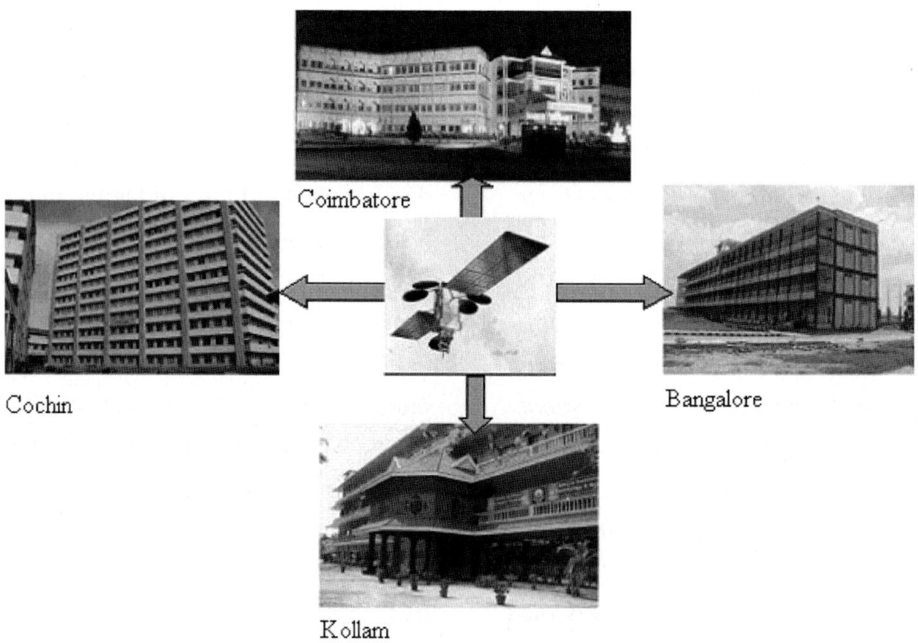


**Figure 1: Amrita E-Learning network**

The system uses a single node to encode the captured video before transmitting them over the network. Encoding, being a computationally intensive process, introduces a time delay. This, being a real-time operation, any delay observed tends to compromise the visual quality.

## 3. Bottlenecks of the existing system

- Overhead due to the usage of a single node for encoding the real time data.
- In addition to transmission delay, encoding contributes a 3 second delay

## 4. Objective and Scope of the Proposed Parallel System

The need for a parallel system arises mainly due to the drawback of using a single node to encode the captured video. The parallel system aims at reducing the time delay incurred during the compression of the captured video, which will then be streamed over a Network. Having a cluster of workstations, each carrying out a part of the encoding task facilitates this process. This follows the SPMD (Single Program Multiple Data) model. To overcome the delay problem, it was decided to implement a multi node encoding system that may reduce the delay to about a manageable 1 - 1.5 seconds. Both uni-processor as well as SMP based nodes can be used to achieve this aim.

The scope of the system is limited to the encoding process. It does not take into account the overhead involved in sending the data over the network nor does it perform traffic management. The role of the scheduler is to split and distribute the video data to the encoding nodes after which the collector nodes collect the data. The two nodes can be used for the encoding process also.

There is no provision for a separate decoder. MPEG-4 is based on the asymmetric complexity. The complexity is present in the encoder, so a separate decoder need not be built. After the encoding process, the data can be transmitted or stored to a file.

## 5. Feasibility Study

One of the main contentions for feasibility analysis would be the hardware involved in the system. The new system would essentially be a cluster. The computers will not be requiring specially configured hardware. Since system operates in a real-time environment, optimization can be achieved largely in the software side. Thus there is a need for the software to be highly customized to this environment. To achieve this purpose, it was decided to use open source libraries, which can be downloaded and updated from their respective web sites. All the nodes in the cluster run on Linux.

In any project, cost is a major issue, when looking at the feasibility of the system. There was no cost involved for the software, as they were freely available and distributed under GPL (GNU Public License). The hardware cost involved buying the required number of systems to build the cluster and other components of the system.

## 6. Architecture of the Parallel System

The video captured from the source (Digital Video Camera, CCTV) is sent to the parallel encoding nodes (cluster or SMP machines), where the video stream is encoded in MPEG-4. The encoded frames are then streamed over the network. At the other end the frames received are decoded and displayed. There could also be a two-way system with provision for encoding and decoding at each station.

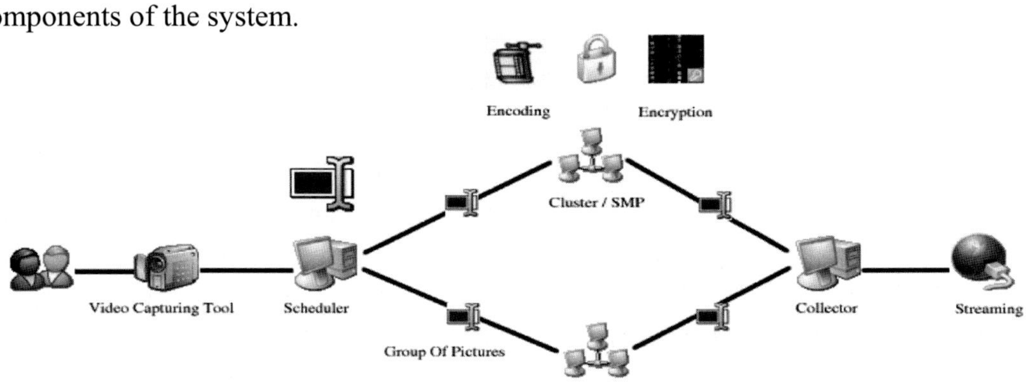


**Figure 2: Detailed Architecture of Parallel System**

Overall, the following components will interact during the process.

- The source of the video – such as a video capturing tool
- A node that will form the front interface to the capturing device
- Nodes that will perform the encoding.
- A node that will collect the encoded frames
- A node (or the same node that will collect the encoded Frames) will then stream the frames over the network in the appropriate format.

For experimental purposes, the source is always assumed to be a raw video file. However, the intended source will be a video capturing device such as a Digital Video Camera. It is expected that the source will emit uncompressed raw video, preferably in YUV format. Thus each frame is uncompressed and "unrelated" to the preceding or succeeding frames.

There is a need to define the unit of video being "treated" by the system. Because the source is expected to deliver individual frames, a frame will be considered to be the unit of video for all practical purposes.

Given the nature of the input and the fact that the encoding libraries expect the input to be in frames, there is no need to split the Video further into Slices, Macro Blocks etc.

Based on our experimentations and study work we arrive at the following graph diagram depicting the delay optimization performed based on the number of workstations used in the cluster.

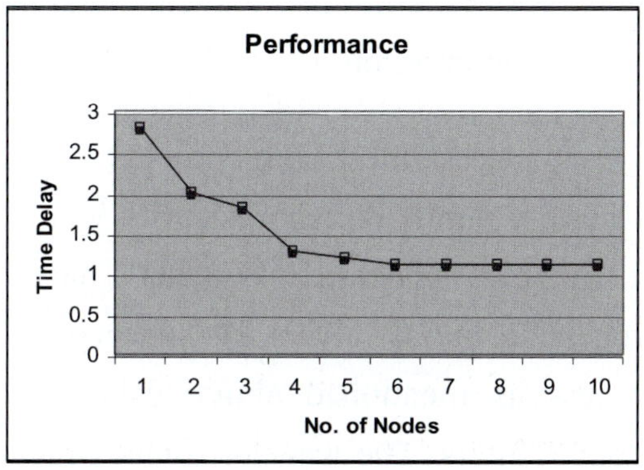


## 6. Future Prospects

The MPEG-4 encoding process usually involves DCT based encoding which involves a lot of complications during implementation.

The general equation for a 2D (*N* by *M* image) DCT is defined by the following equation:

$$F(u,v) = \left(\frac{2}{N}\right)^{\frac{1}{2}} \left(\frac{2}{M}\right)^{\frac{1}{2}} \sum_{i=0}^{N-1} \sum_{j=0}^{M-1} \Lambda(i).\Lambda(j).\cos\left[\frac{\pi.u}{2.N}(2i+1)\right] \cos\left[\frac{\pi.v}{2.M}(2j+1)\right].f(i,j)$$

We propose to replace the DCT based encoding with wavelet based encoding. We can go for Wavelet based implementation for the following reasons:

- It enables better compression with minimal distortion.
- An **encryption** can be performed on the GOPs using the wavelet filters as encryption keys.
- There is no loss of position information in Wavelet Based Compression

If the video conferencing process takes place in a public domain, security would be a big concern. Wavelet based processing enables a very easy encryption process as mentioned above using the wavelet filters. The wavelet filters which are anyway used for the encoding process can be kept secret using them as encryption keys.

SMP Machines to be used to reduce the network overhead. Dual processor machines avoid communication between compute nodes across a network thus overcoming the network communication delay.

Zero copy mechanism increases the network throughput. It enables data transfer of data without using TCP/IP, reducing network traffic thus overcoming the network delay. GAMMA(Genoa Active Message Machines) clusters are to be used to implement the zero copy mechanism.

## 7. References

[1] A. Rodriguez, A. González and M.P. Malumbres, "*Performance evaluation of parallel MPEG-4 video coding algorithms on clusters of workstations*", Technical University of Valencia, Camino de Vera 17, 46071 Valencia, SPAIN, 2004

[2] Chez Skal ‚*SklMP4 –MPEG4 video codec library";* http://www.skal.planet-d.net, 2004

[3] Fabrice Bellard, "*Libavcodec-MPEG4 video codec library";* http://www.ffmpeg.sourceforge.net, 2004

[4] Ajay Gupta, "*MPEG4 Encryption";* http://www.it.iitb.ac.in/~ajay/current.php, 2004

[5]"*A comprehensive index of MPEG resources on the internet",* http://www.mpeg.org, 2004

[6] Peter Symes, "*Digital Video Compression",* http://www.symes.tv, 2004

[8] "*MPEG4 Industry Forum",*http://m4if.org

# Different Strategies for Web Mining

Michael Kernahan
Luiz F. Capretz
Department of Electrical and Computer Engineering
University of Western Ontario
London, Ontario, N6A 5B9
CANADA
{mkernaha, lcapretz}@uwo.ca

*Abstract*-**In the past decade, the amount of information available on the Internet has expanded astronomically. The amount of raw data now available on the Internet presents an opportunity for businesses and researchers to derive useful knowledge out of it by utilising the concepts of data mining. The area of research within data mining, often referred to as web mining, has emerged to extract knowledge from the Internet. Existing algorithms have been applied to this new domain, and new algorithms are being researched to address indexing and knowledge requirements. Three main areas of interest have emerged in web mining: Mining the content of the web pages; mining the structure of the web; and mining the usage patterns of clients. This paper provides an overview of web mining, with an in depth look at each of the three areas just mentioned.**

## I. INTRODUCTION

The rate at which new data is created and stored continues to increase. With the cost of storage continuing to decrease, this rate of increase shows no signs of slowing down. Increasingly, companies are finding themselves with so much data that they do not know how to leverage it effectively. This "explosive growth in data and databases has generated an urgent need for new techniques and tools that can intelligently and automatically transform the processed data into useful information and knowledge" [1]. For this reason, the field of data mining is important to both researchers, who see a new emerging technology, and companies that see a new important revenue source.

With the large diversity of databases and other methods of storing knowledge, no single data mining system is able to address the entire field. Instead, individual data miners must be customised to each application. The diversity of the data has also lead to several different fields of research within data mining itself. Some of the major types of rules and algorithms are described in the next section.

The Internet is quickly becoming one of the largest data sources in the world. "As the popularity of the Internet explodes nowadays, it is expected that how to effectively discover knowledge on the web will be one of the most important data mining issues for years to come" [1].

Due to its rate of growth, it has become nearly impossible for a user to manually traverse the web searching for information. Automated search engines are now a necessity, and "it is believed that a majority of web user sessions now begin by first consulting a search engine for an informational need" [2]. Web mining, or knowledge mining of the World Wide Web, has become an important and necessary part of the web and data mining communities.

This article provides a brief but thorough overview of the field of web mining.

## II. BACKGROUND INFORMATION

An important aspect of data mining is that the search algorithms must be efficient and scalable. For this reason, the optimal solution is rarely found, and a trade off is made between efficiency and accuracy of the results. Reference [1] provides a very valuable and in depth overview of the field of data mining. Different terms referred to throughout this paper require a background in data mining. Some of these terms are explained here briefly.

### *A. Association*

Association algorithms attempt to find the relationships between different items within the dataset. The presence of some items in a given set of data will sometimes imply the presence of other items. This implication is the association rule, and it can be measured in terms of confidence and support. Confidence concerns to the strength of association (percentage of time it holds true), while support is a measure of the frequency of the occurring patterns in the dataset. An example of association with a very simplified dataset is presented in Fig. 1 below.



*T. Sobh and K. Elleithy (eds.), Advances in Systems, Computing Sciences and Software Engineering*, 83–88.

| Dataset |
|---|
| ABC |
| BC |
| AB |
| BD |

Observations:
Every time A appears, B appears
Every time C appears, B appears
Every time D appears, B appears

Evaluation of Associations:
A implies B, 100% confidence (2/2), 50% support (2/4)
C implies B, 100% confidence (2/2), 50% support (2/4)
D implies B, 100% confidence (1/1), 25% support (1/4)

Fig. 1. Examples of Association.

### *B. Generalization*

Depending on the detail level of the items used for finding association rules, it may be difficult to find any relevant rules. By generalizing the data a level or two higher, for example merging *dog* and *cat* into a new class *pet*, new strong associations may be found. The hierarchies used for generalization are generally adjusted dynamically in order to maximize the effectiveness of this rule.

### *C. Classification*

Data classification uses common properties of objects to group them into classes. The classes and the properties are derived from a training set, itself derived from a subset of the dataset, which trains the classifier. Since the classifier is trained, this is technique is known as *supervised learning*.

This is an important field of data mining, as most raw data is not divided into classes. Several fields of research have been applied to this technique, including statistics, machine learning, neural networks, and expert systems.

### *D. Clustering*

Data clustering is similar to classification, except that there are no known classes. This technique is known as unsupervised learning, since it does not use a training set. Clustering is useful for identifying the sparse and crowded places, however it can be expensive to store and compute the clusters.

### *E. Sampling*

Sampling is used to reduce the computational complexity, and can generally be applied to achieve much greater efficiency with little loss of accuracy. Random samples are taken out of the data set, and these samples are examined as if they were the entire data set. This process can be repeated and the results combined in order to improve the accuracy of the results. Combining results like this is sometimes referred to as *bagging*.

## III. WEB MINING

"Web mining is the use of data mining techniques to automatically discover and extract information from Web documents and services" [3]. The size of the web is an elusive number. The moment anyone tries to quantify the size, it has already changed, almost like a Heisenberg principle (The Heisenberg Uncertainty Principle: The more precisely the position is determined, the less precisely the momentum is known) of the Internet. On top of this, not only does the number of resources grow, but also much of the content that does exist is changed from time to time. In order to estimate the current size of the web it is observed that Google (http://www.google.com), a popular search engine, currently indexes 8 billion web pages. While that figure does not encompass the entire web, it does give an idea of just how large the Internet is. Business and e-commerce considerations, combined with the fast growth of information sources and the interest of several fields of research have resulted in a booming research area. There are three main subcategories of web mining: *web content mining*; *web structure mining*; and *web usage mining*. "Web content mining refers to the discovery of useful knowledge from web resources" [4]. Web structure mining refers to studying the hyperlink structure of web pages, both links to and from each page. Web usage mining refers to studying the usage habits of users and other agents as they access web resources. Table 1 [5], gives an overview of the domain of web mining. The columns represent the different domains of web mining, which will be examined in detail in the following subsections, while the rows represent the various concepts and techniques associated with each domain.

### *A. Web Content Mining*

One of the major goals of this field is simply to find and retrieve new resources from the web. Another important goal is to categorize and cluster the resources that have already been found, while still another important goal is to extract the actual information from a web page.

TABLE I.
WEB MANING CATEGORIES [5]

| | Web Mining | | | |
|---|---|---|---|---|
| | Web Content Mining | | Web Structure Mining | Web Usage Mining |
| | **Information Retrieval View** | **Database View** | | |
| **View of Data** | - Unstructured<br>- Semi-structured | - Semi structured<br>- Web site as DB | - Links structure | - Interactivity |
| **Main Data** | - Text Documents<br>- Hypertext documents | - Hypertext documents | - Links structure | - Server logs<br>- Browser logs |
| **Representation** | - Bag of words, n-grams<br>- Terms, phrases<br>- Concepts or ontology<br>- Relational | - Edge-labelled graph (OEM)<br>- Relational | - Graph | - Relational table<br>- Graph |
| **Method** | - TFIDF and variants<br>- Machine learning<br>- Statistical (including NLP) | - Proprietary algorithms<br>- ILP<br>- (Modified) association rules | - Proprietary algorithms | - Machine Learning<br>- Statistical<br>- (Modified) association rules |
| **Application Categories** | - Categorisation<br>- Clustering<br>- Finding extraction rules<br>- Finding patterns in text<br>- User modeling | - Finding frequent sub-structures<br>- Web site schema discovery | - Categorisation<br>- Clustering | - Site construction<br>- Site adaptation<br>- Site management<br>- Marketing<br>- User modeling |

Finding new resources from the Internet generally involves using programs called *spiders*, which *crawl* the Internet by traversing the hyperlinks that are found in web pages. As mentioned in the previous section, there is a lot of data out there. If each web page is estimated as being 12.5 kilobytes of text, then the entire content of the Google's repository could be stored in approximately 100 terabytes. This represents a total cost of only $100,000 to store (at a present cost of $1/GB), which is surprisingly small. However, even if you were able to download the entire contents of the Internet into a massive storage array, not only would it be obsolete before the first few pages were done, but no algorithm could be effectively scaled to be efficient in such a set up. For this reason, "decisions must be made as to how to use the available processing and bandwidth in order to maintain a collection that balances quality, freshness and coverage" [2].

The lack of structure makes any content mining difficult. For this reason, many of the content miners that have been created are limited in their scope or domain. Generally content miners do not interpret the documents themselves, but rather they parse a document and index all the different terms within the document. There are two paradigms for doing this: The Information Retrieval (IR) viewpoint is interested in finding the information and filtering it, while the Database (DB) viewpoint is more interested with structuring the content for sophisticated queries. The IR viewpoint is usually used to index and classify unstructured or semi-structured data. The DB viewpoint generally tries to infer the structure of the web site to transform a web site to become a database.

*B. Web Structure Mining*

Web structure mining is interested in providing a quality rank or a relevancy for each web page, which can be combined with web content mining techniques to return more accurate and complete results to searches. This is accomplished by extracting knowledge from the way documents on the web refer to each other via hyperlinks. Hyperlinks enhance the quantity of information that can be found from a single document, since by following the hyperlinks additional relevant information can usually be found. The problem with web pages is that some are not self-descriptive, while some links are purely for navigational purposes. Web structure mining aims to filter web pages by their structure in terms of their importance.

One of the first major projects undertaken in this field of research was the Hyperlink-Induced Topic Search (HITS), which is an iterative algorithm that mines a graph structure from the web. This graph structure is used to evaluate pages as being either a *hub* or an *authority* on a specific topic. Hubs are pages that refer to many high-ranking authorities on the topic, even though they may have little information about the topic themselves. Authorities are pages that contain information on the specific topic and are referred to by several hubs. Fig. 2 shows the relationship between hubs and authorities. The problem with HITS is that it is a purely link-based algorithm, meaning that all links are considered to have equal weight. This results in HITS having a few problems:

HITS tends to return more general web pages as opposed to actual authorities on a subject. Topic drift may occur when a hub is referenced which has multiple topics since all out-links receive equal weight from the hub.

Popular sites, which may actually be content poor, can receive high marks due the amount of pages that are linked to them.

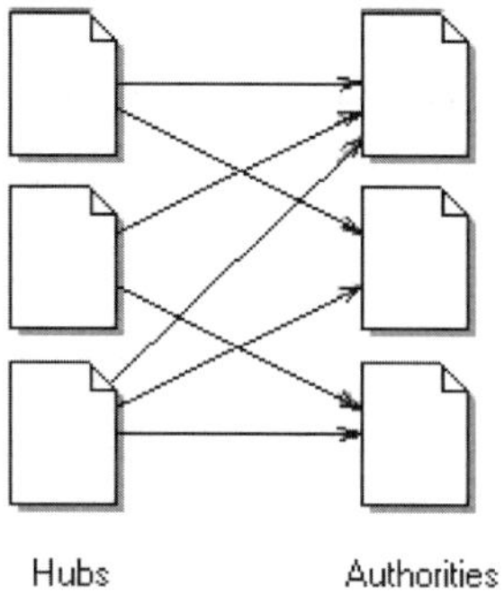


Fig. 2. Hubs and Authorities.

In response to the problems encountered with the HITS algorithm, the CLEVER algorithm was developed. "This algorithm assigns a weight to each link based on the terms of the queries and the end-points of the link" [6]. Hubs are also decomposed by topic (by leveraging content mining techniques) in order to eliminate topic drift by only counting links to authorities that really are related to the searched topic.

A third algorithm which has been developed for this field is the PageRank algorithm, made well known by Google. With this algorithm the web crawler pre-computes page ranks, increasing the speed that ranked search results are returned. The logic is that if a page has important links coming in, then it must have important links going out. PageRank also tries to estimate the probability that a surfer would visit a page without traversing to it through a link (by physically inputting the URL, or via a bookmark in the browser), as this should have some weight to the importance of a page. Currently PageRank distributes the rank equally among all the links, as shown in Fig. 3, however research is being conducted to weight the links based on their relevance [6].

### *C. Web Usage Mining*

Web usage mining tries to make sense of the user's browsing and usage of a set of web pages. This technique can be broken down into three key components. First the pre-processing step converts the raw data into a more concise, usable form. Pattern discovery then tries to find rules and knowledge from the data. Pattern analysis is used to parse these results and determine whether or not the knowledge is actually useful.

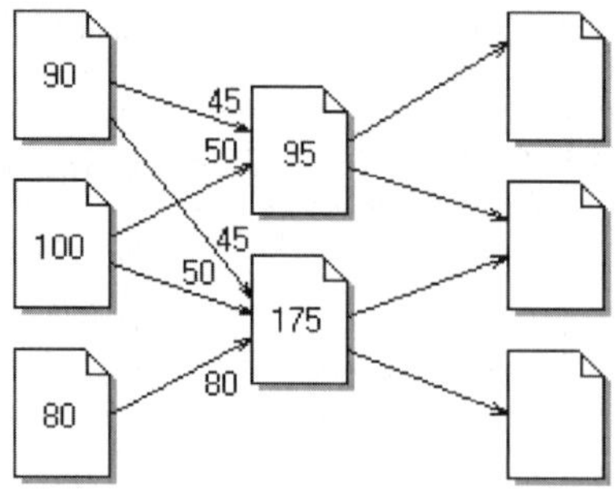


Fig. 3. Simplified example of PageRank [6].

Pre-processing: The amount of information that is available for web usage mining is enormous, as well as noisy. Because of this, data cleaning methods are necessary. The cleaning methods should prepare "data that allows the identification of a particular user's browsing pattern in the form of page views, sessions, and click-streams" [7]. There are three sources of information for web usage mining data: from the server level logs, from the proxy level logs and from the client side level logs.

Mining the server level captures all the external requests for the web page. Since some requests come from proxy servers, it can be difficult to track individual users and sessions. Caching on the client or proxy server can also result in missed statistics since the server is not informed when the site is served to the client this way. Cookies can store client information; however this does raise some privacy issues, which will be discussed later in this article. It is also hard to tell when a user has left a web site. Server side pre-processing attempts to address all these issues by: imposing default timeout restrictions; inferring cached page references; and via sessions representing a single click-stream.

Mining from the proxy level, information about multiple clients visiting multiple servers can be extracted. This information may be too sparsely populated to pull out anything usable; however it is an area where more research could be done.

Mining from the client level fixes the problems of caching and session identification; however such data mining is generally regarded as spy-ware. Today, as privacy concerns are increasing, most users wish to remain anonymous when they are online. Also, even if a special browser is installed to track usage, such as NetZero (http://www.netzero.com) or AllAdvantage (http://www.alladvantage.com) (failed business plans involving paying customers to have targeted banner ads displayed in their browsers), it becomes difficult to convince the user to use the browser for their regular activity.

Pattern Discovery: Once the data has been pre-processed (cleaned), the pattern discovery phase may begin. "Pattern discovery draws upon methods and algorithms developed from several fields such a statistics, data mining, machine learning and pattern recognition" [8]. Pattern discovery applied to web mining requires more prior knowledge, since the Internet is not ordered like a database. This does have a benefit however, since the variety of data and data sources generally adds many dimensions of extra information.

Statistical analysis can be run on the data. This is a simplistic approach to mining, however can yield information such as system performance and security, as well as provide insight on how the site should be modified and help with marketing decisions.

For a more thorough analysis of the structure of the website, association rules can be used to find what pages users frequent regularly. This can find patterns where users have to navigate through several stages to get from one page to another. By identifying these patterns, links can be added or removed in order to facilitate site navigation.

Clustering analysis is used to cluster users and pages that are related. These statistics can make dynamic recommendations to users, such as what Amazon (www.amazon.com) does by recommending books and offering bundles based on what a user's preferences are. These *recommender* systems are used by businesses to convert browsers into buyers, increase cross selling, and build loyalty in their customers.

As a result of this analysis, decisions can be made about groups of users and what they want from the website. "From a business perspective, one of the major goals of web usage mining is the personalization to individual users on a massive scale, often known as *mass customisation*" [8]. By customizing the page to the user's profile, not only does the user get what he/she wants, but marketing and sales advertisements are directed at the correct demographic and thus generate maximum return.

Pattern Analysis: The final stage of web usage mining is pattern analysis. Pattern discovery may not present its results in a readily presentable form; pattern analysis transforms this data into an accessible format. "The overall goal of pattern analysis is to eliminate the irrelevant rules or patterns and to extract the interesting rules or patterns from the output of the pattern discovery process" [4].

## IV. SEMANTIC WEB

A new idea is emerging to have a semantic web that co-exists with the current web. A semantic web is a mesh of linked information that can easily be processed by machines on a global scale. Rather then the current standard of having meta-data, a semantically rich language such as Web Ontology Language (OWL: http://www.w3.org/TR/owl-features) will be embedded in web pages. OWL will enable much more information to be stored compared to the current web-as-database approach with limited meta-data. This semantic web will fit into web mining: web mining will enable the semantic web, and the semantic web will improve web mining effectiveness. Because of the importance of the accuracy of a semantic layer, automation would be required to generate the semantic information. This information would then be available to external web miners, which would allow them to perform their tasks more effectively. Only time will tell if researchers will push the semantic web idea hard enough for it to take hold.

## V. PRIVACY CONCERNS

As stated previously, "Most users want to maintain strict anonymity on the web" [8]. Site administrators however want all the information that they can get on their users so that they can tailor the website to the largest demographic of user. Essentially what is needed is that the analysis on the usage data does not compromise the privacy of any individual user.

Cookies are little packets of information stored on a web user's system by the web sites that they visit. Cookies raise privacy issues because of how they are being used. Originally cookies were designed to allow web advertisers to track what banner ads a user clicked on, so that they could be served similar ads in the future. With time, some companies developed the idea of placing a unique serial key in their cookie and storing all of the user information that they could track on their own server. This information could include real world data such as name, address, phone number, and a financial profile. The prospect is certainly somewhat frightening. Currently most sites that require a user to create an account have a detailed privacy policy, which explains any information that will be collected and how it will be used.

The W3C has an initiative called Platform for Privacy Preferences (P3P: http://www.w3.org/P3P/), which will essentially handshake with a user's computer until a level of privacy is agreed upon between the server and the user, and the connection is made.

## VI. DISCUSSION

This paper has attempted to cover the major aspects of web mining research. Web mining is, like the Internet itself, still in its infancy. The general push is to focus on web usage mining, as businesses try to maximize their online potential, and try to find new ways to use extracted knowledge and information. Web mining still has lots of room for improvement. New types of knowledge are to be found within the web; and extracting and leveraging this knowledge will provide value to both researchers and business. The algorithms used for web mining, and data mining in general, are constantly evolving, and new algorithms will undoubtedly emerge. As the web gets larger and larger, techniques for extracting knowledge incrementally, as well as in a distributed fashion, will help to increase the scope, and the efficiency, of mining for knowledge.

The amount of data available through databases on the web is probably much larger then what spider programs can currently crawl to. While all three of the major fields of web mining are being researched, the emphasis from a business perspective is definitely on web usage mining. When it becomes possible to access all of this information automatically, web mining will become a very important field of research, as it will allow for mining from almost all publicly available information.

Some current research into usage mining includes using Neural Networks to cluster users and pages (Yi [9]), intuitive fuzzy logic (Petrounias *et al.* [10]), session reconstruction (Zhang and Ghorbani [11], and Nadjarbashi-Noghani and Ghorbani [12]), as well as site personalization (Silvestri *et al.* [13] and Eirinaki *et al.* [14]).

Content mining research includes mining data records from web pages (Liu, Grossman and Zhai [15]), indexing pages via key phrases (Hammouda and Kamel [16]), searching for information on people (Harada, Sato and Kazama [17], and Al-Kamha and Embley [18]), and lastly extracting information from the hidden web (Hedley *et al.* [19] and Fontes and Silva [20]).

Structure mining research includes creating smarter web crawlers (Altingovde and Ulusoy [21]) and new extensions of the HITS algorithm (Kao *et al.* [22]).

Privacy will always be an issue with web mining. Since surfers generate most of the data used for web usage mining, measures need to be in place to protect individuals' rights, while at the same time allowing the field of research to grow. The W3 Consortium has made recommendations on how to proceed with this regard, it remains to be seen if they will be implemented.

## REFERENCES


[1] Chen, M., Han, J., Yu, P., "Data Mining: An Overview from Database Perspective." *IEEE Transactions on Knowledge and Data Engineering*, 8(6):866-883, December 1996.

[2] McCurley, K., Tomkins, A., "Mining and Knowledge Discovery from the Web," in *Proceedings of the 7th International Symposium on Parallel Architectures, Algorithms and Networks* (ISPAN '04), 1087-4089/04, 2004.

[3] Etzioni, O., "The World Wide Web: Quagmire or Gold Mine," *Communications of the ACM*, 39(11):65-68, 1996.

[4] Wang, B., Liu, Z., "Web Mining Research," in *Proceedings of the Fifth International Conference on Computational Intelligence and Multimedia Applications* (ICCIMA '03), 0-7695-1957-1/03, 2003.

[5] Kosala, R., Blockeel, H., "Web Mining Research: A Survey," *SIGKDD Explorations*, vol. 2, issue 1, 2000, pp. 1-15.

[6] Xing, W., Ghorbani, A., "Weighted PageRank Algorithm," in *Proceedings of the 2nd Annual Conference on Communication Networks and Services Research* (CNSR '04), 0-7695-2096-0/04, 2004.

[7] Kolari, P., Joshi, A., "Web Mining: Research and Practice," *Computing in Science and Engineering*, vol. 6, issue 4, July-August 2004, pp. 49-53. (1521-9615/04)

[8] Srivastava, J., Cooley, R., Deshpande, M., Tan, P., "Web Usage Mining: Discovery and Applications of Usage Patterns from Web Data," *SIGKDD Explorations*, vol. 1, issue 2, January 2000, pp. 12-23.

[9] Yi, H., "A novel competitive neural network for Web mining," *Proceedings of 2004 International Conference on Machine Learning and Cybernetics*, vol. 6, August 2004, pp. 3417-3422.

[10] Petrounias, I., Tseng, A., Kolev, B., Chountas, P., Kodogiannis, V., "An intuitionistic fuzzy component based approach for identifying Web usage patterns," *Proceedings of the 2nd International IEEE Conference on Intelligent Systems*, vol. 2, June 2004, pp. 430-433.

[11] Zhang, J., Ghorbani, A., "The reconstruction of user sessions from a server log using improved time-oriented heuristics," *Proceedings of the 2nd Annual Conference on Communication Networks and Services Research*, May 2004, pp. 315-322.

[12] Nadjarbashi-Noghani, M., Ghorbani, A., "Improving the referrer-based Web log session reconstruction," *Proceedings of the 2nd Annual Conference on Communication Networks and Services Research*, May 2004, pp. 286-292.

[13] Silvestri, F., Baraglia, R., Palmerini, P., Serrano, M., "On-line generation of suggestions for Web users," *Proceedings of the International Conference on Information Technology: Coding and Computing, 2004*, vol. 1, April 2004, pp. 392-397.

[14] Eirinaki, M., Lampos, C., Paulakis, S., Vazirgiannis, M., "Web personalization integrating content semantics and navigational patterns," *Proceedings of the 6th annual ACM international workshop on Web information and data management*, 2004, pp. 72-79.

[15] Liu, B., Grossman, R., Zhai, Y., "Mining Web Pages for Data Records," *Intelligent Systems, IEEE*, vol. 19, issue 6, November-December 2004, pp. 49-55.

[16] Hammouda, K., Kamel, M., "Efficient phrase-based document indexing for Web document clustering," *IEEE Transactions on Knowledge and Data Engineering*, vol. 16, issue 10, October 2004, pp. 1279-1296.

[17] Harada, M., Sato, S., Kazama, K., "Finding authoritative people from the Web," *Proceedings of the 2004 Joint ACM/IEEE Conference on Digital Libraries*, June 2004, pp. 306-313.

[18] Al-Kamha, R., Embley, D., "Grouping search-engine returned citations for person-name queries," *Proceedings of the 6th annual ACM international workshop on Web information and data management*, 2004, pp. 96-103.

[19] Hedley, Y., Younas, M., James, A., Sanderson, M., "A two-phase sampling technique for information extraction from hidden web databases," *Proceedings of the 6th annual ACM international workshop on Web information and data management*, 2004, pp. 1-8.

[20] Fontes, A., Silva, F., "SmartCrawl: a new strategy for the exploration of the hidden web," *Proceedings of the 6th annual ACM international workshop on Web information and data management*, 2004, pp. 9-15.

[21] Altingovde, I., Ulusoy, O., "Exploiting Interclass Rules for Focused Crawling," *IEEE Intelligent Systems*, vol. 19, issue 6, November-December 2004, pp. 66-73.

[22] Kao, H., Lin, S., Ho, J., Chen, M., "Mining Web informative structures and contents based on entropy analysis," *IEEE Transactions on Knowledge and Data Engineering*, vol. 16, issue 1, January 2004, pp. 41-55.

# Coalesced QoS: A Pragmatic approach to a unified model for KLOS

Ashish Chawla
Sr. Software Consultant, Email: chawlaashish@acm.org

Ramesh Yerraballi[1], Amit Vasudevan
Department of Computer Science and Engineering, The University of Texas at Arlington
Arlington, TX 76019
Email: {ramesh,vasudeva}@cse.uta.edu

*Abstract*— **Advances in software and hardware technologies have given operating systems the ability to process data and handle various concurrent processes. The increased ability has been one of the driving forces which have led to the proliferation of mechanisms in operating systems to satisfy the performance requirements of applications with predictable resource allocation. As different classes of applications require different resource management policies one needs to look into ways to satisfy all classes of applications. Conventional general purpose operating systems have been developed for a single class of best-effort applications, hence, are inadequate to support multiple classes of applications. We present an abstract architecture for the support of Quality of Service (QoS) in Kernel-Less Operating System (KLOS). We propose new semantics for the QoS resource management paradigm, based on the notion of Quality of Service. By virtue of this new semantics, it is possible to provide the support required by KLOS to various components of the operating system such as memory manager, processor time, IO management etc. These mechanisms which are required within an operating system to support this paradigm are described, and the design and implementation of a prototypical kernel which implements them is presented. Various notions of negotiation rules between the application and the operating systems are discussed along-with a feature which allows the user to express its requirements and the fact is that this model assures the user in providing the selected parameters and returns feedback about the way it meets those requirements. This QoS model presents a design paradigm that allows the internal components to be rearranged dynamically, adapting the architecture to the high performance of KLOS.**

**The benefits of our framework are demonstrated by building a simulation model to represent how the various modules of an operating system and the interface between the processes and the operating system can be tailored to provide Quality of Service guarantees.**

***Index Terms***—**Quality of Service, Operating Systems, Resources, Utilization, High Performance, Kernel-Less Architecture.**

## I. Introduction

The rapid growth and coexistence of different application domains, such as multimedia and embedded systems, present a significant challenge to the provision of their Quality of Service (QoS). To solve this challenge, we need a unified QoS framework, which allows flexibility and re-configurability.

Techniques such as over-provisioning of resources work if there is a single application, but cannot easily be extended to realistic scenarios where we have multiple applications competing for the same resources. Therefore, the operating system must allow an application to request some QoS, which is a measure of the resources required by the application. In the following paper, we present an approach which is targeted towards providing quality of service for applications in terms of overall performance which includes both resource (Memory, I/O and Network) usage and timing.

The point to note is that the high performance applications have different QoS requirements and this poses some demands on the resources, like scheduling policies need to emphasize on meeting deadlines, memory management needs to ensure pages are pinned down with bounded access etc. Hence, an operating system which is meant to provide QoS support must have the entire system and each component to be aware of the QoS policies. We thus come up with a Coalesced QoS model which is a unification of the entire resources available in the system and made available to the applications.

KLOS is a Kernel-Less Operating System [1, 2] wherein all processes execute at the unprivileged level having their own private execution space and function independent of all other processes in the system. KLOS is able to achieve high performance due to its architecture and its service call mechanism. In traditional operating systems, a service call in almost all cases involves a ring-transition to the kernel where the required function is then performed. Traditionally this is achieved by using the trap/interrupt which transitions into the kernel to execute the required function. However, in case of KLOS, there is no ring-transition during service calls and this accounts for the High Performance attributed to KLOS. It is

[1] Corresponding Author



*T. Sobh and K. Elleithy (eds.), Advances in Systems, Computing Sciences and Software Engineering*, 89–96.

this feature of KLOS which makes us work on providing Quality of Service to further enhance the performance of the operating system.

Our resource management scheme builds on and significantly extends the established resource allocation model described in [7]. The precise resource requirements of the applications, as well as the number and timing constraints are unknown in advance. These details depend on many factors known only at runtime. The driving principle behind Coalesced QoS is to give applications enough control so as to specify the resources it requires, with the operating system accommodating the request through allocation and negotiation. Till date, research carried out in this domain has not dealt with the management of multiple resource types, concurrent access to different resources, explicit timeliness control, feedback about resource usage, control on resource overruns and management of interactions between resource users and considerations of portability and compatibility.

In our work, we therefore strive for practical and acceptable alternatives which can guarantee access to different resource types. We investigate various resource allocation techniques to provide a unified approach of providing Quality of Service (QoS) to applications. Our methodology takes advantage of the fact the quality of service being provided is not just limited to the processor time but also, all other system components do play a role in the negotiation. We propose integrated architecture for run-time application support, and runtime adaptation to application state variations. Our resource model captures the essential requirements of the applications in a simple, efficient yet general form. Our resource management work significantly extends established real-time scheduling theory, and reservation enforcement mechanisms.

Our approach is in developing a semantically rich model supporting the high-level design/representation of QoS adaptation strategies and concerns. We are designing and implementing mappings to derive runtime resource requirements and for incorporating the information gathered at runtime into refinement of the Quality of Service provided by the operating system.

## II. Related Work

One of the fundamental problems with doing resource allocation in operating systems is that decisions as to which task should receive a given resource. These decisions are based on a measure (e.g. priority) which does not permit control over the actual quantity of a resource to be allocated over a given time period, particularly when this time period is small.

Managing Quality of Service (QoS) in an operating system can be done in a number of ways. At one extreme, one can make hard real time guarantees to applications, refusing them to run if the hard real time guarantees cannot be met at any stage. At the other extreme, one can provide a model which takes into account the fact that the resource reservation decisions are not made only at the beginning of the task but can be recurring to take fluctuations of resource availability and demand into account.

In general, the approach is to provide probabilistic guarantees and to expect applications to monitor their performance and adapt their behavior when resource allocation changes.

### *A. Environment*

The growing popularity of providing QoS in operating systems to support multimedia applications has resulted in several research efforts that have focused on the design of predictable resource allocation mechanisms. Consequently, in the recent past many projects/techniques have attacked the problem of providing quality of service in operating systems in a number of ways [14]. Early work on reservation-based real-time systems focused on the CPU as the resource to be shared. But to make the design of the system more complete, a sufficient model should not only cover CPU but all resources of interest, such as disk bandwidth or network. Hence, systems such as "Resource Kernels" have extended the initial work. They use deterministic reservation times, which lead to over allocation of resources since the worst case must also be covered. The common approach adopted by all of these focused on CPU scheduling mechanisms.

Most of the systems described are intended for more coarse-grained resource allocations. The Q-RAM architecture [7] follows a centralized approach where tasks specify their resource requirements to a central entity responsible for resource management.

In the Eclipse OS [6, 8, 9], applications obtained a desired quality of service by initially acquiring a resource reservation for each required physical resource by incorporating a potential share scheduler for each resource.

Further research resulted in the prediction of response times of the applications before they are executed in order to meet the QoS requirement in the metric of timeliness [5].

Several extensions have been made in existing workstation operating systems, using QoS based Resource Management Model [4], to assist the execution of multimedia systems wherein applications arrive with a request for a certain amount of resources which is provided to the system via the resource demand function and describes the different QoS settings of the application.

Significant information regarding implementing QoS in operating systems could be gathered from the Synthesis Operating System [3] where the kernel combines several techniques to provide high performance, including kernel code synthesis, fine grain scheduling and optimistic synchronization.

Instead of priorities, Synthesis uses fine grain scheduling which assigns larger or smaller quanta to threads based on a "need to execute" criterion. Effective CPU time received by a thread is the determined by the quantum assigned to that thread divided by the sum of quanta assigned to all threads.

Furthermore, guarantees need to be dynamically renegotiated [7] to allow the users to agree upon QoS parameters that satisfy service user(s) and service provider.

The connection request is rejected if no common agreement upon QoS is found. The service provider or called service user should provide the reason for rejection and perhaps additional information to the calling service user so that the request can be modified and resubmitted (with a higher probability of success) or cancelled. Furthermore, a QoS contract might be changed (i.e., renegotiated) during a sessions lifetime.

Providing QoS through Reservation Domains makes use of hierarchical, proportional-share dynamically reconfigurable resource schedulers at the device driver level for the management of disk and network bandwidth and for CPU scheduling.

Reservation domains enable explicit control over the provisioning of system resources among applications in order to achieve the desired levels of predictable performance. In general each reservation domain is assigned a certain fraction of each resource. The basic idea behind reservation domains is to isolate the performance of reservation domains from each other. In particular, time-sensitive applications can coexist with batch processes on the same system. The importance of cumulative service guarantee is discussed which complements other QoS parameters like delay and fairness. Informally, guaranteed cumulative service means that the scheduling delays encountered by a process on various resources do no accumulate over the lifetime of the process.

**Coalesced QoS** addresses the shortcomings of the above mentioned models by accounting for other modules of the operating system also.

Jeffay et al. [15] employ hard real-time scheduling theory in a specialized micro-kernel to address timing issues related to different stages of live video and audio processing. Robin et al. [17] designed a system based on Chorus [16] micro-kernel that addresses both the network and end host QoS control.

*B. Quality of Service in Operating Systems*

The Quality of Service (QoS) refers to the level of commitment provided by the service and the integrity of that commitment.

**Quality of Service** is a general term for an abstraction covering aspects of the non-functional behavior of a system. In particular, it includes not only the specification of non-functional service characteristics, but also necessary data models, operational constraints, and information about data measurement, monitoring and maintenance.

QoS represents a user's expectations for the performance of an application. It can also be defined as the collective measure of the level of service as requested by the user. It has been realized that in order to guarantee any QoS, applications must receive predictable resource allocation from the operating system. Since QoS can change during the execution of the application, the resource allocation from the operating system should also provide flexibility to accommodate the dynamic QoS requirements. The effort of utilizing available resources to satisfy QoS requirements can be largely classified into two types: resource reservation provided by the operating systems, and adaptation from the applications to accommodate resource availability.

In the context of operating systems, on a typical system, multiple applications may contend for the same physical resources such as CPU, memory and disk or network bandwidth. An important goal of an operating system is therefore to schedule requests from different applications so that each application and the system as whole perform well. As an example, let us consider real-time applications, which must have their requests processed within certain performance bounds. To support real time applications correctly under arbitrary system load, the operating system must perform admission control and offer quality of service (QoS) guarantees. Many multimedia applications have timing requirements and other quality of service (QoS) parameters that represent the user's desires and expectations for the performance of the applications. The complexity of providing for these timing requirements at the system level is exacerbated by the fact that the user may change those timing requirements at any time during the execution of the applications; and of course the user may create and terminate multimedia applications at any time.

In dealing with QoS management it is important to realize that there are different types of QoS parameters for different levels of the system. Applications must interact with the user in terms of user-level QoS parameters. Once the user-level QoS parameters are determined, they have to be mapped into system-level QoS parameters that would be meaningful for system-level resource management mechanisms. These system-level QoS parameters would describe how much time is needed on various resources. They depend on the user-level QoS parameters and on detailed computations that the operating system performs on data elements.

To allow the user to specify the QoS parameters desired at the highest level, the application must be able to map from user-level QoS parameters to system-level QoS parameters. The system-level parameters are required for the application to be able to ask for the resources it will need to execute. If the resources are unavailable, the system-level resource management mechanism should be able to communicate the fact that those parameters cannot be guaranteed. It should then initiate a negotiation to arrive at a set of system-level parameters that can be supported by the system. The inverse mapping to user-level QoS parameters should yield a QoS specification that can be tolerated by the user. Thus, the inverse mapping from system-level to user-level QoS parameters is just as important as the forward mapping.

A **resource reservation approach** must allocate the reserve for its computation on behalf of an incoming application request; it must know what the reservation parameters should be. This approach requires the application and OS to enter into a dialogue to allow the application to explicitly request OS specific QoS level, meaning a certain pattern of OS operations to be called with certain timing constraints. The operating system must then map the requested QoS requirements to system resource requirements and decide whether it can acquire the reserves to support that activity. Further, the OS must have the machinery to map those QoS requirements to system resource requirements.

The operating system could store in a persistent preferences database some information about reservation levels used in previous instantiations of the applications. This information would be a good guess as to what reservation levels should be in case the incoming application is memory intensive or IO intensive, and it might be possible to maintain a small database to map prior experience with different QoS parameters to reservation parameters. This approach might get much more complicated as more QoS parameters, reservation parameters, and target system architectures are used. The initial reservation level could be set to zero or some other relatively small value that is known to be smaller than the actual reservation level. This approach requires the mechanisms for reservation level adaptation to quickly acquire the feedback on usage that is necessary to set a reasonable reservation level where desired quality of service parameters can be achieved. Hence, the operating system should support the process of self calibration. Namely the appropriate requirement is calculated from on-the-fly feedback of application performance, and is calibrated when the performance does not meet the QoS requirement. A system that adapts transparently to available platform resources should employ an adaptive QoS-mapping function.

Earlier in this paper, we effectively stated that QoS provides the ability to handle application traffic such that it meets the service needs of certain applications. We recall that, if operating resources were infinite, the service needs of all applications would be trivially met. **It follows that QoS is interesting to us because it enables us to meet the service needs of certain applications when resources are finite.**

A QoS enabled operating system should provide service guarantees appropriate for various application types while making efficient use of available resources. A performance-assured operating system must be able to provide each client or class of clients their desired QoS irrespective of the behavior of other clients. This implies protection of well-behaving clients from those violating their QoS specification.

## III. KERNEL-LESS OPERATING SYSTEM (KLOS)

### *A. Introduction*

The purpose of operating system is to multiplex shared resources between applications. Operating Systems provide services that are accessed by processes via mechanisms that involve a ring-transition to transfer control to the kernel where the required function is performed. This has one significant drawback that every service call involves an overhead of a context-switch where processor state is saved and a protection domain transfer is performed. However, on processor architectures that support segmentation, it is possible to achieve a significant performance gain in accessing the services provided by the operating system by not performing a ring transition. KLOS is a Kernel-Less Operating System, acting as a proof-of-concept vehicle, wherein all processes execute at the unprivileged level having their own private execution space and function independent of all other processes in the system.

Operating systems define the interface between applications and physical resources. Unfortunately, this interface can significantly limit the performance and implementation freedom of applications.

Most, if not all production level operating systems have a dual mode of operation - (1) the privileged level where the kernel resides and, (2) the unprivileged level where application and system processes execute. A ring transition mechanism is used to move from one level to the other. The idea behind this separation has always been protection and stability. However, on processor architectures that support segmentation, it is possible to achieve a significant performance gain by eliminating the ring transition. Further, such gains can be achieved without compromising protection. This is made possible by the use of a subtle trick involving segmentation and Task State Segments (TSS).

To this end, we have designed new operating system architecture, in which –

- There is no kernel as perceived in current operating systems,
- Operating system services are accessed without a ring transition,
- All processes and the operating system execute at the unprivileged level and,
- Each process has its own private address space and virtual memory mappings and functions independent of all other processes in the system.

KLOS, a Kernel-Less Operating System is a realization of this design. Regular runaway processes do not pose a threat to the stability of the operating system built on this design. Processes that are specifically engineered to thwart the stability of the system are contained with a very high probability of success [1, 2]. KLOS is able to achieve high performance due to its architecture and its service call mechanism.

### *B. Design of KLOS*

Traditional operating systems comprise of a kernel, that is responsible for providing the core services of the operating system and a shell or applications that reside on top of the kernel using its services and providing an interface to the user. In KLOS, there is no kernel per se. Instead, the entire operating system is made available to each application as a part of its execution space, running at the same privilege level.

The design of KLOS relies on the segmentation capability of the x86-based processors but is different in its approach with use of the TSS, doing away with the down-call to the kernel. The authors [1, 2] method is dynamic, which means there is no code pre-scanning or other procedures that need to be applied on application code before it is executed.

The architecture, at its heart consists of an event core that is responsible for acting upon external events (hardware interrupts and processor generated exceptions).

Events are the only means of vertical up-calls to the unprivileged level. There are no down-calls to the privileged level eliminating a protection domain transfer during normal program flow. All the components of a typical operating system like the memory manager, scheduler, process manager, device drivers etc. run in the unprivileged level and have a horizontal mode of interaction.

KLOS is able to achieve high performance due to its architecture and its service call mechanism. The service call mechanism of KLOS results in a general 4x improvement over the traditional trap (interrupt) and a 2x improvement over the Intel SYSENTER/SYSEXIT fast system call models.

## IV. COALESCED QOS FRAMEWORK

### A. Design

One of the fundamental problems with conventional operating systems is the decision as to which task should receive a given resource. These decisions are based on a measure (e.g. priority), which does not permit control over actual quantity of resource to be allocated over a given time period, particularly when this time period is small.

By interacting with operating system kernels and application-level hooks, QoS aware operating system has been proved highly effective in supporting high performance applications via run-time probing, and adaptation of applications. Therefore, application configurations and performances can be tailored to different user behavior and characteristics of ubiquitous environments. To be more specific, the QoS aware framework is expected to provide the following critical functions:

**Run-time probing**. Applications are subject to run-time probing with respect to the instantaneous performances of their QoS parameters. Such QoS probing is the responsibility of operating system. Probing is useful in learning about application behavior, so that better run-time adaptation rules may be set accordingly.

**Run-time adaptation**. During application run-time, the operating system may assist the application to adapt to the changes of resource availability or user requirements. Such behavioral changes exist due to the sharing of resources among applications; or lack of resource reservation mechanisms. The operating system changes the application behavior correspondingly when significant variations in these triggering sources are detected.

Our goal is to find adaptation strategies that could be applied on-line during application runtime. We claim that solving sophisticated optimization problems will generate too much overhead and delay. Therefore, we have aimed at rather simple, but efficient heuristics. Therefore, this is the preferred approach for a dynamic "on the fly" adaptation of applications during runtime.

A key problem with the single resource scheduling algorithms is that they are not designed to make use of the information in the additional resource requirements. Our contribution is to come up with a new heuristic for scheduling, wherein our algorithms are *resource-aware*, meaning that they are able to use the additional resource information in the system to intelligently adjust the priority of the execution among the waiting jobs.

Our architecture is based on the notion of feedback so that the period assigned to threads change dynamically as the resource requirements of the processes changes.

Ideally resource allocation should ensure that every process maintains a sufficient rate of progress towards completing its tasks. The progress is determined by how much time does processes take to complete and this progress is relative to its priority in the system. The priority of a process is calculated based on the relative use of all the resources in the system and how much is each process utilizing a particular resource. This feedback mechanism periodically monitors the resource usage of each process and automatically calibrates its priority based on the current usage of the resources.

In our proposed architecture, we use the resource allocation information to come up with a scheduling heuristics so that processes can be sequenced in the queue as per their priority. The priority of the processes is based on a heuristic which takes into account the number of resources the process is using as well as the quantity of the resources. This is done by profiling the system calls for each process.

We modify the system call in KLOS, wherein it is built with QoS Support to include the resource requirements for each process so that we can profile the resource requirement of a process and calculate the priority based on this information.

Run time profiling of the processes is done in the system so that we can raise the priority for processes as time progresses. This is done on the basis of the age of a process and the relative use of all resources in the system by a particular process. Notion is being usable in practice and a process should be able to shed light itself at runtime regarding the resources it will be using.

### B. Stochastic Scheduling

In order to satisfy our needs for scheduling policies that can leverage the performance variability of resources, we come up with a scheduling policy. We describe a methodology that allows our scheduling policy to take advantage of the information to improve application execution performance.

This work on resource-aware scheduling policy is being conducted in a larger context of real-time resource reservations in KLOS as to provide reasonably high Quality of Service. The resource allocation policies are controlled by the scheduler to meet the timeliness requirements, tailoring their behavior in response to the actual resources available. As opposed to systems, where all resource requirements are known well in advance, we have a model where the resource requirements are calculated at run-time and our heuristics support our claim for high performance in such an environment.

TABLE I
SUMMARY FOR EQUAL USAGE OF RESOURCES

| Observed Parameters | Process Intensity = 4 | Process Intensity = 7 | Process Intensity = 10 |
|---|---|---|---|
| CPU Utilization | Coalesced QoS (1.05) | Coalesced QoS (1.10) | Coalesced QoS (1.10) |
| Memory Utilization | Coalesced QoS (1.05) | Coalesced QoS (1.04) | Coalesced QoS (1.10) |
| System Throughput | OSP (1.03) | Coalesced QoS (1.10) | X |
| Avg. Waiting Time | X | X | Coalesced QoS (1.12) |

Process Intensity refers to the Process Creation Intensity. The table tells us which model performs better and the value in bracket is an indicative of how much the performance is increased. An 'X' means that both the models perform equally.

Since our approach is primarily based on real-time scheduling of processes in the presence of multiple resource requirements, it needs to be addressed in a comprehensive manner. Successfully applied scheduling techniques have been based on preemptive fixed priority scheduling in most real-time systems developments since Liu and Layland introduced it in their paper [25]. However, dynamic priority schedulers can achieve higher schedulability than fixed ones, and non-preemptive schedulers incur less run-time overhead.

The fundamental desirable characteristic of a priority scheduler is that the system should always be executing at the highest priority. However, it is not always possible for a system to conform to this requirement. When circumstances within the system force a higher priority task to wait for a lower priority task, **priority inversion** occurs. In our framework, we have priority assigned to each process and this might lead to an uncontrolled priority inversion problem. It is important to note that every operating system experiences incidents of priority inversion; the issue is not the presence of priority inversion but rather the time duration of each source of priority inversion. The magnitude of priority inversion is thus very important in assessing whether an operating system will be suitable for a particular application or system.

We thus know that such problems occur because priorities alone are not expressive enough to capture all the relationships between the resources. Our approach here has an advantage here that the priority is calculated by taking into account the resources a process requires as well as the age of a process and this approach has some clear advantages which can be seen from the results.

## V. PERFORMANCE EVALUATION

In this section, the simulation model designed to evaluate the performance of our resource allocation scheme is described. In order to compare the performance of our algorithm, we simulate a coalesced resource allocation scheme.

In the functional description of a technology, one can describe the different ways of performance, and measure one of the technology performance parameter (application, capability, performance, quality, and safety).

F. Zwicky (Zwicky, 1948) proposed that one can systematically explore technology sources for technical advance in a system by logically constructing all possible combinations of physical alternatives of the system. Thus our performance evaluation technique is based on this where we logically construct combinations of high and low resource utilization with low, medium or high process creation intensity to study the system behavior.

### *A. Performance Metrics*

Once we have built a mathematical model, it must then be examined to see how it can be used to answer the questions of interest about the system it is supposed to represent. If the model is simple enough, it may be possible to work with its relationships and quantities to get an exact, *analytical* solution. If an analytical solution to a mathematical model is available and is computationally efficient, it is desirable to study the model via simulation. However, many systems are highly complex, so that valid mathematical models of them are themselves complex, precluding any possibility of an analytical solution. In this case, the model must be studied by means of *simulation*, i.e., numerically exercising the model for the inputs in question to see how they affect the output measures of performance.

We use simulation to analyze the performance of our proposed resource utilization scheme. Some of the main features in our simulation are:

- The call arrival process is a Poisson Process with rate lambda ($\lambda$). The interarrival times are exponentially distributed with mean interarrival time $1/\lambda$.
- The expected call duration times are exponentially distributed with a mean duration time.

Our algorithm was simulated on a suite of synthetic workload. This allowed us to vary different characteristics of the workload with respect to the resources. While the general shape of this distribution is exponential, the model captures all the characteristics as seen in a production environment.

### *B. Results and Analysis*

In this section, we describe how the components described in the preceding sections can be implemented on a realistic platform. The experimental test-bed and implementation environment is based on OSP [26], an Operating System Project. OSP is a simulated system that provides the illusion of a computer system with a dynamically evolving collection of user processes to be multi-programmed. When simulation commences, OSP uses *simulation parameters* to decide whether the system should be I/O-bound or cpu-bound, how often to generate requests for various system resources, how long the simulation should last, etc.

The OSP simulator accumulates the statistics that reflect the amount of resources consumed by the various modules. The most significant are the following performance indicators:

- CPU Utilization

- Memory Utilization
- System Throughput
- Average Waiting Time per process

Output statistics generated by OSP are used for the performance measures in this simulation. For example, by altering the length of the cpu time quantum, one can see the effect on the overall system throughput and other statistics. Likewise, by varying the degree of prepaging, we can observe the effect of prepaging on the number of page faults, on the total number of pages swapped in or out, and on the average turnaround time.

### *C. Performance Graphs*

As observed from the knee plots in Fig 1 through Fig 8, for the distribution of the various resources, we can say with 95% confidence that the Coalesced QoS based model performs better than a non-QoS based model. The graphs are a clear indication, that the system starts to take advantage of the Coalesced QoS based model when the resource usage in the system is high and also when all the resources are being equally consumed by the processes.

Table I presents the results of the comparison between the performances of a Coalesced QoS based system and a non-QoS based system when the resources are equally distributed in the system.

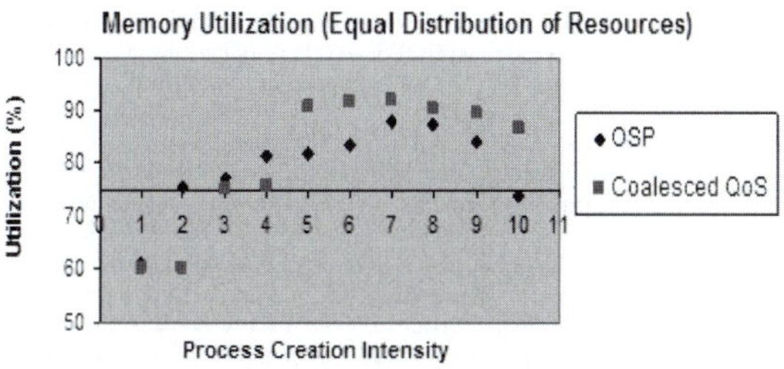


Fig. 1. Knee Plot for Memory Utilization with Equal Distribution of Resources

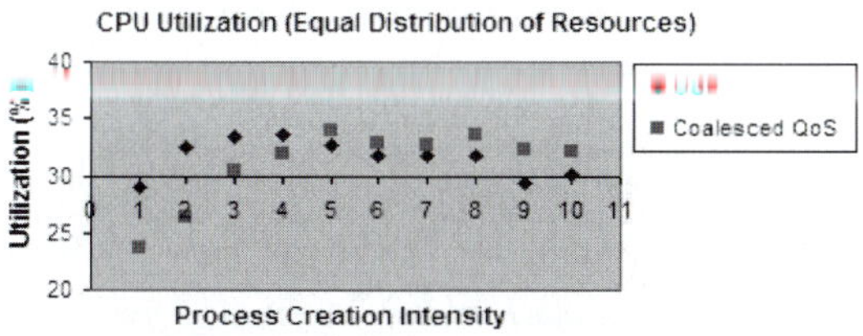


Fig. 2. Knee Plot for CPU Utilization with Equal Distribution of Resources

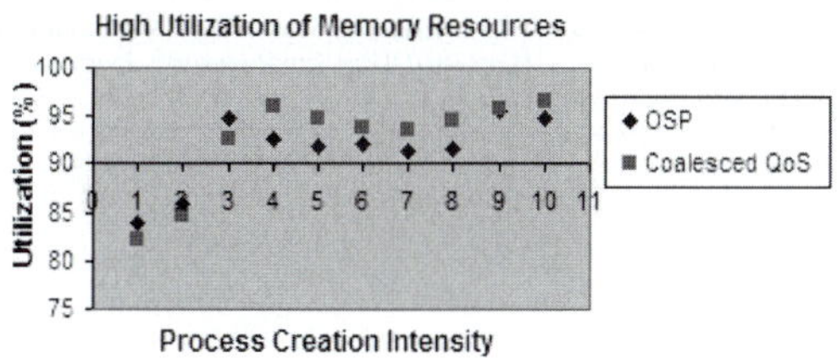


Fig. 3. Knee Plot for High Utilization of Memory Resource

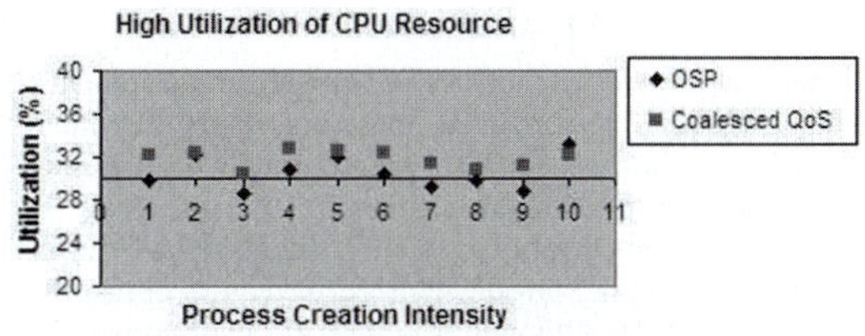


Fig. 4. Knee Plot for High Utilization of CPU Resource

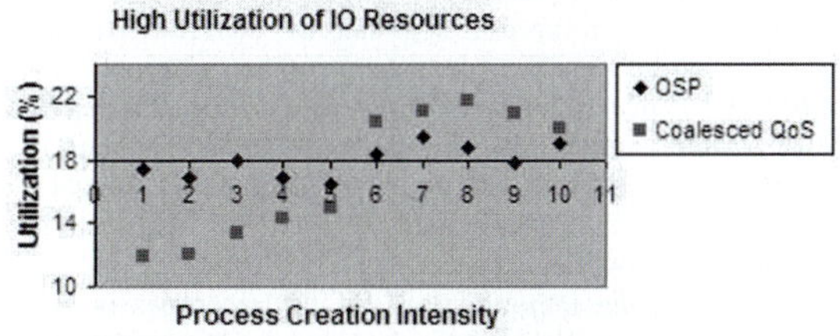


Fig. 5. Knee Plot for High Utilization of IO Resource

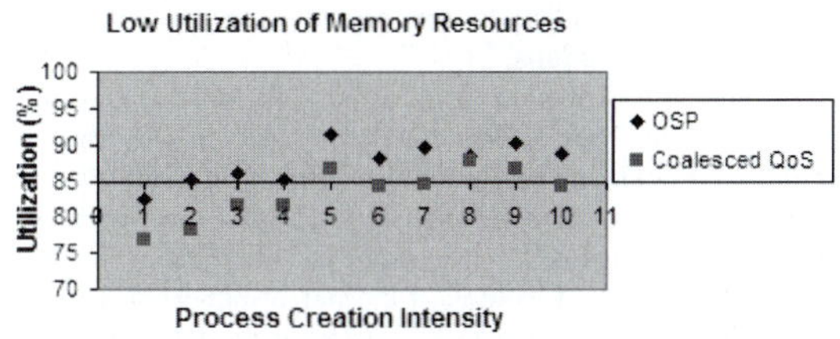


Fig. 6. Knee Plot for Low Utilization of Memory Resource

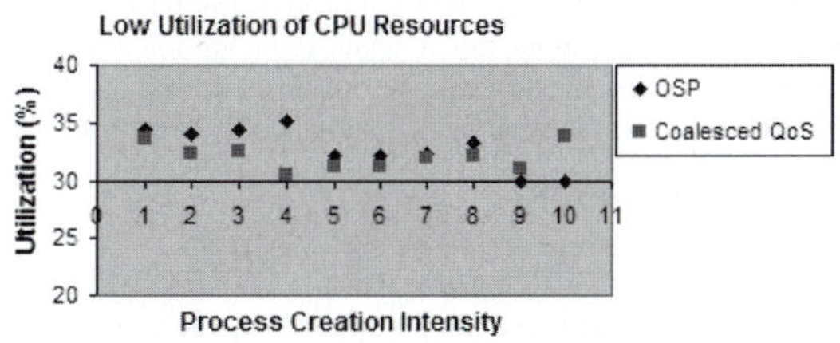


Fig. 7. Knee Plot for Low Utilization of CPU Resource

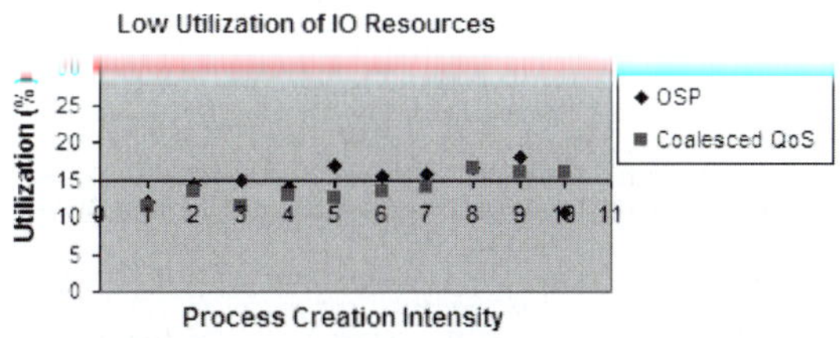


Fig. 8. Knee Plot for Low Utilization of IO Resource

After observing the performance from the graphs, we performed the paired t-test to confirm the authenticity of the above results which is summarized in Table I.

The paired t-test compares two paired groups so one can make inferences about the size of the average treatment effect (average difference between the paired measurements). The most important results are the P value and the confidence interval.

## VI. Conclusion

In this paper, we are concerned with the requirements placed upon high performance operating systems, which are usually, but not necessarily, general-purpose operating systems with soft real-time extensions.

We have attempted to identify an uncontroversial set of requirements that the "ideal" high performance operating system would meet. Although it is unlikely that any single system or scheduling policy will be able to meet all of these requirements for all types of applications, the requirements are important because they describe the space within which high performance systems are designed. A particular set of prioritizations among the requirements will result in a specific set of tradeoffs, and these tradeoffs will constrain the design of the user interface and the application programming model of a system.

Particularly, the importance of all the resources is not reflected in current QoS approaches and we have made an effort to take into account all the resources that a process uses in order to calculate the priority for the efficient execution of applications.

Currently our work has some limitations that are being addressed. Our performance results are based on a simulation model, and hence are preliminary in many ways. We looked into what Kurtzweil (1999) calls the 'knee of the curve' in regard to the resource utilization in the system. We also need to look at the performance measures when this is implemented in the working model of KLOS and that should give us some realistic values as to where our idea stands. Currently, we have used a simulator for our performance evaluation, but it gives us a limited view of the process satisfaction and we need to come up with metrics which would give us this statistics to study process behavior.

## References

[1] Vasudevan, A., Yerraballi, R. and Chawla, A. (2005). A High Performance Kernel-Less Operating System Architecture. In Proc. Twenty-Eighth Australasian Computer Science Conference (ACSC2005), Newcastle, Australia. CRPIT, 38. Estivill-Castro, V., Ed. ACS. 287-296.

[2] Amit Vasudevan, Ramesh Yerraballi, Ashish Chawla, "KLOS: High Performance Kernel Less Operating System", Accepted for publication in the 24th IEEE Real-Time Systems Symposium, WIP Section, Cancun, 2003

[3] Henry Massalin, Calton Pu: "Threads and Input/Output in the Synthesis Kernel", Proceedings of the twelfth ACM symposium on Operating systems principles, 1989

[4] Wonjun Lee, Jaideep Srivastava: "A Market Based Resource Management and QoS Support Framework for Distributed Multimedia Systems", ACM conference on Information and Knowledge Management, USA, 2000

[5] E. Huh, L. R. Welch, B. A. Shirazi, B. Tjaden, and C. D. Cavanaugh, "Accommodating QoS Prediction in an Adaptive Resource Management Framework", IPDPS, Mexico, 2000.

[6] J. Blanquer, J. Bruno, E. Gabber, B. Ozden, and A. Silberschatz: Resource Management for QoS in Eclipse/BSD

[7] Thomas Plagemann, Knut A. Saethre and Vera Goebel: Application Requirements and QoS Negotiation in Multimedia Systems, Second Workshop on Protocols for Multimedia Systems, PROMS'95, Salzburg Austria, October 1995

[8] J. Bruno, E. Gabber, B. Ozden, and A. Silberschatz: The Eclipse Operating System: Providing Quality of Service via Reservation Domains.

[9] J. Bruno, E. Gabber, B. Ozden, and A. Silberschatz: Move-To-Rear List Scheduling: A New Scheduling Algorithm for Providing QoS Guarantees, ACM Multimedia 97 – Electronic Proceedings

[10] R. Rajkumar, C. Lee, J. Lehoczky, and D. Siewiorek: A Resource Allocation Model for QoS Management. In Proc. of the 18th IEEE Real-Time Systems Symposium, San Francisco, USA, December 1997.

[11] B. Bershad, C. Chambers, S. Eggers, C. Maeda, D. McNamee, P. Pardyak, S. Savage, and E. Sirer., SPIN - an extensible microkernel for application specific operating system services. 1994 European SIGOPS Workshop, 1994.

[12] J. Blazewicz, W. Cellary, R. Slowinski and J. Weglarz. Scheduling under Resource Constraints – Deterministic Models. In Annals of Operations Research, Volume 7. Baltzer Science Publishers, 1986

[13] C. W. Mercer, S. Savage and H. Tokuda. Processor Capacity Reserves for Multimedia Operating Systems. In Proceedings of the IEEE International Conference on Multimedia Computing and Systems. May 1994

[14] Raj Rajkumar, K. Juvva, A. Molano and S. Oikawa. Resource Kernels: A Resource-Centric Approach to Real-Time and Multimedia Systems. In Proceedings of the SPIE/ACM Conference on Multimedia Computing and Networking, January 1998.

[15] K. Jeffay, D. L. Stone, and F. D. Smith. "Kernel Support for Live Digital Audio and Video". (Nov 1991), 10-21.

[16] Bricker, M. Gien, M. Guilllemont, J. Lipskis, D. Orr and M. Rozier. "Architectural Issues in Microkernel-base Operating Systems: the CHORUS Experience". Computer Communication 14, 6 (July 1991), 347-357.

[17] P. Robin, G. Coulson, A. Campbell, G. Blair and M. Papathomas. Implementing a QoS Controlled ATM Based Communications System in Chorus. Technical Report MPG-94-05, Dept. of Computing, Lancaster University, March, 1994.

[18] D. Engler, M. Kaashoek, and J. O'Toole. Exokernel: An operating system architecture for application-level resource management. In Proceedings of the 15th ACM Symposium on Operating System Principles, pages 251 - 266, 1995.

[19] H. Hartig, M. Hohmuth, J. Liedtke, S. Schonberg, and J. Wolter. The performance of μKernel-based systems. In Proceedings of the 16th ACM Symposium on Operating Systems Principles, pages 66 - 77, 1997.

[20] K. M. Zuberi, P. Pillai, and K. G. Shin. EMERALDS: a small-memory real-time microkernel. In Proceedings of the 17th ACM Symposium on Operating System Principles, pages 277–291, Kiawah Island, SC, 1999.

[21] T. Shinagawa, K. Kono, T. Masuda. Fine-grained Protection Domain based on Segmentation Mechanism. Japan Society for Software Science and Technology, 2000.

[22] Tao Li, Lizy Kurian John, Anand Sivasubramaniam, N. Vijaykrishnan, Juan Rubio. Understanding and improving operating system effects in control flow prediction. In Proceedings of the 10th international conference on architectural support for programming languages and operating systems, pages 68-80, San Jose, CA, 2002.

[23] R. Sekar, V.N. Venkatakrishnan, Samik Basu, Sandeep Bhatkar, Daniel C. DuVarney. Model-carrying code: a practical approach for safe execution of untrusted applications. In Proceedings of the nineteenth ACM symposium on Operating systems principles, pages 15–28, Bolton Landing, NY, 2003.

[24] David Lie, Chandramohan A. Thekkath, Mark Horowitz. Implementing an untrusted operating system on trusted hardware. In Proceedings of the nineteenth ACM symposium on Operating systems principles, pages 178-192, Bolton Landing, NY, 2003.

[25] C. L. Liu and J. W. Layland, "Scheduling Algorithms for Multiprogramming in a Hard Real-Time Environment", Journal of the ACM, V20, N1, 1973, pp. 46-61.

[26] Michael Kifer, Scott A. Smolka: OSP: An Environment for Operating System Projects. *Operating Systems Review 26(4): 98-100 (1992)*

# Method of Key Vectors Extraction Using R-cloud Classifiers

A. A. Bougaev[1], A. M. Urmanov[2], K. C. Gross[2] and L. H. Tsoukalas[1]
[1]School of Nuclear Engineering, Purdue University, 400 Central Dr., W. Lafayette, IN, 47907, USA
[2]Sun Microsystems, 9515 Towne Centre Dr., San Diego, CA, 92121, USA

***Abstract*— A novel method for reducing a training data set in the context of nonparametric classification is proposed. The new method is based on the method of R-clouds. The advantages of the R-cloud classification method introduced recently are being investigated. A data set reduction approach using Rvachev functions-based representation of the separating boundary is proposed. The R-cloud method was found instructive and practical in a number of engineering problems related to pattern classification. The method of key vectors extraction uses the property of the normal R-cloud boundary to evaluate the distance from the sample to the separating boundary.**

## I. Introduction

The computational load of classification of input data by pattern recognition algorithms, which are used in diagnostics, increases linearly (or even quadratically for some algorithms) with the number of training patterns. This limits the applicability of pattern classification for online diagnostics and also for offline mining of large (from 10,000 patterns and up) data bases.

If one can systematically decrease the size of the collection of training patterns it is possible to significantly reduce the compute load for subsequent classification analyses. One cannot arbitrarily eliminate training patterns. If the training patterns are not pruned out reasonably, one introduces bias and inconsistency in the subsequent classification analyses.

Conventional approaches for training set reduction suffer from one of the following deficiencies:

- prohibitively long running time (3-rd and higher order polynomial in the number and in the dimension of training patterns analyzed)
- inconsistency of the resultant decision boundary obtained on the reduced set (i.e. the decision boundary is different than would have been obtained with the complete set of training patterns)
- suboptimal size for the reduced training set (i.e. there exists a smaller subset that results in the same decision boundary as obtained with the complete set of training patterns)

Hence, there is a need for a method that is fast, preserves the original decision boundary, and generates a minimal reduced training set.

The proposed key vectors extraction (KVE) algorithm provides a new and unique training set reduction technique to shorten significantly the computation time required for two-group pattern classification of input data.

The KVE algorithm realizes a method which:

- takes advantage of R-cloud classifier (defined below), which permits the evaluation of the shortest distance from any point to the boundary
- requires $O(dn^3)$ operations ($d$ is the dimension of the input space, $n$ is the number of training samples), which is significantly faster than conventional decision boundary consistent methods
- preserves the decision boundary locally in the region of interest
- allows for further reduction of the training set in the cases where an admissible decision boundary change (specified by the user as an allowable tolerance) is specified

## II. Previous Work

Automated systems for pattern classification applications such as, for example, component or system fault identification, computer network intrusion detection, and denial of service attack detection, process sets of input data by dividing the input data into more readily processable subsets. Such data processing employs pattern classification to divide the available data into two (and sometimes multiple) subsets.

The computation time of classification of input data by pattern recognition algorithms that are used in diagnostics, such as k-Nearest Neighbor (kNN) classifiers, Radial Basis Function (RBF) networks, Least-Squares Support Vector Machines (LSSVM), Multivariate State Estimation Techniques (MSET), and other algorithms, increases linearly (or quadratically for some algorithms) with the number of training patterns. This compute cost limits the applicability of classification pattern recognition for online diagnostics and also for offline mining of large data bases.

As known from the literature, reduction of the computation time can be achieved by reducing the training set size. In one example, as described by Hart in [1], the condensed nearest neighbor rule, which is one of a class of ad hoc decisions rules, employs a procedure that starts with a one-pattern reduced set and sequentially examines the remaining training patterns, discarding the ones that are correctly classified by the current reduced set and adding the ones that are classified incorrectly to the reduced set. The algorithm then iterates through the discarded patterns until all the remained discarded patterns are classified correctly by the reduced set. This algorithm is not decision boundary consistent and does not always find a minimal set.



*T. Sobh and K. Elleithy (eds.), Advances in Systems, Computing Sciences and Software Engineering*, 97–100.

One of the decision-boundary consistent algorithms known is the one introduced by Toussaint in [2]. The algorithm uses the Voronoi diagrams and, hence, its name the Voronoi editing algorithm. The Voronoi diagram partitions the input space into regions that are the loci of points of space closer to each data point than to any other data point. The Voronoi editing algorithm maintains exactly the original decision boundary of the nearest neighbor decision rule; however, the reduced set produced by the algorithm is not minimal, and the algorithm requires $O(n^{d/2})$ operations what makes it impractical for dimensions higher than 4.

An improvement over the Voronoi editing algorithm is the Gabriel graph condensing algorithm proposed in [3]. The Gabriel graph condensing algorithm constructs the Gabriel graph - a set of edges joining pairs of points that form the diameter of an empty sphere - of the training set. This method is much faster; it requires $O(dn^3)$ operations. However, the Gabriel graph condensing does not preserve the decision boundary.

Another iterative training set reduction algorithm is described by Brighton and Mellish [4]. This algorithm applies a rule that identifies patterns to be removed, removes the identified patterns, and applies the rule again to the reduced set until no more patterns can be removed. The deletion rule is as follows. For each point **x**, if the number of other points that classify **x** correctly is greater than the number of points classified by **x**, then discard point **x**. This algorithm does not preserve the decision boundary and may require excessively long execution times due to its iterative nature.

## III. R-CLOUD CLASSIFIERS

A novel pattern classification method based on Rvachev functions (termed R-functions) was introduced recently by Bougaev and Urmanov [5]. The method is termed the R-cloud classifier and uses the notions of the separating primitive and the separating bundle. The R-cloud classification is a non-parametric approach. It makes no assumptions about the data model and does not require class density estimates.

Real valued functions, whose sign is completely determined by the signs of their arguments, belong to the class of R-functions. One of the important applications of R-functions is the representation of geometric objects. An object, which can be defined as a set of geometric primitives can be represented by an R-function. A brief introduction into the theory of R-functions is given by Shapiro in [6].

The construction of R-cloud classifiers utilizes the methodological apparatus of R-functions. The R-cloud classification allows for the implicit analytical representation of the decision surface. This makes such type of classifiers attractive from an application viewpoint. The separating surfaces are represented by R-clouds, which are built on the notions of the separating primitive and the separating bundle.

R-cloud classifiers utilize the R-function $R(\mathbf{x})$ constructed of separating primitives and separating bundles.

In a setting of the two-group pattern classification problem $C_1$, $C_2$, and $\mathbf{D}=\{C_1,C_2\}$ denote the collection of data patterns in $\mathbf{R}^d$ that belong to class (group) 1, the collection of data patterns in $\mathbf{R}^d$ that belong to class (group) 2, and the collection of all data patterns in the training data set, respectively. In this case, the R-cloud function associated with class 1 is formulated as

$$R(\mathbf{x}) = \bigvee_{\mathbf{u}\in C_1} \bigwedge_{\mathbf{v}\in C_2} \rho(\mathbf{x},\mathbf{u},\mathbf{v}) \qquad (1)$$

where $\vee$ and $\wedge$ are the operations of R-disjunction and R-conjunction defined in [6], $\rho(\mathbf{x},\mathbf{u},\mathbf{v})$ is the separating primitive function defined in [5], and **x** is an unlabeled vector from the input space, which needs to be classified.

The R-clouds are R-functions by design. Therefore, the value of R-cloud function, $R(\mathbf{x})$ evaluated on a previously unseen data vector **x** gives the shortest distance from **x** to the decision boundary [7]. This property of the R-cloud is utilized in the algorithms described below.

## IV. KEY VECTORS EXTRACTION METHOD

The new training set reduction algorithm employs a new Key Vectors Extraction (KVE) procedure described as follows. A nonparametric pattern classifier uses the training set, **D**, which comprises $n$ samples of both classes to assign new unknown-class patterns into one of the two classes, based on the decision rule the classifier implements. An example would be mining a large data base of telemetry signals from a population of servers that either suffered a particular failure mechanism (class 1), or did not (class 2). If one can apply pattern recognition to successfully classify the data into class 1 versus class 2 (i.e. the 'failed' vs. 'not failed' categories), then one can take that trained classification algorithm and processes telemetry data sets from the field at some frequency. This process will flag servers that are considered at elevated risk of failure.

Typically, known types of nonparametric classifiers require $O(n)$ to $O(n^2)$ operations. Such a computation complexity limits their usage for applications involving online diagnostics and also for offline mining of large data bases. Therefore, the goal of a training set reduction algorithm is to thin the training data set by deleting patterns that are not representative.

The pruning of data patterns must of course be done carefully. The reduction is required to preserve exactly or change insignificantly the original decision boundary of the classifier. The proposed key vectors extraction algorithm exploits the R-cloud classifiers and comprises the following steps:

1. Construct the original decision boundary, $R(\mathbf{x})$, with R-cloud classifier, using the full training set **D**.
2. For each data pattern **x** in **D**:
3. Compute the multidimensional Euclidian distance, $d_1$, to the original decision boundary via evaluating $R(\mathbf{x})$.
4. Construct the decision boundary $R^{\mathbf{x}}(\mathbf{x})$ using the training set without **x**, represented as $\mathbf{D}\backslash\mathbf{x}$.
5. Compute the distance, $d_1=R^{\mathbf{x}}(\mathbf{x})$, to the new decision boundary.
6. If $d_1$-$d_2$ (the distance did not changed), mark the vector as a *key vector*.
7. Remove all non-key data patterns from **D**; the resultant reduced training set, **D'**, comprises only the key vectors.

This algorithm requires $O(dn^3)$ operations and preserves the original decision boundary of the classifier in the region of interest.

Advantageously, in the cases where an allowable level of decision-boundary change is specified, the key vector extraction algorithm reduces the size of the training set even further. The modified algorithm that accounts for the allowable level of decision-boundary change, $\tau$, consists of the following steps:

1. Construct the original decision boundary $R(\mathbf{x})$ using the full training set **D**.
2. Specify the threshold $\tau$.
3. For each data pattern **x** in **D**:
4. Compute the shortest distance, $d_1$-$R(\mathbf{x})$, to the original decision boundary.
5. Using the training set without **x**, **D\x**, compute the distance, $d_2$=$R(\mathbf{x})$, to the new decision boundary.
6. If $|d_1-d_2|>\tau$ (the change in the shortest distance is greater than the allowed threshold), mark the pattern as a key pattern.
7. Remove all non-key vectors from **D**; the resultant reduced training set, **D'**, comprises only the key patterns.

This modification of the KVE algorithm produces a reduced training set in $O(dn^3)$ operations. The reduced training set changes the decision boundary locally by no more than the user-specified tolerance value, $\tau$. In addition, for small values of the tolerance, the new decision boundary differs from the original decision boundary mainly in the region outside of the convex hull of the training set, which is usually not of interest for the classification problems (i.e. the data outside of the convex hull of the training set has minimal, if any, influence on the decision boundary).

The KVE algorithm maintains the level of the deviation of the decision boundary from the original decision boundary below a specified tolerance value, which is beneficial in engineering applications of this approach for proactive health monitoring of complex servers.

## V. Example Application

The proposed KVE algorithm is illustrated using a synthetic data set in dimensions ranging from 2 to 1024. For illustration of the linear dependence of the computation time on the data dimension, the training data set comprises two classes of patterns. Each class has 150 data patterns. The data patterns of each class are generated from two multidimensional Gaussian distributions with different means. The Bayes error of the training set is 5% (this reflects the fraction of patterns that are misclassified by the most accurate pattern classifier theoretically possible).

Fig. 1 demonstrates the computation time required by the algorithm for fixed $n=150$ and different dimensions. As evident from the figure, the dependence is linear as a theoretical analysis of the algorithm predicts.

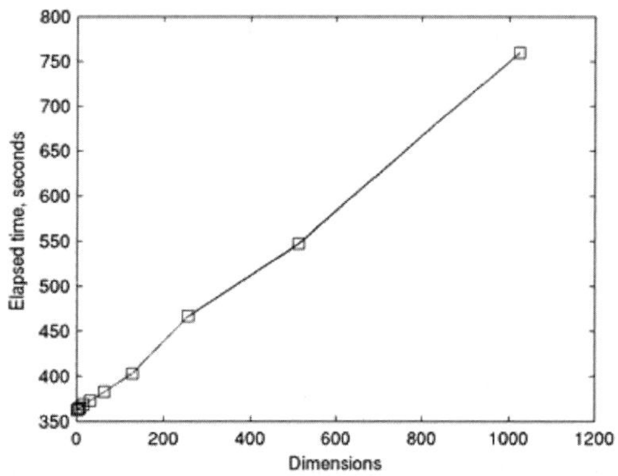


Fig. 1. Linear dependence of the computation time on the dimension of input data space of the key vectors extraction algorithm. A synthetic data set is used with 150 patterns in each class. The patterns of each class are generated from two multivariate Gaussian distributions in $\mathbf{R}^d$ with different means. The theoretical linear dependence on the dimension $O(dn^3)$ and the simulation results agree.

Fig. 2 and Fig. 3 illustrate the case of the reduction of a double horse-shoe training set in two dimensions. A double horse-shoe configuration often represents a challenge for a pattern classification algorithm. The reduced data patterns are marked with filled circles (class 1 data patterns) and filled triangles (class 2 data patterns). The shaded area represents the region of class 1. The decision boundary is preserved in Fig. 2. Approximately 25% of data patterns are removed by the KVE algorithm. This reduces the computational cost of the classification algorithm, yet the decision boundary computed from the KVE-processed reduced training data set is nearly identical to that obtained from the original training data set, as desired.

The decision boundary is preserved in the region of interest in Fig. 3. Approximately 75% of data patterns are removed by the KVE algorithm with the tolerance value of $\tau$=$10^{-6}$. This

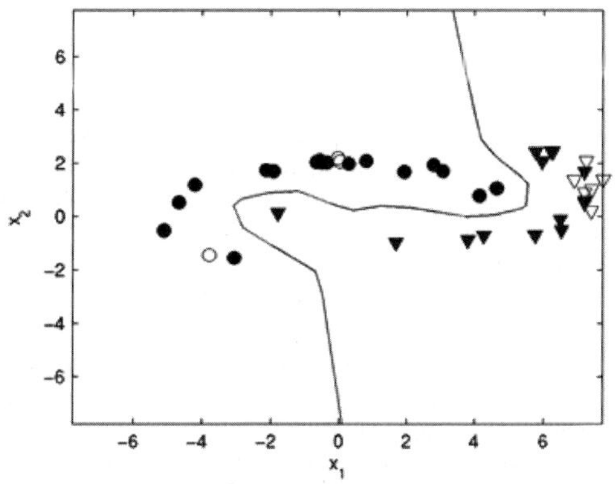


Fig. 2. The reduced training set (the filled circles and triangles) obtained by the key vectors extraction algorithm on a synthetic data set. 25% of the patterns are removed from the data set; the decision boundary is preserved. The black line that represents the original decision boundary coincides with the gray line that represents the new decision boundary.

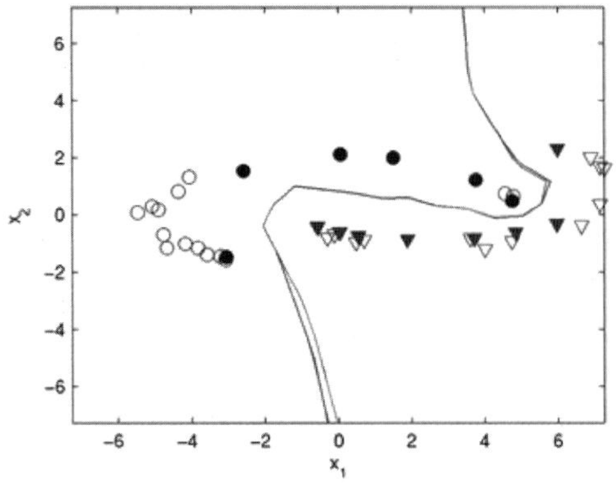


Fig. 3. The reduced training set (the filled circles and triangles) obtained by the key vectors extraction algorithm with tolerance. 75% of the patterns are removed from the data set; the decision boundary is preserved in the region of interest. The black line represents the original decision boundary; the gray line represents the new decision boundary.

significantly reduces the computational cost of the classification algorithm, yet the decision boundary computed from the KVE-processed reduced training data set is nearly identical to that obtained from the original training data set in the region of interest inside of the convex hull of the training set.

## VI. CONCLUSION

The proposed key vectors extraction algorithm provides a new training set reduction technique to shorten the computation time required for two-group classification of input data. The KVE algorithm realizes a method which:

- takes advantage of the R-functions-based representation
- requires $O(dn^3)$ operations
- preserves the decision boundary in the region of interest
- allows for further reduction of the training set

The advantage of the R-functions-based representation of the decision boundary is the ability to evaluate the shortest distance from any point **x** to the decision boundary. The computational complexity of $O(dn^3)$ operations is significantly faster than other decision boundary consistent methods for dimensions greater than 3. The KVE algorithm preserves the decision boundary in the region of interest inside of the convex hull of the training set and allows for further reduction of the training set in the cases where the admissible decision-boundary change is specified.

The reduced training set changes the decision boundary locally by no more than the specified threshold value, which is unique in the field. The usage of this algorithm results in earlier detection of component degradation and better avoidance of system failures which is crucial for achieving higher availability of computer systems.


## ACKNOWLEDGMENT

The authors would like to thank Joshua Walter for his help during the preparation of this manuscript.


## REFERENCES


[1] P. E. Hart, "The condensed nearest neighbor rule," *IEEE Transactions on Information Theory*, vol. 14 (3), pp. 515-516, 1968.

[2] G. T. Toussaint and R. S. Poulsen, "Some new algorithms and software implementation methods for pattern recognition research," *In Proceedings of the IEEE International Computer Software Applications Conference*, Chicago, 1979, pp. 55-63.

[3] G. T. Toussaint, B. K. Bhattacharya and R. S. Poulsen, "The application of Voronoi diagrams to non-parametric decision rules," *In Proceedings of 16th Symposium on Computer Science and Statistics: The Interface*, Atlanta, Georgia, March 14-16, pp. 97-108, 1984.

[4] H. Brighton and C. Mellish "Advances in instance selection for instance-based learning," *Algorithms,Data Mining and Knowledge Discovery*, vol. 6(2), 2002, pp. 153-172.

[5] A. Bougaev and A. Urmanov, "R-functions based classification for abnormal software process detection," *In Proceedings of The International Conference on Computational Intelligence and Security (CIS05),* Xian, China, 2005.

[6] V. Shapiro, "Theory of R-functions and applications: A primer," Cornell University, Computer Science Department, Ithaca, NY Tech. Rep. TR91-1219,1991.

[7] A. Bougaev, A. Urmanov, K. Gross, and L. Tsoukalas, "Pattern recognition method based on Rvachev functions," Tech. Rep. PU/NE-05-15, Purdue University, West Lafayette, IN, USA, 2005.

# Semantic Web Knowledge Management

Malik F Saleh
Nova Southeastern University
Saleh@nova.edu

## ABSTARCT

*The aim of the Semantic Web is to allow access to more advanced knowledge management by organizing knowledge in conceptual spaces according to its meaning. Semantic Web agents will use the technologies of the Semantic Web to identify and extract information from Web sources. However, the implementation of the Semantic Web has a problem, namely, that the Semantic Web ignores the different types of already-existing resources in the current implementation of the Web. As a result, many of the resources will not be included in the conceptual spaces of the Semantic Web. This research introduces a framework that creates a catalog of Internet resources. This Internet catalog allows multi-access points to different resources, and it allows agents to discover the sought-after knowledge. This catalog can be established through the creation of a framework that helps computer-to-computer communication, knowledge management, and information retrieval in general.*

## I. INTRODUCTION

The Semantic Web, the next generation of the Web, aims to make better use of the Web as it stands. The Semantic Web is built on markup languages and document descriptions that will let machines understand the nature of a page's content. It will take the Web beyond the era of HTML, which lets machines understand the nature of a page's appearance, and it will bring order to the loosely connected networks of digital documents that make up the Web [1, 2].

The key to integrating information in a reusable way is the use of semantics, which describe the meaning of a word or concept. Semantics ensure that resources appearing in different domains in different forms with different names can be described as truly equivalent. The ability to distinguish among these resources is essential when integrating data from different domains [3].

One of the technologies that the Semantic Web needs is an ontology, a truly semantic representation of knowledge that is scalable, flexible, and better able to support multiple existing and future applications. An ontology contains a representation of all of the concepts present in a domain and of all of the relations between them. These associations among concepts are captured in the form of assertions that relate two concepts by a given relation. Ontologies have a broad range of relations among concepts that cover all variant forms of binding relations among resources. An ontology enables information from one resource to be mapped accurately at an extremely granular level to information from another source. When aggregated, these assertions can be built into a hierarchy or taxonomy. The taxonomies of concepts are extremely useful when the concepts are annotated with properties such as synonyms. This concept enables users to specify high-level concepts when performing a search or selecting data for analysis. [3].

The Semantic Web introduces the concept of agents roaming the Internet. Agents are using domain specific ontologies to discover information. An ontology, a shared and machine-executable vocabulary for a specific domain, is increasingly regarded as the key to making possible semantic-driven data access and processing. There are many applications for such an approach, such as document and content management, information integration, and knowledge management. Within the Semantic Web, ontologies play central roles in enabling agents to navigate the Web. The future vision of the Web is based on services that can be discovered and utilized by anyone. However, such an open and flexible approach to the Web has many obstacles before it becomes reality. One obstacle is the numerous and heterogeneous data formats [4].

## II. A PROBLEM WITH THE SEMANTIC WEB

The aim of the Semantic Web is to allow access to more advanced knowledge management systems by organizing knowledge in conceptual spaces according to its meaning. Semantic Web agents will use the technologies of the Semantic Web to identify and extract information from Web sources [5]. However, the implementation of the Semantic Web has a problem, namely, that the Semantic Web ignores the different types of already-existing resources in the current implementation of the Web. As a result, many of the resources will not be included in the conceptual spaces of the Semantic Web.

The Semantic Web has a multi-faceted problem, and this research envisions three related problems with the current



*T. Sobh and K. Elleithy (eds.), Advances in Systems, Computing Sciences and Software Engineering*, 101–106.

implementation of the Web. First, diverse access points to Web resources cannot be achieved without creating catalogs of Web resources. Agent, or computer-to-computer communication, is part of the new implementation of the Web that needs catalogs for fast access to information. Second, a feasible implementation and a framework that allows the creation of the catalog is needed. Finally, different formats existed without any structure control. A standard format is needed for all resources, but this standard format should not take away the openness of the Web.

Why create a catalog for Web resources? Most of the Web's content today is designed for humans to read, not for computer programs to interpret. The Semantic Web, the next generation of the Web, aims at making better use of the Web as it stands. The Semantic Web will be built on markup languages and document descriptions that will let machines understand the nature of a page's content. Creating a catalog that can be used by human and interpreted by agents will provide more knowledge about all Web resources and it will enhance information retrieval.

The ultimate goal of this research is to create a catalog of Internet resources that allows multi-access points to different resources and allows agents to discover the sought-after knowledge. This catalog can be established through the creation of a framework in order to help computer-to-computer communication, knowledge management, and information retrieval in general.

Different domains, which have different conceptualizations of their specific data sets, will understand the data by understanding the data representation [6]. Over the past few years, new sources of information in the form of multimedia, computer software, graphic files, and other file types, have become accessible via the Internet, and their access is rapidly growing. Metadata facilitates the cataloging of resources such as non-textual objects. The objects themselves may not be able to provide the information and semantic dependencies, and will likely be computationally too expensive to automate the process of discovering data about the objects. Metadata becomes important when the resource itself is not as easily accessible as the index. The easiest way to help organize these resources is to provide a system that helps contributors catalog their own resources—a system in which every contributor to the Web provides keywords along with his contributions to enhance their retrieval. This approach has been advocated by Tim Berners-Lee, the man credited with creating the World Wide Web. This approach was also suggested by the popular and sometimes controversial columnist for PC Magazine, John Dvorak. In addition, this approach stresses the fact that a catalog is needed to help knowledge management and information retrieval [7,8,9].

Accessing and extracting semantic information from Web documents is crucial for the realization of the Semantic Web. Manual extraction of semantic information is impractical, and fully automated tools are still in a very early stage of development. Therefore, the use of specialized domain knowledge in the form of an ontology is a practical short-term solution to searching for and extracting semantic information from unstructured text on the Web [10]. However, an ontology is not available for all resources on Web. Designing tools to assist in extracting such content is a challenging task, because the knowledge embedded in the Web is in a textual form, unstructured, and lacking in semantic tags. The solution is to create a Web catalog for all resources on the Web. The Semantic Web will use this catalog to discover information about any resource.

## III. THE FRAMEWORK

Implementing the new Internet catalog requires a framework that defines the rules for publishing resources on the Web. A resource may be anything from a Web page to a document, a Web-based printer, a Web Service, and many other items. This research proposes a framework for cataloging Internet resources that can integrate the components for resource discovery into the Semantic Web. There are four main components in the framework. First, resources must have consistent structures without taking away the openness of the Web. This part of the framework will be addressed by creating a resource cover page that gives information about the resource. Second, the Web contains different types of resources, but not all of them can have the agents infer knowledge from them; therefore, the publisher must publish information about the resource in a cover page. Third, the cover page must contain metadata that is common among all cover pages; therefore, controlled vocabulary must be established in order to describe the information. Finally, the cover page metadata must be validated; therefore, validation rules must be enforced.

Creating the cover page requires using specialized vocabulary that describes the resource and the meaning of the entries in the cover page. To create the resource file for each published document, this research introduces a Resource Descriptor XML (RDXML) as an extension to XML. The RDXML resource file, or cover page, will contain XML tags corresponding to the catalog fields used in retrieving information.

## IV. CREATING THE RESOURCE DESCRIPTOR XML (RDXML)

Extensible Markup Language (XML) is a text-based markup language that can be used in any discipline to create specific tags that serve the unique needs of that discipline. The XML schema is used to validate the XML resource file. Having a schema ensures that each resource file is compliant with the RDXML extension and the document structure.

RDXML creates definitions for tags to describe a file's contents. For example, a Title tag would describe the title of the resource, which, just like a book or an article, must have a title to be cataloged. A Keyword tag can be repeated multiple times to accommodate multiple keywords. All these tags will be necessary in order to describe a resource in a way that would make retrieving and finding the resource easy.

The assignment of these values for the creation of a resource cover page will be performed by the user who created the resource. It will, therefore, be necessary to control user input, by using a schema to check the cover page for the right format, tags, and valid data. A cover page that fails any of these tests would be invalid and rejected by the catalog.

The tags created for RDXML are listed in Table I. Multiple entries are possible for each tag in order to accommodate all resource configurations. For example, if a document has two authors, there will be two author tags.

**Table I. RDXML Tags**

| XML Tags |
| --- |
| 1- <Title> |
| 2- <Keyword> |
| 3- <Subject> |
| 4- <URL> |
| 5- <Author> |
| 6- <Date> |
| 7- <LastModified> |

## V. CREATING THE XML SCHEMA

The schema consists of validation rules for each tag. For example, the keyword fields can be a text field and can be repeated multiple times. Each field has its own validation rules. A sample schema is listed in table II

## VI. CREATING THE COVER PAGE

The cover page file is given the same name as the resource, with the extension .rdxml, and it is created using the tags defined in RDXML. A validation reference to an XML schema is also added to the cover page in order to check the validity of the XML. Most XML tags will be mandatory in the cover page to ensure consistent structure.

Creating the cover page is the most important step in the process of cataloging the Internet and in knowledge management because the cover page needs to serve as a standardized source of information describing each published resource. Providing a standard format, the cover page maintains the Web's universality while making Web resources uniform, manageable, and retrievable. Although the cover page adds an extra step for the user publishing on the Web, it will make information retrieval more accurate.

Metadata is information that describes data. Metadata services allow users to record information about the creation, transformation, the semantics of data items, and other data attributes. The purpose of cover page's metadata is to give potential users, search engines, and agents a variety of ways to discover and locate information that might meet their needs. Cataloging data provides diverse access points such as names, subjects, places, languages, and physical characteristics of the data item. The cataloging of material and the maintenance of catalogs is a complex task, but with the advent of the Internet and the Web, information of different types, and different formats existed without any structure control. The cover page would provide control over locating and understanding different resources without taking away the openness of the Web.

**Table II. A sample schema**

**XML Schema**

```
<xsd:schema
xmlns:xsd=
"http://www.w3.org/2001/XMLSchema">
 <xsd:annotation>
  <xsd:documentation xml:lang="en">
           Resource Cover Page
  </xsd:documentation>
 </xsd:annotation>
 <xsd:element name="CoverPage"
type="CoverPageType"/>
<xsd:attribute  name="id" type="xsd:string"
use="required"/>
<xsd:complexType
name="CoverPageType">
  <xsd:sequence>
   <xsd:element name="Category"
type="xsd:string" minOccurs="1"
           maxOccurs="unbounded"/>
   <xsd:element name="Subject"
type="xsd:string" minOccurs="1"
           maxOccurs="unbounded"/>
<xsd:element name="Author" type="xsd:string"
minOccurs="1"
maxOccurs="unbounded"/>
 <xsd:element name="Keyword"
type="xsd:string" minOccurs="1"
maxOccurs="unbounded"/>
  <xsd:element name="Rating" type="xsd:string"
minOccurs="1"
maxOccurs="unbounded"/>
<xsd:element name="URL"
type="xsd:string" minOccurs="1"
maxOccurs="1"/>
<xsd:element name="Date"
type="xsd:string" minOccurs="1"
           maxOccurs="1"/>
<xsd:element name="ModifiedDate"
type="xsd:string"
           minOccurs="1" maxOccurs="1"/>
</xsd:sequence>
</xsd:complexType>
</xsd:schema>
```

## VII. CREATING THE CATALOG

The next generation of the Web aims to alleviate the lack of consistent structure by adding metadata, which describes Web content in a machine-accessible fashion. The meanings of the metadata must be understood by all agents in order to provide indexing to Web resources [11]. Therefore, there must be some common ways of providing meaning for metadata. The cover page can serve as the uniform structure providing meaning to Web resources. In the absence of cover pages, Web crawlers—applications that search the Web for new and updated pages to catalog—process Web pages in their entirety in order to find relevant information.

A crawler is created to gather all the cover pages, and a new site catalog is created from all the resources that have cover pages. The process of creating the site catalog is simplified if all the resources have cover pages to describe them. The crawler validates that the resource cover page conforms to the resource schema and that it can be added to the catalog. The site catalog is created as a resource Descriptor Framework (RDF) file from all resources' cover pages. The crawler acts as a wrapper that maps the RDXML tags into RDF tags in the site catalog. A wrapper normally consists of a set of extraction rules which can precisely identify the text fragments to be extracted from Web pages. RDF is the language that agents on the Semantic Web understand. RDF supports the exchange of knowledge on the Web and it addresses the global meanings of metadata and offers extensibility and interoperability among the different frameworks.

Discovering the meanings of resources requires the use of specialized knowledge about the resource. This knowledge requires representation before agents can discover it, but the Web is currently structured to support humans, not agents. For the Semantic Web to realize its goal of enabling the agent to discover information, it must provide the means for agents to discover and understand the meanings of information (Jacob, 2003). Considering the important role of agents on the Semantic Web, the interoperation issue between the Web and the Semantic Web needs to be addressed from the agents' point of view. Agents are expected to comprehend the resources, process the information, and understand the knowledge in a human-like manner, and consequently reduce the user's workload in knowledge processing (Huang & Webster, 2004). Therefore, the catalog will act as a source of information about Web resources. It is used as a first point in acquiring information about any published Web resource, but before a resource can be discovered, it must to be categorized. Ontologies are the categories of items that exist or may exist; therefore, any catalog of types is an ontology [12].

Different arguments about ontologies suggest that ontologies are not new to the Web. According to [12], any metadata schema is, in effect, an ontology specifying the set of characteristics of resources that have been deemed relevant for a particular community of users. Therefore, ontologies are implemented as domain-specific. There are various approaches to dealing with the challenges that organizations face in order to share knowledge between different ontologies and this catalog simplifies the entire process.

The site catalog serves many purposes. It simplifies the search and makes it multifaceted. The catalog is used in navigational searches and research searches. In the navigational search case, the user is using the catalog as a navigation tool to navigate to a particular intended resource. In the research search, the user specifies a search query and the catalog will generate the search result. Agents on the Semantic Web can use the same catalog to locate resources and navigate to them.

## VIII. BARRIERS AND ISSUES

Because cataloging Web resources is not a trivial process, agreeing on and drafting a standard is a lengthy process that requires a lot of effort and the involvement of major organizations in order to test any proposed approach. Therefore, this research created a prototype to be used as an Internet cataloging standard. The idea of the model is to test the process of creating the catalog; not to test the validity of the proposed approach. To create the catalog, Web crawlers were used to gather the information from Web sites. The Web crawling process was outside the scope of this research, but the functionality of a Web crawler was utilized. Because of limitations, a sample set of resources were used.

Testing distributed applications involves using automation in the testing process, and tools to automate test planning, test preparation, and test execution. Also test tools are evolving to include tools for testing data management and for configuration management. Because of limitations and the expenses associated with acquiring these tools, the testing process was conducted manually without testing tools.

## IX. DISCUSSION

The implementation of the Semantic Web ignores the different types of already-existing resources in the current implementation of the Web. As a result, many of the resources will not be included in the conceptual spaces of the Semantic Web. Data intensive applications produce and analyze large data sets that may span millions of files or data objects. Metadata services are required to support these data intensive applications. Metadata provides query for data items based on these descriptive attributes; therefore, accurate metadata tagging is essential for correct analysis of results and for understanding the semantic of the data [13]. Current semantic techniques usually do not have a global mechanism to automatically infer semantic features available in the data being mined. The semantics are either embedded in the document or domain-specific semantics are automatically extracted. Both of these techniques have proved effective but require a lot of human effort [14]. This research simplifies inferring information from Web resources by providing global meaning about Web resources in an Internet catalog. The catalog serves as the first point of access for discovering published Web resources.

The Semantic Web is being designed to enable automated reasoning to be used by agents. In order for an agent to accept and trust a result produced by unfamiliar software, the result needs to be accompanied by a justification that is understandable and usable by the agent. This justification must be expressed using a standard ontology and a standard Semantic Web language. One way for producing justifiable result is by using inference. This framework is used for explaining answers produced from Semantic Web Services and applications. The question answering components of the Semantic Web may generate answers and justifications for their answers using proof markup languages. Proof markup languages is an ontology added to W3C's OWL Semantic Web representation language so that the justification is expressed in OWL and is therefore exchangeable among Semantic Web services and clients using the RDF/XML syntax [15]. In order for an agent to use automated reasoning and to accept and trust a result produced by unfamiliar software, the meaning must be used by agents. This research simplifies inferring by providing global meaning and justifiable answers can be defined from the RDF/XML syntax of the catalog.

The multi-faceted problem of the Semantic Web has been addressed in this research by creating a framework that sets rules for publishing resources on the Web. The framework consists of different parts. First, the cover page creates a standard structure for all Web resources. The standard structure is accessible by crawlers to create the site catalog. The cover page has the metadata that describe the resource and it provides the needed classification. In the Semantic Web, classification must be accomplished on the basis of knowledge extracted from the documents because there is no catalog to rely on. Relying only on endogenous knowledge means classifying a document based solely on its semantics, with the understanding that the semantics of a document is a subjective notion. In particular, this means that metadata such as publication date, document type, publication source, etc., is not assumed to be available. The endogenous knowledge limits the capabilities of the agents and the extracted knowledge may misrepresent the publisher

Different domains have different conceptualizations of their specific data sets. Different domains will understand the data by understanding the data representation. The aim of the Semantic Web is to allow more advance knowledge management by organizing knowledge in conceptual spaces according to its meaning. The catalog organizing knowledge according to its meaning and it provides meaning that can be interpreted by all Web agents.

## X. CONCLUSION AND FUTURE WORK

This paper has presented a problem within the Semantic Web in building knowledge management systems by organizing knowledge in conceptual spaces according to its meaning. While not all Web resources has meaning that can be extracted by agents for the purpose of classifying resources. A framework that builds a catalog of cooperative knowledge management was introduced as a site catalog that has user-created classification. This paper has outlined the motivation and system design principles, and the technological realization of the proposed solution.

The main contribution of this research is a concise framework that combines the key technologies of the Semantic Web into a system that supports resource discovery within the Semantic Web. The framework also contributes a resource management system by creating a conceptual space for organizing resources according to their meaning.

Unlike other solutions, this framework solved different problems, namely, Web resource classification, creating a catalog, and creating consistence and controlled structure of the different Web resources by leveraging the Web technologies while minimizing implementation time and cost.

The aim for future development is to expand the site catalog to be used as a general Internet catalog and as an Open Directory for managing all published Web resources.

## ACKNOWLEDGMENT

We would like to thank the faculty of the Graduate School of Computer and Information Sciences at Nova Southeastern University for their support. Special thanks for Dr. Easwar Nyshadham, Dr. Amon Seagull, and Dr. Junping Sun for all their support and input to the work presented here.

## REFERENCES

[1] R. Hellman, "Semantic approach adds meaning to the web," in Computer, 1999, pp. 13-16.

[2] C. C. Marshall and F. M. Shipman, "Which semantic web," presented at Proceedings of the Fourteenth ACM Conference on Hypertext and Hypermedia, Nottingham, UK, 2003.

[3] S. P. Gardner, "Ontologies and semantic data integration," Drug Discovery Today, vol. 10, pp. 1001-1007, 2005.

[4] C. Bussler, D. Fensel, and A. Maedche, "A Conceptual Architecture for Semantic Web Enabled Web Services," ACM SIGMOD Record, vol. 31, pp. 24-29, 2002.

[5] G. Antonio and F. Harmelen, A Semantic Web Primer. Cambridge, Massachusetts: The MIT Press, 2004.

[6] D. Bell, C. Bussler, and J. Yang, "The Semantic Web and Web Services," Information Systems, vol. In Press, 2005.

[7] T. Berners-Lee, J. Hendler, and O. Lassila, "The Semantic Web," in Scientific American, 2001.

[8] J. Dvorak, "Fixing the Internet," in PC Magazine, 2001.

[9] B. Desai, "Cover Page aka Semantic Header," http://www.cs.concordia.ca/~faculty/bcdesai/web-publ/semantic-header-R.html, 2003.

[10] S. A. Noah, L. Zakaria, A. Alhadi, T. Sembok, and S. Saad, "Towards Building Semantic Rich Model for Web Documents Using Domain Ontology," presented at IEEE/WIC/ACM International Conference on Web Intelligence (WI'04), Beijing, China, 2004.

[11] I. Horrocks and P. Patel-Schneider, "Three Theses of Representation in the Semantic Web," presented at The Twelfth International Conference on World Wide Web, Budapest, Hungary, 2003.

[12] E. K. Jacob, "Ontologies and the Semantic Web," Bulletin of the American Society for Information Science and Technology, vol. 29, pp. 19-22, 2003.

[13] G. Singh, S. Bharathi, A. Chervenak, E. Deelman, C. Kesselman, M. Manohar, S. Patil, and L. Pearlman, "A Metadata Catalog Service for Data Intensive Applications," presented at Proceedings of the 2003 ACM/IEEE conference on Supercomputing, Washington, DC, 2003.

[14] R. Ghani and n. Fano, "Using text Mining to Infer Semantic Attributes for Retail Data Mining," presented at 2002 IEEE International Conference on Data Mining (ICDM'02), Maebashi City, Japan, 2002.

[15] P. P. d. Silva, D. L. McGuinness, and R. Fikes, "A Proof Markup Language for Semantic Web Services," Information Systems, in press, 2005.

# Augmented Color Recognition by Applying Erasure Capability of Reed-Solomon Algorithm

Jaewon Ahn, Cheolho Cheong, Tack-Don Han
Dept. of Computer Science, Yonsei University
134 Shinchon, Seodaemoon
Seoul, 120-749 KOREA (ROK)

***Abstract*-Color-image data is generally more complex than black and white version in terms of the number of variables in the channels and "color constancy" issues. Most of two-dimensional (2D) codes are dominant in a black and white domain such as QR code, PDF417, Data Matrix and Maxicode rather than in color domain, affected usually by such technical difficulties of handling color variables. Conventional 2D codes adopt Reed-Solomon (RS) algorithm for their error control codes mainly to recover from damages such as stains, scratches and spots, while highlight and shadow issues in color codes are much more complex. In this paper we propose a new decoding approach by applying recoverable erasures by RS algorithm in color codes to resolve color recognition issue effectively. Using erasure concept ultimately for color constancy, marking erasure strategy during color recognition processing steps is very crucial. Our algorithm ultimately mitigates color recognition load overall in decoding color codes. Consequently our new erasure decision algorithm prevents color recognition failures within erasure correction capability of RS algorithm, and it can be applied along with other color constancy algorithms easily to increase overall decoding performance just using RS erasure.**

## I. Introduction

In modern digital communication, storage systems design and numerous applications, the performance of *error detection and correction* function is becoming increasingly important. It is widely-accepted understanding that the importance of not only speed but also accuracy in the storage, retrieval and transmission of data is essential. Reed-Solomon (RS) code has played a significant role, practically everywhere.

Numerous applications are used in situations, in which the exact identification of an object is necessary. This identification can be done either automatically or by humans. For example, bar codes have been applied in various businesses and applications and are a part of our everyday lives. They are an automatic identification (Auto ID) technology that streamlines product identification and data collection, and are subsequently found in a category called "*Information technology -- Automatic identification and data capture techniques*" within the International Standards Organization (ISO) standards [7]. Bar codes have an apparent weakness in terms of information density: The vertical dimension does not carry any information. It only provides redundancy that enables decoding of partially damaged symbols (caused by stains, scratches and spots) and also allows for imperfect scans when the user is not careful about certain orientation and registration bounds [8]. To comply with industrial and commercial requirements, new forms of automatic identification technology applications have been introduced such as "two-dimensional codes (2D codes)," RFID (Radio Frequency Identification), "smart cards," and GPS (Global Positioning System) [10]. Among these, 2D codes can replace bar codes directly and easily since its similarity with bar codes in functionality and domain than any other techniques. RS codes are widely used in most 2D codes for error correction algorithm [5].

This work was supported in part by the Korea Science & Engineering Foundation (KOSEF) under the Basic Research program (No. R01-2005-000-10898-0) and by the Grant BK21 (Brain Korea 21) Project in 2003-2005.

ColorCode$^{TM}$ [12] is another form of 2D codes that is made up of blocks of cells, more than one bit per cell, and mainly used in an online code [13]. Most conventional black and white 2D codes are for "portable database" in an offline version. More bits per cell means higher information density and it is very good advantage of color codes. One of most serious issues is, however, color constancy issue when migrating from the domain of black and white codes into that of color codes. In this paper, our research was derived from the fact that major 2D codes in ISO standards adopt RS codes for their error control. The recovery performance of RS erasure is outstanding and we demonstrated that RS erasure can be marked on the occasion of color constancy failure in early decoding process, then RS decoding in post-processing recovers from the marked erasures back to original color values to the full extent of RS erasure correction capability.

## II. Reed-Solomon Codes

A five-page paper appeared in 1960 in the Journal of the Society for Industrial and Applied Mathematics where the paper, "Polynomial Codes over Certain Finite Fields," by Irving S. Reed and Gustave Solomon, then staff members at MIT's Lincoln Laboratory, introduced ideas that form the core of current error-correcting techniques. RS correcting codes have



*T. Sobh and K. Elleithy (eds.), Advances in Systems, Computing Sciences and Software Engineering*, 107–112.

become commonplace in modern digital communications. RS codes made possible the stunning pictures of the outer planets sent back by Voyager II [1]. It can safely be claimed that RS codes are the most frequently used digital error control codes in the world [3]. Versions of RS codes are now used in error correction systems found just about everywhere as follows.

- Storage devices (hard disks, compact disks, DVD, barcodes);
- Wireless communication (mobile phones, microwave links)
- Digital audio/video (digital television, digital audio system)
- Satellite communications (including deep space missions like Voyager)
- Broadband modems (ADSL, xDSL)
- Two-dimensional codes [5] (QR Code, PDF417, Data Matrix, MaxiCode).

*A. The Polynomial Approach for Generic Reed-Solomon Codes*

The generator polynomial construction for RS codes is the approach most commonly used today in error control literature [3]. If an $(n, k)$ code is cyclic, a generator polynomial $g(x) = g_0 + g_1x + g_2x^2 + \cdots + g_{n-k}x^{n-k}$. In this definition each code word is interpreted as a *code polynomial.*

$$(c_0, c_1, c_2, \ldots, c_{n-1}) => c_0 + c_1x + c_2x^2 + \cdots + c_{n-1}x^{n-1} \quad (1)$$

Let $\mathbf{m} = (m_0, m_1, \ldots, m_{k-1})$ be a block of $k$ information symbols. These symbols can be associated with an information polynomial $m(x) = m_0 + m_1x + \cdots + m_{k-1}x^{k-1}$, which is encoded through multiplication by using the generator polynomial g(x).

$$c(x) = m(x)g(x) \quad (2)$$

Suppose that we want to build a $t$-error-correcting RS code of length $q$-1 with symbols in GF($q$). Recall that the nonzero elements in the Galois field GF($q$) can be represented as ($q$-1) powers of some primitive element α.

$$g(x) = \prod_{j=1}^{2t}(x - \alpha^j) \quad (3)$$

It follows that the dimension of a code with a degree-$2t$ generator polynomial is $k = q - 2t - 1$. Once again we see the MDS (Maximum distance separable) relationship [3]:

$$\text{Error correction capability} = \frac{\text{Length - Dimension}}{2} \quad (4)$$

*B. Properties of Reed-Solomon Codes*

RS codes are a *systematic linear block code.* It is defined as a block code because the code is put together by splitting the original message into fixed length blocks. Each block is further subdivided into m-bit symbols. Each symbol has a fixed width, usually 3 to 8 bits wide [4].

If we exclude the erasure parameter, we can obtain the following parameters for any positive integer $t \leq 2^m - 1$ and the symbol width $m$: (expanded from [20])

$$\begin{aligned}
&\text{Block Length}: n = 2^m - 1\ (symbols) = m(2^m - 1)\ (bits)\\
&\text{Number of parity-check symbols}: n - k = 2t\ (symbols) = 2mt\ (bits)\\
&\text{Dimension}: k = 2^m - 1 - 2t\ (symbols) = m(2^m - 1 - 2t)\ (bits)\\
&\text{Minimum distance}: d_{\min} = 2t + 1 = n - k + 1\ (symbols)\\
&\text{Code rate} = k / n
\end{aligned} \quad (5)$$

*C. Erasure*

There are many situations in which a decision on a received symbol is not considered reliable. It may be more advantageous, from the viewpoint of minimizing the probability of a decoding error, to declare a "no-decision." In such a case, the received symbol is "erased" and it is called an *erasure* [14].

The Binary Erasure Channel (BEC) is a good model to explain such "no-decision" state while general approaches adopt typical Binary Symmetric Channel (BSC) model (see detain in [15]). Variable '$v$' means (unknown) random errors and '$e$' means (known) erasures in (6). The word 'error' normally means '$v$' otherwise described differently. As a non-binary codes like RS, errors should be searched for their locations first and evaluated their magnitude second. By the definition, erasure requires second step only. Therefore erasure performance is double than that of error in correction capability of RS as described in (6).

$$\begin{aligned}
&2v + e + 1 \leq d_{\min}\\
&v + e/2 \leq t
\end{aligned} \quad (6)$$

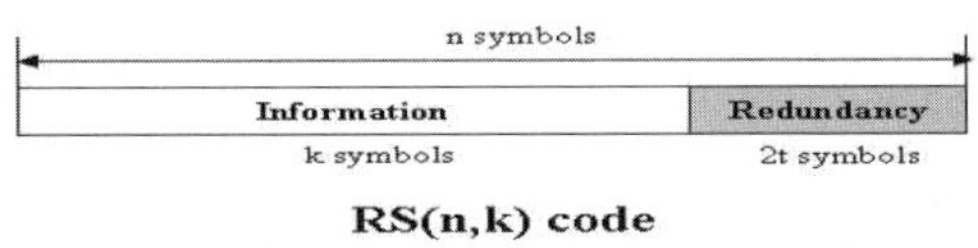

Figure 1. Systematic block encoding for error correction

This is the very impressive feature of RS code itself, naturally for user to consider erasure more than random error correction if permitted.

## III. TWO-DIMENSIONAL CODES

Use 2D codes (symbologies) first appeared in 1988 when Code 49 was introduced by Intermec. Since that time, there are well over 20 different 2D codes available. 2D codes can be classified into several types, with stacked and matrix being the most prevalent. Some of the advantages of 2D over one-dimensional (1D) bar codes are the physical size, storage capability and data accuracy [5]. A comparison between 1D bar codes and 2D codes is presented in Table 1.

2D codes are largely divided into two groups: black and white codes and color codes.

### *A. Black and White 2D Codes*

Fig. 2 shows black and white 2D codes certified in ISO/IEC standards [7]. 2D matrix codes (e.g. Figure 2(a), (c),(d)) usually offer higher data densities than stacked codes (e.g. Fig. 2(b)) in most cases, as well as for use orientation independent scanning applications [6]. A matrix code is made up of a pattern of cells that can be square, hexagonal or circular in shape. While the image of bar codes and stacked codes are obtained by using a laser scanner as a scanning method, the image of matrix codes are usually obtained through the use of a charge-coupled device camera [8]. All of the four 2D codes presented above adopt a RS code as their error correction technique (BCH code may apply in QR) [5].

### *B. Erasure Concept for Black and White Codes*

As mentioned in the Introduction section, all major ISO-certified black and white 2D codes adopt RS code for their error correction and detection. RS code can be a watchdog in a back-end for securing decoding process as well as correcting and detecting errors to recover information as expected. To discriminate 'white' from 'black' from the captured image is relatively simple and easy in black and white codes, thus they are more focused on soiled or damaged occasions such as stains, scratches and spots, by applying RS algorithm. Most of such occasions can be possibly recovered by RS erasure correction capability if definable. 2D code in the right side of Fig. 4 is a good example to explain the mechanism to work with RS erasure. If some part of 2D code is missing, such lost area is not definable and all cell values in the missing area can be categorized as erasure, for surely we may predict how many cells (black or white) of 2D code should be read during decoding procedure. Once erasures are marked during early decoding, RS can recover marked (located) erasures within the erasure boundary specified in (6) where '$e$' can be maximized as '$v$' is minimized. It is also true in some cases that decision-making could be technically difficult between 'random error' and 'erasure' based on the condition of captured image data, and decoding algorithm design only.

Table 1. Comparison between 1D and 2D Codes (modified from [6])

| Item | 1D Bar Codes | 2D Codes |
|---|---|---|
| Information Capacity | Approx. 20 characters | Approx. 2000 characters** |
| Information Type | Alphanumeric | Alphanumeric, other languages |
| Storage Density* | 1 | 20 to 40 |
| Data Restoration Ability | No | Yes |

* Storage comparison for an area with identical size, with the bar codes taken as a criterion "1."

** Typical 2D codes in case of an off-line use.

Figure 2. 2D Codes in ISO/IEC Standards: (From left) (a) QR Code (ISO/IEC 18004); (b) PDF417 (ISO/IEC 15438); (c) Data Matrix (ISO/IEC 16022); (d) Maxicode (ISO/IEC 16023)

## IV. ERASURE METHOD FOR COLOR RECOGNITION

### *A. Color Constancy*

Color constancy is a phenomena of representing (visualizing) the reflectance properties of the scene independent of the illumination spectrum [16].

Ordinary human vision system will generate responses that the objects in both the images (Fig. 3(a) and (b)) are the 'same', and the difference between the two images is due to the difference in the color of the illumination under which images were captured. However, machine vision system is more subject to the influence of illumination on color image formation and high correlation between the surface reflectance and spectral properties of the illumination. If machine vision system is unable to cope with changes in color due to the different illumination conditions it may result in misclassification, segmentation, recognition and incorrect target detection. In order to resolve color constancy issue in machine vision system, many algorithms have been researched [16], [18]

Figure 3. Possible erasure candidates

Figure 4. Macbeth color checker under different illuminant. (a) Phillips Ultralume fluorescent tube (b) Solux 3500K +Roscolux 3202 blue filter [17].

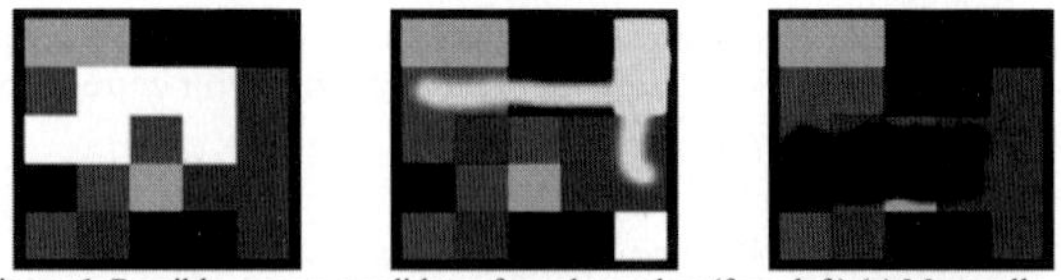

Figure 6. Possible erasure candidates for color codes: (from left) (a) Most cells in second and third codeword symbols are to be decoded as 'white'; (b) Most cells in second and sixth codeword symbols are to be decoded as 'yellow'; (c) Most cells in third and fourth codeword symbols are to be decoded as 'black'.

, [19]: *Retinex algorithm; Gray world algorithm; Gamut mapping algorithm; Bayesian algorithm; Neural network algorithm.* Such algorithms can be collaborated with our proposed algorithm to secure their own color constancy performance.

*B. Erasure Concept for Color Codes*

A color code is another type of 2D matrix codes applied with colors, such as ColorCode [9], Ultracode [5], and Hue Code [5]. Color constancy is one of the most difficult issues to resolve for decoding color codes. Such reason hinders in the increase of the number of applying colors each color cell. These codes are made up of blocks of cells containing more than one bit per cell, particularly in this example, two bits per cell as shown in Fig. 6, while black and white codes use one bit per cell in design. Therefore, the characteristic of errors in color codes look burst errors in domain (multi-bit errors).

## V. AUGMENTED COLOR RECOGNITION BY REED-SOLOMON ERASURE MODEL

*A. Encoding Color Code Using Reed-Solomon Algorithm*

Among color codes, we conduct simulation with ColorCode™ [9], [12]. The basic rule of codeword encoding layout is "left-to-right" and "top-to-bottom" scheme in our design. For the suggested algorithm, a *reserved cell*[1] at the corner is not in use. In the example in Fig. 5, particularly, each cell contains two bits encoding as a shorten RS code (6, 4). Originally from the (1) and (2), information vector (message) **m** = (5, 245, 238, 114) and encoded vector (codewords) **c** = (5, 245, 238, 114, 229,106) as a systematic form [5] is presented where each codeword is a symbol composed in a 8-bits byte and four

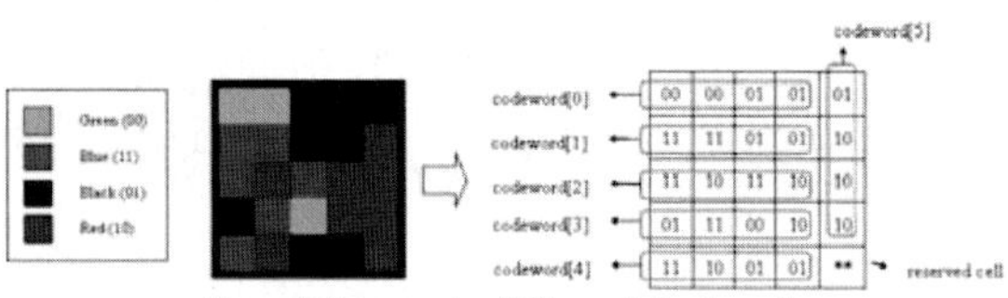


Figure 5. An example of RS encoded color code

[1] Sometimes used as a finder pattern [2].

information symbols ($k$=4) and two redundancy symbols ($n$-$k$=2). Such encoded codewords are mapped as shown in Figure 3. From the equation (6), we can calculate error correcting parameter $t = (n-k)/2$. For the simulation example in Fig. 3, we obtain $t = 1$. Consequently this RS code (6, 4) is a *1-error –correcting code*[2], that is, $v = 1$. If $v = 0$, then $e = 2$, which means *2-erasure –correcting code*[3].

*B. Decoding Color Code Using Reed-Solomon Algorithm*

The erasure test cases are especially created to test how to cope with color constancy and recognition issue in Fig. 6.

ColorCode™ example here uses four colors (R, G, B, K) only. Other than four colors should be returned invalid and need to be filtered into "no-decision" state, resulted in marked erasures. In our simulation, we did not conduct with actual illuminants; instead we designed corrupted samples in Fig. 6, to test with for three major situations by illuminants. For 'white', 'yellow' and 'black' cells in Fig. 6(a)-(c), we assumed that original color value of those regions are influenced by an unknown illuminant each case: (a) for "highlighted regionally"; (b) for "wrong color regionally"; (c) for "shadowed regionally." Pre-processing in our simulation remained the same each test, where pre-processing is in our simulation:

- Binarization
- Search the boundary rectangle of 2D code from the captured image
- Edge detection

If there were no erasure consideration in 2D code design, we could easily anticipate situations in following points:

1) Threshold for determining color value should be almost perfect to determine all color data to classify each into suitable color groups (R, G, B, K for this example); Ideally, it might be possible, however not actually applicable for ordinary color codes once if chromaticity of color code is shifted to "no-decision" state by an unknown illuminant.
2) We should set a bypass from threshold failures

[2] Any single-symbol (one-byte here) random error can be corrected.
[3] Any two-symbol erasures can be corrected.

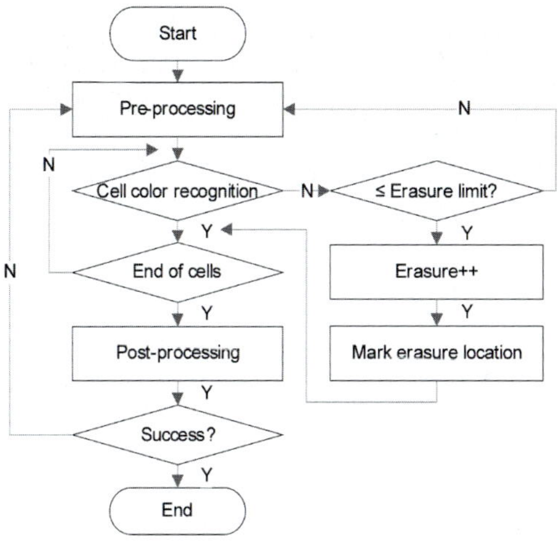


Figure 7. Simplified decoding routine with erasure marking technique.

(indecisive conditions): we may acquire to recapture color code; this is simple but it may request many iterative recaptures unavoidably once if chromaticity of color code is shifted by an unknown illuminant.

3) Such corrupted two-codeword each case in Fig. 6 exceeds *v* boundary (decoding fail) while all is within *e* boundary (decoding pass) otherwise.

Third case in Fig. 6(c) was conducted to see the case that the original colors of cells are partially corrupted as shown. This case depends on how many pixels are taken from each cell to decode. If retrieved pixels are not equivalent one another in a cell, decoding engine should put 'erasure' mark on the cell location. In our test, this case was also successfully decoded.

Erasure marking strategy is guided in the following flowchart in Fig 7. Erasure performance is determined by the key parameter *t* in (6), and it is possibly increased when more redundancy is added to 2D codes if required. The post-processing steps for RS algorithm in our simulation,

- Codeword formulation
- Initialization for RS
- Syndrome calculation
- Check syndrome
- Error or erasure correction.

We encountered several fundamental issues and consideration with our new algorithm in the course of defining a usage of erasure compared with random error of RS decoding for color constancy and recognition:

1) Use erasure more than random error correction to utilize the maximum performance of RS decoding;
2) Erasure adoption in decoding reinforces the accuracy of threshold, while this alleviates the decision load of an applied color constancy algorithm;
3) The performance of erasure as well as random error correction can be flexible by simply changing variable *t* in the encoding step of 2D codes.

## VI. Conclusion

Although black and white 2D codes are still widely used these years, 2D color codes will probably catch up their position gradually. One of reasons is information density in a limited space. Color itself is another dimension, thus 2D color codes may be considered as three-dimensional (3D) codes since multi-bits apply instead of single-bit in a cell. Color constancy issues will be resolved a lot as new techniques from numerous researches emerge more and more in the near future. Most of all, human are more subject to perceive color than black and white, and colors are more appealing to our vision system.

While RS code is mainly applied for soiled and damaged occasions over most of black and white codes, we found that RS code can even improve color recognition correlated to constancy issues of color codes, using RS erasure correction capability without any additional overhead. RS erasure recovery is very powerful for location-known errors, and also effectively covers "no-decision" state of decoding processes if there is any. It is obvious that threshold optimization and pre-processing method is still very critical to determine overall decoding quality and performance. If conventional color constancy algorithms are combined with our new algorithm, augmented color recognition with RS erasure, overall performance of color recognition would be increased tremendously than just using color constancy algorithm only. For our future work, we are planning to perform simulation with color codes under various and actual illuminants.

## Acknowledgment

The authors are grateful to Emanuel Taropa for suggestions.

## References

[1] B. A. Cipra, "The Ubiquitous Reed-Solomon Codes," Society for Industrial and Applied Mathematics (SIAM) News, Vol. 26, No. 1, Jan.1993.
[2] Denso Wave Inc. [illegible] Sep. 2005.
[3] S.B. Wicker and V.K. Bhargava, "An introduction to Reed-Solomon Codes," Reed-Solomon codes and their applications, pp. 1-16, IEEE Press, 1994.
[4] J. Sylvester, "Reed Solomon Codes," Available: http://www.elektrobit.co.uk, Jan. 2001.
[5] D. Barnes, J. Bradshaw, L. Day, T. Schott, and R. Wilson, "Two Dimensional Bar Coding," Tech 621, Purdue University, spring 1999.
[6] T. Soon, "An Introduction to Bar Coding," Available:http://www.itsc.org.sg, Sep. 2005.
[7] ISO Standards, Available: http://www.iso.org/iso/en/CatalogueListPage.CatalogueList, Sep. 2005.
[8] T. Pavlidis, J. Swatz, and Y. Wang, "Information Encoding with

Two-Dimensional Bar Codes," *IEEE Computer Magazine*, Vol. 25, No. 6, pp. 18-28, 1992.

[9] Tack-Don Han, et al., "Machine readable code and method and device of encoding and decoding the same," Japan Patent 3336311, Aug. 2, 2002.

[10] M. Karkkainen, T. Ala-Risku, and P. Kiianlinna, "Item Identification / Applications and Technologies," TAI Research Center, Helsinki University of Technology, 2001.

[11] G.C. Clark, Jr. and J. Bibb Cain, *Error-Correction Coding for Digital Communications*, Plenum Press, 2001, pp. 188.

[12] ColorZip Media Inc., Available: http://www.colorzip.com, Sep. 2005.

[13] J. Rekimoto and Y. Ayatsuka, "yberCode: Designing Augmented Reality Environments with Visual Tags," Proceedings of DARE 2000 on Designing augmented reality environments, 2000, pp. 1-10.

[14] R.H. Morelos-Zaragoza, *The Art of Error Correcting Coding*, Wiley, 2002, pp. 55.

[15] J.L. Massey, "Applied Digital Information Theory I," Lecture Notes, unpublished. Available: http://www.isi.ee.ethz.ch/education/public/pdfs/aditI.pdf

[16] V. Agarwal, "Ridge Regression Approach to Color Constancy," M.S. thesis, Dept. Electron. Eng., The University of Tennessee, Knoxville, May 2005.

[17] K. Barnard, L. Martin, B. Funt, and A. Coath, "A Data Set for Color Research," *Color Research and Application*, vol. 27, no. 3, 2002, pp. 147-151. Available: http://www.cs.sfu.ca/~colour/data/colour_constancy_test_images/

[18] K. Barnard, V. Cardei, and B. Furt, "A Comparison of Computational Color Constancy Algorithms-Part I: Methodology and Experiments With Synthesized Data," *IEEE Trans. Image Processing*, vol. 11, pp. 972-983., Sep. 2002.

[19] K. Barnard, L. Martin, A. Coath, and B. Funt, "A Comparison of Computational Color Constancy Algorithms-Part II: Experiments With Image Data," *IEEE Trans. Image Processing*, vol. 11, pp. 985-996., Sep. 2002.

[20] S. Lin, and D. J. Costello, *Error Control Coding: Fundamentals and Applications,* 2nd ed., Pearson Prentice Hall, 2004, pp. 217, 237.

# Decoupling of Collaboration-Based Designs

Dr. Osama Izzat Salameh, Dima Jamal Damen
Emails: osama_salameh@uop.edu.jo, ddamen@vitamin.com.jo
Computer Science Department
University of Petra, Amman, Jordan

*Abstract* - **Collaboration-based design is a well known method for constructing complex software systems [1, 12, 13]. A collaboration implements one feature of the system. Because of the independent development of collaborations, collaborations might easily produce methods with identical signatures though no intention of overriding [7, 10]. This paper differentiates between accidental and intended overriding and proposes a solution to the problem generated from overriding method signatures between collaborations. Our solution is based on method renaming at the compilation level. The predominant goal is the clarity, measured by the ease-of-use by developers.**

**Keywords :** collaboration-based design, mixin, mixin layer, method decoupling.

## I. Introduction

In object oriented design, it is well known that objects are rarely self sufficient. Each software incorporates multi-features. A feature could be understood as a distinctive aspect or job of the system. Practice has shown that most of the modifications (adding, removing or changing implemented features) in an object-oriented system will affect more than one class [2, 6].

This is because objects depend on each other to implement the same feature. These interdependencies can be expressed in collaborations. Collaboration is a set of objects and a protocol that encodes one feature of the system [1], so feature modification affects only the corresponding collaboration. The importance of collaboration-based designs in Component-Based Software Engineering is indicated in many papers [1, 6, 8].

## II. Collaboration-Based Design Implementation

As a result, it has been necessary to separate different collaborations, so each collaboration can stand on its own. Such a separation can be implemented using a programming pattern called mixin layer [12]. Each mixin layer is a class (outer class) that consists of a number of nested subclasses (inner classes), each of which encodes an object role in the collaboration. Mixin is a class whose super class is specified by a parameter [1]. A mixin layer's super class is also specified as a parameter. This parameter would be used to determine the super classes of all subclasses in that layer as well. So there are two levels of inheritance: one for the outer class and one for inner classes. The inheritance of the outer class determines which collaboration comes before (i.e. depends on) this collaboration, while the subclasses add behavior to the *same* subject of the previous collaboration.

Several researchers have worked on enhancing well-known OOP languages like Java to enable such Mixin Layers and Mixins. One of the well-known successes is JL (Java Layers) developed by Cordone et. al [5]. Using JL syntax , a mixin layer is specified as follows:

```
class MixinLayer <SuperCLass> extends
SuperCLass {
   // outer class
   class role1 extends   SuperClass.role1
   {…} //nested inner class
   …
   class roleN extends SuperClass.roleN
   {…}
}
```

To illustrate the basic concepts mentioned above, we provide a simple but revealing example.

### A. An Example

In a simple registration system, three classes could be identified: Student, Course, and Attendance. We would present three features of such a system. The first feature represents the enrollment in classes, the second is for grading and the third is for attendance control. Enrollment expresses the basic interaction between the Course and Student classes, where the student enrolls or withdraws from a course. Grading collaboration is used to alter the student's average based on course grades. Finally, attendance collaboration would be used to give warnings to students who have exceeded their maximum allowance of absenteeism. Three collaborations would be employed for those features. Figures 1-3 depict, using UML interaction diagrams, how each feature functions:

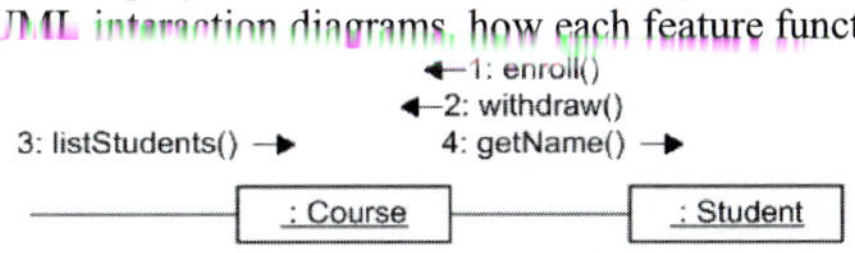


Fig. 1. Enrollment Collaboration

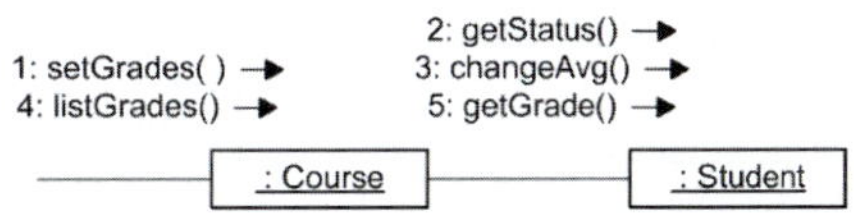


Fig. 2. Student Grades Collaboration



*T. Sobh and K. Elleithy (eds.), Advances in Systems, Computing Sciences and Software Engineering*, 113–117.

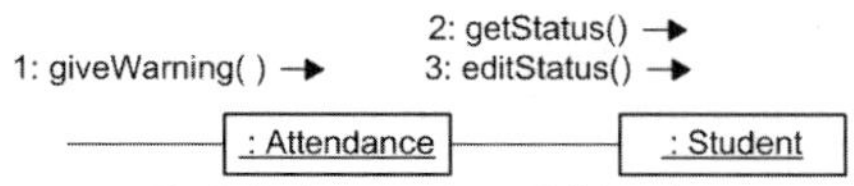


Fig. 3. Student Attendance Collaboration

In the second interaction diagram the grading feature would be responsible for assigning grades to students, and then enlisting all those grades. setGrade() method of the Course object would trigger getStatus() and changeAvg() methods for each enrolled student.

Notice the duplication of the method "getStatus()." The method is used in the second collaboration to decide the student's grading, while it's used in the third to get the number of absent classes.

Moreover, each feature of the system can be implemented in a variety of interpretations. For example, the grading could follow the typical percentage grading, or a symbol grading. In a symbol-based grading, symbols (A, B, …) substitute the ordinary quantified representation of student averages. Another collaboration, labeled 'Symbol Grading', may be introduced to the same registration system. getGrade() method would be used in both collaborations to return different representations of the student grade.

Figure 4 presents the collaborations and classes of our example application (We follow the notation in [2] where Ovals represent collaborations and rectangles represent classes). The intersection of a class and a collaboration illustrates the role prescribed by the class in that collaboration. A role encodes the part of an object in a collaboration.

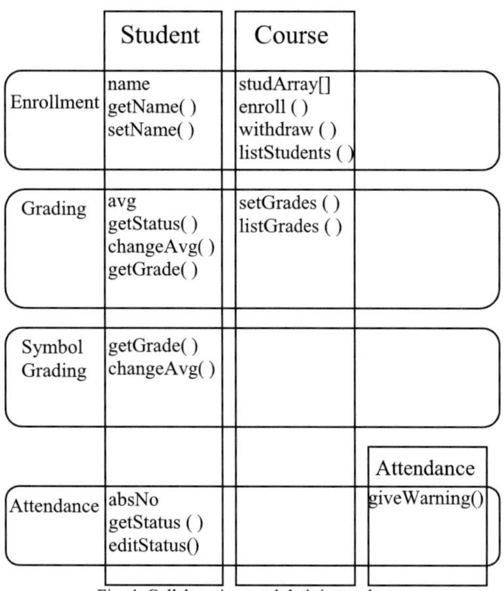


Fig. 4. Collaborations and their inner classes

Via studying the system, one would perceive that the second and the fourth collaborations are independent of each other, but they both depend on the first one. Grading and Attendance won't be valid unless the student is enrolled in the class. This is why the second and fourth collaboration mixin layers depend on the Enrollment mixin layer. Any added collaboration should therefore build upon one of the already available collaborations, or it would stand alone as a separate system.

The skeleton of the Enrollment collaboration implementation in JL is as follows:

```
class Enrollment {
class student {
String name;
void setName(String s) {…}
String getName() {…}
…
}
class Course {
Student studArray[];
void enroll(Student s){…}
void withdraw(Student s){…}
void listStudents () {…}
…
}// class Course
} //class Enrollment
```

The implementation of the other dependent mixin layers differs slightly. As an example we provide the skeleton of Grading collaboration. Notice that the Enrollment class wasn't hard coded into the application. It's merely left to run-time where the Enrollment/parent class is clarified.

```
class Grading <T> extends T {
class Student extends T.Student {…}
class Course extends T.Course {…}
}
```

Several different compositions are now possible resulting in different systems. These compositions are expressed in so called type equations [2, 12]. The type equation specifies the order of inheritance of collaborations in a composition. In mixin layers, collaborations stand on top of each other to utilize the data stored in underlying features. The grading feature would definitely depend on the enrollment of students in courses. For example, the type equation of a registration system with grading (the second collaboration) in JL is:

```
Grading<Enrollment> a;
```

An object of that type is instantiated as follows:

```
Grading<Enrollment> a = new
Grading<Enrollment>();
```

The type equations for the remainder collaborations in the above-given example are expected to be:

```
SymbolGrading <Grading <Enrollment>>
```

and

```
Attendance <Enrollment>
```

After instantiations, it is possible to work with the subclass objects (for example, Student class) using the standard Java conventions.

## III. Decoupling Of Mixin Layers

As indicated in [6], two important questions have to be answered about any newly introduced collaboration:

1. Does the new collaboration break the properties of the existing system?
2. Does the existing system invalidate the local properties of new collaboration?

To resolve such questions, a thorough study should be conducted to discover such situations that may arise from the programming point of view.

One situation that would violate both rules is method overriding. In method overriding, the properties of the existing collaborations would be affected by the newly introduced duplicate signature methods. Moreover, the properties of the existing collaborations would invalidate the new collaboration as method duplication would cause confusion when adding a new collaboration. The main issue behind solving the decoupling of methods is whether this overriding is required or not. This paper proposes the term "accidental overriding" to refer to methods of different layers with equal signatures but perform dependent jobs. Typical overriding is thus referred to using the phrase "intended overriding." To differentiate between accidental and intended overriding, it is important to understand the reciprocal relationship between collaborations. For the example above, the relationships are represented using Figure 5.

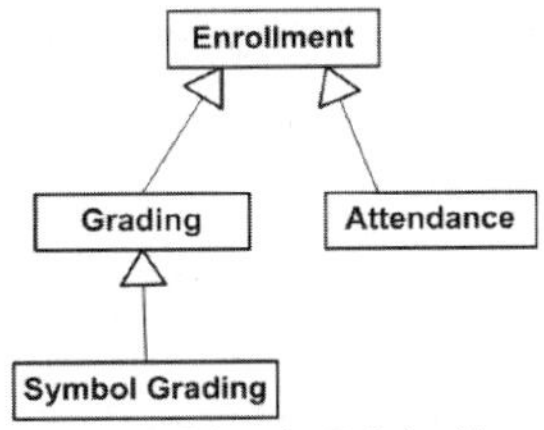


Fig. 5. Collaboration Relationships

Thus, we would clearly differentiate between intended and accidental overriding based on the collaborations dependencies. Intended decoupling is identified where collaborations could be related via an inheritance chain such as in the case of the two methods: getGrade() and changeAvg(). Those methods actually intended to override the methods in their parent collaboration. getStatus(), on the other hand, was accidentally overridden.

### A. The proposed solution

A resolution to accidental decoupling of methods is renaming methods differently so that they would have unique signatures. Intended decoupling would either re-write an already implemented method or provide an implementation to an otherwise abstract method. Below is a pseudo-code to assure proper renaming. The proposed below-written pseudo-code utilizes the compilation stage to solve the decoupling problem. The implementation of the pseudo-code could be added to any JL or other language compilers.

```
Let n : number of layers
Let i : layer number ( i = 1..n) ;
Let cls(i) : number of classes in layer i
Let cls(i, j) : class number j in layer i;
Let m(i, j) : number of methods in class j layer i;
Let m(i, j, r) : method number r in class j layer i;
/* scan layers starting from new ones to minimize changes to riginal layers */
for i = n to 1 (step -1)
{ /* p goes through all previous layers */
for p = i-1 to 1 (step -1)
{ /* j goes through all classes of layer i */
for j = 1 to cls( i )
{ /* Is there an inheriting class for class j layer i in layer p */
if cls(p, j) exists and cls(i, j) exists
{ /* Compare the method signature of all methods in layer i with corresponding
   methods in layer p */
for r = 1 to m(i, j )
{ for r1 = 1 to m(p, j )
{ /* Is there a method with the same signature */
if m(i, j, r) = m(p, j, r1)
{ assign m(i, j, r) a unique name;
for s = 1 to cls(i)
{ update all references to m(i, j, r) from classes of the same layer i;
}
/* check that there's an inheritance relationship between i and p layers */
if (p recursively extends i)
```

```
/* abstract class implementation should be updated to enable overriding */
if (m(p, j, r1) is abstract)
{ change m(p, j, r1) to non- abstract
}
add to m(p, j, r1) the following code
 { if calling method belongs to layer i or any layer that extends i
      m(i, j, r);
 }
for all children of layer i
{ update the references to m(i ,j ,r) from any inheriting class if any;
}
exclude the keyword super used to reference m(i, j, r) from layer p if used;
} /* end if statement comparing methods*/
break; /* compare the next method*/
} /* end r1 loop*/
} /* end r loop*/
} /* end if statement checking class existence */
} /* end j loop*/
} /* end p loop*/
} /* end i loop*/
```

The code above assumes that new layers have higher index numbers than already available ones. This assumption is justified as new layers could not be added to the system prior to their ancestor mixin layers.

To illustrate how abstract method overriding is resolved using the previously-laid pseudo-code, figure 6 and the code below show how abstract method m() at layer A is changed after compilation and re-generated as follows:

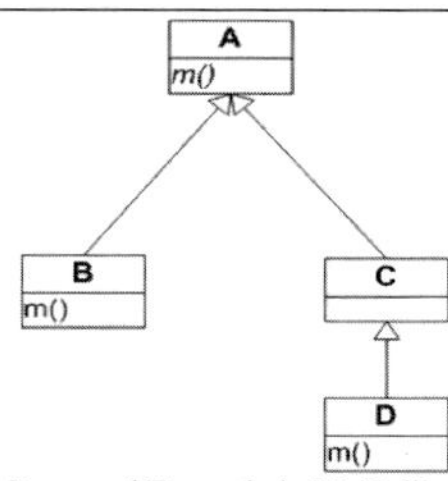


Fig. 6. Proposed Example in UML-like notation

```
public void m()
{  if (calling method from layer B or any layer that inherits B)
       /* this is done by referring to stack of functions */
        call B_m();
        /* this is the renamed function */
    else if (calling method from layer D or any layer that inherits D)
        call D_m();
    else
        ERROR! Method is abstract.
}
```

*B. Inheritance Information Initialization*

It has been perceived that the inheritance information is only provided at run-time. Such information should be collected when apparent at run-time to be utilized for decoupling. This is a technical issue that could be solved using attribute adding or simply storing this information in a special deployment file.

## IV. Future Work And Conclusion

Several other situations would be explored where the new collaboration may affect the existing system. One condition proposed for further research is the improper implementation of an interface method that the new collaboration requires. Since usually interface specification information is available in the form of method signatures, no constraints are placed on the collaborations implementation. This may lead to unexpected collaboration behavior when integrated to the system (For example, there are different ways to handle exceptions and possible errors). So, more collaboration specification information has to be developed.

In this paper we explored some implementation issues of collaboration-based designs. We put the focus on layer decoupling, clarifying the difference between accidental and intended decoupling. A re-naming methodology was proposed that would resolve decoupling while reserving the effects of intended overriding.

## References

[1] Batory D., Cardone R., Smaragdakis Y. ***Object-Oriented Frameworks and Product-Lines***, 1st Software Product-Line Conference, Denver, Colorado, 1999.

[2] Batory D., Geraci B. ***Validating Component Compositions and Subjectivity***, GenVoca Generators. IEEE Transactions on Software Engineering, p. 67-82, 1997.

[3] Batory D., Singhal V., Thomas J. ***Scalable Software libraries***, ACM SIGSOFT 1993.

[4] Bracha and Cook, ***Mixin-Based Inheritance***, ECOOP/OOPSLA 90,303-311, 1990.

[5] Cardone R., Brown A., McDirmid S., Lin C. ***Using Mixins to build flexible Widgets***, 1st International Conference on Aspect Oriented Software Enginering, The NetherLands, 2002.

[6] Fisler K., Krishnamurthi S., Batory D. ***Verifying Component-based Collaboration designs***, 4th ICSE Workshop on Component based software engineering, Canada, 2001.

[7] Herrmann, S. ***Object Teams: Improving Modularity for Crosscutting Collaborations***, Proc. Of Net. Object Days, Erfurt, 2002.

[8] Holland I. ***Specifying reusable components using contracts***, ECOOP, p. 287-308, 1992.

[9] ***Java Layers Home Page*** at http://www.cs.utexas.edu/users/richcar/JavaLayers.html

[10] Ovlinger J. ***Combining Aspects and Modules***. PhD Dissertation, College of Computer and Information Science, Northeastern University, 2004.

[11] Roberts S., Heller P., Ernest M. ***Complete Java 2 Certification study guide.*** 2nd ed. Sybex , 2000.

[12] Smaragdakis Y. ***Implementing Large Scale Object Oriented Components***. PhD Dissertation, CS Dept. University of Texas at Austin, 1999.

[13] Smaragdakis Y. and Batory D. ***Implementing Layered designs with Mixin Layers***, in Proc. of ECOOP'98, 1998.

[14] Szyperski, ***Component Software: Beyond Object-Oriented programming***, Addison-Wesley, 1998

# A Location Service for Pervasive Grids

M. Ciampi
ICAR-CNR
Via Castellino 111, 80131
Napoli, Italy
mario.ciampi@na.icar.cnr.it

A. Coronato
DRR-CNR
Via Castellino 111, 80131
Napoli, Italy
coronato.a@na.drr.cnr.it

G. De Pietro
ICAR-CNR
Via Castellino 111, 80131
Napoli, Italy
giuseppe.depietro@na.icar.cnr.it

M. Esposito
ICAR-CNR
Via Castellino 111, 80131
Napoli, Italy
massimo.esposito@na.icar.cnr.it

***Abstract*-Grid computing environments are being extended in order to present some features that are typically found in pervasive computing environments. In particular, Grid environments have to allow mobile users to get access to their services and resources, as well as they have to adapt services depending on mobile user location and context. This paper presents a 2-layers location service that locates mobile users both in intra-Grid and extra-Grid architectures. The lower layer of the location service locates mobile users within a single intra-Grid environment, i.e. mobile users can move among different areas of the physical environment and the Grid can provide its services accordingly to the user position. The upper layer locates mobile users in an extra-Grid, which is composed by a distributed federation of intra-Grids, by collecting location information coming from basis layers. The location service has been developed at the top of the standard OGSA architecture.**

## I. INTRODUCTION

In the late 1990s, the Grid computing model has emerged and rapidly affirmed as the new computing paradigm for large-scale resource sharing and high-performance distributed applications.

The term "The Grid" was primarily introduced by Foster and Kesselman to indicate a distributed computing infrastructure for advanced science and engineering [1]. Successively, it has been extended to denote the virtualization of distributed computing and data resources such as processing, network bandwidth and storage capacity to create a single system image, granting users and applications seamless access to vast IT capabilities.

As a result, Grids are geographically distributed environments, equipped with shared heterogeneous services and resources accessible by users and applications to solve complex computational problems and to access to big storage spaces.

Recently, Grid computing environments are being extended in order to present some characteristics that are typically found in pervasive computing environments.

The goal for Pervasive Computing is the development of environments where highly heterogeneous hardware and software components can seamlessly and spontaneously interoperate, in order to provide a variety of services to users independently of the specific characteristics of the environment and of the client devices [2]. Therefore, mobile devices should come into the environment in a natural way, as their owner moves, and transparently, that is owner will not have to carry out manual configuration operations for being able to approach the services and the resources.

The conjunction of the two paradigms is leading towards Pervasive Grid environments [3].

A key feature for Pervasive Grids is the possibility of making mobile users able to get access to the Grid and to move both among different areas within the same Grid environment and among different Grid environments. In this scenario, Pervasive Grid environments should customize services and resources access depending on mobile user location and context.

This requires that such environments be supported by advanced location and tracking services, which have to localize mobile users and to track their movements in the physical environment.

In this paper we propose a location and tracking service for Pervasive Grid environments. In addition, the service offers functionalities to mobile users for supporting transparent and spontaneous wireless connectivity, as well as reliable disconnections.

The service has been developed at the top of the standard OGSA (Open Grid Services Architecture) and so may easily extend traditional OGSA-compliant Grids.

The rest of the paper is organized as follows. Section 2 presents some motivations and related works. The architecture of the Pervasive Grid environment that we built is shown in section 3. Section 4 describes the location and tracking service. In section 5 the implementation details are outlined. Finally, section 6 concludes the paper.

## II. RELATED WORKS AND MOTIVATIONS

Current Grid architectures and algorithms do not take into account the mobile computing environment since mobile devices have not been considered as valid resources or interfaces in Grid community. However, in the last few years mobile devices have substantially been enhanced and have got a large diffusion in the market. As a consequence, they can no longer be discriminated by Grid community. Differently, they can be effectively incorporated into the Grid either as service consumer or as service provider [4].

Such new potentialities have been attracting the interest of several researchers. In the following we report some remarkable works that face some aspects related to combining Grid and Mobile computing.

In Reference [5] new challenges stemming from implicit requirements are outlined. In particular, authors emphasize some characteristics of mobile devices and their impact in the Grid environment.



*T. Sobh and K. Elleithy (eds.), Advances in Systems, Computing Sciences and Software Engineering*, 119–123.

In Reference [6] a new scheduling algorithm for Grid environments have been proposed. Such algorithm takes into account the intermittent connectivity of mobile nodes, which are interfaced with the Grid by specific proxy components.

In Reference [7] mobile devices are considered as active resources for the Grid. In particular, authors propose an architecture for deploying Grid services on mobile nodes by making them active actors in the Grid.

In this paper we focus on the effect of user mobility in the Grid. This has two main aspects to concern.

First, services can be customized depending on user context and location. This is a typical scenario in pervasive environments. For an example, if a mobile user moves in a multimedia room equipped with a large display after having required to view a video presentation, the output results will be redirected to the room display rather than to the mobile device.

Second, the set of Grid services can be extended with services that relies on user location. This is the case of an E-Testing service, for which the access could be restricted to users located in a specific classroom rather than an indiscriminate access from everywhere.

## III. A PERVASIVE GRID ENVIRONMENT

Grid applications can be physically placed in different sites, which are topologically organized to compose intra-Grids, extra-Grids or inter-Grids [8,9].

Intra-Grids pool resources offered by different divisions existing within a single organization, which defines policies to share, access and use such resources. Computers of a single organization use a common security domain and share data internally on a private/LAN network.

Extra-Grids bring together more intra-Grids using a remote/WAN connectivity. They are based on multiple organizations, characterized by more than one security domain, which cooperate among them through external partnerships in order to address about business, academic and other topics. For this reason, these partners have to agree common policies, in addition to or extending the existing single organization ones, to enable the access to services and resources of the extra-Grid.

Finally, inter-Grids are based on many organizations distributed in the world, which pool their own resources using an Internet/WAN connectivity in order to sell, for example, computational or storage capacity. Because of the great number of organizations involved in this type of Grid, a partner agreement is very difficult to realize and so a more complex management about the offered services and their integration and collaboration in a secure context is required.

The architecture that we implemented consists of a Pervasive Grid environment, which is an extra-Grid composed by two intra-Grids. These latter are deployed in different physical sites, are equipped with wireless access point and are both independently connected to the internet.

The intra-Grid 1 is equipped with a Linux cluster composed of five computers, a network printer and other resources, while in the intra-Grid 2 there are computational and other resources, a multimedia laboratory and some multimedia rooms. In particular, intra-Grid 2 is made up by the following resources:

- Rendering Station – This is a workstation for rendering row motion data in 3D graphic applications; it is a Silicon Graphics workstation with IRIX as operating system.
- Projector - This is a projector, driven by a pc, to project multimedia presentations.
- E-Testing Server - This server hosts an E-Testing application.
- Streaming Server - This server hosts a video streaming application.

The services of our Pervasive Grid environment are the following:

- RenderingService - This service enables users to submit row motion data and to build 3D graphic applications. This is a computational service that executes processes on the Rendering Station.
- PresentationService - This service enables a user to project its presentation and to control the presentation flow in a multimedia room. For this reason, this service must be available only in the multimedia rooms of the intra-Grid 2.
- VideoConferenceService - This service enables attendees to follow a presentation on their personal mobile device. A video server captures a presentation with its videocam and streams it over the network.
- E-TestingService - This service performs on-line evaluation tests for courseware activities. When a session test starts, students must be in a multimedia room of the intra-Grid 2. Evaluation tests are synchronized and students have a predefined period of time for completing each test section. Students can interrupt their test by explicitly closing the service or by leaving the multimedia room. This service must be available only in the multimedia rooms of the intra-Grid 2.

In the Pervasive Grid, mobile users can dynamically enter in a physical site and leave it. They can also move among sites. Every time a mobile user enters in a site, the environment has to locate him and to provide him with the list of services available at that location.

A possible scenario can be represented by the execution of an evaluation test. It has to be performed in the multimedia room of the intra-Grid 2. As a consequence, the access to the E-TestingService has to be allowed only to students that are physically located in the multimedia room.. Moreover, if a student leaves the room during the test, its session has to be interrupted and any resources associated to it have to be released.

A different scenario is represented by a mobile user that launches a rendering operation to the RenderingService. When the results are ready, the RenderingService returns them to the user depending on his position, that is, if the user is located in the intra-Grid 2, which is equipped with multimedia displays, the results are presented on them rather than on his mobile device.

It is then important to note that both scenarios require the presence of a location and tracking service able to provide information about mobile user position in Pervasive Grid environments.

## IV. THE LOCATION AND TRACKING SERVICE

Traditional Grid environments are not equipped with mechanisms for handling mobile users. In particular, they are typically not able to recognize user disconnections and when they do that, they always threat disconnections as systems failures. Moreover, they are not equipped with location and tracking mechanisms.

Differently, modern Pervasive Grid environments must be equipped with location discovery systems, which have to i) discover and connect incoming mobile device; ii) localize them and track their movements; and iii) recognize implicit disconnections, that is the mobile user leaves the environment without a logout operation.

The service here proposed is able to locate active mobile users in any physical site of the extra-Grid, and to track their movements. Moreover, it has been augmented with connecting facilities, which enable mobile users to get transparent access, as well as disconnection mechanism that reliably handles implicit disconnection.

The location service implements the following strategies.

*Connecting incoming mobile users*

The mechanism we chose for providing network connectivity on-the-fly is based on the Dynamic Host Configuration Protocol (DHCP) [10], which is also a well known solution for the implementation of basic location functions [11]. DHCP dynamically assigns an IP address to an incoming device, which is, then, able to access to the network. It can be used both for 802.11b and for Bluetooth enabled devices in the PAN profile or in the LAN Access Profile [12,13]. No particular assumptions about the mobile device must be taken except that it must be configured as a DHCP client. It's worth noting that standard DHCP protocol has not been devised for highly dynamic environments. As a matter of fact, the IP address assigned to a device is locked until the LEASE_TIME expires. The LEASE_TIME is a parameter that typically varies from 12 to 24 hours. During this time, standard DHCP protocol doesn't take care about possible early user disconnections. Some limitations of the standard DHCP protocol for mobile computing environments were already pointed out in [14] and are not faced by the forthcoming DHCP RFC [15], which only introduces the possibility of forcing the mobile device to renew the IP request once the lease time expires. Pervasive Grid environments have further requirements. Indeed, when a mobile device disappears from the environment, it would continue to have an associated IP address until its LEASE_TIME expires, making such a network resource unavailable for new incoming devices. For this reason, some additional functions must be developed. In particular, it has been built a DHCP server that is able to release IP addresses on demand. By doing so, the environment can require to release network resources (IP addresses) when mobile users leave. It's also important to note that such additional functionalities do not affect the DHCP protocol ar DHCP clients are just standard clients.

*Locating mobile users*

In the environment, at any time, a variable numbe devices may be active. These devices can be located different sites and even in different locations within the sam site. The environment has to locate them in order to provic customized services. In general, in a Pervasive Gr environment both location and locating functions must b available. A location function is a mechanism for identifyir objects active at a specific location, whereas a locatir function is a mechanism for identifying the location of specif objects [16]. As an example, the environment has to use th location function for determining active devices in th multimedia room and providing them with access to the I TestingService. Differently, when a mobile user requires render a multimedia file, the environment has to invoke th locating function to locate him in order to present results to th better and nearest display.

In our Pervasive extra-Grid, each intra-Grid is composed b different locations that correspond to physical areas covere by distinct wireless Access Points (APs), i.e. each A identifies a physical area in the site. As a consequenc knowing objects location is a two level problem. First lev consists in determining the physical site, whereas second lev consists in determining the physical area within the site. T locate devices, wireless APs can periodically be interrogate Indeed, each AP writes an event into a log file whenever device becomes active into its area. By comparing such log and by handling global states, it is possible to detect locatic changes. A similar approach has been realized in [18].

It is finally worth to note that a more effective locatic mechanism can be obtained by equipping environments wi active location systems like those described in [17]. Ou environment has been equipped with RFID [20] systems ar we are now integrating them in the architecture.

*Detecting user disconnections*

In order to recognize user implicit disconnections it possible to adopt a strategy based on checkpoints. As a matt of fact, the environment can periodically detect each mobi device with a ping operation. After having issued a pir message, the environment waits for a response or for timeout. A mobile device is declared inactive after havir missed a certain number of consecutive ping messages. Pir intervals, as well as the number of missing ping responses declare inactive a mobile device, have to be parameterized.

*Tracking user movements*

The service can catch user movements by collecting even related to i) the entrance of a new mobile device, ii) th movement of the device within the environment, and iii) th exit of the device. Such events have to be produced b functions that implements previous strategies.

## V. IMPLEMENTATION DETAILS

The service architecture is shown in figure 1. It consists the following components:

- DHCPComponent – This component implements a DHCP service. It provides network connectivity to the incoming devices as a standard DHCP, but it has additional functionalities. In particular, it communicates to the SiteLocationComponent when a new device enters the intra-Grid. Moreover, it releases allocated IP addresses when the SiteLocationComponent requires to do that.
- EcoComponent – This component sends ping messages towards mobile devices in order to detect implicit disconnections. When an implicit disconnection is detected, the component communicates such an event to the SiteLocationComponent.
- LocatingComponent – This component is in charge of locating mobile devices active in the area covered by a wireless AP. In the intra-Grid, one LocatingComponent is deployed per each wireless AP. Currently, we are also developing LocatingComponents for the RFID technology.
- SiteLocationComponent – This component locates mobile devices in the intra-Grid. To perform its task, it collects location information coming from different LocatingComponents. Moreover, it gets messages from DHCPComponents and EcoComponents for incoming and leaving mobile users. As a consequence, it forces DHCPComponents and LocatingComponents to release and allocate resources respectively. Finally, any change of state of a mobile user is communicated to the GlobalLocationComponent.
- GlobalLocationComponent – This component locates mobile devices in the extra-Grid. It interacts with the SiteLocationComponents and with the TrackingComponent as well.
- TrackingComponent – This component tracks users' movements within the extra-Grid.

The location service has been realized as an OGSA-compliant Grid Service. It has been developed and integrated in the Globus Toolkit 4.0 [19], extending the open source collection of OGSA-based Grid Services offered by it.

It basically exposes the following functions:

- LocalizeDevice – This function returns the position of a specific device.
- ActiveDevices – This function returns the list of devices active at a specific position.
- UserStory – This function returns the list of movements that a specific device has performed within the environment.

When a new mobile device comes into the extra-Grid, it dynamically obtains an IP address from the DHCPComponent. The DHCPComponent communicates to the SiteLocationComponent that a new device is active. The SiteLocationComponent activates an Eco function for the new device, updates its internal data structures, and communicates that a new device is active in its site to the GlobalLocationConponent, which propagates it to the TrackingComponent.

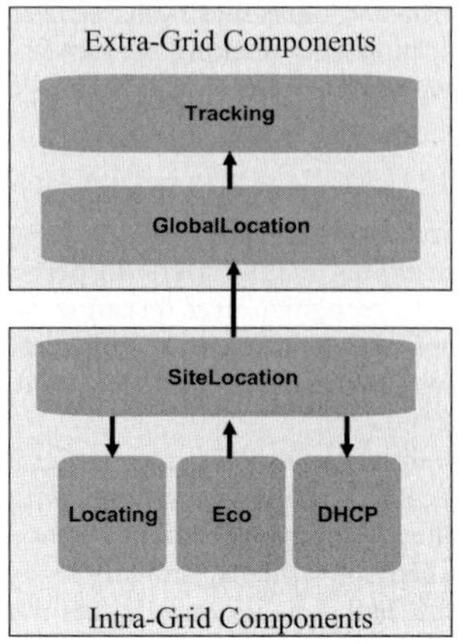


Fig. 1. Services architecture

When the device becomes inactive, the EcoComponent communicates such a condition to the SiteLocationComponent, which requires to free allocated IP resources to the DHCPComponent.

While the mobile device is active in the extra-Grid (i.e. in any intra-Grid), every location change is caught by a LocatingComponent and propagated to upper components.

## VI. CONCLUSIONS AND DIRECTIONS FOR FUTURE WORKS

In this paper we presented a location and tracking service that make Pervasive Grid environments enable to locate and track active mobile devices in a federation of intra-Grids, that is an extra-Grid. This facility provides the Grid with support for customizing services depending on the user location, as well as enabling mobile users to get access.

Future work will aim to implement mechanisms for handling user sessions. In particular, the environment has to distinguish between computational and interactive services. As a matter of fact, if a user leaves after having launched services, the environment has to free resources allocated to interactive services (the user is supposed to no longer be interested in such services), whereas computational services, which do not require interactions but may take long execution time, have to continue until results are produced even if the user is no longer active (the user is supposed to return back in the environment to pick up results).

## REFERENCES

[1] I. Foster, C. Kesselman, "The Grid: Blueprint for a New Computing Infrastructure". Morgan Kaufmann, 1999.

[2] D. Saha and A. Murkrjee, "Pervasive Computing: A Paradigm for the 21st Century", IEEE Computer, March 2003.

[3] V. Hingne, A. Joshi, T. Finin, H. Kargupta, E. Houstis, "Towards a Pervasive Grid", International Parallel and Distributed Processing Symposium, IPDPS 2003.

[4] B. Clarke and M. Humphrey, "Beyond the 'Device as Portal': Meeting the Requirements of Wireless and Mobile Devices in the Legion of Grid Computing System", International Parallel and Distributed Processing Symposium, IPDPS 2002.

[5] T. Phan, L. Huang and C. Dulan, "Challenge: Integrating Mobile Devices Into the Computational Grid", International Conference on Mobile Computing and Networking, MobiCom 2002.

[6] S. M. Park, Y. B. Ko and J. H. Kim, "Disconnected Operation Service in Mobile Grid Computing", International Conference on Service Oriented Computing, ICSOC 2003.

[7] D. C. Chu and M. Humphrey, "Mobile OGSI.NET: Grid Computing on Mobile Devices", International Workshop on Grid Computing, GRID 2004.

[8] L. Ferreira, V. Berstis, J. Armstrong, M. Kendzierski, A. Neukoetter, M. Takagi, R. Bing-Wo, A. Amir, R. Murakawa, O. Hernandez, J. Magowan, N. Bieberstein, "Introduction to Grid Computing with Globus", IBM RedBooks, September, 2003.

[9] J. Joseph, M. Ernest, C. Fellenstein, "Evolution of grid computing architecture and grid adoption models", IBM Systems Journal, December, 2004.

[10] R. Droms, "Dynamic Host Configuration Protocol", RFC 2131, Internet Engineering Task Force, www.ietf.org.

[11] G. Banavar, J. Beck, E. Gluzberg, J. Munson, J. Sussman, and D. Zukowski, "Challenges: An Application Model for Pervasive Computing", in the proc. of the 6th ACM/IEEE Int. Conference on Mobile Computing and Networking, MOBICOM2000.

[12] S. Berger, S. McFaddin, N. Narayanaswami, M. Raghunath, "Web Services on Mobile Devices - Implementation and Experienced", in the proc. of The Fifth IEEE Workshop on Mobile Computing Systems and Applications, WMCSA 2003.

[13] M. Albrecht, M. Frank, P. Martini, M. Schetelig, A. Vilavaar, A. Wenzel, "IP Services over Bluetooth: Leading the Way to a New Mobility", in the proc. of The 24th conference on Local Computer Networks, LCN 1999.

[14] C. E. Perkins and K. Luo, "Using DHCP with computers that move", Wireless Networks, 1995, pp 341-353, Baltzer AG, Science Publisher.

[15] Y. T'Joens, et all, "DHCP Reconfigure Extention", RFC 3203, Network Working Group, www.ietf.org.

[16] S. Fischmeister, G. Menkhaus and A. Stumpfl, "Location-Detection Strategies in Pervasive Computing Environments", in the proc. of the 1st international conference on Pervasive Computing, PERCOM03.

[17] J. Hightower and G. Borriello, "Location Systems for Ubiquitous Computing", IEEE Computer, August 2001.

[18] S. G. M. Koo, C. Rosenberg, H. H. Chan, Y. C. Lee, A. Vilavaar, A. Wenzel, "Location-Based E-Campus Web Services: From Design to Deployment", in the proc. of The first IEEE International Conference on Pervasive Computing and Communications, PERCOM 2003.

[19] I. Foster, C. Kesselman, J. Nick, S. Tuecke, "The Physiology of the Grid: An Open Grid Services Architecture for Distributed Systems Integration", Open Grid Service Infrastructure WG, Global Grid Forum, June 22, 2002.

[20] Ron Weinstein and Johns Hopkins University, "RFID: A Technical Overview and Its Application to the Enterprise", ITProfessional, May/June 2005 (Vol. 7, No. 3) pp. 27-33.

# Extending an existing IDE to create non visual interfaces

Amina Bouraoui
ESTI-University of Carthage
45 rue des entrepreneurs,
Charguia II 2035, Tunisia
hannibal.a@wanadoo.tn
+21698200097

**Abstract-This paper exposes how to modify an existing programming workbench to make it useful and efficient for developing interfaces for the blind and partially sighted people. This work is based on abstract data types and components.**

## I. INTRODUCTION

Non visual interfaces (interfaces for the visually handicapped persons) represent a specific type of software development, because all the existing tools : architectures, interaction description methods, developing tools, and technologies are designed and developed for standard user interfaces only.

Creating software for people with visual deficiencies can be achieved using one of these options:

- adapting existing software using screen-memory information for essentially non-graphical user interfaces, or to enlarge graphical user interfaces; the screen enlargers are used only by partially sighted people;
- filtering the messages circulating in GUIs, and adapting them in order to present information on adapted peripherals (Off Screen Models); the major inconvenient of this method is that it makes original applications very slow, and its use very complex [1];
- encouraging manufacturers such as Microsoft to provide possibilities for better accessibility, by building [illegible] features into their products in the very early phases of conception [2];
- using the built-in features offered by manufacturers, for example the Accessibility module existing in Windows versions since Windows95;
- developing new and specific interfaces using specific tools and interactions : this is necessary when existing applications do not respond to the handicapped "specific" needs.

This latter is the best option because it takes into account the blind and the partially sighted people at the same time; and gives the developer the possibility to create interfaces optimized for the visually handicapped persons.

Let's take the case of a software developer in front of the task of creating an application for the visually handicapped persons. He has to face two major problems.

First of all the use of different and complex peripherals such as Braille terminals, speech synthesizers, speech recognisers, sound cards, sensitive keyboards...etc. The communication with the peripherals must be established.

The second point is related to information presentation and ways of interaction with the computer. The interaction objects and principles used in graphical user interfaces are not useful in interfaces targeted to visually handicapped persons, unless for creating dual interfaces (for both sighted and blind people). So, they have to be replaced by other interaction principles such as multimodality which permit to replace the deficient sense of the user [3]. The developer has to design and build up these interactions.

For these reasons, development of a non visual interface takes much longer time than the development of a standard user interface : there is no existing tool to make non visual interfaces development as easy as for the classic interfaces. I aimed to find a solution to this problem.

With the emergence of "interfaces for all" [2], I think that the most urgent thing to do is to provide the developer's community with tools that help them in creating non visual interfaces, as well as they dispose of tools and environments for creating GUI's.

There are two interesting approaches that inspired my work :

- the extended User Interface Managing System proposed in Reference [4]. It consists in adding a specific toolkit to an existing toolkit. The principle is summarized in Figure 1;
- the use of generic tools to create non visual interfaces, recommended in Reference [5]. These tools may permit to manage input-output, peripheral constraints, basic callbacks, and specific objects.

Instead of adding one more specific application builder, it is better to improve an existing one. Besides to its classic functionalities, an Integrated Development Environment (IDE) may be optimized by adding all the components needed for the developing of non visual interfaces.



*T. Sobh and K. Elleithy (eds.), Advances in Systems, Computing Sciences and Software Engineering*, 125–130.

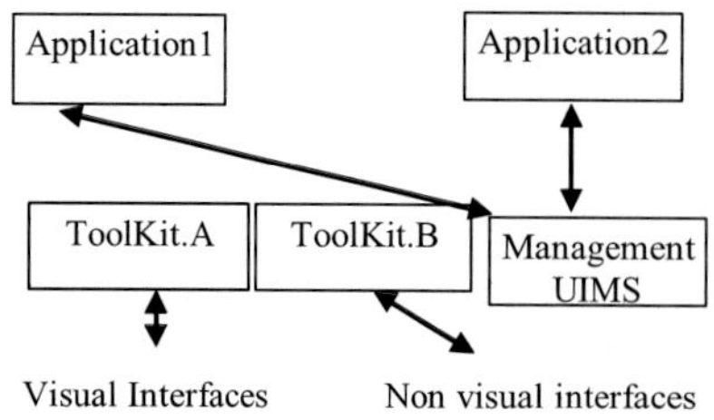


**Figure 1 : Extension of the UIMS principle**

The advantage of this method is to facilitate the developer's task, to integrate the developing of specific interfaces among the classic developing tools, and to make it faster and less fastidious.

## II. SPECIFICATION OF THE NON VISUAL WORKBENCH

After developing non visual interfaces mainly in the field of education [6], it came out that a standard platform must be completed with the following components :

- a simple interface to allow the developer to choose between existing peripherals;
- a system to deal with the communication protocols of the specific input/output peripherals;
- a tool to help the developer in configuring the peripherals and defining simple events;
- a tool that allows the definition of multimodal events;
- a tool to define time constraints existing on events;
- a library of adapted interactive objects.

All these components are integrated in a development environment which has to generate some of the application code.

The IDE chosen for our experimentation is Microsoft Visual C++ because it is complete, and the language used is known by a large scale of people: students, researchers, and engineers.

MSVC++ is made out from four components :

- the application framework called AppWizard: defines and generates the general aspect and behaviour of the application;
- the MFC toolkit : combined with Windows SDK toolkit, furnishes a large amount of necessary functions;
- the resource editor or AppStudio which permits creating interface objects and designing the application interface;
- the dialog controller or ClassWizard which associates callbacks to interface objects and creates classes for interface objects (different controls like edit boxes, buttons, windows, dialog boxes…etc.).

These components work with an event manager integrated into Windows environment. This manager deals with all the standard peripherals and their associated events. The programming language is C++.

It is possible to use interface objects and function libraries issued from the DLL, ActiveX, OLE, OCX or COM technologies in order to enrich the possibilities furnished by MSVC++. All these controls constitute the component gallery.

This original IDE must be extended as shown in **Figure 2** and the method is exposed in the following paragraphs.

Comparatively to Reference [4] recommendation, my work does not end to the function toolkit, but it is extended to all the components of the original IDE. And the extended components are ***integrated*** among the original components.

### *A. The extended toolkit*

Standard libraries and toolkits offer various functions to the developer. They have to be completed with functions permitting for example to communicate with a non standard peripheral, to write a text on Braille terminal display, to read a text written using a Braille terminal keyboard, to pronounce a text by a speech synthesizer, or to make a sound.

The needed functions are divided in two great categories:

- basic functions necessary to establish communication with peripherals; these functions when packaged as components constitute platform independent device drivers [7];
- basic functions necessary to use non standard interaction objects.

For these two categories I designed abstract data types composed by data, operations, conditions and axioms. In object oriented modelling and programming, these data types are represented and finally implemented using the class concept.

#### *Input/Output peripherals managing*

The extended function toolkit needs the creation of a library of non visual classes, which take place in elaborating the non visual interaction modalities. We need abstract classes to represent each of the possible peripheral devices, plus classes defining communication protocols.

For example, I designed BrailleTerminalType, an abstract data type on which I based the implementation of CBrailleTerminal : an abstract class representing all types of Braille terminals.

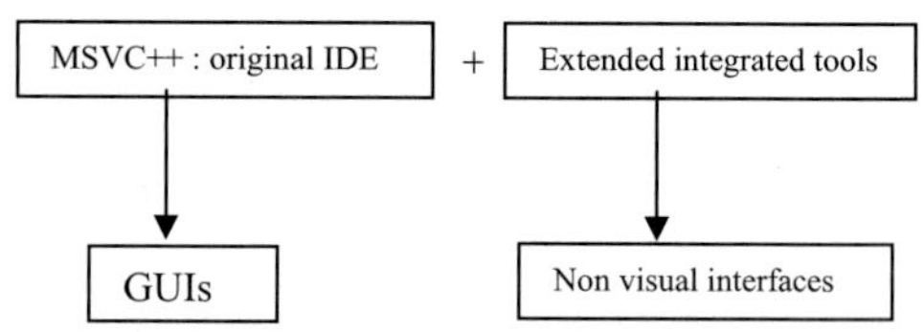


**Figure 2 : The extension of MSVC++**

General attributes and methods are defined in this abstract class. I have assumed that the Braille terminal may use function keys, interactive keys, and a Braille keyboard besides the Braille display.

A small part of the class specification is given here :

**Class Name** : CBrailleTerminal
**Inherits** : CSerial, CParallel, CUsb
**Description** : defines the Braille terminal functionalities
**Constraints** : only one Braille terminal used per application

**Class attributes**

| **Name** | **Type** | **Initial value** | **Description** |
|---|---|---|---|
| m_displaysize | integer | 20 | Display size |
| m_celltype | integer | 8 | Braille cell type |
| m_currentpos | integer | 0 | Display position |
| m_cursor | Ascii | | Cursor aspect |
| m_error | integer | | Error managing |

**Class methods**

- activate, deactivate the Braille device,
- initialize communication with the Braille device,
- display a line on the Braille display,
- display a character string at a given position on the Braille display,
- display a single character at a given position of the display,
- read a line on the display,
- read a string beginning at a given position of the display,
- reads a single character on the display,
- Etc…

Specific models of Braille terminals inherit general characteristics from the abstract class CBrailleTerminal, and implement their specific characteristics. The class CBrailleTerminal has a multiple inheritance feature since we can find different communication protocols for Braille terminals.

Each of the peripheral classes developed has a basic event manager function.

*Adapted interactive objects*

Once the developer finds object galleries in the existing environments, such as buttons, list boxes, dialog boxes, that are very easy to use, and help in the design of the graphical user interface, it would be interesting to find matching non visual objects. For example a list box using the Braille and the speech modalities.

Several woks have been done in this field, and I have proposed adaptations for a major number of graphical interactive objects [8] based on these studies. An example of adapting an edit box using the Braille modality is given in Figure 3 .

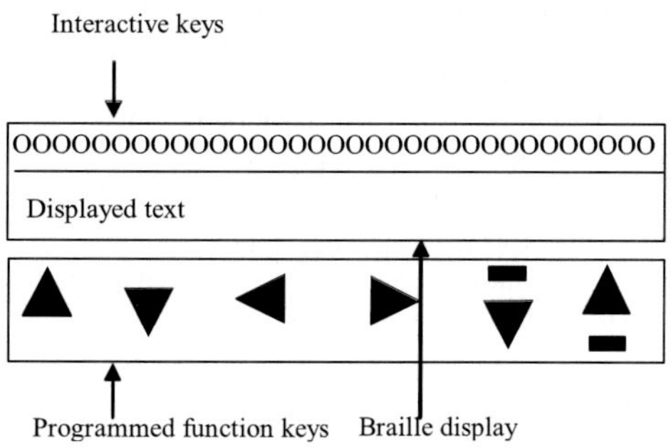


**Figure 3 : Non visual Braille text box.**

A line of Braille text is displayed in the edit box, and the Braille terminal function keys are programmed to navigate in the whole text, while interactive Braille keys allow to point on a single character or to make a word selection.

The sound modality can be used in the text box adaptation, in order to give some extra information to the user. In this case we obtain a multimodal edit box.

Design and implementation of the adapted controls goes through the same steps as for the I/O communication. Abstract data types are defined and then classes are implemented to represent the non visual controls. For example a Braille text editing box called CBrailleEdit, or a multimodal edit box using speech recognition input and Braille input/output called CMultimodalEdit.

In order to implement these controls we have two propositions :

- creating a library of classes implementing the functionality of a non visual control, MFC classes are used as mother classes for the new controls; these controls can be modified easily by the developer;
- creating reusable controls using OLE, ActiveX or Component Object Model concepts. Reusable controls can activate events, they have their own attributes and methods and are portable between applications.

For the two cases the I/O classes developed are used to implement the interactions. The two solutions are exposed for the Braille text box.

a) CBrailleEdit inherits from the MFC class CEdit, then we :

- redefine the creation method of CEdit,
- add non visual attributes in CBrailleEdit,
- define and implement the events to which the control responds;
- implement the non visual behaviour, response to the events, such as the navigation into the Braille control.

b) CBrailleEdit is an OCX reusable control :

- we design the graphic aspect of the control (if needed, it's a screen adaptation of the control);

- we implement code sources, the control is implemented using a class definition and implementation, with a possibility of subclassing;
- we define the events that can be activated;
- we implement the responses of the control to the different events arriving from the peripherals.

The Figure 4 shows the functioning of a reusable control.
In this case, the new control will be integrated among classic controls of the IDE's resource editor.

After the completion of this step we will be in presence of a personalized and specific version of the MFC and the component gallery.

*B. The application framework*

By following a few steps and answering some questions, the developer gives the AppWizard tool the basis on which it will build up the general application framework. The structure consists in header and source files, containing a minimum of source code and defining a default behaviour.

The questions asked are for example : is the application Single Document Interface or Multiple Document Interface, does the application use a database...etc.

I propose to implement a new application framework called *NonVisualAppWizard* for Non Visual Interfaces.

*NonVisualAppWizard* permits to start a new application by giving some options to the developer. The options in a sequence of dialog boxes, must concern the types of specific I/O devices, the model, the communication protocol, the configuration of each previously chosen device, the simple events to take into consideration (monomodal events), the combined events associated to two or more different devices (multimodal events), and finally the time constraints for some events (for example the elapsed time between two clicks for a double-click).

Once the choices made by the developer, *NonVisualAppWizard* creates and compiles a new application framework containing header and source files composing the application, and a basic event manager.

The non visual application framework designing goes trough three major steps :

- designing the general options of the application,
- designing the general event manager,
- creating the template files.

*General options*

In order to define the general options of the application framework we have to design the dialog boxes allowing the different choices stated in the previous paragraph.

The developer has only to check the appropriate options in the dialog boxes. Each dialog box updates some variables I will describe in the coming paragraphs. The figures below show some examples of these dialog boxes.

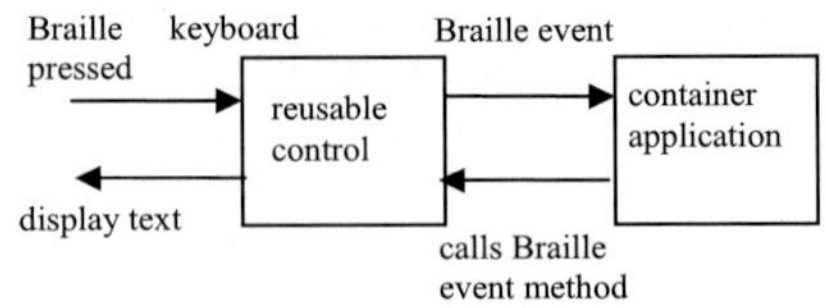


**Figure 4 : A reusable control.**

The number of dialog boxes proposed to the developer depends on the peripheral devices he chooses in the first step. There is a dialog box per device permitting to define its specific features.

For instance the Figure 6 represents the dialog box proposed after the developer has validated his choices illustrated in Figure 5. Then, other dialog boxes are proposed to configure the speech recognition and the speech synthesis devices.

*The event manager*

Dialog boxes are designed to describe the different combinations of events allowed depending on the peripherals chosen in the initial steps :

- simple events coming from input peripherals, for example Braille keyboard input, Braille function key input;
- events on output peripherals, for example displaying a text on the Braille display, or pronouncing a text by speech synthesizer;
- multimodal input events, some realistic multimodal events[9] are proposed, for example speech recognition and click on the Braille interactive keys;
- time constraints on multimodal events and on some simple events will be entered by the developer, for example : a multimodal event composed by a click on a Braille interactive key and a speech recognition event, with time elapsing between events not exceeding 350 milliseconds.

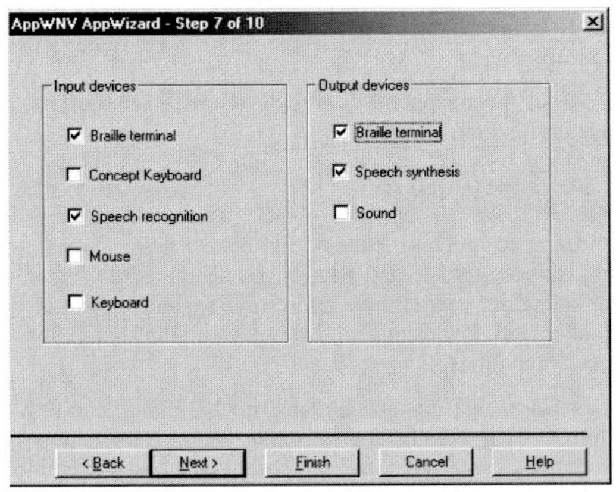


**Figure 5 : Choosing the application I/O devices.**

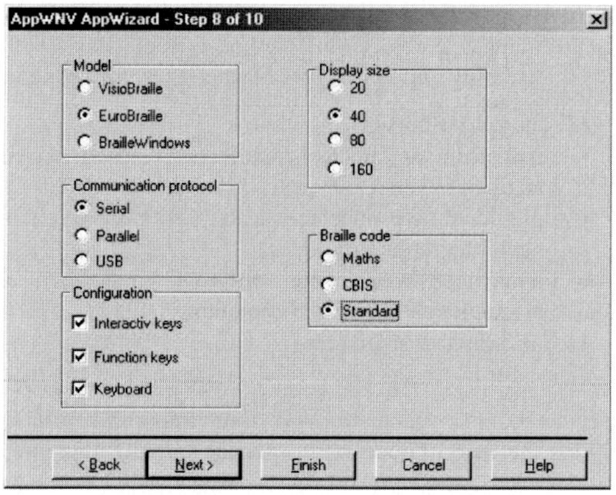


**Figure 6 : Configuration of the Braille terminal.**

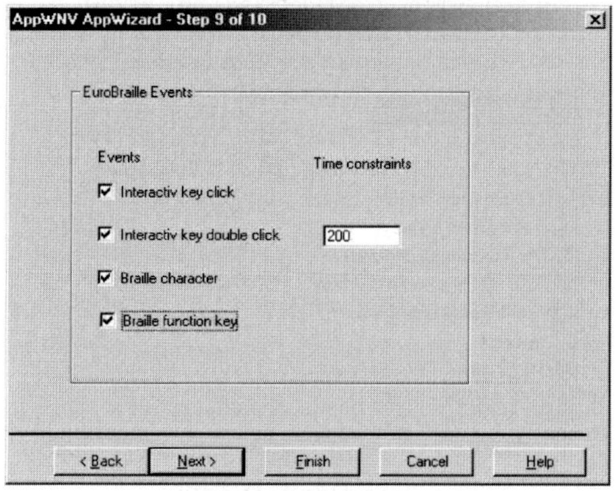


**Figure 7 : Defining Braille terminal simple events**

The Figure 7 shows the definition of events that may come from the chosen Braille terminal.

*The template files*

The template files are used by the application framework as models to generate the application source files. There are template files for each one of the application header and source files, together with a main template file, containing the description of the application framework, and controlling the generation of the final source files.

Template files consist in conditional instructions, source code and keywords. The keywords (also called macros) are stocked in a dictionary; assimilated to variables (Boolean or numeric), their final values depend on the choices made by the developer in the sequential dialog boxes proposed above. Each dialog box permits to update one or more keywords in the dictionary. This latter is then used to generate the source files.

For example by choosing the "Braille terminal" check button in Figure 5, the associated keyword (BRAILLE_TERMINAL) in the dictionary is set to TRUE. This will instantiate every instruction containing the keyword BRALLE_TERMINAL in the template files to TRUE, and due to this notion, a source code is generated that takes into account the presence of the Braille terminal. This is a piece of code contained in the main template file :

```
$$IF(BRAILLE_TERMINAL)
+tbraille.cpp      tbraille.cpp
tbraille.h  tbraille.h
$$ENDIF // BRAILLE_TERMINAL
```

These instructions mean that if the BRAILLE_TERMINAL macro is true, then the application source files must contain two more files : "tbraille.h" and "tbraille.cpp" which will also be generated using their template files. These two files will be included in the project and compiled.

The source files are instantiated after replacing each keyword with its value. For example if the Braille keyboard is not selected for input events, then all the related methods will not be generated, they will remain as a commentary in the source file. And if the developer chooses Braille terminal but not speech recognition as I/O peripheral, the source files of Braille classes are included into the project but not the speech recognition ones.

*The template event manager*

The template event manager consists in a simple algorithm that scrutinizes all peripherals. Each peripheral class (developed in the toolkit extension) contains a local event manager which waits for events, and puts them in a local queue.

The principal event manager, reads all events in the queues, interprets them in order to combine some events into multimodal events, and puts them in a general queue.

A rule base is used by the algorithm to help classification and interpretation of the events. The rule base is generated after the definition of simple and multimodal events and temporal constraints. Only the chosen peripherals are involved in the final event manager.

After the events are recognised and put in the general event queue, they are dispatched adequately by the interface to the controls or to the application that will process them.

*Conclusion*

The result of these implementations is a tool I called NonVisualAppWizard which is completely integrated in MSVC++, and is found among the other wizards furnished by Microsoft as shown in Figure 8.

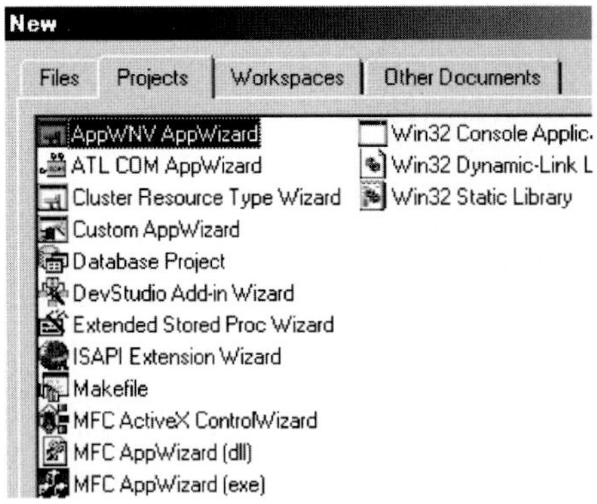


**Figure 8 : The non visual AppWizard**

The developer of a non visual interface has to complete the generated code, in case he deals with peripherals and device models we did not have the time or the opportunity to consider. But he can always rely on the abstract classes offered, because they are complete and give a model for implementing specific classes.

### *C. The resource editor*

The resource editor is composed of controls developed for the generic toolkit, such as the Braille edit box given for example. These controls are naturally integrated into the *AppStudio* editor.

In case we develop controls inheriting from MFC features, they will not be proposed as objects by the resource editor, and the developer has to manage between their appearance on the interface and their behaviour at user inputs.

### *D. The dialog controller*

The dialog controller has a tight relationship with the application framework and the toolkit components. It will work normally with the controls developed as reusable controls, integrate new classes defined for peripherals or adapted objects, and assimilate the connection between events and their callbacks.

## III. CONCLUSION

The work presented can be exploited in different ways. I have experimented with MSVC++ environment, but it is possible to do similar work with other IDEs.

Technologies based on binary code components, allow to implement modules in a chosen language and to use them from a different language (COM/Dcom components). So it is possible to develop libraries, DLLs and controls in COM technology and use them from VB or Visual Java and on different platforms.

This experimentation may lead to a project allowing the use of an existing interface manager to implement specific user interfaces. I have based my study on non visual interfaces, but by developing libraries which command specific peripherals or transactions, by designing and creating adapted interactive objects, the method is applicable to other kinds of disability and other types of interfaces, for example :

- motor disabilities;
- industrial or medical applications using specific engines;
- graphical user interfaces using new peripherals and new interaction principles;
- new interfaces using different metaphors and principles as for mobile computing;
- multilingual interfaces;
- ...etc.

My work has gone through four important phases : studying the field of accessibility for the visually impaired an defining needs, specifying and designing abstract data types, implementing libraries and objects and finally realising a solid documentation targeted to the developers.

The general purpose is to make specific developments as classic as possible, by making IDE as generic as possible [10]. The final aim is to take an IDE, to design a specific or a dual interface as easily as adding windows and buttons on a GUI, and to obtain a valid application in a reasonably short time.

## REFERENCES

[1] Becker, S, Landman, D, Improving Access to computers for blind and visually impaired persons, In *Proceedings of the TIDE 98 conference*, Helsinki, 1998.

[2] Stephanidis, C, User interfaces for all : new perspectives into Human-Computer Interaction, In User interfaces for all concepts-methods, and tools, C Stephanidis (Ed), 2001.

[3] Burger, D, La multimodalité : un moyen d'améliorer l'accéssibilité des systèmes informatiques pour les personnes handicapées. L'exemple des interfaces non visuelles, In *Proceedings ERGO IA 92,* Biarritz, France 1992, 262-290.

[4] Stephanidis, C, Savidis, A, Chomatas, G, Spyridou, N, Sfyrakis, M, Weber, G, Concerted action on technology and blindness, *Medical and Health research programme of the European community CEE,* Bruxelles, 1991.

[5] Burger, D (1994) Improved access to computers for the visually handicapped : New prospects and principles. In : *IEEE transactions on rehabilitation engineering,* vol.2, N°3, Septembre 1994.

[6] Bouraoui, A, Burger D., Tactison : a multimedia learning tool for blind children , In *Computers for Handicapped Persons, ICCHP 94, Lectures Notes in Computer Science 860, Springer Verlag,* Zagler W.L, Busby G. and Wagner R.R. (Eds), Vienna, Austria 1994, , 471-478.

[7] Wang A.J.A, Diaz Herrera, J.L, Device drivers as reusable components In Software Engineering and Applications, Hamza (Ed), Marina Del Rey, USA, 2003.

[8] Bouraoui A, Etude et réalisation d'un éditeur d'interfaces non visuelles multimédias, Thesis, University of Paris XI, Paris, 1998.

[9] IHM 92 « Quatrièmes journées sur l'ingénierie des interfaces Homme-Machine », Compte rendu des ateliers, Paris, Septembre 1992.

[10] Stephanidis, C, Emiliani , P..L , Universal Access to Information Society Technologies: Opportunities for People with Disabilities, In *Computer Helping People with Special Needs : 8th International Conference, ICCHP 2002*, Miesenberger, Klaus, Zagler (Eds.), Linz, Austria, July 15-20, 2002.

# Performance Comparison of Two Identification Methods for Analysis of Head Related Impulse Responses

Kenneth John Faller II[1], Armando Barreto[1], Navarun Gupta[2] and Naphtali Rishe[3]

Electrical and Computer Engineering Department[1] and School of Computer Science[3]
Florida International University
Miami, FL 33174 USA

Department of Electrical and Computer Enginering[2]
University of Bridgeport
Bridgeport, CT 06604 USA

***Abstract*-Head-Related Impulse Responses (HRIRs) are used in signal processing to model the synthesis of spatialized audio which is used in a wide variety of applications, from computer games to aids for the vision impaired. They represent the modification to sound due to the listener's torso, shoulders, head and pinnae, or outer ears. As such, HRIRs are somewhat different for each listener and require expensive specialized equipment for their measurement. Therefore, the development of a method to obtain customized HRIRs without specialized equipment is extremely desirable. In previous research on this topic, Prony's modeling method was used to obtain an appropriate set of time delays and a resonant frequency to approximate measured HRIRs. During several recent experimental attempts to improve on this previous method, a noticeable increase in percent fit was obtained using the Steiglitz-McBride iterative approximation method. In this paper we report on the comparison between these two methods and the statistically significant advantage found in using the Steiglitz-McBride method for the modeling of most HRIRs.**

## I. INTRODUCTION

Humans have the remarkable ability to determine the location and distance of a sound source. How we are able to do this has been a topic of research for some time now. Some aspects of this topic are well understood while other aspects still elude researchers. For example, it is known that the time difference between the arrival of a sound to each ear provides a strong cue for the localization of the sound source in azimuth, while elevation is primarily determined by the perceived modification of sound that takes place in the pinnae or outer ear [1]. Many modern technologies benefit from generating synthetic sounds that have a simulated source location. Currently there are two approaches to synthetic spatial audio: multi-channel and two-channel approaches. The multi-channel approach consists of physically positioning speakers around the listener (e.g., Dolby 5.1 array). This is an effective solution but impractical for the majority of applications that utilize spatial audio. The two-channel approach is more practical because it can be implemented using digital signal processing (DSP) techniques and delivered to the user through headphones.

One such technique is the use of Head-Related Impulse Responses (HRIRs). HRIRs capture the location-dependent spectral changes that occur due to environmental (walls, chairs, etc.) and anatomical (torso, head, and outer ears or pinnae) factors [1]. This approach requires the availability of an HRIR for each ear and each position (elevation, azimuth) of the sound source. The sound signal is then convolved with the HRIR for each ear, to create a binaural sound (left channel, right channel), which gives the listener the sensation that the sound source is located at a specific point in space (Fig. 1). This ability to emulate spatial audio with only two channels has broadened its uses in several important areas: human/computer interfaces for workstations and wearable computers, sound output for computer games, aids for the vision impaired, virtual reality systems, "eyes-free" displays for pilots and air-traffic controllers, spatial audio for teleconferencing and shared electronic workspaces, and auditory displays of scientific or business data [1].

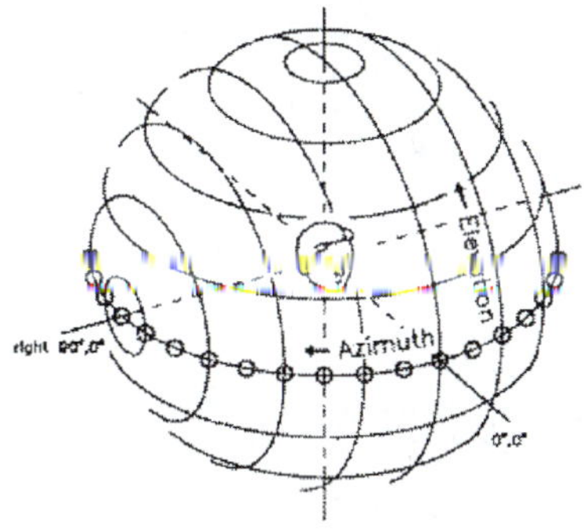


Fig. 1. Diagram of spherical coordinate system [2]

At present, the HRIRs that are used for the synthesis of spatialized audio are either generic or individual. Generic HRIRs are measured using a manikin head (e.g., M.I.T.'s measurements of a KEMAR Dummy-Head Microphone [3]) or using a limited number of subjects to represent the general population (e.g., the CIPIC Database [4]). Individual HRIRs



*T. Sobh and K. Elleithy (eds.), Advances in Systems, Computing Sciences and Software Engineering*, 131–136.

require the subject to undergo time consuming measurements with specialized equipment. Furthermore, a trained and experienced technician is necessary to operate the equipment. Unfortunately, access to the equipment necessary to measure HRIRs is limited for the general public. As a consequence, many spatialized audio systems rely on generic HRIRs, although these are known to reduce the fidelity of the spatialization and increase phenomena such as front to back reversals [5]. These reversals occur when a sound simulated in the front hemisphere is actually perceived in a symmetrical position of the back hemisphere, or vice versa.

Previous research by our group has sought to create a model to generate customized HRIRs with only a few simple measurements. The basic model that resulted from previous research comprises a single resonance feeding its output to a set of parallel paths, each with a magnification and a delay factor, which could be obtained from measurements of the head and pinnae and the use of Prony's method (Fig. 2) [5][6]. Prony's method is an algorithm for finding the coefficients for an IIR filter with a prescribed time domain impulse response. The algorithm implemented is the method described in reference [7].

During recent experimentation on this topic, Prony's method ("Prony") was substituted by the Steiglitz-McBride iteration method ("STMCB"). The STMCB method is similar to Prony in that it also tries to find an IIR filter with a prescribed time domain impulse response. The only difference is that the STMCB method attempts to minimize the squared error between the impulse response and the input signal. A noticeable improvement was observed after the substitution of Prony with STMCB for HRIR modeling. The algorithm for the STMCB method implemented is the method described in reference [8].

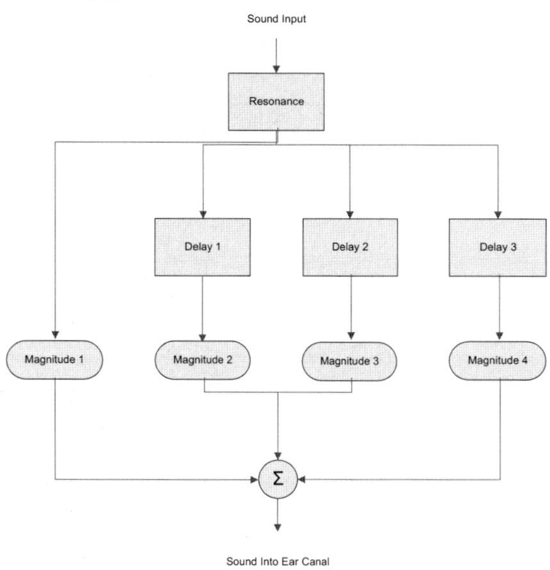


Fig. 2. Block diagram of pinna model

## II. Methodology

The following subsections describe the methodology used to compare STMCB and Prony for HRIR modeling.

### *A. Best Fit Iteration Algorithm*

The purpose of this experiment is to show that there is a statistically significant modeling improvement when STMCB is used for HRIR analysis instead of Prony. In order to do this, a sample population of HRIRs is necessary. Fortunately the CIPIC database, which is a database that contains HRIRs recorded at 44.1 kHz. from 45 subjects for various azimuths and elevations, is available from [1]. This database contains a large number of HRIRs and is impractical to analyze all azimuths and elevations for both ears. Hence, only HRIRs for the right ear at 0° elevation and 25 different azimuths ranging from -80° to 80° were involved in this comparison.

A Matlab® script was created to iterate through each of the CIPIC HRIRs described above. The script attempts to discover the best fit between a measured HRIR and the HRIR that can be reconstructed by adding the partial 2nd order responses (equivalent to a full path from top to bottom in Fig. 2) extracted from the HRIR using both Prony and STMCB. Both of these methods can estimate a full signal with a smaller segment of the original signal. Furthermore, considering that the original HRIR is believed to consist of a primary resonance and at least two delayed echoes [5], processing the entire HRIR with Prony or STMCB at once would result in a large approximation error sequence, as defined in equation (1). Therefore, data "windows" of increasing sizes have to be tried iteratively, to define each of the 2nd order "echoes" that make up the HRIR, as indicated in Fig. 3. The sizes of the windows to use are determined by iteration, subject to the constraints found in previous work in this area [5]: The first window is at least 5 samples which results in window1 in Fig. 3 starting at 5. Additionally, the windows are not allowed to grow wider than 10 samples.

In this comparison study, the reconstructed HRIRs will only consist of three 2nd order responses that are obtained from Prony or STMCB. These are the "primary" response and two delayed responses, referred to as "echoes." While there may be other late components in the HRIRs, such as the third echo recovered in [5], it is clear that these first three components contain most of the power in the HRIR and were selected as the basis of comparison to keep the number of iterations manageable. Once the primary response and echoes are determined, the reconstructed HRIR is created by adding the extracted responses at the determined delays and comparing the resulting sequence to the original HRIR, in terms of mean square (MS) value:

$$\text{Error} = \text{Original HRIR} - \text{Reconstructed HRIR}, \quad (1)$$

$$\text{Fit} = [1 - \{\text{MS(Error)}/\text{MS(Original HRIR)}\}]. \quad (2)$$

The percentage fit ("fit") between the original HRIR and the reconstructed HRIR was calculated for every subject and every azimuth, and used as the figure of merit to compare the performance of STMCB and Prony for this modeling task.

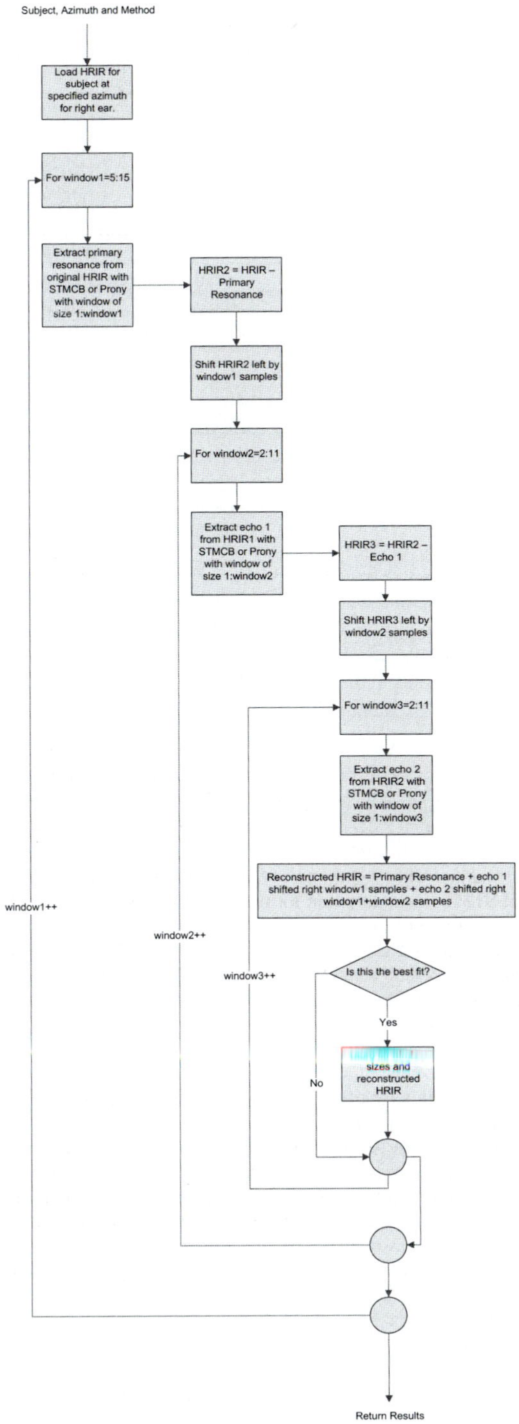


Fig. 3. Flow chart for the iterative process that determines best fit.

### *B. Statistical Analysis Algorithm*

Additional Matlab® scripts were created to statistically analyze the results of the previous section. Matched-*t* tests were utilized in order to determine statistical significance of performance differences observed when the modeling task used Prony or STMCB, for each given source azimuth. The fit obtained through STMCB was subtracted from the fit obtained through Prony, for each azimuth. The 45 differences for one azimuth form a single sample and there were 25 samples (i.e., 25 azimuths) in total.

To assess whether the STMCB significantly improved the fit percentage, the following hypotheses were tested:

$$H_0\colon \mu = 0. \tag{3}$$

$$H_a\colon \mu > 0. \tag{4}$$

Here $\mu$ is the mean improvement that would be achieved by using STMCB over Prony in the modeling process. The null hypothesis says that no improvement occurs, and $H_a$ says that the fit from STMCB is higher on average.

In this case, the one-sample *t* statistic is:

$$t = \frac{\overline{x} - 0}{s / \sqrt{n}}, \tag{5}$$

were $\overline{x}$ is the sample mean, $s$ is the standard deviation and $n$ is the sample size.

The results of the significance test will determine if STMCB outperformed the Prony method for HRIR analysis. Unfortunately, the size of the improvement cannot be determined from these results. A statistically significant but very small improvement would not be sufficient to claim that STMCB is a superior method. A confidence interval is used to remedy this problem. The confidence interval will display how much STMCB improved over Prony with a margin of error:

$$\overline{x} \pm t * \frac{s}{\sqrt{n}}. \tag{6}$$

The procedure followed and a complete example implementation is available in [9].

## III. Results And Disscusion

The following section will overview and discuss the results obtained. Table 1 displays the mean fits for both STMCB and Prony. The "Gain" column is calculated by subtracting the Prony column from the STMCB column. For example, at azimuth -80° the fit improved from 81.20% (with Prony) to 87.57% (with STMCB), which results in a 6.36% gain.

TABLE 1
MEAN FIT OF PRONY AND STMCB

| Azimuth (°) | Prony | STMCB | Gain |
|---|---|---|---|
| -80 | 81.20% | 87.57% | 6.36% |
| -65 | 75.80% | 80.86% | 5.05% |
| -55 | 70.83% | 77.97% | 7.14% |
| -45 | 69.42% | 76.04% | 6.61% |
| -40 | 68.17% | 75.05% | 6.88% |
| -35 | 70.15% | 76.61% | 6.45% |
| -30 | 68.09% | 73.50% | 5.41% |
| -25 | 68.48% | 73.53% | 5.05% |
| -20 | 69.35% | 73.82% | 4.46% |
| -15 | 66.90% | 71.48% | 4.58% |
| -10 | 65.72% | 70.49% | 4.77% |
| -5 | 61.78% | 68.48% | 6.70% |
| 0 | 61.20% | 66.52% | 5.33% |
| 5 | 59.98% | 65.87% | 5.89% |
| 10 | 58.79% | 63.22% | 4.43% |
| 15 | 60.21% | 63.49% | 3.28% |
| 20 | 60.07% | 62.08% | 2.01% |
| 25 | 60.18% | 66.71% | 6.53% |
| 30 | 63.31% | 66.96% | 3.65% |
| 35 | 63.04% | 72.46% | 9.42% |
| 40 | 68.84% | 75.00% | 6.15% |
| 45 | 67.71% | 75.92% | 8.21% |
| 55 | 74.76% | 82.13% | 7.38% |
| 65 | 77.05% | 85.49% | 8.44% |
| 80 | 82.73% | 88.66% | 5.93% |

TABLE 2
RESULTS OF MATCHED *t* PAIR PROCEDURE

| Azimuth (°) | Null Hypothesis | p | t |
|---|---|---|---|
| -80 | 1 | 9.356E-11 | 8.445E+00 |
| -65 | 1 | 9.716E-03 | 2.704E+00 |
| -55 | 1 | 1.092E-03 | 3.496E+00 |
| -45 | 1 | 3.337E-06 | 5.319E+00 |
| -40 | 1 | 4.020E-09 | 7.311E+00 |
| -35 | 1 | 2.176E-11 | 8.895E+00 |
| -30 | 1 | 7.245E-06 | 5.086E+00 |
| -25 | 1 | 3.127E-06 | 5.339E+00 |
| -20 | 1 | 2.527E-04 | 3.982E+00 |
| -15 | 1 | 2.440E-04 | 3.993E+00 |
| -10 | 1 | 3.970E-05 | 4.567E+00 |
| -5 | 1 | 5.826E-06 | 5.152E+00 |
| 0 | 1 | 7.957E-04 | 3.603E+00 |
| 5 | 1 | 2.299E-04 | 4.013E+00 |
| 10 | 0 | 5.191E-02 | 1.998E+00 |
| 15 | 0 | 1.717E-01 | 1.390E+00 |
| 20 | 0 | 3.411E-01 | 9.624E-01 |
| 25 | 1 | 7.808E-04 | 3.610E+00 |
| 30 | 1 | 2.388E-02 | 2.340E+00 |
| 35 | 1 | 2.726E-07 | 6.063E+00 |
| 40 | 1 | 7.363E-05 | 4.375E+00 |
| 45 | 1 | 4.562E-08 | 6.591E+00 |
| 55 | 1 | 4.130E-10 | 7.993E+00 |
| 65 | 1 | 8.665E-09 | 7.082E+00 |
| 80 | 1 | 9.702E-06 | 4.998E+00 |

* Degrees of freedom (df) is 44

To investigate the statistical significance of this apparent improvement achieved by using STMCB, the fit values associated with the HRIRs from each of the azimuth values studied were processed with the "ttest" command in Matlab®. This command performs a t-test of the hypothesis that the data submitted to it (in this case, the fit differences between STMCB and Prony) comes from a distribution with a pre-specified mean (in this case 0). The command provides the values of the t-statistic, as well as the associated p-value, i.e., the probability that the value of the t-statistic is equal to or more extreme than the observed value by chance, under the null hypothesis (mean difference = 0). Additionally, the command provides both limits (CI1 and CI2) of a 95% confidence interval on the mean [10]. Table 2 summarizes the p-value and t-statistic results, for each population of fit differences, by azimuth. The second column of this table ("Null Hypothesis") displays a flag that summarizes the result of the test, in terms of significance. If the flag is "0", it means that the null hypothesis cannot be rejected in those cases, since the difference is not significant ($p > 0.05$). If the flag is "1", it means that null hypothesis is rejected, with $p < 0.05$, i.e., for these azimuths the use of STMCB resulted in a significant improvement over the use of Prony.

As seen in Table 2, the improvement in percent fit with the use of STMCB is significant for many of the azimuths studied. In fact there were only 3 azimuths in which that was not the case: 10°, 15° and 20°. For these azimuths the null hypothesis cannot be rejected, which says that no statistically significant improvement in performance has occurred. However, the vast majority of the results support the view that the use of the Steiglitz-McBride approximation methods within the iterative process outlined in Figure 3 results in improved performance, as opposed to the use of the traditional Prony method [10].

From a different point of view, a statistically significant but very small improvement could be insufficient to prefer the use of an iterative method, such as STMCB, over a single-pass method, such as the traditional Prony algorithm. To illuminate this point, Table 3 displays the improvement of fit observed for each studied azimuth in terms not only of the mean improvement, but also indicating its standard deviation, and, most importantly a 95% confidence interval ([CI1, CI2]) for this improvement.

TABLE 3
CONFIDENCE INTERVAL AND STANDARD DEVIATION OF RESULTS

| Azimuth (°) | CI 2 | CI 1 | Mean | SD |
|---|---|---|---|---|
| -80 | 4.845% | 7.882% | 6.363% | 5.055E-02 |
| -65 | 1.287% | 8.822% | 5.054% | 1.254E-01 |
| -55 | 3.025% | 11.259% | 7.142% | 1.371E-01 |
| -45 | 4.107% | 9.117% | 6.612% | 8.338E-02 |
| -40 | 4.984% | 8.778% | 6.881% | 6.314E-02 |
| -35 | 4.990% | 7.914% | 6.452% | 4.866E-02 |
| -30 | 3.265% | 7.550% | 5.408% | 7.132E-02 |
| -25 | 3.142% | 6.952% | 5.047% | 6.341E-02 |
| -20 | 2.204% | 6.721% | 4.463% | 7.518E-02 |
| -15 | 2.268% | 6.889% | 4.578% | 7.691E-02 |
| -10 | 2.662% | 6.868% | 4.765% | 6.999E-02 |
| -5 | 4.080% | 9.323% | 6.702% | 8.726E-02 |
| 0 | 2.347% | 8.303% | 5.325% | 9.914E-02 |
| 5 | 2.929% | 8.841% | 5.885% | 9.839E-02 |
| 10 | -0.038% | 8.892% | 4.427% | 1.486E-01 |
| 15 | -1.476% | 8.028% | 3.276% | 1.582E-01 |
| 20 | -2.201% | 6.225% | 2.012% | 1.402E-01 |
| 25 | 2.885% | 10.180% | 6.533% | 1.214E-01 |
| 30 | 0.507% | 6.797% | 3.652% | 1.047E-01 |
| 35 | 6.289% | 12.552% | 9.421% | 1.042E-01 |
| 40 | 3.318% | 8.988% | 6.153% | 9.435E-02 |
| 45 | 5.699% | 10.720% | 8.210% | 8.356E-02 |
| 55 | 5.516% | 9.236% | 7.376% | 6.191E-02 |
| 65 | 6.038% | 10.842% | 8.440% | 7.994E-02 |
| 80 | 3.538% | 8.320% | 5.929% | 7.957E-02 |

In order to verify the validity of the percentages of fit found by the automated script employed for the comparison, a few individual modeling results were inspected. Two of these individual results are used for illustration. Figure 4 shows one original (measured) HRIR sequence (subject 24, 35° azimuth) in the top panel, as well as the reconstructed HRIRs obtained through STMCB (middle panel) and Prony (bottom panel). This figure confirms that the main morphology of the measured HRIR sequence has been preserved when the three 2nd order responses found by either STMCB or Prony were assembled together. This is in agreement with the high numerical values found by our comparison script in this case (approximately 94% for both STMCB and Prony). These results, in turn, confirm that the limitation to the modeling of just two "echoes" was not too restrictive.

In contrast, Figure 5 displays the results of approximating a different measured HRIR (subject 27, 20° azimuth). The original and reconstructed HRIR sequences appear in the same order as for Figure 4: original at the top, STMCB reconstruction in the middle, and Prony reconstruction at the bottom.

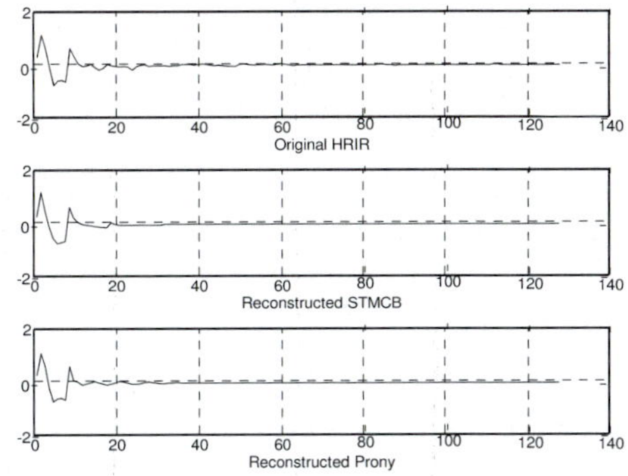


Fig. 4. Plot of the original and reconstructed HRIRs for subject 24 at 35° azimuth.

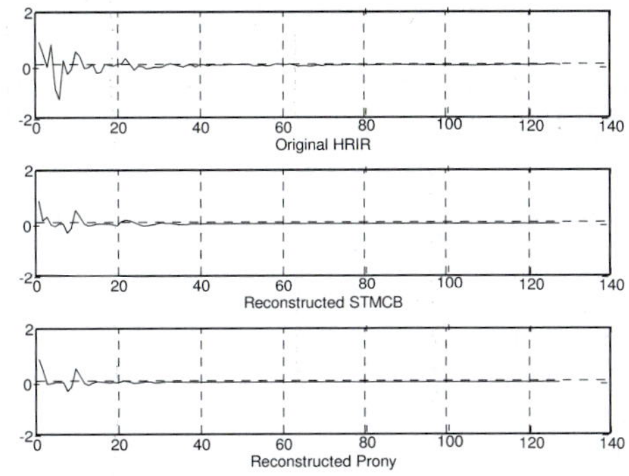


Fig. 5. Plot of the original and reconstructed HRIRs for subject 27 at 20° azimuth.

The fit for this particular case was about 28%, for both methods. As seen in the figure, the reconstructed HRIRs do not resemble the original. It would seem that both methods were able to approximate the second positive "peak" in the HRIR, appearing at a latency of about 12 sampling intervals. On the other hand, it is apparent that both STMCB and Prony minimized the error in the approximation of the first positive peak and the negative peak that immediately follows it by substituting both with a data segment that hovers around zero, which is clearly inappropriate. It is possible that the separation of these two echoes in HRIRs such as this might be very small, particularly considering the limited temporal resolution afforded by the 44.1 kHz sampling rate employed in the development of the CIPIC Database, as compared to the 96 kHz sampling rate used in other previous studies that have attempted this kind of HRIR decomposition [5][6]. However, further research is needed to ultimately pinpoint the reasons for the degradation of this technique for some azimuth values.

## IV. Conclusion

We have implemented a semi-automated comparison of the modeling of measured HRIRs as triads of 2nd order responses. The extraction of these responses was achieved by the Stieglitz-McBride and Prony sequence approximation methods. The fit of reconstructed HRIRs obtained by re-assembling the 2nd order responses extracted to the original measured HRIRs was used as the figure of merit to compare the advantage of using one approximation method over the other. According to the analysis of our results, it has been shown that there is a statistically significant increase in percent fit when STMCB is used rather then Prony for the modeling of most of the HRIRs studied. On the other hand, while the STMCB decomposition of HRIRs at 10°, 15° and 20° had also a better average fit than the corresponding Prony decomposition, the statistical significance of the superiority of STMCB at these three azimuths was not confirmed.

Since STMCB was significantly better than Prony for most of the azimuth angles studied, and it still had a better average fit for the three exception cases, it seems reasonable to recommend the use of STMCB signal approximation methods for HRIR modeling.

## Acknowledgment

This work was partially sponsored by NSF grants IIS-0308155, CNS-0520811, HRD-0317692 and CNS-0426125.

## References

[1] "Spatial Sound." CIPIC Interface Laboratory. University of California, Davis. 23 Aug. 2005 <http://interface.cipic.ucdavis.edu>.

[2] J. C. Makous, J. C. Middlebrooks, and D. M. Green. "Directional Sensitivity of Sound-Pressure Levels in the Human Ear Canal." Journal of the Acoustical Society of America 86 (1989): 89-108.

[3] W. Gardner, and K. Martin. HRTF Measurements of a KEMAR Dummy-Head Microphone. 18 May 1994. Massachusetts Institute of Technology. 24 Aug. 2005 <http://sound.media.mit.edu/KEMAR.html>.

[4] V. R. Algazi, R. O. Duda, D. M. Thompson and C. Avendano. Workshop on Applications of Signal Processing to Audio and Acoustics, 21-24 Oct. 2001, Audio & Electroacoustics committee of the IEEE Signal Processing Society. New Paltz, NY: IEEE, 2001.

[5] N. Gupta, "Structure–Based Modeling of Head-Related Transfer Functions Towards Interactive Customization of Binaural Sounds Systems." Ph.D. Dissertation, Florida International Univ., 2003.

[6] A. Barreto, and N. Gupta, "Dynamic Modeling of the Pinna for Audio Spatialization", WSEAS Transactions on Acoustics and Music, 1 (1), January 2004, pp. 77 - 82.

[7] T.W. Parks, and C.S. Burrus, Digital Filter Design, John Wiley & Sons, 1987, pp. 226-228.

[8] K. Steiglitz, and L.E. McBride, "A Technique for the Identification of Linear Systems," IEEE Trans. Automatic Control, Vol. AC-10 (1965), pp. 461-464.

[9] D. S. Moore, and G. P. McCabe. Introduction to the Practice of Statistics. 4th ed. New York: W. H. Freeman and Company, 2003. 501-504.

[10] "Ttest - Statistics Toolbox." Matlab Version 7 Release 14. Mathworks Inc. 1 Sept. 2005 <http://www.mathworks.com/>.

# Reversers — A programming language construct for reversing out of code

Raphael Finkel
Department of Computer Science
University of Kentucky
raphael@cs.uky.edu

***Abstract*-This paper proposes a new programming language construct called a reverser for situations in which a subroutine performs actions that it must reverse if it encounters a failure. The reversal code stacks up as more actions are performed. A failure invokes all the reversals in LIFO order; success invokes none of them. The reverser construct avoids a common situation in the Linux source code that is currently programmed by goto statements.*** [*]

## I. THE PROBLEM

The Linux source code has a surprisingly large number of **goto** statements [1]. In the past, Linux used **goto** extensively in an effort to keep the code for the more likely branch of a conditional in line; these usages have been superseded by the **likely** and **unlikely** macros (which become compiler directives). The remaining uses of **goto** mostly handle cases in which an error is discovered and a subroutine must exit without fulfilling its purpose, but it must first reverse actions that it has taken. These actions are often memory allocations, which must be reversed by a memory release, and counter increments, which must be reversed by counter decrements. The code snippet in Figure 1 shows an excerpt of the Linux code for the subroutine `copy_process()`.

The invocation of `dup_task_struct()` in line 5 might fail, in which case the routine should return a failure code. If it succeeds, though, it has begun to "dig a pit", having created a task structure that needs to be destroyed if `copy_process()` fails later. In particular, the check `tooMany_Threads()` in line 12 might succeed, in which case the code jumps to "internal" code at the end. The deeper the pit, the more reversal code the program needs to execute. The Linux style is to place labels for reversal code at the end of routines that dig pits and to jump to the appropriate label upon failure. Each reversal falls through to ones pertaining to earlier actions.

This use of **goto** statements follows a pattern, which gives us hope that we can replace it by a more structured statement, in keeping with the continuing effort started by Dijkstra in 1968 [2].

```
1  // simplified from kernel/fork.c
2  task_t *copy_process() {
3      if (paramsLookWrong())
           return ERR_PTR(-EINVAL);
4      retval = -ENOMEM;
5      p = dup_task_struct(current);
6      if (!p) goto fork_out;
7      retval = -EAGAIN;
8      if (someLimitReached())
           goto bad_fork_free;
9      atomic_inc(&p->user->__count);
10     atomic_inc(&p->user->processes);
11     get_group_info(p->group_info);
12     if (tooManyThreads())
           goto bad_fork_cleanup_count;
13     if (!try_module_get())
           goto bad_fork_cleanup_count;
14     ...
15     return p;
16 bad_fork_cleanup_count:
17     put_group_info(p->group_info);
18     atomic_dec(&p->user->processes);
19     free_uid(p->user);
20 bad_fork_free:
21     free_task(p);
22 fork_out:
23     return retval;
24 }
```

Fig. 1. Extract of Linux code for copy_process()

```
task_t *copy_process() {
    reverser return ERR_PTR(retval);
    retval = -EINVAL;
    if (paramsLookWrong()) punt;
    retval = -ENOMEM;
    p = dup_task_struct(current);
    if (!p) punt;
    reverser free_task(p);
    retval = -EAGAIN;
    if (someLimitReached()) punt;
    atomic_inc(&p->user->__count);
    reverser free_uid(p->user);
    atomic_inc(&p->user->processes);
    reverser atomic_dec
        (&p->user->processes);
    get_group_info(p->group_info);
    reverser put_group_info
        (p->group_info);
    if (tooManyThreads()) punt;
    if (!try_module_get()) punt;
    ...
    return p;
}
```

Fig. 2. Modified Linux code for `copy_process()`

[*]This work was partially supported by the National Science Foundation under Grant IIS-0325063. Any opinions, findings, conclusions or recommendations expressed in this material are those of the author and do not necessarily reflect the views of the funding agency.



*T. Sobh and K. Elleithy (eds.), Advances in Systems, Computing Sciences and Software Engineering*, 137–140.

As we do so, we want to associate each reversible activity with its reversal code, so the code becomes more readable, more maintainable, and more likely to be correct.

We also want to preserve two good `properties` of the current pattern: It places infrequently executed code out of line (the Linux community values this optimization), and it emits each reversal only once, to avoid bloating code space.

## II. THE SOLUTION

We suggest two new keywords, **reverser** and **punt**. We use **reverser** to register code that must be executed to reverse an action that has just succeeded. We use **punt** to indicate that a failure has occurred; it executes all the reversers that have been registered and exits the subroutine. Figure 2 recasts `copy_process()` with these keywords.

Figure 2 has no **goto** statements. The first reverser (line 2) gives the default **return** statement in case of a failure. Without that reverser, the default action would be simply to return, which does not satisfy the procedure signature. The **punt** statement in line 4 replaces a simple **return** statement in Figure 1, but it makes all the error cases follow a regular pattern. The first reversible action is duplicating the task structure (line 6), with its associated reverser in line 8. Each of lines 11, 13, and 15 is reversible. None of them can fail. The original code of Figure 1 considers them as a single action, with a single reversal method at `bad_fork_cleanup_count` (Figure 1, line 16). We have chosen to explicitly show the reverser for each of these three statements in order to clarify how each reverser correlates with its statement. As we introduce explicit reversers, we notice a coding anomaly in line 12: The reverser is not what we expect, a call to `atomic_dec()`, but rather a call to `free_uid()`. Careful scrutiny of Figure 1 might also notice the anomaly, but in the original code of Linux (version 2.6.11.11), 242 lines of code separate the call to `atomic_inc()` and the reversal call to `free_uid()`[1].

## III. DISCUSSION

Although the **reverser** and **punt** constructs appear straightforward, we must deal with several complexities.

### *A. Static versus dynamic semantics*

We need to define the semantics of **reverser** and **punt** carefully. The most straightforward interpretation is static: The keyword **reverser** is syntactic sugar for a sequence of labeled statements at the end of the subroutine, and the keyword **punt** is syntactic sugar for a **goto** into that sequence. This interpretation meets all our goals: It replaces **goto** statements, it keeps infrequently executed code out of line, and it emits each reversal only once.

Furthermore, the static interpretation is easy to implement.

```
int getSomething() {
    reverser return -1;
    if (!getResource1()) punt;
    reverser putResource1();
    lock();
    if (!getResource2()) {
        unlock();
        punt;
    }
    reverser putResource2();
    unlock();
    if (!getResource3()) punt;
    reverser putResource3();
    return(0);
}
```

Fig. 3. Releasing a lock

One can easily build a script that converts **reverser** and **punt** to the necessary **goto** and labeled statements. Similarly, a compiler could easily generate code for the static interpretation. There is no runtime cost associated with **reverser**, and the cost of **punt** is a single jump, followed by executing all the reversal code.

On the other hand, reversible actions within loops and conditionals argue for a dynamic interpretation: Each execution of **reverser** pushes another piece of code onto a stack, and an execution of **punt** executes the code on that stack. A loop or a conditional can dig a pit of depth unpredictable at compile time, but the runtime execution of **punt** reverses whatever actions have actually occurred. Conditional cases are quite common in the Linux kernel; one such actually occurs in `copy_process()`. The Linux reversal code repeats the conditional. Iterative cases also occur in the Linux kernel, although they are rare; the Linux reversal code then also contains a loop.

Implementation of a dynamic interpretation requires a runtime data structure that records the current stack of reversal code. Executing **reverser** has a small cost: pushing the address of the reversal code onto that stack. Executing **punt** requires looping across each reverser on the stack, with at least one jump to the reverser and one jump back to the loop. The runtime environment must allocate a reversal stack for each activation record of a subroutine that uses reversers. Placing the reversal stack at the end of the activation record makes this allocation inexpensive, but it might collide with other requirements, such as space for dynamically allocated temporaries.

### *B. Scope issues and exceptions*

The scope of **reverser** and **punt** is the subroutine in which they are placed. The compiler (and runtime, for the dynamic interpretation) can forget all knowledge of the reverser stack at the end of the procedure.

Unlike exceptions, we do not want **punt** to implicitly pass failure to the calling subroutine. The purpose of **reverser** and **punt** is not to propagate failures, but rather to reverse actions

---

[1] It turns out that `free_uid()`, although capable of performing other actions as well, in this case has the sole effect of `atomic_dec()`.

cleanly. A programming language that includes exceptions might also profit from reversers; raising an exception would implicitly also invoke all the reversers. But then we need to associate reversers with a finer scope than the whole subroutine, because exception scopes (**try—catch** blocks in C++) are smaller than full subroutines.

Dealing with the interplay between reversers and exceptions might be fertile ground for further research.

### *C. Pure stack order?*

Situations arise that might be coded by introducing temporary reversers and removing them in a non-stack order. Consider the example of Figure 3. The program acquires resource 2 (line 6) while holding a lock (line 5). Any **punt** while the lock is held, such as at line 8, must also release the lock. On the other hand, the **punt** following line 11 must not release the lock, so we must not introduce an ordinary reverser that releases the lock.

We need to violate the stack order of reversers. A temporary reverser that releases the lock needs to be in force before the program acquires resource 2, but that reverser must stop being in force after line 11. We might suggest the code in Figure 4. The **climb** statement in line 9 invokes the reverser it names (here, `unlocker`) and removes it from the stack of reversers.

Implementing **climb** under the static interpretation is intricate. The fall-through behavior of the stack of reversers changes dynamically based on whether the temporary reverser is currently in force. The compiler must introduce a conditional at the end of the next reverser, either to fall through to the temporary reverser or to jump around it. Under the dynamic interpretation, **climb** is easier to implement: It excises one entry from the middle of a runtime stack.

### *D. Unexpected change in expression values*

The body of the **reverser** construct is an arbitrary statement, which may refer to arbitrary variables. These variables might have different values between the time the reverser is stacked and the time a **punt** statement invokes it. For example, line 8 of Figure 2 refers to variable `p`. The programmer must be careful not to modify the value of `p` from this point forward in the subroutine; otherwise, the reversal code misbehaves. This warning applies to the original code as well, of course. It is not in general possible for a compiler to determine that a statement at some point after **reverser** modifies a value that the reversal code needs. However, a compiler might be able to present warnings in simple cases. For example, it might flag changes to variable `p` and to `p->user`, although it would not be able to flag changes to `t->user->__count` where `t` has the same value as `p`.

```
1   int getSomething() {
2       reverser return -1;
3       if (!getResource1()) punt;
4       reverser putResource1();
5       lock();
6       reverser "unlocker" unlock();
7       if (!getResource2()) punt;
8       reverser putResource2();
9       climb unlocker;
10      if (!getResource3()) punt;
11      reverser putResource3();
12      return(0);
13  }
```

Fig. 4. Climb

### *E. The default reverser*

One of the goals of the **reverser** and **punt** mechanism is to cleanly exit the subroutine. A clean exit means returning a value of the type specified in the subroutine's signature. One way to ensure such a value is to require that the first **reverser** in the subroutine must end with a **return** statement. A compiler can verify this condition in the same way it verifies that the subroutine as a whole ends with a **return** statement. This suggestion requires no additional syntax. Good programming practice might suggest that the first **reverser** be nothing but such a **return** statement.

### *F. More comprehensive syntax*

One helpful feature of the general C **for** loop is that the header specifies all the details up front: the initialization, the test for remaining in the loop, and the modification to control variables between iterations. Similarly, we might like our syntax to show a closer connection between an action and its reverser. Between those components, however, the program often needs to verify that the action has in fact succeeded. As we see in Figure 1, many reversible actions can themselves fail. An extended syntax that combines all these components needs to have a slot for the action, for the test for success (along with an exit path upon failure), and the reverser. We now rewrite Figure 1 using such a threefold syntax, as shown in Figure 5.

The new **action** keyword introduces a reversible action. It has an optional **punt if** clause, showing the exit path on failure, and an associated **reverser** clause. The default reverser (line 2) does not have either the **action** or the **punt if** clause. The syntax must also allow more complex punt methods that perform some local fixup, such as releasing locks, before invoking **punt**.

This syntax flags reversible actions so the compiler can enforce a rule that all reversible actions must be associated with a reverser. Other than this rule, the syntax adds nothing new.

### *G. Conditional compilation*

One advantage of the **reverser** and **punt** syntax is that conditional compilation can simultaneously surround both an action and its reverser. If the condition is not met, then the compiler generates code for neither part. In the absence of reversers, the programmer needs to replicate the conditional tests, leading to error-prone and less readable code.

## IV. RECOMMENDATIONS

Is a new language construct needed?

```
task_t *copy_process() {
    reverser return ERR_PTR(retval);
    retval = -EINVAL;
    if (paramsLookWrong()) punt;
    retval = -ENOMEM;
    action p =
            dup_task_struct(current);
        punt if (!p);
        reverser free_task(p);
    retval = -EAGAIN;
    if (someLimitReached()) punt;
    action atomic_inc
            (&p->user->__count);
        reverser free_uid(p->user);
    action atomic_inc
            (&p->user->processes);
        reverser atomic_dec
            (&p->user->processes);
    action get_group_info
            (p->group_info);
        reverser put_group_info
            (p->group_info);
    if (tooManyThreads()) punt;
    if (!try_module_get()) punt;
    ...
    return p;
}
```

Fig. 5. Alternative modified Linux code for `copy_process()`

The Linux kernel uses **goto** extensively, mostly to encode failure handling. (The other principal use is retrying.) In many cases, the code only needs a default reverser. However, it is quite common to need multiple reversers within a single subroutine. The `copy_process()` case is extreme, with 19 separate reversers. In most cases, converting existing code to using reversers is straightforward. Some cases (such as `copy_mm()`) require some thought.

On the other hand, situations requiring reversers are rare in general practice. It is not wise to clutter a programming language with features that are unlikely to see wide use. As Tony Hoare put it, "You include only those features which you know to be needed for *every* single application of the language." [3]

There are alternatives to placing this facility in the compiler. One might imagine that a simple set of preprocessor macros would suffice to handle the static interpretation. Unfortunately, the C preprocessor is apparently insufficient, because it cannot construct the stack of reversers. Instead, one might use a script. A 70-line Perl [4] script, available at `http://www.cs.uky.edu/~raphael/fixup.pl`, does a fairly good job of translating these constructs into native C. A script that has a less rudimentary parser would do far better.

Another alternative is to consider code reversal a "crosscutting concern" in the terminology of Aspect-Oriented Programming (AOP) [5]. The techniques currently available for AOP, such as those of AspectJ [6], involve weaving optional advice code into target code at well-defined places. Unfortunately, these techniques don't fit our need. First, reversal code is not optional[2]. Second, each reverser is special-purpose, tuned to the activity it needs to reverse. Even though reversal cuts across many routines, the exact method of reversal is particular to each reversible action. One cannot invoke a generic reverser. Third, languages like AspectJ do not insert inline code; they insert procedure calls. They are therefore unable to insert jumps and returns, which our reversal code needs. Still, AOP might eventually be able to capture code reversal.

My recommendation is to take a first step and gain experience with reversers. That first step is to use a language-specific preprocessor that converts **reverser** and **punt** to the code dictated by the static interpretation. I would avoid the **climb** construct; it looks error-prone and it makes the static implementation much harder to implement. I expect that reversers will be a welcome programming technique in the realm of system programming but will find few applications elsewhere. They will most likely never find a place in general-purpose programming languages.

## REFERENCES


[1] Linux source code at `http://lxr.linux.no/source/`, for example.

[2] Dijkstra, E. W., "GOTO statements considered harmful," *Comm. ACM*, 11:147 148, 1968.

[3] C. A. R. Hoare., "The emperor's old clothes," *CACM*, vol. 24 no. 2, pp. 75-83, 1981.

[4] Larry Wall, Tom Christiansen, and Jon Orwant, *Programming Perl*, O'Reilly, 3rd edition, 2000.

[5] Gregor Kiczales, John Lamping, Anurag Mendhekar, Chris Maeda, Cristina Videira Lopes, Jean Marc Loingtier, and John Irwin, "Aspect oriented programming," in Mehmet Aksit and Satoshi Matsuoka, Eds, *Proceedings European Conference on Object Oriented Programming*, pages 220-242, Berlin, June 1997.

[6] Gregor Kiczales, Erik Hilsdale, Jim Hugunin, Mik Kersten, Jeffrey Palm, and William G. Griswold. "An overview of AspectJ," in J. Lindskov Knudsen, Ed, *Proceedings European Conference on Object Oriented Programming*, pages 327-353, Berlin, June 2001.


[2] Some in the AOP community believe that advice code should add new capability to legacy code; the advice code is therefore optional. Others say that advice code, whose purpose is to modularize crosscutting concerns, may be mandatory.

# Hand-written Character Recognition Using Layered Abduction

Richard Fox, William Hartmann
Department of Computer Science
Northern Kentucky University
Nunn Drive
Highland Heights, KY 41099

**Abstract – Even though automated hand-written character recognition can be highly accurate, most of these systems are unable to apply context to improve results, unlike human readers. This paper describes an approach to automated hand-written character recognition that seeks explicit features amongst the input data, and applies layered abduction to derive an explanation to account for the input in terms of English characters. Layered abduction is used because it can provide top-down guidance to improve accuracy. Such an approach has been taken here resulting in more than 96 % accuracy for hand-written printed character recognition in a limited domain.**

## I. Introduction

Automated optical character recognition dates back to the 1960s and has been applied to numerous fields, notably postal mail sorting [1]. When the characters are handwritten, performance often degrades. Modern approaches to solving optical character recognition revolve around neural networks, stochastic approaches, genetic algorithms, and search-based Artificial Intelligence approaches [2, 3, 4, 5, 6]. For instance, automated character recognition used in flat-bed scanners apply trained neural networks to perform the recognition task, and can achieve impressive (accurate) results when the text being scanned is computer-generated. However, hand-written character recognition often results in errors even when the hand-writing is clear (typical performance for printed handwriting is 80-90% accuracy with some results being as high as 97-99.8%) [7]. Human readers will often bring context into consideration when deciphering hand-writing and thereby have additional knowledge to use to improve accuracy.

Unlike trained perceptrons or neural networks which, once training is over, are unable to apply other knowledge sources, a feature-based pattern matching approach can bring in variety of knowledge sources. Through the use of layered abduction, a problem solver can apply knowledge in a bottom-up fashion (that is, analyze data and infer characters) but also top-down knowledge (once some of the characters have been recognized, these can be used as "clues" to further identification). Thus, context can be applied. A layered abduction approach is being researched and has thus far achieved over a 96% recognition accuracy for hand-written printed character recognition in a limited domain.

This paper describes this approach, and is laid out as follows. First, a brief introduction to layered abduction is offered, including a discussion of the use of top-down guidance. Next, the CHREC (CHaracter RECognizer) system is introduced. This system uses layered abduction to perform printed hand-written character recognition and uses both bottom-up and top-down processing. Some examples and experimental results are offered to provide an indication of the performance of this approach. Finally, conclusions and future work are mentioned. It should be noted that CHREC is currently under construction and there is a good deal of more work planned.

## II. Layered Abduction

Abduction is defined as "inference to the best explanation," a task of generating an explanation to account for the appearance of a given set of findings or data [8]. Within the context of handwritten character recognition, the task is one of hypothesizing the characters/symbols/words responsible for the features found in scanned bitmaps and attempting to find the best collection of characters to explain these features. Abduction has previously been applied to a wide range of interpretation problems including speech recognition [9], diagnosis [10, 11, 12], medical test interpretation [13], story understanding [14], natural language understanding [9, 15], theory formation and legal reasoning [16].

One specific strategy for abduction, discussed in [8] and applied in numerous prior systems, attempts to form an explanatory hypothesis through a process of hypothesis generation and instantiation of elementary hypotheses, followed by assembling a subset of these hypotheses into the best possible coherent explanation. Hypothesis generation comprises a search through some source of domain hypotheses; each elementary hypothesis being able to explain some of the given findings (symptoms, data, perceptions, signals, etc). Instantiation comprises evaluating each generated hypothesis in terms of its utility toward the given



*T. Sobh and K. Elleithy (eds.), Advances in Systems, Computing Sciences and Software Engineering*, 141–147.

data (how likely is the hypothesis? what can it explain from the data? does it have any form of interaction with other hypotheses such as mutual exclusiveness or expectations that might lead one to believe that if one hypothesis is true, the other might be true?) Hypothesis assembly selects data to explain and seeks the best available hypothesis to explain it. Explained data are removed from consideration and the selected hypothesis is then built upon by considering whether this hypothesis has expectations or incompatibilities with other hypotheses. Thus, the assembly algorithm forms "islands of certainty" from which the explanation can "grow." The process of assembly continues until all of the data are explained, or there are no more plausible hypotheses left to continue building the explanation.

The term "best," as used in the definition for abduction, is subjective but includes factors such as parsimony (i.e. no superfluous parts), consistency, completeness (in terms of explanatory coverage), terseness, and high likelihood (plausibility). Under good conditions, hypothesis assembly will terminate with a complete explanation for all the findings. Under less favorable conditions, hypothesis assembly will terminate with a partial explanation, explaining as much as can be plausibly explained consistent within any constraints.

Figure 1 illustrates an abstract abduction problem. A collection of explanatory hypotheses {H1, H2, H3, H4, H5, H6} have been generated to explain a collection of data {D1, D2, D3, D4, D5}. The lines between hypotheses and data indicate what data each hypothesis can explain. The dotted line between hypotheses H1 and H4 indicate incompatibility (if one hypothesis is used as part of the explanation, the other cannot be used). The numbers above each hypothesis indicate the individually assessed likelihoods (on a scale of 0 to 1).

The assembly process generates an explanation that can account for all of the data (or as many as possible). The assembly process is one of composing the best subset of hypotheses, focusing one what each hypothesis can explain and how plausible that hypothesis is. From figure 1, a complete explanation is the set {H1, H2, H3}. The explanation {H1, H4} will not work because H1 and H4 are incompatible hypotheses. An alternative explanation is {H1, H5} which is simpler than {H1, H2, H3} but not as plausible since H5 was more poorly evaluated than H2 and H3. By using the explanatory power of the individual hypotheses, knowledge of incompatibilities and expectations (absent in this example), and the individual hypothesis likelihoods, the explanation {H4, H6} might be deemed the best (simplest complete explanation with the highest overall rating). As the reader might infer from this brief example, the assembly task becomes intractable if one attempts to generate *all* possible explanations. The strategy used here takes shortcuts to ensure tractability (see [8] for details).

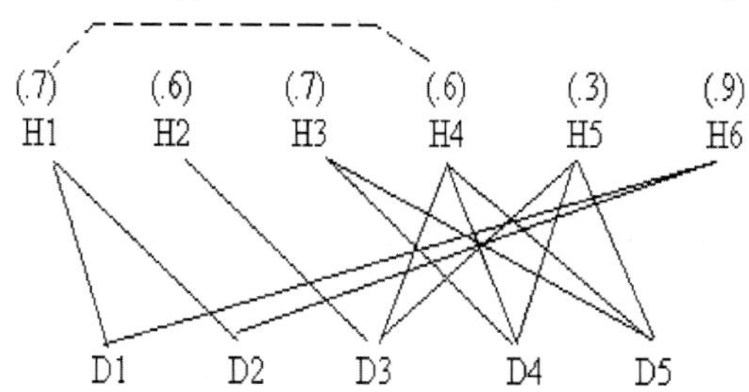


Figure 1: Explanatory Hypotheses

For some perceptual and perception-like interpretation problems, a *layered* approach is taken where hypotheses accepted at one level of description become data to be explained at a higher level of description. This cascaded inference process allows for an explanation to be generated using a variety of different knowledge types that are appropriate for the problem. For instance, in a visual recognition task, one might use low-level knowledge to recognize such features as edges and shapes, and then use knowledge of objects to explain those shapes. Once a shape has been identified, expectations regarding that shape can be used to reprocess edge and shape detection in an attempt to explain data that was either unexplained originally or was poorly explained.

The control strategy for layered abduction includes both bottom-up and top-down elements, with top-down processing used to allow accepted hypotheses at a higher level of abstraction to remove ambiguities and generate expectations at lower levels. The ultimate explanation must be at a high enough level of abstraction and be of a proper vocabulary to be useful. The process of layered abduction is one in which (possibly independent) abductive problem solvers generate their explanations to explain previously accepted composite explanations from lower levels as the data to be explained. Cascaded inferences of this type seem to be used in many types of reasoning dealing with perception, diagnosis and natural language understanding where there is a need to abstract types of data into other types (for instance, in natural language understanding, a syntactic parse of a sentence might be explained in terms of semantic roles for each sentence constituent, and in turn the semantic explanation can be used as data to be explained in terms of a pragmatic explanation) [17].

One of the advantages to layered abduction is the ability to employ a variety of hypotheses types. These include "noise" hypotheses, which can signify that a datum is not worth explaining by a higher level hypothesis because the datum itself is noise. This permits a problem solver to distinguish between significant and insignificant data without having to spend much attention on the insignificant data. [18]

## III. The CHREC System

The layered abduction strategy described above has been used to implement a modest hand-written character recognition system for printed characters. This system currently can recognize the capital letters A-F, the digits 0-9 and the = sign. This will eventually be expanded to include all upper case letters as well as other punctuation symbols. For now, however, the system attempts to properly recognize an equation that specifies a decimal (base 10) number and a hexadecimal (base 16) number. The rationale for this will be explained later.

The CHREC system consists of three primary components, as shown in figure 2. First is the "parser" which takes a bitmap of the scanned input and breaks it into individual regions to be explained. The parser includes a collection of feature detectors. Feature detectors scan the region of the bitmap to identify lines and curves. Some features sought include "diagonal line," "vertical line," and "upward curving open arc." Each feature detector will provide information such as the approximate location of the feature (top, middle, bottom, left, center, right), a slope or acuity for a line or curve respectively, and a rating of how certain a given feature is, which will aid the character recognizers. Features are the output generated from feature recognizers, and are used as the data to be explained.

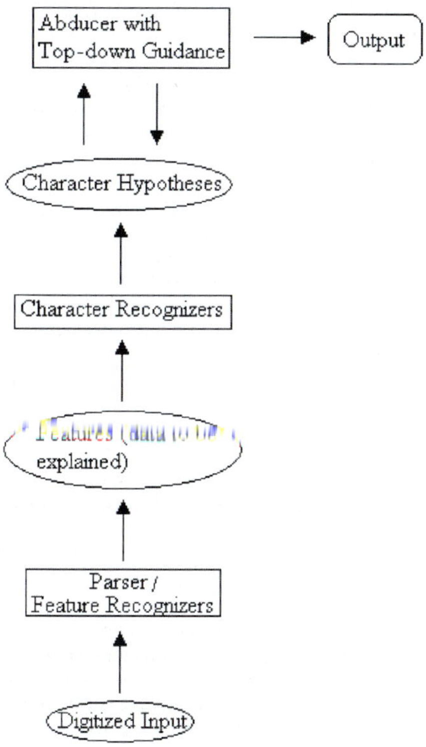


Figure 2: CHREC Architecture

The next component is a collection of character recognizers. Each character that can be recognized has a recognizer. A recognizer seeks the presence of certain features and the absence of other features. As an example, the "7" detector looks for a roughly horizontal line near the top and a diagonal line from left to right, but no curves or other lines. A character recognizer returns a hypothesis stating the likelihood that the features in the given region can be explained by this character. The likelihood is based on what expected features are found, what expected features not found, and what unexpected features are found.

The final component performs abductive assembly to form the best explanation of the digitized input in terms of a collection of highly evaluated characters. This component, once it has formed an explanation, critiques the explanation by making sure that it forms a legal decimal to hexadecimal equation. If the equation is not legal (the left side and right side are not equivalent), then the abducer uses the information regarding the most accurately recognized portions of the explanation as top-down guidance. This permits the abducer to reconsider the explanation and attempt to correct the explanation.

CHREC contains 9 feature detectors, listed in Table 1. These have been fine-tuned through a process of experimentation and seem to work very well for upper-case printed characters. They are robust enough to be able to handle writing from 5 different people and from very neatly written to poorly written characters. These features are applied to each grouping of pixels found, which are anticipated to correspond to a character. In the case that pixels are found that don't form a line or curve or have sufficient width or height, then they are labeled as "noise" to be disregarded by the character recognizers.

There is one character recognizer for each character in the system. Table 2 (at the top of the next page) illustrates the rules for the "F" character recognizer. In this table, there are several features being sought and some features that should

TABLE 1: Features

| Feature | Descriptor |
|---|---|
| Curve | Angle (line segments < 170 degree angle) |
| Line | Angle (line segment > 160 degree angle) |
| Width | Large, Medium, Small |
| Height | Large, Medium, Small |
| Slope | Vertical, Horizontal, +/- Diagonal |
| Curve Opening | N, NE, E, SE, S, SW, W, NW |
| Connections | None, Partial, Fully |
| Location | Top, Top Left, Top right, Left, Right, Bottom left, Bottom right |
| Region | Top, Middle, Bottom, Left, Center, Right |

TABLE 2: EXAMPLE CHARACTER RECOGNIZER FOR "F"

| Feature | Specification | Present/Absent |
|---|---|---|
| Line | Vertical, Top Left, Medium | + |
| Line | Vertical, Middle Left, Medium | + |
| Line | Horizontal, Top Left, Small | + |
| Line | Horizontal, Middle Left, Small | + |
| Line | Vertical, Bottom Left, Medium | - |
| Line or Curve | Right | - |

not appear. Features expected are denoted with a + and those not expected are denoted with a -. Based on how many of the expected features are found and how well they match, and how many of the unexpected features are not found, the character recognizer generates a likelihood score (a value between 0 and 1). To recognize an "F" the character recognizer expects two horizontal lines across the top and middle, one vertical line across the left, which is broken into two separate line segments to make detection easier, and no line on the bottom and nothing (line or curve) on the right.

Some of the recognizers have multiple sets of features. For instance, "7" is sometimes written with a horizontal line in the middle. The "Seven Recognizer" has one set of features that includes this second horizontal line and one set that does not. The "0" recognizer also has two sets of rule to represent the number without and with a diagonal line going up from left to right. Some effort must go into fine tuning each character recognizer, but this work is straight-forward and can be derived largely by following rules of penmanship. It should be noted that expanding the system to recognize a larger number of characters requires adding more character recognizers but not any additional feature detectors.

Abduction works in CHREC as follows. Features have been identified to be explained. Features are grouped based on their location within the original bitmap (that is, features are divided into what is thought of as the first character, the second character, etc). Pixels that could not be identified as features are annotated as "noise." Next, all of the character recognizers are invoked for each region of features, and a score for every possible character is generated. Now, the abducer selects the best character to explain the features. The best character is the one that accounts for all (or the most) features while having the highest likelihood as generated by its character recognizer.

For initial testing, the CHREC system has been implemented to recognize mathematical equations where one side represents a number in decimal (base 10) and the other represents a number in hexadecimal. In spite of this domain as being impractical (i.e., useless), it was chosen to provide a domain that used both letters and digits, and contained some form of top-down knowledge. A more useful application is planned and will be discussed in the conclusions. In this given domain, 17 characters are represented (10 digits, 6 upper case letters, 1 equal sign) whereas typical hand-writing would consist of 10 digits, 52 upper and lower case letters and perhaps 20 punctuation marks. Therefore, this domain provides characters for only approximately one-fourth of the total needed when the system is expanded to a more realistic domain.

At this point in the processing, the abducer has put together an explanation which is a mathematical equation. The equation contains an equal sign, with a decimal number on one side of the equal sign and a hexadecimal number on the other side. Top-down guidance can now take place. In should be noted that top-down guidance is domain dependent. For this system, top-down guidance will involve ensuring that both sides of the equation are equivalent. This allows CHREK to test the two numbers that it generated to make up the explanation.

The top-down guidance works as follows. The collected characters on both sides of the equal sign are converted from a string of characters into two numbers. For this to work accurately, the equal sign must be clearly identified since this character is where the *break point* is. Both numbers are then converted to both bases. For instance, if the number 31 was generated as one of the two numbers, it is converted into decimal as 49 and into hexadecimal as 15. These generated numbers (49 and 15 in this example) are used as expectations for the number on the other side of the equal sign. Additionally, since the number "3" is somewhat similar to "8," "B" and possibly "5," the characters "8," "B," and "5" are somewhat more plausible and may be added as further top-down expectations later.

One of three cases may arise. First, one of the two numbers is found on the other side of the equal sign, and similarly, one of the two numbers converted from the number on the other side of the equals sign is found. In this case, the abducer has confirmed that its explanation is accurate. CHREK reports the explanation with high confidence.

In the second case, one of the numbers derived by conversion matches only some of the characters of the other number. In such a situation, one of the numbers seems to have been accurately identified and the other number is erroneous because some characters do not match exactly. Note that if one of the two numbers has been identified accurately, the converted number should still match some of the characters of the other number. Therefore, the error should lie with the other number.

With this mismatch, the abducer reconsiders the incorrect number. Specifically, the abducer examines those characters that were evaluated at least somewhat likely from their recognizers. It builds a collection of plausible numbers. If any of these numbers match the expected number, then CHREK attempts to correct its explanation. This is done by increasing an individual character's likelihood based on the expected number. Additionally, if any of the mismatched characters has a character that is similar in appearance, then those character hypotheses have their likelihoods increased. For instance, a “0” is similar to an “8.” If the “8” character recognizer thought “8” was at least somewhat likely but did not give it a high score, and a “0” was recognized previously, then “8” is reconsidered by having its likelihood increased a small amount. Other similarities currently applied include “B” and “8,” “0” and “D,” and “C” and “6.”

At this point, applying the expectations in this top-down form will have hopefully altered some of the likelihoods of characters. The abducer runs again to see if it can form a better explanation. If the abducer creates a new number, then the top-down process repeats by converting the numbers and checking to see if there is a match. If a match is found, CHREK reports the new explanation and provides the user with a statement that this explanation is a corrected version, that the original explanation, while (possibly) being more likely, was less consistent. If there are still errors, CHREK outputs both generated explanations and states that both answers are uncertain at best.

In the final situation, errors have arisen on both sides of the equal sign. This is the most problematic of situations that CHREK faces. In this case, once both numbers have been converted, neither converted value matches the other number, and so the expectations can not simply be used to correct the mistakes. Instead, CHREK will go back to the character recognizers and request all characters that generated at least a somewhat likely result. CHREK then forms all possible numbers from these. For instance, if a number was thought to be “38” but it is possible that the first character was a “B” and the second was either a “5” or “0”, then CHREK generates “38,” “B8,” “35,” “B5,” “30” and “B0.” The same takes place with the other number. Now, all numbers are converted and from both lists, comparisons are made to see if there are any that match exactly. If so, CHREK increases the likelihoods of those characters that match and the abducer generates a new explanation. If the explanation derived is coherent, it is reported, but CHREK reports the doubt that this is a correct explanation because of the errors found.

It should be noted that the top-down strategies employed for the equation checker rely on accurately identifying the “=” among the data. To date, the “=” has been recognized with 100% accuracy. Also, CHREK is currently only doing one layer of layered abduction because there is no explanation generated for an equation. Future versions that are implemented for more relevant domains will have more layers for additional top-down guidance.

## IV. Examples And Analysis

In this section, three examples are demonstrated, followed by some brief statistical analysis of the system’s performance. The inputs for all three example inputs are shown in figure 3.

The first example demonstrates a case where CHREC correctly identified all of the characters initially and so no top-down guidance was required. This example is of the equation 652 = 28C. Based on the features detected for the first character, only “5” and “6” were generated as plausible, with “6” getting a very high likelihood of 0.91 and “5” scoring 0.75. Aside from the higher score, a curve was found in the lower-left area of the character and therefore “5” was not as good an explainer, leading “6” to be selected. For the second character, “5” was the only character that received a plausible score (0.5) and so was selected. Similarly, “2,” “=,” “2,” “8” and “C” were the only characters generated with reasonable likelihoods for the remaining input characters and so they were all selected to explain the features. At this point, 652 and 28C were both converted, with 28C only be converted to decimal (since it had a hexadecimal character, there was no need to convert it to hexadecimal). CHREC confirmed that the answers corresponded, so processing terminated.

The second example is of the equation 285=11D. CHREC originally inferred that the equation was 285=11B. The problem with this example, which you might infer in looking at the figure, is that the right arc of “D” is somewhat unclear in the scan. This problem arose because of the ink used to write the character. CHREC had difficulty with this character’s recognition because it miss-parsed the pixels and thereby missed two distinguishing features between “D” and “B” when performing its feature detection. These features are that of a horizontal line in the middle (for “B”) and a single leftward-opening arc (for “D”). What was inferred from the pixels located was that there were two curves, one in the lower-right and one in the upper-right. These two features are sought for the letter “B.” Therefore, the character recognizer for “B” found more evidence than the recognizer for “D” resulting in a higher likelihood (0.8 for “B” and 0.6 for “D”). In spite of this error, top-down guidance was able to resolve this problem by converting 285 into hexadecimal and finding that “B” should be “D.” In converting 11B into decimal, 285 is converted into 11D, the result should be 283, but there was no evidence at all that “5” should be “3,” and so the best available explanation was that “B” was incorrect. With this correction, CHREC is able to infer the correct answer, 285=11D.

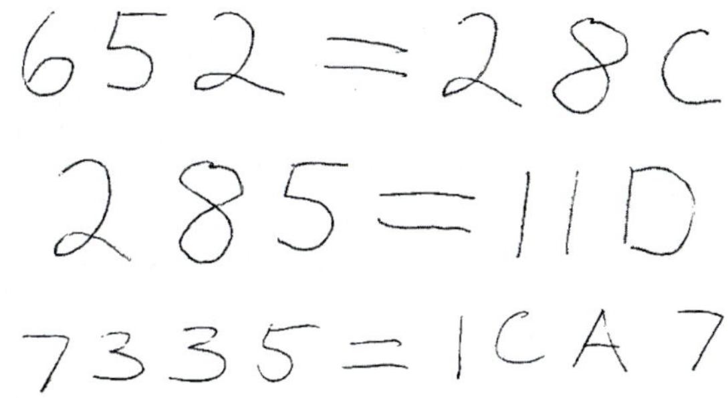


Figure 3: Examples

The third example presents the equation 7335=1CA7. In this case, there are two errors in CHREK's original attempt at an explanation, 733F=16A7. The selection of "5" over "F" was made because "5" scored 0.5 while "F" scored 0.8. This is not a substantial difference and so top-down guidance proved to help. Similarly, in the case of "6" instead of "C," "6" scored a 0.6 and "C" scored a 0.5.

In this case, the abducer detects that the left-side and right-side numbers do not cohere and attempts to use top-down guidance to resolve the problem. First, CHREK attempts a simple solution – by replacing the somewhat highly scored "5" with "F," would that resolve the matter? No, the converted number would still not be 16A7. Similarly, using 1CA7 for the right-hand side number does not convert into 733F. So, top-down guidance needs to be more involved.

CHREK now generates all numbers formed from any plausibly recognized characters from both sides of the equal sign. These numbers are converted to decimal and hexadecimal (for those that did not contain hexadecimal digits). Each of these converted numbers are considered as *mild* expectations. Now it is a matter of attempting to infer what *could* be the correct answer. CHREC will notice that 7335, which is somewhat plausible because three of the four characters are correct and the incorrect character still rated fairly well, matches 1CA7, which itself is also plausible since three of its four characters are correct. Through this top-down guidance, "5" and "C" are both given increased likelihoods. CHREK then regenerates its explanation and finds that the results cohere. Therefore, CHREC provides this explanation, but tempers it with uncertainty because neither "5" nor "C" was highly rated to begin with.

The CHREC system was first tested out on individual characters. From 738 characters, written by 5 different people, the system accurately recognized 93%. The system was then tested on 75 equations from 2 people. Each equation consists of a decimal value on one side of the equation, and a hexadecimal value on the other. There were a total of 526 characters in the 75 equations. Without top-down guidance, the system achieved 87% accuracy, but with top-down guidance, accuracy improved to 96%. Testing equations of just one person resulted in 92.5% accuracy without top-down guidance and 100% accuracy with top-down guidance. The other writer's "E" caused problems because he wrote the "E" like a backward "3" (that is, with curves) while the character recognizers had not been programmed to recognize that form of "E." That is being fixed and it is anticipated that overall accuracy with top-down guidance will improve to over 98%.

The hand-written characters used as test data were scanned into bitmaps using an HP Scanjet 3970. This scanner has its own "intelligent scanner" optical character recognition program. As a test, several of the inputs were run through this OCR program. Figure 4 contains the output from the scanner for the three examples from this section (652=28C, 285=11D and 7335=1CA7 respectively). The Scanjet's recognition program had tremendous difficulty with the recognition of hand-written characters, particularly of the second example. The scanner is primarily set up to handle machine-prepared text and far better character recognition results have been achieved, but it is interesting to compare against CHREC's performance.

```
b 5 ~~ ~ 8 (

:185, ~/IO }

733S-=- jCA 7
```

Figure 4: HP Scanner's Recognition

## V. Conclusions And Future Work

The CHREC System performs hand-written character recognition by using layered abduction. First, features are sought from the digitized input. Character recognizers, each responsible for identifying one character (such as "2" or "A") examines the features for a given character and based on the presence or absence of expected features, assigned a likelihood. An abductive assembler then attempts to explain the data by selecting those characters that best explain the pixels in the region of the character. Noise hypotheses are also available to explain spurious or irrelevant pixels found. Top-down guidance is applied to ensure that the explanation is coherent. In the current case, "coherence" means a mathematical equation of a decimal value and a hexadecimal value that must be equal to each other.

The layered abduction approach differs greatly from other recent approaches to hand-written character recognition that might use neural networks or genetic algorithms for example. By using layered abduction, knowledge can be applied to assist in a top-down fashion to improve on the accuracy of a merely bottom-up approach. This provides CHREK with numerous advantages over neural network and stochastic approach. First, CHREK is able to express uncertainty in its solutions when there is sufficient uncertainty. This allows the user to know just how confidently an answer should be taken. The ability to apply a variety of types of knowledge permits the use of indirect reasoning, which helps correct errors. By explicitly generating an explanation, noise hypotheses are available, which can improve accuracy.

While the domain currently implemented for CHREC is somewhat of a toy domain, it demonstrates the utility of top-down guidance as well as the capabilities of this approach. Current accuracy is in the 96% range. CHREC is currently under construction. It is limited to the domain of decimal and hexadecimal equations and therefore only recognizes 17 characters. However, plans are to expand the system to include all 52 alphabetical characters and to perform postal address recognition. In this new domain, a greater amount of top-down guidance can be applied by ensuring that city, state and zip code cohere although street addresses and names may be more problematic. With features detection mechanisms already in place, expanding the system is merely a matter of adding character recognizers for new characters. It is hoped that the system will achieve similarly high performance once expanded to the domain of postal addresses.

## REFERENCES


[1] H. F. Herbert. *The History of OCR, Optical Character Recognition.* Manchester Center, VT: Recognition Technologies Users Association, 1982.

[2] M. A. Otair and W. A. Salameh, "An improved back-propagation neural network using a modified non-linear function," *Proc. of the IASTED Intl. Conf.*, 2004, pp. 442-447.

[3] G. Mayraz and G. E. Hinton, "Recognizing handwritten digits using hierarchical products of experts," *IEEE Trans. on Pattern Analysis and Machine Intelligence*, vol 24, no. 2, Feb 2002, pp. 189-197.

[4] L. S. Oliveira, R. Sabourin, F. Bortolozzi, and C. Y. Suen, "Feature selection using multi-objective genetic algorithms for handwritten digit recognition," *16th Int'l. Conf. on Pattern Recognition*, vol 1. 2002, pp. 572-575.

[5] K-F. Chan and D-Y. Yeung, "An efficient syntactic approach to structural analysis of on-line handwritten mathematical expressions," *Pattern Recognition*, vol 33. no. 3, 2000, pp. 375-384.

[6] E. H. Ratzlaff, "Methods, report and survey for the comparison of diverse isolated character recognition results on the UNIPEN database," *Int'l. Conf. on Document Analysis and Recognition*, vol. 2, 2003, pp. 623-628.

[7] S. Tanner, *Deciding whether Optical Character Recognition is feasible.* London: King's Digital Consultancy Services, 2004.

[8] J. Josephson and S. Josephson, eds., *Abductive Inference: Computation, Philosophy, Technology*, Cambridge University Press, New York, 1994.

[9] R. Fox and J. Hartigan, "An algorithm for abductive inference in artificial intelligence," *Encyc. of Library and Information Science*, vol 64, A. Kent, Ed., Marcel Dekker, Inc: New York, 1999, pp. 22-38.

[10] H. Pople, "The formation of composite hypotheses in diagnostic problem solving: An exercise in synthetic reasoning," *Proc. of IJCAI Five*, San Franscisco, 1977, pp. 1030-1037.

[11] J. Pearl, *Probabilistic reasoning in intelligent systems: Networks of plausible inference*, Morgan Kaufmann, 1988.

[12] J. Reggia, "Diagnostic expert systems based on a set covering model." Internat. *J. Man-Machine Stud.*, vol. 19, 437-460, Nov 1983.

[13] J. Josephson, M. Tanner, J. Svirbely, and P. Strohm, "Red: Integrating generic tasks to identify red-cell antibodies," *Proc. of the Expert Systems in Government Symposium*, K. N. Karna, Ed, 1985, pp. 524-531, IEEE Computer Society Press.

[14] J. R. Hobbs, M. Stickel, D. Appelt and P. Martin, "Interpretation as abduction," *Artificial Intelligence J.*, vol. 63, issue 1-2, pp. 69-142.

[15] V. Dasigi, R. C. Mann, V. A. Protopopescu, "Information fusion for text classification - an experimental comparison," *Pattern Recognition*, vol. 34, issue 12, pp. 2413-2425, 2001.

[16] P. Thagard, "Explanatory Coherence," *Behavioral and Brain Sciences*, vol 12, issue 3, 1989.

[17] J. Josephson, "A layered abduction model of perception: Integrating bottom-up and top-down processing in a multi-sense agent" in the *Proc. of the NASA conf. on Space Telerobotics*, 1989, pp. 197-206, Pasadena: JPL publication.

[18] R. Fox and J. Josephson, "Explicit noise hypotheses in speech recognition" *Proc. of SPIE*, vol 2032 (Neural and Stochastic Methods in Image and Signal Processing II), San Diego, California, July 1993), pp 245-255.

# Dynamic Pricing Algorithm for E-Commerce

Samuel B. Hwang
*Cafedvd.com*
*Sunnyvale, CA, USA*
*samuel@cafedvd.com*

Sungho Kim
*Cafedvd.com*
*Sunnyvale, CA, USA*
*sungho@cafedvd.com*

***Abstract*** **– E-Commerce has developed into a major business arena during the past decade, and many of the sales activities are handled by computers. Intelligent algorithms are being developed to further automate the sales process, which reduces labor costs as well as business operational expenses. This paper describes an automatic sales-price determination algorithm for online markets in a competitive environment. It tries to dynamically adjust the sales price to maximize the profit and minimize the sales time. This algorithm gathers sales prices of competing online stores, and optimally adjusts the advertising price. It is particularly useful when a store carries numerous items so manual price adjustments are laborious and costly. The algorithm has been applied to DVD movie sales in the online market, and shown to shorten the sales time and increase the profit.**

***Index Terms – Dynamic Pricing, Online Sales, Pricing Strategy, E-Commerce Automation***

## I. INTRODUCTION

Dynamic pricing responds to market fluctuations in a real-time basis to achieve specific sale objectives such as maximize profit, maximize sales volume and minimize sales time. This often results in charging different prices for the same goods to different customers. The airline industry is often cited as a success story for dynamic pricing, as customers are used to paying different prices for the same ticket [1]. In other sectors, however, people feel dynamic pricing have serious ethical and fairness problems. When Amazom.com charged different prices for the same DVD movies to people in different physical locations or with different purchase history, many people were upset [2]. In the e-commerce arena where both the suppliers and the purchasers have real-time, global information on product prices and availability, dynamic pricing is not a luxury, but a must-have tool for sales activities. In this paper, we are more interested in real-time adjustment of price by comparing the prices of our competitors, and not in charging different prices to different buyers.

Many online stores carry a variety of products and compete with other online stores. It is prudent to gather competitors' prices and then set your price to optimize profit and sales time. This, however, requires searching the internet to get competitors' prices, which is labor extensive. If you are selling DVD movies, for example, the labor cost of searching competitors' prices of all the movies offsets the additional profits you may gain by optimally adjusting the prices. The same is true for marketing lodging facilities for vacations, DVD movies, or used cars. In another situation, you may want to sell your new or used items by using well known sales portals such as amazon.com, ebay.com or half.com. If you regularly use such sales portals, monitoring your price against competitors' is also laborious. Whether you operate your own online store or use existing sales portals, you will greatly benefit from a software package that has a web crawler which gathers competitors' prices and an intelligent algorithm that sets your price based on them. This enables you to optimally set your sales price with little labor costs.

## II. PREVIOUS WORK

Throughout the history of commerce, charging different prices for identical goods has been a common practice in markets separated by geography or defined by distinguishable customer types [3]. Among many works on dynamic pricing, a classical work can be found in [4, 5] and an overview in [6]. In [4] it classified the pricing strategies into three categories: skimming, neutral and penetration (Fig. 1), whereas in [7] the high-value/high-price and low-value/low-price are categorized as *premium* and *economy*, respectively (Fig. 2). We will use this classification in the development of our dynamic pricing algorithm. The pricing is part art and part science, and pricing strategies based on human nature and psychology can be found in [8]. An example of such is the "charge more, and they think it is a better product" strategy. Questions to consider for pricing and different strategies such as maximize short term profit, gain market share, survive and help society can be found in [9, 10, 11].

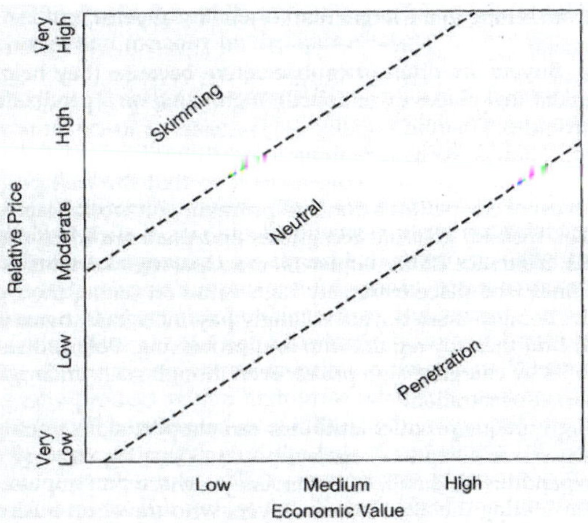


Fig. 1 Relationship between price and economic value in Strategy selection [4].



*T. Sobh and K. Elleithy (eds.), Advances in Systems, Computing Sciences and Software Engineering*, 149–155.

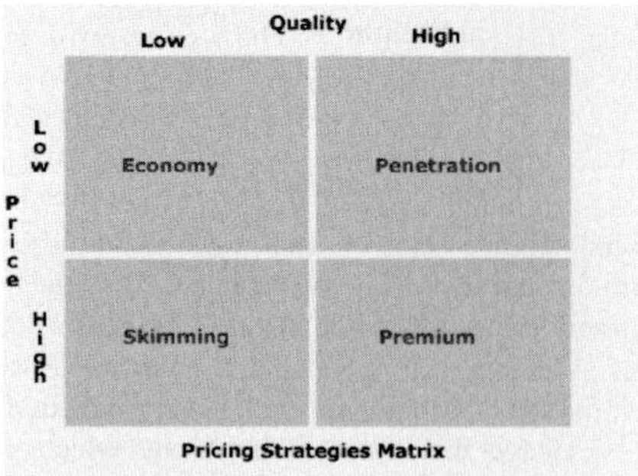


Fig. 2 Different pricing situations [7].

There are several companies that develop dynamic pricing software. Netpricing-solutions.com [12] has developed *AutoPricer* pricing system that is designed to allow tour operators to price any combination of airline (scheduled and chartered), hotel, cruise, car rental and coach travel. It loads actual cost data and pricing assumptions, generates prices and margins, models different pricing scenarios and then outputs the prices to the reservation system.

Finally, internet user behavior for search engines is reported in [13, 14]. It includes user population, loyalty to particular search engines, reaction to search failure, toolbar usage and paid vs. free. These pieces of information on user behavior are useful when developing dynamic pricing algorithms for e-commerce as the web-page browser is the user interface.

## III. Dynamic Pricing Algorithm

We follow the steps in [4] in developing our dynamic pricing algorithm shown in Fig. 3. There are three phases: data collection (Steps 1, 2, 3), strategic analysis (Steps 4, 5, 6) and strategic formulation (Step 7). Our algorithm is general in that it can be applied to most market sectors given we have correct statistics of the corresponding sales process. When we give examples, however, we will use the DVD movie market since it is our first application of our dynamic pricing algorithm. The main contributions of our paper are the use of probability models of sales process and the introduction of benchmark prices that reflect the characteristics of the online store's primary user interface, namely, the web pages.

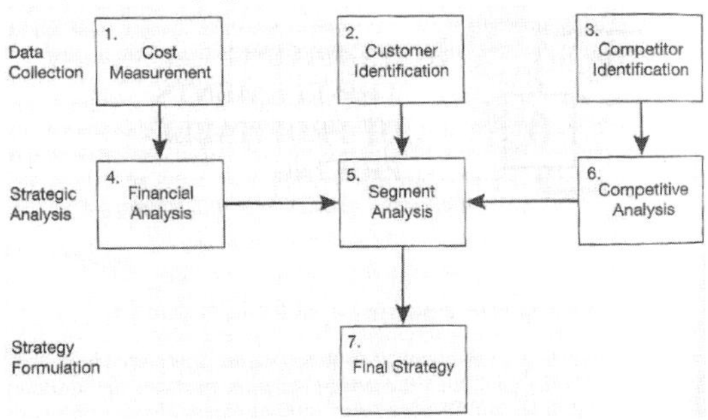


Fig. 3 Steps for more profitable pricing.

### A. Data Collection

The first step of the data collection is the cost measurements. This can be calculated in a straightforward manner using the traditional accounting method. For DVD movie sales, it is the sum of the purchase price, the operational cost of labor and maintenance of online store infrastructure, and the capital cost of equipment and buildings. The second step is the identification of customers. For online DVD movie sales, the customers are the movie collectors who are globally located.

The third step is the competitor identification. For online stores, they are any companies or individuals who try to sell the same or substitute products via internet. They are easily identifiable through a search engine, e.g., search for "DVD movie sale" in the google.com or yahoo.com business section. By using a web crawler, the competitors' names, products and prices are available 24 hours a day, 7 days a week. If you use a sales portal such as half.com, it emails you a weekly report of your price and statistics on competitors' prices in your product category.

### B. Strategic Analysis

Step 4 of the pricing process is the financial analysis. The cost of goods and the expected revenue are used to calculate the profit. The relationship between the price and the expected sales volume is identified here so it can be used in strategy formulation step (Step 7). Correctly estimating the sales volume is a challenge. Companies have to rely on statistics from market surveys or recent sales records. For the case of selling a product on a sales portal such as DVD movies at half.com, one can have samples of price/sales-time/sales-volume by monitoring how long the listed items stay in the sales portal. Given enough such data samples, the relationship between price and expected sales volume can be estimated.

Segment analysis (Step 5) classifies customers into different classes by geographical locations, socioeconomic status, purchase characteristics, and other factors. If online stores give prices after the customer is identified, it can give a personalized price as practiced by the airline industry and amazon.com [2]. The emphasis of our paper is, however, on the cases where we do not discriminate customers, but on the determining our price by comparing with competitors' prices. As a result the resulting price is advertised on the web before customers are identified. We do, however, identify the market phase over the product life cycle so we can maximize the profit which is the product of the sales volume and the profit per unit. Fig. 4 shows the product life cycle for a typical product. The DVD movie market, however, has no development and very, very short growth phase since movies are typically shown in theaters three to six months before being sold in DVD form. Most of movie introduction and marketing are done before and during the time movies are played in theaters. We will use for a typical movie the right-hand side of a Gaussian curve,

$$SalesRate = \frac{2}{\sigma\sqrt{2\pi}} e^{-t^2/2\sigma^2} \quad (1)$$

for the sales rate as a function of time as shown in Fig. 5

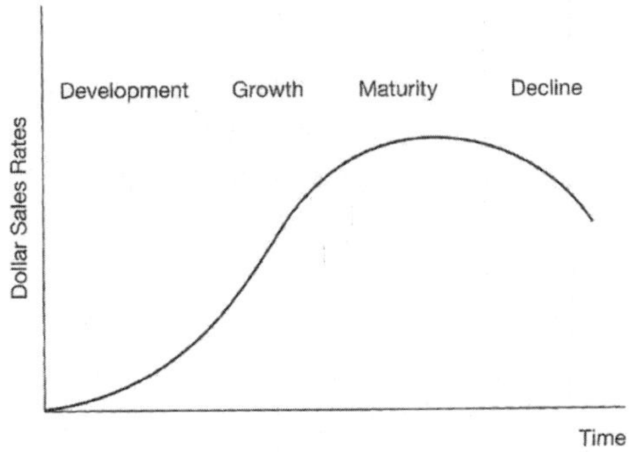


Fig. 4 Phases of the market over the product life cycle.

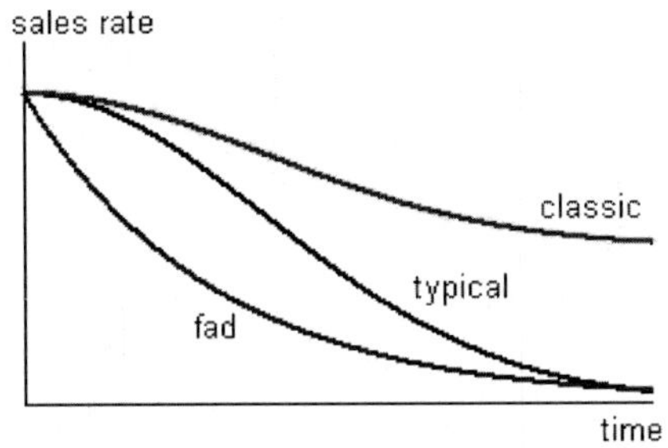


Fig. 5 Product life cycle of DVD movie market.

For classic movies, however, the sales rates stay above certain levels. For fad-type of movies the sales rates drop more rapidly, and we model them with an exponential decay function. They are also shown in Figure 5. We note here that the sales rate is the inverse of the sales time and that the sales rate and the time value of the product are positively correlated. The exact shapes of these curves are unknown in advance, and the curves of similar products from the past have to be used in the pricing algorithm.

The last step of strategic analysis is the competitive analysis (Step 6). After the web crawler gathers the prices of competitors, the following steps are used to calculate statistics on competitors' prices. In the first step, the prices are normalized according to the quality of the items from their descriptions in the posting in case there are differences in product qualities among competitors. The resulting prices are further normalized by the seller's rating, brand name and online exposure amount. Secondly, the standard *Q-test* is used to eliminate all outliers in the price set. Outliers correspond to either 1) sellers who try to sell an item quickly and advertise it at "dumping" prices, or 2) sellers who try to skim novice buyers with high prices. These are eliminated and not used in determining our price.

In the third step, several benchmark prices are computed from the remaining price data. The benchmark prices incorporate the characteristics of the internet and its users' behavior. In internet browsing, the most attention is given to the items on the first page of the search result. The probability of users visiting the subsequent pages decreases dramatically with the number of clicks required to visit them. It is thus crucial to get your item listed on the first page for the maximum amount of exposure. The minimum, average and maximum prices of the first-page-item prices (P1mn, P1ave, P1max, respectively) are included in the benchmark price set. (For the first-page maximum price, we actually use the price less one cent to actually list our product in the first page of the sales portal.) To maximize profit, however, the average and maximum prices of all the price set (Paave, Pamax) are also included in the benchmark price set. Lastly, the regular prices of the product in big-name stores, e.g., amazon.com (Pamazon), walmart.com (Pwalmart), etc., complete the benchmark price. All the prices in the benchmark set are sorted in decreasing order to form the price hierarchy. If there is a large price jump between adjacent prices in the price hierarchy, some intermediate prices can be inserted if the user chooses so.

### C. *Strategy Formulation - Dynamic Pricing Algorithm*

Online stores have different selling situations for each of its products at different times. There are numerous pricing objectives such as: maximize long-run or short-run profits, increase sales volume or return on investment, build store traffic, or price for market survival [9]. In terms of the product's economic value and the price matrix, the pricing objectives are classified as skimming/neutral/penetration in [4] and skimming/economy/premium/penetration in [7] as mentioned earlier. The economic value and pricing policy can be illustrated as in Fig. 6 in a graphical form. Suppose the value of an item decays exponentially in time. If the store lowers its price following the staircase function above the exponential curve, it is trying to gain a large profit from this item and its policy is skimming and maximize-short-term-profit. If the store's pricing follows the staircase below the curve, its selling objective is maximize-sales-volume and market penetration. The actual pricing typically crisscrosses the curve, staying above the curve at times and below it at other times.

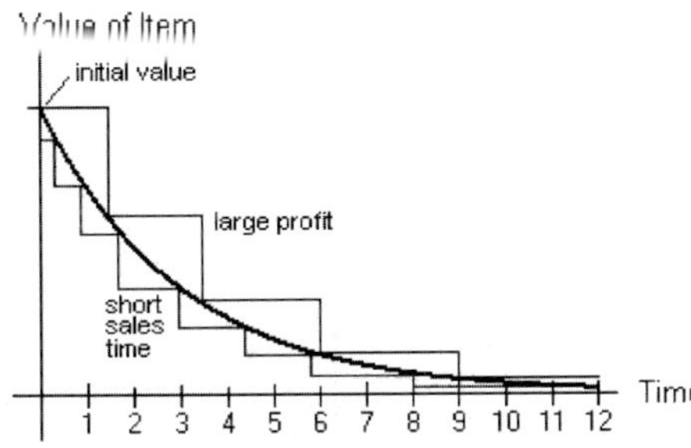


Fig. 6 Item value verses two different pricing policy.

Using the price statistics obtained from the data on the key competitor sites, our dynamic pricing algorithm generates different staircase functions depending on the pricing policy set by the store. The price levels in the staircase functions come from the benchmark prices computed in the competitive analysis, and the length of each staircase level is set either manually or automatically using the mathematical model we develop below.

**Mathematical model of sales process**

The sales rate curves in Fig. 5 shifts toward the y-axis and the peaks at the initial time get higher as we lower the price. For the Gaussian curve in Eqn. (1), the lower the price, the smaller $\sigma$ becomes and the overall sales volume (the total area under the curve) increases. The sales rate as a function of time and price is then

$$SalesRate = \frac{A}{p}\frac{2}{\sigma\sqrt{2\pi}}e^{-t^2/(2(Bp)^2)} \quad , \quad (2)$$

where $p$ is the price, $A/p$ is factored in to show volume increase, and $\sigma = Bp$ shows a smaller spread for a smaller price. The sales-rate graph reflecting only the smaller $\sigma$ effect is shown in Fig. 7(a), while that reflecting both the smaller $\sigma$ and increased volume is shown in Fig. 7(b). We note that similar equations and graphs can be worked out for the other sales-rate curves in Fig. 5.

If we know the graph shown in Fig. 7(b) in advance, we can determine the price that maximize the profit, which is the product of (price minus cost of goods) and the overall sales volume. Fig. 8 shows the profit rate as a function of price and time, and the overall profit for a particular price can be computed by integrating the curve at that price over the time axis. But all we have are the graphs corresponding to similar goods we sold in the past. We can start with one

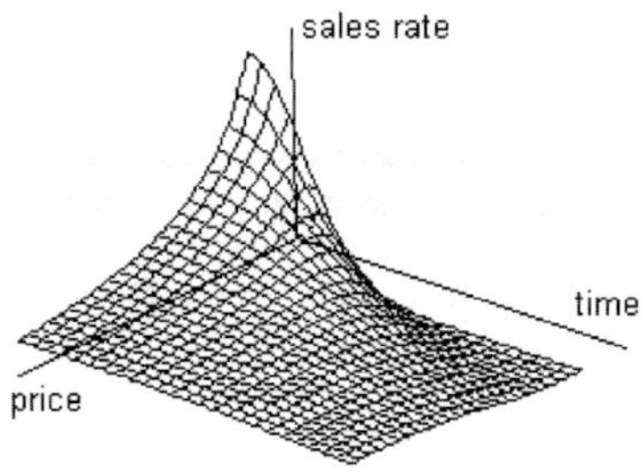


(a) Sales rate showing only the effect of smaller spread $\sigma$, plotted for price ranging from 0.5 ~3.

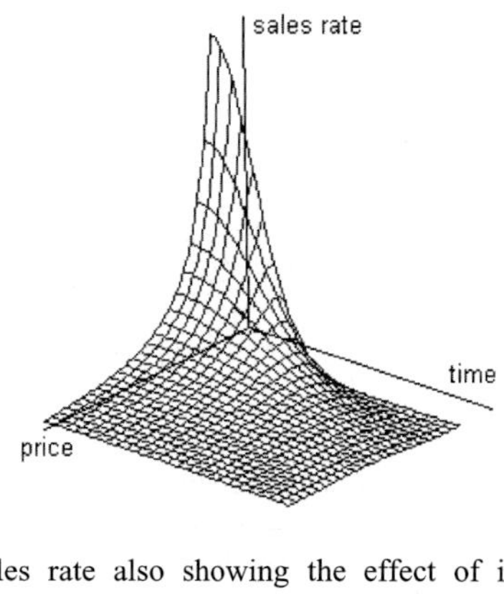


(b) Sales rate also showing the effect of increased overall sales volume

Fig. 7 Sales rate as a function of time and price.

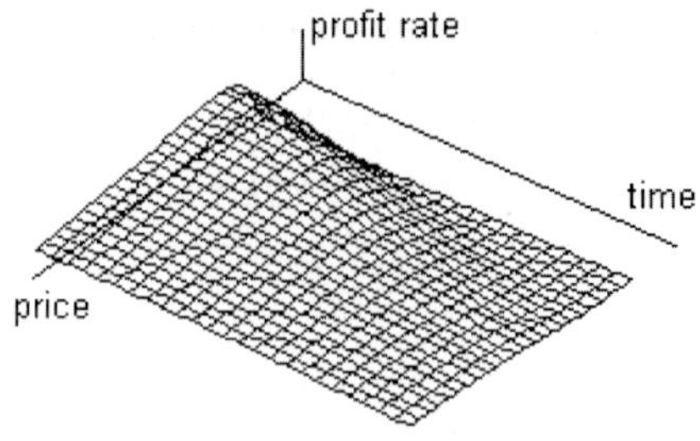


Fig. 8 Profit rate as a function of time and price.

of such graphs, but a better way is to incrementally construct the graph for the item current being sold. Since we have a function with 2 parameters, $A$ and $B$, in Eqn. (2), all we need is 2 data points to construct this graph. Of course, to minimize statistical errors, many more data points are needed. By having the mathematical model of the sales rate, it is possible to construct a graph of the profit rate on the fly. (Note that we have to sell the product at different prices initially to estimate the sales-rate graph.) As time goes on we get a better estimate of this graph, so we can better set the price to maximize the profit or the sales volume. This is one of the main contributions of this paper. The minimum-error curve fitting the Eqn. (2) is discussed in the appendix.

For the DVD movie market, this graph is valid for 3 to 6 months as the movie production companies reprice DVD movies after such periods, often lowering the prices by five to ten dollars from twenty to twenty five dollars. When this happens, our algorithm needs to rebuild the sales-rate graph from the start.

**Dynamic pricing algorithm**

Now that we have the mathematical model of the sales rate and the data on competitors' prices, we are ready to formulate our dynamic pricing algorithm. There are three steps in this formulation: pricing-policy selection and optimal price computation from the sales/profit rate curves, conversion of the selected price to one of the benchmark prices, and determination of the time interval to keep the selected benchmark price.

In the first step, the store chooses for a particular item the pricing policy. Modifying the classification in Fig. 1 and 2, we have the following three pricing objectives that constitute a pricing policy: maximize-profit, maximize-sales-rate and premium. For maximize-profit, we select the price that maximizes the time-integral of the profit-rate curve in Fig. 8. For the maximize-sales-volume policy, we are trying to increase the market share (market penetration) by sacrificing the profit. For this objective, we specify two parameters: the minimum sales volume and the maximum loss. Our algorithm then chooses the maximum price at which the time-integral of the sales-rate curve in Fig. 7(b) meets our total sales volume target while keeping the price such that the total loss is less than the specified maximum loss. If no such price is found, it asks the user to modify the two parameters. When we notice that the demand far exceeds the supply for an item, we set the price to the highest of the benchmark prices. This is the case of the premium pricing objective. We will not use the skimming pricing strategy, as there are buyer ratings and comments for most online stores.

In the second step, the price selected in the first step is converted to one of the benchmark prices. If the nearest lower benchmark price to the selected price is the first-page maximum price, then the price is converted to the first-page maximum price to increase the online exposure of our item. In all other cases, the selected price is converted to the nearest benchmark price in the price hierarchy. The rationale for this conversion is that human mind is sensitive to the least expensive, the average and the best (most expensive) items for sale. These prices are reflected in the benchmark-price set. Then again considering the human psychology on numbers, the cent digit of the converted price is changed to 9 given this does not bump our item out of the first page of the sales portal.

Finally, as the length of time an item is put for sale increases, the sale price is lowered to the next level on the benchmark-price hierarchy. The time intervals between the price adjustments are crucial to balance between profit and sales time. In the manual mode of our algorithm they are set by the store manager, while in the automatic mode they are determined by our algorithm as follows. When we put an item for sale at a certain price, we expect a certain sales rate as defined by the curve in Fig. 7(b). If the actual sales rate is higher or lower than this expected rate, the current sales-rate surface is not an accurate enough estimate. By incorporating the currently measured sales-rate data into Eqn. (2), a more accurate sales-rate surface can be obtained. The new surface is used to update the price selected in the first step, and consequently in the second step described above. The result is the price jumping among benchmark prices when the time on the market prolongs without being sold (typically a price reduction)

There is a question of how often our algorithm should get the competitors' prices, daily, weekly or monthly. Since it does not cost any expense to get this data, we believe it is prudent to run our algorithm daily if you are selling an item in a sales portal. If one operates one's own online store, a weekly update is better to show price stability to consumers.

## IV. Application to Online Sales of DVD Movies

Our algorithm is general in that it can be applied to set the product prices for any online sales activities. Our first application has been to the sale of DVD movies at the sales portal half.com. The half.com classifies each DVD movie by its condition into four categories: New, Like New, Very Good, and Good. It then lists items in each category in ascending order of their prices. Cafedvd.com is an online movie rental store, and regularly sells new and used DVD movies that are overstocked or becoming old. Although half.com emails weekly statistics of competitors' prices on all the DVD movie titles you are selling in their portal, it is a laborious task to monitor them and dynamically change your prices as Cafedvd.com sells hundreds of titles at any time. We have, therefore, customized our dynamic pricing algorithm to this application.

We have implemented our algorithm in Perl for web crawling and simple computations, and in C for the mathematically intensive statistics part. Fig. 9 shows the user interfaces of our dynamic pricing algorithm customized for sales of DVD movies on half.com. Fig. 9(a) is the policy editor with which one can generate a customized pricing policy, and this is used to modify our dynamic pricing algorithm manually if desired. Fig 9(b) is the price setter which is used to set the price of individual DVD movies according to the pricing policy of the user's choice.

The pricing policy editor allows a user to select a price from the benchmark set presented in a pop-up menu (shown on the right side) for each of the DVD conditions and for each of the sales objectives. The table is initially comprised of the benchmark prices from the default policy that is described in Sec. III.D. Below the table are the time periods after which a mandatory price reductions are required. This specifies for each of the benchmark prices the time period after which the price should be lowered to the next price in the benchmark price hierarchy. The intervals suggested by our algorithm are displayed as the default values, and the user can manually override them. Note that we update our prices on daily basis by collecting competitors' prices every day, but these time intervals represents the times at which a mandatory price reduction to the next lower benchmark price occurs. The user can generate any number of pricing policies, e.g., one for typical DVD movies, another for classic movies, etc.

In the price setter (Fig. 9(b)) for individual DVD movies, the user enters either the movie title or the UPC code at the top. The quantity to be sold and the original

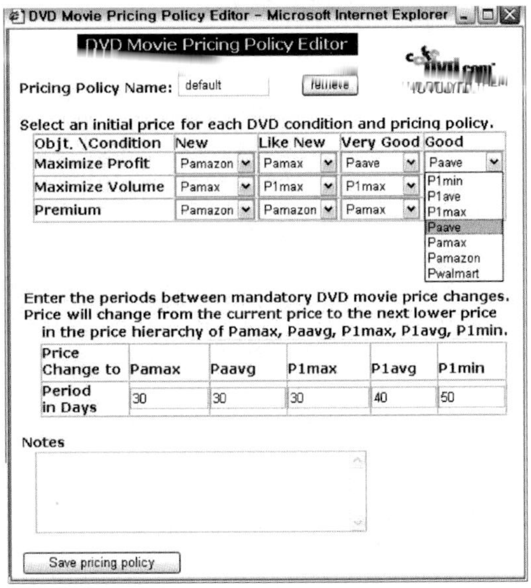


Fig. 9(a) Dynamic pricing policy editor.

cost of purchase are also entered. After that, the DVD condition, the pricing policy to use and the sales objective are entered by clicking radio buttons. Note even if we sell only one copy of a particular DVD movie, maximize-sales-volume is a valid objective as it translates into minimize-sales-time. Then the benchmark prices from the competitive analysis are displayed automatically by our algorithm. For this particular movie, 45 competitors listed their items for sale. For maximize-volume objective, the user has to enter both the maximum loss and the minimum sales volume desired. The maximum-profit and maximum-volume prices are either computed from either the sales statistics of the similar items we sold in the past, or commonsense prices for each sales objective if no such statistics exists. The price of the DVD movie determined by the user's choice and our algorithm is highlighted with a green box color. Note that in this particular example, the price maximizing sales volume subject to $100 maximum loss and minimum volume of 10 copies is $4.99, and the benchmark price nearest to it is $4.48 corresponding to P1max minus one cent. Finally, clicking the set-price button sets the price.

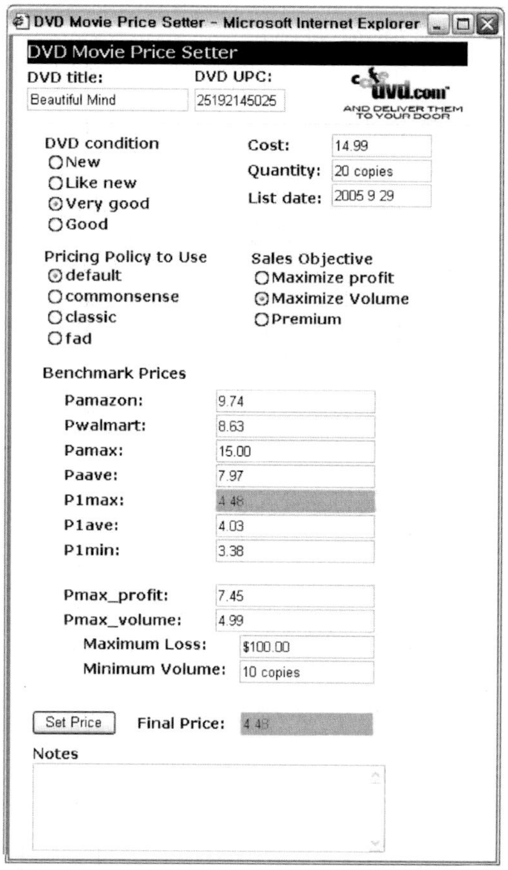


Fig. 9(b) User interface of our item price setter.

Here are some details and happenings of our application to DVD movie sales. Half.com shows four items for each condition on the first web page in the listing. For brand-name stores such as amazom.com or walmart.com, we have decided to give them a 10% edge, i.e., the price normalization factor is 1.11. In another words, our price is 10% less than their prices for the same item. Since we are at the beginning stage of using our algorithm, the initial prices are set using commonsense prices. In the first day of production, i.e., go-alive, we ended up selling several DVD movies in a day. This was because the price normalization was not done correctly. Some of the lowest-priced competitors' items had descriptions like "have different cover" or "have Asian cover." These DVD movies were actually produced for Australian or Asian markets, and were likely have lower video quality. So when our items of good qualities were listed at comparable prices to theirs, ours were sold in a short time. We believe we could have sold one of them at $11.99 instead of $9.99. What we need are more comprehensive keyword sets to judge item quality/condition so the price normalization can be done accurately.

The most quantity of a single item we sell at half.com is a few tens of copies, and it has proved to be difficult to estimate sales-rate surface accurately. For sufficient accuracy we need either to sell at least hundreds of copies of the same item or have sales-rate surfaces for a similar movie from the past sales. As it stands now, we do not have assurance that the prices for maximum profit are accurate. Most of our DVD movies that were sold were priced between P1ave and P1max, since the ones we sell are overstocked movies that are not popular (thus not being rented often).

## V. CONCLUSIONS AND FUTURE WORK

The dynamic pricing algorithm has shown its value in application to the DVD movie sales market. Although we need more time to estimate how much profit has increased by using the dynamic pricing algorithm, it is definitely enabling dynamic pricing. Before this implementation, the labor cost prohibited the dynamic pricing of the DVD movies for Cafedvd.com. There are numerous paths for improvements and new research following our work in this paper. One simple improvement will be extending the benchmark prices to the second, third, etc., web-page prices, and also including the impact of whether the user has to scroll down to see the sales items shown toward the bottom of a web page. The most important and difficult research, however, is the understanding products' and their condition descriptions and correctly evaluating their economic values. This will result in a better price normalization (Step 6 of competitive analysis). Another research topic is the employment of more elaborate machine-learning techniques [15, 16] for updating the sales-rate surface on the fly.

The main contributions of our paper is the mathematical formulation of the "experience" and "black-box" art of dynamic pricing, and the introduction of benchmark prices that reflect the human mindset and web-page specific characteristics of online stores' primary user interface. Our dynamic pricing algorithm will be particularly more useful for markets with 1) small to medium sales volumes, 2) items with a variety of qualities and conditions resulting in frequent changes in competitors' price statistics, and 3) rapid product advances mandating frequent re-pricing such as the high-tech electronic industry. The dynamic pricing has not quite become a mainstream pricing strategy yet due to the complexity of integrating it with the enterprise resource planning (ERP) software, e.g., the accounting system has to get the correct price for each particular instance of the item, the customer, and the time of sales for each transaction [3]. We believe, however, that the dynamic pricing will become a must-have tool and not a luxury item for running a business in the near future.

## APPENDIX

There are extensive literatures on minimum-error curve or surface fitting. Substituting $\sigma = Bp$ into Eqn. (2) yields

$$s \equiv SalesRate = \frac{A}{p}\frac{2}{Bp\sqrt{2\pi}}e^{-t^2/(2(Bp)^2)} \quad (3)$$

We take natural log of both sides to get

$$\ln s = \ln(\frac{A}{p^2}) + \frac{1}{2}\ln(\frac{2}{\pi}) - \ln B - \frac{t^2}{2p^2}\frac{1}{B^2} \quad (4)$$

which is linear in *lnA* but nonlinear in *B*. Given the data set of the triples (*p, t, s*), any one of the numerical surface-fitting methods, can be used to compute *A* and *B* with a minimum error measure of user's choice such as the least square error [17-21].

## ACKNOWLEDGMENT

Samuel Hwang would like to thank Dr. Sungho Kim for providing Samuel with an opportunity and guiding him in his participation in a state-of-the-art project on the internet research frontier.

## REFERENCES

[1] Elizabeth Millard, "Dynamic Pricing for E-Commerce," www.EcommerceTimes.com, Part of the ECT News Network, August 6, 2003.

[2] -, Amazon.com Varies Prices of Identical Items for Test, Wall St. J., Sept. 7, 2000, at B19.

[3] Robert M. Weiss and Ajay K. Mehrotra, "Online Dynamic Pricing: Efficiency, Equity and the Future of E-commerce," *Virginia Journal of Law and Technology*, Summer 2001.

[4] Thomas T. Nagle, Reed K. Holden and Reed Holden, *The Strategy and Tactics of Pricing: A Guide to Profitable Decision Making,3rd Edition,* Upper Saddle River, New Jersey: Prentice Hall, 2002.

[5] Morris Engelson, *Pricing Strategy: An Interdisciplinary Approach*, London, England: Joint Management Strategy, 1995.

[6] -, "Dynamic pricing overview," http://www.managingchange.com/dynamic/overview.htm

[7] -, Pricing strategies, http://www.marketingteacher.com/Lessons/lesson_pricing.htm, Sept. 2005.

[8] Malaney Smith, "Perplexing Pricing Strategies," http://www.clickz.com/experts/crm/analyze_data/article.php/1582081, Feb. 2003.

[9] -, Pricing objectives, http://en.wikipedia.org/wiki/Pricing_objectives, Sept 2005.

[10] Joel R. Evans and Barry Berman, "Pricing and small retailers: questions to consider," http://retailindustry.about.com/library/uc/be/uc_be_pricing1.htm

[11] Ralph F. Wilson, E-Commerce Consultant, "P4: Pricing Strategy as Part of Your Internet Marketing Plan," Web Marketing Today, May 9, 2000.

[12] -, http://www.netpricing-solutions.com/

[13] -, Search engine ratings and reviews, http://searchenginewatch.com/reports/article.php/2156451

[14] -, Search engine user attitude, http://searchenginewatch.com/searchday/article.php/3357771

[15] Tom M. Mitchell, *Machine Learning,* New York, NY: McGraw Hill, 1997.

[16] Stuart J. Russell and Peter Norvig, *Artificial Intelligence: A Modern Approach, 2nd Edition,* New York, New Jersey: Prentice Hall, 2002.

[17] Charles L. Lawson, Richard J. Hanson, *Solving Least Squares Problems (Classics in Applied Mathematics, No 15),* Philadelphia, PA: Society for Industrial & Applied Mathematics 1995.

[18] Ake Bjorck, *Numerical Methods for Least Squares Problems*, Philadelphia, PA: Soc for Industrial & Applied Mathematics, 1996.

[19] Harvey Motulsky, Arthur Christopoulos, *Fitting Models to Biological Data Using Linear and Nonlinear Regression: A Practical Guide to Curve Fitting*, New York, NY: Oxford University Press, 2004.

[20] Albert Cohen (Editor), et al, Curve *and Surface Fitting: Saint-Malo 2002*, Brentwood, TN: Nashboro Pr Inc., 2003.

[21] Sung Joon Ahn, *Least Squares Orthogonal Distance Fitting of Curves and Surfaces in Space (Lecture Notes in Computer Science), 1 edition*, New York, NY: Springer, 2005.

# Runtime support for Self-Evolving Software

cisterni@di.unipi.it
Dipartimento di Informatica
L.go Bruno Pontecorvo, 3
I-56127 Pisa – Italy

ambriola@di.unipi.it
Dipartimento di Informatica
L.go Bruno Pontecorvo, 3
I-56127 Pisa - Italy

***Abstract* - Software development and deployment is based on a well established two stage model: program development and program execution. Since the beginning of software development, meta-programming has proven to be useful in several applications. Even though interpreted languages have successfully exploited the ability of generating programs, meta-programming is still far from being main stream. The two stage model is largely due to the computing metaphors adopted for defining programs, and deeply entrenched within the whole software management infrastructure. In this paper we explore how a runtime, multi-staged, self-contained, meta-programming system based on the notion of partial application can provide a suitable support for programs capable of evolve along time. To make the discussion more concrete we discuss two scenarios where the suggested infrastructure is used: software installation and robot control by means of programs embedding knowledge into their code rather than into data structures.**

## I. INTRODUCTION

Software development is facing new challenges due to the [illegible] dimension of source code. Time is mature for a leap comparable to the move from machine language to structured programming. Programming languages, despite their increasing expressivity, seem unable to dominate the complexity of the programs being developed; this trend is also strengthened by the need for generic software, capable of adapting to users and different operating environments.

The generative programming approach [1] suggests that a possible way to dominate software complexity without loosing performance is by raising the abstraction level, shifting the focus from writing programs to write programs generating other programs. This approach has been successfully exploited in several domains and is quickly becoming part of the mainstream programming.

A fundamental issue in having programs specializing themselves into other programs is that the computation model does not fit the current software deployment model. This is especially true for compiled languages, because at runtime a program is expressed in a different language, and it relies on different abstractions than those provided by the programming language used to its development.

Programming languages are trying to provide support for this kind of manipulation in several ways: C++ template metaprogramming [1, 2] is slowly becoming part of the language; Aspect Oriented Programming systems focus on specific program transformations; MetaML [3, 4], MetaOCaml [5], or Jumbo [6] are interested in studying a suitable notion of staged computation inside programming languages. These attempts are all in the right direction, though they focus on different aspects of the problem, with little effort in providing a complete programming infrastructure capable of providing the appropriate support to program manipulation.

The advent of virtual machines, like Sun JVM and Microsoft CLR, have improved the state of the art by proposing a mixed solution where the compilation model is greatly enriched by a significant set of dynamic services available at runtime. In these environments it is easier to combine the advantages of compilation and interpretation. Recent advances in computer systems like XML and Web services have benefit from these execution environments.

Web technologies are almost entirely built upon program generation: the Web server is responsible for generating programs that dynamically [illegible]. Web services infrastructure relies on code generation for generating a WSDL and SOAP mapping to classes and methods.

Despite this large field of application for program generation techniques, there is not a well established model and programming system capable of supporting it. The main-stream focuses on a string-based source code generation approach. Although effective, this approach suffers of several problems: programs are difficult to debug; there is very little support for program manipulation provided from the programming toolbox; for compilation based systems program manipulation is not homogeneous because the compiled program generates a source program that should be compiled in order to become executable.

Another approach to program manipulation is based on some form of module selection. In this case the transformation is performed outside of the program object of the transformation; this



*T. Sobh and K. Elleithy (eds.), Advances in Systems, Computing Sciences and Software Engineering*, 157–163.

is the case, for instance, of program configuration, where program modules are chosen from a repository and the program is unaware of this process of selection performed by the installation procedure.

The contribution of the paper is twofold: first we discuss the impact that a runtime with multi-stage and meta-programming support would have on the current software deployment infrastructure; in the second part of the paper, to make the discussion more concrete, we sketch two scenarios that would benefit from the infrastructure.

## II. VISION

Imagine a world where programs are able to define all their transformations in themselves. The entire deployment infrastructure in place today would become obsolete because programs would be free to evolve by changing their structure over time. In the beginning we can simply assume that the newer infrastructure would permit mimicking the current model for software deployment and execution.

All the current stages at which a program gets manipulated could be seen as instances of this general model: installation is simply a matter of specializing a program against a set of install-time known parameters; software adaptation driven by user monitoring involves code specialization rather than data interpretation. The overall infrastructure would be more efficient and scalable. A hint that this approach can be successful has been given by the IronPython experiment has shown how an appropriate use of runtime code generation on a virtual machine may lead to faster code than the standard C interpreted approach [7].

After an initial phase where the current software deployment infrastructure would be mapped onto the new one, there will be a growing interest for the facilities provided by this infrastructure. The mere possibility for a program to change its own structure at runtime may lead to programs more autonomous and capable of evolve over time. We call these *self-evolving programs*, and we believe that this will be a new abstraction level into the computer system infrastructure.

How a program can successfully evolve over time is still to be figured out; we can imagine borrowing models from nature, as it is suggested by Peter Bentley, for instance, in [9]. Computer science has already been inspired by nature, and the genetic programming community [10] is already attempting to find such techniques. However they miss the appropriate infrastructure capable of support their work.

## III. RUNTIME CODE TRANSFORMATIONS

Self evolving programs require a program infrastructure capable of homogeneous program transformations. Unless we assume only interpreted languages, where source to source program transformations are naturally possible because of the *eval* function, we have to face the problem of having a homogenous transformation system within the runtime support of programming languages.

In a compiled scenario homogeneous program transformations should be performed at runtime level: the target program is expressed in object language rather than source language, unless we assume that the whole compilation infrastructure is available at runtime. For instance a C++ program would face machine code manipulation for changing its own structure at runtime. Although feasible, in practice this approach is not viable because it assumes too much knowledge about compilers on the programmer side.

If we raise the abstraction level we find the increasingly popular world of virtual machines of Java and .NET. In this case binary code is expressed in an intermediate form, therefore runtime code generators may avoid machine code generation in favor of byte code generation. This greatly simplifies the code generation task, that however is still based on a language far from the one used by developers. It is a fact that the *Reflection.Emit* library provided by the .NET base class library, supporting byte-code generation at runtime, is not very popular among programmers.

Recent research on meta-programming [3, 4, 6, 15] has focused on how to compile not only programs, but also program transformations. All these approaches are based on some form of generation of pre-compiled fragments that are mixed together by little interpreters at runtime. These transformations are implicitly homogeneous and can be easily iterated allowing the definition of multi-staged computations.

Although this ability of combining code fragments may be restricted to a single programming language's runtime support, it seems that the possibilities would grow richer if programs can combine code fragments at runtime no matter of the programming language used to write them. In this respect Microsoft CLI [8] is a good platform for providing such support because CLR is a runtime shared among several languages, and the runtime is dynamic and rich enough to provide support for code manipulation.

The CodeBricks [11, 12] project has proven that there exists a homogeneous, type-safe, self-contained, program transformation system that allows programs to combine program fragments at runtime. The system allows program evaluation in stages; though at the moment without persistence (all transformations must take place at runtime).

A significant difference from the other approaches is that CodeBricks is focused on providing runtime support for meta-programming languages. Therefore the library defines a general mapping from programming languages to runtime objects that allows programmers to perceive program transformations as they were at language level, rather than at the virtual machine level.

In the rest of this section we introduce the CodeBricks generation model. And we briefly discuss the issue of persisting program changes when the evaluation of a program is done in different stages of execution.

*A. CodeBricks*

The library provides a simple abstraction for code manipulation, based on the well-known abstraction of *partial application*.

**Definition:** Let $T_r\ m(a_1 : T_1, \ldots, a_n : T_n)$ be a function (or a method) with a given signature. A *partial application p* of *m* is an application where only *k* arguments are applied: $T_r\ p(a'_1 : T'_1, \ldots, a'_{n-k} : T'_{n-k}) = m(e_0, \ldots, e_h, a'_1, e_{h+1}, \ldots, e_k, a'_2, e_{k+1}, \ldots, a'_{n-k}, \ldots, e_k)$ where $e_i$ are expressions such that if $e_i$ is in position *j* in the arguments list then $e_i \rightarrow T_j$, assuming $\rightarrow$ to be the typing judgment of the language.

CodeBricks is a code generation system capable of generating code equivalent to partial applications. Therefore methods are bricks that can be composed by means of partial application.

In order to be able to manipulate code, the library provides a way to lift methods into *code bricks*. Partial application is the operation available on code bricks, whose result is a new brick representing the code equivalent to the application. Generated code is made executable by the inverse of *lift* operator, which converts a code brick into a callable function.

In this approach methods correspond to pre-compiled code fragments that are composed at runtime. The system rely on the fact that runtimes like Sun JVM and Microsoft CLR preserve enough information within their binaries that types and methods are still available at runtime.

The library assumes four possible kinds of values that can specified as arguments in the partial application:

1. a special value *free* to indicate that an argument should not be bound;
2. [illegible] values;
3. a code brick returning the value [illegible] compatible with the one expected as argument;
4. a code brick to be used as high order value inside the method

A simple example of CodeBricks application could be the following (expressed in a pseudo-language):

*add(x, y) = x + y;*
*addb = lift(add);*
*incb = addb.apply(1, free);*
*inc = makeExecutable(incb);*
*i = inc(2); // assign 3 to i*
*add3b = addb.apply(addb.apply(free, free), free);*
*add3 = makeExecutable(add3b); // x + y + z*

In [11] it is shown that there exists a code transformation that combines code fragments while preserving the partial application semantics. The code generation strategy is, in its essence, a sort of inlining of bricks with an appropriate renaming of arguments and local variables.

CodeBricks is an object oriented library designed to be part of the base class library of a modern runtime execution environment. Therefore there is no notion of "source language" in the program transformations based on it. Nevertheless there is a natural one on one mapping between the semantic objects used by the library, and those provided by programming languages: therefore programmers will reason on the common subset of concepts that is shared between a programming language and its runtime.

*B. Persistent Program Transformations*

Runtime support for program manipulation like that provided by CodeBricks is necessary for enabling programs to modify themselves. Meta-programming facilities are, however, not sufficient to have programs capable of modifying their own structure: we need some form of persistence of the transformations performed so that the program can terminate without loosing the information stored into the code.

Code persistence can be easily achieved in runtime environments where dynamic loading is available: the program can generate a module that becomes part of the program definition and that will be load and accessed through generic interfaces. However this approach suffers of relevant problems:

- the model is coarse grain: even small changes to a program must be represented as modules (i.e. types or DLLs);
- the approach is not declarative, therefore the programmer must explicitly code module generation and the related configuration
- [illegible] program are not part of the program itself; some form of configuration data, [illegible] part of the program, should keep track of generated modules

It is our belief that a runtime should provide better support to orchestrate this transformation process. For this reason we are currently developing a system capable of loading a program and evaluate portions of it.

The transformation will be driven by annotations present within the executable; these annotations will be in the form of custom attributes: meta-data extensions provided by the .NET runtime. The annotation model provided by the runtime allows decorating only objects accessible through reflection (roughly speaking classes and class members). Using techniques like those presented in [13], it is possible to extend this model to allow annotations on code blocks inside methods. Annotations will be also used to keep track of the transformation used for generating the various code fragments.

The evaluation of a program takes place in distinct stages, each stage characterized by a *label* indicating the *name* of the current stage. At each stage the system loads the program; it looks for annotations with a label matching the name of the stage, and executes a corresponding method that generates a code fragment that will be stored in the resulting executable.

Let us consider the following example written in C#:

```
[InstallTime(Method("InstallDir"))]
public string InstallationDir() {
  return null; // this ensures typesafety
}
//...
Code InstallDir(Context c) {
  // Generates the brick of the constant
  // function specialized with the actual
  // value of the parameter
}
//...
File.Open(InstallationDir() + @"\myf.txt");
```

In this example we use a method to indicate where a string should be placed within the code. When the stage named *InstallTime* is evaluated the *InstallationDir* method should be evaluated. This method simply defines the signature that a code brick returned by the *InstallDir* will have. During evaluation each invocation of the *InstallationDir* method causes the invocation of the corresponding *InstallDir* method, and the resulting brick will replace the method invocation in a type-safe way. Presumably the *InstallDir* method will prompt the user a dialog asking for the installation dir and it will simply generate the brick representing th constant function returning the value provided by the user.

When the execution of the computations marked with the name of the stage is completed, a new program will be generated ready for the next computation stage. The whole transformation process is based on the CodeBricks library.

## IV. SELF-EVOLVING PROGRAMS

A program may want to change its own definition mainly for two reasons:

1. to improve its performance by specializing itself once part of its input become known;
2. to adapt its behavior to a particular environment

Several decades of research in the field of partial evaluation and program specialization make the first motivation worthless of further discussion. Besides, the second motivation is less evident; a program may adapt, and usually do, by producing an appropriate set of data structures that are interpreted to determine its behavior. We have plenty of these examples in our everyday life: program configuration provides the user with the opportunity to decide how a program should behave.

Autonomous programs, like software agents, may have to cope with very complex data structures in order to adapt their behavior to very dynamic environments. A self-evolving program may reduce this complexity by storing part of the acquired knowledge into its code.

A *self-evolving* program is a meta-program which is capable of rewriting its own structure to evolve its behavior along time. Program manipulation is performed by mixing existing code fragments (usually present within the program itself) to derive new behavior or improving the existing ones.

This class of meta-programs is very relevant because it allows describing the current software deployment infrastructure in a single and coherent framework. Nowadays programs pass through a number of systems responsible for configuration: domain specific generators, setup generators, installers, are but few examples. The bad thing is that the program is unaware of most of the transformations performed on it; this greatly affects the maintainability of the software, introducing several issues in the configuration of programs. With respect to this particular domain, recent systems have introduced some embryonic form of evolution by adding the check for newer version of the program on a Web site at startup time; if the new version is available the program exits and the newer version is installed.

Self-evolving software paradigm permits to define the whole lifetime of a program within the program itself. Applications would obtain a homogenous framework for describing how the program should change over time, without having to resort to some external program that manipulates the application from outside.

## V. CASE STUDY: PROGRAM CONFIGURATION

Program configuration is a relevant area of software development, deployment and maintenance. In its most abstract form, a configuration system defines a set of parameters determining the code to be executed by the configured program. Usually parameters are defined at different times in the program's lifetime.

In the actual software infrastructure configuration is handled by several systems in the different moment of program life cycle. At compile time code selection is driven by the value of symbols and coordinated by a build system like *make*.

At install time there often is a further selection of code performed by module selection. In this case compiled units are copied in the installation location depending on the preferences expressed by the user.

The rest of the configuration is usually provided as a set of values stored in some form of database. In the Windows system, for instance, most of the configuration is stored in the system registry. These values are retrieved by the program through a set of methods incapsulating the access to the configuration database.

A self-evolving program can perform code selection by synthesizing the appropriate code once the value of the parameter controlling the selection is known.

The program do whatever is needed for obtaining the value of the parameter (perhaps by prompting the user with a dialog) and then generates the required code.

Let us consider the problem of selecting an algorithm depending on the fact that the system has single or multiple processors. Using our runtime this can be expressed using the following schema:

```
void AlgorithmSP() {
  // Single processor version
}

void AlgorithmSMP() {
  // Symmetric multiprocessor code
}

[InstallTime(Method("SelectAlgorithm")]
void Algorithm() {
  // Do nothing: it's just a placeholder
}

Code SelectAlgorithm(Context c) {
  return ProcessorNumber() > 1 ?
    new Code(Method("AlgorithmSMP") :
    new Code(Method("AlgorithmSP");
}
```

In this case we select between two alternatives, though it is easy to imagine how to perform other kinds of code selection.

Using the same approach, as already discussed in section 3.2, we can insert into the target program the values instead having to store them into an external database (though in some cases it can make sense to still use it). In this case we still use the method responsible for retrieving the value, though this method will simply act as a placeholder that will be replaced with the parameter value.

Stage names allow determining when configuration parameters must be set. At the end of the process the resulting program will be specialized to the desired configuration.

Software configuration can be expressed as a simple program evolution. Using simple transformations it is possible to express code and parameter selection without having to rely on external tools. The main advantage of our approach is evident: the whole program definition is self-contained, rather being spread across several files and systems. This will simplify the maintenance process because we get rid of a number of tools; moreover when the program changes it is easier to ensure that the transformations remain consistent with it.

## VI. CASE STUDY: ADAPTIVE ROBOT CONTROL

In the previous section we have discussed how self-evolving software infrastructure supports the configuration problem in a cleaner and simpler way than the current one. Now we would like to consider a more challenging scenario where a program represents the knowledge in the form of code rather than data structures.

Software controlling autonomous robots should adapt the robot behavior to a particular environment. The robot should keep information about its past experience in the environment in order to adapt its behavior to be more effective.

A very abstract way to think of the control software of a robot is the perception/action loop as discussed, for instance, in AIMA by Russel and Norvig [14]:

*while (true) {*
*p = ReadPerception();*
*a = SelectAction(p, $a_1$, ..., $a_n$);*
*Do(a);*
*}*

In a robot actions $a_i$ are commands for the actuators. The *SelectAction* method determines the behavior of the robot, and it is the method that should adjust its behavior as time passes.

In a robot the actions available are very small, and even simple behaviors require many of them. Therefore it is not infrequent that the control software issues the same sequence more than once. There are also techniques, like hierarchical planning, that tend to build upon these sequences more abstract actions. In this case the program should book-keep in some data structure these aggregate actions.

A self-evolving program can rely on the ability of combining code at runtime for generating new actions by composition of existing ones. Let us suppose that the sequence $a_3$, $a_1$, $a_7$ is frequently applied by the robot. We can synthesize a new operation as follows:

*a3b = lift($a_3$);*
*a1b = lift($a_1$);*

*a7b = lift($a_7$);*
*pairb = lift(ActionSequence);*
*na = pairb.apply(pairb.apply(a3b, a1b), a7b);*
*$a_{n+1}$ = makeExecutable(na);*

The *ActionSequence* method, as suggested by its name, simply applies two actions in sequence. We partial apply the sequence brick *pairb* using the action bricks as arguments. At the end of the composition we make the new action executable.

Once the control program has generated a new action, the application can select among $n+1$ actions at each step, forgetting that $a_{n+1}$ is just a composition of pre-existing operators. This saves the application from having to keep track that $a_{n+1}$ is just the sequence of the three actions $a_7$, $a_1$, and $a_3$.

The ability of synthesizing new actions may help the program to simplify the data structure needed to make decisions. Moreover the procedure can be iterated, allowing the robot to generate new operators combining either basic or derived actions.

Although it is convenient to treat robot operators in the same way, no matter if they are primitive or obtained by composition, there are situations where the program may need to know the true nature of an operator.

In order to be able to provide this information the composition used to generate a given operator is stored inside the generated code in the form of custom annotations. As it is done for annotated C# [13], the reflection API will be used to retrieve the structure, in terms of partial applications.

We are currently implementing the control software for an experimental robotic platform, built at our department, in order to prove the effectiveness of self-evolving software paradigm in this application domain.

## VII. PROGRAMMER'S PERSPECTIVE

So far we have discussed self-evolving software from the infrastructure perspective. But what a programmer should do to use the facilities provided by this richer infrastructure? The runtime exposes its facilities through a set of API, however it is plain that the syntax is rather clumsy and not really effective. We expect that programming languages will rely on the infrastructure provided by the runtime to compile programs with meta-programming elements.

It is also worth notice that self-evolving programs stress a problem that is already present in standard software: when a program adapts its behavior over time, the experience acquired cannot be neglected; therefore a program is not just a functional piece of software anymore. The problem of how to preserve the knowledge retained by a program is important because otherwise simple operations like installation may lead to an unwanted loss of data.

## VIII. CONCLUSIONS

The idea of self-evolving programs is indeed a fascinating one. We believe it is also a useful one: under this broader model of computation falls the traditional one. Therefore the existing software infrastructure may greatly benefit from a model in which existing tasks can be performed in an easiest and cost-effective way. Nevertheless the proposed model would make easier to explore new models of program development based on program transformation.

The ability of transforming programs by combining fragments of existing ones strongly remind natural process, based on the recombination of existing elements. Thus this new software infrastructure will encourage further investigation of new computation models based on programs that evolve over time. The infrastructure will also be validated against application domain. Robotics, for instance, is an application domain where the self-evolving programs model may contribute to realize control programs capable of adapt to the surrounding environment.

Microsoft .NET seems good platform to be used to realize this software infrastructure: it is a standard platform, targeted by several programming languages; moreover the availability of fundamental mechanisms as customizable reflection provides the appropriate support for realizing the transformation system.

## REFERENCES

[1] Czarnecki, K., Eisenecker, "U.W.: Generative Programming – Methods, Tools, and Applications." Addison Wesley, Reading, MA (2000)

[2] G. Attardi, A. Cisternino, "Self Reflection for Adaptive Programming", Proceedings of Generative Programming and Component Engineering Conference (GPCE), 50-65, LNCS 2487, October 6-8, 2002

[3] Taha, W., "Multistage Programming: Its Theory and Applications", PhD thesis, Oregon Graduate Institute of Science and Technology, July 1999. Available at ftp://cse.ogi.edu/pub/tech-reports/1999/99-TH-002.ps.gz.

[4] Taha, W., Sheard, T. "Multi-stage programming with explicit annotations." In Proceedings of the ACMSIGPLAN Symposium on Partial Evaluation and semantic based program manipulations PEPM' 97, Amsterdam, p. 203-217. ACM, 1997.

[5] http://www.metaocaml.org, MetaOCaml Web site

[6] Kamin, S., Clausen, L., Jarvis, A., "Jumbo: run-time code generation for Java and its applications", Proceedings of the international symposium on Code generation and optimization: feedback-directed and runtime optimization

[7] Hugunin, J., "IronPython: A fast Python implementation for .NET and Mono", PyCON March 24, 2004, available at http://www.python.org/pycon/dc2004/papers/9/.

[8] ECMA 335, "Common Language Infrastructure (CLI)", http://www.ecma.ch/ecma1/STAND/ecma-335.htm.

[9] "The garden where perfect software grows", NewScientist, 6 March, 2004, available at http://www.cs.ucl.ac.uk/staff/p.bentley/seedsarticle.html.

[10] Genetic programming community, http://www.genetic-programming.org/

[11] Cisternino, A. "Multi-stage and Meta-programming support in strongly typed execution engines." PhD Thesis, TD-5/03, Dipartimento di Informatica, Università di Pisa, May 2003, available at http://www.di.unipi.it/phd/tesi/tesi_2003/PhDthesis_Cisternino.ps.gz.

[12] Attardi, G., Cisternino, A., Kennedy, A., "Code Bricks: Code Fragments as Building Blocks", in proceedings of 2003 SIGPLAN Workshop on Partial Evaluation and Semantic-Based Program Manipulation (PEPM), 66-74, San Diego, CA, USA, 2003.

[13] Cazzola, W., Cisternino, A., Colombo, D., "[a]C#: C# with a Customizable Code Annotation Mechanism", Proceedings of Symposium of Applied Computing, Santa Fe (NM), March 2005

[14] Consel, C., Noël, F., "A general approach for run-time specialization and its application to C", In POPL'96: 23rd Principles of Programming Languages, St. Petersburg Beach, Florida, January 1996, pages 145–156, 1996.

[15] Russle, S. J., and Norvig, P., "Artificial Intelligence. A Modern Approach" (AIMA), Prentice-Hall, Englewood, 1995.

# A Mobile Location Algorithm Using Clustering Technique for NLoS Environments

Cha-Hwa Lin and Juin-Yi Cheng
Department of Computer Science and Engineering
National Sun Yat-sen University
Kaohsiung, 80424, Taiwan

***Abstract*—For the mass demands of wireless communication application services, the mobile location technologies have drawn much attention of the governments, academia, and industries around the world. In wireless communication, one of the main problems facing accurate location is nonline of sight (NLoS) propagation. To solve the problem, we present a new location algorithm with clustering technology by utilizing the geometrical feature of cell layout, time of arrival (ToA) range measurements, and three base stations. The mobile location is estimated by solving the optimal solution of the objective function based on the high density cluster. Simulations study was conducted to evaluate the performance of the algorithm for different NLoS error distributions and various upper bound of NLoS error. The results of our experiments demonstrate that the proposed algorithm is significantly more effective in location accuracy than linear line of position algorithm and Taylor series algorithm, and also satisfies the location accuracy demand of E-911.**

***Keywords-Mobile location, Nonline of sight (NLoS), Clustering, Time of arrival (ToA).***

## I. INTRODUCTION

The location estimation of a mobile station (MS) in a wireless system has drawn much attention of the researchers in recent years, especially since the U.S. Federal Communication Commission (FCC) mandated cellular providers to generate location estimates for Enhanced-911 (E-911) services [1]. The accuracy requirement of E-911 for phase II is 100m for 67% of the time and 300m for 95% of the time for network-based location system. Many location-based applications include fleet management, package and personnel tracking, location-sensitive billing, intelligent transportation system applications, and so on, are relied on the location technology.

The most widely employed location technology is radio location system that attempts to locate an MS by measuring information between the MS and a set of base stations (BSs). Radio location system can be based on signal strength, angle of arrival (AoA), time of arrival (ToA), or time difference of arrival (TDoA), and can be network or terminal based [2][3]. The accuracy of mobile location schemes depend on the propagation conditions of wireless channels. If line of sight (LoS) propagation exists between the MS and all BSs, a high location accuracy can be achieved. However, the direct path from the MS to a BS is always blocked by buildings and other obstacles in wireless communication systems, so the signal measurements include an error due to the excess path length traveled because of reflection or diffraction, which is termed the NLoS error [4]. Fig. 1 shows the NLoS propagation

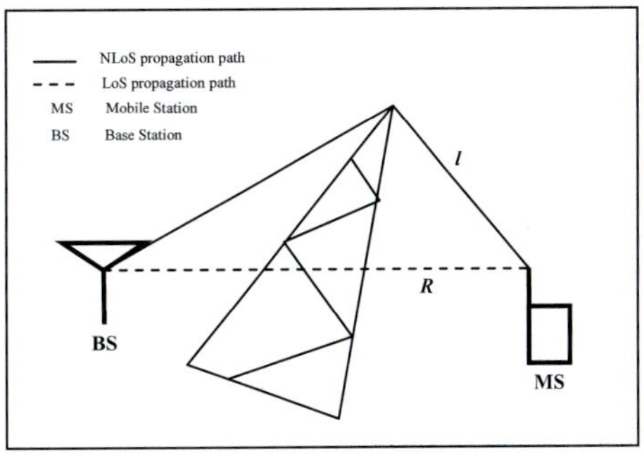


Fig. 1. NLoS propagation condition.

between the MS and a BS. The measured distance form MS to BS is denoted as $l$ and the true distance is $R$, which can be expressed as

$$l = R + \text{NLoS error}$$

The NLoS error is quite common in all environments, except for rural areas, and causes considerable degradation of location accuracy [5]. Hence many location algorithms focus on identifying and mitigating the NLoS error. Some of these approaches require certain assumptions to be satisfied in order for them to be applicable. NLoS mitigation techniques [4] [6] require distance measurements from MS in a time series and are based on the assumption that the standard deviation of NLoS distance measurements is greater than that for LoS measurements. These techniques may be useful when a mixed measurements set including both LoS and NLoS measurements in the time series. However these algorithms cannot be expected to be effective when the signal propagation is only through NLoS path. The algorithm in [7] uses signal measurements at a set of participating BSs and weighting techniques to mitigate NLoS effects. But if the propagations at all BSs are NLoS, this algorithm cannot improve the location accuracy.

A new linear line of position method (LLoP) is presented in [8] which make it easier estimating the unknown MS location than traditional geometrical approach calculating the intersection of the circular lines of position (LoP) [9]. LLoP algorithm can mitigate the NLoS error as well as the measurement noise, but it needs at least four BSs to achieve better location accuracy, and its performance highly depends on the relative position of MS and BSs. Recently, a constrained



*T. Sobh and K. Elleithy (eds.), Advances in Systems, Computing Sciences and Software Engineering*, 165–169.

optimization approach is presented in [10] named RSA that utilizes bounds on the NLoS rang error inferred from the geometry of the cell layout and range circles for three BSs. RSA indeed improved the location accuracy in the NLoS environment, but cannot satisfy FCC E-911 demand.

We propose a new location algorithm named density-based clustering algorithm (DCA), which only needs three range measurements, and does not discriminate between NLoS and LoS BSs. DCA analyses the geometrical feature of cell layout, time of arrival (ToA) range measurements, and three base stations, and estimates the mobile location by solving the optimal solution of the objective function modified by the high density cluster. DCA location algorithm can apply to any system based on ToA range measurements or ranging techniques which measured range is greater than true range. The location accuracy of DCA has been greatly improved over previous algorithms.

The remainder of this paper is organized as follows. The proposed algorithm is outlined in Section II. The simulation and result are discussed in Section III, followed by some concluding remarks in Section IV.

## II. Density-Based Clustering Algorithm (DCA)

For the range measurements using ToA method at the MS from the *i*th BS, the range equation can be expressed as

$$R_i = \sqrt{(x-x_i)^2 + (y-y_i)^2}\ , i = 1, 2, 3 \tag{1}$$

where $(x_i, y_i)$ is the *i*th BS location and $(x, y)$ is the true location of MS. Let $l_i$ and $R_i$ indicate the measured ranges and true ranges, respectively. Then $R_i$ can be written in terms of $l_i$ as

$$R_i = \delta_i l_i \tag{2}$$

Because the NLoS error is positive, the measured range is greater than the true range. Hence, the value of $\delta_i$ is bounded by $0 < \delta_i \le 1$. If and only if there is no NLoS error, $\delta_i$ equals 1. It is assumed that the measurement noise is a zero-mean Gaussian random distribution with relatively small standard deviation and is negligible as compared to the NLoS error.

Squaring the range in (1) and substituting (2) results in

$$(x-x_i)^2 + (y-y_i)^2 = \delta_i^2 l_i^2\ , i = 1, 2, 3 \tag{3}$$

For simplifying equation, we define

$$[\alpha, \beta, \gamma] = [\delta_1^2, \delta_2^2, \delta_3^2],\ 0 < \alpha, \beta, \gamma \le 1 \tag{4}$$

The proposed DCA algorithm utilizes the circle equations and the boundaries of α, β, and γ to calculate all the possible locations of MS, defined as candidate points (CPs). The optimal MS location is estimated by applying an objective function which is modified by the high density cluster of CPs.

### A. Calculation of Candidate Points

The measured ranges by ToA method expressed as (3) and (4) are circle equations as follows.

$$(x-x_1)^2 + (y-y_1)^2 = \alpha\ l_1^2 \tag{5}$$

$$(x-x_2)^2 + (y-y_2)^2 = \beta\ l_2^2 \tag{6}$$

$$(x-x_3)^2 + (y-y_3)^2 = \gamma\ l_3^2 \tag{7}$$

Subtracting (6) and (7) from (5), we can respectively obtain the linear equations of $L_{12}$ and $L_{13}$, as Fig. 2 showed. Calculating the intersection point (*x*, *y*) of these two linear equations results in

$$x = A\alpha + B\beta + C\gamma + D \tag{8}$$

$$y = A'\alpha + B'\beta + C'\gamma + D' \tag{9}$$

We define

$$k = 2\ [(x_2 - x_1)(y_3 - y_1) - (x_3 - x_1)(y_2 - y_1)]$$

$$k' = 2\ [(y_2 - y_1)(x_3 - x_1) - (y_3 - y_1)(x_2 - x_1)]$$

and

$$A = \frac{(y_3 - y_2)\ l_1^2}{k}$$

$$B = \frac{(y_1 - y_3)\ l_2^2}{k}$$

$$C = \frac{(y_2 - y_1)\ l_3^2}{k}$$

$$D = \frac{(y_2 - y_3)(x_1^2 + y_1^2) + (y_3 - y_1)(x_2^2 + y_2^2) + (y_1 - y_2)(x_3^2 + y_3^2)}{k}$$

$$A' = \frac{(x_3 - x_2)\ l_1^2}{k'}$$

$$B' = \frac{(x_1 - x_3)\ l_2^2}{k'}$$

$$C' = \frac{(x_2 - x_1)\ l_3^2}{k'}$$

$$D' = \frac{(x_2 - x_3)(x_1^2 + y_1^2) + (x_3 - x_1)(x_2^2 + y_2^2) + (x_1 - x_2)(x_3^2 + y_3^2)}{k'}$$

If the intersection point of $L_{12}$ and $L_{13}$ linear equations is on the circle formed by $BS_i$ and $\delta_i l_i$ (*i*=1, 2, 3), we define this point as candidate point (CP). The intersection of the three dotted line circles is the possible location of MS, as shown in Fig. 2.

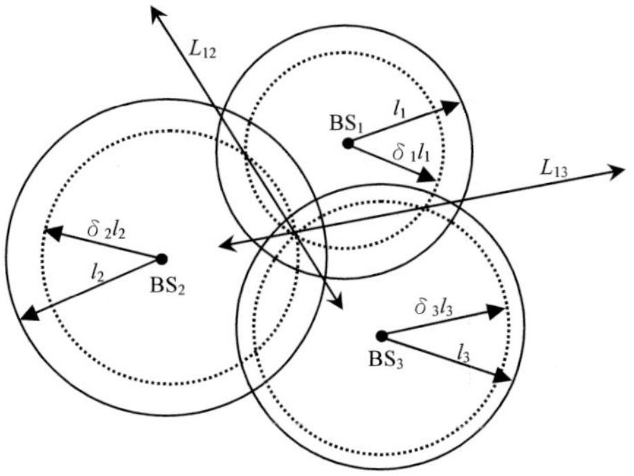


Fig. 2. The geometry of the three circular equations formed by $BS_i$ and $\delta_i l_i$ ($i$=1, 2, 3).

## B. *Boundaries of α, β, and γ*

When the virtual circles (as the three dotted line circles in Fig. 2) are too small to intersect, the $(x, y)$ calculated by (8) and (9) is no significance. In order to decrease the computing time of candidate points, we bound α, β, and γ to omit the nonsignificant calculations. The technique of finding the boundaries of α, β, and γ adopted here is identical to the method outlined in [10].

Because the NLoS error is always positive, the measured ranges are greater than the true ranges and the MS location must lie in the region of overlap of the range circles (region enclosed by U, V, W) as shown in Fig. 3.

Let the NLoS range error of the $BS_i$ be $\eta_i$. Assuming the measured range of $BS_2$ is LoS, it can be seen that if the true range from $BS_1$, namely, $R_1$ is less than $l_1 - \overline{AB}$, then the true range circles of $BS_1$ and $BS_2$ will not overlap or intersect. But the true range circles should intersect at the MS location, which is impossible, and $\eta_1$, the NLoS error of $BS_1$, cannot be larger than $\overline{AB}$. Applying the same argument to the ranges from $BS_2$ and $BS_3$, the value of $\eta_1$ cannot be larger than $\overline{EF}$. Thus, the upper bound of $\eta_1$ is

$$\max \eta_1 = \min\{\overline{AB}, \overline{EF}\}$$

Similarly, the upper bounds of $\eta_2$ and $\eta_3$ are

$$\max \eta_2 = \min\{\overline{AB}, \overline{CD}\}$$

and

$$\max \eta_3 = \min\{\overline{CD}, \overline{EF}\}$$

We know that $\eta_1 = l_1 - R_1 = (1-\delta_1)\, l_1$ so that $\delta_1 = 1 - \eta_1 / l_1$. Thus the minimum value that $\delta_1$ can take is given by

$$\min \delta_1 = 1 - \frac{\max \eta_1}{l_1} = 1 - \frac{\min\{\overline{AB}, \overline{EF}\}}{l_1}$$

From Fig. 2, we can see that

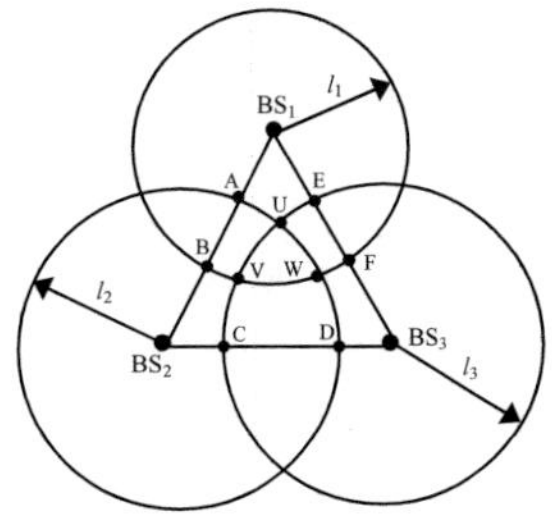


Fig. 3. Geometry of ToA-based location showing measured range circles and the region of overlap in which the MS lies.

$$\overline{AB} = l_1 + l_2 - L_{12}$$

$$\overline{CD} = l_2 + l_3 - L_{23}$$

$$\overline{EF} = l_1 + l_3 - L_{13}$$

Since $\overline{AB}$, $\overline{CD}$ and $\overline{EF}$ are positive, $\min\delta_1$ can be written as

$$\min \delta_1 = 1 - \frac{\min\{\overline{AB}, \overline{EF}\}}{l_1}$$

$$= 1 - \frac{\min\{l_1 + l_2 - L_{12}, l_1 + l_3 - L_{13}\}}{l_1}$$

$$= \max\left\{\frac{L_{12} - l_2}{l_1}, \frac{L_{13} - l_3}{l_1}\right\}$$

Similarly, the lower bounds of $\delta_2$ and $\delta_3$ also can be written as

$$\min \delta_2 = \max\left\{\frac{L_{12} - l_1}{l_2}, \frac{L_{23} - l_3}{l_2}\right\}$$

$$\min \delta_3 = \max\left\{\frac{L_{13} - l_1}{l_3}, \frac{L_{23} - l_2}{l_3}\right\}$$

When the range of the MS from a serving BS is small, it is possible that the NLoS error is large enough that the range circle of the serving BS lies fully within the range circle of the other BS, as illustrated in Fig. 4. In this scenario, $\delta_i$ calculated by the previous equations may be negative. So the equations are modified as

$$\min \delta_1 = \max\left\{\frac{L_{12} - l_2}{l_1}, \frac{L_{13} - l_3}{l_1}, \rho\right\}$$

$$\min \delta_2 = \max\left\{\frac{L_{12} - l_1}{l_2}, \frac{L_{23} - l_3}{l_2}, \rho\right\}$$

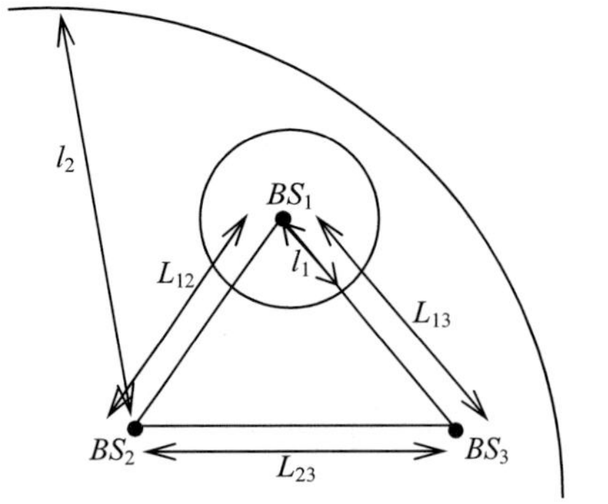


Fig. 4. Range circles of $BS_1$ and $BS_2$ that do not intersect.

$$\min \delta_3 = \max\left\{ \frac{L_{13}-l_1}{l_3}, \frac{L_{23}-l_2}{l_3}, \rho \right\}$$

where $0 < \rho << 0.1$, and $0 < \delta_i \le 1$. Therefore, the lower bounds of α, β and γ can be expressed as

$$[\min \alpha, \min \beta, \min \gamma] = [\min \delta_1^2, \min \delta_2^2, \min \delta_3^2].$$

### C. *Modification of The Objective Function*

The location estimation problem can be formulated as a nonlinear optimization problem. The cost function to be minimized is taken to be the sum of the square of the distances from the MS location to the points of intersections of the range circles closest to it (i.e., points U, V, and W in Fig. 3). The coordinates of U, V, and W are ($U_x$, $U_y$), ($V_x$, $V_y$), and ($W_x$, $W_y$) , respectively. The objective function to be minimized for the nonlinear optimization problem is, therefore [10]

$$f(x,y) = (x-U_x)^2 + (y-U_y)^2 + (x-V_x)^2 + (y-V_y)^2 + (x-W_x)^2 + (y-W_y)^2 \qquad (10)$$

The candidate points are all in the region enclosed by U, V, and W. However, these points are usually not uniformly distributed. The optimal MS location would probably more likely to appear in a more density distributed subregion within the region U, V, and W. This desired subregion enclosed by U', V', and W' (as shown in Fig. 5) is determined by the density-based clustering algorithm. Assume that the set of all candidate points for MS is N, and the set of candidate points in subregion is M. The DCA algorithm first randomly assigns |M| candidate points to Cluster-A, where |M| is the size of M candidate points. The rest |N|-|M| candidate points would be assigned to Cluster-B. The area formed by the candidate points of Cluster-A is then calculated. If a candidate point P in Cluster-A replaced by a candidate point Q in Cluster-B would narrow the area of Cluster-A, then P and Q are swapped to form new Cluster-A and Cluster-B. Repeat this process until the area of Cluster-A could no longer be narrowed. Thus, Cluster-A forms the smallest subregion which would have |M| candidate points, as Fig. 5 shows the region enclosed by U', V', and W'.

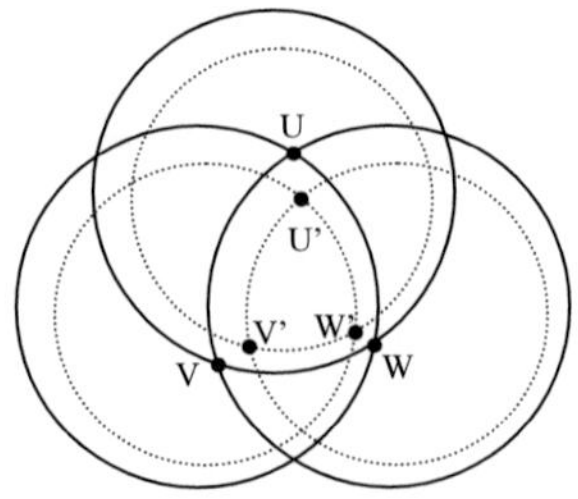


Fig. 5. The overlapping region of the three dotted line circles is the smallest region with most dense candidate points by clustering.

The coordinates of U', V' and W' are ($U_x$', $U_y$'), ($V_x$', $V_y$') and ($W_x$', $W_y$'), respectively. The objective function is then modified as

$$f(x,y) = (x-U'_x)^2 + (y-U'_y)^2 + (x-V'_x)^2 + (y-V'_y)^2 + (x-W'_x)^2 + (y-W'_y)^2 \qquad (11)$$

### D. *Estimation of The Mobile Location*

Since the candidate points are all the possible locations of a mobile station, the location estimation problem can be formulated as a nonlinear optimization problem. The optimal candidate point which minimizes the objective function (11) is the mobile location estimated by the proposed DCA algorithm.

## III. Performance And Simulation Results

We use the cell layout as shown in Fig. 3 to examine the performance of the DCA location algorithm. The MS location is uniformly distributed within the area covered by the triangle formed by points $BS_1$, $BS_2$, and $BS_3$. The reason for restricting the MS location in this area is that the range measurements are considered only from the three nearest BSs. If the MS location is outside this area, a neighboring BS, not in the set considered, will be closer. All the numerical quantities are expressed in kilometers. The NLoS range errors are modeled as positive random variables having support over [0, 0.4] km, generated according to different probability density functions, such as CDSM [11] model, uniform distribution, and normal distribution.

Simulations are performed for cellular environments with cells of radii 1 km. The coordinates of the three BSs are $BS_1$: (0, 0), $BS_2$: (0.866, 1.5), and $BS_3$: (1.732, 0). The performance of DCA location algorithm are compared to LLoP algorithm [8] and the Taylor series algorithm (TSA) [9].

Fig. 6 shows the cumulative distribution functions (CDFs) of the average location error of the algorithm when the range errors are generated using CDSM model. The location error of DCA is less than 0.2 km for 98% of the time. Whereas LLoP and TSA are for 70 % and 50 % of the time for the same location error, respectively. It can be seen that the DCA performs better than both LLoP and TSA for the error model considered. Note that DCA performs well regardless of the NLoS error distribution as shown in Fig. 7, where the CDFs for DCA are nearly identical whether the probability of large NLoS error is high or low.

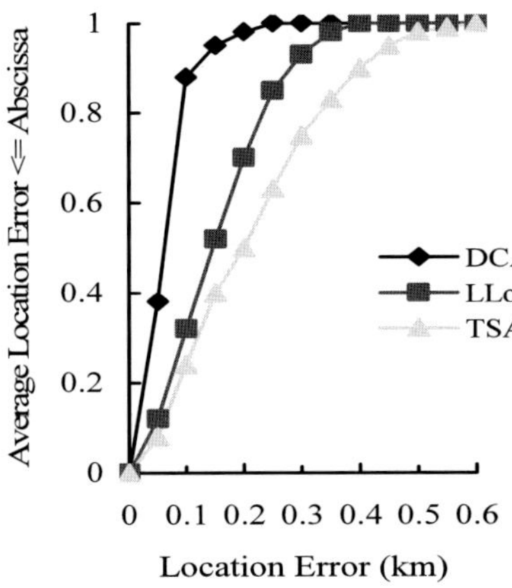


Fig. 6. The CDF of location error of the DCA, LLoP, TSA algorithms on CDSM NLoS model.

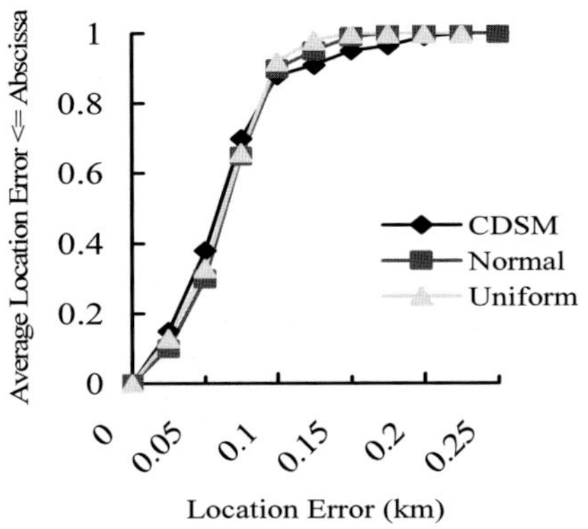


Fig. 7. The CDF of location error of the DCA on CDSM, uniform, and normal NLoS models.

Fig. 8 shows the average location error versus the upper bound on uniform NLoS error. This simulation is performed to examine how the proposed algorithm compared with the other algorithms when the upper bound of NLoS error is varied. It can be seen that the sensitivity of the DCA increases in maximum NLoS magnitude is much less than that for the LLoP and TSA.

## IV. CONCLUSION

In this paper, a new location algorithm using clustering technique is presented for location estimation using range measurements from only three BSs in the NLoS environments. DCA location algorithm utilizes the geometrical feature of cell layout, ToA range measurements, and three base stations without requiring discrimination between LoS and NLoS. Simulation results show that the location accuracy of DCA is much better than LLoP and TSA algorithms. The location error

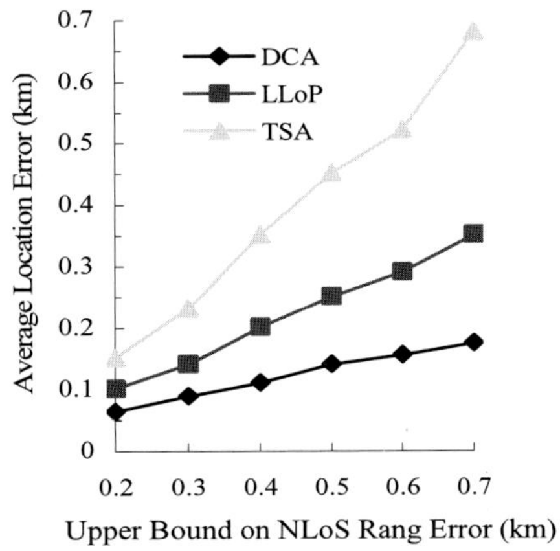


Fig. 8. The average location error versus the upper bound on uniform NLoS error.

of DCA is less than 0.08 km for 67% of the time, and less than 0.15 km for 95% of the time. The results of the experiments satisfy the location accuracy demand of E-911.


## ACKNOWLEDGMENT

This study was supported by the National Science Council, Taiwan, under grant NSC 93-2745-E-110-001.


## REFERENCES


[1] *Revision of the Commissions Rules to Insure Compatibility with Enhanced 911 Emergency Calling Systems,* Fed. Commun. Commission (FCC), Washington, DC, Tech. Rep. RM-8143, 1996.

[2] J. Caffery and G. Stuber, "Overview of radiolocation in CDMA cellular systems," *IEEE Commun. Mag.*, vol. 36, pp. 38–45, Apr. 1998.

[3] J. Caffery, *Wireless Location in CDMA Cellular Radio Systems*. Norwell, MA: Kluwer, 1999.

[4] M.P. Wylie and J. Holtzman, "The nonline of sight problem in mobile location estimation," in *Proc. IEEE Int. Conf. Universal Personal Communications (ICUPC'96)*, vol. 2, pp. 827–831, 1996.

[5] M.I. Silventoinen and T. Rantalainen, "Mobile station emergency locating in GSM," in *Proc. IEEE Int. Conf. Personal Wireless Communications*, pp. 232–238, 1996.

[6] S. S. Woo, H. You, and J. S. Koh, "The NLOS mitigation technique for position location using IS-95 CDMA networks," in *Proc. IEEE Vehicular Technology Conf. (VTC'00)*, vol. 6, pp. 2556–2560, 2000.

[7] P.-C. Chen, "A nonline-of-sight error mitigation algorithm in location estimation," in *Proc. IEEE Wireless Communication and Networking Conf. (WCNC'99)*, pp. 316–320, 1999.

[8] J. Caffery, "A new approach to the geometry of TOA location," in *Proc. IEEE Vehicular Technology Conf. (VTC'00)*, pp. 1943–1949, 2000.

[9] W. Foy, "Position-location solutions by Taylor series estimation," *IEEE Trans. Aerosp. Electron. Syst.*, vol. AES-12, pp. 187–193, Mar. 1976.

[10] S. Venkatraman, J. Caffery, and H.-R. You "A novel ToA location algorithm using LoS range estimation for NLoS environments," *IEEE Transactions on Vehicular Technology*, vol. 53, pp.1515-1524, Sep. 2004.

[11] P. van Rooyen, M. Lotter, and D. van Wyk, *Space-Time Processing for CDMA Mobile Communications*. New York: Kluwer, 2000.

# A Simplified and Systematic Technique to Develop and Implement PLC Program for a Control Process

Aamir Hanif, Muhammad Ahmad Choudhry, Senior Member IEEE and Tahir Mahmood
Department of Electrical Engineering
University of Engineering and Technology
Taxila, Pakistan.

*Abstract*— **Programmable logic controllers (PLCs) have made it possible to precisely control large process machines and driven equipment with less physical wiring and lower installation costs than required with standard electromechanical relays, pneumatic timers, drum switches, and so on. The programmability allows for fast and easy changes in relay ladder logic to meet the changing needs of the process or driven equipment without the need for expensive and time consuming rewiring.**

**In this paper a target to develop, simulate and, implement PLC software for automation of a water treatment plant is achieved. The algorithm developed in this paper may be used to develop PLC based software for any control process.**

## I. INTRODUCTION

Automatic control of a system requires that output of the system may be maintained at a desired magnitude or changed to some desired value without human interference. The field of automatic control system is under rapid development and has a promising future [1]. PLC based automation is a replacement for the hardwired relay and timer logic found in traditional control panels. PLCs provide ease and flexibility of control based on programming and executing simple logic instructions. PLCs have internal functions such as timers; counters and shift registers making sophisticated control possible with even small size PLCs.

A PLC is modular, reconfigurable and programmable controller which is a user-friendly electronic computer that carries out control functions of many types and levels of complexity [2]. The programmable controller operates by examining the input signals from a process and carry out logic instructions on these input signals, producing output signals to drive the process. Standard interfaces built in to PLCs allow them to be directly connected to process actuators and transducers (e.g. Pumps and valves) without the need for intermediate circuitry or relays.

Using PLCs, it is possible to modify a control system without having to disconnect or re-route a single wire. The requirement is to just change the control program using a keypad or Human Machine Interface (HMI) terminal. Programmable controllers also require shorter installation and commissioning times than do hardwired systems.

Although PLCs are similar to conventional computers in terms of hardware technology, however the specific features suite to industrial control.

The main features that suit industrial environment include:

1. Rugged, noise immune equipment.
2. Modular plug-in construction.
3. Standard input/output connection and signal levels.
4. Easily understood programming languages.
5. Ease of programming and reprogramming the PLC as per requirement.

These features make programmable controllers highly desirable in a wide variety of industrial situations.

Unfortunately the lack of standardization coupled with continually changing technology has made PLC common nightmare of incompatible protocols and physical networks. During 1980s an attempt was made to standardize communications with General Motors manufacturing automotive protocol. It was a time for reducing the size of PLC and making them programmable through symbolic programming on personal computers, instead of dedicated programming terminals.

IEC has recently standardized programmable controllers and their associated peripherals to facilitate automation industry. The commission has developed a series of standards (IEC-61131 series [7]-[12]) to address salient issues and problems confronting the automation industry. It also provides guidelines for the implementation of programming languages in programmable controller and their programming support environment.

## II. PROBLEM STATEMENT

In Pakistan, the Karachi Development Authority has to supplement its resources for drinking water from two new projects. These projects are located at Hub and Pepri near Karachi. This research work is related to one of many applications of Programmable Logic Controllers (PLCs) for the water treatment plants in these locations. PLC based software for backwash water treatment process used for the above mentioned projects is developed.

## III. METHODOLOGY

The concept of controlling a plant is a simplistic in nature. It involves a systematic approach by following the operation procedure as shown in Fig. 1. The algorithm involves the following steps.



*T. Sobh and K. Elleithy (eds.), Advances in Systems, Computing Sciences and Software Engineering*, 171–178.

*A. Determine the Sequence of Operation*

We have to identify the equipment or the system to be controlled. The ultimate purpose of the programmable controller will be to control an external system. The system to be controlled may be machine, equipment, or process to be controlled [4].

The movement of the controlled system is monitored in real time by the input devices. It gives a specified condition and sends a signal to the programmable controller. The programmable controller outputs a signal to the external output devices which actually controls the movement of the controlled system as specified and thus achieves the extended control action [3]. In simple, one needs to determine the sequence of operation by developing flowchart.

*B. Assignment of Inputs and Outputs*

All external input and output devices to be connected to the programmable controllers must be determined. The input devices are the various switches, sensors and timers. The output devices are the solenoids, electromagnetic valves, motor indicators and likewise.

After identifying all the INPUT and OUTPUT devices, the numbers corresponding to the INPUT and OUTPUT of the particular programmable controller will be assigned. The actual wiring will follow the numbers of the programmable controller.

The assigning of INPUT and OUTPUT numbers must be carried out before writing the ladder diagram because the number dictate what is the precise meaning of the contacts in the ladder diagram [5].

C. *Develop Timing Diagram*

Timing diagram is developed to show the length of occurrence of signals (I/O signals) of control process.

*D. Develop Flow Chart*

Flow chart is drawn considering the sequence of operation of control process.

*E. Writing of the Program*

Write the ladder program by following the control system sequence of operation as determined by step A.

## IV. PROGRAMMING SEQUENCES

A programmable controller has the following software design procedure.

1. The process is described.
2. Input, Outputs are defined for control process.
3. This description by considering I/Os is translated into a flow chart or function diagram.
4. Finally this flowchart is converted into ladder logic format and then programmed into the PLC.

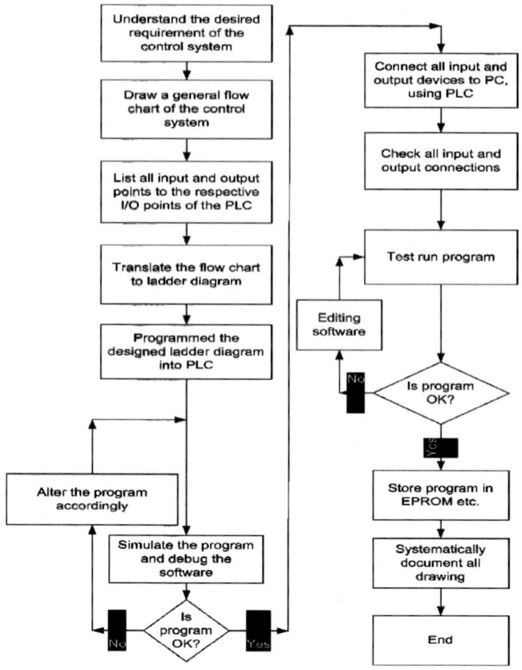


Fig. 1. A systematic approach to develop PLC based software

The verbal or written description of an automatic process is usually long and difficult to follow. The complete process is more easily understood when divided into a number of self-contained sub-units or sub-processes. Each sub-unit can then be constructed from sequences and interlocks to create the required function.

Several options exists e.g. relay logic diagrams, logic schematics, flowcharts and function charts. Choice of a particular method depends mainly on previous experience and the nature of the application [6].

## V. SEQUENCE OF OPERATION FOR BACKWASH WATER TREATMENT PLANT

After fulfilling the pre-requisites for automatic operation and the start up sequence, we converted detailed theoretical description of sequence of operation for backwashing a filter in steps, so that flow chart and hence software can be easily developed.

*A. Initiation of a Backwash*

Three methods for initiating backwashing sequence will be available.

*1) Timed wash/automatic initiation:* The filter will be washed in sequence by a timer, which will be adjusted from 12 to 36 hours.

TABLE I
MASTER PLC ALL INPUT DETAIL

| S.No | WIRE COLOR | WIRE# | PLC ADDRESS | VOLTAGEE | INPUT FROM |
|---|---|---|---|---|---|
| 01 | BLUE | MX0100 | 00000 | 24VDC | RUNNIG SIGNAL FOR F-200/1 |
| 02 | BLUE | MX0101 | 00001 | 24VDC | RUNNIG SIGNAL FOR F-200/2 |
| 03 | BLUE | MX0102 | 00002 | 24VDC | RUNNIG SIGNAL FOR F-200/3 |
| 04 | BLUE | MX0103 | 00003 | 24VDC | RUNNIG SIGNAL FOR F-200/4 |
| 05 | BLUE | MX0104 | 00004 | 24VDC | RUNNIG SIGNAL FOR F-200/5 |
| 06 | BLUE | MX0105 | 00005 | 24VDC | RUNNIG SIGNAL FOR F-200/6 |
| 07 | BLUE | MX0106 | 00006 | 24VDC | RUNNIG SIGNAL FOR F-200/7 |
| 08 | BLUE | MX0107 | 00007 | 24VDC | RUNNIG SIGNAL FOR F-200/8 |
| 09 | BLUE | MX0108 | 00008 | 24VDC | RUNNIG SIGNAL FOR F-200/9 |
| 10 | BLUE | MX0109 | 00009 | 24VDC | AIR DUMP VAL V-228(OPEN) |
| 11 | BLUE | MX0110 | 00010 | 24VDC | AIR DUMP VAL V-228(CLOSE) |
| 12 | BLUE | MX0111 | 00011 | 24VDC | START |
| 13 | BLUE | MX0112 | 00012 | 24VDC | STOP/ EMERGENCY STOP |
| 14 | BLUE | MX0113 | 00013 | 24VDC | RESTART |
| 15 | BLUE | MX0114 | 00014 | 24VDC | FAULT ACKNOWLEDGE |
| 16 | BLUE | MX0115 | 00015 | 24VDC | LS-203(LEVEL SWITCH)LOW WATER LEVEL |
| 17 | BLUE | MX0200 | 00100 | 24VDC | LS-203(LEVEL SWITCH)HIGH WATER LEVEL |
| 18 | BLUE | MX0201 | 00101 | 24VDC | BLOWER DUTY SELECTOR SWITCH(POSITON 1) |
| 19 | BLUE | MX0202 | 00102 | 24VDC | BLOWER DUTY SELECTOR SWITCH(POSITON 2) |
| 20 | BLUE | MX0203 | 00103 | 24VDC | BLOWER DUTY SELECTOR SWITCH(POSITON 3) |
| 21 | BLUE | MX0204 | 00104 | 24VDC | PUMP DUTY SELECTOR SWITCH(POSITON 1) |
| 22 | BLUE | MX0205 | 00105 | 24VDC | PUMP DUTY SELECTOR SWITCH(POSITON 2) |
| 23 | BLUE | MX0206 | 00106 | 24VDC | PUMP DUTY SELECTOR SWITCH(POSITON 3) |
| 24 | BLUE | MX0207 | 00107 | 24VDC | AIR SCOUR BLOWER B-200/1 SWITCHED REMOTE POSITION |
| 25 | BLUE | MX0208 | 00108 | 24VDC | AIR SCOUR BLOWER B-200/2 SWITCHED REMOTE POSITION |
| 26 | BLUE | MX0209 | 00109 | 24VDC | AIR SCOUR BLOWER B-200/3 SWITCHED REMOTE POSITION |
| 27 | BLUE | MX0210 | 00110 | 24VDC | BACK WASH PUMP P-200/1 SWITCHED REMOTE POSITION |
| 28 | BLUE | MX0211 | 00111 | 24VDC | BACK WASH PUMP P-200/2 SWITCHED REMOTE POSITION |
| 29 | BLUE | MX0212 | 00112 | 24VDC | BACK WASH PUMP P-200/3 SWITCHED REMOTE POSITION |
| 30 | BLUE | MX0213 | 00113 | 24VDC | AIR SCOUR BLOWER B-200/1 SWITCHED AUTO POSITION |
| 31 | BLUE | MX0214 | 00114 | 24VDC | AIR SCOUR BLOWER B-200/1 SWITCHED AUTO POSITION |
| 32 | BLUE | MX0215 | 00115 | 24VDC | AIR SCOUR BLOWER B-200/1 SWITCHED AUTO POSITION |
| 33 | BLUE | MX0300 | 00200 | 24VDC | BACK WASH PUMP P-200/1 SWITCHED AUTO POSITION |
| 34 | BLUE | MX0301 | 00201 | 24VDC | BACK WASH PUMP P-200/2 SWITCHED AUTO POSITION |
| 35 | BLUE | MX0302 | 00202 | 24VDC | BACK WASH PUMP P-200/3 SWITCHED AUTO POSITION |
| 36 | BLUE | MX0303 | 00203 | 24VDC | V-228 AIR DUMP VALVE SWITCHED AUTO |
| 37 | BLUE | MX0304 | 00204 | 24VDC | V-228 AIR DUMP VALVE SWITCHED REMOTE |
| 38 | BLUE | MX0305 | 00205 | 24VDC | TRIPPED SIGNAL FOR BACK WASH PUMP P-200/1 |
| 39 | BLUE | MX0306 | 00206 | 24VDC | TRIPPED SIGNAL FOR BACK WASH PUMP P-200/2 |
| 40 | BLUE | MX0307 | 00207 | 24VDC | TRIPPED SIGNAL FOR BACK WASH PUMP P-200/3 |
| 41 | BLUE | MX0308 | 00208 | 24VDC | TRIPPED SIGNAL FOR AIR SCOUR BLOWER B-200/1 |
| 42 | BLUE | MX0309 | 00209 | 24VDC | TRIPPED SIGNAL FOR AIR SCOUR BLOWER B-200/2 |
| 43 | BLUE | MX0310 | 00210 | 24VDC | TRIPPED SIGNAL FOR AIR SCOUR BLOWER B-200/3 |
| 44 | BLUE | MX0311 | 00211 | 24VDC | STOP SIGNAL FOR BACK WASH PUMP200/1 |
| 45 | BLUE | MX0312 | 00212 | 24VDC | STOP SIGNAL FOR BACK WASH PUMP200/2 |
| 46 | BLUE | MX0313 | 00213 | 24VDC | STOP SIGNAL FOR BACK WASH PUMP200/3 |
| 47 | BLUE | MX0314 | 00214 | 24VDC | STOP SIGNAL FOR AIR SCOUR BLOWER B-200/1 |
| 48 | BLUE | MX0315 | 00215 | 24VDC | STOP SIGNAL FOR AIR SCOUR BLOWER B-200/2 |
| 49 | BLUE | MX0400 | 00300 | 24VDC | STOP SIGNAL FOR AIR SCOUR BLOWER B-200/3 |
| 50 | BLUE | MX0401 | 00301 | 24VDC | AIR DUMP VALVE V-228 FAULT SIGNAL |
| 51 | BLUE | MX0402 | 00302 | 24VDC | SPARE |
| 52 | BLUE | MX0403 | 00303 | 24VDC | SPARE |
| 53 | BLUE | MX0404 | 00304 | 24VDC | SPARE |
| 54 | BLUE | MX0405 | 00305 | 24VDC | SPARE |
| 55 | BLUE | MX0406 | 00306 | 24VDC | SPARE |
| 56 | BLUE | MX0407 | 00307 | 24VDC | SPARE |
| 57 | BLUE | MX0408 | 00308 | 24VDC | SPARE |
| 58 | BLUE | MX0409 | 00309 | 24VDC | SPARE |
| 59 | BLUE | MX0410 | 00310 | 24VDC | SPARE |
| 60 | BLUE | MX0411 | 00311 | 24VDC | SPARE |
| 61 | BLUE | MX0412 | 00312 | 24VDC | SPARE |
| 62 | BLUE | MX0413 | 00313 | 24VDC | SPARE |
| 63 | BLUE | MX0414 | 00314 | 24VDC | SPARE |
| 64 | BLUE | MX0415 | 00315 | 24VDC | SPARE |

TABLE II
SLAVE PLC/X ALL INPUT DETAIL

| S.NO | WIRE COLOUR | WIRE# | PLC ADDRESS | VOLTAGE | INPUT FORM |
|---|---|---|---|---|---|
| 01 | BLUE | SXN00 | 0000 | 24VDC | S/S KEY OPERATED POSITION 1 (AUTO) |
| 02 | BLUE | SXN01 | 0001 | 24VDC | S/S KEY OPERATED POSITION 2 (SEMI AUTOMAIC) |
| 03 | BLUE | SXNO2 | 0002 | 24VDC | S/SKEY PERATED POSITION 3 (MANUAL) |
| 04 | BLUE | SXNO3 | 0004 | 24VDC | COLSE CONFIRMATION OF PENSTOCK V-202/X |
| 05 | BLUE | SXN04 | 0005 | 24VDC | OPEN CONFIRMATION OF PENSTOCK V-203/X |
| 06 | BLUE | SXN05 | 0006 | 24VDC | PEN CONFIRMATION OF PENSTOCK V-203/X |
| 07 | BLUE | SXNO6 | 0010 | 24VDC | CLOSE CONFIRMATION OF VALVE V-204/X |
| 08 | BLUE | SXN07 | 0008 | 24VDC | OPEN CONFIRMATION OF VALVE V-204/X |
| 09 | BLUE | SXN08 | 00011 | 24VDC | CLOSE CONFIRMATION OF VALVE V-206/X |
| 10 | BLUE | SXN09 | 00100 | 24VDC | OPEN CONFIRMATION OF VALVE V-206/X |
| 11 | BLUE | SXN10 | 00101 | 24VDC | CLOSE CONFIRMATION OF VALVE V-207/X |
| 12 | BLUE | SXN11 | 00102 | 24VDC | OPEN COFIRMATION OF VALVE V-207 |
| 13 | BLUE | SXN12 | 0013 | 24VDC | AUTO(LOOP INPUT)FOR ALL VALVES &PENSTOCK |
| 14 | BLUE | SXN13 | 0014 | 24VDC | REMOTE CONTROL(LOOP I/P) FOR ALL VALVES&PENSTOCKS |
| 15 | BLUE | SXN14 | 0015 | 24VDC | START KEY OPERATED P.B |
| 16 | BLUE | SXN14 | 00103 | 24VDC | RESTART |
| 17 | BLUE | SXN15 | 00104 | 24VDC | V-202/X FAULT INDICATION |
| 18 | BLUE | SXN16 | 00105 | 24VDC | V-203/X FAULT INDICATION |
| 19 | BLUE | SXN17 | 00106 | 24VDC | V-204/X FAULT INDICATION |
| 20 | BLUE | SXN18 | 00107 | 24VDC | V-206/X FAULT INDICATION |
| 21 | BLUE | SXN19 | 00108 | 24VDC | V-207/X FAULT INDICATION |
| 22 | BLUE | SXN20 | 00109 | 24VDC | SPARE |
| 23 | BLUE | SXN21 | 00110 | 24VDC | SRARE |
| 24 | BLUE | SXN22 | 00111 | 24VDC | SPARE |
| 25 | BLUE | SXN23 | 01200 | 24VDC | SPARE |
| 26 | BLUE | SXN24 | 01201 | 24VDC | SPARE |
| 27 | BLUE | SXN25 | 01202 | 24VDC | SPARE |
| 28 | BLUE | SXN26 | 01203 | 24VDC | SPARE |
| 29 | BLUE | SXN27 | 01204 | 24VDC | SPARE |
| 30 | BLUE | SXN28 | 01205 | 24VDC | SPARE |
| 31 | BLUE | SXN29 | 01206 | 24VDC | SPARE |
| 32 | BLUE | SXN30 | 01207 | 24VDC | SPARE |
| 33 | BLUE | SXN31 | 01208 | 24VDC | SPARE |
| 35 | BLUE | SXN32 | 01209 | 24VDC | SPARE |
| 36 | BLUE | SXN33 | 01210 | 24VDC | SPARE |

N=SLAVE#1 TO SLAVE #20
V- ♦♦♦ /X WHERE ♦=1------------20 (FILTERNUMBER) X= VALVES & PENSTOCK NUMBER.

TABLE III
SLAVE PLC/X ALL OUTPUT DETAIL

| S.NO | WIRE COLOURE | WIRE# | PLC ADDRESS | VOLTAGE | OUT PUT TO |
|---|---|---|---|---|---|
| 01 | BROWN | SYN00 | 01000 | 110VAC | FILTER WASHING LAMP AT FCC/X& MIMICS |
| 02 | BROWN | SYN01 | 01001 | 110VAC | FILTER IN SERVICE LAMP AT FCC/X & MIMICS |
| 03 | BROWN | SYN02 | 01002 | 110VAC | V-202/X (CLOSE) |
| 04 | BROWN | SYN03 | O1003 | 110VAC | V- 202/X( OPEN ) |
| 05 | BROWN | SYN04 | 01004 | 110VAC | V-203 /X( CLOSE ) |
| 06 | BROWN | SYN05 | 01005 | 110VAC | V-203/X( OPEN) |
| 07 | BROWN | SYN06 | 01006 | 110VAC | V-204 /X (CLOSE ) |
| 08 | BROWN | SYN07 | 01007 | 110VAC | V-204/X (OPEN) |
| 09 | BROWN | SYN08 | 01100 | 110VAC | V-206/X (CLOSE) |
| 10 | BROWN | SYN09 | 01101 | 110VAC | V-206 /X (OPEN) |
| 11 | BROWN | SYN10 | 01102 | 110VAC | V-207/X (CLOSE) |
| 12 | BROWN | SYN11 | 01103 | 110VAC | V-207/X (OPEN) |
| 13 | BROWN | SYN12 | 01104 | 110VAC | BACK WASH FALIURE INDICATIN |
| 14 | BROWN | SYN13 | 01105 | 110VAC | FILTER OUT OF SERVICE OF FCC/X TO MIMICS |
| 15 | BROWN | SYN14 | 01106 | 110VAC | KALAXOL ACTIVATE |
| 16 | BROWN | SYN15 | 01107 | 110VAC | SIGNAL THE 1$^{ST}$ AIR BLOWER TO RUN IN SEQUENCE |
| 17 | BROWN | SYN16 | 01200 | 110VAC | SIGNAL TO 2$^{ND}$ DUTY AIR BLOWER TO RUN |
| 18 | BROWN | SYN17 | 01201 | 110VAC | SIGNAL THE 1$^{ST}$ DUTY BACKWASH PUMP TO RUN |
| 19 | BROWN | SYN18 | 01202 | 110VAC | V-228(OPEN) |
| 20 | BROWN | SYN19 | 01203 | 110VAC | V-228(CLOSED) |
| 21 | BROWN | SYN20 | 01204 | 110VAC | SIGNAL 2$^{ND}$ DUTY BACKWASH TO RUN |
| 22 | BROWN | SYN21 | 01205 | 110VAC | SPARE |
| 23 | BROWN | SYN22 | 01206 | 110VAC | SPARE |

# Development of Mathematical Model of Blast Furnace Smelting

Andrey N. Dmitriev
*Institute of Metallurgy of Ural Branch of Russian Academy of Sciences,*
*101 Amundsen st., Ekaterinburg, 620016, Russia*
*dmi_imet@r66.ru*

## Abstract

*The solution of a problem of mathematical description of heat exchange, gas dynamics and the physicochemical phenomena taking place in blast furnace in their correlations, and some its application for study of processes, defining reduction of metals from multicomponent iron ores are considered.*

## 1. Introduction

Exploration of heat and mass transfer phenomena in a blast furnace with the purpose of determination of reserves of effectiveness of its work - drop of a coke rate and a raise of efficiency – is connecting with major financial, technological and engineering difficulties. Therefore use of mathematical models on study of blast furnace smelting, system engineering monitoring and control of a blast-furnace smelting operation is great importance. The role of mathematical models will increase at a shortage of the information on the phenomena happening in the blast furnace, such as temperatures of charge (material) and gas, pressure and composition of gas, a degree of reduction of iron in volume of the blast furnace.

## 2. Mathematical model

### 2.1. Formulating of a problem

The physical formulation of a problem consists in the following [1] (fig. 1). In a shaft furnace in radius $R_0$ and height $H_F$ a continuous stream of gas and a material towards each other along streamlines with initial temp eratures $t_G'$ and $t_M'$, accordingly are move. Is supposed the tuyere fireplace located on distance $L_{Tu}$ from a wall of the furnace, serves as a point source of gas and a drain of material. Heat capacity of gas $W_G$, an integral transfer coefficient of mass exchange $K_{\Sigma V}$ and integral coefficient of heat-away $\alpha_{\Sigma V}$ are functions of length of streamlines $h$, i.e. velocities of gas. Temperatures of the beginning of softening and melting of a material, dependent on chemical and mineralogical composition, are function of a degree of reduction.

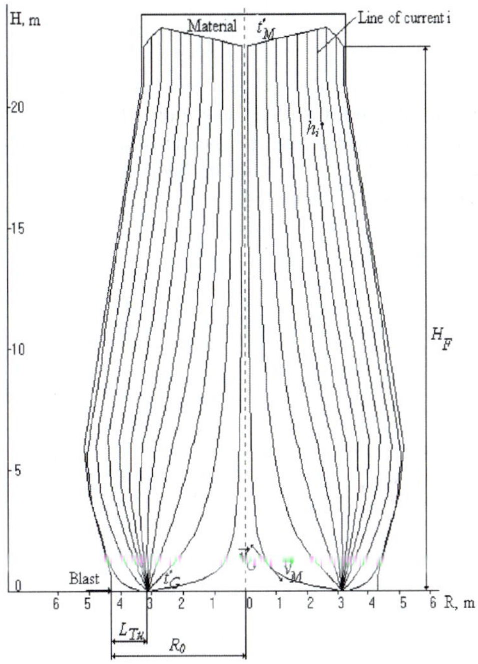


**Figure 1.** Mathematical formulating of a problem of making of mathematical model of blast furnace process: $H_F$ and $R_0$ – height of a charge layer and radius of the furnace; $L_{Tu}$ – distance from a wall of the furnace up to the center of tuyere fireplace; $\vec{V}_M$ and $\vec{V}_G$ - vectors of speeds of material (charge) and gas; $t_M'$ and $t_G'$ - temperatures of material and gas on an input in a layer; $h_i$ – distance from a charge level up to a calculation point lengthways *i*-streamline.



*T. Sobh and K. Elleithy (eds.), Advances in Systems, Computing Sciences and Software Engineering*, 179–183.

## 2.2. Mathematical model of gas dynamics

The velocity vector of gas in any point is defined on equation [1]

$$\vec{V}_G = \frac{2k_P V_{GEg} K_2 \sqrt{1-\delta^2 k_2^2}\,\mathrm{sn}\left(\frac{K_1}{H_F}\vec{Z},k_1\right)}{\pi\left[\delta^2+\left(1-\delta^2\right)\right]\mathrm{sn}^2\left(\frac{K_1}{H_F}\vec{Z},k_1\right)}. \quad (1)$$

Here $K_1$ is full elliptic integral of the first sort with the module $k_1$; $K_2$ is full elliptic integral of the first sort with the module $k_2$; $V_{GEg}$ is average speed of gas in the horizontal section of furnace shaft, m/s; $k_P$ is the factor which is taking into account a nonlinear dependence of pressure drop on height of a blast furnace;

$$\delta = sn\left(\frac{K_1}{R_0}L_{Tu},k_2\right). \quad (2)$$

Results of calculation are images on display, namely gas dynamics grids of movement (fig. 2) which is non-uniform, and a field of speeds of gas, i.e. value of speeds in units of a grid on the basis of which lines of equal speeds are calculated.

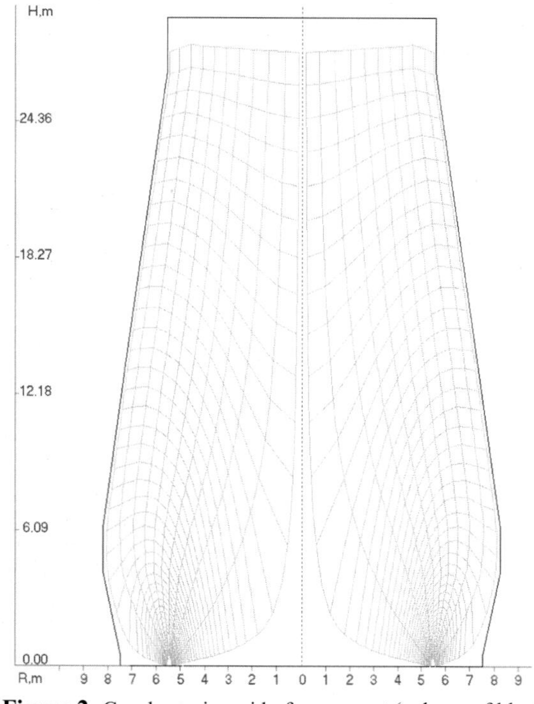


**Figure 2.** Gas dynamics grid of movement (volume of blast furnace is 5500 $m^3$).

## 2.3. Model of heat exchange

Values of temperatures of material $t_M$ and gas $t_G$ at any point of blast furnace are calculated [1]:

$$\left.\begin{aligned} t_G(h) &= B + A\int_0^h \frac{\alpha_{sv}(h)S(h)}{W_\Gamma(h)} e^{-f(h)}dh, \ {}^oC; \\ t_M(h) &= t_M' + A\int_0^h \frac{\alpha_{sv}(h)S(h)}{m(h)W_\Gamma(h)} e^{-f(h)}dh, \ {}^oC. \end{aligned}\right\} \quad (3)$$

Here

$$\left.\begin{aligned} A &= \frac{t_G' - t_M'}{e^{-f(H)} + \int_0^H \frac{\alpha_{sv}(h)\cdot S(h)}{m(h)\cdot W_\Gamma(h)} e^{-f(H)}dh}; \\ B &= t_M' + A; \\ f(h) &= \int_0^h \frac{\alpha_{sv}(h)\cdot S(h)}{m(h)\cdot W_\Gamma(h)\cdot}\left[1-m(h)\right]dh. \end{aligned}\right\} \quad (4)$$

Results of calculations on the equations (3) - (4) are isotherms of material and gas (fig. 3), and also distribution of temperatures of material and gas in anyone horizontal or a vertical section, used for model adaptation and for the assaying of appearances.

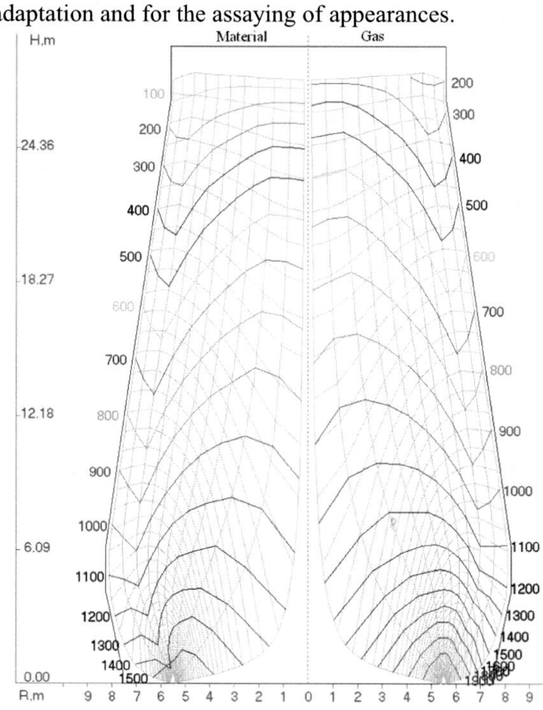


**Figure 3.** Isotherms of material (to the left) and gas (to the right) (volume of blast furnace is 5500 $m^3$), °C.

## 2.4. Model of mass exchange

Values of relative concentration potential of a gas stream and degree of reduction of iron oxides are calculated by decision of the system of the differential equations of mass transfer and reduction [1]

$$\left.\begin{aligned} V_G(h)dC &= -K_{\Sigma V0}e^{-(E/RT)}(1-\varphi_{Fe})Cdh; \\ m^*(h)\frac{d\varphi_{Fe}}{dh} &= \frac{dC}{dh}; \end{aligned}\right\} \quad (5)$$

where $C$ - a relative concentration potential of a gas stream (or a relative potential of mass transfer), part of a unit; $\varphi_{Fe}$ - a degree of reduction of iron oxides, part of a unit; $m^*$ - the ration mass capacity of streams of material and gas, part of a unit; $K_{\Sigma V0}$ - an integral transfer coefficient at $\varphi_{Fe}$ = 0, sm/s; $E$ - critical increment of energy, the J/mole; $R$ - gas constant, J/mole·K.

This set of equations is solved by a numerical method, with pre-award carrying out of interpolating of variables included in a system. For this purpose applied a method of splines-functions - interpolation by the generalized cubic splines, namely, the rational spline, permitting to interpolate functions with heavy gradients.

Outcomes of calculation are fields of degrees of reduction of iron and concentration potentials of gas (fig. 4).

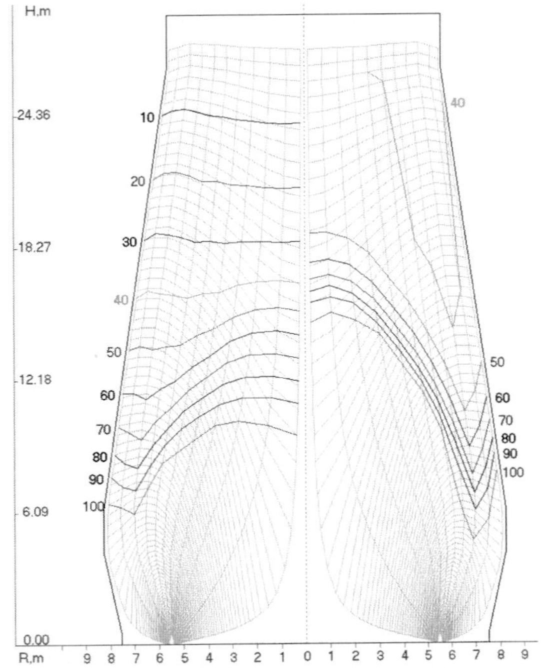


**Figure 4.** Fields of degrees of reduction of iron (on the left) and concentration potentials of gas (on the right) (volume of blast furnace is 5500 m$^3$), %.

## 3. Complex of mathematical models

The block-diagram of a complex of mathematical models is shown on fig. 5 [2].

The computer program is written on Fortran language and compiled with the help of the compiler Compaq Visual Fortran v.6.6B.

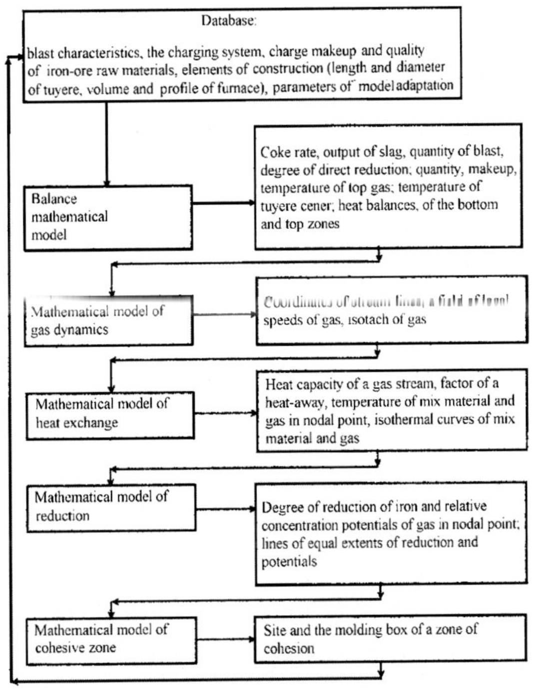


**Figure 5.** The block-diagram of a complex of two-dimensional mathematical models.

## 4. Use of mathematical models

This model is used for decision of practical problems of blast furnace smelting [3]. For example, on fig.6 is shown the influence of shaft profile on two-dimensional temperature fields of material and gas.

On fig. 7 two-dimensional fields of speeds of gas are resulted at various pressure differences $\Delta P = P_{Tu} - P_{To}$ at tuyeres $P_{Tu}$ and top $P_{To}$.

On fig. 8 - 10 two-dimensional fields of degrees of reduction of iron are shown at various values $K_{\Sigma V0}$ and $\Delta P$ .

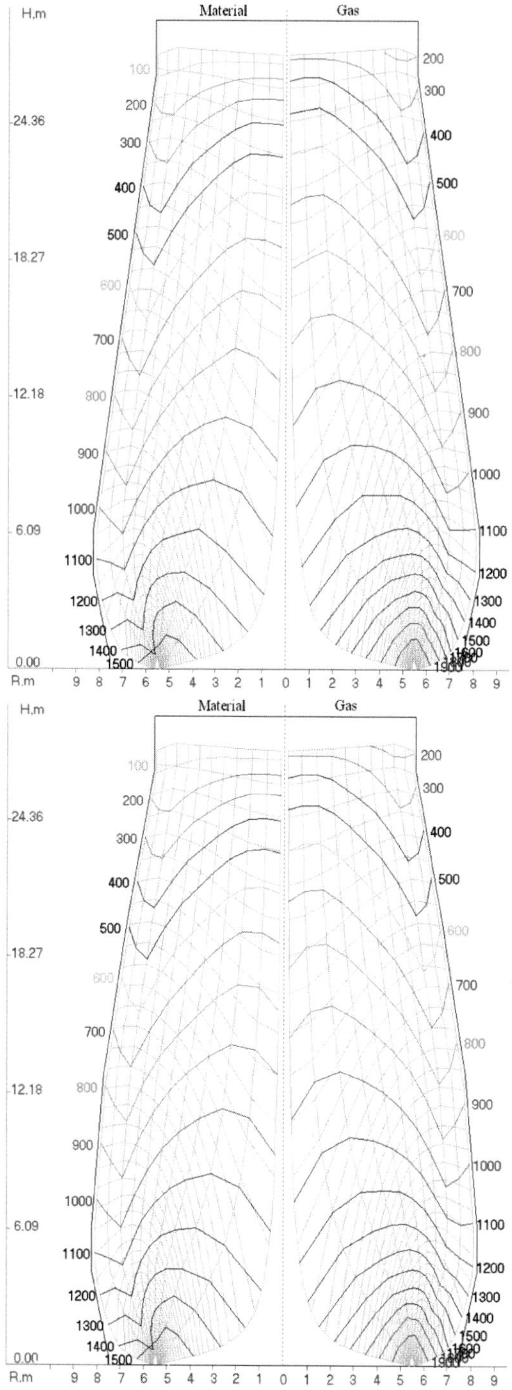


**Figure 6.** Two-dimensional temperatures fields: constant angle of shaft (from above) and variable angle of shaft (from below) (volume of blast furnace is 5500 m$^3$).

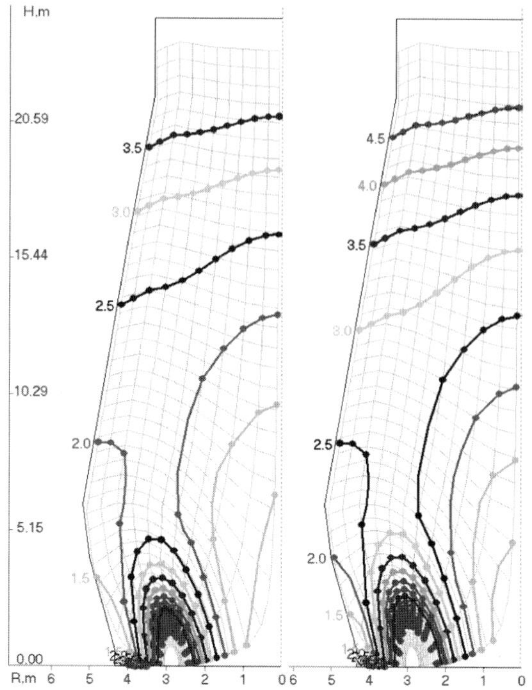


**Figure 7.** Fields of speeds of gas (m/s) at $\Delta P$ =*1,2* gauge atmosphere (on the left) and $\Delta P$ =*1,5* gauge atmosphere (on the right) (volume of blast furnace is 1500 m$^3$).

## 5. Conclusion

Thus, the mathematical model of blast furnace smelting is developed. It allows to analyze the processes of reduction of iron on height of a blast furnace in conditions of non-uniform movement of a material and gas on radius of the furnace.

## 6. Acknowledgments

Work is executed at support:
- Council under Grants for Leading Scientific Schools of Russia (School № 1997.2003.3);
- Regional Ural Branch of Academy of Engineering Sciences of the Russian Federation of name A.M. Prokhorov.

## 7. References

[1] A.N. Dmitriev. "Heat and mass transfer in blast furnace in conditions of non-uniform movement of material and gas", *Proceedings of the First International Conference on Diffusion in Solids and Liquids DSL-2005,* University of

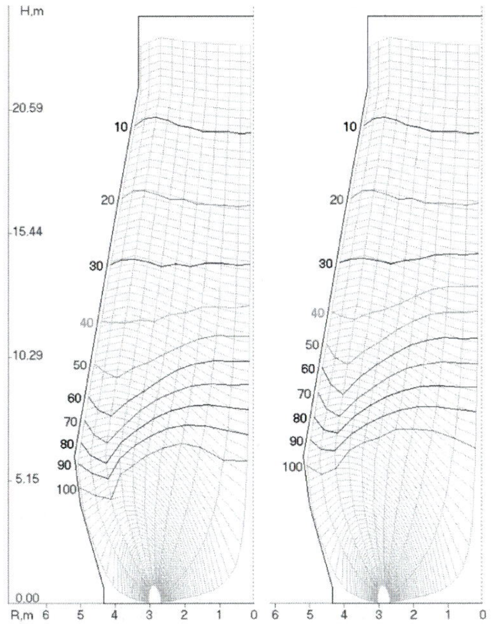


**Figure 8.** Fields of degrees of reduction of iron (%) at $K_{\Sigma V0}$ = 0.2 sm/s (to the left) and $K_{\Sigma V0}$ = 0.5 sm/s (to the right) at $\Delta P$ =1.2 gauge atmosphere; $P_{Tu}$=2.7 gauge atmosphere (volume of blast furnace is 1500 $m^3$).

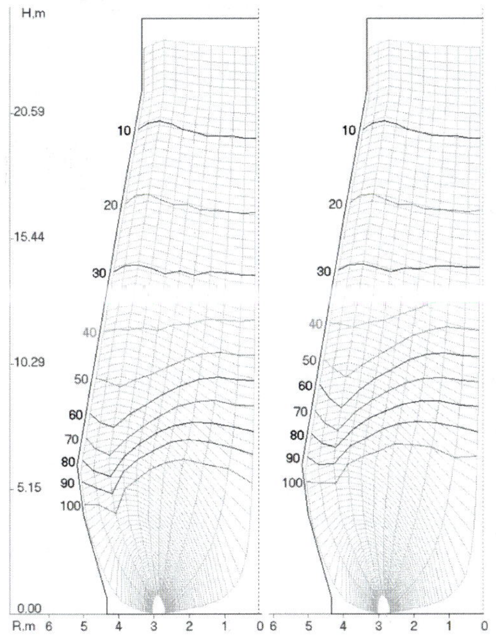


**Figure 9.** Fields of degrees of reduction of iron (%) at $K_{\Sigma V0}$ = 0.2 sm/s (to the left) and $K_{\Sigma V0}$ = 0.5 sm/s (to the right) at $\Delta P$ =1.5 gauge atmosphere. $P_{Tu}$=2.7 gauge atmosphere (volume of blast furnace is 1500 $m^3$).

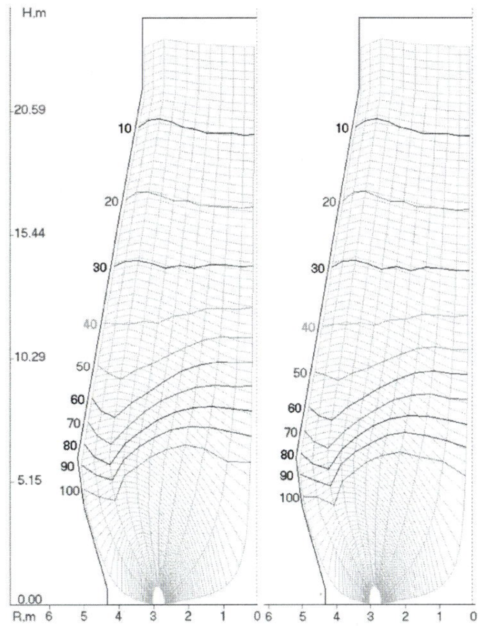


**Figure 10.** Fields of degrees of reduction of iron (%) at $K_{\Sigma V0}$ = 0.2 sm/s and $P_{Tu}$=2.7 gauge atmosphere: at $\Delta P$ =1.2 gauge atmosphere (to the left) and at $\Delta P$ =1.5 gauge atmosphere (to the right) (volume of blast furnace is 1500 $m^3$).

Aveiro, Aveiro, Portugal. 2005, July 6-8, 2005, Volume I, pp.181-186. (http://event.ua.pt/dsl2005/page11.html).

[2] A.N. Dmitriev, and S.V. Shavrin, "Development of mathematical methods of the analysis and modelling of the phenomena in a blast furnace", *Proceedings of the Third International Conference on Mathematical Modeling and Computer Simulation of Material Technologies (MMT-2004)*, College of Judea and Samaria, Ariel, Israel. September 06-10, 2004, pp. 2-24 – 2-33. (http://www.yosh.ac.il/research/mmt/MMT-2004/mmt-2004.htm).

[3] A.N. Dmitriev, "Mathematical modeling of two-dimensional processes in a blast furnace", *Computing methods and programming*, 2004, V.5, Section 1, pp. 252-267. (http://www.srcc.msu.su/num-meth).

# Research and researcher implications in sustainable development projects: Muti-agent Systems (MAS) and Social Sciences applied to Senegalese examples

Alassane Bah, Jean-Max Estay, Christine Fourage and Ibra Touré

**A new approach to problems of land allocation for livestock grazing, combining both computer science and social science tools has been developed since 1998 in Senegal, especially in the Ferlo area. Here we examine the implications for the different research centres such as the CIRAD[1], the Pôle Pastoral Zones Sèches (PPZS), the ESP and the laboratories of the UCO.**

**A computer simulation (MAS), based on sociological data, has been introduced in order to obtain a more neutral evaluation of the possible approaches to the problem, for example the approach of the local people involved, together with that of regional land development and allocation policy. This is preferable to practice that consists of experimenting policy in real-life situations with potentially dire consequences for the population and the environment. This initiative aims both to stabilise the social position of shepherds/herders and to preserve the production potential of ecosystems used for grazing. Its approach is one of sustainable development and the underlying theory must be questioned in order to clarify the empirical scope and pertinence.**

**This applied research (which, owing to the questions it asks, is rooted "in the most immediate reality" [1]), should not be naïve concerning its implications. Is it possible, with theoretical knowledge and an objective approach, to give research results to local management especially when the latter are separated from the decision making process?**

**The project's aim is to modify social behaviour, to rationalise it or to induce new behaviour. Groups or individuals that have to modify their behaviour as a result will naturally give rise to questioning concerning the social benefits of such changes.**

**Who is destined to benefit from the MAS platform? From the perspective of increased well-being to local populations a long-term follow-up is required together with an evaluation of social relations resulting from its use. It will certainly be necessary to implement a large-scale survey of people concerned both by the need for the platform (survey carried out during the phase of elaboration) and by its consequences (survey of the results and possible improvements). This obviously appears to be a long and expensive use of financial and intellectual resources but at the same time reflects the cost of working towards sustainable social and economic innovation in, as far as is permitted, a democratic manner. However what is more important than the principle is the fact that continuous follow-up and control seem to be a guarantee of success for such projects.**

## I. Introduction

For several years, Multi-Agent Systems have been used in the field of natural and renewable resource management and also to simulate competitive behaviour [2] [3]. They can be compared to micro-programs running in an environment exchanging information, skills and services between themselves and their environment. Applications exist in the fields of industry, computer assisted geography but also in the field of human sciences.

MAS and social sciences can beneficially combine their complementary approaches when confronting the realities of fieldwork with modelling. This collaboration has taken place in the field of sustainable and/or local development (and to a lesser degree in the field of medicine, notably in relation to home-care for serious pathologies in Quebec).

The simulations can be concerned with the impact of technical transformations, renewable energy or water management, organisational change or new judicial systems but in each situation the problems/issues concerning the decision maker (agent) are central to the computer specialist. These different aspects of social change are tributary to the representations from which they [1].

In this respect the social science contribution is double:

- Theoretical: by enabling an interpretation of the social reality that is to be modelled. It should be remembered that sociology is a cumulative discipline in so far as it possesses analytical theory and social mechanisms that can be tested and thus guide the work of the programmer. It is also an approach based on axiological neutrality, the external analytical

[1] Centre de coopération Internationale en Recherche Agronomique pour le Développement (Montpellier-France)



*T. Sobh and K. Elleithy (eds.), Advances in Systems, Computing Sciences and Software Engineering*, 185–191.

viewpoint that recommends, in the Durkheim tradition, that social phenomena must be observed as things and that they are recognizable by the coercive power that they exert on individuals without the latter being aware of the coercive process.
- Methodological: as an empirical discipline, sociology has over time equipped itself with data gathering methods that profess a certain objectivity and that allow, in a sense, to interrogate fieldwork.

On the other hand, thanks to MAS, sociology will have to accept the challenge of experimentation, something that it mistrusts both by tradition and habitus! Sociology will have to confront the analysis and the description of social mechanisms with the perspectives generated by the simulation that it has helped to build. It offers a unique opportunity to test its hypotheses.

The computer specialist must provide the evaluation tools that the sociologist is cruelly lacking, however we are forced to note that the connection between the two disciplines remains marginal.

## II. MULTI-AGENT SYSTEMS (MAS)

A multi-agent system is a computer system, which enables a virtual universe to be constructed for the purpose of simulation [4] [5]. The models are organised around entities, called objects or agents, which are placed in a dynamic environment/universe. These entities evolve with time both as a result of their own characteristics and of their interactions with their environment. The agents are capable of influencing that which surrounds them, both other agents and objects.

The action of the agents is defined by decomposing their behaviour mechanism into three stages: perception-deliberation-action. Generally the agent is given a limited field of perception, which enables it to obtain information about its environment. During deliberation it chooses between a number of predetermined actions based on the knowledge and objectives that it has received. Upon taking action it will change its environment and consequently the perception that it had will have to be reconsidered in light of this change leading to the choice of new actions.

This iterative process is all the more complex in that any one agent is not the only source of change for the environment and at each time step the universe is modified by the action of all the agents simultaneously. There are a number of ways in which agents can influence one another: by changing the others, by communicating (sending messages which change knowledge or objectives) and by modifying the common environment (which changes perception) [2].

In order to define a system from which to conduct simulations, it is necessary to describe the characteristics of the objects and agents, the dynamics of their evolution and the action method of the agents. The system is launched from an initial or starting state and repeats at each time step a succession of transformations and actions. The universe and the agents evolve until states of equilibrium appear in their characteristics or until regularities can be seen in their actions. The term emergence is generally used to describe these global phenomena that can be found in the universe after a large amount of local dynamics has taken place [5] [6] [7].

Given the importance attributed to the interaction when multi-agent modelling the latter is often used to describe societies whether these are animal or human. Most of the time the capacity for self-organization is examined and the supposed processes tested. When describing the appearance of hierarchy in human societies the most common simulation is based on the quest for natural resources with the appearance of leader dependant groups, that either survive or do not, a common action [5] [8]. The multi-agent framework thus enables the observation of consequences of local actions, relative to explicitly described logic, on the collective form.

Faced with these simulations of emergence where the agent 'understands' nothing about the collective process in which he participates through his actions other users of multi-agent systems base themselves on the existence of a predefined social group and the ability of each agent to perceive and in a sense to make commitments to the others [9] [10]. It is also possible to define a group of agents that represent collective characteristics and that interact with individual agents [11]. In this way the influence of social groups on the behaviour of individuals is possible.

Similarly different hypotheses concerning the information received by the agent, its individual objectives, together with diverse forms of collective influence on these criteria can be tested. It is this definition of the model via the rules of local behaviour that makes multi-agent system simulation so interesting for the social sciences [12].

## III. SUSTAINABLE DEVELOPMENT

The 1950s and 70s concept of development can be summarized as being growth, self-maintained through the development of natural resources and unexploited land resources, through the domination development or mastering of nature [1].

The concept of sustainable development appeared in the mid 70s. It consists of a development that 'satisfies the needs of the present generation without compromising the capacity of future generations to satisfy theirs' [13]. Nature is perceived as a stock to be optimally managed to equilibrium. It consists of preserving the environment and maintaining or restoring equilibriums.

In the 1990s, 'the intrusion of variability, of uncertainty and irreversibility into system dynamics led to viewing development in terms of management of the interactions between economic and social variability and natural variability, in both space and time' [1]. It is what we call viable development. The concept is different, an optimum is even more required but the elaboration of adaptive strategies is preferred to natural or economic variability. It consists of managing for the best, with management based on stable long-term objectives and the interaction between the different sources of natural and social variability. To reason in terms of viable development is to affirm that both the rules for equity and the long-term objectives result from political debate and not from analytical definitions. A shift is made from models where either economic or ecological optimisation is the sine qua non of development, to a model where ethical norms rank higher than the others [14]. From now on, sustainable development implies respecting simultaneously more than one objective [14]: economic development; preservation of the natural resource base and the implied ecological constraints; social equity both inter and intra-generational (especially between southern and northern countries). Sustainable development is a development trajectory that enables the co-evolution of economic, social and ecological systems [14].

## IV. Applied Research Dynamics: Sustainable Development in Senegal

The MAS-sociology partnership, given the specificities outlined in the introduction, is particularly adapted to research interventions in the field of sustainable development.

The role of the computer specialist is to provide tools for the sociologist. The sociologist's needs are similar to the computer modelling of sociological data and events. It is interdisciplinary methodology, 'accompanying and modelling' [2] which is at the centre of the research. From this perspective, the bridge between the two disciplines is a two-way road between perceived reality and the virtual model. Observation and formulation are respectively questioned in order to attain a better understanding of reality.

A new approach combining the two disciplines has been developed since 2001 within the Pôle Pastoral Zones Sèches (PPZS) based in Dakar with a view to preserving sustainable livestock grazing.

Livestock grazing is an important way of life and method of food production in arid areas of Africa, in spite of the general crisis that threatens the ecosystems. The question concerning better integration of sustainable livestock grazing into national society and linked to other arid zone production methods is at the heart of regional, national and international development programmes. The aim of the research development is twofold: to provide information for political and economic decision making via a global approach and to identify, together with the local populations, practical solutions that not only meet their immediate needs but also allow them to secure both social projects and the production capacity of the livestock grazing ecosystems.

A computer simulated MAS has been introduced in order to obtain a more neutral evaluation of the different approaches to the problem, for example the approach of the local people involved together with that of the regional land development and allocation policy. This is preferable to practice that consists of experimenting policy in real-life situations with potentially dire consequences for the population and the environment.

The experimentation has two objectives:

- One research orientated: to succeed in modelling the rationalities of the different approaches in compliance with reality and to clarify the interdisciplinary analysis (computer science and sociology) for a good conceptualization of livestock grazing. The user-friendliness, the flexibility of the hypotheses and the interactivity of the simulation enable integration of information from both these disciplines and from the different approaches (land ment, livestock breeding).

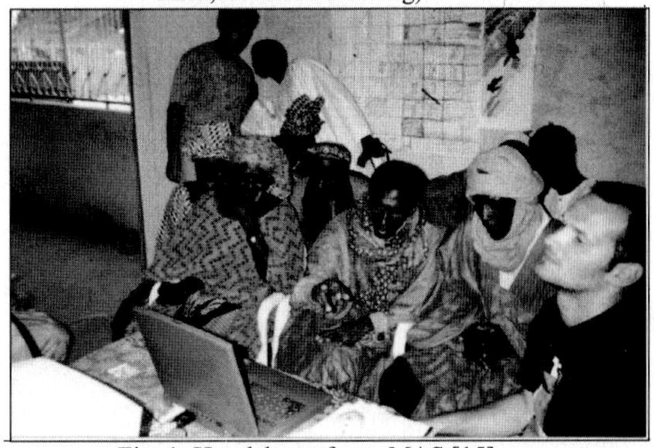

Fig. 1: Hot debates from MAS [15]

- The other is to evaluate the impact of the different possible types of livestock breeding/grazing and land development (including usage and ownership practice) on the current situation (state of the environment, degree of mobility and territorialization, socio-economic impact).

Two case studies carried out in Senegal will also be outlined. The first, known as 'SelfCormas' [15] is concerned mainly with support for the decentralized management of the northern Senegalese territory. The methodology involves the endogenous development of a management tool for the different decision makers linking Geographical Information Systems (GIS), Multi-Agent Systems (MAS) and role-plays. The second study entitled 'Thieul' [16] is concerned with the dynamics of land allocation in a zone used for agriculture, forestry and grazing in the mid-east of Senegal.

The models coming from these two experiments are of practical interest as a result of their development involving not only many specialists (sociologists, modellers, herders, geographers and others) but also many local decision makers (farmers, herders and representatives of local government).

### *A. SelfCormas experiment*

The experiment called 'SelfCormas' comprised four three-day test workshops organised in the delta of the river Senegal at different levels and with target populations. The first workshop, in French, was concerned with a group of 'people-resources' of the local community (school teacher, young person returning to the village, etc.) directly chosen by the community so that it could afterwards assist them in territorial development. The three other workshops were tested in Wolof, the local language, with representatives from the local illiterate population.

After the first stages of the supportive approach, information and dialogue, the participants were required to test the resource management rules and to imagine possible evolutions.

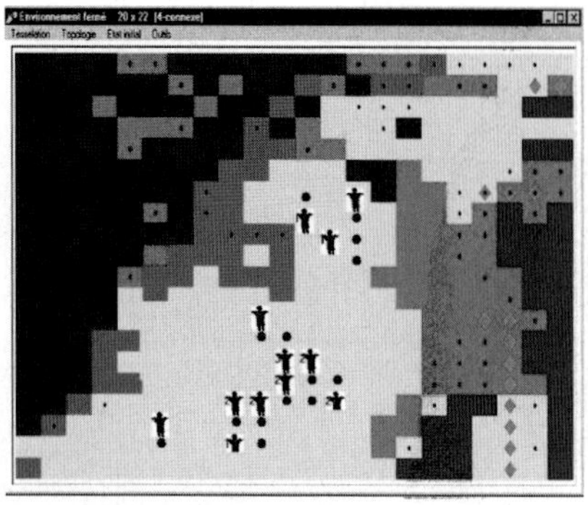

Fig. 2: Simulation interface[2] (SelfCormas MAS model).

The simulations resulting from these different experiments helped the participants work together and enriched their thinking at each step of the decision making process, that is, from the identification of access rules for a given type of land (agricultural, grazing) up to the evaluation of the social and environmental impact of different types of land allocation.

*B. Thieul experiment*

The primary objective of this experiment, carried out within the Thieul grazing unit[3], is to construct a model using the knowledge acquired from the multiple use of land and resources around the Thieul drilling area. It consists of interdisciplinary research on the feasibility of grazing in difficult bio-climatic conditions and in a particularly sensitive environment (social, ethnic, cultural, economic and political).

The multi-agent modelling that was used has generated an innovative way of approaching modelling and simulation in the environmental sciences by enabling simultaneously the representation of individuals, their behaviour, and their interactions [3]. A few authors use this approach, with some success, to try and understand the interactions between social and ecological phenomena [11] on irrigated areas in the valley of the river Senegal and [16] on grazing in the Sahelian zone.

To enable diverse decision makers and experts to benefit from the development process of the MAS, the methodology is based on a participative approach from the initial analysis up to the implementation of the model.

Fig. 3: Hot debates for the installation of an threshold at Thieul village.

The elaboration of the endogenous knowledge base, the understanding of activities, and the logic of the decision makers are all based on the active participation of the populations. This participative approach, based on the design of maps and computer tools using the knowledge and techniques of the local populations, is divided into four steps: 1) analysis of the external situation; 2) reinforcement of endogenous skills; 3) design of maps by the decision makers; 4) participative design of a multi agent simulator which uses the data (maps etc.) coming from the different field workshops.

The current set of results shows that the local population have adapted well to the use of the maps. The restitution of the results of the simulation is being prepared.

Fig. 4: Decision makers debating (photo I Touré).

---

[2] This Model is implemented using CORMAS platform

[3] Surface area: 1,031.46 $km^2$. The grazing unit is situated in the grazing/forestry zone of Ferlo at about 60 km southeast of Dahra in Senegal.

Fig. 5: Simulation interface[4] (Thieul MAS Model)

## V. Positioning of the Researcher with Respect to the Socio-Economic Demands of Sustainable Development – What Is at Stake?

Sociology outlines an initial analysis of social conditions concerning agricultural activity. It does this from social and environmental data (population characteristics, herds, grazing zones, farming or transhumance…) and by taking into account the cultural forms of practice. It also takes into account natural or man-made factors that are susceptible to transform socio-economic relations and production methods through technical innovation (change, rationalization). The sociologist and computer specialist share the objective of local populations mastering change with as little upheaval as possible in their fundamental social dynamics.

However, research, especially applied research, when 'the questions asked are rooted in the most immediate reality' [1], should not be naive concerning its implications. It serves a social project, which like any project, has theoretical and/or social and political foundations. Sociology's impartiality is frequently questioned [17], its approach reveals the hidden mechanisms of domination and constitutes a critical analysis of society. Here it is necessary to assume an intellectual heritage and to say that it is possible, with theoretical knowledge and an objective approach, to give research results to local management even when the latter are separated from the decision making process. They are free to crosscheck results with social protest movements that reject growth as the only model of economic and human development. The sociologist does not posses a better knowledge of what is 'just' or 'good' for the local decision makers. Nevertheless, basing himself on methodological options, and accepting his implication in the situation, he seeks to contribute to the building of practical material susceptible to be of use to his interlocutors [18].

There is nothing to prevent the computer specialist and the sociologist, within the framework of applied research, answering economic or political questions and measuring the risks of a critical positioning towards free-market nations. It is thus possible to examine what 'theory' (from sustainable development to bearable decline) can bring to the operational aspect of the research and to the deprived (symbolically, socially, economically, culturally, politically [19]) decision makers enabling them to gain legitimacy in politics and economics. This means adopting a 'more respectful attitude' towards those for whom the different disciplinary approaches are destined. Researcher commitment implies behaviour based on an awareness of one's involvement in the world and therefore responsibility for the consequences of one's actions in the world [18].

Pierre Bourdieu, throughout his career, insisted on sociology's ability to conduct critical analysis. This is why it appears as dangerous and relativistic to people in positions of dominance (no matter what the social domain). Indeed, by making explicit and dismantling social relationships, it reveals the mechanisms of power. It is in the interest of those who profit from this power that these mechanisms remain hidden or be considered as self-evident, that they cannot or should not be questioned coming as they do from a 'higher order' (divine, social, or political). Bourdieu also thought that a sociologist's work was finished when he was in a position to provide those submitted to symbolic, social, economic or political (…) violence, with the means by which they could understand and escape their 'negative privileges'. By doing so the sociologist is responsible for the theories he/she produces, and is necessarily socially committed. It is not possible to lose interest in the use(s) of one's scientific output. The researcher is inevitably committed to producing and structuring the world he is studying. The sociologist has a permanent dialogue with the constituents, and builds society whilst trying to understand it. Neutrality is not therefore possible [18].

Consequently, when social science researchers participate in applied research whose purpose is to modify social behaviour, to rationalize it or to induce new behaviour (as is the case when collaborating with computer science specialists on MAS project development), essential questions concerning the benefits to those who will undergo such changes arise quite naturally.

Who is destined to benefit from the MAS platform? From the perspective of increased well-being to local populations, a long-term follow-up is required together with an evaluation of social relations resulting from its use. Have the planned benefits been obtained? Does this have a social cost? If so, is it accepted or acceptable? It is then necessary to evaluate the impact of the MAS on the behaviour and practice of the users.

In data gathering, from a methodology point of view, it seems that the approach is based on privileged information sources, chosen because of their ability to act as relays for the local population or because of their position within the

[4] This Model is implemented using CORMAS platform

community (an economic, political or symbolical function). As a result it is necessary to analyse the implied effects of these positions as well as the implied social relationships to be sure of the acceptability and logic of the installed procedures. If such aspects are overlooked the researcher runs the risk of only using one of the parties concerned to the detriment of those who will have to modify their behaviour on a long-term basis. It is easy to see the limitations of such projects: the researcher looking for 'competent' interlocutors will be perceived as being 'in the service of', his/her implication places he/she 'naturally' on the side of the powerful.

In other respects the political desire, the reform or the social project which constitutes the starting point of the research tool must also be investigated. The question asked is therefore one of 'fit' or appropriateness between what the platform offers (why it is developed) and its social use.

If as is hoped, the platform serves to improve living conditions of individuals, it is necessary to be able to verify this in the long run. It is also necessary to be sure that not only the political or administrative structures are reinforced by the platform and that a new split does not appear between those who have access to the platform and promote it and those who, for whatever reasons (which need to be understood), lack the means to do so. A utility analysis is required in order to determine how and which decision makers actually benefit: end users, institutional decision makers, scientists, planners, other decision makers, NGOs, etc.

Moreover, the researcher makes an important contribution when working in the field of technical innovation. Methods have to be proved to be effective and useful. The reputation of the academic discipline is at stake. If the researcher loses sight of the fact that the project is perhaps also an attempt to increase legitimacy, that it is indeed possible to use the project for symbolically gratifying reasons (participating in international programmes, working with NGOs), then it is probable that the original intended contribution to scientific knowledge be betrayed. The risk is all the greater when the project is perceived as 'exciting' or when exploring new territory. Does the reality in the field justify taking such risks? Or is the researcher going to use the 'urgent need for research' as an excuse to satisfy his own needs?

## VI. Conclusion: Is It Possible to Avoid Permanent Project Evaluation?

It will certainly be necessary to implement a large-scale survey of people concerned both by the need for the platform (survey carried out during the phase of elaboration) and by its consequences (survey of the results and possible improvements). This obviously appears to be a long and expensive use of financial and intellectual resources but at the same time reflects the cost of working towards sustainable social and economic innovation in, as far as is permitted, a democratic manner. However what is more important than the principle is the fact that continuous follow-up and control seem to be a guarantee of success for such projects.

An essential question addresses the possible research spin off and its effect on local populations. This could be seen as being primarily to the advantage of the researchers and one can seriously ask whether proven classical social science methodologies (as opposed to the highly specialized computer simulations) would not do the same job at a lower cost. The researcher's intervention would be perceived as being more 'human' and would minimize both the risks associated with imposing problematics and the risks accompanying symbolic violence (stereotyped responses when faced with an order).

Finally, in an approach linking sociologists and developers, it is necessary, within the team, to evaluate the interdisciplinary aspect itself. Researchers need to confront their approaches and develop a culture of cross disciplinary implication that will enable/require a good conceptual and methodological understanding of each others fields and thus overcome the obstacle of identical terminology referring to very different realities. Rational MAS handle beliefs, a fact held as true or verified by past history, but for the sociologist they are no more than representations.

If the researcher's implication mobilises a form of theoretical and practical intelligence linked to different paradigms [18], then cross disciplinary implication is a tacit mutual agreement based on the continuous questioning both of disciplinary 'truths' and of the way work is usually carried out. In this situation scientific collaboration bears the mark of modesty and relativism, a relativism which does not believe that everything is possible bur rather that everything can be challenged in order to gain a better understanding of reality, of the social decision makers and of the aim of a sustainable development project. From this moment the quarrel of exact science verses social science is abandoned in favour of research that corresponds to the objectives of development and to the professed social benefit to the target population.

## Bibliography


[1] Weber, J., 1995 *"Gestion des ressources renouvelables : fondements théoriques d'un programme de recherche."* http://cormas.cirad.fr/pdf/green.pdf

[2] Bousquet F., 1996, "Systèmes multi-agents et action sur l'environnement", *Actes du colloque "Mémoires, inscriptions, actions, individuelles et collectives" 22-26/01/96* Centre de recherche de Royallieu, Compiègne.

[3] Ferber J., 1995: *Les systèmes multi-agents, vers une intelligence collective.* InterEditions.

[4] Lenay C., 1994, "Intelligence Artificielle Distribuée : modèle ou métaphore des phénomènes sociaux" In *Revue Internationale de systémique, vol 8(1).* p1-11.

[5] Doran J., Palmer M., Gilbert N., Mellars P.,1994, "The EOS project: modelling Upper Paleolithic social change", *Simulating societies. The computer simulation of social phenomena*

[6] Cariani P., 1991, "Emergence and artificial life", In *Artificial Life II*, vol. X, Langton C.G. ed., AddisonWesley, pp 775-.

[7] Baas. 1994 “Emergence, hierarchies and hyperstructures” In *Artificial Life*, numéro III, vol XVII, Langton C.G, ed, Addison-Wesley, pp 515-537

[8] Doran J., Palmer M., 1993, "The EOS project: Integrating two models of Paleolithic social change", *Artificial Societies*, N. Gilbert & R. Conte ed., UCL Press

[9] Rao A.S., Georgeff M.P., 1995, "BDI", *Actes du colloque ICMAS 95,* V.Lesser ed., MIT Press, pp

[10] Castelfranchi C., 1995, "Commitment: from individual intention to groups and organizations", *Actes du colloque ICMAS'95,* V.Lesser ed., MIT Press, pp 41-48.

[11] Barreteau, O., 1998. *Un système Multi-Agent pour explorer la viabilité des systèmes irrigués: dynamique des interactions et modes d'organisation.* [A Multi-Agent System for Exploring the Viability of Irrigated Systems: Dynamics of Interactions and Organisational Modes]. Montpellier, National University for Rural, Water and Forestry Engineering. 260p

[12] Gilbert N., 1993, Emergence in social simulation, prepared for Sim Soc 93, Cartosa di Pontignano, Siena.

[13] Bruntland, 1987 COMMISSION MONDIALE SUR L'ENVIRONNEMENT ET LE DÉVELOPPEMENT (CMED), 1987. *Notre avenir à tous,* aka Bruntland report.

[14] Torres E., 2000 "Développement durable et territoire", Presses universitaires Septentrion, Villeneuve d'Ascq.

[15] D'Aquino, P., Le Page, C., Bousquet, F. and Bah, A. 2003. "Using self-designed role-playing games and a multi-agent system to empower a local decision-making process for land use management: The SelfCormas experiment in Senegal”. In Journal of Artificial Societies and Social Simulation 6(3) http://jasss.soc.surrey.ac.uk/6/3/5.html.

[16] Bah, A., Canal, R., D'Aquino, P. and Bousquet, F., 1998. "Application des systèmes multi-agents et des algorithmes génétiques à l'étude du milieu pastoral Sahélien. [Application of Multi-Agent Systems and Genetic Algorithms for the Study of the Pastoral Environment in the Sahel]", p. 207-220. N. Ferrand (ed.), *Modèles et systèmes multi-agents pour la gestion de l'environnement* [Multi-Agent Models and Systems for Environmental Management]. *Proceedings from the SMAGET colloquium.* Cemagref Editions.

[17] Bourdieu P., 1980. *Questions de Sociologie*, Ed de minuit,

[18] Herreros G., 2002. *Pour une sociologie d'intervention* Ed Érès.

[19] Bourdieu P., Wacquant L. 1992. *Réponses : pour une anthropologie réflexive.* - Paris Seuil, - 267 p.

## AUTHORS BIOGRAPHIES

**Mr Alassane Bah** is an engineer in computer science from the Ecole Supérieure Polytechnique (ESP-UCAD) in Dakar. He has been a lecturer at the same department since December 2000. He is working on multi-agent simulation and genetic algorithms: their application to the management problems of natural resources in Sahel (Africa). Within this context, he has been carrying out several research projects on desertification problems, on space modelling problem, on the distributed multi-agent simulation theme.

LGLIA, Ecole Supérieure Polytechnique (ESP), Département Génie Informatique, Université Cheikh Anta Diop, BP 5085, DAKAR Fann, Sénégal, bah@ucad.sn

**Dr Jean-Max Estay** is Assistant Professor in Computer Sciences at Université Catholique de l’Ouest (Angers, France).

CREAM, Institut de Mathématiques Appliquées, Université Catholique de l’Ouest, BP 10808, 49008 Angers cedex 01 France, Jean-Max.Estay@uco.fr

**Dr Christine Fourage** is Assistant Professor in Sociology at Université Catholique de l’Ouest (Angers, France).

CERIPSA, Institut de Psychologie et Sociologie Appliquées
Université Catholique de l’Ouest, BP 10808, 49008 Angers cedex 01 France, Christine.Fourage@uco.fr

**Dr Ibra Touré** is the co-ordinator of Pôle Pastoral Zones Sèches (Dakar, Senegal)

Cirad-Emvt, Pôle Pastoral Zones Sèches (PPZS) Isra-Dakar, Sénégal B.P.: 2057 Dakar-Hann, Sénégal, ibra.toure@cirad.fr

# The importance of modeling metadata for Hyperbook

**Abdoulmajid Hakki**
Lappeenranta University of Technology
Department of Information technology
ahakki@lut.fi

***Abstract* - This paper provides a review of modelling metadata for adaptive Hyperbook. We analyse the adaptivity and conceptual model for representing and sorting a Hyperbook, the language and tools for metadata and the architecture of the Hyperbook.**

## 1. INTRODUCTION

Hyperbook document consist of different types of objects (text, pictures, sounds, etc.), interrelated by a link structure for navigation. In cases that the Hyperbook consists of different sections, systematic metadata management is needed to guarantee consistency, maintainability and extendibility. The types mentioned in [1] of metadata have to be managed.

Hyperbook document is made of fragments or piece of information, which can be assembled to constitute directly readable hypertext document read by any navigation [2]. From this perspective we will consider that the information contents of Hyperbook consist of a set of fragments which have content, a description and relation between them. The access of information is to be done through views which are hyper documents generated from documents.

The concept of cross media Hyperbook is a particular case of electronic book. In our case the Hyperbook is understood as a kind of monograph, since it is limited to the subject matters of a book and has no encyclopaedic purpose. In contrast with ordinary documents the adaptivity aspect of the Hyperbook is the challenging features of our studies.

This paper describes a meta-modeling approach to Hyperbook design which is rigorously database-oriented and supports maintenance and reuse. The set of current paper is organised as follows: The next two sections will present adaptivity and a conceptual model for representing and storing a Hyperbook and finally we will discuss on how the Hyperbook system can be implemented.

## 2. ADAPTATION COMPONENT OF THE HYPERBOOK SYSTEM

The goal of our study is to increase the functionality of Hyperbook by making it personalised. Adaptive Hyperbook will build a model of preferences [3] of individual user and use this throughout the interaction of the content based on the needs of that user.

For our studies we will follow a constructivist learning approach, building on project-based learning. The critical feature of Hyperbook system is possibility of providing Hyperbook adaptation on the base of the user model as shown in Fig. 1.

The project based learning approach will rich the practical knowledge of developing Hyperbook publishing and will help the scientific contribution. The adaptive Hyperbook has to implement the following functionality:

- adaptive presentation
- adaptive navigation support
- adaptive information resource

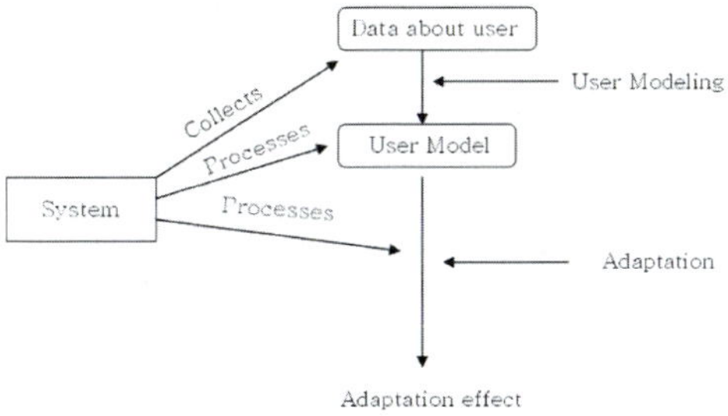


Fig. 1. User model adaptation in adaptive system [3]

The aim of the Adaptive presentation is to adapt the content of Hyperbook view to the user goals: device profile and other information stored in model (Fig. 1). In a system with adaptive presentation the views are not static, but adaptively generated or assembled from pieces for each user [3]. The idea of adaptive navigation support is to guide the user in reading the Hyperbook by changing the appearance of visible link. Adaptive navigation as defined in [3] can be considered as a generalisation of curriculum sequencing technology in Hyperbook context.

Adaptive information resource gives the authors and publisher editors appropriate information while developing the content of the book.

## 3. CONCEPTUAL MODEL FOR REPRESENTING AND STORING A HYPERBOOK

Semantic annotations and metadata are seen as a crucial technique for transforming the content [4] of Hyperbook to give meaning to amorphous pages interlinked and their associations.

The semantic web is concerned with adding a semantic level to resources that are accessible over the Internet in order to enable sophisticated forms of use and reuse.

The content of Hyperbook data faces same challenges as outlined in the Semantic Web Proposal of the W3Consortium [5]. This allows us to use to some extent the same constructs to denote the metadata.

As in [6] argued there is no consensus about a common virtual document model. Nevertheless most of the proposed models are comprised of domain ontology and a fragment base [6, 7]. In [8], the architecture of the Semantic Web is outlined in the following three layers:

1. **The metadata layer:** The data model at this layer contains just the concepts of resource and properties. The RDF (Resource Description Framework) is believed the most popular data model for the metadata layer [9].
2. **The schema layer:** Web ontology languages are introduced at this layer to define a hierarchical description of concepts and properties. RDF Schema is considered as a candidate schema layer language [10].

*T. Sobh and K. Elleithy (eds.), Advances in Systems, Computing Sciences and Software Engineering*, 193–195.

3. **The logical layer:** More powerful web ontology languages are introduced at this layer. This language provides a richer set of modelling primitives that can be mapped to the well-known expressive Description Logics. OIL (Ontology Inference Layer, 2000) and DAML-OIL (Darpa Agent Markup Language-Ontology Inference Layer, 2001) are to popular logical layer languages [7]

Like other publishing technologies, the interest in creating and developing the Semantic Hyperbook is motivated by opportunities that might bring to publishing industry in the information age. Hyperbook based on web-services, agent based distributed computing, semantic based content search engines and semantic based Hyperbook libraries.

For our study there is a need to develop a conceptual modelling tool for construction of Hyperbook. The need for a Hyperbook conceptual modelling tool is essential for the effectiveness of Hyperbook, because of the importance of the fully understanding and exploiting the semantics of the pages of Hyperbook and subject index terms. Different Hyperbook models have been proposed in the past [6, 11, 12]. Most of these models employ different types of data structure at higher abstraction levels to describe the informative content of the document collection placed on the lowest level of the abstraction [11].

[13] propose two models of virtual documents, which use domain ontologies for indexing informational fragments. [14] propose a comprehensive and detailed model of virtual documents, which is based on four ontologies for modelling the domain, the metadata, the applications and the user.

Our approach to the Hyperbook model is comprised of an information fragments, domain ontology and interface specification. The fragments and the ontology with their consistent links form the structural part of the Hyperbook (Fig. 2.).

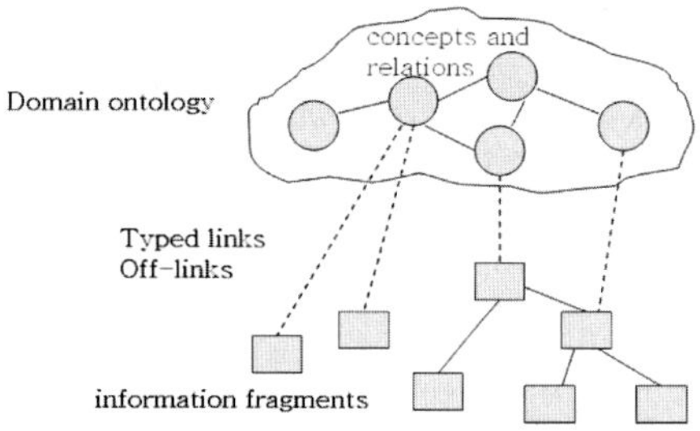


Fig. 2: The Hyperbook principle structure

A semantic data model defines a number of semantic entities (categories, concepts, interrelated by means of semantic relations between such entities), which model a particular subject domain [15].

Such a model provides a conceptualisation of that particular subject domain. Semantic data modelling is similar to different knowledge representation techniques [15]. Hyperbook characterised as a collections of hypertext documents [16]. We generalise the approach to Hyperbook system by decoupling metadata/conceptual models, which explicitly model all relevant units of the Hyperbook and data referencing the actual data and implementing a metadata-base system using several sets of abstractions to visualise document units and their semantic relationships. As product model we recommend to use XML for the implementation of actual Hyperbook. The use of XML and the focus of the current method on the specific domain of Hyperbook, which sets certain limitations in the structure of applications, make it different from the other methods [17, 18].

Semantic web technologies like RDF or RDF Schema [4, 19] provide us interesting possibilities. An ontology formally defines the relations among terms, which, following the specific terminology, are referred to ass classes of objects [20].

In RDF everything is expressed through statements: simple triples made up of resources, namespace and literals [21].

## 4. ARCHITECTURE

The next challenge in establishing Hyperbook system is to decide on how the data will be organised and stored. In our study the architecture of Hyperbook system follows the commonly used application development that is shown in Fig. 3.

As shown in Fig. 3. The reader browses the Hyperbook with any browsers, while all necessary processing is done on the server side. The Hyperbook architecture will consist of three modules: Storage Module, User Adaptation Module and Bookmark Module. The main purpose of the storage module is to manage the persistent storable objects that as a whole constitute the Hyperbook [22]. The User Adaptation Module stores, describes and infers information, knowledge, preferences etc. about an individual user. The user adaptation module expresses, derives and draws conclusions about the characteristics of users [23]. Unlike the cluttering or defacing of a paper book, a hypertext system lets users mark nodes, by putting a name in a list of bookmarks, doing no damage to the document itself. This bookmark name can be a system-defined name or a name the user can choose. Most hypertext systems do not display the list of bookmarks unless you ask for it, so the bookmarks do not hinder the reading process. The ability to add a bookmark and to jump to the indicated node at any time is beneficial to the online learner. Browsers for the World Wide Web save bookmarks as an html file, so that the bookmark node can be used like an ordinary node in the hypertext document [24]. Storage Module contains the data of the specific Hyperbook, e.g. the instances of the concepts from the conceptual model of the Hyperbook in question and implements a query interface. Thus it serves as a data repository and is called by the other modules [25].

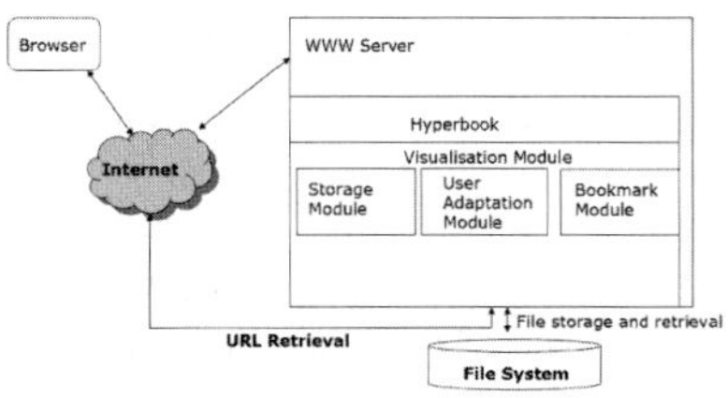


Fig. 3: General system architecture of Hyperbook.

## 6. CONCLUSION

We have analysed a methodology and modelling language for describing major aspects of Hyperbook, including architecture modules. These modules specify all design decisions needed or adaptive Hyperbook and also the basis for creation of personalised pages which are displayed using a browser. The next work shall extend the current work by introducing an application neutral layer which allows describing knowledge about aspects of the conceptual and the modular structure of Hyperbook. We will present a XML binding to our model specifications and discuss how they serve data schema that applies the content of Hyperbook in cross media environment.

## ACKNOWLEDGEMENT

I thank my supervisor Professor Arto Kaarna and Dr. Kari Heikkinen for their helpful support, suggestions and comments.

## REFERENCES

1. Peter Fröhlich, N. Henze and W. Nejdl, Meta Modeling for Hypermedia, University of Hannover – Germany (1996)
2. G. Falquet, J. Hurni, J. Guyot, L. Nerima, Learning by creating multipoint of view scientific Hyperbooks, Universit de Gen ve (2001)
3. Peter Brusilovsky, Methods and techniques of adaptive hypermedia, User Modeling and User Adapted Interaction, 1996, v 6, n 2-3, pp 87 – 129
4. Wolfgang and Martin Wolpers Christian Capelle, RDF Metadata, Semantic Data Models and Structured Hypertext, Institut für Rechnergestützte Wissensverarbeitung, University of Hannover, 1999
5. Tim Berners-Lee, James Hendler and Ora Lassila, The Semantic Web, Scientific American, May 2001
6. Gilles Falquet, Luka Nerima, Jean-Claude Ziswiler, Adative Mechanisms in Virtual Hyperbook, Centre Universitaire d´Informatique, University of Geneva, 2004 IEEE
7. Shiyong Lu, Ming Dong and Farshad Fotouhi, The Semantic Web, Opportunities and challenges for next-generation Web application, Department of Computer science, Wayne State University, Information Research, Vol. 7, No, 4, July 2002
8. Tim Berners Lee, Semantic Web Road, [W3C Design Issues, Cambridge, MA: W3C, Cambridge, MA: W3C, available at: http://www.w3.org/DesignIssues/Semantic.html September 1998. [site visited 30.8.2005]
9. Jeen Broekstra, Michel Klein, Stefan Decker, Dieter Fensel and Ian Horrocks, Adding formal semantics to the Web: building on top of RDF Schema, 2002
10. Dan Brickley, W3C , R.V. Guha, IBM , RDF Vocabulary Description Language 1.0: RDF Schema,W3C Recommendation 10 February 2004, http://www.w3.org/TR/rdf-schema/ [site visited 30.08.2005]
11. Fabio Crestani, Massimo Melucci, Automatic construction of hypertextsfor self-referensing: The hyper-textBook project, Information Systems 28, 2003
12. Peter Fröhlich and Wolfgang Nejdl, A Database-Oriented Approach to the Deisgn of Educational Hyperbooks, 8th World Conference of the AIED Society, Japan 18 – 22 August 1997
13. Sylvie Ranwez and Michel Crampes, Conceptual Documents and Hypertext Documents are two Different Forms of Virtual Document, Laboratoire de Génie Informatique et d'Ingénierie de Production, EMA - EERIE, Parc Scientifique Georges Besse, 1999
14. Serge Garlatti and Sébastien Iksal, A Semantic Web Approach for Adaptive Hypermedia, Department of Artificial Intelligence and Cognitive Sciences, 2003
15. Denis Helic, Aspects of Semantic Data Modeling in Hypermedia Systems, Institute of Information Processing and Computer Supported New Media (IICM), Graz University of Technology, Austria, July 2001
16. Nicola Henze, Kabil Naceur, Wolfgang Nejdl and Martin Wolpers, Adaptive Hyperbooks for Constructivist teaching, Institut für Technische Informatick, Universität Hannover, 1999
17. Symeon Retalis and Andreas Papasalouros, Yet Another Approach to Support the Design Process of Educational Adaptive Hypermedia Applications, Eindhoven University of Technology, 2004
18. Changtao Qu, Wolfgang Nejdl, Bringing O-Telos and XML with XML schema: The Authoring Environment for KBS Adaptive Hyperbook, IEEE 2002
19. Peter Dolog, Rita Gavriloaie, Wolfgandg Nejdl, Jan Brase, Integrating Adaptive Hypermedia Techniques and Open RDF-Based Environments, WWW2003, Budapest, Hungary, 2003
20. Dieter Pfoser, Evaggelia Pitoura, Necatari Tryfona, Metadata Modeling in a Global Computing Environment, GIS´02, November 8-9, 2002, Virginia, USA
21. Wolfgang Nejdl, Hadhami hraief, Boris Wolf, Building up AI Resources as an AI Testbed, Computer Engineering/Knowledge Based Systems, University of Hannover, 2002
22. Antonina Dattolo, Vincenzo Loia, Active distributed framework for adaptive hypermedia, Dipartimento di Informatica ed Applicazioni, Universit_a di Salerno , 1997
23. Nicola Henze and Wolfgang Nejd, Logically Characterizing Adaptive Educational Hypermedia Systems, University of Hannover, 2003
24. Mark David Mann, Using The Adaptive navigation support technique of Link hiding in an educational hypermedia system, an experimental study, Oklahoma State University, 1999
25. Nicola Henze and Wolfgang Nejdl, Adaptation in Open Corpus Hypermedia, International Journal fo Artificial Intelligence in Education, 2001

# Cluster-Based Mining of Microarray Data in PHP/ MYSQL Environment

E. Udoh, S. Bhuiyan
Department of Computer Science
Indiana University-Purdue University
2101 E. Coliseum Blvd, Fort Wayne, IN 46805 USA

*Abstract* - **Extracting biological significance from a large microarray dataset using data mining clustering technique is an important process in bioinformatics. In this paper, a microarray dataset (matrix 504 x 227) made available by SAMSI institute, was used as the base sample to develop a new demo web-based clustering system that exploits the improved efficiency and functionality of PHP/MYSQL technology. The clustering algorithms and robustness of PHP/MYSQL produced categorized microarray data that can be associated with diseases with improved visualizations.**

*Keywords* - **Data Mining, Microarray, Clustering Dendrogram, and PHP/MYSQL.**

## I. INTRODUCTION

Microarray technology generates extremely large gene expression dataset (hundreds of rows and columns) in a single biological experiment (Fig. 1). The gene expression patterns provide unprecedented information on human disease, aging, drug, mental illness, diet and many other clinical matters, because they correlate strongly with function. In the medical world, microarray has paved the way for a new era of genetic screening, testing and diagnostics [7]. However, the dataset often contains missing values, and exhibits high dimensional attributes, i.e. large number of genes with relatively small number of samples [1]. It is cumbersome to manually examine these large dataset for biological significance, hence the need for automation (e.g. using clustering techniques) to reduce the quantity of data to a manageable level [2, 3].

Clustering techniques can determine intrinsic grouping in a set of microarray data using distance or conceptual measures. To determine membership in a cluster, clustering algorithms evaluate distance between a point and the cluster centroid. The output is basically a statistical description of the cluster centroid with the number of components in each cluster. However, domain knowledge is useful to formulate appropriate measure in a clustering algorithm, which may be exclusive, overlapping, hierarchical or probabilistic.

There are several clustering algorithms to process and establish relationships in large dataset generated by microarray experiments [4]. These algorithms can be used to determine what group a particular genetic sample belongs to and the tendency for certain clusters to be associated with certain characteristics. This helps to eliminate less relevant dimensions or genetic characteristics. In this paper, six algorithms were applied, e.g. un-weighted pair group centroid, weighted pair group centroid or ward's method to achieve clustering of the samples [5]. These different algorithms can be applied individually or compared, and a suitable one chosen for the investigation.

The microarray data in Fig. 1, made available by SAMSI institute, was used for this project. The data consist of a series of genetic profiles for breast, prostrate, liver, colon, kidney, lung, ovary and testis cancer (matrix 504 x 227). Some of the cell samples are malignant while others are normal. The data are log-transformed, with row and column means subtracted (with positive values showing well expressed gene, while negative values indicate non-expression of gene). The used data have been subjected to singular value decomposition by authors in [2] as an effective way to provide dimensionality reduction, and eventual elimination of less relevant data [7, 10].

The main thrust of this project is to cluster microarray samples with the rich visualization features in the PHP/MYSQl development environment [6] as the displays in this project attest to. PHP, an open source, server-side scripting language, allows user to easily develop a robust and dynamically generated page quickly. PHP is cross-platform and easy to learn with well balanced memory and server load.



*T. Sobh and K. Elleithy (eds.), Advances in Systems, Computing Sciences and Software Engineering*, 197–200.

The improved visualization features are attractive to any bioinformatics programmer, since the representations are intuitive.

## II. ANALYSIS and DESIGN

In the first phase to develop bioinformatics software in PHP/MYSQl environment, this clustering system was developed in a Linux server environment hosting Tomcat 5.0, PHP 5.0, MySQL 4.1 and Ghostscript 8.15. The base technology used for the analysis and design are the server-side scripting language (PHP) and MySQL database for persistence storage. On execution, the PHP code retrieves the microarray data from the MYSQL database. It converts the data into a usable format, and then passes the output to the clustering software, which in turn sends the result to a dendrogram and ghostscript programs for visualization. Some of the clustering algorithms programmed, include single link, complete link, group average, weighted average, weighted centroid and ward's method (Fig. 2).

As can be gleaned from Fig. 2, the PHP code calls the mathematical algorithms to perform the clustering, as well as provide an easy to use interface to the system. The user will be able to specify a variety of parameters or algorithms using the web based interface, while the results of the clustering can be presented by showing which clusters are included in particular groups (Fig. 2). The program shows the percentages of each cluster with or without the cancer malignancy. For example, few clusters may be used and the percentages for cancer within those groups may be too mixed with normal samples (such as a 30% cancerous 70% normal) within a cluster. In such a scenario, the number of clusters may be increased to allow a better level of granularity when clustering. Attribute grouping and associative clustering explain similar dependencies and also offer improved classification of such genes [8, 9]. Below is a dendrogram produced for the analyzed sample (Fig. 3).

Cancerous samples are indicated by arrows. These cancerous samples are all in the first group, with some few normal samples. It is clear in this example that almost all of the cancer samples are present in the first group, with a very high probability that any sample in that group will be cancerous (Fig. 3). A variation of this dendrogram can be obtained if another algorithm is selected (Fig. 4).

| Id | Sample_Tissue_Site | Sample_General_Pathologic_Category | 34449_at CASP2 | 35150_at TNFRS |
|---|---|---|---|---|
| 72 | LIVER, NOS | NORMAL | -0.64565 | -0.71759 |
| 77 | COLON, NOS | NORMAL | 0.38159 | -0.32849 |
| 79 | KIDNEY, NOS | MALIGNANT | 0.83026 | -0.2525 |
| 83 | LIVER, NOS | MALIGNANT | 0.047638 | -0.81221 |
| 91 | KIDNEY, NOS | MALIGNANT | -0.22587 | -0.45274 |
| 96 | KIDNEY, NOS | NORMAL | -0.0528 | -0.21131 |
| 98 | LUNG, NOS | MALIGNANT | -0.02914 | -0.2353 |
| 100 | COLON, NOS | MALIGNANT | 0.65866 | 0.78431 |
| 101 | LUNG, NOS | NORMAL | 0.23735 | 0.003307 |
| 109 | LIVER, NOS | MALIGNANT | -0.3751 | -0.10789 |
| 117 | LIVER, NOS | NORMAL | -1.0638 | -0.84062 |
| 118 | COLON, NOS | NORMAL | -0.88117 | -0.7941 |
| 124 | LIVER, NOS | NORMAL | 0.073264 | -0.29181 |

Fig. 1: A cross-section of a microarray dataset used for the design of the system (**http://www.samsi.info/200304/dmml/web-internal/bio/data.html**)

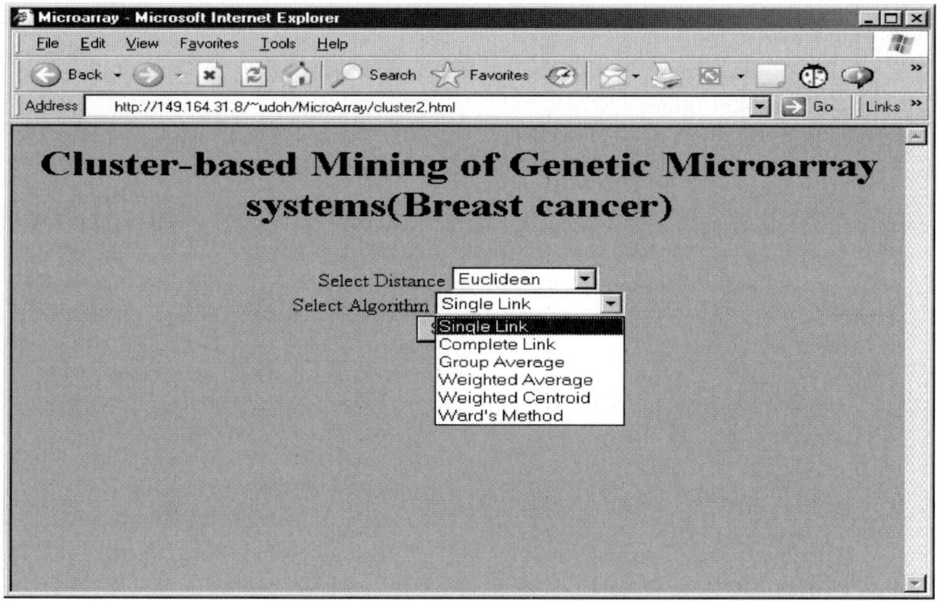


Fig. 2: Web interface to the cluster program.

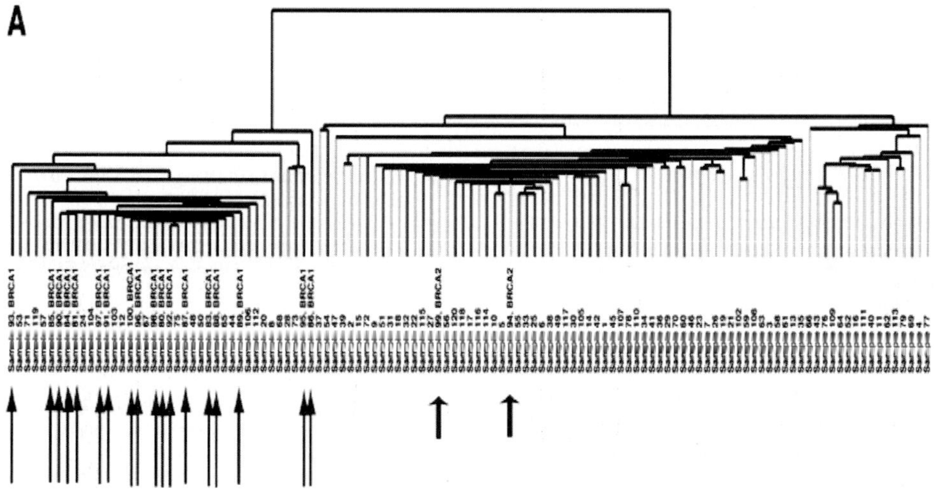


Fig. 3: A dendrogram of microarray data (Arrow indicates cancerous sample).

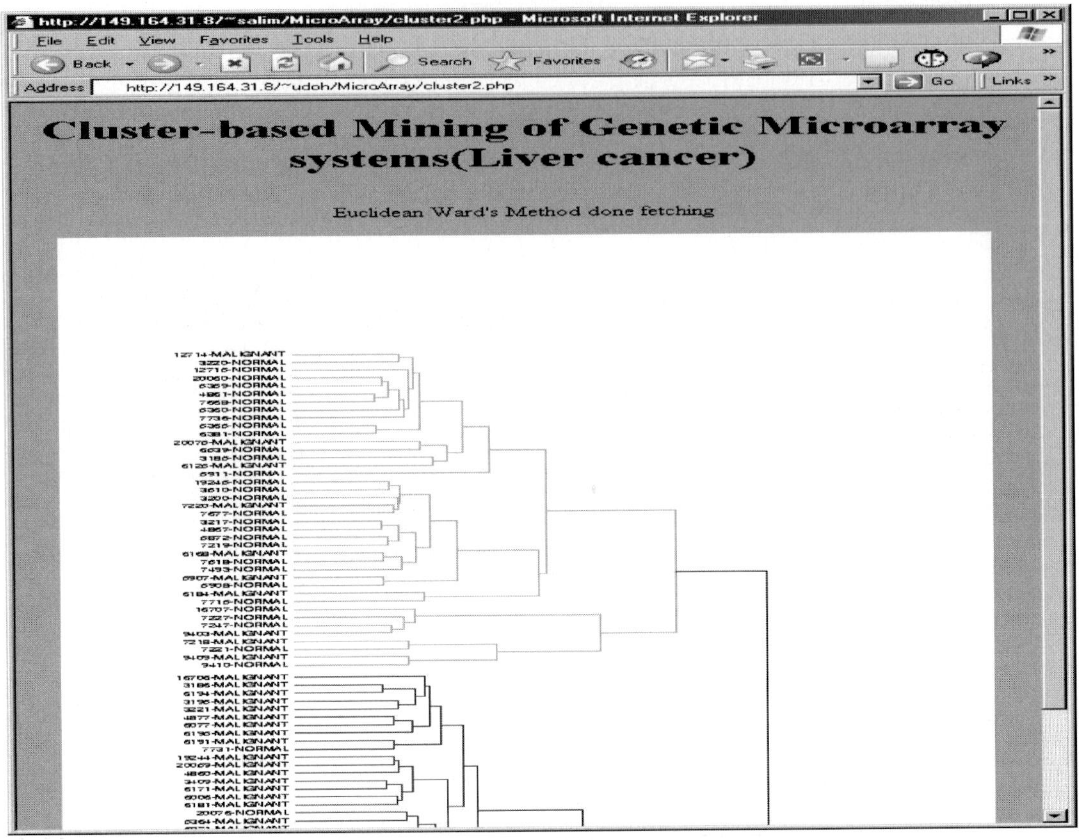


## III. CONCLUSION

The process of finding relevant genetic markers in a large microarray dataset is a difficult task. By reducing the amount of data using data mining technique like clustering, the task is made a lot easier. This paper exploited the improved functionality of PHP/MYSQL programming environment to design a new demo cluster program. It will be refined in the future to serve the bioinformatics community better. The PHP/MYSQL programming is certainly attractive to the bioinformatics programming community.

## REFERENCES

[1] A. D. Baxevanis & F. Ouellette, *Bioinformatics: A practical guide to the analysis of genes and proteins* (New York, NY: Wiley-Interscience, 2001).

[2] L. Liu, D. Hawkins, S. Ghosh & S. Young, Robust singular value decomposition analysis of microarray data, *Proceedings of the National Academy of Sciences PNAS*, USA, 2003, 100( 23), 13167-13172.

[3] G. Piatetsky-Shapiro & P. Tamayo, Microarray data mining: Facing the challenges, *ACM-SIGKDD Explorations*, 5(2), 2003, 1-5.

[4] P. Glenisson, J. Mathys & B. De Moor, Meta-clustering of gene expression data and literature based information. *SIGKDD Explorations*, 5(2), 2003, 101-112.

[5] M. Scheena, *Microarray analysis* (New Jersey: Wiley, 2003).

[6] J. Meloni, *PHP 5*, (Boston, MA: Thomson Course Technology, 2004).

[7] N. Bolshakova, F. Azuage & P. Cunnigham, An integrated tool for microarray data clustering and cluster validity assessment, *Proceedings of ACM Symposium on Applied Computing*, Nicosia, Cyprus, 2004: 133-137.

[8] W. Au, K. C. Chan, A. Wong & Y. Wang, Attribute of grouping, selection, and classification of gene expression data. IEEE/ACM Transactions on Computational Biology and bioinformatics, 2(2): 2005, 83 – 101.

[9] S. Kaski, J. Nikkila, J Sinktonen, L. Lahti, J. Knuuttila, C. Roos, Associative clustering for explaining dependencies between functional genomics data sets. IEEE/ACM Transactions on Computational Biology and bioinformatics, 2(3): 2005, 203 – 216.

[10] L. Parsons, E. Haque & H. Liu, Subspace clustering for high dimensional data: a review, ACM SIGKDD Explorations newsletter, 6(1): 2004, 90-105.

# Grid and Agent based Open DSS Model*

Zhiwu Wang [1,2]
[1] *Department of Control Science and Engineering, Huazhong University of Science and Technology, Wuhan, 430074*
[2] *Zhengzhou Economic Management Institute, Zhengzhou, 450052*
zhiwuwang@mail.hust.edu.cn

Qianqian Wei [1]
[1] *Department of Control Science and Engineering, Huazhong University of Science and Technology, Wuhan, 430074*
treeleave0724@tom.com

Xueguang Chen [1]
[1] *Department of Control Science and Engineering, Huazhong University of Science and Technology, Wuhan, 430074*
xgchen9@mail.hust.edu.cn

*Abstract* - **The paper discusses an open DSS model based on grid and agent technologies for decision resources sharing and reusing. The model faces decision objects and organizes decision resources in the form of e-market, so the model improves the timeliness, openness and intelligent capacities of DSS, also enhances its ability to handle distribution issues. The structure and the operating process of the model are presented in detail and its advantages compared with the traditional DSS are also discussed.**

## I. INTRODUCTION

It seemed that DSS has some new opportunities recently with the development of the correlative scientific fields. D.N. Davis et al. developed a Multiple Agent Decision Support System, which not only remains the multiple base architecture of the traditional DSS, but introduces a flexible architecture of the Agent organization, [1]. Manfred et al. discussed problems such as using distributed DSS to solve the complicated decision problems, and searching proper DSS components in the Internet space, [2]. J. Y. Chi et al. referred to the new concept of Grid-Based Open Decision Support Systems and presented concept model, [3]. Specialists have investigated DSS from different aspects utilizing new technologies. The research outcome has greatly promoted DSS. This paper presents a grid and agent based open DSS model, analyses the architecture and operation mechanism of this model in details, and finally discusses the differences between this model and the traditional DSS.

## II. A GRID AND AGENT BASED OPEN DSS MODEL

Grid technology originated in grid computing, whose purpose is to combine many computers, huge databases, precious research equipments, communication facilities and diversified sensors in different places to a gigantic super computing system which can support high performance computing and scientific research, [4]. With the research going on, The function and application of grid is also expanded, many new types of grid come out such as computing grid, information grid, data grid and service grid. Grid has many excellent characteristics such as resources share in large scale, parallel processing, cooperation with others in different places, open standards support and dynamic services. All these characteristics can solve the supporting incapacities of the traditional DSS in distributed and dynamic environment, improve the share and reuse of decision resources, enhance the flexibility of DSS, reduce the difficulty and cost of the development of DSS, and provide DSS with a brand new platform.

Meanwhile, the new concepts, new methods and new tools appearance in the correlative scientific fields also provided new support for the development of DSS. For example, the research on the solution of problems in distributed system brought new challenges to DSS; the corporation work mechanism of multi-agent system opened up a new perspective for data obtaining, modeling DSS, [5]; Mobile Agent is autonomic, mobile and intelligent which provides new solution for the distributed problems of grid based DSS, [6]; agent-oriented programming also lays the foundation for the constitution and soft-integration of the new DSS. When the new DSS is constructed, we must consider the influence of the new concepts, new methods and new tools on DSS. The new constructed system model is shown as Figure1.

## III. THE ARCHITECTURE ANALYSIS OF GRID AND AGENT BASED OPEN DSS MODEL

The new constructed DSS consists of three components: DSS control center, interior system and decision service e-market. The three components are analyzed as follows.

### *A. DSS control center*

DSS control center, the interface between users and the system, is the core of the new constructed DSS. It functions as problem-analyzing, solution-planning, resource arrangement, task assignment, task-monitoring, result-synthesizing and so on. Multi-Agent and Mobile Agent, which are the latest technologies in the distributed AI fields, are applied to function as DSS control center, to gain excellent human computer interaction, strong distributedness, the ability to solve complicated problems, good intelligence, dynamics and openness.

*1) User Interface Agent*: User Interface Agent helps users' and DSS's mutual cooperation with each other with its decision knowledge, self-knowledge and domain knowledge. It is intelligent that it can do feedback and initiative

* This work is supported by SRFDP Grant #20040487076



*T. Sobh and K. Elleithy (eds.), Advances in Systems, Computing Sciences and Software Engineering*, 201–204.

adjustment according to users' reaction, and direct users' operation to lighten burden.

*2) Problem Disassembling Agent (PD-Agent)*: Problem Disassembling Agent disassembles decision problems that are gained from the User Interface Agent with its decision knowledge, self-knowledge and domain knowledge. A complicated decision problem can be disassembled into a series of sub-problems that can be solved individually. All results of the sub-problems can be synthesized into the solution of the decision problem.

*3) Solution Planning Agent (SP-Agent)*: The sub-problems are obtained by disassembling the complicated decision problem. So there are certain logic and time relations between the sub-problems definitely. Solution Planning Agent arranges the sequence of sub-problems according to the logic and time relations between them. Therefore, the sub-problem solving chain is formed which can make the problem-solving process effective and rapid.

*4) Resource Arrangement Agent (RA-Agent)*: The decision resources used in the decision problem solving process have complex relations with each other. The key to solve sub-problems is deciding the combination relation of all the models, the utilization relation between the models and data, and the coordination relation between model computing and knowledge reasoning, which form the decision resource arrangement chain. Resource Arrangement Agent does decision resource planning and allocating in the system or outer resource market according to the need of problem-solving, thus to optimize the indexes such as time and costs of resource utilization, and the system stability.

*5) Task Assignment Agent (TA-Agent)*: Task Assignment Agent assigns problem-solving tasks and relative resource flow and resource address to proper Mobile Agent. Task Assignment Agent does the assignment according to the problem-solving chain and decision resource arrangement chain that are provided by the Solution Planning Agent and Resource Arrangement Agent. Mobile Agents carry decision tasks to the server, which has the appointed decision resources to do the problem-solving.

*6) Result Synthesizing Agent (RS-Agent)*: Result Synthesizing Agent collects results returned by each mobile agent. It processes from bottom level tasks and gets the result of the super level task according to the logic relation of the bottom level task. It finally obtains the result of the decision problem level by level, and then submits the result to the user by User Interface Agent.

*7) Mobile Agent Library*: First of all, Mobile Agent is a software Agent that satisfies the target-driven characteristic of Agent and is intelligent and autonomic. Secondly, Mobile Agent can move to different address space to execute tasks among network nodes. The executing status can be saved when Mobile Agent is transferring. The execution process persists when Mobile Agent moves to the destination. Mobile Agent can not only carry on the codes and data needed in computing,

but also the Mobile Agent`s status, transferring information among diverse and various servers in the network. We can assort Mobile Agents into three types: resource management mobile agent, resource trade mobile agent, task mobile agent by the task of every Mobile Agent in order to gather them into a library to do good management.

*8) Agent Manager*: Agent Manage is the core component of the new constructed DSS control center. It takes charge of Agent life cycle management, provides services for communication and information share among agents, and adjusts the relations of Agents.

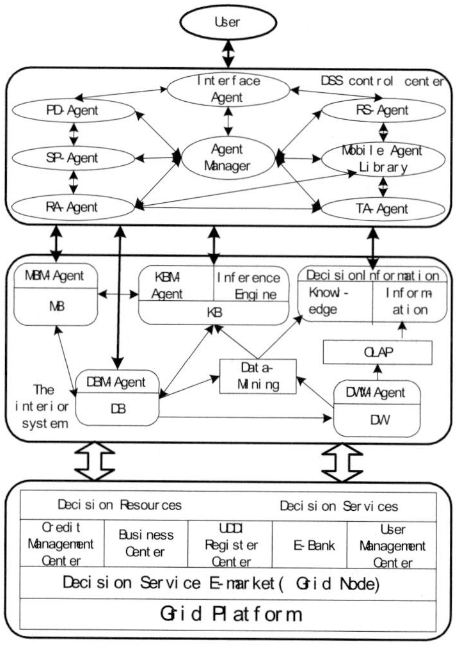


Figure 1. A grid and agent based open DSS model

## *B. The interior system*

Traditional DSS is model-driven; IDSS combines the qualitative analysis of the expert knowledge and quantitative computing ability of DSS model to do the intelligent decision support; data warehouse, data-mining and OLAP based DSS focuses on data, and provides decision support to users by analyzing, and processing the data to obtain information and knowledge. These three different types of DSS can support users to do decision at diverse aspects. With the development of the society, the decision problems and decision environment that users are facing more and more complicated. In order to provide users with more comprehensive and intelligent decision support, model, data and knowledge need to be integrated together to construct compositive and intelligent DSS that is the trend of DSS development and research.

### C. *Decision Service E-market*

Grid based decision service e-market is an environment where decision resources are shared, published and traded. It provides a decision resource sharing platform for decision resource creators and users, which is the foundation of flexibility and dynamic expanding of DSS based on grid and agent. Decision service e-market provides decision resources such as models, data, algorithms, knowledge and decision toolkits, as well as decision service which is the operation environment of decision problem long-distance and distributed solving. Decision service is a compounding system of decision resources and resource operation platform. It is an environment that consists of software and hardware set up by decision resource provider to solve certain decision problems. Countless commercial decision services offer a large number of alternatives for users when buying decision resources and leasing decision services, which is the embodiment of the distributedness, dynamics and openness of grid and Agent based DSS.

The new resource and service organization style, which makes decision resources and decision services as electronic commodity, changes the development and operation mechanisms of DSS. The market-oriented and cooperative development mode of DSS substituted for the traditional independent and close mode. The operation mechanism of DSS does not limit to the centralized mode as well. DSS can operate in an open and distributed environment according to its needs, which promotes resource sharing, technology advancing, cost reducing, fields expanding of DSS.

## IV. Analysis of the Operation Mechanism of Grid and Agent Based Open DSS

Grid and Agent based open DSS is constructed on the new platform and it has a brand new structure whose components and relations of the components have all been changed. The analysis of the operation mechanism of grid and agent based open DSS checks the foundation of every component, and lays out a complete and transparent view to the users. Figure 2 shows the operation flow of the system.

In figure 2, the resource arrangement Agent manages decision resources dynamicly that bespeaks the openness and dynamics of the system. The following text analyzes the three situations the resource arrangement Agent will face.

1) Resource Management Mobile Agent searches all the decision resources needed in solving the decision problem in the interior system. Resource Arrangement Agent arranges the decision resources according to the relative information returned by the Resource Management Agent, and creates the decision resource arrangement chain.

2) Resource Management Mobile Agent finds part of needed decision resources in the interior system. Resource Arrangement Agent sends Resource Management Agent to decision service e-market to search the lacking resources. The information of the needed decision resources which are found

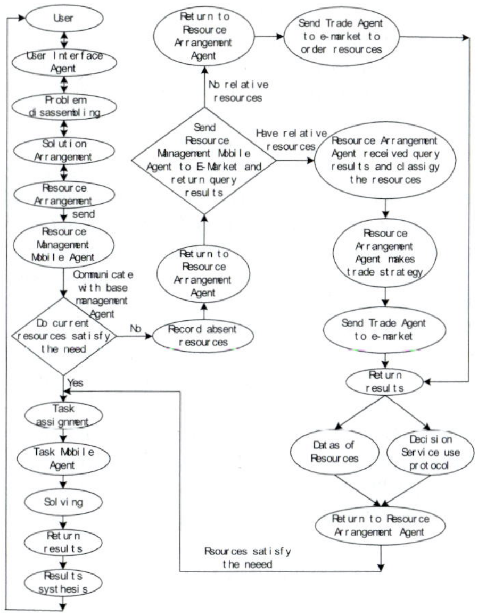


Figure 2. The flow chart of operation mechanism of grid and agent based open DSS

at the resource register center will be returned to Resource Arrangement Agent, that will decide whether to buy or lease the decision service so as to achieve the optimization according to the information of the resources in the interior system. Then, Resource Arrangement Agent makes the trade strategy and sends trade Mobile Agent to the e-maket to do the resource trade. The resource data and specifications of the leased service will be given to Resource Arrangement Agent. Finally, Resource Arrangement Agent completes resource arrangement and makes the resource arrangement chain.

3) If Resource Management Mobile Agent does not find all the decision resources at the interior system or the e-market resource register center, Resource Arrangement Agent will describe the function and interface parameters of the resources on order, make the resource order strategy, send trade mobile agent to the e-market to negotiate with the resource provider brokers and order the resources.

## V. The Differences Between the Grid and Agent Based Open DSS and the Traditional DSS

Analyzing the operation mechanism of grid and agent based open DSS, we can conclude:

1) Grid and agent based open DSS is an open and dynamic system. The organization of decision resources is open, dynamic and flexible. Therefore, it can dynamically expand decision resources via the e-market that is on the grid platform according to the practical needs.

2) Grid and agent based open DSS is a decision problem driven system. Though the traditional model-driven DSS has been advanced, the model-driven theory impedes its openness,

flexibility and the capability of solving dynamic problems at real time. Grid and Agent based DSS orients decision problems and organizes the decision resources based on the problem-solving, so it has excellent openness and flexibility.

3) Grid and agent based open DSS focuses on the openness, dynamics and commercialization of decision resources and decision services. It transforms the mode of DSS development by utilizing the new resource organization, which makes decision resources and decision services as information commodities and trading them on the e-market. The development of DSS transform from the independent and separated mode to the market-oriented open and cooperated mode, which brings DSS to a new era.

4) Grid and agent based open DSS is more intelligent. In this system, intelligent Agent can accomplish communicating with users, disassembling problems, problem solving arrangement, decision resource arrangement, decision task assignment, result synthesis and so on. The intelligence and mobility of Agent enhance the intelligence and the capability of solving distributed problems of the DSS.

## VI. Summary and Prospect

Grid and agent based open DSS is the production of decision support technology combining grid and intelligent agent technology. Utilizing grid's distributed resource sharing and cooperating ability, it improves the distributed cooperating ability of traditional DSS. At the same time, by using multi-agent and mobile agent technology, it enhances the intelligence and distributed problem solving ability. Analyzing the architecture and operation mechanism, the openness and flexibility of the new constructed system is manifested. Decision resources shared in the e-market is the foundation of the system's dynamic and open working. To publish decision resources in the e-market is our further research subject.

## References


[1] D. N. Davis. Agent-based decision support framework for water supply infrastructure rehabilitation and development. Computers, Environment and Urban Systems, vol. 24, no.3, pp. 173-190, 2000

[2] A. J. Manfred. Distributed decision support and organizational connectivity: a case study. Decision Support System, vol. 19, no. 3, pp. 215-225, 1997.

[3] J. Y. Chi, X. G. Chen. A Model of Grid Based Decision Support System. Computer Science, vol. 33, no. 3, pp. 121-124, 2004.

[4] I. Foster. The Grid: A New Infrastructure for 21st Century Science. Physics Today, vol. 55, no. 2, pp. 42-47, 2002.

[5] R. Vahidov, B. Fazlollahi, Pluralistic multi-agent decision support system: a framework and an empirical test. Information and Management, vol. 41, no. 7, pp. 883-898, 2004.

[6] D.B. Lange, and M. Oshima, Seven Good Reasons for Mobile Agents *Comm. ACM,* vol. 42, no. 3, pp. 88-89, Mar. 1999.

# The Design and Implementation of Electronic Made-to-Measure System

*SHI Xiu-jin, SUN Li & CHEN Jia-xun*
College of Computer Science & Technology, Donghua University
1882, Yan An Xi Rd.
Shanghai, 200051 China

***Abstract*-Electronic Made to Measure is a new mode of garment production. This paper analyses the state of eMTM, and introduces the functional components of the eMTM infrastructures, designs it's workflow, then gives one feasible solution. The research work of this paper will play a good role for future research and popularization of eMTM in China.**

## I. INTRODUCTION

There are two modes of traditional garment customization. One is manual customization, where designer (namely tailor) uses such tools as a measuring tape , etc., to measure body shape data for customers, then customizes clothing based on the data. Manual customization has existed for many years in the world. Since its customized period is too long, the precision relies on the tailor's experience and judgment, and it lacks digitized standard, and it needs numerous modification, it makes against mass-customized clothing production and garment customization for remote customer. The other mode is customization according to the size serial standards of clothing, confirming the corresponding garment size based on the customer's body shape first, and then the second stage is to customize clothing by garment size. This mode takes less time, and at the same time it benefits mass-production. But there are many problems in it: First, because personal body shape is different from the standard body shape and people with different areas, different age brackets, even the crowds of different jobs have different body shape feature, depending on a set of unified standard simply can't meet the different people's need of garment fitting. Second, because the size standard of China was formulated relatively early, and the categorized method is simple, it can't reflect accurately the body shape feature of Chinese people at present. And the third is that it is unfavorable to the clothing industry carrying on the dress designing and production accurately based on body shape feature and requirement of the different consumer groups.

Made to Measure (MTM) is a mode which regards the consumer as the center totally. The mode which organically combines body measurement, style choice, body shape analysis, dress design, and clothing order, etc. achieves high-efficient and fast garment customization production [1]. Therefore, MTM can meet the demand of garment fitting, and it can meet the need of not only the group but also the individual. However, for every segment in MTM production is relatively independent, it is unable to form a intact industry roller chain. And some segments are still at manual stage, e.g. the critical step in MTM ----finding the difference in control segment and manually revising the selective template, where the decision depends on the worker's own knowledge structure and working experience, so it limits the industrialized application. As a result, the industrialized MTM production has become the important problem which should be solved urgently with the digitized development of current garment manufacture.

With the rapid development of computer and network communication technology, especially the emergence of the 3D body scanner that serves for garment manufacture makes possible using computer technology to achieve the digitized eMTM production, and it aims to achieve mass-customization utilizing the digitized and networked technology to combine the digitized body size scanning and measurement, digitized body shape analysis, automatic mapping between body shape and size, CAD design, garment customization and vendition through network etc. to, then it makes possible individual clothing and service for every customer.

## II. RESEARCH BACKGROUND

At present, more than twenty academies and a dozen large-scale clothing manufacturing enterprises have already participated in the research of eMTM system. The European textile clothing organization (EURATEX) proposed a plan that aims to achieve a new electronic commerce E-Tailor which uses 3D body scanner with modern network technology and clothing CAD technology to establish eMTM production [2]. Japan and America have already exploited their own eMTM system respectively too, and the systems have been applied in market for test. Such as Baird Menswear suit company of Britain, 80 percent of suits which are sold to internal and external market finished by eMTM system, and the clothing series contain thousands upon thousands combinations with different styles, color and specification. For another instance, Europe is building the body shape databank, England is establishing size serial standards—SizeUK, America is also establishing a new SizeUA and so on. They as the basic technology in eMTM are developing more rapidly. Now EURATEX has made the digitized garment customization as a important direction of studying at the development of clothing industry in the future. EMTM system is also used in



*T. Sobh and K. Elleithy (eds.), Advances in Systems, Computing Sciences and Software Engineering*, 205–208.

mass-customization widely. In 2002,the American navy advanced "Navy Uniform Project". And this project utilizing eMTM system has succeeded in customizing altogether forty shapes of body for 40,000 soldiers [3].

In China, the research of eMTM technology still quite lags behind, and the key technology enumerated above hasn't been solved. One suit of digital garment customization system which fit Chinese garment manufacture hasn't emerged to date. Therefore, researching the key technology of eMTM has very important academic and practical meaning for the digitized development of Chinese garment manufacture. Donghua University has begun to cooperate with European Union in launching "The New Generation Digitized Clothing Technology Research in Eurasia (Eurasia Tex)", and has begun to research eMTM system based on BMS system of the American textile technology corporation($[TC]^2$).

BMS system of $[TC]^2$ is noncontact body scanning and measurement system [4]. It utilizes equipment to cast grid towards body surface, then it arises echo on body surface according to the optical principle, and it is shot by CCD camera, and it conveys the information to the imaging software system through sensor. We can use 3D body scanner to obtain the picture of a human body as well as body model, and its detailed size.

The scanned file format provided by BMS system has many categories which contain image format (.tif), 3D lattice format (.bin, .url), body model format (.rbd) and size format (.ord). Among them size format can obtain more than 200 body size parameters directly, but 3D lattice format can gain more parameters. Figure 1 illustrates some file formats.

## III. OUR RESEARCH

E-Tailor that EURATEX realized has proposed the main functional components of E_TAILOR infrastructure [5], as Fig. 2 shows.

Customer stores his personal information including body data scanned by 3D body scanner to multi-application smartcard, and sends to remote human body measurement databank through Internet. Then customer can remote order garments. There are some key problems need to be tackled.

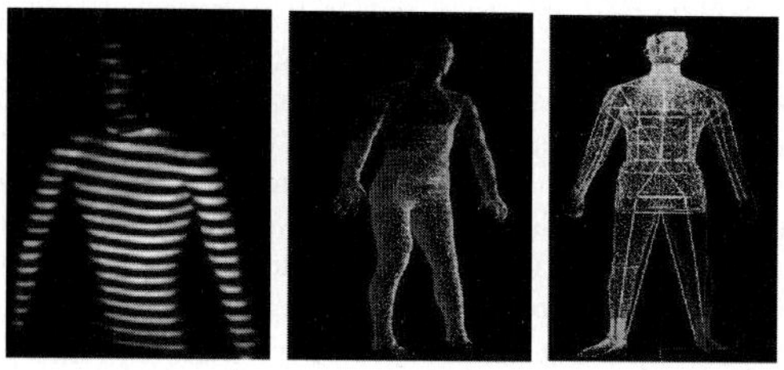

a. Image format b. Lattice format c. Size format

Fig.1. Formats of 3D body scanning image

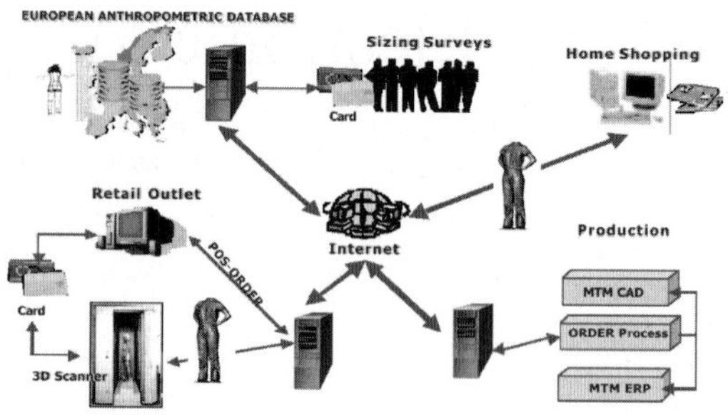


Fig.2. The functional components of the E-TAILOR infrastructure

### A. *Modeling 3D human body*

We can use the VRML file of body by 3D body scanner to create 3D human body model. One simple way to display 3D body model is reading VRML file directly through the browser plug-in of VRML, but the model can't be rotated with the scale freely, and can only show the whole human body, can't show one special part of body alone, and the display mode is single , can't show by way of some clouds or nets. It is inconvenient to further application, so we need to deal with VRML file further by means of programming.

Java3D, as an advanced 3D graphics programming API, nearly includes all function offered by VRML2.0, and have the power of Java at the same time, can use to write out complicated 3D application program. Java3D encapsulates OpenGL and DirectX, adopts scene picture which can compiled to raise the efficiency as data structure. So we adopt Java3D technology to deal with VRML file and create 3D body model. Fig. 3 is the scene picture of 3D body model.

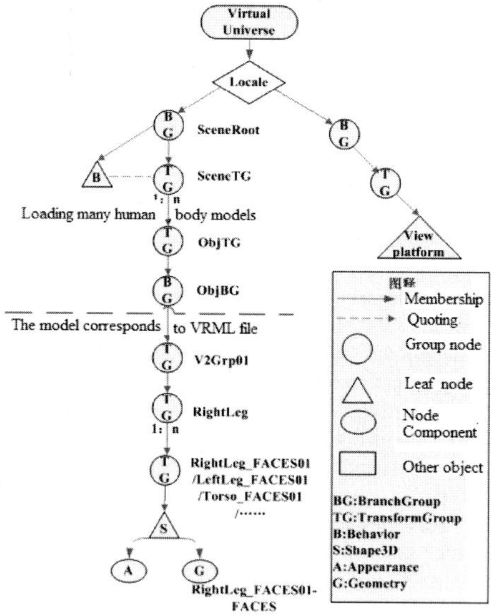


Fig.3 The scene picture of 3D body model

*B. Data exchange*

EMTM system allows users to upload local 3D human body data remotely, can offer various kinds of search information to user. XML as the standard language can help system become open. Customer can upload their data which mapped into RDB on the server in XML. At the same time eMTM deals with the data to store into RDB. Enterprise can query some information in XML, the result from RDB will be converted into XML format.

One XML-RDB displayed in Fig.4 lies in the intermediate tier of three tiers application can realize the data exchange between XML and RDB.

## IV EMTM SYSTEM SOFTWARE

*A. Key Workflow*

The kernel workflow of the eMTM system is shown in fig.5. Donghua University has already cooperated with Wacoal Corporation in the survey of woman's body shape in Jiangsu province, Zhejiang province and Shanghai. 850 groups of the woman's measurement data where the age ranges between 20 years old and 50 years old and the job involve each trade have been obtained using 3D body scanner. The measurement data was made a body shape analysis firstly, then standard body shape & size database was built using clustering analysis method, and the standard 3D human body visualization database was created utilizing Java 3D. After the customer stored the 3D measurement data in body databank with 3D body scanner of BMS system, the corresponding body shape & size can be found in the standard body shape & size databank using matching arithmetic, and the virtual try-on will be achieved after the 3D human body visualization is acquired through individualized processing using standard 3D human body visualization database.

*B. Logic structure*

We use J2EE technology combined with MVC design pattern to design the logic structure of eMTM system. In the construction of the system we consulted the WAF（Web Application Framework） used by BluePrints samples which were issued by SUN corporation [6]. The systemic logic structure is shown in fig.6.

*C. Functional module*

The main functional module is divided into several parts as follows according to the structure of eMTM system:

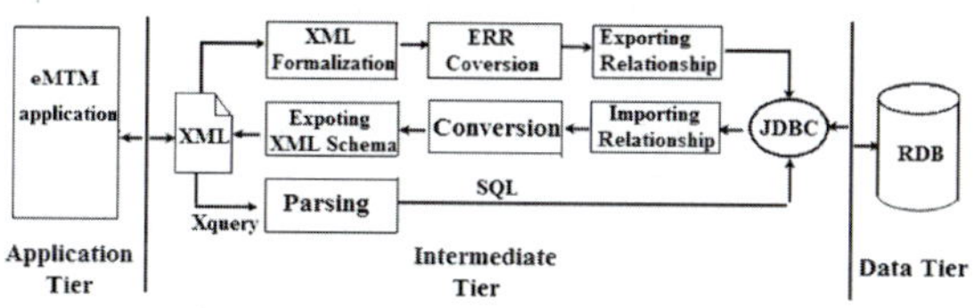


Fig.4. XML-RDB Prototype structure

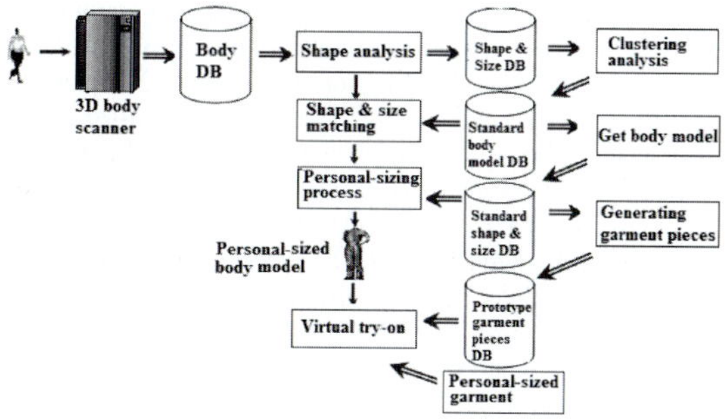


Fig.5. eMTM system workflow

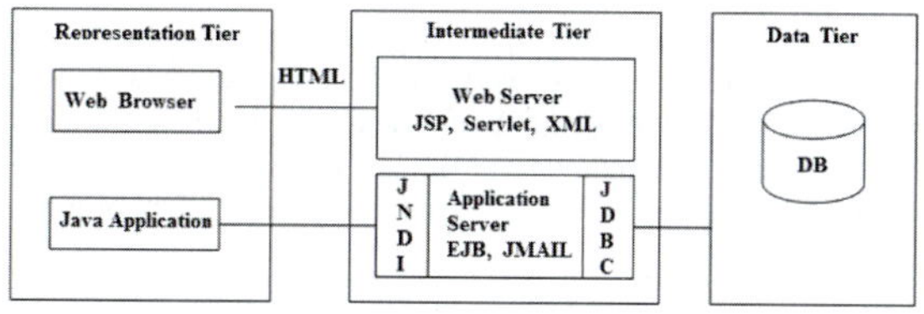


Fig.6. Logic structure of eMTM system

Data support and preparing module: it provides integrated relevant data support for describing body shape, building body measurement database, analyzing body shape and application of eMTM oriented. And the main function of it is collecting and uploading the user's body measurement data and other relevant data, and providing the database access interface for body shape analysis and 3D human body visualization.

Body shape & size analyzing module: the main function of it is according to mass body measurement data using clustering analysis to confirm the classified body shape reference standard, building standard body shape databank, and then with the combination of the former Chinese apparel size serial standard building the apparel size reference standard according with the current Chinese body shape feature, building standard size database. After received the user's body shape data, eMTM system will analyze the user's body shape feature and send the result to the clothing design system. Then the clothing design system will calculate the apparel size according with the user's body shape based on the user's body shape feature, and will automatically chalk out and do template. At the same time, the apparel CAD system will generate 3D models which contains outer swear, business suit, trousers, skirt, shirt etc. style based on the user's need. And these 3D models will be stored in system server with VRML file format.

3D human body visualization module: its main function is generating corresponding 3D human body visualization based on the user's 3D body measurement data, and implementing virtual try-on when the user is online.

Garment customization module: its main function is to achieve the user's garment customization. The user chooses clothing style he needs from the clothing sample catalogue, then downloads the 3D apparel model created by apparel CAD system to the client, shows the model at the client's browser. The system provides virtual-try-on that shows 3D clothing model for user with this method. After he has confirmed the custom-made clothing, the user can make a clothing order online pay for it through network. The eMTM system sends the user's order to the clothing industry which will begin the clothing production and complete the procedure of eMTM.

Smartcard module: it offers multi-application smartcard platform divided into two parts: making cards and retailer clients. Smartcard stores the user's social information and body shape information. Customers send their information stored in smartcards to remote eMTM system through POS machine on Internet, then order garments remotely.

Data exchange module: the main function is describing in XML standard language and establishing the exchange reference standard about 3D body measurement data, thereby the exchange and integration of 3D measurement data is achieved among the different area and operation system with XML document.

In this system the representation tier takes charge of representing the data to the user and interacting with user and communicating with other tiers. For example, it deals with the user's input, shows the analysis result for user, provides service with server and so on. The representation tier communicates with other layers with defining explicit interface, and it is implemented by static HTML page and dynamic JSP page. The intermediate tier is composed of web server and application server. Web server adopts the form of JSP+JavaBean+Servlet. The business agent component at application server is implemented by SessionBean which links the preceding and the following and reduces the network burden. The data service tier connects with database closely. We utilize EntityBean to encapsulate the operation of the relevant table in database and implement the encapsulation of the bottom and the access to the lower tier.

## V. CONCLUSIONS

As a new mode of digitizing clothing production, eMTM production meet clothing production's need of varying from mass but little varieties to few but much varieties and single production development. And it will be widely applied at the garment production in the future. Some functional module in eMTM system designed at this paper have already put into practice, e.g. it provides the service of body shape analysis in woman's underwear production of Wacoal Corporation and has made better result. The impact of the research result of this paper concluded by this paper will play positive guidance and support function in applying and spreading eMTM production in China. And the result offers very useful reference from theory and practice for the digitized development of clothing industry in China.

## ACKNOWLEDGEMENTS

The body scan data used in this research was provided by Fashion Institute of Design Of Donghua University. This project has been supported by the Shanghai Science & Technology Committee of Defense, under contract number 045107026.

## REFERENCES

[1] WAN Zhi-qin, "Garment Production Managemnt", Shanghai, Chinese Textile Press, 2001, 57－80

[2] G.A.Kartsounis, N.Magnenat-Thalmann, Euratex (2001-5), The European Textile/Clothing Industry on the eve of the New Millennium, Brussels, Euratex, 2001

[3] http://bupersed.technology.navy.mil/bup_updt/upd_CD/BUPERS /Unireg/4.PDF

[4] Tailored Clothing Technology Corporation. 3D Body Scanner Specifications. http://www.tc2.com/products/body_scanner.html

[5] G.A.Kartsounis,N.Magnenat-Thalmann,Hans-Christian Rodrian. E-tailor:Integration of 3D Scanners,CAD and Virtual-Try-on Technologies for Online Retailing of Made-to-Measure Carments. http://www.atc.gr/e-tailor/e-Tailor_Paper.PDF

[6] Sun Microsystems, Inc. Designing Enterprise Applications with the J2EE Platform, Second Edition. http://java.sun.com/ blueprints/guidelines

# Ants in text documents clustering

Łukasz Machnik
L.Machnik@ii.pw.edu.pl
Department of Computer Science, Warsaw University of Technology,
ul. Nowowiejska 15/19, 00-665 Warsaw, Poland

*Abstract* - **Ant systems are flexible to implement and give possibility to scale because they are based on multi agent cooperation. The aim of this publications is to show the universal character of that solution and potentiality to implement it in wide areas of applications. The increase of demand for effective methods of big document collections management is sufficient stimulus to place the research on the new application of ant based systems in the area of text document processing. The author will define the ACO (Ant Colony Optimization) meta-heuristic, which was the basis of method developed by him. Presentation of the details of the ant based documents clustering method will be the main part of publication.**

## I. The Origin Of Aco (Ant Colony Optimization)

One of the topics that was deeply explored in the past by ethnologist was the understanding of mechanism how almost blind animals were able to find the shortest way from nest to a food. Comprehension of the way to achieve this task by nature was the first step to implement that solution in algorithms area. Main inspiration to craete ACO metaheuristic were researches and experiments done by Goss and Deneubourg [6].

Ants (*Linepithaema humile*) are the insects that live in the community called colony. The primary goal of ants is the survival of the whole colony. A single specimen is not essential, only bigger community may efficiently cooperate.

Ants possess the ability of such efficient cooperation. It bases on work of many creatures, who evaluate one solution as a colony of cooperative agents. Individuals do not communicate directly. Each ant creates its own solution that contributes to the whole colony's solution [8]. The ability to find the shortest way between the source of the food and the ant-heel is a very important and interesting behavior of the ant colony. It has been observed that ants use the specific substance called pheromone to mark the route they have already gone through. When the first ant randomly chooses one route it leaves the specific amount of pheromone, which gradually evaporates. Next ants which are looking for the way, will, with greater probability, choose the route where they feel more pheromone and after that they leave their own pheromone there. This process is autocatalic – the more ants choose a specific way, the more attractive it stays for the others. Above information bases mainly on Marco Dorigo publications. He is the one who most of all contributed to develop the research in the ant systems area. His publications are the bigest repository of ACO information [7] [8].

## II. ACO-Based Clusterig Method

Noticed analogy between finding the shortest way by ants and finding documents most alike (the shortest way between documents), and in addition ability to use agents who construct their individual solutions as an element of the general solution, became the stimulus to begin research on using the ant based algorithms in the documents clustering process.

### A. *Details of processing*

The method of document clustering which is introduced here, is based on artificial ant system. Application of such solution will be used as a method of finding the shortest path between the documents, which is the goal of the first phase (trial phase) of considered method. The second phase (dividing phase) will have a task to actually separate a group of documents alike.

The aim of trial phase is to find the shortest path connecting every document in the set using ACO algorithm. That is equivalent to building a graph, whose nods would make up a set of analyzed documents. The probability of choosing next document *j* by ant *k* occupying document *i* is calculated by following function (1).

$$p_{ij}^{k}(t) = \frac{[\tau_{ij}(t)]^{\alpha} * [s_{ij}]^{\beta}}{\sum_{k \in Z_k} [\tau_{ik}(t)]^{\alpha} * [s_{ik}]^{\beta}} \qquad (1)$$

In the above formula, $Z_k$ represents list of documents not visited by ant *k*, $\tau_{ij}(t)$ represents the amount of pheromone trail between documents *i,j*, $\alpha$ is intensity of pheromone trail parameter, $\beta$ is visibility of documents parameter, however $s_{ij}$ is cosine distance between documents *i* and *j*. After ants complete their trace the pheromone trail is evaporated and new amount of pheromone is left between every pair of documents. The amount of pheromone that is left by the ants is dependent on quality the constructed solution (length of the path). In practice, adding the new portion of pheromone to trail and its evaporating is implemented by formula presented below. This formula (2) is adapted to every pair of documents (*i,j*).

$$\tau_{ij}(t) \leftarrow (1 - \rho) * \tau_{ij}(t) + \Delta\tau_{ij}(t) \qquad (2)$$



*T. Sobh and K. Elleithy (eds.), Advances in Systems, Computing Sciences and Software Engineering*, 209–212.

In the above formula, $\rho \in (0,1>$ stands for the pheromone trail decay coefficient, while $\Delta\tau_{ij}(t)$ is an increment of pheromone between documents (i,j). Below the dependence (3) that controls the amount of pheromone left by ant k between pair of documents (i,j) is presented.

$$\Delta\tau_{ij}^{k}(t) = \begin{cases} n / L_k(t) & \text{, for (i,j)} \in T^k(t) \\ 0 & \text{, for (i,j)} \in T^k(t) \end{cases} \quad (3)$$

In the above formula, $T^k(t)$ means a set of document pairs that belong to path constructed by ant k, $L_k(t)$ is length of path constructed by ant k, while n is the amount of all documents. Finding the shortest path connecting every document in the set will be equivalent to building a graph, which nods would make up a set of analyzed documents. Documents alike would be neighboring nods in the graph, considering that the rank of the individual nods will fulfill the condition of being smaller or equal to 2, which means that in the final solution one of the documents would be connected to only two others (similar) – each document in the designed solution would appear only ones. Gaining of such solution would mean the end of the first phase, known as *preparing*.

The code below represents the trial phase.

```
Procedure sequence_preparation()
{
        reset_pheromone();
        inialize_ants(number_of_ants);
              for(number_of_ants)
              {
                reset_ant();
                build_solution();
                update_best_document_sequence();
              {
         distribute_pheromone();
 }
```

```
Procedure build_solution()
{
     while  (available_documents)
            {
            update_ant_memory(current_document);
compute_transition_probablities(current_doc,ant_mem);
            choose_document();
            move_to_next_document();
            }
     record_document_sequence();
}
```

The Fig. 1 represents the result of trial phase.

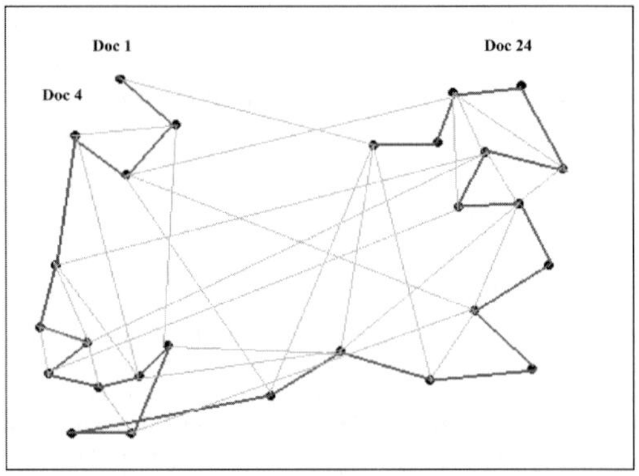


Fig. 1. The shortest path between documents.

In the following stage of the process it is necessary to separate a group of documents alike in a sequence obtained in the first phase. The separation of groups is obtained by appropriate processing of the sequence of documents (the shortest path) received in the preparing phase. Following individual steps of that process are described. The vector that represent the first document in sequence is recognized as centroid $\mu$ of the first group that is separated. In the next step we calculate the sum of all elements (positions) of the centroid vector. After that we calculate the cosine distance between centroid vector $\mu$ and vector $D$ that represent the next element of documents sequence. Next, we check the condition (4). If it is true, then the considered element permanently becomes the member of first group. We recalculate the value of centroid and try to wide this group by adding the next element from sequence.

$$\delta * \sum_{k=1}^{n} t_{\mu k} < \cos(\mu, D) \quad (4)$$

The δ parameter is called attachment coefficient and its range is (0,1>. However, if the condition is false, then the separation of first group is finished and the separation of the next (second) group begins. Vector of considered document that couldn't be added to the first group becomes initial centroid of the new group. The whole process is repeated from the beginning. Processing is finished when whole sequence of documents is done.

The Fig. 2 represents groups separation.

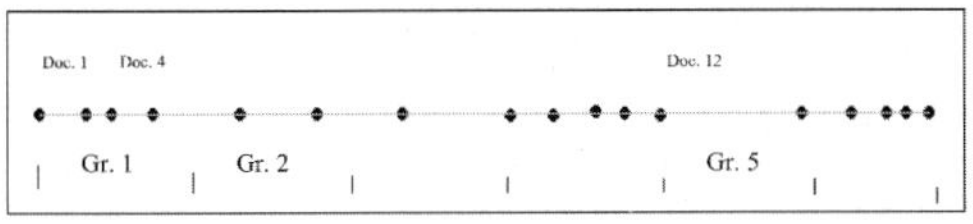


Fig. 2. Groups separation.

The code below represents the dividing phase.

```
Procedure groups_separation()
{
while  (available_documents)
     {
            if (current_document==first_document)
            {
             new_group_creation();
             add_document_to_group(current_document);
             centroid_calculation(current_group);
             }
       else
         {
          if (check_attachment_condition)
               {
                add_document_to_group(current_document);
                centroid_calculation(current_group);
               }
          else
               {
               new_group_creation();
               add_document_to_group(current_document);
               centroid_calculation(current_group);
               }
            }
     }
}
```

B. *Variants of method*

The amount of separated groups depends precisely on attachment coefficient. When we use a big value (close to 1) of δ parameter as a result of processing we received a big amount of groups with high degree of cohesion. The decrease of δ value causes receiving smaller amount of groups with less cohesion. In connection with above conclusion there is a possibility to propose two variants of considered method.

The first variant called by author – single pass, is based on very precise execution of the trial phase - a lot of ants. The duration of the first phase increases, however this activity permits to accept smaller value of attachment coefficient during dividing phase and finishing processing after single pass of algorithm – single trial phase and single dividing phase.

The clustering method that uses the single pass variant is the example of non-hierarchical clustering method. The main advantage of that method is that operator does not have to set the expected number of clusters at the beginning of processing. Results received in this variant are less precise than results from second variant, however the time of processing is much shorter than time of a second proposed variant. This type of considered method can also act as trial phase for other clustering algorithms. The example can be separations of centroids for K-means method.

Second variant called by author – periodic, differs a little bit from variant proposed earlier. It assumes periodic processing of both phases: trial and dividing. In every iteration of dividing phase the small numbers of neighbors are connected into small groups. The value of attachment coefficient is very high in initial phases and is gradually decreased to allow group creation in next iterations. Each group during processing is represented by centroid. After group creation and centroids calculations the next iteration can be started – finding the shortest path between centroids and documents. The whole process is finished when all documents are connected as a single cluster or when the stop criterion is reached.

This variant is an example of agglomerative hierarchical clustering method that begins from a set of individual elements which are then connected to the most similar elements forming bigger and bigger clusters. The result of hierarchical technique processing is creating nested sequence of partitions. The main partition is placed at the top of hierarchy. It includes all elements from considered collections. The base of hierarchy creates individual elements. Every middle level can be represented as combination of clusters that are at the lower level in hierarchy. User can choose any level that satisfied him as solution.

C. *Optimization*

The second variant proposed by author is the dynamic one. It means that during each iteration optimal solution (the shortest path) is changed. It is indicated to use optimization method that adopts solution to changing optimum. The key aspect is to use solution that was received in previous phases – previous iterations; to find a solutions to the changing problem. Till now one of the dynamic problems that was solved by using ant algorithms was the problem of finding route in telecommunication network was [9] [10]. In presented method (periodic variant), a change (adding new calculated centroids) takes place in the exact point of time (next iteration) and it is required that algorithm should adopt to the change. In the basic version of presented method after the problem is changed (adding new centroids and erasing early grouped documents) algorithm is reset. If we assume that the change of problem is relatively small, it is probable that the new optimum will be connected with the old one. It can be useful to transfer a knowledge that was discovered during creating the old solution to build the new one.

To reach the strategy described above, author proposes to use the modification of pheromone trail between documents as a response for changing the problem: adding new centroid and erasing document. During pheromone trail modification the problem is to keep right balance between resetting the correct amount of pheromone, to make process of finding new optimal solution flexible, and keeping enough knowledge to accelerate searching process.

The strategies of pheromone modification were presented inter alia in publications [11] [12]. Modifications that were described in those publications can be called - global, but their disadvantage is the fact that they do not include place where the change took place. According to that, to calculate initial amount of pheromone trail for iterations <>1, author proposes using strategy that is called η-startegy, described in [13]. The „η-startegy" uses heuristic information, distance between documents, to define a degree of compensation that should be performed on a value of pheromone trail. This method is based on implementing the function that is presented below to calculate pheromone trail for every couple of documents/centroids (*i,j*):

$$\tau_{ij} \leftarrow (1 - \gamma_i) * \tau_{ij} + \gamma_i*(n-1)^{-1} . \quad (5)$$

Parameter $\gamma_i \in <1,0>$ is called the reset value and for every document/centroid its value is proportional to distance between document/centroid *i* and new added element *j*. The value of reset parameter:

$$\gamma_i = \max( 0 , d_{ij}^s ) , \quad (6)$$

where

$$d_{ij}^s = 1 - (s_{avg}/\lambda*sij), \quad (7)$$

$$s_{avg}=[n*(n-1)]^{-1} \Sigma^n_{i=1} \Sigma_{k<>i} s_{ki} , \quad (8)$$

$$\lambda \in <1,\infty). \quad (9)$$

Parameter *n* define the number of elements that take part in processing.

## III. SUMMARY

Clustering of big documents sets may be classified as a complicated computing problem. Ant systems are flexible to implement and give possibility to scale because they are based on multi agent cooperation. Experiments have shown that idea of using ACO meta-heuristic in described way to solve document clustering problem is useful and functional. Indication the place of that method in the document clustering area needs comparison with other available methods. The research is in progress and detailed results of tests will be soon published.

## REFERENCES


[1] J.-L. Deneubourg, S. Goss, N. Franks, A. Sendova-Franks, C. Detrain, L. Chretien, *The dynamics of collective sorting: Robot-like ants and ant-like robots*, First International Conference on Simulation of Adaptive Behaviour: From Animals to Animats 1, 356-365, MIT Press, MA, 1991.

[2] E. Lumer, B. Faieta, *Diversity and adaptation in populations of clustering ants*, Third International Conference on Simulation of Adaptive Behaviour: From Animals to Animats 3, 501-508, MIT Press, 1994.

[3] Ł. Machnik, *Documents Clustering Techniques*, IBIZA 2004, Poland, 2004.

[4] H. Azzag, G. Venturini, *A clustering model using artificial ants*, Universite Francois-Rabelais, France, 2004.

[5] A. Lioni, C. Sauwens, G. Theraulaz, J.-L. Deneubourg, *The dynamice of chain formation In Oecophylla longinoda*, Journal of Insect Behavior, vol. 14, 2001.

[6] J.-L. Deneubourg, J. M. Pasteels, J.C. Verhaeghe, *Probabilistic behaviour in Ants: a strategy of errors*, Journal of Theoretical Biology, 259-271,1983.

[7] M. Dorigo, *Optimization, Learning and Natura Algorithms* (In Italia), PhD thesis Dipartimento di Elettronica e Informazione, Politecnico di Milano,IT, 1992.

[8] M. Dorigo, V. Maniezzo, A. Colorni, *The ant systems: optimization by colony of cooperating agents*, IEEE Transactions on Systems, Man, and Cybernetics-PartB, 1996.

[9] G. Di Caro, M. Dorigo, „*AntNet: Distributed Stigmergetic Control for Communications Networks*", Journal of Artificial Intelligence, 1998.

[10] R. Schoonderwoerd, O. Holland, J.Bruten, L. Rothkrantz, „*Ant-based Load Balancing in Telecommunications Networks*", Adaptive Behavior, 1996.

[11] L.-M. Gambardella, E. D. Taillard, M. Dorigo, „*Ant Colonies for the Quadratic Assignment Problem*", Journal of the Operational Research Society, 1999.

[12] T. Stützle, H. Hoos, „*Improvements on the ant system: Introducing MAX(MIN) ant system*", Proc. Of the International Conf. On Artificial Neural Networks and Genetic Algorithms, Springer-Verlag, 1997.

[13] M. Guntsch, M. Middendorf, „*Pheromone Modification Strategies for Ant Algorithms applied to Dynamic TSP*", Proceedings of EvoWorkshops, Italy, 2001.

# Optimal Control in a Monetary Union: An Application of Dynamic Game Theory to a Macroeconomic Policy Problem

Reinhard Neck and Doris A. Behrens
Department of Economics,
University of Klagenfurt
Universitaetsstrasse 65–67, A-9020 Klagenfurt, Austria

***Abstract*—We develop a dynamic game model to study the optimal control of the economies in a two-country monetary union under strategic interactions between macroeconomic policy-makers. In this union, governments of participating countries pursue national goals when deciding on fiscal policies, whereas the common central bank's monetary policy aims at union-wide objective variables. For a symmetric demand shock, we derive numerical solutions of the dynamic game between the governments and the central bank. The different solution concepts for this game serve as models of a conflict between national and supra-national institutions (non-cooperative Nash equilibrium) on the one hand and of coordinated policy-making (cooperative Pareto solutions) on the other. We show that there is a trade-off between instruments' and targets' deviations from desired paths; moreover, the volatility of output and inflation increase when private agents react more strongly to changes in actual inflation.**

## I. Introduction

DYNAMIC macroeconomic policy-making has usually been regarded as a problem of optimal control of a national economy under a policy-maker's objective function in the theory of quantitative economic policy. However, conflicts between decision-makers with different objectives constitute an essential element of the policy-making process. In particular, different policy-making institutions, which are responsible for specific policy instruments and/or areas, may differ with respect to their preferences. For example, central banks are often highly adverse against inflation, while governments frequently put more emphasis on goals like full employment or high GDP growth. Another possible conflict may arise between policy-makers of different countries, who often pursue primarily their national interests and do not care about effects of their actions to other countries. There may be even a conflict of interest between the policy-makers of a country and the preferences of (a majority of) their own citizens. For modeling these and similar potential conflicts, dynamic game theory has proved to be a valuable analytical tool, and several macroeconomic policy applications of this theory can be found in the literature (see, e.g., [4], [5]).

This research was financially supported by the Jubilaeumsfonds of the Austrian National Bank (project no. 9152) and the Ludwig Boltzmann Institute for Economic Analyses, Vienna.

Dynamic game models are usually more complex than optimum control problems, and only in rare cases analytical solutions for these models are available (see, e.g., [1]). Therefore, even for small macroeconomic models, numerical solutions or approximations to them are the best one can hope for. In this paper, we use the OPTGAME algorithm [3] to analyze a macroeconomic policy problem for a two-country monetary union. OPTGAME is a numerical algorithm designed for calculating solutions of dynamic games with a finite planning horizon. It solves discrete-time LQ (linear-quadratic) games, and approximates the solutions of nonlinear-quadratic difference games by iteration. At present, the algorithm calculates the open-loop and the feedback Nash equilibrium solution and the cooperative Pareto-optimal solutions for an arbitrary number of players.

## II. The Model

In a monetary union, national currencies and national central banks are completely replaced by a common currency and a common central bank, respectively. Here we consider the simple case of a monetary union consisting of two symmetric countries (countries of equal size with identical model parameters).

In the following description of the macroeconomic model, capital letters indicate nominal values, while lower case letters correspond to real values. The superscripts $^d$ and $^s$ denote demand and supply, respectively. Model parameters are denoted by Greek letters. The model consists basically of a (mostly exogenous) supply side showing long-run equilibrium growth and short-run deviations from the long-run growth path due to Keynesian features of goods and financial markets. The two countries under consideration are linked both through goods markets (exports and imports of goods and services) and through the integrated money markets. Three active policy-makers are considered: the governments of the two countries and the common central bank of the monetary union.

The goods market for each country is modeled by a short-run income-expenditure equilibrium relation superimposed



*T. Sobh and K. Elleithy (eds.), Advances in Systems, Computing Sciences and Software Engineering*, 213–218.

on an exogenous natural growth path. For $t = 1,...,T$, real output in country $i$ ($i = 1,2$) at time $t$ is given as the sum of the long-run equilibrium level of real output, $\overline{y}_{it}$, and the short-term deviation there from, $\widetilde{y}_{it}$, i.e.,

$$y_{it} = \overline{y}_{it} + \widetilde{y}_{it}, \tag{1}$$

where

$$\overline{y}_{it} = (1+\theta)\overline{y}_{i(t-1)}, \ \overline{y}_{i0} \text{ given,} \tag{2}$$

$$\widetilde{y}_{it} = \frac{\delta_i (P_{jt} - P_{it})}{P_{it}} - \gamma_i (r_{it} - \theta) + \rho_i \widetilde{y}_{jt} - \eta_i \widetilde{f}_{it} + z_{it}, \tag{3}$$

for $i \neq j$ ($i, j = 1,2$). The variable $P_{it}$ ($i = 1,2$) denotes country $i$'s general price level, $r_{it}$ ($i = 1,2$) represents country $i$'s real interest rate, and $\widetilde{f}_{it}$ ($i = 1,2$) denotes country $i$'s real fiscal surplus (if negative, its real fiscal deficit). $\widetilde{f}_{it}$ ($i = 1,2$) in (3) is country $i$'s fiscal policy instrument, i.e., its control variable. The natural real growth rate, $\theta \in [0,1]$, is assumed to be equal to the natural real rate of interest. The parameters $\delta_i$, $\gamma_i$, $\rho_i$, $\eta_i$, $i = 1,2$, in (3) are assumed to be positive. The variables $z_{1t}$ and $z_{2t}$ are non-controlled exogenous variables and represent exogenous demand-side shocks on the goods market.

For $t = 1,...,T$, the current real rate of interest for country $i$ ($i = 1,2$) is given by

$$r_{it} = R_{Et} - X_{it}, \tag{4}$$

where $R_{Et}$ denotes the common nominal rate of interest determined by the common central bank of the monetary union, and $X_{it}$ ($i = 1,2$) represents country $i$'s rate of inflation. The long-run equilibrium or natural (nominal and real) interest rate, $\overline{R}_{Et} = \overline{r}_{it} = \theta$, is "inflation-free", i.e. $\overline{X}_{it} = 0$ for $i = 1,2$.

The general price levels and inflation rates for $i = 1,2$ and $t = 1,...,T$ are determined according to an expectations-augmented Phillips curve, i.e., the rate of inflation depends positively on expected inflation and on goods market excess demand:

$$P_{it} = (1 + X_{it})P_{i(t-1)}, \quad P_{i0} \text{ given,} \tag{5}$$

$$X_{it} = X_{it}^e + \xi_i \widetilde{y}_{it}, \tag{6}$$

where $\xi_1$ and $\xi_2$ are positive parameters. $X_{it}^e$ ($i = 1,2$) is the rate of inflation of country $i$ ($i = 1,2$) expected to prevail during time period $t$, which is formed at the end of time period $t - 1$, $t = 1,...,T$. Inflationary expectations are formed according to the hypothesis of adaptive expectations:

$$X_{it}^e = \varepsilon_i X_{i(t-1)} + (1 - \varepsilon_i)X_{i(t-1)}^e, \tag{7}$$

where $\varepsilon_i \in [0,1]$ for $i = 1,2$ are positive parameters determining the speed of adjustment of expected to actual inflation.

We also define average variables for output and inflation in the monetary union as

$$y_{Et} = \omega y_{1t} + (1-\omega) y_{2t}, \quad \omega \in [0,1], \tag{8}$$

$$X_{Et} = \omega X_{1t} + (1-\omega) X_{2t}, \quad \omega \in [0,1]. \tag{9}$$

Real money demand in country $i$ ($i = 1,2$) is the sum of long-run and short-run real money demand:

$$m_{it}^d = \overline{m}_{it}^d + \widetilde{m}_{it}^d. \tag{10}$$

Short-run real money demand is determined by a Keynesian money demand function involving both the transactions and the speculative motive:

$$\widetilde{m}_{it}^d = \kappa_i \widetilde{y}_{it} - \lambda_i (R_{Et} - \theta). \tag{11}$$

Here $\kappa_i$, $\lambda_i$ ($i = 1,2$) are positive parameters, $\theta$ is the natural rate of interest, and $R_{Et}$ denotes the common nominal interest rate. In accordance with the long-run equilibrium relations, $\overline{y}_{it} = y_{it}$, $\widetilde{y}_{it} = 0$, $\overline{X}_{it} = 0$ and $\overline{r}_{it} = \theta$ ($i = 1,2$), and long-run equilibrium money demand is given by

$$\overline{m}_{it}^d = \kappa_i \overline{y}_{it}. \tag{12}$$

This leaves us with the following relationship for the long-run demand for money in country $i$ ($i = 1,2$):

$$\overline{M}_{it}^d = P_{it}\overline{m}_{it}^d = P_{it}\kappa_i (1+\theta)\overline{y}_{i(t-1)}. \tag{13}$$

In a monetary union, the sum of the countries' money demands has to be equal to the monetary union's money supply. Here we assume the money market always to clear in the short-run, too, and hence money supply to be equal to the sum of short-run money demands in countries 1 and 2,

$$M_{Et}^s = M_{1t}^d + M_{2t}^d. \tag{14}$$

This leads to the money market equilibrium condition for the monetary union:

$$M_{Et}^s = \kappa_1 y_{1t} P_{1t} + \kappa_2 y_{2t} P_{2t} - (\lambda_1 P_{1t} + \lambda_2 P_{2t})(R_{Et} - \theta) \tag{15}$$

This implies that in both countries the price level will stay constant in the long run if money supply, $\overline{M}_{Et}^s$, grows at the natural rate $\theta$.

The government budget constraint is given as an equation for government debt of country $i$ ($i = 1,2$),

$$D_{it} = (1 + R_{E(t-1)})D_{i(t-1)} - F_{it} - \beta_i \widetilde{B}_{Et}, \quad D_{i0} \text{ given,} \tag{16}$$

where the nominal fiscal surplus of country $i$ ($i = 1,2$) is determined by the identity

$$F_{it} = P_{it} f_{it} = P_{it} \widetilde{f}_{it}. \tag{17}$$

$\widetilde{B}_{Et}$ denotes the short-term deviation of high-powered money, $B_{Et}$, from its long-run equilibrium level, $\overline{B}_{Et}$. The long-run (equilibrium) stock of high-powered money is assumed to grow at the constant natural rate $\theta$. Hence,

$$B_{Et} = \overline{B}_{Et} + \widetilde{B}_{Et} = (1+\theta)\overline{B}_{E(t-1)} + \widetilde{B}_{Et}. \tag{18}$$

$\widetilde{B}_{Et}$ represents the control variable of the monetary union's common central bank. The change in high-powered money is distributed as seigniorage to the two countries according to given positive parameters $\beta_1 \in [0,1]$ and $\beta_2 := 1 - \beta_1$. Assuming a constant money multiplier, $\psi$, the broad money supply of the monetary union is given by

$$M_{Et}^s = \psi B_{Et}. \tag{19}$$

Both national fiscal authorities are assumed to care about stabilizing inflation, output, debt, and fiscal deficits of their own countries, i.e., they aim at zero inflation, natural output growth, zero government debt and a balanced budget at each time $t$. The common central bank is interested in stabilizing inflation and output in the monetary union and in a low variability of its supply of high-powered money. Hence, the individual objective functions (loss functions, to be minimized) of the national governments ($i = 1,2$) and of the common central bank are given by

$$J_i = \frac{1}{2}\sum_{t=1}^{T}\left(\frac{1}{1+\theta}\right)^t \Big(\alpha_{iy}\left(y_{it} - \bar{y}_{it}\right)^2 + \alpha_{iX}\left(X_{it} - \bar{X}_{it}\right)^2 + \alpha_{iD}\left(D_{it} - \bar{D}_{it}\right)^2\Big) + \frac{1}{2}\sum_{t=1}^{T}\left(\frac{1}{1+\theta}\right)^t \alpha_{if}\tilde{f}_{it}^2, \tag{20}$$

$$J_E = \frac{1}{2}\sum_{t=1}^{T}\left(\frac{1}{1+\theta}\right)^t \Big(\alpha_{Ey}\left(y_{Et} - \bar{y}_{Et}\right)^2 + \alpha_{EX}\left(X_{Et} - \bar{X}_{Et}\right)^2 + \alpha_{EB}\tilde{B}_{Et}^2\Big), \tag{21}$$

where all weights are positive numbers $\in [0,1]$. The joint objective function for the calculation of the cooperative Pareto-optimal solution is given by $J = \mu_1 J_1 + \mu_2 J_2 + \mu_E J_E$, ($\mu_1, \mu_2, \mu_E \geq 0$, $\mu_1 + \mu_2 + \mu_E = 1$).

The parameters of the model are specified numerically in the simplest possible way, leaving us with a symmetric monetary union (see Table I). For the parameter $\varepsilon_i$, Table I gives the value for our benchmark case (Section III); it will be varied later on (Section IV). The target values assumed for the objective variables of the players are given in Table II; they are basically the long-run equilibrium values of the respective variables. The initial values of the state variables of the dynamic game model are shown in Table III.

Equations (1) – (21) constitute a nonlinear dynamic game with a finite planning horizon, where the objective functions are quadratic in the paths of deviations of state and control variables from their respective desired values.

TABLE I
PARAMETER VALUES FOR THE SYMMETRIC MONETARY UNION

| $T$ | $\theta$ | $\delta_i$, $\gamma_i$, $\rho_i$, $\varepsilon_i$, $\omega$, $\mu_i$ | $\xi_i$ | $\lambda_i$ | $\psi$ | $\eta_i$, $\kappa_i$, $\alpha$'s | $\mu_i$ |
|---|---|---|---|---|---|---|---|
| 20 | 0.03 | 0.5 | 0.25 | 0.15 | 2.0 | 1.0 | 0.33 |

TABLE II
TARGET VALUES FOR THE SYMMETRIC MONETARY UNION

| $\bar{y}_{it}$ | $\bar{y}_{Et}$ | $\bar{X}_{it}$ | $\bar{X}_{Et}$ | $\bar{D}_{it}$ | $\bar{\tilde{f}}_{it}$ | $\bar{\tilde{B}}_{Et}$ |
|---|---|---|---|---|---|---|
| $(1+\theta)^t$ | $(1+\theta)^t$ | 0 | 0 | 0 | 0 | 0 |

TABLE III
INITIAL VALUES AT TIME 0 FOR THE SYMMETRIC MONETARY UNION

| $\bar{y}_i$ | $\tilde{y}_i$ | $P_i$ | $X_i$ | $D_i$ | $R_E$ | $\bar{B}_E$ | $\tilde{f}_i$ | $\tilde{B}_E$ |
|---|---|---|---|---|---|---|---|---|
| 1 | 0 | 1 | 0 | 0 | θ | 1 | 0 | 0 |

## III. Optimal Fiscal and Monetary Policies: A Benchmark Case

Next, we study the behavior of the model under different assumptions about the shocks acting on the monetary union, i.e., about the paths of the exogenous non-controlled variables. We consider a temporary positive symmetric demand shock influencing the economies of the two countries in the same way. In particular, we assume that autonomous real output (GDP) in both economies rises by 1.5 % of GDP above the long-run equilibrium path for the first four periods and less (declining) for the next three periods: $z_{i0} = 0$, $z_{i1} = z_{i2} = z_{i3} = z_{i4} = 0.015$, $z_{i5} = 0.01$, $z_{i6} = 0.005$, $z_{i7} = 0.0025$, and $z_{it} = 0$ for $t \geq 8$, $i=1,2$. We start with the benchmark case of $\varepsilon_i = 0.5$.

Without policy intervention, this demand side shock leads to higher output during the first four periods and to higher inflation during the first five periods (compared to the long-run equilibrium path), but lower output and inflation afterwards (see the path denoted as "uncontrolled" in Figs. 3 and 4). The uncontrolled dynamic system adjusts in dampened oscillations, approaching the long-run path only slowly (not within the planning horizon of twenty periods). Deviations of output from its equilibrium path amount to less than 0.33 % (in the first period), of actual inflation less than 0.11 %; the expected rate of inflation remains below 0.1 %. This shows that, even without policy intervention, there is sufficient negative feedback in the system to reduce the impact of the shock on output to not more than one fifth of the original shock in the case of a temporary symmetrical shock. The price level rises cumulatively by less than 4.4 % until period five. Due to the symmetry of the economies and of the shock, the reactions of all variables are identical in both economies.

When policy-makers are assumed to react on this shock according to their preferences as expressed in their objective functions, outcomes depend on the assumptions made about the behavior of the respective other policy-makers. Here we consider two non-cooperative equilibrium solutions of the resulting dynamic game, namely the open loop Nash and the feedback Nash equilibrium solution, and one cooperative solution, the Pareto-optimal collusive solution (where all players get the same weight $\mu_i = 1/3$, $i = 1,2,E$). The feedback Nash equilibrium solution is subgame perfect or Markov perfect, while the open-loop Nash equilibrium solution requires assuming that all policy-makers commit themselves unilaterally and decide upon trajectories of their instrument variables once for all at $t = 0$.

The trajectories of the control variables – real fiscal surplus for either country and additional high-powered money for the central bank – under the three solution concepts considered are shown in Figs. 1 and 2, respectively, those of the state (and target) variables' deviations from long-run equilibrium output and inflation in Figs. 3 and 4, respectively. The common nominal rate of interest exhibits a behavior very similar to the uncontrolled case. All country-specific variables show exactly the same time paths for both countries.

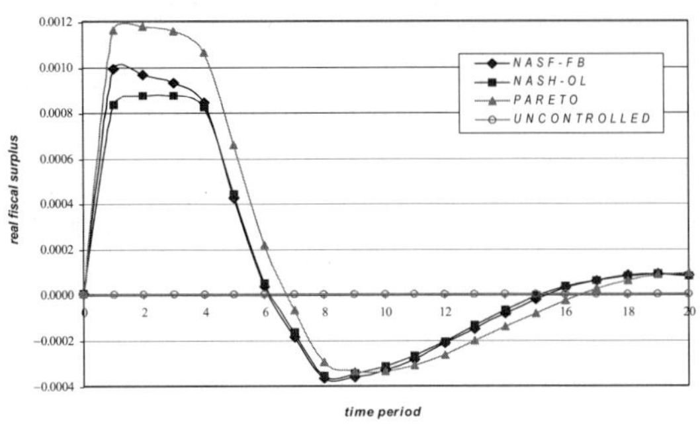


Fig. 1. Country $i$'s fiscal surplus for $i = 1, 2$ and $\varepsilon = 0.5$

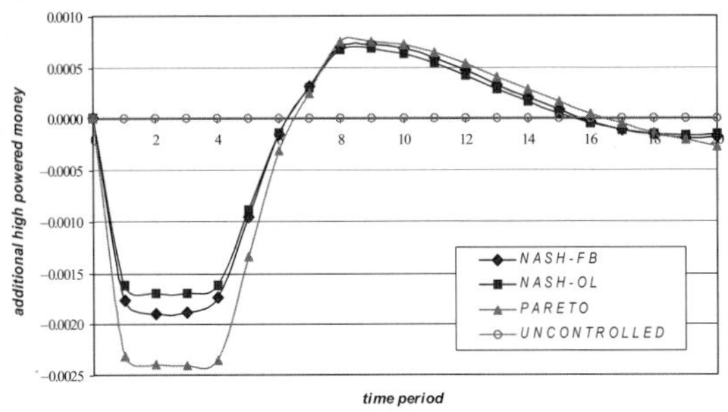


Fig.2. Additional high-powered money for $\varepsilon = 0.5$

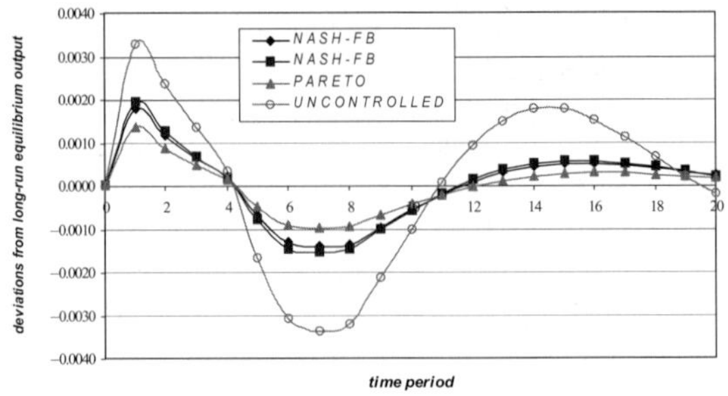


Fig. 3. Country $i$'s output-deviation from its long-run equilibrium level for $i = 1, 2$ and $\varepsilon = 0.5$

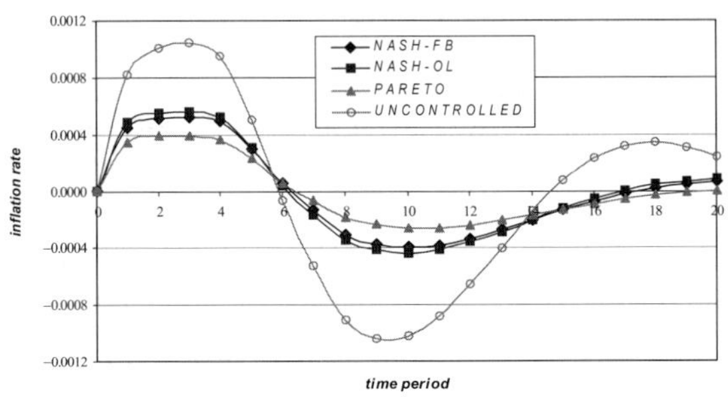


Fig. 4. Country $i$'s inflation rate for $i = 1, 2$ and $\varepsilon = 0.5$

As can be seen from these figures, both fiscal and monetary policies react on the positive demand shock in a restrictive and hence counter-cyclical way: both countries create a fiscal surplus during the first six periods and alternate between periods of fiscal deficit and surplus afterwards, and the central bank decreases its supply of high-powered money during the first five years and increases it afterwards. This results in less additional output (and hence less excess demand loss) and lower inflation than in the uncontrolled solution during years 1 to 4; in fact, in the cooperative solution inflation is nearly reduced to one half of the uncontrolled values. Oscillations of these variables are dampened more strongly than in the uncontrolled solution also in later periods.

Note the small magnitude of the (absolute) values of the instruments involved: the fiscal surplus/deficit created is to the order of one tenth (or less) of one percentage point of GDP, for example. Also changes of the monetary base in periods 1 to 4 amount to only about 0.25 percent of its stock. These small policy reactions are due to the strong self-stabilizing forces in the model used, acting especially through the interest rate channel. As there is not much need for counter-cyclical action, it is not surprising that optimal (equilibrium) policies entail only cautious activities.

When we compare the non-cooperative equilibrium solutions and the cooperative solution, another interesting observation can be made: All show qualitatively the same behavior, and the two non-cooperative Nash equilibrium solutions are rather close together in terms of all control and state variables. The Pareto-optimal collusive solution, although not too distant from the other two, entails more active control: higher fiscal surplus and money reduction in the first periods. This different policy-mix does not change the path of the rate of interest, but does so for the paths of output and inflation: both are closer to their long-run values and hence contribute more to reaching the common goal in the cooperative than in the non-cooperative solutions.

## IV. Optimal Policies and Expectations Formation

Next, we investigate how the dynamics of the model and the results of the policy game depend on the way inflationary expectations are formed. For this purpose, we retain the assumption of adaptive expectations, i.e., equation (7), but vary the parameter $\varepsilon_i$ between 0 and 1 by considering the different solutions for $\varepsilon_i = 0$, $\varepsilon_i = 0.25$, $\varepsilon_i = 0.5$ (the benchmark case of Section III), $\varepsilon_i = 0.75$, and $\varepsilon_i = 1$. Again, we assume the same value of $\varepsilon_i$ for both countries $i = 1, 2$ in each of these cases.

A value of $\varepsilon_i = 0$ means that private agents do not revise their inflationary expectations when new information about actual inflation gets known; the expected inflation is completely unaffected by news. With the values of parameters and initial conditions assumed in Section II, this implies that the expected rate of inflation is always zero, irrespective of the actual rate of inflation. The other extreme, $\varepsilon_i = 1$, implies that the expected rate of inflation is always equal to the previous period's actual rate of inflation. In this case, the public believes that current inflation will persist in the next period.

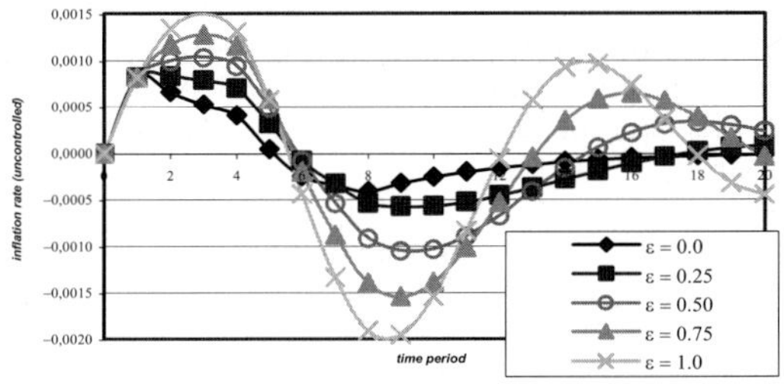


Fig. 5. Country $i$'s uncontrolled inflation rate for $i$ = 1, 2 and different values of $\varepsilon$

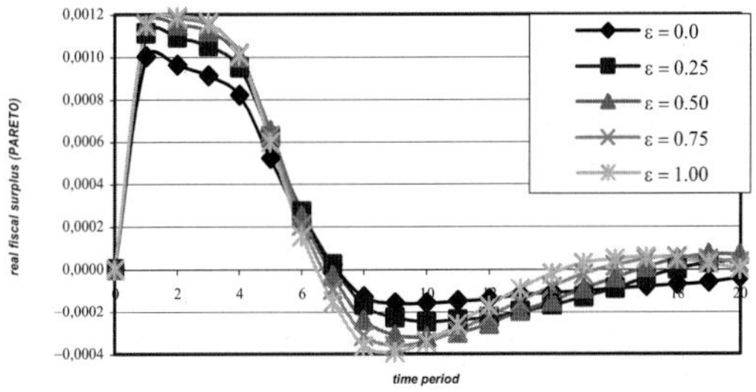


Fig. 7. Country $i$'s fiscal surplus (Pareto-optimal solution) for $i$ = 1, 2 and different values of $\varepsilon$

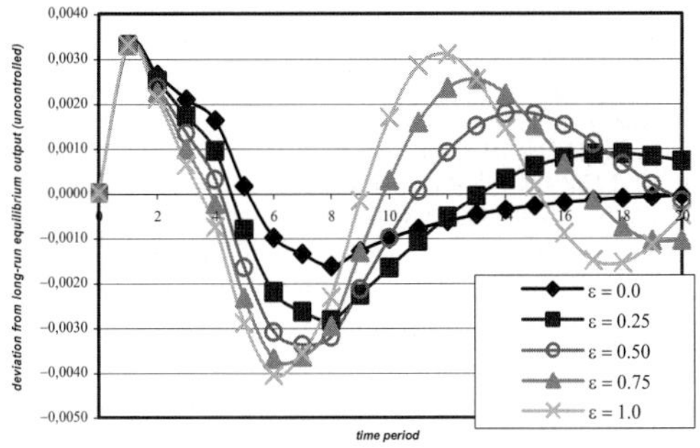


Fig. 6. Country $i$'s uncontrolled deviations from its long-run equilibrium output level for $i$ = 1, 2 and different values of $\varepsilon$

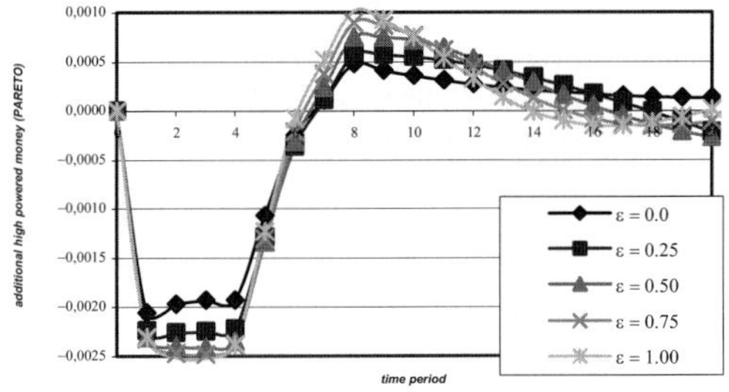


Fig. 8. Additional high powered money (Pareto-optimal solution) for different values of $\varepsilon$

Figs. 5 and 6 show the development of the rate of inflation and the deviation of output from its long-run equilibrium path (excess demand), respectively, without intervention of the monetary union's policy-makers (the uncontrolled solution). For both the rate of inflation and excess demand, the following pattern of dependence upon $\varepsilon_i$ can be observed: as $\varepsilon_i$ increases, oscillations of these variables become wider (larger amplitude) and faster (shorter period). Expected inflation lags behind actual inflation, except for the case of $\varepsilon_i$ = 0, where inflation is just determined by excess demand. The latter case corresponds to a similar model without inflationary expectations analyzed previously [2]. When inflationary expectations react on actual inflation, output, price level and inflation converge more slowly towards the long-run equilibrium path. Inflation lags behind output in the present model, instead of the parallel movement of these two variables prevalent in the model without inflationary expectations.

Figs. 7 and 8 show the optimal reaction of the control variables fiscal surplus and additional high-powered money, respectively. Here we show the time paths of these variables from the Pareto-optimal solution; the other solution concepts yield similar trajectories for these variables with even smaller differences between paths from models with different values of $\varepsilon_i$. The counter-cyclical behavior of all these policy instruments is shown for different values of $\varepsilon_i$. They are applied more vigorously for increasing values of $\varepsilon_i$, due to larger deviations of the target variables from their desired values. There is a trade-off between instrument and target variables: higher deviations of the latter in the uncontrolled solution call for more active policies, which in turn imply larger instrument costs. This trade-off is less favorable if inflationary expectations react more strongly to actual inflation.

For the time paths of the target variables output and inflation in the dynamic game solutions resulting from these policies, excess demand and inflation are still oscillating, hence these are not fully counteracted by policies, due to the trade-off between control and target variables. The qualitative pattern of dependence of oscillations' amplitude and frequency is the same as in the uncontrolled case. But policy actions smooth the time paths of these variables. This is particularly true for output deviations, which are reduced rather quickly after the peak in year 1. For the rates of inflation, the amplitudes of the first oscillations are reduced to roughly one half of those in the uncontrolled case, and even more later on. Expected inflation again lags behind actual inflation when $\varepsilon_i$ is positive. A higher value of $\varepsilon_i$ makes stabilization of inflation and output more costly and the equilibrium and optimal time paths of the target variables less smooth than a lower value or even $\varepsilon_i$ = 0. The differences between the non-cooperative and the cooperative solutions are minor; in particular, the qualitative behavior is similar in all cases considered.

## V. Concluding Remarks

Applying dynamic game theory and the OPTGAME algorithm to a simple macroeconomic model of fiscal and monetary policies in a two-country monetary union, we obtained several insights into the control of the economies of the union under a symmetric excess demand shock. In particular, optimal policies of both the governments and the common central bank are counter-cyclical but not very active for the model under consideration. The outcomes of the different solution concepts of dynamic game theory are rather close to each other. In particular, a periodic update of information and related reduction of commitment (a change from an open-loop to a feedback Nash equilibrium solution) does not cause benefits or costs to either decision-maker. Cooperative economic policies (both fiscal and monetary ones) are more active or "aggressive" than non-cooperative ones, resulting in a somewhat different policy-mix with higher stabilization effects. In all cases, there are trade-offs between the vigor of policy actions and the smoothing effect on target variables. If private agents' inflationary expectations react more strongly to actual inflation, this complicates the stabilization task of macroeconomic policies in the monetary union. Further research will have to show how sensitive these results are with respect to the assumptions about the model and the shock.

## References

[1] T. Başar and G. J. Olsder, *Dynamic Noncooperative Game Theory*, 2nd edition. Philadelphia, PA: SIAM, 1999.

[2] D. A. Behrens and R. Neck, "Optimal decision rules in a monetary union," in *Operations Research Proceedings 2002*, U. Leopold-Wildburger, F. Rendl, G. Wäscher, eds., Berlin: Springer, 2003, pp. 437–445.

[3] D. A. Behrens and R. Neck, "Approximating Equilibrium Solutions for Multi-Player Difference Games Using the OPTGAME Algorithm," unpublished, Working Paper, Department of Economics, University of Klagenfurt, 2003.

[4] E. Dockner, S. Jorgensen, N. v. Long and G. Sorger, *Differential Games in Economics and Management Science*. Cambridge: Cambridge University Press, 2000.

[5] M. L. Petit, *Control Theory and Dynamic Games in Economic Policy Analysis*. Cambridge: Cambridge University Press, 1990.

# Schema Matching in the Context of Model Driven Engineering: From Theory to Practice

Denivaldo Lopes(1) Slimane Hammoudi(2) Zair Abdelouahab(1)

(1) Federal University of Maranhão (UFMA)
CCET - Department of Electrical Engineering
65080-040 São Luís - MA, Brazil

(2) ESEO
4, rue Merlet de la Boulaye, BP 30926
49009 cedex 01 Angers, France

denivaldo.lopes@gmail.com, shammoudi@eseo.fr and zair@dee.ufma.br

***Abstract* - Recently, software engineering is required to make face to the development, maintenance and evolution complexity of software systems. Among the proposed solutions for managing the complexity, Model Driven Engineering (MDE) has been accepted and implemented as one of the most promising solutions. In this approach, models become the hub of development, separating platform independent aspects from platform dependent aspects. Among the more important issues in MDE, we remark *model transformation* and *model matching* (or *schema matching*). In this paper, we propose an approach to take into account *schema matching* in the context of MDE. A *schema matching algorithm* is provided and implemented as a *plug-in* for *Eclipse*. We discuss an illustrative example to validate our approach.**

## I. Introduction

Recently, software engineering has once been requested to make face to the development, maintenance and evolution complexity of software systems. In order to provide rational solutions to enable the management of this complexity, several organizations and enterprises have proposed approaches based on models. This new trend has resulted in Model Driven Engineering (MDE). OMG's Model Driven Architecture (MDA™1) [17], Microsoft's Software Factories [14], and the initiative of the *Eclipse Project* denominated Eclipse Modeling Framework (EMF) [3] are examples of MDE.

Models have been used for a long time ago in the software development. Booch, OMT, UML and CASE tools based on models were proposed in the 80's and 90's years to improve the software development process. However, in this period, models were often used in initial phases of software development such as requirements, analysis and design. What is different with MDE? In this new trend, models are used in all phases of software development process. Thus, each model that is used in a phase can be projected in another model in the subjacent phase.

In MDE, models become the hub of software development process, separating platform independent aspects from platform dependent aspects. Several concepts and techniques are important in MDE approach such as *metamodels*, *transformation language*, *mapping* and *methodologies* to apply MDE. Unified Modeling Languages (UML) [16] and Enterprise Distributed Object Computing (EDOC) [18] are examples of metamodels used to create models. Atlas Transformation Language (ATL) [4], Yet Another transformation Language (YATL) [20] and QVT-Merge transformation language [22] are examples of transformation languages. The specification of relationships between the UML metamodel and a WSDL[2] metamodel is an example of mapping [13]. Several methodologies conform to an MDE approach were proposed in the literature [5] [8] [9]. Among these fundamental concepts and techniques for MDE, *transformation language* and *mapping* have been studied for a long time ago in other domains, e.g. database domain. In [12] [13], an MDA approach separating *mapping specification* from *transformation definition* was introduced and implemented in a tool denominated Mapping Modeling Tool (MMT) that enables the manual creation of *mapping specifications*. However, the manual creation of *mapping specifications* is a fastidious and error-prone task. Generally, this task implies in the search of *equivalent* or *similar elements* between two metamodels. In database domain, this task is called *schema matching* [23].

In this paper, we propose: a new tool for *mapping modeling* called Mapping Tool for MDE (MT4MDE), an approach to take into account *schema matching* and an implementation of this approach called Semi-Automatic Matching Tool for MDE (SAMT4MDE). This paper is organized in the following way. Section II is an overview of MDE and *schema matching* approaches. Section III shows our proposition to take into account *semi-automatic schema matching* in the context of MDE. Section IV presents the implementation of our proposed approach for *schema matching* as a *plug-in* for *Eclipse* and an illustrative example. Section V contains conclusions and presents the future directions of our research.

## II. Overview

Model Driven Engineering (MDE) is an approach to develop, maintain and evolve software systems driven by models. In

1 MDA™ is a trademark of the Object Management Group (OMG).

2 Web Services Description Language (WSDL) [26].



*T. Sobh and K. Elleithy (eds.), Advances in Systems, Computing Sciences and Software Engineering*, 219–227.

the literature, some aspects of MDE were identified, for example:

- "*An important aspect of MDE is its emphasis on bridges between technological spaces, and on the integration of bodies of knowledge developed by different research communities*" [7].
- "*MDE is wider in scope than MDA. MDE combines process and analysis with architecture*" [11].

Some benefits of MDE were identified: it saves time and resources, and it avoids error-prone factors [15]. However, it is still evolving and need more time to be stable. MDE aims to manage the development, maintenance and evolution complexity of software system through a rational utilization of models. Thus, *models* are the impetus to manage this complexity, and *model transformation languages* constitute an important issue for manipulating models. However, the task of creating *model transformation programs* is a task of program codifying. So, the move from *traditional programming languages* (e.g. Java, C++ and C#) to *model transformation languages* (e.g. ATL and YATL) still results in program codifying. In [13], we propose an approach separating *mapping specification* from *transformation definition*. In this approach, a *mapping specification* is a model that represents the relationships between metamodels, while *transformation definition* contains the operational description of the transformation between models. In addition, a *transformation definition* can be generated from a *mapping specification*. Thus, a *transformation program* can be generated from a *transformation definition*. Figure 1 depicts the *mapping specification* as a `mapping model`, and *transformation definition* as a `transformation model`.

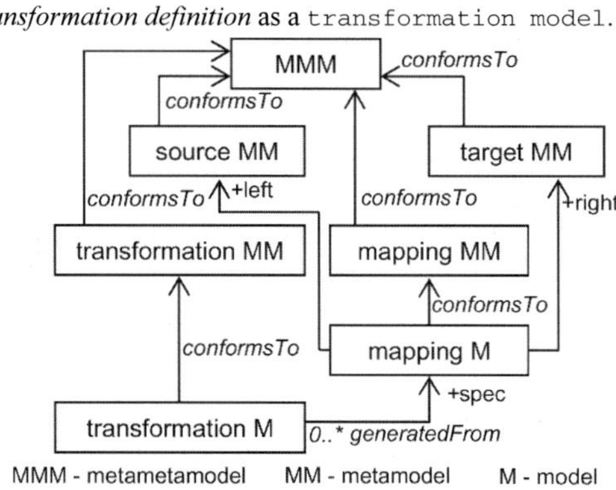


Fig. 1. Mapping Specification and Transformation Definition as different models [13].

A `transformation model` (considered as a PSM) can be generated from a `mapping model` (considered as a PIM) through transformations. Thus, the important issue here is the elaboration of `mapping models`, which is not yet an easy task. However, this task can be semi-automatized[3] using *schema matching algorithms*[4] that have been studied for a long time in database domain [1] [21] [23].

*A. Approaches for Schema Matching*

In general, metamodels are created with a specific purpose[5] and by different groups of persons. Each purpose is determined in function of the domain, and each group of persons models a system in different ways. In the modeling task, each group abstracts, classifies and generalizes the reality based on its own knowledge. Consequently, metamodels that are created in the same context but by different groups can have different structure and terminology [23] causing the *semantic distance* among them [2] [10] [24].

A model can be transformed in another model, only if the metamodel of the former can be mapped in the metamodel of the later. In order to map metamodels, the *equivalent* or *similar elements* must be identified, and the *semantic distance* should be minimized. The first step is denominated *schema matching* [23]. And "*the notion of semantic distance was developed to cover the notion of how close is close enough*" [10]. A dual for *semantic distance* is *schema similarity* that is defined as "*the ratio between the number of matching elements and the number of all elements from both input schemas*" [6] ($SS = N_m/N_t$, where *SS* is the *schema similarity*, $N_m$ is the number of matching elements and $N_t$ is the number of all elements). *Semantic distance* can also be quantified as a numeral value (like *schema similarity*) or as a subset of a metamodel [13]. In the literature, several *schema matching approaches* have been proposed [25]. Each *schema matching approach* has its own characteristics that were grouped in taxonomy [23]. In addition, each approach has been evaluated through *match quality measures* [6].

## III. Schema Matching in MDE

An approach to take into account *schema matching* in the context of MDE should provide answers for the following questions:

- What is a *mapping*?
- How is a *mapping model* (or *mapping specification*)?
- What are *equivalent* or *similar elements* from two metamodels?
- How can I find *equivalent* or *similar elements* from two metamodels?
- How can I create a general *algorithm for schema matching*?

In the next paragraphs, we provide some insights to answer these questions.

*A. Foundation for Mapping*

We can define a *transformation function* as follows [12]:

3 In general, a *schema matching algorithm* is not so powerful to find all correspondences between two schemas. So, the human intervention is often requested.

4 We can think in *model matching algorithms*. However, this new terminology will only increase the vocabulary in MDE without providing real contribution. Thus, we prefer to use the well know terminology, *schema matching algorithm*. Moreover, a schema is the equivalent of metamodel in the XML technological space.

5 UML is a general-purpose modeling language, but it provides profiles as extension mechanism in order to be adapted to a domain.

$$Transf(M_1(s)/M_a, C_{Ma \rightarrow Mb}/M_c) \rightarrow M_2(s)/M_b$$

where:

$M_1$ is a model of a system *s* created using the metamodel $M_a$.

$M_2$ is a model of the same system s created using the metamodel $M_b$.

$C_{Ma \rightarrow Mb}$ is the mapping between $M_a$ and $M_b$ created using the metamodel $M_c$.

In general, $M_a$, $M_b$ and $M_c$ conform to the same metametamodel, simplifying the *transformation process* [13]. A mapping model in the context of MDE must be conforming to a metamodel. In [12] [13], an initial *mapping metamodel* was proposed and implemented in a tool called MMT. However, this initial *mapping metamodel* mixes information about mapping and versioning and it is a fixed metamodel. Our new approach separates the mapping metamodel in two complementary parts: *mapping* and *versioning*. Moreover, we introduce the notion of properties as a predefined mechanism to extend this mapping metamodel.

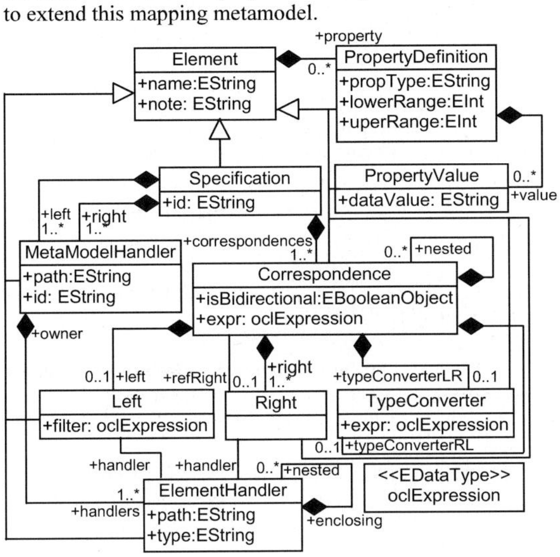


Fig. 2. A mapping metamodel: basic

Figure 2 illustrates our new mapping metamodel that presents the following elements:

- `Element` is a generalization for the other elements. The attribute `name` is an identification of an element, and `note` has a remark about this element.
- `Specification` contains references for two or more metamodels (one or more `+left`, and one or more `+right`), and it has `+correspondences` among these metamodels.
- `Correspondence` defines the interrelationships between two or more metamodel elements. A `correspondence` can relate one `Left` with one or more `Right`, or it relates one `Right` with another `Right` of another `Correspondence` (through the relationship `+refRight`). If a correspondence is bidirectional, then `isBidirectional` is `true`, else it is `false`. If the correspondence is complex (i.e. it needs defining a query), then a `TypeConverter` must be defined.
- `MetaModelHandler` allows the navigation into a metamodel. A specification can reference two or more `MetaModelHandlers`.
- `Left` identifies the left element of a mapping. It has an OCL expression that is a filter.
- `Right` identifies the right elements of a mapping.
- `ElementHandler` allows the navigation into the elements being mapped without changing them.
- `TypeConverter` contains an expression in OCL (`+expr`) that allows the navigation in a metamodel element. This expression in OCL is a query to obtain a specific element, attribute or reference from the metamodel `+left` to be related with an element, attribute or reference from the metamodel `+right`.
- `PropertyDefinition` allows the creation of properties attached to an `Element`. They are an extension mechanism to add additional information to a mapping model.
- `PropertyValue` is an element used to store the values of a property (in `dataValue`).

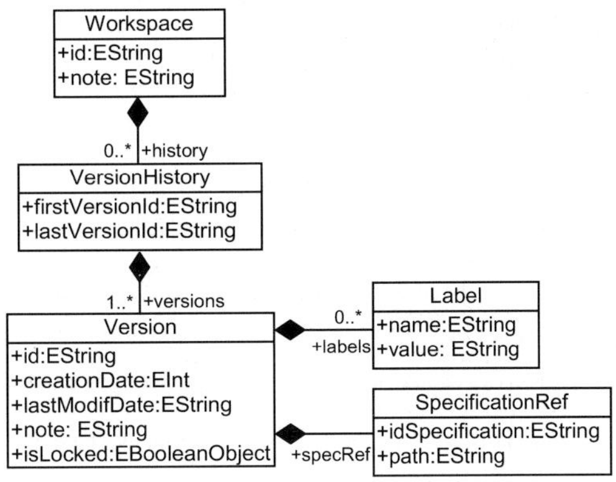


Fig. 3. A mapping metamodel: versioning

Some tools have supported the management of multiple versions of a same program such as Concurrent Versions System (CVS). Recently, OMG has adopted the MOF 2.0 Versioning and Development Lifecycle Specification [19] for managing the co-existence of multiple versions of a metadata, i.e. metamodels, models and information. Versioning seems to be a frequent issue in the development of software systems. So, we have studied, proposed and implemented a metamodel to support versioning. Figure 3 presents the second part of our mapping metamodel that takes into account versioning. This metamodel contains the following elements:

- `Workspace` contains information about versions of mapping models (zero or more `+history`). It has an unique `id` and a general `note`.

- `VersionHistory` is a container for each history of only one mapping model. It points for the first version (`+firstVersionId`) and for the last version (`+lastVersionId`) of a mapping model.
- `Version` identifies each version of a mapping model. It has an unique identifier (`id`), creation and modification date, a note and a lock.
- `Label` is attached to a `Version` for adding more information about this version.
- `SpecificationRef` is a reference for a specification of a mapping model. It has the identifier of a mapping model (`idSpecification`) and the location where this mapping model is stored (`path`).

In [1], by *similar*, "*we mean that they are related but we do not express exactly how*". In [21], two models are defined equivalent "*if they are identical after all implied relationships (i.e. relationship kinds and cross-kind-relationship implications) are added to each of them until a fix point is reached*".

*B. A Schema Matching Algorithm*

This section describes a *schema matching algorithm* defined here as the operator $Match'(M_a,M_b) = C_{Ma \to Mb}/M_c$. It takes two metamodels as inputs and produces a mapping model. In our approach, this mapping model conforms to our mapping metamodel (cf. figure 2). In fact, we have considered models and metamodels as sets and vice versa, but we must understand that they are complex and heterogeneous sets. So, we must define the sets $M_a$, $M_b$ and $C_{Ma \to Mb}$.

- $M_a$ can be defined as follows:

$$M_a = \bigcup_{i=1}^{m'} a_{1i} \bigcup \bigcup_{j=1}^{n'} a_{2j} \bigcup \bigcup_{k=1}^{o'} a_{3k}$$

Elements $a_{1i}$ are classes (i.e. meta-classes), $a_{2j}$ are data types and $a_{3k}$ are enumerations.

Classes can have attributes and relationships, thus:

$$a_{1i} = \bigcup_{\phi=1}^{d'} t_\phi \bigcup \bigcup_{\psi=1}^{e'} r_\psi$$

Elements $t_\phi$ are attributes and $r_\psi$ are relationships between classes.

Enumerations have literals, thus $a_{3k}$ can be defined as:

$$a_{3k} = \bigcup_{\sigma=1}^{r'} l_\sigma$$

Elements $l_\sigma$ are literals.

- Similarly, we define $M_b$ as follows:

$$M_b = \bigcup_{x=1}^{s'} b_{1x} \bigcup \bigcup_{y=1}^{t'} b_{2y} \bigcup \bigcup_{z=1}^{u'} b_{3z}$$

Elements $b_{1x}$ are classes (i.e. meta-classes), $b_{2y}$ are data types and $b_{3z}$ are enumerations.

Classes can have attributes and relationships, thus:

$$b_{1x} = \bigcup_{\eta=1}^{w'} g_\eta \bigcup \bigcup_{\theta=1}^{f'} h_\theta$$

Elements $g_\eta$ are attributes and $h_\theta$ are relationships between classes.

Enumerations have literals, thus $b3z$ can be defined as:

$$b_{3z} = \bigcup_{\tau=1}^{q'} v_\tau$$

Elements $v_\tau$ are literals of an enumeration.

The correspondence can be defined as follows:

$$C = \bigcup_{p=1}^{a'} c_{1p} \bigcup \bigcup_{q=1}^{b'} c_{2q} \bigcup \bigcup_{s=1}^{c'} c_{3s}$$

Elements $c_{1p}$ constitute one set of matched classes, elements $c_{2q}$ constitute one set of matched data types, and elements $c_{3s}$ constitute one set of matched enumerations.

Elements $c_{1p}$ contain the classes belonging to $M_a$ and $M_b$ that are equal or similar, thus:

$$c_{1p} = \{a_{1i}, b_{1x}, e_p\}$$

Such that:

$a_{1i} \in M_a$, $b_{1x} \in M_b$, and $e_p$ is a set constituted of matched attributes and matched relationships. So, we define:

$$e_p = \bigcup_{\alpha=1}^{x'} m_{p\alpha} \bigcup \bigcup_{\beta=1}^{z'} n_{p\beta}$$

Where, we can define:

$$m_{p\alpha} = \{\{t_\phi, g_\eta\} \mid t_\phi \in M_a \text{ and } g_\eta \in M_b\}, \text{ where:}$$

$$\alpha = \{\alpha \in N \mid 1 \le \alpha \le x'\}$$

and

$$n_{p\beta} = \{\{r_\psi, h_\theta\} \mid r_\psi \in M_a \text{ and } h_\theta \in M_b\}, \text{ where:}$$

$$\beta = \{\beta \in N \mid 1 \le \beta \le z'\}.$$

Elements $m_{p\alpha}$ constitute the set of matched attributes, and elements $n_{p\beta}$ constitute the set of matched relationships.

Elements $c_{2q}$ contain the data types from $M_a$ equal or similar to others in $M_b$. Thus:

$$c_{2q} = \{\{a_{2j}, b_{2y}\} \mid a_{2j} \in M_a \text{ and } b_{2y} \in M_b\}$$

Elements $c_{3s}$ contain the enumerations in $M_a$ that are equal or similar to others in $M_b$. Thus:

$$c_{3s} = \{a_{3k}, b_{3z}, o_s\}$$

Such that:

$a_{3k} \in M_a$, $b_{3z} \in M_b$, and $o_s$ is a set constituted of matched literals. So, we define:

$$o_s = \bigcup_{\delta=1}^{j'} w_{s\delta}$$

$$w_{s\delta} = \{\{l_\sigma, v_\tau\} \mid l_\sigma \in M_a \text{ and } v_\tau \in M_b\}, \text{ where:}$$

$$\delta = \{\delta \in N \mid 1 \le \delta \le j'\}.$$

After the definition of the sets $M_a$, $M_b$ and $C$, we can present our algorithm for matching two metamodels:

1. **Create C:** create the set $C$.
2. **Initialize:** assign $\varnothing$ to $C$.
3. **Find leaf classes**[6]**:** select $a_{1i}$ from $M_a$ that are leafs, and select $b_{1x}$ from $M_b$ that are leafs.
4. **Select equal or similar classes:** For each pair of leaf classes $a_{1i}$ from $M_a$ and $b_{1x}$ from $M_b$, the function φ classifies the elements of this pair in *equal*, *similar* or *different*.

The function φ mapping two classes to an integer can be defined as follows:

6 Leaf classes are classes that do not have child classes.

$$\varphi(a_{1i}, b_{1x}) = \begin{cases} 1 \text{ if } a_{1i} = b_{1x} \\ 0 \text{ if } a_{1i} \cong b_{1x} \\ -1 \text{ if } a_{1i} \neq b_{1x} \end{cases}$$

Where:

= symbolizes the equality between two classes.

≅ symbolizes the similarity between two classes.

≠ symbolizes the difference between two classes.

In order to determine if $a_{1i}$ and $b_{1x}$ can be matched, the function φ compares the attributes $t_\phi$ with $g_\eta$ and the references $r_\psi$ with $h_\theta$. Moreover, the relationship kinds and cross-kind-relationship implications (see section III-A) are used to find *similarities* and *equivalences*. If the number of matched attributes and relationships is bigger than a predetermined average (threshold), then an element $c_{1p}=\{a_{1i}, b_{1x}, e_p\}$ is created, initialized with $a_{1i}$, $b_{1x}$, and $e_p$, else nothing is created.

5. **Put each $c_{1p}$ into *C*:** after iterating (4) to a fixpoint, put each $c_{1p}$ in *C*.
6. **Select equal or similar data types:** For each pair $\{a_{2j}, b_{2y}\}$, the function φ' classifies the elements of this pair in *equal*, *similar* or *different*.

The function φ' mapping two data types to an integer can be defined as follows:

$$\varphi'(a_{2j}, b_{2y}) = \begin{cases} 1 \text{ if } a_{2j} = b_{2y} \\ 0 \text{ if } a_{2j} \cong b_{2y} \\ -1 \text{ if } a_{2j} \neq b_{2y} \end{cases}$$

An element $a_{2j}$ is equal to $b_{2y}$ if they are of the same data type. An element $a_{2j}$ is similar to $b_{2y}$ if $a_{2j} \subset b_{2y}$. For example, an *int* can be represented by a *float*, but a *float* cannot be represented by an *int*. For complex data types, e.g. an *array*, this principle can be used too.

7. **Put each $c_{2q}$ into C:** after iterating (6) to a fixpoint, put each $c_{2q}$ into *C*.
8. **Select equal or similar enumerations:** For each pair $\{a_{3k}, b_{3z}\}$, the function φ'' classifies the elements of this pair in *equal*, *similar* or *different*.

The function φ'' mapping two enumerations to an integer can be defined as follows:

$$\varphi''(a_{3k}, b_{3z}) = \begin{cases} 1 \text{ if } a_{3k} = b_{3z} \\ 0 \text{ if } a_{3k} \cong b_{3z} \\ -1 \text{ if } a_{3k} \neq b_{3z} \end{cases}$$

An element $a_{3k}$ is equal to $b_{3z}$ if all literals are matched. An element $a_{3k}$ is similar to $b_{3z}$ if a part of literals are matched. An element $a_{3k}$ is different to $b_{3z}$ if literals are not matched.

9. **Put each $c_{3s}$ into *C*:** after iterating (8) to a fixpoint, put each $c_{3s}$ into *C*.

In this algorithm, we presented the functions $\varphi(a_{1i}, b_{1x})$, $\varphi'(a_{2j}, b_{2y})$ and $\varphi''(a_{3k}, b_{3z})$ explaining the meaning of inputs and outputs. However, the detailed description of the bodies of these functions is out of the scope of this paper. In our experimentations presented hereafter, we have used the following criteria to build these functions: *name similarity*, *type similarity* and *graph matching*. So, we have adopted a *hybrid matching approach*. Moreover, $\varphi(a_{1i}, b_{1x})$ and $\varphi'(a_{2j}, b_{2y})$ were created considering the relationship kinds and a part of cross-kind-relationship implications [21].

## IV. A PLUG-IN FOR SCHEMA MATCHING

We have implemented a *plug-in* for *schema matching* denominated Semi-Automatic Matching Tool for MDE (SAMT4MDE). This *plug-in* is based on the algorithm presented in section III-B and it was implemented using EMF [3]. We have considered the sets $M_a$, $M_b$ and *C* as metamodels (conform to Ecore) that are illustrated in figure 4 (we omitted $M_b$ because it is similar to $M_a$).

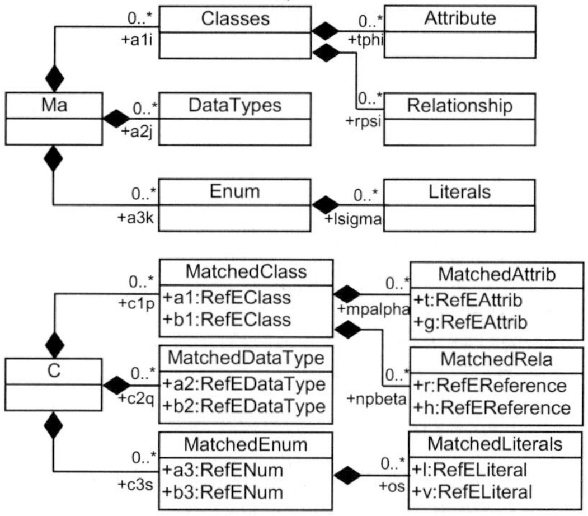


Fig. 4. Modeling the schema matching algorithm: class diagrams of *Ma* and *C*

SAMT4MDE extends MT4MDE with *semi-automatic schema matching*. MT4MDE was designed to be easily extensible, e.g. through another *plug-in* for generating a *transformation definition* from a *mapping specification*, or through a *plug-in* for matching two metamodels. In order to interact with MT4MDE, a plug-in for schema matching must have a class realizing the interface *ITFMatchEngine.java*. This class receives the two metamodels to be matched and returns a mapping model.

The file `plugin.properties` of MT4MDE has two variables `_UI_MatchEngines` and `_UI_MatchEngine_classes` employed to locate and to load the class used to match two metamodels. The first variable indicates the name of a *schema matching algorithm*, and the second variable indicates the package and class implementing this algorithm, as follows:

```
_UI_MatchEngines= Default,
_UI_MatchEngine_classes=match.engine.Match
```

In this case, the algorithm name is `Default`, the package is `match.engine` and the class is `Match`.

SAMT4MDE is a semi-automatic matching tool because the user must validate the matched elements or may complete the generated mapping model. In other words, the object instantiated of the class `Match` determines the matched elements between two metamodels, and afterwards the user can validate or refuse them. Later, a *mapping model* can be generated from these matched elements, but it may be incomplete, thus the user must manually put the other elements in correspondence.

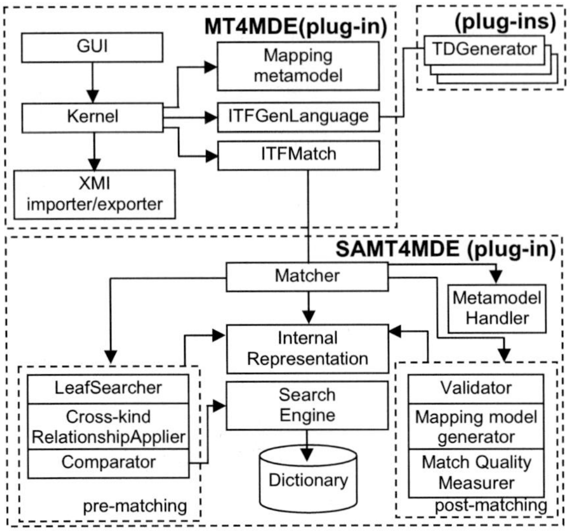

Fig. 5. Plug-in for schema matching: architecture

Figure 5 presents the MT4MDE architecture (MT4MDE), the SAMT4MDE architecture, and the transformation definition generators (`TDGenerator`) that were implemented as *plug-ins* for *Eclipse*. The MT4MDE architecture is composed of:

- `GUI` is the graphical user interface of MT4MDE (it is hereafter presented in figure 6).
- `Kernel` executes all the basic functionalities of MT4MDE.
- `XMI importer/exporter` takes a metamodel in the XMI format and translates it in the same metamodel conforms to *Ecore*. It also takes a metamodel conforms to *Ecore* and translates it in the same metamodel in the XMI format.
- `Mapping metamodel` is the metamodel used to create mapping models. It was detailed in section III-A.
- `ITFGenLanguage` is an interface to handle transformation definition generators that create *transformation definitions* from *mapping specifications*. The MT4MDE kernel uses this interface to interact with the `TDGenerators`.
- `ITFMatch` is an interface to handle matchers.

The SAMT4MDE architecture is composed of:

- `Matcher` implements the interface `ITFMatch` and it coordinates the matching of two metamodels.
- `Internal Representation` is a representation more adapted to create matches (i.e. matching elements) between metamodels. In our *plug-in*, the representation for the matching elements and metamodels were illustrated in figure 4. We remark that our *internal representation* (cf. figure 4) is different from the *mapping metamodel* (cf. figure 2). The former aims to describe the matched elements from two metamodels, while the later aims to describe the correspondences between two metamodels, including the queries needed to navigate inside a metamodel.
- `MetamodelHandler` allows the navigation into the metamodels.
- `LeafSearch` searches for metaclasses that are leafs such as discussed in our proposed algorithm for schema matching.
- `Cross-kind relationship applier` searches and applies cross-kind relationships.
- `Comparator` implements part of the functions $\varphi(a_{1i}, b_{1x})$, $\varphi'(a_{2j}, b_{2y})$ and $\varphi''(a_{3k}, b_{3z})$.
- `Search Engine` searches for name synonyms.
- `Dictionary` is a database and stores domain dictionaries.
- `Validator` receives the matching elements and interacts with the user that validates the matching elements or refuses them.
- `Mapping model generator` creates a mapping model (cf. figure 2) from the validated matching elements (cf. figure 4).
- `Match quality measurer` evaluates the results and provides the values of match quality measures [6]: *Schema Similarity*, *Precision*, *Recall*, *F-Measure* and *Overall*.

### *A. An Illustrative Example*

We choose the UML metamodel 1.4 [16] and Java metamodel [12] for illustrating our *plug-in* for *schema matching*. Figure 6 presents both metamodels in the form of trees. The steps to use our *plug-in* for *schema matching* (i.e. SAMT4MDE) with MT4MDE can be illustrated as follows:

- **Import left and right metamodel**: MT4MDE loads the UML and Java metamodel.
- **Select a schema matching algorithm**: in MT4MDE, we choose a *plug-in* that implements a *schema matching algorithm*.
- **Run the selected schema matcher**: the selected *plug-in* is executed for finding pairs of matched elements, i.e. derived matches [6].
- **Validate the pairs of matched elements**: the user can validate or refuse the pairs of the matched elements (cf. figure 7).

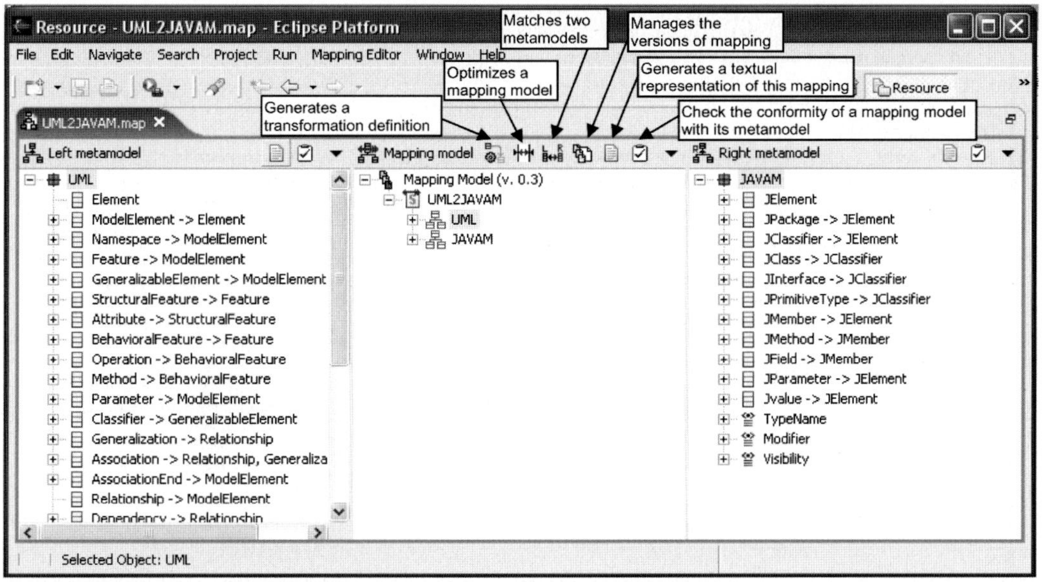


Fig. 6. MT4MDE: UML and Java metamodel (fragment)

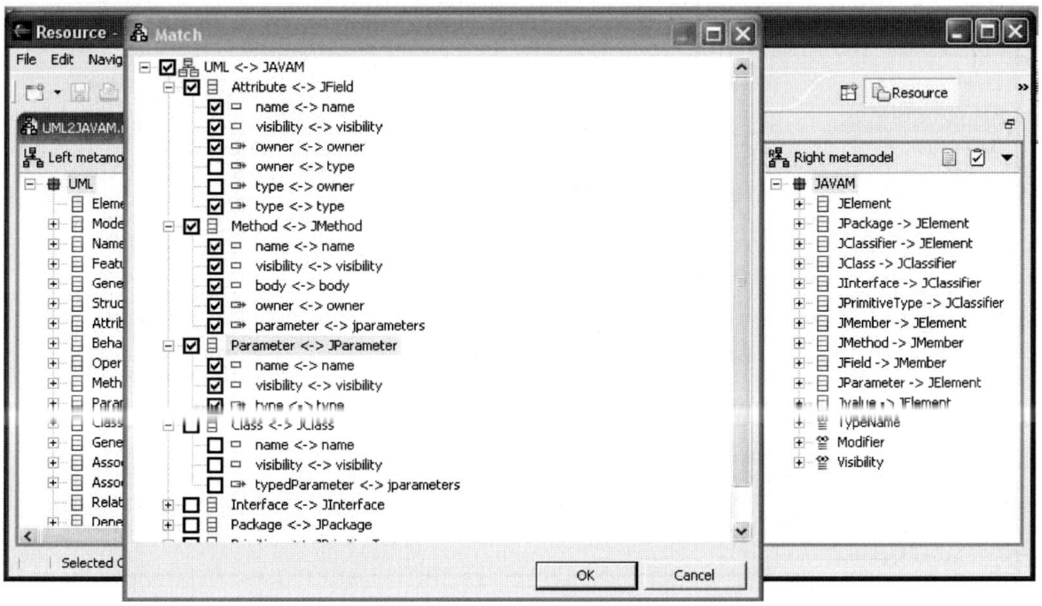


Fig. 7. Validating matched elements (fragment)

- **Generate a mapping model:** `Mapping model generator` uses the validated pairs of matched elements to generate a mapping model (cf. figure 8).
- **Complete the mapping model**: if the mapping model is complete, then we can move to the next step, else the mapping model must be completed.
- **Select a transformation language**: the user chooses a transformation language such as ATL.
- **Generate the transformation definition**: a *transformation definition* is generated from the *mapping specification* (i.e. mapping model). This *transformation definition* is written in the selected transformation language (cf. figure 9).

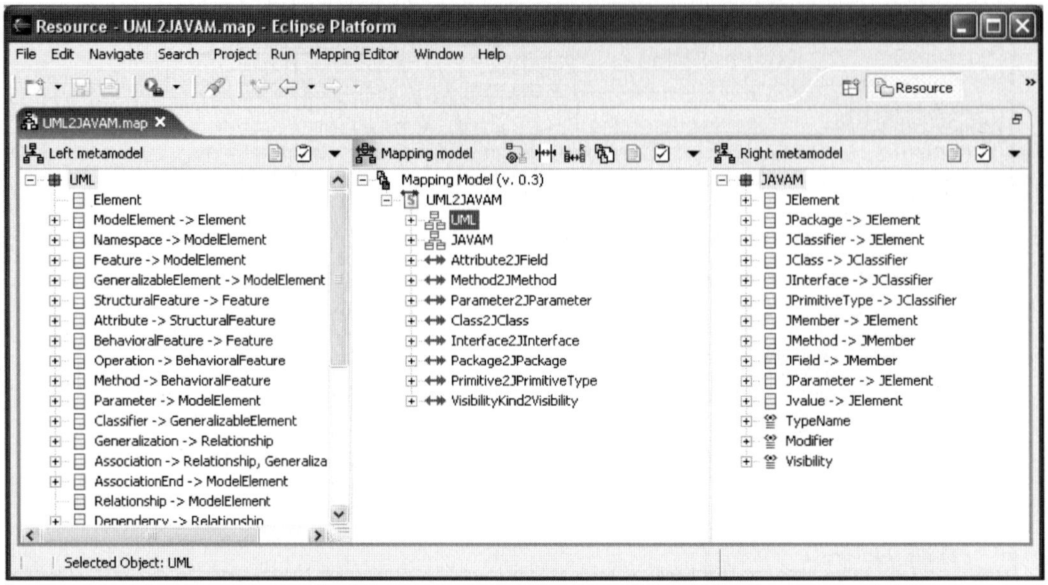


Fig. 8. Mapping specification created applying the match algorithm (fragment)

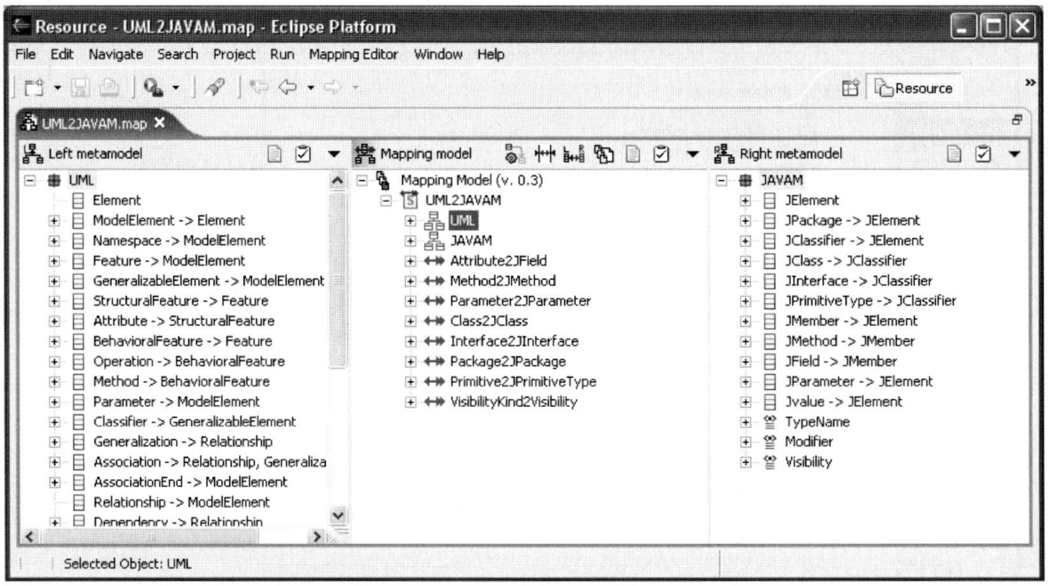


Fig. 9. Transformation definition generated from mapping specification (fragment)

- **Complete the transformation definition**: if the *transformation definition* is complete, then we can move to the next step, else the *transformation definition* must be completed before passing to the next step.
- **Apply the transformation definition**: the t*ransformation definition* can be applied to transform a UML model into a Java model.

In order to evaluate our approach and tool, we have used *match quality measures* [6]. SAMT4MDE in conjunction with MT4MDE provided the following values for *match quality measures*:

*Precision* = 0.86
*Recall* = 0.68
*F-Measure* = 0.76
*Overall* = 0.57

In the ideal case, *Precision*=*Recall*=1.0, i.e. when the number of *false negatives* and *false positives* are both zero [6]. In our experimentation, *Precision*=0.86 demonstrates that 86% of *derived matches* were correctly determined using our *schema matching algorithm*. And *Recall*=0.68 demonstrates that 68% of *real matches* were automatically found.

When 0.5 < *Precision* < 1, *Recall*≠1 and *Recall*≠*Precision*, the *match quality measures* are interrelated as follows: "*the value of F-Measure is within the range determined by Precision and Recall, Overall is smaller than both Precision and Recall*" [6]. According to this proposition, the results of our experimentation are correct, because *Recall* < *F-Measure* < *Precision* (*Recall* =0.68, *F-Measure* = 0.76 and *Precision* = 0.86). In addition, if *Recall* < *Precision*, then *Overall* < *Precision*, else *Overall* < *Recall* (*Overall* = 0.57).

In this illustrative example, we provided a proof of concept with preliminary experimental results that demonstrate the utility of our approach and tool to match two metamodels.

## V. Conclusion

In this paper, we have presented our approach to take into account *schema matching* in the context of MDE. The study of *schema matching* in MDE is a promising trend to improve the creation of *mapping specifications* and, consequently, *transformation definitions*. Tools for *schema matching* are necessary to avoid error-prone factors linked to the manual creation of *transformation definitions* and to evolve *mapping specifications* when metamodels change. In fact, we have used the principles of MDE to develop MDE. First, the *schema matching algorithm* helps in the creation of *mapping specifications*. Afterwards, a *mapping specification* is transformed in a *transformation definition*. Thus, a *mapping specification* is a PIM, and a *transformation definition* is a PSM.

The main contributions of this work are a *schema matching algorithm*, a *plug-in* for *schema matching* (SAMT4MDE) and an illustrative example. A *schema matching algorithm* was proposed and implemented as a *plug-in* for *Eclipse*. An illustrative example using UML and Java metamodels was presented to validate our approach.

In future work, we will make qualitative and quantitative comparison (through taxonomies and match quality measures) of our approach with other approaches supporting *schema matching*. We will study and implement other *schema matching algorithms*, e.g. algorithms based on *machine learning* and other *heuristics*. In addition, we envisage studying the optimization of mapping models which seems to be another important issue in MDE.

## Acknowledgments

The work described in this paper was financed by **Fundo Setorial de Tecnologia da Informação (CT-Info), MCT, CNPq (CT-Info/MCT/CNPq)**.

We thank Jean Bézivin for many useful discussions.

## References

[1] P. A. Bernstein, "Applying Model Management to Classical Meta Data Problems", Proceedings of the 2003 CIDR, pages 209–220, January 2003.

[2] G. Booch, A. Brown, S. Iyengar, J. Rumbaugh, and B. Selic, "An MDA Manifest", MDA Journal, May 2004.

[3] F. Budinsky, D. Steinberg, E. Merks, R. Ellersick, and T. J. Grose, *Eclipse Modeling Framework: A Developer's Guide*, Addison-Wesley Pub Co, 1st edition, August 2003.

[4] J. Bézivin, G. Dupé, F. Jouault, G. Pitette, and J. E. Rougui, "First Experiments with the ATL Model Transformation Language: Transforming XSLT into XQuery", 2nd OOPSLA Workhop on Generative Techniques in the context of Model Driven Architecture, October 2003.

[5] P. Cáceres, E. Marcos, and B. Vela, "A MDA-Based Approach for Web Information System Development", Workshop in Software Model Engineering, October 2003.

[6] H.-H. Do, S. Melnik, and E. Rahm, "Comparison of Schema Matching Evaluations", Revised Papers from the NODe 2002 Web and Database-Related Workshops on Web, Web-Services, and Database Systems, pages 221–237, 2003.

[7] J. M. Favre, "Towards a Basic Theory to Model Driven Engineering", UML 2004 - Workshop in Software Model Engineering (WISME 2004), 2004.

[8] F. Fondement and R. Silaghi, "Defining Model Driven Engineering Processes", WiSME@UML 2004, October 2004.

[9] A. Gavras, M. Belaunde, L. F. Pires, and J. P. A. Almeida, "Towards an MDA-based Development Methodology", First European Workshop on Software Architecture (EWSA 2004), May 2004.

[10] H. T. Goranson, "Semantic Distance and Enterprise Integration", International Conference on Enterprise Integration and Modelling Technology (ICEIMT2004), October 2004.

[11] S. Kent, "Model Driven Engineering", Integrated Formal Methods - IFM, pages 286–298, May 2002.

[12] D. Lopes, "Study and Applications of the MDA Approach in Web Service Platforms", Ph.D. thesis (written in French), University of Nantes, 2005.

[13] D. Lopes, S. Hammoudi, J. Bézivin, and F. Jouault, "Generating Transformation Definition from Mapping Specification: Application to Web Service Platform", The 17th Conference on Advanced Information Systems Engineering (CAiSE'05), (LNCS 3520):309–325, June 2005.

[14] Microsoft. "Software Factories". Available at http://msdn.microsoft.com/architecture/overview/softwarefactories/.

[15] Middleware Company. "Model Driven Development for J2EE Utilizing a Model Driven Architecture (MDA) Approach", Technical report, The Middleware Company, June 2003. Available at http://www.middleware-company.com/casestudy.

[16] OMG, "Unified Modeling Language Specification", Version 1.4, September 2001.

[17] OMG, "MDA Guide Version 1.0.1", Document Number: omg/2003-06-01. OMG, June 2003.

[18] OMG, "Enterprise Collaboration Architecture (ECA) Specification", OMG formal/04-02-01, February 2004.

[19] OMG, "MOF 2.0 Versioning and Development Lifecyle Specification" - ad/2005-05-011, May 2005.

[20] O. Patrascoiu, "Mapping EDOC to Web Services using YATL", 8th IEEE International Enterprise Distributed Object Computing Conference (EDOC 2004), pages 286–297, September 2004.

[21] R. A. Pottinger and P. A. Bernstein, "Merging Models Based on Given Correspondences", Proceedings of the 29th VLDB Conference, pages 826–873, 2003.

[22] QVT-Merge Group, "Revised submission for MOF 2.0 Query/Views/Transformations RFP (ad/2005-07-01) version 2.1", July 2005. Available at http://www.omg.org/docs/ad/05-07-01.pdf.

[23] E. Rahm and P. A. Bernstein, "A Survey of Approaches to Automatic Schema Matching", VLDB Journal, 10(4):334–350, 2001.

[24] O. Sims, "Enterprise MDA or How Enterprise Systems Will Be Built", MDA Journal, September 2004. Available at http://www.bptrends.com-/search.cfm?keyword=MDA+journal&gogo=1&go.x=68&go.y=4.

[25] X. L. Sun and E. Rose, "Automated Schema Matching Techniques: An Exploratory Study", Research Letters in the Information and Mathematical Sciences, 4:113–136, 2003.

[26] W3C, "Web Services Description Language (WSDL) 1.1", March 2001. Available at http://www.w3c.org/tr/wsdl.

# FOXI - Hierarchical Structure Visualization

Robert Chudý[1]
Faculty of Information Technology
Brno University of Technology

Jaroslav Kadlec[2]
Faculty of Information Technology
Brno University of Technology

**Abstract**

This paper presents a novel approach in hierarchical structure visualization. The main goal of the presented approach is to achieve an ability to display infinite hierarchy size in a limited display area, maintaining high level of orientation and fast navigation in the resulting hierarchical structures. High level of orientation is achieved by specific hierarchy visualization techniques where each level of hierarchy is abstracted from its substructures and limited visualization area leads to high speed navigation engaging hierarchy substructure zooming.

**CR Categories**: H.5.2 [User Interfaces] – Graphical user interface; E.2 [Data Storage Representations] – Composite structures

**Additional Keywords:** Hierarchical structure visualization, zooming, limited display size

## 1 Introduction

Hierarchies are one of the most commonly used data structures. The principle of hierarchical organization is very well known and understood by most of the typical computer users. While a strong emphasis exists on the automatic retrieval of information through search engines that have already migrated to personal computers, the hierarchical structures still seem to keep their importance in situations where the context is not recorded along with the information it relates to or where the information structure is too simple to be searched using a search engine. In the end, the hierarchical structure needs not to be perceived as a contradictory approach to automated searching and may serve as its effective supplement.

The major challenges in the area of visualization of hierarchical structures include making efficient use of the available display-area; prevent visual clutter and information overload, and the development of effective navigation approaches. Over the past years a lot of research work has been done in the field of effective display and interaction with hierarchically organized data. This research resulted in many different approaches like Fisheye Views [1], SemNet [2], Cone Trees [3], Tree Maps [4], Hyperbolic Browser [5] Disk Trees [6], Goldleaf [7] and others. These techniques usually provide a better perception of attributes of a hierarchical tree or present novel navigational strategies that go beyond the traditional scrolling and panning either in 2D or 3D space.

The visualization techniques are usually classified as overview+detail or focus+context approaches [8]. The principal difference between these two classes of interfaces is that the overview+detail techniques display the general overview of the rendered structure and the detail of the structure in focus in visibly separated areas, whereas the focus+context techniques use an integrated view utilizing special effects to enhance the perception of the area in focus.

The FOXI technique is based upon assumptions similar to Bubble Tree [9] and can be classified as overview+detail approach. Just as the Bubble Tree, it is based on the property of trees to recursively sub-categorize themselves into sub-trees while utilizing structure based method of abstraction of detail. The major difference between FOXI and Bubble Tree is that the sub-categorizing is used not only as a method for abstraction but also automatically creates navigational points in the layout of the interface.

The basic advantage of FOXI when compared to previously mentioned techniques is its ease of use, fairly straightforward concept that is easily understood by typical computer users, and stability of appearance during navigation even if the underlying structures are gradually changing. This enables effective exploitation of user's spatial memory and fast learning capabilities.

This paper describes the FOXI concept, its implementation into functional prototype, preliminary user testing results, discovered advantages, and disadvantages with corresponding enhancements and future work related to FOXI development.

### 1.1 FOXI

The primary objective for FOXI was to allow rendering of hierarchical structures with unlimited size in limited display area. This ability makes FOXI ideal for navigation over structures growing in space, like virtual cities and data landscapes.

FOXI is usable in cases where the displayed hierarchies reach a moderate complexity. For dealing with highly complex hierarchies, enhancements need to be applied both over the hierarchy and the interface utilizing FOXI technique. This situation is described under the Usability review section of this paper.

The layout of the resulting interface structure is created by inscription of circles representing sub-trees into a base circle. The sub-circles are touching their neighbor sub-circles and the base circle.

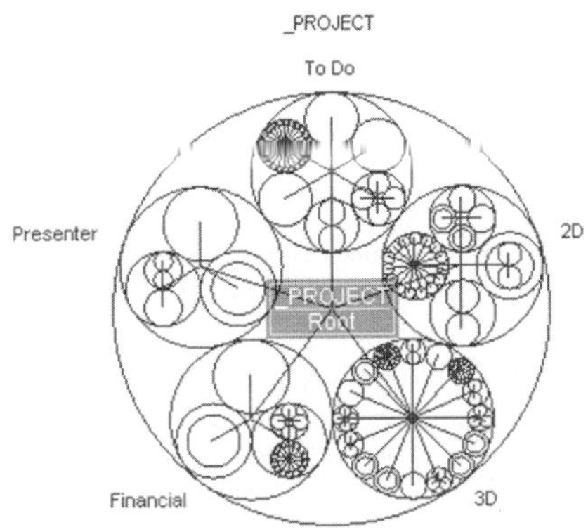


Figure 1. FOXI Layout

[1]email: chudy@fit.vutbr.cz
[2]email: kadlecj@fit.vutbr.cz



*T. Sobh and K. Elleithy (eds.), Advances in Systems, Computing Sciences and Software Engineering*, 229–233.

This technique allows intuitive recognition of spatial relations among displayed sub-trees. Despite the usage of the same building elements, different number of nodes in the sub-trees results in automatic creation of easily recognizable navigational points. These navigational points are simple to read and bring stability to the appearance of the structure even if the underlying hierarchy is gradually changing.

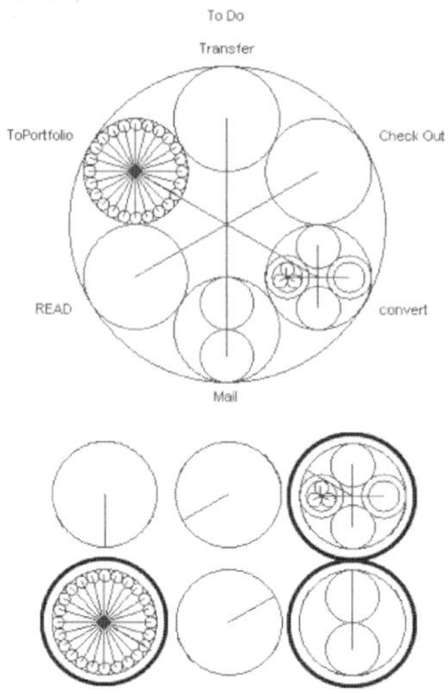


Figure 2. Navigational points

Increased density of nodes in the displayed sub-trees results in creation of higher number of visual signs in the area. FOXI allows free rotation of the sub-trees around the axis of the base circle what gives variability to possible setups of navigational strategies.

FOXI technique tends to preserve the most stable inner distribution of visual elements to make the best use of user's spatial memory and to allow reading of the context of the rendered structures thus allowing not only their perception but also understanding.

## 2 Implementation

The prototype of FOXI described in this paper is a zooming interface that works with the hierarchy provided by the file-system in common computer system. In this case FOXI displays the folder structure only to limit the number of displayed elements and to avoid visual clutter as much as possible without modifying the construction of the hierarchy it displays. To display the files a separate window is used.

### 2.1 Basic layout

FOXI interface visuals implemented in the prototype include:

1. Display area with navigational points
2. Textual description of folders on the current level – these descriptions work as supplements to navigational points. In case of properly structured and simple hierarchies it is possible to make them display only temporarily and over the display area.
3. Back button allowing moving one step up in the hierarchy – the placement of the button in the middle of the structure is strategic and allows very fast movement up in the hierarchy. The positioning in the same place throughout the whole navigation process requires no movement of mouse when moving up the hierarchical structure.

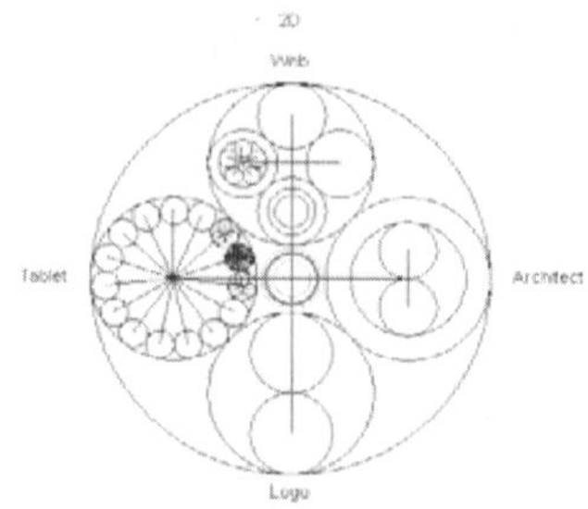


Figure 3. Basic layout with back button

### 2.2 Enhanced layout and interactions

FOXI was originally developed as navigational gizmo concept for a spatial interface controlled with hand-gestures. The prototype described in this paper is controlled by standard 3-button mouse; therefore, minor changes in visuals of the interface and control interactions had to be implemented:

1. *On mouse-over highlight* – the highlighting of current selectable circle allows user to easily identify his current options for movement in the structure. This feature is especially usable when jumping directly to a sub-folder.

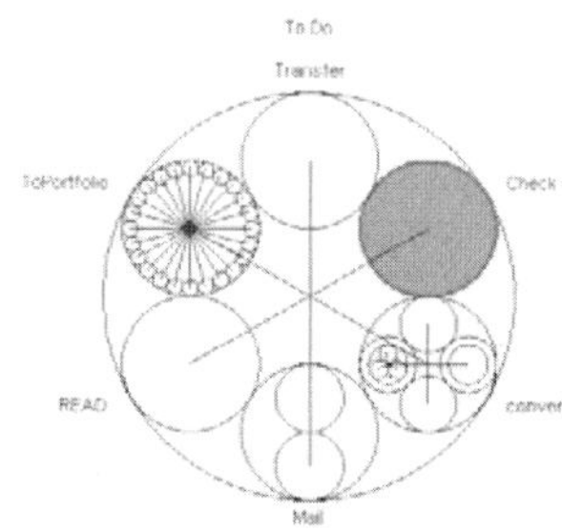


Figure 4. On mouse over highlight

2. *Direct jump* – allows user to jump directly into a sub-folder skipping a number of folders on the way. With this feature active, both *On mouse-over highlight* and name display of the current selectable folder are rendered.

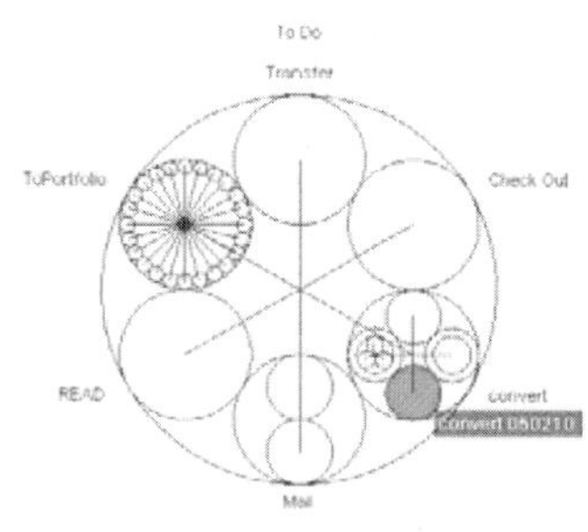


Figure 5. Direct jump highlight with folder name

3. *File window open* – the prototype was developed as a functional application for general usage to show performance in real-life situations. This limited the options for effective and precise evaluation of the interface. Opening of file window helped to identify that the user has reached his desired destination folder. Double click used for this operation seemed to slow down interaction with the interface. Instead a middle mouse button click was used.

4. *Enhanced back* – in cases where the direct jump to sub-folder was used, the right click on mouse provides jump back to the original folder skipping the folders on the way. In normal situations, left click in the middle of the base circle results in a move up in the rendered structure.

### 2.3 Basic Visualization

As mentioned before, the main task of the FOXI system is to visualize hierarchical structure in small area limited by circular shape. With exact knowledge of a hierarchy structure – its sub-structures – and radius of limiting circular shape a task of inscribed circles must be evaluated (fig. 6).

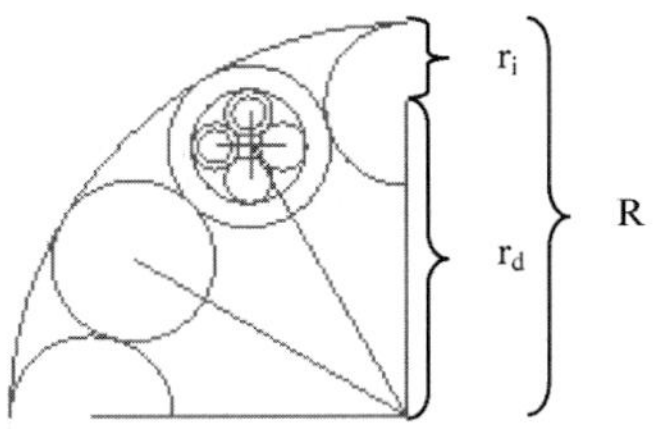


Figure 6. Inscribed circles

Knowing radius of the limiting circle R and number of inscribed circles N, representing number of sub-structures in the hierarchy, a function computing radius of inscribed circle can be evaluated. First a sector size in radians for each of the inscribed circles is computed (1)

$$\alpha = \frac{\pi}{N} \tag{1}$$

From knowledge of inscribed circle sector size a radius of inscribed circle can be evaluated (2)

$$r_i = \sin(\alpha) * \frac{R}{1 + \sin(\alpha)} \tag{2}$$

Radius R represents radius of the limiting circle, α sector size in radians, and $r_i$ final radius of the inscribed circle. The radius of the virtual circle where the inscribed circles are visualized can be computed from (3)

$$r_d = R - r_i \tag{3}$$

Further visualization of the next level of hierarchy is done similarly. Limiting radius is taken from the inscribed circle and the task of the new inscribed circles must be evaluated. The visualization of a hierarchy can be rather slow, especially in case of huge hierarchy with high structural complexity. In such cases, a limiting inscribed radius can be set to stop recursive visualizing at a level with radius that is too small to be rendered. The visualization process can benefit from utilization of this attribute as the physical visualization display has limited resolution and displaying of inscribed circles with too small radius has no positive effect on the visualization itself, because the displayed radius can be eventually smaller than 1 screen pixel.

### 2.4 Zooming In and Zooming Out

FOXI interface is using zooming technique for navigating in the hierarchy. Zooming involves limiting circle and one selected inscribed circle for zooming in or parent circle for zooming out. With these two circles and their positions and radii, a function of radius is computed to obtain appropriate position relative to the limiting circle. May $l$ be a distance from the center of the limiting circle to the center of inscribed circle. Inscribed circle is resized by small delta radius $\Delta_r$ (dependent on the zooming speed) and original radius before resize is saved to $old_r$. May radius of limiting circle be $R$ and its position in zero coordinates (inscribed circle is positioned relatively to the limiting circle). Points initializing the final function are [$old_r$, $l$] and [$R$, $0$] defining function of radius where distance from the beginning is decreasing with increasing radius. Final function computing position based on initial points is a linear function (4).

$$ax + by + c = 0 \tag{4}$$

Where $y$ is the final distance from the zero and $x$ is radius after resize for which the distance is applicable. Constants $a$, $b$, $c$ are predefined as follows:

$$\begin{aligned} a &= 0 - l \\ b &= -(R - old_r) \\ c &= (-b) * l - a * old_r \end{aligned} \tag{5}$$

With final distance after inscribed circle radius resized, the proper position can be computed. Similarly zooming out can be computed from current limiting circle and parent circle resizing radius down to the limiting circle radius.

A problem arises with the $\Delta_r$ increment. Too small increment sizes lead to slow zooming speed and are vastly slowing the whole interface operations. Too big increment leads to quick zooming making user feel little dizzy and disoriented with a question of "what happened?" and "where I am?". The best results with increment coefficient have been reached with custom increment specifications as various users enjoyed various zooming speeds.

### 2.5 Hierarchy caching

Directory structure, as a source of hierarchy, works fine as the test subjects most usually know contents of their hard drive well and learning time to get familiar with the new interface has therefore been shorted to minimum. In case of artificial hierarchy, visualized by FOXI, users would have to learn and get familiar with this unknown hierarchy. This may extend the "get used to" time significantly.

However, usage of directory structure on a hard drive brings several problems. The main problem is the drive speed and seeking speed. If the seeking speed of a hard drive would be about

10 ms, reading of a hundred directories would last over 1 second. Typical hard drive containing several hundreds of thousands of directories would take a very long time to load. Even visualization of current directory with 50 directories visible in depth would get stuck for almost a half of a second, what is very inconvenient.

A caching system has been proposed to solve this loading problem raised from the hard drive speeds. External cache file has been created to hold hierarchy data. Main benefit of external file is that FOXI can be brought to virtually any environment with its own structure representing menu with actions, database with images and so on. Test users got, thanks to this external cache generation, very fast tool for browsing contents of their hard drives.

## 3 Efficiency Measurement

As stated before, with FOXI prototype users got a tool for browsing the contents of their hard drives. Several approaches were proposed to test the efficiency of the proposed FOXI structure.

Approach with predefined hierarchy structure and set of tasks to do seemed to be a good idea as it would measure browsing speed in distinct tasks. However, this approach required users to learn to control not only the user interface but also to learn the structure of the test hierarchy. With too simple hierarchy structure there would be only small differences between tree browsing and FOXI browsing so another approach has been proposed.

Users working with well known hierarchies would have to learn the new user interface only and the total learning time would be much shorter. Using FOXI with their own hierarchies with no specific tasks to read and perform was for most of the users very engaging.

In this case the goal of measurement was to acquire all possible data from user's browsing actions. The problem arose because of the need to distinguish the moments when the user started to browse and he stopped to browse in the hierarchy. The user typically stops browsing when he finds certain information and displays it. In this case a window is opened. Opening of the window takes focus from the testing and measuring application. This action is considered stopped browsing. Similarly, when the user opens FOXI application, he starts browsing or navigating through the structure for some kind of information. This moment of gained focus is considered the start of browsing. Between these two moments, application measures number of levels that the user has browsed and the number of clicks she made to reach her goal. The time measured between focus gain and focus loss indicates how long was the user browsing through the structure. However, this time is not precise because the user could spend some time to open the application and the time to the first click could be very long. That is why a time from the first click is also measured to eliminate these problematic parts of browsing time.

As no other possibility exists how to compare the FOXI approach, a secondary application has been developed. This application was similar to the tree browser, featuring similar set of functionalities like FOXI and measuring the same actions like time of browsing, number of clicks and number of browsed levels. These values can be compared, giving approximate level of differences between these two approaches.

Table 1. Table with FOXI measured data by novice user

| Levels | 2 | 3 | 4 |
|---|---|---|---|
| Clicks | 2 | 3 | 4 |
| Total Time (ms) | 3355 | 6022 | 8805 |
| Time from first click (ms) | 3345 | 5835 | 8091 |

Table 2. Table with FOXI measured data by skilled user

| Levels | 2 | 3 | 4 |
|---|---|---|---|
| Clicks | 1 | 2 | 2 |
| Total Time (ms) | 1823 | 2364 | 2828 |
| Time from first click (ms) | 1012 | 1512 | 1802 |

Table 3. Table with Tree measured data by novice user

| Levels | 2 | 3 | 4 |
|---|---|---|---|
| Clicks | 2 | 3 | 4 |
| Total Time (ms) | 3145 | 4956 | 6732 |
| Time from first click (ms) | 3124 | 4812 | 6830 |

Table 4. Table with Tree measured data by skilled user

| Levels | 2 | 3 | 4 |
|---|---|---|---|
| Clicks | 2 | 3 | 4 |
| Total Time (ms) | 2183 | 2933 | 3835 |
| Time from first click (ms) | 2183 | 2489 | 2864 |

### 3.1 Efficiency measurement results

The efficiency measurement was focused on the speed of browsing only. The results indicate that the FOXI technique is slightly slower than the common tree browser when used by novice users. However the learning curve for FOXI interface is very steep, what makes the initial speed unimportant.

In case of skilled users, FOXI usually performs better than the tree browser. In the best observed situations, FOXI performs more than twice as good as the tree browser. The slightly worse results between the total time and time from first click, even if in favor of the FOXI, indicate a slower speed of user orientation in the radial layout of the interface. After the initial click, the spatial memory is on work and browsing times improve significantly.

## 4 Usability Review

### 4.1 Positive usability features

1. Ease of use – if the direct jump feature is not used, browsing of the FOXI structure requires no precise aiming with mouse. In most cases, the circles representing nodes are large enough to be hit with low accuracy of aiming when compared to the tree browser.

2. Mouse moves – the structure of FOXI allows navigation with very little mouse movements. This enhances the ergonomic aspects of usage significantly.

3. Developing of movement routines – the users of the FOXI interface usually develop a routine of clicks and movements that allow intuitive browsing in the interface. Along with the stability of FOXI in situations, where the underlying structure is changing gradually and with the help of navigational points, this feature is a powerful rendering of user's ability to employ spatial memory with the FOXI interface. This behavior is not forced and develops automatically, allowing users to browse at extremely high speeds.

5. Stability in gradually changing environment – FOXI proved to be stable in cases where the underlying structure got changed during the process of learning. However, the changes must have been gradual and not too complex. In these cases even some of the movement routines were easily adopted.

4. Unforced learning of the structure – the users reported an intuitive learning of the hierarchical structure rendered by FOXI. FOXI provides both general overview and the learning experience of details of the structure.

5. Orientation in unknown structures – after getting used to the radial layout even orientation in previously unknown structures became for most users easier.

### 4.2 Problematic usability areas

1. Navigation points differentiation – the ability of FOXI to create navigational points is based on the differences in number of nodes among the sub-trees. A uniform distribution of nodes among the sub-trees may decrease the usability of this feature. However, the navigational points are not defined only by their structure but also by their position and corresponding description. This characteristic preserves user's ability to navigate effectively with the help of his spatial memory.

2. Visual clutter – the ability of FOXI to abstract detail from the hierarchical structure is based on a complying distribution of complexity in the hierarchy. FOXI is primarily intended for shallow hierarchies, however with a proper method of abstraction it may be applied universally.

If no abstraction method is available, FOXI may require interface enhancements to provide easy navigation.

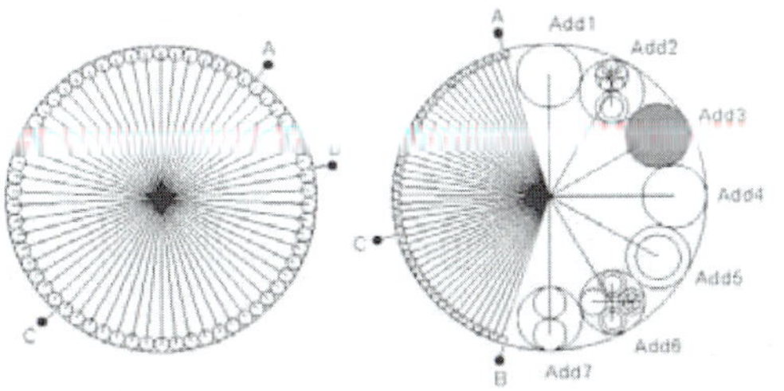


Figure 7. Interface enhancement option

3. Central node – FOXI, when compared to the tree browser, allows no jumps over nodes that accidentally happen to be very close to each other during browsing of the hierarchical structure. FOXI always displays a wide spread of nodes and allows no selection of nodes that will be displayed, whereas the tree browser may allow to jump from a node that is very deep in the hierarchy to a node high in the hierarchy because its sub-nodes have not been displayed. In this situation nodes from very different levels in the hierarchy can get very close and make the jump easy to perform; however, this feature is not controllable, happens mostly by accident and tends to diminish throughout the browsing actions. FOXI compensates for this characteristic with its fast back browsing and spatial memory features.

## 5 CONCLUSION

The usage of the interface proved that the FOXI interface is a very powerful navigation and structure-learning tool. The prototype was created to prove usability of the FOXI technique. Results of efficiency measurement indicate that the FOXI technique is not only an easy to use but also an efficient way to visualize and browse hierarchical structures.

Future improvements of the prototype will include color coding to further enhance the appearance of navigational points in the structure, visual clutter reduction enhancements, operations and interface for adding and removing folders, bookmarks, and tracing of browsing history.

## ACKNOWLEDGEMENT

This work has been supported by research grant User Interface with Hierarchical Structures from FRVŠ MŠMT FR200/2005/G1.

## REFERENCES

[1] Furnas G.W.: Generalized fisheye views, Proceedings of CHI'86 (Boston,MA),ACM,16-23.

[2] Fairchild K.M.,Poltrock S.E.,Furnas G.W: SemNet:Three-dimensional graphic representation of large knowledge bases, Cognitive Science and its Application for Human-Computer Interface, Lawrence Erlbaum, NewJersey, 1988.

[3] Robertson G.,Mackinlay J.,Card S.: ConeTrees: Animated 3D visualizations of hierarchical information, Proceedings of CHI'91 (NewOrleans,LA), ACM, 189-194.

[4] Johnson B.,Shnedierman B.: Tree-maps: A space-filling approach to the visualization of hierarchical information, Visualization 1991, IEEE, 284-291.

[5] Lamping J.,Rao R.: Laying out and visualizing large trees using a hyperbolic space, Proceedings of UIST'94, ACM, 13-14.

[6] Chi E., Pitkow J., Mackinlay J., Pirolli P., Gossweiler R., Card S.: Visualizing the evolution of web ecologies, Proceedings of CHI'98, ACM, 400-407.

[7] Faichney J., Gonzalez R.: Goldleaf hierarchical document browser, Proceedings of the 2nd Australasian conference on User interface, IEEE Computer Society, 2001

[8] Boardman R.: Bubble trees the visualization of hierarchical information structures, Conference on Human Factors in Computing Systems, ACM, 2000

[9] Cava R.A., Luzzardi P. R. G., Freitas C. M. D. S.: The Bifocal Tree: a Technique for the Visualization of Hierarchical Information Structures, 2002

# A Hands-Free Non-Invasive Human Computer Interaction System

F. Frangeskides and A. Lanitis
School of Computer Science and Engineering,
Cyprus College, P.O. Box 22006, Nicosia, Cyprus.
ffranges@cycollege.ac.cy, alanitis@cycollege.ac.cy

**Abstract - Conventional Human Computer Interaction requires the use of hands for moving the mouse and pressing keys on the keyboard. As a result paraplegics are not able to use computer systems unless they acquire special Human Computer Interaction equipment. In this paper we describe a system that aims to provide paraplegics the opportunity to use computer systems without the need for additional invasive hardware. Our system uses a standard web camera for capturing face images showing the user of the computer. Image processing techniques are used for tracking head movements, making it possible to use head motion in order to interact with a computer. The performance of the proposed system was evaluated using a number of specially designed test applications. According to the quantitative results, it is possible to perform most HCI tasks with the same ease and accuracy as in the case that a touch pad of a portable computer is used. Currently our system is being used by a number of paraplegic users.**

## I. Introduction

In several occasions it may be difficult to use a computer mouse in order to interact with a computer system. When using a conventional mouse the user needs to be able to hold the mouse, move it in a controlled fashion and also there is need to press the mouse buttons. Such actions require the use of hands and fingers, making the process of human machine interaction difficult (and in some cases impossible) for paraplegics. In this paper we describe a system for hands-free control of a computer based on head movements. A standard web camera is used for capturing images of the computer user and image-processing techniques are used for tracking his/her head movements. The face tracker activates cursor movements consistent with the detected head motion. The proposed system also allows the user to activate mouse clicks and enter text using a virtual keyboard so that users of the system are able to achieve full control of the system, in a hands-free fashion.

Our system is based on image processing algorithms that process images captured by an ordinary web camera mounted at the monitor of the system. At each frame of the image sequences captured, the system locates the position of the eyes of the computer user, enabling in that way the definition of his/her head movements. Those movements are translated into cursor movements on a computer screen so that the user of the system can use his/her head for moving the cursor. Figure 1 shows users using a computer system based on the system developed in this project.

The algorithms developed as part of the project, formed the basis for developing an integrated software package. The package contains a program that enables the user to control his/her computer using the proposed system. The package also includes training and test applications that enable users to become familiar with the system before they use it in real applications. Test applications enable the quantitative assessment of the performance of users when using our system. A number of volunteers tested our system and provided feedback related to the performance of the system. Both the feedback received and the quantitative assessment of the performance of volunteers, who tested the system, prove the potential of using our system in real applications. Currently a number of paraplegic computer users use our system on regular basis.

Fig. 1. Hands free human-computer interaction based on the proposed system



*T. Sobh and K. Elleithy (eds.), Advances in Systems, Computing Sciences and Software Engineering*, 235–242.

The remainder of the paper is organized as follows: In section 2 we present a brief overview of the relevant bibliography, in section 3 we describe the face-tracking algorithm. In section 4 we describe the functionality offered by the proposed system and in section 5 we describe the familiarization and test applications developed for testing the performance of the system. In section 6 we present quantitative system evaluation results and in section 7 we present our conclusions.

## II. LITERATURE REVIEW

In this section we provide a brief description of related work on face tracking – a more comprehensive treatment on related topics can be found in [9, 13, 15]. In this review we provide both a general overview on face tracking and we then focus on specific tracking systems used for hands free computer control.

### *A. Face Tracking*

Early attempts of locating faces in images relied on background subtraction and frame differencing [11]. Many researchers developed face-tracking systems based on colour information [4]. In this context the distributions of RGB intensities for the human skin are defined, enabling in this way the estimation of the probability that an image region contains skin. In several occasions [7] skin color information is used in conjunction with other visual information, such as edge maps, in order to improve the accuracy of face location.

The approaches mentioned above rely mainly on the distribution of image intensities of the facial regions to be tracked. Methods that utilize constraints related to the spatial relationships between facial features were also investigated [3]. Cootes et al [3] describe how Active Appearance Models can be used for tracking faces in sequences. During the tracking process Active Appearance Models deform, move and change intensity in order to match the movements and variations of faces in sequences, achieving in this way accurate face tracking. However, the process of tracking using Active Appearance Models is time consuming making this method inappropriate for real time human computer interaction tasks. Recently almost real time Active Appearance Model fitting algorithms were reported [12], but still the computation load required is higher than other tracking methods.

Chellapa and Zhou [2] use a probabilistic time series state space model for recovering both the motion and the identity of faces in sequences. With this approach the processes of tracking faces and recognizing faces are integrated, improving in that way the robustness of the overall system.

Template matching using either rigid [1] or flexible [14] templates has also been widely used for locating facial features. Instead of using the traditional template matching methodology, template matching can be done based on integral projections [1] in an attempt to reduce the computational cost and also make the method invariant to facial appearance variations. Recently Mateos [8] developed a system for tracking facial features using horizontal and vertical integral projections of the facial area. The difference in the projections, between two image frames, is used to estimate the movement parameters, which are applied in order to locate the face and facial features in the current image frame. The tracking algorithm used in our system is related to the framework reported by Mateos [8].

### *B. Hands-Free Computing*

Toyama [10] describes a face-tracking algorithm that uses Incremental Focus of Attention. In this approach they perform tracking incrementally starting with a layer that just detects skin color and through an incremental approach they introduce more capabilities into the tracker. Motion information, information related to the shape of the face and information related to the appearance of specific facial features are eventually used in the tracking process. Based on this approach they achieve real time robust tracking of facial features and also determine the facial pose in each frame. Information related to the face position and pose is used for moving the cursor on the screen.

Gorodnichy and Roth [6] describe a template matching based method for tracking the nose tip in image sequences captured by a web camera. Because the intensities around the nose tip are invariant to changes in facial pose they argue that the nose tip provides a suitable target for face tracking algorithms. In the final implementation cursor movements are controlled by nose movements, thus the user is able to perform mouse operations using nose movements. In addition to tracking using a single camera, Gorodnichy and Roth also implement and use a stereo face tracking method. Authors have used the system for several applications like drawing and gaming but they do not provide a quantitative evaluation of the proposed system.

Darrel et al [5] investigate different ways in which face-tracking technologies can be used in Human Machine Interaction tasks. Specifically they investigate the use of facial trackers for direct manipulation of cursor position, an agent based interaction paradigm and a perceptive presence example. In their experiments they use face trackers based on stereo information.

Several commercial hands-free computer control systems are available [16]. In most cases head tracking is done by using special hardware such as infrared detectors and reflectors [19] or special helmets[18, 21]. Hands free non-invasive systems are also available in the market [17 ,20].

## III. FACE TRACKING

Our system relies on an eye tracking algorithm used for determining the position of the eyes of subjects using the system. In this section we describe the basic integral projection tracking algorithm and then present the overall face tracking system.

### *A. Integral Projections*

An integral projection [1 ,8] is an one-dimensional pattern, whose elements are defined as the average of a given set of pixels along a specific direction. Each element of a vertical projection is defined by the sum of intensity values of a set of pixels along the horizontal axis divided by the number of pixels considered. Similarly, each element in a horizontal projection is given by the sum of the intensities of a given set of pixels along the vertical axis divided by the number of pixels.

Integral projections represent two-dimensional structures in image regions using only two one-dimensional patterns, providing in that way a compact method for region representation. Since during the calculation of integral projections an averaging process takes place, spurious responses in the original image data are eliminated, resulting in a noise free region representation.

### *B. Tracking Using Integral Projections*

In order to perform tracking based on this methodology, we calculate the horizontal and vertical integral projections of the image region to be tracked. Given a new image frame we find the best match between the reference projections and the ones representing image regions located within a predefined search area. The center of the region where the best match is obtained, defines the location of the region to be tracked in the current frame. This procedure is repeated on each new frame in an image sequence.

The process of tracking using integral projections is similar to the process of template matching with the difference that in the case of projection matching one-dimensional patterns instead of two-dimensional templates are utilized.

### *C. Face Tracker*

The method described in the previous section formed the basis of the face tracking system employed in our system. Instead of tracking a single facial region using integral projections, we track two facial regions – the eye region and the nose region. The nose region and eye region are primarily used for estimating the vertical and horizontal face movement respectively.

During the training process the vertical projection of the nose region and the horizontal projection of the eye region are calculated and used as the reference projections during tracking.

Once the position of the two regions in an image frame is established, the exact location of the eyes is determined by performing local search in the eye region. In particular the area of the eye region is divided into three columns and the center of gravities of the two side columns indicate the approximate location of the eyes. Based on the initial estimate of the locations of the eyes, we perform local search in order to refine the initial approximation. The local search is based once again on integral projections for each eye.

In order to improve the robustness of the system to changes in lighting, we employ a normalization algorithm that aims to remove global differences in gray-level intensities between integral projections derived from successive frames.

In order to deal with rotated faces at each frame we estimate the rotation angle of the face, based on the location of the eyes. On the next frame prior to the calculation of the integral projections we rotate accordingly the nose and eye region, so that the projections obtained are invariant to changes in rotation angle.

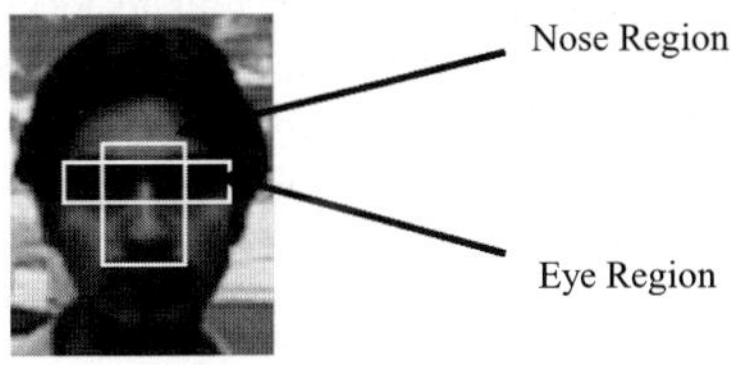


Fig. 2. Example of face and eye regions

In order to deal with occlusion and significant changes in pose we apply constraints related to the relative position of the two regions tracked (the nose and eye regions) so that their relative positions comply with the expected positions on face images. Based on this approach, during the process of tracking the two boxes tracked are only allowed to move to positions that do not violate the statistical constraints pertaining to their relative positioning.

### D. Experimental evaluation

The completed tracking system was tested on a portable Intel Pentium III 800 MHz computer with 256 MB ram. A low quality consumer web camera has been used in our experiments. The distance between the camera and the user was between 70 and 130 cm. Testing was done using seven pre-recorded image sequences containing about 200 frames each. Test sequences aim to evaluate the tracking process under different conditions such as changes in lighting, changes in speed of face movement, changes in the expression of the face, tracking different individuals, changes in the distance between the user and the camera, tracking in the presence of occlusion and 3D pose variation. Typical examples of face images from the test sequences are shown in figure 3. For each test sequence the tracker is manually initialized on the first frame and the tracking process is carried out automatically for the rest of the sequence.

The tracking results are encouraging. In a 160 x 120 pixel frame sequence, the application was able to track accurately the exact position of the eyes in real time with an average error of 0.73 pixels. Failures in tracking correctly the eyes of the user were recorded in the cases that a great part of the face was not visible in the camera and in the cases that the horizontal rotation of the face is greater than 70°. The tracking process has been proved to be robust to usual destructors encountered during tracking such as sudden changes in lighting conditions, reasonable occlusion, changes in facial expressions and variations in the background.

Even in the cases that the tracker fails to locate the eyes correctly, the system usually recovers and re-assumes accurate eye-tracking.

Fig. 3. Typical images from the test sequences. (In the first row the eye positions recovered by the eye tracker are overlaid)

## IV. COMPLETE SYSTEM DESCRIPTION

In this section we describe how various functions are implemented in the proposed non-invasive human computer interaction system. Those actions refer to system initialization, system training and simulation of click operations.

### A. System Initialization

The first time that a user uses the system he/she is required to keep his/her face still and perform blink actions. Based on a frame-differencing algorithm the positions of the eyes and nose regions are determined and integral projections for those areas are computed. Those projections are saved in the system so that in the case that the same user uses the system the same projections are used. Once the projections are estimated the tracking process starts.

### *B. Simulation of Mouse Operations*

In this section we describe how mouse operations are implemented in our system.

*Moving the cursor:* The divergence of the face location from the initial location is translated in cursor movement speed, towards the direction of the movement. Based on this approach only minor face movements are required in order to initiate substantial cursor movement. The sensitivity of the cursor movement can be customized according to the abilities of different users.

*Mouse Click actions:* Mouse click operations can be activated in three ways. According to the first way, clicks are activated by the stabilization of the cursor to a certain location for a time period longer than a pre-selected threshold (usually around one second). In this mode, users select the required click action to be activated when the cursor is stabilized. The predefined options include: left click, right click, double click, drag and drop and scroll.

As an alternative, click actions can be performed using an external switch attached to a computer. In this mode the user directs the mouse to the required location using head movements and the appropriate click action is activated based on the external switch. The switch can be activated either by hand, foot or by voice.

Click actions can also be activated by sudden movements in the area of the eyes, such as eye blinking or eyebrows movement. However, users of the system demonstrated strong preference to the first two methods, which are less tiring, hence in the quantitative evaluation we only consider the first two click methods.

*Text Entry:* Text entry is carried out by using the "On-Screen Keyboard" - a utility provided by the Microsoft Windows Operating System (see figure 4). Once the On-Screen Keyboard is activated it allows the user to move the cursor on any of the keys of the keyboard and by clicking actions activate any key. As a result it is possible to use head movements in order to write text or trigger any operation that is usually triggered from the keyboard, achieving in that way full hands-free control over the computer.

Fig. 4. Screen shot of the "On Screen Keyboard"

## V. HANDS-FREE COMPUTING APPLICATIONS

Although the proposed hands-free HCI system can be used for any task where the mouse and/or keyboard is currently used, we have developed dedicated computer applications that can be used by perspective users of the system for familiarization purposes. Also test applications were developed and used for obtaining quantitative evaluation results of the overall system operations. In this section we briefly describe the familiarization and test applications.

### *A. Familiarization applications*

The following familiarization applications have been developed:

*Paint-tool Application:* The user is able to draw on the screen various shapes based on corresponding head movements in order to get familiar with cursor movement.

*Car Driving Game:* The user directs a visual vehicle with head movements. This game is designed to train the user how to move the cursor over a predefined pattern.

*Piano:* With this program the user can move the cursor over a visual piano and click on the notes in order to hear the corresponding tune. This application aims to train perspective users to direct the cursor to a specific location and activate click actions.

Screen shots of the familiarization applications are shown in figure 5.

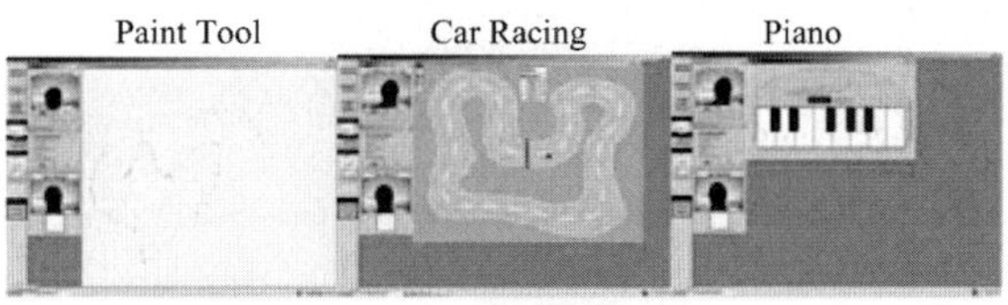


Fig. 5. Screen shots of the familiarization applications

### B. Test Applications

Test applications are used as a test bench for obtaining quantitative measurements related to the performance of the users of the system. The following test applications have been developed.

*Click Test:* The user is presented with four squares on the screen. At any time one of those squares is blinking and the user should direct the cursor and click on the blinking square. This process is repeated several times and at the end of the experiment the average time required to direct the cursor and click on a correct square is quoted.

*Draw Test:* The user is presented with different shapes on the screen (square, triangle and circle) and he/she is asked to move the cursor on the periphery of each shape. The divergence between the actual shape periphery and the periphery drawn by the user is quoted and used for assessing the ability of the user to move the cursor on a predefined trajectory.

*Typing Test:* The user is presented with a word and he/she is asked to type in the word presented. This procedure is repeated for a number of different randomly selected words. The average time required for typing a correct character is quoted and used for assessing the ability of the user to type text.

Screen shots of the three test applications are shown in figure 6.

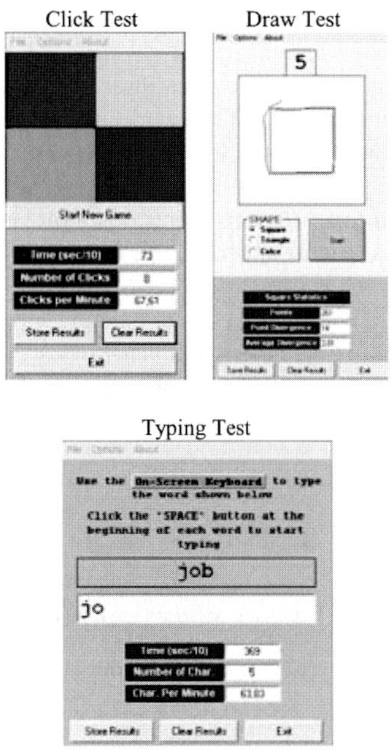


Fig. 6. Screen shots of the test applications

It is important to note that all test applications can run either using any conventional pointing device or the hands-free human computer interaction system described in this paper.

## VI. SYSTEM EVALUATION

The test applications presented in the previous section were used for assessing the usefulness of the proposed system. In this section we describe the experiments carried out and present the results.

### A. Experimental Procedure

Twenty volunteers tested our system in order to obtain quantitative results related to the performance of the system. The test procedure for each subject is as follows:

*Familiarization stage:* The subject is instructed how to use the hands-free computing system and he/she is allowed to get familiar with the system by using the familiarization programs. On average the duration of the familiarization stage was about 15 minutes.

*Benchmark performance:* The benchmark performance for each volunteer is obtained by allowing the user to complete the test applications using a conventional mouse and a typical touch pad of a portable PC. The performance of the user is assessed on the following tests:

- Click Test: The average time required for five clicks was recorded.
- Draw Test: The subject is asked to draw a square, a triangle and a circle and the average discrepancy between the actual and the drawn shape is quoted.
- Type test: The user is asked to type five randomly selected 3-letter words and the average time for typing a correct letter is recorded (In this test text input was carried out by using the "On Screen Keyboard").

*Hands-Free test:* The user is asked to repeat the procedure used for obtaining the benchmark performance, but in this case he/she runs the test programs using the hands-free computing system.

*Hands-Free with an external switch test:* The test procedure is repeated, but in this case the user is allowed to use the Hands Free system in conjunction with an external switch for activating mouse clicks.

The 20 volunteers who tested the system were separated into two groups according to their prior expertise in using the hands-free computing system.

Group A contains subjects with more than five hours prior experience in using the hands free system. Subjects from group B used the system only as part of the familiarization stage (for about 15 minutes).

### *B. Results - Discussion*

The results of our experiments are summarized in Table I.

TABLE I
QUANTITATIVE EVALUATION RESULTS

| Test | Test Method | Group A | Group B |
|---|---|---|---|
| Click test (Units: seconds/click) | Mouse | 0.76 | 0.86 |
| | Touch Pad | 1.45 | 2.18 |
| | Hands Free | 3.58 | 4.84 |
| | Hands Free + switch | 1.41 | 2.5 |
| Draw Test (Units: Divergence in Pixels) | Mouse | 2.62 | 2.05 |
| | Touch Pad | 3.01 | 4.01 |
| | Hands Free | 3.07 | 5.93 |
| | Hands Free + switch | N/A | N/A |
| Typing Test (Units: seconds/click) | Mouse | 0.89 | 0.86 |
| | Touch Pad | 1.73 | 3.41 |
| | Hands Free | 4.39 | 5.99 |
| | Hands Free + switch | 2.37 | 3.70 |

Based on the results shown above the following conclusions are derived:

*Click test:* In all occasions the results obtained by using a conventional mouse are better. When the hands free system is combined with a switch for performing click actions, the performance of the system is comparable with the performance achieved when using a touch pad. In the case that the hands free system is not used with a switch, the performance of the users decreases. The additional delay introduced in this case is mainly due to the requirement for stabilizing the cursor for some time (1 second according to the default setting) in order to activate a click action.

*Draw Test:* For experienced users of the system (Group A) the performance achieved using the free hand mouse is comparable with the performance achieved when using a touch pad. Once again subjects achieved the best performance when using a conventional mouse. Subjects from group B (inexperienced users) produced an inferior performance when using the hands free system. The main reason is the reduced ability to control precisely cursor movements due to the limited prior exposure to the system.

*Typing Test:* In this test the use of mouse or touch pad for typing text is significantly superior to the performance of users using the hands-free system, indicating that the proposed system is not the best alternative for typing applications. However, the performance obtained when using the hands-free system in conjunction with the external switch, is once again comparable to the performance obtained with the touch pad. The main reason for the inferior performance obtained when using the hands free system, is the small size of the keys on the "On Screen Keyboard" that requires precise and well-controlled cursor movements. The ability to precisely move the cursor requires extensive training. Instead of using the "On Screen Keyboard", provided by the Windows operating system, it is possible to use dedicated virtual keyboards with large buttons in order to improve the typing performance achieved when using the hands-free computing system.

*User Expertise:* The abilities of users to use the hands free system increase significantly with increased practice. Based on the results we can conclude that subjects with increased prior experience in using the hands-free system (from group A) can perform all usual HCI tasks efficiently. It is expected that with increased exposure to the system, users will be able to improve even more their performance. It is worth mentioning that even for volunteers from Group A the practice time for using a computer with a mouse or a touch pad is much longer than the total prior practice time with the hands free system.

*External Switch:* The introduction of an external switch that can be activated either by foot or hand enhances significantly the performance of the system.

## VII. CONCLUSIONS

We presented a hands-free computing system that relies on head movements for performing all Human Computer Interaction tasks. The proposed system caters for common HCI tasks such as mouse movement, click actions and text entry (in conjunction with the "On Screen Keyboard").

Based on the quantitative results presented, head based HCI cannot be regarded as a substitute for the conventional mouse, since the speed and accuracy of performing most HCI tasks is below the standards achieved when using a conventional mouse. However, in most cases the performance of the proposed system is comparable to the performance obtained when using touch pads of portable computer systems. Even though the accuracy and speed of accomplishing various HCI tasks with a touch pad is less than in the case of using a conventional mouse, a significant number of computer users use regularly touch pads, since they provide flexibility to portable computing. We are convinced that computer users will also find the proposed hands free computing approach useful.

The proposed system is particularly useful for paraplegics with limited (or without) hand mobility. Such users are able to use a computer system based only on head movements. Currently a number of paraplegic computer users are using the system – the feedback received so far proves the usefulness of the system for allowing disabled users to use computer systems.

In the future we plan to upgrade the system to a multi-modal hands-free system that combines both head movement and speech input in order to perform HCI tasks more efficiently. Initial experimentation in this area proved the potential of this direction.

Once the multi-modal HCI system is completed we plan to stage a large-scale evaluation test is order to obtain concrete conclusions related to the performance of the proposed system. Since the hands-free system is primarily directed towards paraplegic computer users, we plan to include evaluation results from paraplegics in our quantitative evaluation results.

## ACKNOWLEDGEMENTS

The work described in this paper was supported by a Cyprus College Research Grant and a Research Promotion Foundation Grant (project 26/2001). We are grateful to members of the Cyprus Paraplegics Organization for their valuable feedback and suggestions.

## REFERENCES

[1] R. Brunelli, T. Poggio, T. 1993. "Face Recognition: Features versus Templates". IEEE Transactions on Pattern Analysis and Machine Intelligence, Vol 15, No 10, 1993 pp, 1042-1052.

[2] R. Chellappa and S. Zhou. "Face tracking and recognition from video". Handbook of Face Recognition, S. Li and A. K. Jain (Eds.), Springer, 2004.

[3] T.F. Cootes, G.J. Edwards, C.J. Taylor. "Active Appearance Models". IEEE Transactions of Pattern Analysis and Machine Intelligence, Vol 23, 2001, pp 681-685.

[4] J. Crowley, K. Schwerdt. "Robust Tracking and Compression for Video Communication". Procs. Of the International Workshop on Recognition, Analysis and Tracking of Faces and Gestures in Real Time Systems, 1999, pp 2-9.

[5] T. Darrell, K. Tollmar, F. Bentley, N. Checka , L. Morency, A. Rahimi and A. Oh. "Face-responsive interfaces: from direct manipulation to perceptive presence", Procs. of the International Conference of Ubiquitous Computing, 2002, pp. 135-151.

[6] D.O Gorodnichy and G.Roth, "Nouse 'Use Your Nose as a Mouse' – Perceptual Vision Technology for Hands-Free Games and Interfaces", Image and Vision Computing, Vol. 22, No 12 , 2004, pp 931-942.

[7] K. Huang and M. Trivedi, "Robust Real-Time Detection, Tracking, and Pose Estimation of Faces in Video Streams" Procs of International Conference on Pattern Recognition ,2004.

[8] G.G. Mateos. "Refining Face Tracking With Integral Projections". Procs. Of the 4th International Conference on Audio and Video-Based Biometric Person Identification, Lecture Notes in Computer Science, Vol 2688, 2003, pp 360-368.

[9] A. Pentland. "Looking at People: Sensing the Ubiquitous and Wearable Computing". IEEE Transactions of Pattern Analysis and Machine Intelligence, Vol 22, No 1, 2000, pp 107-119.

[10] K. Toyama, "Look, Ma – No Hands! – Hands Free Cursor Control with Real Time 3D Face Tracking", Procs. Of Workshop on Perceptual User Interfaces, 1998, pp. 49-54.

[11] M. Turk, A. Pentland. "Eigenfaces for Recognition". Journal of Cognitive Neuroscience, Vol 3, No 1, 1991, pp 71-86.

[12] J.Xiao, S.Baker, I.Matthews, and T.Kanade. "Real-Time Combined 2D+3D Active Appearance Models" Proceedings of the IEEE Conference on Computer Vision and Pattern Recognition, 2004.

[13] M.H. Yang, D.J. Kriegman and A. Ahuja. "Detecting Faces in Images: A Survey". IEEE Transactions of Pattern Analysis and Machine Intelligence, vol 24, no 1, 2002, pp 34-58.

[14] A. Yuille, D. Cohen, P. Hallian. "Feature extraction from faces Using Deformable Templates". Procs. International Journal of Computer Vision, Vol 8, 1992, pp 99-112.

[15] W. Zhao, R. Chellapa, P.J. Philips, A. Rosenfeld, "Face Recognition: A Literature Survey". ACM Computing Surveys, Vol 35, No 4, 2003, pp 399-458.

[16] Assistive Technology Solutions. http://www.abilityhub.com/mouse/

[17] CameraMouse: Hands Free Computer Mouse. http://cameramouse.com/

[18] EyeTech Digital Systems-Eye Tracking Device. http://www.eyetechds.com/

[19] Hands Free Mouse – Assistive Technology Solution. http://eyecontrol.com/smartnav/

[20] Mouse Vision Assistive Technologies, http://mousevision.com/

[21] Origin Instruments Corporation http://orin.com/index

# A Model for Anomalies of Software Engineering

Gertrude N. Levine
Fairleigh Dickinson University
1000 River Road
Teaneck, NJ 07666

**Abstract- We present a model for the anomalies of software engineering that cause system failures. Our model enables us to develop a classification scheme in terms of the types of errors that are responsible, as well as to identify the layers of a system architecture at which control mechanisms should be applied.**

## I. Introduction

Software Engineering has been defined as the "application of a systematic, disciplined, quantifiable approach to the development, operation, and maintenance of software …"[11]. A development team needs to assure that it has adequate resources and specifications [10]. A procedure must be followed during which requirements are systematically and thoroughly gathered and a software system is designed, implemented, and tested. If there are errors in the procedure or in the time allocated, work may be lost during execution. If these anomalies are not detected and corrected within an acceptable time interval, they cause software failures. Cost overruns, time overruns, and malfunctions have plagued the software industry since the early days of computers. This paper considers anomalies of software systems, which we define as errors that result in the loss of work and potentially cause system failure. We establish a classification scheme for anomalies and the mechanisms that are used in their control. Classes are defined by the type of interaction between processes; subclasses are determined by the characteristics of the error that causes each anomaly. We provide examples of anomalies in each class and subclass in order to illustrate the comprehensiveness of our study.

A model is defined, which has previously been studied for other applications [13, 14, 15] and which is also applicable to anomalies of software engineering. The same mechanisms that are the basis of the model are shown to be the causes of anomalies. This approach brings together topics from different areas of software engineering into one unifying analysis.

The word "process" is frequently applied to a software engineering project, but we limit the term to denote an entity with a single thread of control that is executed and allocated service as a unit. A software engineering project is then a set of cooperating and potentially concurrent processes that are bound together with control mechanisms to ensure that they are developed, integrated, tested, and provided completed service as a cohesive unit. Independent projects may be combined to compose a system of systems [10].

## II. A Model for Software Systems

### A.. The process

$\mathcal{P}$ is a finite, nonempty set of processes in a software system. A process, $p \varepsilon \mathcal{P}$, is a set of requests that are bound together by control mechanisms in order to complete service as a unit. In addition, requests of the same process are restricted so that they are serviced in sequential order. A concurrent process has additional restrictions at coordination points. Each process has a unique key, which includes a project and user identification. Processes of the same project and/or user may share buffer elements.

### B. The Resource

$\mathcal{R}$ is a finite, nonempty set of resources in a software system. Each resource, $r \varepsilon \mathcal{R}$, has a unique key, which consists of several fields. Fungible elements[1] share most of their resource's key for user identification, although this key has an additional field for identification of the individual elements by the resource system. Resources of the same system may share buffer elements. Each resource element has a finite number of linearly ordered units corresponding to instances of Time.

### C. Time

$\mathcal{T}$ is a set of units of Time represented by an initial subset of the natural numbers that is bounded by the lifetime of the Software System.

### D. The Layers

All requests attempt to traverse a set, $\mathcal{M}$, of ordered layers in a system's architecture (Fig. 1):

1. A Process Conception layer, **PC**, is a set of requests that are conceived by a user and continually reconceived until they can be submitted to the next higher layer. If requests move down to this layer, they are requesting to be reconceived, i.e., corrected.
2. A Process Buffer layer, **PB**, is a set of requests that have been conceived and then postponed (buffered in the user system) while requesting delivery to resource layers.
3. A Delivery Buffer layer, **DB**, is a set of requests that are buffered in an independent delivery system requesting access to resource layers.

[1] We borrow the term fungible from law to denote resource units that are interchangeable to the user.



*T. Sobh and K. Elleithy (eds.), Advances in Systems, Computing Sciences and Software Engineering*, 243–250.

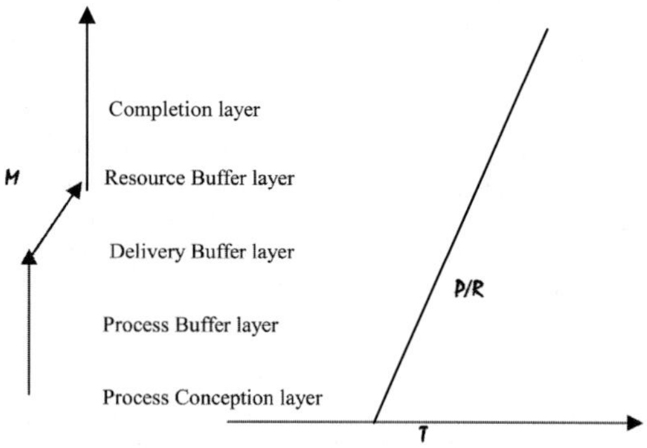


**Fig. 1. The Layers of a Software System**

4. A Resource Buffer layer, **RB**, is a set of requests that are stored in the resource system requesting service.
5. A Resource Service layer, **RS**, is a set of requests that are in a state of execution requesting completion of service.
6. A Completion layer, **C**, is a set of requests that have completed service.

We define an ordering relation, >, on $\mathcal{M}$, such that **C** > **RS** > **RB** > **DB** > **PB** > **PC**. For M and M' $\varepsilon$ $\mathcal{M}$, where M > M', with m, m' subsets of M and M' respectively, we say that m > m' and m is in a higher layer. If m and m' are subsets of the same layer, then m == m'.

The requirement of six layers is a simplification. Software systems routinely add additional layers, as in the waterfall model of software engineering. For systems with fewer layers, such as a 10Base-2 Ethernet, we model the absence of layers by the movement of requests through them without delay. The layers are conceptually circular, so that requests that complete service can be reconceived for reuse [15].

For peer-to-peer systems, the above diagram "folds over," (Fig. 2) so that both processes and resources initiate and receive requests, acting as both source and destination. The ordering relation above is thus defined for the direction of the request mapping in its attempt at service, through source conception, source buffering, delivery buffering, destination buffering, destination service, completion, and potential reuse. $\mathcal{T}$ is not shown in Fig. 2. We need a 3-dimensional diagram for a peer-to-peer system, in which requests can be shown initiating and terminating from left to right during the increment of time.

*E. The Request*

Each request is a triple (m, r, t), where:

m $\varepsilon$ $\mathcal{M}$ identifies the current position of the request in the system layers in its request for completed service. It also identifies a specific fungible element where appropriate. A request, (m, r, t), requests to be mapped to a layer m' > m.

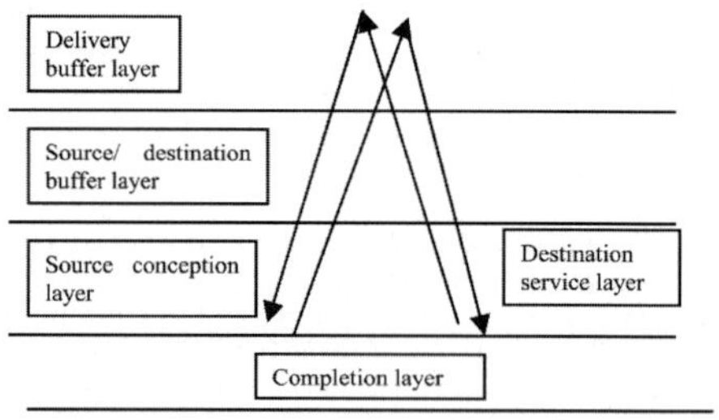


**Fig. 2. Architecture for Peer-to Peer-Systems**

r $\varepsilon$ $\mathcal{R}$ identifies the resource element at which service is requested.

t $\varepsilon$ $\mathcal{T}$ identifies a discrete unit of Time. A request is always requesting movement to a higher layer at its current unit of T, but may be postponed to a t' > t.

All requests are either **Input** or **Output** operations, specified by a predicate on the request that also contains data values to be input or output, together with process and data format keys. An input request that contains data values other than a wild card is a match request and is serviced only if its data values matches the values currently being stored at the requested location. An input request with a wild card in place of data returns the value being stored. If such an input request also matches the format key of the stored data (in contrast to undecipherable input, for example), the input is called a read.

*F. The Software System*

A Software System is a quintuple {$\mathcal{M}$, $\mathcal{P}$, $\mathcal{R}$, $\mathcal{T}$, $\mathcal{F}$ }, where $\mathcal{M}$ is a set of ordered layers, $\mathcal{P}$ is a set of processes, $\mathcal{R}$ is a set of resources, $\mathcal{T}$ is slotted time, and $\mathcal{F}$ is a function that controls the movement of requests through the single and composite layers of $\mathcal{M}$ . $\mathcal{F}$ assigns a mapping, $g_P$: $\mathcal{P} \rightarrow \mathcal{P}'$ for each p $\varepsilon\mathcal{P}$, $g_P$ recursively defined, such that:

$g_P$(m,r,t)= (m', r, t), m' > m, if restrictions permit this mapping (called promotion), else

$g_P$(m,r,t)= $g_P$ (m', r', t'), m == m', t' > t, if restrictions permit the mapping (called rescheduling), else

$g_P$(m,r,t)= $g_P$ (m', r', t), m > m' (called demotion).

A request is mapped from the process conception layer, in order, to the completion layer as restrictions allow. If restrictions prevent mapping to a higher layer at some specific time unit, the request attempts rescheduling to the same layer for a later time unit; if restrictions prevent promotion and rescheduling, the request is demoted and again requests promotion. All requests that do not complete service by the end of Time are demoted to **PC**. Requests that are demoted to **PC** and never reconceived and promoted are in a dead state [14].

*G. The Control Mechanisms*

Each request is assigned three types of restrictions that, together with input and output operations, match the subclasses of errors to be defined:

1) Promotion and demotion **dependency** predicates specify a set of process mappings (events) that effect a request's upward or downward movement. A request cannot be promoted if any process event is contained in its promotion dependency. Its process must wait for these events to complete, at which time they are removed from the dependency. A request cannot continue to be rescheduled if an event in its demotion dependency occurs.

2) Promotion and demotion **times** predicates specify the minimum/ maximum number of times that a request must be/ can be rescheduled to a particular element of $H$. These values are decremented, down to 0, each time a request is rescheduled. A restriction of a minimum number of rescheduling times delays the request from moving to a higher layer until that number is decremented to 0; that of a maximum number of rescheduling times sets a threshold for how long a request can remain at its present layer (e.g., a hard deadline). The maximum rescheduling time assigned to all requests is limited by the lifetime of the system, causing any process that has not completed service at the end of Time to be demoted to **PC**. Times predicates control rescheduling at both single and composite layers.

3)Promotion and demotion **priority** predicates determine request mappings when conflict occurs. A request is enabled for promotion iff its promotion dependency set is empty and its promotion times predicate is 0. A request is enabled for rescheduling to the same layer iff its demotion times predicate is not 0 and an event in its demotion dependency does not occur. **Conflict** occurs if an output request and at least one other request with a different key value are both enabled for mapping to the same resource or buffer unit. (For fungible resources, conflict occurs if there are more enabled output requests with different key values than there are fungible units.) Requested service or buffering is provided at each resource unit to at most one conflicting request.

*H. Definitions*

We have introduced the term, anomaly, which is not widely used in the software engineering field, to denote an execution time error that results in loss of service, and may or may not cause a failure. One benefit derived from the study of the model is the enabling of definitions of these terms.

**Service** is rescheduling at **RS** with a reduction of the request's promotion time predicate. (Depending on the subject of analysis, a buffer layer may be considered a service layer.)

**Completed service** is a request's mapping to **C**.

**User requirements** is a nonempty set of process requests that are conceived, bound together, and repeatedly rescheduled in **PC** by a user and/or by a development team, and then submitted to resource layers to request completed service as a unit.

An **error** is a state containing a nonempty set of processes that conflict with user requirements.

An **anomaly** is an error that contains at least one process that is demoted from **RS** and causes a loss of (partial) service.[2] The process must repeat previous upward movement to **RS** as well as service at **RS** before it can complete service. Note that a process may complete service even if an anomaly occurs. If an anomaly is caused by a user error, the user may provide for effective error recovery. If an anomaly is caused by a system error, the resource system may detect the anomaly and recover from it.[3]

A **failure** is an anomaly in which process output conflicts with user requirements and causes the demotion of the process to **PC**.

## II. THE CLASSIFICATION SCHEME

Anomalies are divided into three classes: scheduling, security, and development. These categories are then subdivided into five subclasses for the errors that cause them: errors of output, input, dependency, priority, and timing.

A process or set of cooperating processes always requests to complete a service. In development anomalies, the process set is developed incorrectly and cannot complete service unless it is returned to its maintenance team for correction. Yet, even a process set that has been correctly developed has no guarantee that it will complete service unless the resource system prevents interference from other processes, either that of scheduling errors (errors in the resource system) or of security errors (errors in the violating process and in the resource system). The resource system can prevent scheduling and security anomalies, although prevention mechanisms may be onerous. Alternatively, it may rely on detection, using key matches or time-based heuristics, and recovery of valid processes, typically with roll backs and restarts. The resource system cannot prevent development anomalies; in particular, it cannot achieve recovery for a security violator.

Scheduling is the assignment of independent processes to different units of shared resources. For example, an operating system's scheduler chooses which of competing processes to assign to the CPU and a network router schedules packets onto positions in output queues. A process or a set of cooperating processes in a scheduling anomaly would have completed its service if it had been serially scheduled, assuming that security and development anomalies had not occurred and that time constraints were satisfied. To enhance throughput and minimize response time, the scheduler had assigned independent processes to concurrent intervals of time, but did not secure the order of their competing requests.

Security is the prevention of unauthorized access to a resource system, while authorizing a set of valid users to share resources. A malicious insider or an intruder with an insider's access key is still an invalid user, since some of the fields of the keys in its data requests will not match the stored data

[2] An anomaly is an error in execution, with a relationship similar to that of a process and a program.

[3] The demotion of an intruder process is also an anomaly, since the intruder should have been denied access, preventing the resultant work loss.

keys. Although most security anomalies can be prevented, control mechanisms have become expensive as systems have evolved and violators have become more sophisticated. A process or set of cooperating processes in a security anomaly would have achieved its service without the interference of malicious processes, assuming that scheduling and development anomalies had not occurred. The security system, while providing access for independent processes during concurrent intervals of time, inadvertently permitted the access of violators.

Development is the production of a software system: understanding, designing and implementing the processes that cooperate in comprising the system. Development consists of multiple activities, including requirements gathering, system design, project management, and implementation. Development errors can be subdivided into those that occur during the development process and those that remain in the resultant system, but both of these types display the same characteristics. Development anomalies are due to errors that are introduced at these stages, before a project is promoted to buffer or service layers. Such anomalies require abortion and correction by the production or maintenance team for completed service. Scheduling and security anomalies, on the other hand, occur asynchronously at execution time, and are due to errors in the software system that controls them. These latter anomalies can be corrected by the system at higher layers without changing the processes involved.

Control mechanisms are applied for the prevention of all types of anomalies. In competition synchronization, mechanisms secure resource allocation; in cooperation synchronization, they secure process interaction. We consider two types of orders for control mechanisms: Process/Resource (P/R)-ordered and Times (T)-ordered. P/R-ordered mechanisms prevent anomalies by establishing procedures that follow orders of resource and/or process keys. Examples of such mechanisms include linearly ordering resources (to prevent resource deadlock), serializing transactions (to prevent inconsistent data), and priority scheduling (to satisfy real-time constraints); law-governed interaction [16] uses P/R-ordered methods in security systems, for example, in exchanging key values during authentication. T-ordered policies make decisions based on the number of time units that processes are rescheduled to a resource or buffer unit. Examples of such mechanisms include aging (to prevent unbounded waits), timestamps (to prevent duplicate service following man-in-the-middle attacks), time-outs (to prevent unbounded waits), FIFO queues (to prevent unfair service), and quantifiable approaches (to prevent time overruns); budgetary controls may be applied to establish a threshold on the entry of incorrect passwords [1] and statistical decision theory may be applied to achieve a threshold for assuming that a watermark is present [17]. P/R-ordered algorithms frequently incorporate T-orders, such as sequence numbers or timestamps, into key values to order requests with the same process key, as is used in ARQ [20]. T-ordered schemes may incorporate P/R orders to allocate alternating turns to ordered processes, as is used in TDM [20] and round robin scheduling. For efficient resource utilization during non-uniform traffic conditions, protocols alternate between P/R-orders during heavy traffic and T-orders for light traffic [13], as is used in right turns on red, Binder's protocol [3], and Dekkar's algorithm [6].

The model presented in Section 2 introduced dependency, times, and priority predicates. Dependency predicates are P/R-ordered; times predicates are T-ordered; priority predicates alternate between T-ordered service during no conflict and P/R-ordered service when conflict occurs.

We next introduce the subclasses: errors in outputs, inputs, dependencies, priorities, and mapping times.

### *A. Output Errors*

Scheduling output anomalies result from the scheduling of outputs in an erroneous order. Security output anomalies result from intruders outputting erroneous data into a system. Development output anomalies result from erroneous output decisions reached during development. In each case, key values of requests conflict and cause failures unless the conflict is controlled. P/R orders are useful to either prevent or detect output anomalies, although large systems rely on automated T-ordered tools to identify anomalous behavior.

If scheduling mechanisms are incomplete, outputs from independent processes may be scheduled to the same resource unit and interfere with each other. P/R-ordered prevention schemes include network collision-free protocols, such as a Token Bus in Local Area Networks that orders media access by station addresses, and transaction serialization in Database Management Systems, which prevents inconsistent data. Alternatively, collisions may be allowed and detected, perhaps by checksums or time-out heuristics that wait for acknowledgments, followed by recovery via restarts. For example, Aloha's collision-based media access protocol dampens repeated collisions by retransmitting the colliding broadcast at random intervals [20], as a heuristic for linearly ordering rebroadcasts. Unless control measures completely order subsequent access, outputs may continue to collide. An example of such an anomaly is the repeated alternating of reads and arm cylinder moves in disk sharing [7].

If security mechanisms are incomplete, an intruder's output can interfere with valid processes. An intruder's key will not match that of the process that is being serviced at the accessed resource and its output will overwrite any data stored there, canceling the valid process's continuing service. Or a process may repudiate a message, and thus the key that identifies the sender of the message, canceling the message's service. An intruder who obtains a user's access key may be able to output invalid data; these contain the intruder's key value and cancel data requests currently being serviced. Examples of such anomalies are virus propagation, cracking, and message replays. (A duplicate transaction cancels the valid process's quantity specification.) Law-governed interaction [16], including mechanisms such as authorization, encryption, and third party authentication, is effective in preventing many security output anomalies. Virus detection software matches known virus signatures to values in attempted accesses. Alternatively, systems may use heuristics to flag anomalous behavior as evidence of violators. If these mechanisms are insufficient, the system may also rely on integrity check values and audits to detect anomalies. Recovery involves roll back, restoration, and restart. Similar to scheduling anomalies, recovery from output anomalies is achieved in the resource layers.

If development procedures are incomplete, a process's outputs may cause system failure. Data keys do not match and data requests that are currently being serviced are overwritten. Examples of these errors are incorrect assignments, version inconsistencies, and pointer arithmetic errors. Development output anomalies are frequently prevented by program verification and validation or by testing procedures, configuration management tools, and other automated tools. Such methods define a series of cooperating processes that must be followed in an established order. As systems have expanded, methods to prevent anomalies with verification tools [4] or comprehensive test suites [19] have become unmanageable. Developers utilize heuristics to test systems, with the understanding that not all errors are detected, relying on users and exceptions to notify them of conflicts with outputs after system release. Unlike scheduling and security output anomalies, resource systems cannot recover from development errors, but must rely on maintenance teams to achieve recovery at the process conception layer.

*B. Input Errors*

Scheduling input anomalies result from scheduling inputs in an erroneous order. Security input anomalies result from intruders obtaining unauthorized information. Development input anomalies result from erroneous inputs during requirements gathering and/or implementation. Key values of input requests conflict with output requests being serviced and cause anomalies.[4] P/R orders are used to prevent or detect input anomalies. Recovery is more difficult than that of output anomalies.

If a process's inputs are scheduled incorrectly, they may be interleaved with another process's updates, so that an inputting process may conflict with the outputting process and be denied service. An example of such a scheduling anomaly is inconsistent retrieval [2]. P/R ordered mechanisms, similar to those effective for output errors, can prevent scheduling input errors; examples of such mechanisms are critical regions and resource pre-allocation. Erroneous inputs cannot be corrected unless they result in outputs that are detected by key matches. Detection and correction of invalid scheduling inputs are generally more difficult than that of output anomalies, since dependent results may be output at multiple and/or distant locations.

If security mechanisms are incomplete, a violator may read data for which it does not have a valid key. Such input cancels a user's request for restricted access. Examples of such failures are invasion of privacy, traffic analysis, and identity theft. The same methods of law-governed interaction that are appropriate to prevent security output anomalies, such as authorization and encryption, are effective for most security input anomalies. Detection and recovery are generally complex, if at all possible.

If a process's inputs are not validated during development, data with non-matching keys may be input. Examples of resulting anomalies include reading from uninitialized data, incorrect data gathering, and incomplete comprehensive of requirements or implementations. Prevention methods include initializing storage with invalid data keys that conflict with the variable type and raise an exception during testing and verifying the range of input values. Development input anomalies, similar to development output anomalies, may be prevented by verification and validation procedures and by automated testing tools. Detection and recovery can be complex.

*C. Dependency Errors*

Processes are frequently dependent upon actions of other processes. A scheduling dependency error occurs if processes are scheduled to an incorrect dependency, such as a circular wait. A security dependency error occurs if a violator corrupts an assigned dependency. A development dependency error occurs if a process's progress is dependent upon an incorrectly programmed event.

If a scheduling protocol is incomplete, processes may be scheduled to wait forever. Four cars at stop signs on four corners of an intersection will be in a dead state if they each wait for the vehicle on the left to proceed into the intersection. A cyclical dependency can also occur among processes if each process's release of a held resource is dependent upon its later request for a resource held by another process in a cycle, causing a resource deadlock. P/R-ordered policies will prevent circular waits with uniquely identifiable resource units and also detect them but are less effective for tangible resources [14]. Automated tools, serving as heuristics of banker's algorithm, set a threshold on resource use, throttling new processes, in order to prevent resource deadlock. Alternatively, the system may assume an erroneous scheduling wait if a process exceeds some time-out period without making progress, and roll back and restart the process for attempted recovery.

If security measures are incomplete, an attacker can corrupt an event upon which a process is dependent. For example, a TCP/IP system establishes a half-open connection following a connection request. If an attacker fills a server's backlog queue without completing the connections, new users cannot be serviced. To prevent the possibility of a server waiting forever, half-open connections are discarded after a time-out

4 Suppose that an output to a process is received incorrectly. What type of error has occurred? If a value is incorrectly output or corrupted during delivery, an output error occurs. If a value is sent correctly, but the receiving process does not have the proper format key for the input, an input error occurs. If data are delayed in transit, a times error may occur.

period (about 75 seconds), but an attacker can continue to flood the server with connection requests. A relay (SYNDefender) may be placed between connection requests and the server, so that only completed connections reach the server. Processes are required to follow an ordered sequence of requests in order to establish connections through the relay.

An erroneous development dependency occurs if a process's progress is dependent upon an incorrect event. Examples of such anomalies include an infinite loop [14] and an infinite wait for a process that does not exist. These anomalies may be prevented by formal specification or by testing procedures. Alternatively, they may be detected by time-outs, which are heuristics that rely on the detection of a wait that exceeds a threshold, but which do not identify the nature of the wait.

*D. Priority Errors*

A scheduling priority anomaly can occur if an incorrect priority is assigned due to an incomplete scheduling policy. A security priority anomaly can occur if violators corrupt scheduling priorities. A development priority anomaly can occur if the development team assigns incorrect priorities to development or execution processes. Recovery from priority anomalies is typically ineffective.

A scheduling priority anomaly occurs if a process is chosen for execution in an incorrect priority order. For example, in priority inversion, a high priority process waits for service while a lower priority process is allocated the CPU [12]. Priority inversion can be prevented by priority inheritance, where a server's priority is ordered based on the priority of its waiting processes. If a wait resulting from a priority anomaly is unacceptable, recovery is no longer possible [14].

A security anomaly can occur if violators corrupt the scheduling order. For example, a set of web pages may link to each other repeatedly, causing a counting spider to assign these pages higher priority for a search engine's pages than their popularity warrants. If a search engine can detect this misuse, perhaps after reaching a threshold on the number of hits within a small set of nodes, it can change the links' priorities, but this will be too late for the users who have missed choosing the more popular web pages.

During development, erroneous priorities are frequently assigned to the allocation of resources, causing such failures as cost overruns and time overruns. A development policy establishes ordered steps in system development, so that delays can be identified and higher priority assigned to processes that are bottlenecks in system development. Erroneously assigned priorities are usually detected when established milestones are missed, although reprioritizing of manpower during production generally does not enable recovery. Erroneous priorities that are assigned to processes during development may not be detected during software testing, since results are dependent upon execution orders.

*E. Times Errors*

A scheduling times anomaly occurs if the number of time units allocated for a process to be serviced is insufficient due to an error in the scheduling policy. A security times anomaly occurs if the time units are insufficient due to an error in the security policy. A development times anomaly occurs if there is an overrun of time units due to the miscomprehension or implementation of a project. Times errors are widely prevented by automated tools. If these errors cause anomalies, recovery is not possible since the processes have already exceeded their acceptable allocated time.

A scheduling times anomaly occurs if a scheduler is overwhelmed by traffic, causing the demotion of process as traffic increases. The scheduler has accepted more traffic than can be serviced within the available time. In networks, packets are discarded. In virtual memory systems, page-thrashing causes repeated preemption and context switching. Such congestion must be prevented by budgetary controls, such as maintaining metrics of resource use and throttling new processes when a threshold is reached.

A security times anomaly occurs if an attacker prevents valid processes from obtaining service within their required time. Examples of such service threats include distributed denial of service and internet worms. Firewalls at clients can deny the access of many service threats, but the processing burden may prevent the service of valid processes. Budgetary controls to identify and abort such intruders at routers or other intermediate switches before they swamp clients are frequently effective prevention mechanisms.

A development times anomaly occurs if insufficient time units are allocated to complete required tasks. T-based orders are the basis for stress tests, volume tests, timing tests and timelines for project milestones. These measures are frequently effective for preventing production time overruns, but, as in prevention methods for scheduling and security timing errors, they are only heuristics.

*F. Causes of the Classes and Subclasses*

The following table outlines the above classification, providing examples of anomalies in each category, as well as appropriate control mechanisms for prevention, detection and recovery (Table 1). Characteristics are shown to apply across subclasses. Although this classification scheme was developed before specific anomalies were considered, examples in each subclass were obtained.

**Table 1. Classification Scheme for Anomalies of Software Systems**

| | | Scheduling anomalies | Security anomalies | Development anomalies |
|---|---|---|---|---|
| Caused by | | Incorrect competition mechanisms allowing interference between independent processes. | Incorrect cooperation mechanisms allowing interference from an (independent) attacker. | Incorrect competition or cooperation mechanisms within a set of cooperating processes. |
| | Recovery by (where possible) | Restoration of processes by resource system. | Restoration of victim(s) by resource system. Abortion of violator. | Abortion by resource system. Correction by the development team. |
| Output errors | Examples of anomalies: | Lost update, collisions in Aloha or Ethernet | Viruses, hacking, man-in-the-middle attack, buffer overflow | Range errors, buffer overflow, memory leaks |
| | Prevention mechanisms: P/R orders | Ex: resource pre-allocation, locking mechanisms, collision free protocols | Ex: authorization, authentication, encryption, virus detection software, law-governed interaction | Ex: program verification and validation, testing procedures, automated tools |
| | Detected by key matches | Ex: frame check sequences; Recovery by backup and restart | Ex: integrity check values, stack overflow registers; Recovery by backup and restart | Ex: exception handlers, memory bounds registers; Recovery by correction and resubmission |
| Input errors | Examples of anomalies: | Inconsistent retrieval with interleaved read requests | Invasion of privacy, traffic analysis, sniffing, identity theft | Uninitialized data, incorrect data or requirements gathering |
| | Prevented by P/R- orders | Ex: resource pre-allocation, locking mechanisms | Ex: encryption, law-governed interaction | Ex: program verification, test suites, automated tools |
| | Detected by key matches | Ex: conflict between users or data; Recovery is difficult | Ex: conflicts between users or data; Recovery is difficult | Ex: conflicts between users or data; Recovery is difficult |
| Dependency errors | Examples of anomalies: | Resource deadlock, circular scheduling | Filling a TCP/IP backlog queue with half-open connections | Infinite loop, wait for a signal from a nonexistent process |
| | Prevented by P/R-orders | Ex: linear ordering of processes or resources | Ex: gateway filtering of connection requests | Ex: program verification, tests of process paths |
| | Detected by T-orders (heuristics) | Ex: budgetary controls; time-outs; Recovery by back ups and restarts | Ex: budgetary controls, time-outs; Recovery by backups and restarts | Ex: statistical tools; time-outs; Recovery by correction and resubmission |
| Priority errors | Examples of anomalies: | Priority inversion where lower priority process is served first | Incorrectly raising priorities via circular URL links | Incorrect allocation of manpower, resources |
| | Prevented by P/R-orders | Ex: priority inheritance used to raise server's priority | Ex: tree-based traversal to prevent circular links | Ex: high priorities for potential bottlenecks |
| | Detected by T-orders | Ex: failure to meet time goals; Recovery is impractical | Ex: failure to meet time goals; Recovery is impractical | Ex: failure to meet time-lines; Recovery is impractical |
| Times errors | Examples of anomalies: | Congestion, page thrashing, system overload | Service threats – Internet worms, DDOS, spam | Time overruns, failure to meet time constraints |
| | Prevented by T-orders (heuristics) | Ex: thresholds on resource usage; monitoring usage to throttle new processes during overloads | Ex: budgetary controls, such as thresholds, for early detection and denial of attacks | Ex: statistical tools during development, such as timing tests, stress tests, and volume tests |
| | Detected by T-orders | Recovery is not possible once a process exceeds its allocated time | Recovery is not possible once a process exceeds its allocated time | Recovery is not possible once a process exceeds its allocated time |

## IV. STRUCTURED ANALYSIS

We consider three types of data-modeling tools, state-transition diagrams [9, 18], data flow diagrams [8] and entity-relationship diagrams [5]. Each of these tools is preferred over our model in providing focused clarity for the design of software systems. To illustrate the power of our model, however, we show that data-modeling constructs can be expressed within the model.

### A. State-Transition Diagrams

In state-transition diagrams, a system state is represented by values of stored data at each instant of time. Then an event causes a transition of the system from one state to another, similar to the mappings from state to state in our model.

States are represented in our model by sets of requests at each unit of Time. Transitions are represented by mappings to other request sets. Events that cause mappings are combinations of completed mappings contained in dependencies, expirations of times restrictions, and priority restrictions following conflicts. Inputs or outputs that cause transitions are represented by input and output requests that are serviced at RS. When states cycle back on themselves, requests are mapped to a later element of T so that $g_p(m, r, t) = g_p(m, r, t'), t' > t$.

*B. Data Flow Diagrams*

Data flow diagrams illustrate the movement of data from one entity to another. This data flow is cleaner than the movement of data contained in the requests of our model.

In our model, all output requests transmit data and all input requests (except match requests) receive data. Sources of external data conceive their data in **PC** and transmit their data to **RS**. (Thus, input from external sources in data flow diagrams are outputs to the resource system in terms of our model.) Destinations input their data through input requests that are serviced at **RS**. (Thus, outputs to users and other external sources in data flow diagrams are inputs at these external entities in terms of our model.) Requests can originate from the **PC** layer at any time, allowing users to output data interactively. Each data store/ database is represented by resource buffers at a specific sublayer. Data flows are represented by mapping of input or output requests. Transformations are the mechanisms by which inputs yield outputs. In terms of our model, transformations occur when output requests are repeatedly rescheduled, perhaps with their values altered by other output requests to the requested resource.

As an example, consider a student process that requests to be registered in a course. Some of the process's output requests containing initial data values (name, id) are mapped to a resource buffer. Other process requests are mapped to data stores, requesting information on seats available, course hours, etc. These activate dependencies in the data store, that then outputs additional values to be mapped to the reserved buffer. The final process request, for the output of a completed registration form, is only serviced after all the other requests have been serviced.

*C. Entity-Relationship Diagrams*

Entity-relationship diagrams express relationships between objects. These relationships are implicit in our model, which primarily focuses on mappings (transitions, transformations).

In our model, entities are sets of requests that are bound together by control mechanisms. Sets of requests are assigned dependencies so that they can be identified and accessed as a unit. Data fields of the same entity each contain the same identifier as part of their key. Relationships between entities are best expressed by rescheduling the entity that is represented by one request set to a request set with additional characteristics.

## V. SUMMARY

Software engineering is concerned with how digital information is produced, organized, and delivered. Of major concern is the control of anomalies that are caused by software errors, whether by prevention or by detection and recovery, and what kind of control mechanisms are effective. This paper proposes a model and a classification scheme for errors of software systems that lead to the loss of work and potential system failure, offering a unifying view of the control of anomalies.

The model is expressed in terms of requests being mapped through different stages in the software lifecycle. These mappings are controlled by a small set of restrictions, which are also the causes of software errors. The constructs of the model are shown to be equivalent to those of some data-modeling tools.

## REFERENCES


[1] X. Ao and N.H. Minsky, "On the role of roles: from role-based to role-sensitive access control," *SACMAT,* June 2004, pp. 51-60.

[2] P. Bernstein and N. Goodman, "Concurrency control in distributed database systems," *ACM Computer Surveys*, vol. 13, 2, June 1981, pp. 185-211.

[3] R. Binder, "A dynamic packet-switching system for satellite broadcast channels," *Proc. ICC,* 1975, pp. 41.1-.5.

[4] B. Bruegge and A. Dutoit, A. *Object-Oriented Software Engineering*, Prentice-Hall, Inc., NJ, 2000.

[5] P. Chen, "The entity relationship model – towards a unified view of data." *ACM Transactions of Database Systems*, vol. 1, 1, 1976, pp. 9-36.

[6] E.W. Dijkstra, "Cooperating sequential processes," *Programming Languages*, Academic Press, London, 1965.

[7] M. Flynn, and A.M. McHoes, *Understanding Operating Systems*, Brooks/Cole, Australia, 2001.

[8] C.P. Gane and T. Sarson, *Structured Systems Analysis: Tools and Techniques*. Prentice-Hall, NJ, 1979.

[9] D. Harel, "Statecharts: a visual formalism for complex systems," *Science of Computer Programming*, vol. 8, 1987, pp. 231-274.

[10] I. Hooks, "Managing requirements for a system of systems," *Crosstalk*, vol. 17, 8, Aug. 2004, pp. 4-7.

[11] *IEEE Standards Collection: Software Engineering*, IEEE Standard 610.12-1990, IEEE, 1993.

[12] G. Levine, "The control of priority inversion in Ada," *Ada Letters*, vol. 8, 6, Nov., Dec. 1988, pp. 53-56.

[13] G.N. Levine, "The control of starvation," *International Journal of General Systems*, vol. 15, 1989, pp. 113-127.

[14] G.N. Levine, "Defining deadlock," *Operating Systems Review*, ACM Press, vol. 37, 1, Jan. 2003, pp. 54-64.

[15] G.N. Levine, "A model for software reuse.," *Proceedings of the 5th Workshop of Specification of Behavior Semantics, OOPSLA '96,* Oct. 1996, pp. 71-87.

[16] N.H. Minsky, and V. Ungureanu, "Law-governed interaction: a coordination and control mechanism for heterogeneous distributed systems," *ACM Transaction on Software Engineering and Methodology*, vol. 9, 3,June 2000, pp. 273-305.

[17] A. Piva, F. Bartolini, and N. Barni, "Managing copyright in open networks," *IEEE Internet Computing*, vol. 6,3, May-June 2002, pp. 18-26.

[18] K.G. Salter, "A methodology for decomposing system requirements into data processing requirements," *Proceeding of the 2nd Int. Conference on Software Engineering*, San Francisco, Calif., 1976, 91-101.

[19] K. Stobie, "Too darned big to test," *Queue*, vol.3, 1, Feb. 2005, pp. 30-37.

[20] A. Tanenbaum, Computer Networks, Prentice-Hall, Inc., New Jersey, 2003.

# A Qualitative Analysis of the Critical's Path of Communication Models for Next Performant Implementations of High-speed Interfaces

Ouissem Ben Fredj
GET / INT
Évry, France
Ouissem.BenFredj@int-evry.fr

Éric Renault
GET / INT
Évry, France
Eric.Renault@int-evry.fr

***Abstract*— Recent high-speed networks provide new features such as DMA and programmable network cards. However standard network protocols, like TCP/IP, still consider a more classical network architecture usually adapted to the ethernet network. In order to use the high-speed networks efficiently, protocol implementors should use the new features provided by recent networks. This article provides an advanced study of both hardware and software requirements for high-speed network protocols. A focus is made on the programming model and the steps involved in the transfer's critical path.**

## I. INTRODUCTION

High-speed network interconnects that offer low latency and high bandwidth have been one of the main reasons attributed to the success of commodity cluster systems. Some of the leading high-speed networking interconnects include Gigabit-Ethernet, InfiniBand [1], Myrinet [2] and Quadrics. Two common features shared by these interconnects are User-level networking and Direct Memory Access (DMA). Gigabit and 10-Gigabit Ethernet offer an excellent opportunity to build multi-gigabit per second networks over existing Ethernet installations due to their backward compatibility with Ethernet. InfiniBand Architecture (IBA) is a newly defined industry standard that defines a System Area Network (SAN) to enable a low-latency and high-bandwidth cluster interconnect.

The Transmission Control Protocol (TCP) is one of the universally accepted transport layer protocols in today's networks. The introduction of high-speed networks a few years ago has challenged the traditional TCP/IP implementation in three aspects, namely performance, CPU requirements and memory traffic. In order to allow TCP/IP based applications achieve the performance provided by these networks while demanding less CPU resources, researchers came up with solutions in two broad directions: user-level sockets and TCP Offload Engines (TOE). User-level socket implementations rely on zero-copy OS-bypass high-performance protocols built on top of high-speed interconnects. The basic idea of such implementations is to create a socket-like interface on top of these high-performance TCP Offload Engines, on the other hand, offload the TCP stack on to hardware in part or in whole. Earlier Gigabit-Ethernet adapters offloaded TCP and IP checksum computations on to hardware. Both these approaches concentrate on optimizing the protocol stack but they do not rely on memory traffic which becomes the bottleneck.

This paper is divided into two parts. First one analyses the existing programming models in order to choose the model that use new network hardware features. The second part analyses the communication's critical path to determine the best way to take advantage of this new hardware. A comparaison between existing high-speed communication protocols and libraries come with all steps of the study.

The next section II of the article analyses existing communication models. The section III choose the best programming model suited with the communication model. The next section IV is an overview of the one-sided communication model. The second part of the article analyses the communication critical's path. We concentrate on issues that determine the performance and the semantics of a communication system: memory management (section V), host memory - NI (Network Interface) data transfer (section VI), send and receive queue management (section VII), data transfer (section VIII), and communication control (section IX).

## II. THE COMMUNICATION MODELS

One way to compare communication models is to classify them according to the sender-receiver synchronization mechanism required to perform data exchanges. There are three synchronization modes: the full synchronization mode, the rendez-vous mode, and the asynchronous mode.

With the full synchronization mode, the sender has to ensure that the receiver is ready to receive incomming data. This means that a flow control is required. FM [3] and FM/MC [4] both implement flow control using a host-level credit scheme. Before a host sends a packet, it checks for credits regarding the receiver; a credit represents a packet buffer in the receiver's memory. Credits can be handed out in advance by pre-allocating buffers for specific senders, but if a sender runs out of credits it must block until the receiver sends new credits. LFC [5], specifically designed for Myrinet clusters, implements two levels of point-to-point synchronization: the



*T. Sobh and K. Elleithy (eds.), Advances in Systems, Computing Sciences and Software Engineering*, 251–256.

NI-level and the host-level. At the host level, when the NI control program receives a network packet, it tries to fetch a receive descriptor. If the receive queue is empty, the control program defers the transfer until the queue is refilled by the host library. At NI-level, the protocol garantees that no NI sends a packet to another NI before the receiving NI is able to store the packet. In order to achieve this, each NI assigns a number of its receive buffers to each other NI in the system. An NI can transmit a packet to another one if there is at least one credit for the receiver. Each credit matches a free receive buffer for the sender. Once the sender has consumed all its credits for this receiver, it must wait until the receiver frees some of its receive buffers for this sender and returns new credits. Credits are returned by means of explicit acknowledgements or by piggybacking them at the application-level return traffic. This mechanism is set up to all communication node's pair, so that they are very expensive in both NI memory resource and synchronization time. Indeed, for applications using a lot of small messages, NI buffers could overflow quickly and synchronization time may exceed the latency.

The rendez-vous mode discharge the duty of flow control to the application. For example, BIP [6], VIA [7], BDM [8] and GM require that a receive request is posted before the message enters the destination node. Otherwise, the message is dropped and/or NACKed. VMMC [9] uses a *transfer redirection* that consists in pre-allocating a default redirectable receive buffer whenever a sender does not know the final receive buffer address. Later, when the receiver posts the receive buffer, it copies the data from the default buffer to the receive buffer. To implement such a model, a middleware must be added between the user application and the implementation ensuring the flow control. Similar to VMMC, QNIX [10] program moves incoming data into a buffer in the NIC memory if incomming data arrive before the receiver creates the corresponding *Receiver Context*.

The asynchronous mode breaks all synchronization constraints between senders and receivers. The completion of the send operation does not require the intervention of the receiver process to take a complementary action. This mode allows an overlapping between computation and communication, a zero-copy without synchronization, a deadlock avoidance, and an efficient use of the network (since messages do not block on switches waiting for the receive operation). As a consequence, the asynchronous mode provides a high throughtput and a low latency, in addition to a flexibility (as the synchronized mode can be implemented using the asynchronous mode).

AM [11] and PM2 [12] (Parallel Multithreaded Machine) are using the later mode to perform RPC-like communications. Each AM message contains the address of a user-level handler which is executed on message arrival with the message body as an argument. Unlike RPC, the role of the handler is to get the message out of the network and integrate it into the receiver process space. The problem with this scheme is that, for each message, a process handler is created (as with PM2) or an interrupt is generated (as with Genoa Active Message MAchine(GAMMA) [13]) which is very expensive (for both time and resource).

Other libraries (like VIA, AMII and DP [14]) require a startup connection to be executed before any communications. Such a connection consists in creating a *channel* that allows communication between a sender and a receiver. This step is often used to exchange capabilities (reliability level, quality of service...) and restrictions (maximum transfer size ...) of the process. This can be usefull for dynamic and heterogenous topologies.

As discussed previously, each synchronization mode has advantages and drawbacks. The rest of the article focuses on the asynchronous mode for both its simplicity and efficiency.

## III. The programming model

There are many programming models that use the asynchronous mode. The best suited for the RDMA is the one-sided model. It means that the completion of a send (resp. receive) operation does not require the intervention of the receiver (resp. sender) process to take a complementary action. RDMA should be used to copy data to (from) the remote user space directly. Suppose that the receiver process have allocated a buffer to room incoming data and the sender have allocated a send buffer. Prior to the data transfer, the receiver must have sent its buffer address to the sender. Once the sender owns the destination address, it initiates a direct-deposit data sending. This task does not interfere with the receiver process. On the receiver side, it keeps on doing computation tasks, testing if new messages have arrived, or blocking until an incoming message event arises.

There are several classes of applications that are easier to write with one-sided communication. They include:

- remote paging;
- adaptative codes where communication requirements are not known in advance;
- codes that perform random access to distributed data. In this case the process owning the data does not know the data to access;
- asynchronous parallel algorithms;
- symetric machines programming;
- data storage algorithms...

The one-sided programming model is simple, flexible and can be used as a high-level interface, or as a middleware between a high-level library such as MPI and the network level. A recent study proved that all MPI-2 routines can be implemented on top of a one-sided interface easily and efficiently [15]. This means that any message-passing algorithms may be implemented using this programming model.

The one-sided scheme can be acheived either by using one-sided read or by a one-sided write. The remote read requires at least two messages, the first to inform the remote DMA engine (the remote network interface, the remote OS, or the remote process) about the requesting data and the second to send affectively the data. The remote write operation requires one message.

## IV. THE ONE-SIDED COMMUNICATION PROTOCOL

The need for a one-sided communication protocol has been recognized for many years. Some of these issues were initially addressed by the POrtable Run-Time Systems (PORTS) consortium [16]. One of the missions of PORTS was to define a standard API for one-sided communications. During the development, several different approches were taken towards the one-sided API. The first one is the *thread-to-thread* communication paradigm which is supported by CHANT [17]. The second one is the *remote service request* (RSR) communication approach supported by libraries such NEXUS and DMCS. The third approach is a *hybrid* communication (that combines both prior paradigms) supported by the TULIP [18] project. These paradigms are widely used. For example, NEXUS supports the grid computing software infrastructure GLOBUS. MOL [19] extends DMCS with an object migration mechanism and a global namespace forthe system. DMCS/MOL is used both in Parallel Runtime Environment for Multi-computer Applications (PREMA) [20] and in the Dynamic Load Balancing Library (DLBL).

In 1997, MPI-2 [21] (a new MPI standard) have been including some basic one-sided functionalities. Although, many studies have integrated one-sided communications to optimize MPI [22]. In 1999, a new communication library called Aggregate Remote Memory Copy Interface (ARMCI) [23] has been released. ARMCI is a high-level library designed for distributed array libraries and compiler run-time systems. IBM have maintained a low-level API, named LAPI [24], implementing the one-sided protocol and runing on IBM SP systems only. Similarly, Cray SHMEM [25] provides direct send routines.

At the network layer, many factories have built RDMA features that ease the implementation of one-sided paradigms. For example, the HSL [26] network uses the PCI-Direct Deposit Component (PCI-DDC) [27] to offer a message-passing multiprocessor architecture based on a one-sided protocol. InfiniBand [1] proposes native one-sided communications. Myrinet [2], [28] and QNIX [10] do not provide native one-sided communications. But these features may be added (as for example in GM [29] with Myrinet since Myrinet NICs are programmable).

The arrival of these kind of networks has imposed common message-passing libraries to support RDMA (RW, GM , VIA [7]...). Most of these libraries have extended with one-sided communications to exploit RDMA features. But they do not use these functionalities as a base for their programming model.

Machine hardware have provided some kind of one-sided communication. Thinking machines [30] (CM1 in 1980, CM2 in 1988, and CM5 in 1992) are representative examples. CM5 uses two primitives, called PUT and GET, to allow thousands of simple processors to communicate over a the TeraFLOPS.

The rest of the article analyzes each step of the communication's critical path.

## V. MEMORY MANAGEMENT

Memory allocation preceeds any data transfers. It consists in reserving memory areas to store data to send or to receive. The way allocated areas are managed influences the performance.

DMA should be used to transfer data. The main constraint for DMA operations is that physical addresses are required rather virtual ones. Therefore, transfer routines must provide physical addresses of all pages of a message to the DMA engine. This is a tricky task because a contiguous message in the virtual address space is not necessarily contiguous in physical memory. A virtual-to-physical translation table built at allocation time can be used. Later, at the send (resp. receive) step, the translation table is used to gather (resp. scatter) data from (resp. to) memory to (resp. from) the network interface.

GM and PM2 add some optimizations to use the translation table: it stores the table in the user's memory to be able to translate the whole memory and it creates a small cache table in the NIC memory. The cache table contains a virtual-to-physical translation of most used pages. To avoid page swapping, allocated buffer have to be locked.

Another solution used by the network layer of MPC-OS [31] consists in splitting the message to send into several smaller messages which size is less than the size of a page.

Yet another solution consists in managing physical addresses directly, without operating system intervention. The idea is to allocate a physical contiguous buffer and to map it into the virtual contiguous address space of the user process. Thus, just one translation is needed. Its most important advantage is the avoidance of scatter/gather operations at transfer time. FreeBSD provides a kernel function that allows to allocate physical contiguous buffers. Linux do not provide such an operation. However, there are two methods to allocate physical contiguous memory. The first one is to change the kernel policy by changing the source code of Linux. The second one consists in allocating memory at boot-time. A driver maps the whole physical contiguous memory into a virtual contiguous area. Then, a function is used to search for a contiguous memory area that fits the requested size in the set of free spaces. Note that this function can be executed in user space without any call to the operating system.

Memory allocation is not a step of the communication's critical path, but the policy used to manage memory has an important impact on data transfers. Our goal is to reduce the time spent in the virtual-to-physical translation by using physical contiguous memory allocations.

## VI. HOST MEMORY - NI DATA TRANSFER

According to the one-sided scheme, the NI must communicate with the host memory in three cases. The first case is when the user process informs the NI for a new send, ie, when the user process sets up a send descriptor to be used by the NI to send message. Both the second and the third cases are when sending and receiving messages. For traditional message-passing systems, the user process must provide a receive descriptor to the NI. There are three methods to communicate between the host memory and the NI: PIO,

WC and DMA. With the Programmed IO (PIO), the host processor writes (resp. reads) data to (resp. from) the I/O bus. This method is extremely fast. However, only one or two words can be transferred at a time resulting in lots of bus transactions. Throughtput is different for writes and reads, mainly because writing across a bus is usually a little bit faster than reading. Write combining (WC) enhances write PIO performance by enabling a write buffer for uncached writes, so that affected data transfers can occur at cache line size instead of word size. Write Combining is a hardware feature initially introduced on the Intel Pentium Pro and now available on recent AMD processors;

A Direct Memory Access (DMA) engine can transfer entire packets in large bursts an proceed in parallel with host computation. Because a DMA engine works asynchronously, the host memory being the source or the destination of a DMA transfer must not be swapped out by the operating system. Some communication systems pin buffers before starting a DMA transfer. Moreover, DMA engines must know the physical addresses of the memory pages they access.

Choosing the suitable type of data transfer depends on the host CPU, the DMA engine, the transfer direction, and the packet size. A solution consists in classifying messages into three types: small messages, medium messages, and large messages. PIO suits small messages, write combining (when supported) suits medium messages, and DMA suits large messages. The definition of a medium message (and then the definition of both short and large messages) changes according to the CPU, the DMA engine, and the transfer direction. Since DMA-related operations (initialization, transfer, virtual-to-physical translation) can be done by the NI or the user process, a set of performance tests is the best way to define medium messages.

## VII. Send and receive queue management

In order to ensure an asynchronous exection of communication routines, two queues are used: a *send queue* and a *receive event queue*. Unlike synchronous libraries, the asynchronous mode does not need a *receive queue* to specify receive buffers. Although, it needs a *receive event queue* which contains a list of incoming messages.

Queues allow asynchronous operations. In fact, the user process just appends a descriptor to the *send queue* to send a message. Once the operation is finished, the sender continues with the next send or with the computing task. *Receive event queue* is used to probe or poll for a new receive event.

The *send queue* contains a set of send descriptors provided by user processes and read by the NI at send time. A send descriptor determines the outgoing buffer address, its size, the receiver node and the receiver buffer address. Additional attributes can be specified to personalize the send operation (security level, the reliability level). The NI uses send descriptors to initiate the send operation. Three steps are required:

The first one is the initialization of the send descriptor. This step is a part of the transfer's critical path if the send is a point-to-point communication. For collective operations (multicast, broadcast...), this step can be done once for multiple send requests.

The second step consists in appending the send descriptor to the *send queue*. This step depends on the send queue management. In fact, according to the NI type and the memory size, the *send queue* can be stored either in the NI memory or in the host memory. The first case (used by FM and GM typically) avoids the NI from polling on the host memory queue. The second case (defined in the VIA specifications and implemented in Berkeley VIA [32]) allows a larger queue. MyVIA [33], an implementation of VIA over Myrinet, uses two queues (the *small ring* in host memory and the *big ring* on the NIC). If the *small ring* is not full, the send descriptor is written directly. Otherwise, it is written in the *big ring* . If the number of unused descriptors in the *small ring* reaches a lower limit and if there are unprocessed descriptors in the *big ring*, the NI requests a driver agent to move *big ring* descriptors to the NI.

The third step of the critical path is the polling performed by the NI on the *send queue*. This step depends on the previous one. A comparison between MyVIA and Berkeley VIA proved that storing the *send queue* in the NI memory ensures a more efficient management of the send queue especially for small messages. In fact, Berkeley VIA requires two transactions between the NI and the host memory (the first one to inform the NI about the send and the second one to read the host memory descriptor) whereas MyVIA needs only one transaction. As for the *send queue*, the *receive event queue* should be stored on the host memory to allow easy polling by user processes.

Since the size of the send descriptor is only several byte long, PIO or write combining techniques should be used to update the NI *send queue* or to inform the NI about a new send.

## VIII. Data transfer

As introduced in the section III, RDMA-write is the sufficient and the efficient way to send data. The receive operation is detailed in the next section.

In order to avoid bottlenecks and use available resources efficiently, a data transfer should take into account the message size, the host architecture (processor speed, PCI bus speed, DMA characteristics), NIC properties (transfer mode, memory size), and the network characteristics (routing policy, route dispertion...).

Many studies have tried to measure network traffics to determine the size of message. However, they mainly focused either on a set of applications [34], a set of protocols [35], a set of networks [36] or a specific environnement (a single combination of networks, protocols, machines and applications) [37]. All these studies show that small messages are prominent (about 80% of the messages have a size smaller than 200 bytes). Moreover, the one-sided scheme requires an extra use of small messages to send receive buffer addresses. Thus, it is interesting to distinguish between small and large messages. As discussed earlier, the maximum size of small messages should be determined using performance evaluation.

For the transfer of small messages, no send buffer address nor receive buffer address are required. Therefore, it is possible to store the content of small message in the send descriptor. To send such a message, seven operations are performed: (1) the sender sets up the send descriptor (including data); (2) the sender informs the NI about the send descriptor; (3) the NI copies necessary data from the host memory, (4) the NI sends the message to the network; (5) the remote NI receives the message and appends it to the *receive event queue*; finally, (6) the receiver process reads the data from the receive descriptor and (7) informs the NI that the receive is done successfully.

For the transfer of large messages, the sender has to specify both the send buffer address and the remote buffer address and ten steps are involved : the sender prepares the send buffer; (2) the sender sets up the send descriptor and writes it to the NI memory using PIO; (3) the sender informs the NI about the send descriptor; (4) the NI reads the send descriptor; (5) the NI copies data from the host memory according to the send descriptor; (6) the NI sends the message to the network; (7) the remote NI receives the message and writes incoming data to its final destination; (8) the remote NI appends a receive descriptor to the *receive event queue*; (9) the receiver process reads the receive descriptor and (10) informs the NI that the receive is done successfully.

Network policy can also affect data transfers. Adaptative routing, which allows multiple routes for each destination, may cause buffer overwriting due to unordered arrival of messages. This problem doesn't exist with synchronous transfer.

It is benefit to distinguish between small and large messages in order to use efficiently the hardware and to implement a performant one-sided scheme. Finally, Data tranfer is the most important step of the communication. So care should be taken when writing its routines.

## IX. Communication control

The communication control focuses on how to retreive messages from the network device. How should the NI inform the user process about the completion of the receive? With user-level access to the network, an implementor can choose between using interrupts or polling.

The interrupt-driven approach lets the network device signals the arrival of a message by raising up an interrupt. This is a familiar approach, in which the kernel is involved in dispatching the interrupt to the application in user space.

The alternative model for message handling is polling. In this case, network device does not actively interrupt the CPU, but merely sets some status bits to indicate the arrival of a message. The application is required to poll the device status regularly; when a message is detected, the polling function returns the receive descriptor describing the message.

Quantifying the difference between using interrupts and polling is difficult because of the large number of parameters involved: hardware (cache sizes, register windows, network adapters), operating system (interrupt handling), runtime support (thread packages, communication interfaces), and application (polling policy, message arrival rate, communication patterns). First, executing a single poll is typically much cheaper than taking an interrupt, because a poll executes entirely in user space without any context switching. Recent operating systems decrease the interrupt cost by saving minimal process state, but interrupts remain expensive. Second, comparing the cost of a single poll to the cost of a single interrupt does not provide a sufficient basis for statements about application performance. Each time a poll fails, the user program wastes a few cycles. Thus, coarse-grain parallel computing favors interrupts, while fine-grain parallelism favors polling.

For application containing unprotected critical sections, interrupts lead to nondeterministic bugs, while polling leads to safe run. Moreover, for asynchronous communication, polling can lead to substantial overhead if the frequency of arrivals is low enough that the vast majority of polls fail to find a message. With interrupts, overhead only occurs when there are arrivals.

Most of high-speed communication libraries (AM, FM, PM [38], GM) use polling and let interrupt to signal exceptions like queue overflow or invalid receive buffer address. FM/MC, LFC, and PANDA [39] use a system that integrates automatically polling and interrupts through a thread scheduler. Since there is no incompatibility between both polling and interrupt, users can use both depending on the application context. Note that, interrupts may be imposed by some libraries to garantee a forward progress for system communications.

## X. Conclusion

This paper presented several design issues for high-speed communication protocol. First, we classified communication models into three synchronization modes: full synchronization mode, rendez-vous mode and asynchronous mode. The asynchronous mode removes all synchronization constraints between the sender and the receiver. Moreover, this mode is suited with one-sided programming model which offers a simple programming interface and high performance. In addition, both the asynchronous mode and the one-sided scheme take advantage of the DMA feature to acheive a RDMA communication. Note that the one-sided programming model insure the implementation of all message-passing application. Finally, the RDMA write is sufficient to implement both these two modes.

The memory allocation step preceed any data transfer but it affects the implementation way and the performance of the data transfer routine. Network protocol implementors should avoid the virtual-to-physical translation of memory page addresses while transfering data. The communication routines should take into account that a big number of small message may be exchanged. Theses messages correspond to memory address which are used in the RDMA communication. Thus, a distinction between small message and large message is interresting. For small message, PIO or write-combining method is usable and DMA is suitable for large message.

In order to implement an asynchronous communication, the send request should be stored in a send queue. The send routine has just to append a send descriptor to the send queue.

It is preferable to store the send queue in the network interface memory and thus avoiding the network interface to poll send descriptor from the host memory. Similar to the send message, the receive completion descriptor should be stored in a host receive queue to avoid the process to poll from the network interface.

Finally, to acheive a good communication control, the programmer should use both interruption and polling. Polling is suited with fine-grain applications and interrupt is used with coarse-grain one. The programmer should avoid all unprotected critical section and shoul use the interrupt to signal exceptions such as the overflow of the buffer or a security problem.

## REFERENCES


[1] I. T. Association, *The InfiniBand Architecture, Specification Volume 1 & 2*, June 2001, release 1.0.a.

[2] N. J. Boden, D. Cohen, R. E. Felderman, A. E. Kulawik, C. L. Seitz, J. N. Seizovic, and W.-K. Su, "Myrinet: A gigabit-per-second local area network," *IEEE Micro*, vol. 15, no. 1, pp. 29–36, 1995. [Online]. Available: citeseer.ist.psu.edu/boden95myrinet.html

[3] S. Pakin, M. Lauria, and A. Chien, "High performance messaging on workstations: Illinois Fast Messages (FM) for Myrinet," 1995, pp. ??–?? [Online]. Available: citeseer.ist.psu.edu/pakin95high.html

[4] "(r) efficient reliable multicast on myrinet," in *ICPP '96: Proceedings of the Proceedings of the 1996 International Conference on Parallel Processing (ICPP '96)-Volume 3*. IEEE Computer Society, 1996, p. 156.

[5] R. Bhoedjang, T. Ruhl, and H. Bal, "Lfc: A communication substrate for myrinet," 1998. [Online]. Available: citeseer.ist.psu.edu/bhoedjang98lfc.html

[6] L. Prylli and B. Tourancheau, "BIP messages user manual." [Online]. Available: citeseer.ist.psu.edu/prylli97bip.html

[7] "Virtual Interface Architecture Specification, Version 1.0, published by Compaq, Intel, and Microsoft," December 1997. [Online]. Available: http://www.viarch.org

[8] G. Henley, N. Doss, T. McMahon, and A. Skjellum, "Bdm: A multiprotocol myrinet control program and host application programmer interface," 1997. [Online]. Available: citeseer.ist.psu.edu/henley97bdm.html

[9] M. A. Blumrich, K. Li, R. Alpert, C. Dubnicki, E. W. Felten, and J. Sandberg, "Virtual memory mapped network interface for the shrimp multicomputer," in *ISCA '98: 25 years of the international symposia on Computer architecture (selected papers)*. ACM Press, 1998, pp. 473–484.

[10] A. D. Vivo, "A light-weight communication system for a high performance system area network," Ph.D. dissertation, Università di Salerno - Italy, November 2001.

[11] T. von Eicken, D. E. Culler, S. C. Goldstein, and K. E. Schauser, "Active Messages: A mechanism for integrated communication and computation," in *19th International Symposium on Computer Architecture*, Gold Coast, Australia, 1992, pp. 256–266. [Online]. Available: citeseer.ist.psu.edu/eicken92active.html

[12] R. Namyst and J.-F. Méhaut, *Parallel Computing: State-of-the-Art and Perspectives. Proceedings of the Intl. Conference ParCo '95, Ghent, Belgium, 19–22 September 1995*, ser. Advances in Parallel Computing. Elsevier, Feb. 1996, vol. 11, ch. PM²: Parallel Multithreaded Machine. A Computing Environment for Distributed Architectures, pp. 279–285. [Online]. Available: citeseer.ist.psu.edu/27887.html

[13] G. Chiola and G. Ciaccio, "Gamma: a low cost network of workstations based on active messages," 1997. [Online]. Available: citeseer.ist.psu.edu/chiola97gamma.html

[14] C.-L. Wang, A. T. C. Tam, B. W. L. Cheung, W. Zhu, and D. C. M. Lee, "Directed Point: a communication subsystem for commodity supercomputing with Gigabit Ethernet," *Future Generation Computer Systems*, vol. 18, no. 3, pp. 401–420, 2002. [Online]. Available: citeseer.ist.psu.edu/417848.html

[15] O. Gluck, "Optimisations de la bibliothèque de communication mpi pour machines parallèles de type grappe de pcs sur une primitive d'écriture distante," Ph.D. dissertation, Université Paris VI, juillet 2002.

[16] T. P. Consortium, "The PORTS0 Interface," Mathematics and Computer Science Division, Argonne National Laboratory, Technical Report ANL/MCS-TM-203, February 1995. [Online]. Available: ftp://ftp.globus.org/pub/globus/papers/ports0_spec_v0.3_ps.gz

[17] M. Haines, P. Mehrotra, and D. Cronk, "Chant: Lightweight threads in a distributed memory environment," 1995. [Online]. Available: citeseer.ist.psu.edu/haines95chant.html

[18] P. Beckman and D. Gannon, "Tulip: Parallel run-time support system for pc++."

[19] N. Chrisochoides, K. Barker, D. Nave, and C. Hawblitzel, "Mobile object layer: A runtime substrate for parallel adaptive and irregular computations," *Advances in Engineering Software*, vol. 31, no. 8–9, pp. 621–637, August 2000.

[20] K. J. Barker, "Runtime support for load balancing of parallel adaptive and irregular applications," Ph.D. dissertation, 2004, adviser-Nikos Chrisochoides.

[21] M. P. I. F. MPIF, "MPI-2: Extensions to the Message-Passing Interface," Technical Report, University of Tennessee, Knoxville, 1996. [Online]. Available: citeseer.ist.psu.edu/517818.html

[22] J. Dobbelaere and N. Chrisochoides, "One-sided communication over MPI-1."

[23] J. Nieplocha and B. Carpenter, "ARMCI: A portable remote memory copy library for distributed array libraries and compiler run-time systems," *Lecture Notes in Computer Science*, vol. 1586, pp. 533–??, 1999. [Online]. Available: citeseer.ist.psu.edu/nieplocha99armci.html

[24] I. Corporation, Ed., *LAPI Programming Guide*, ser. IBM Reliable Scalable Cluster Technology for AIX L5. Poughkeepsie, NY: First Edition, September 2004, no. IBM Document Number: SA22-7936-00.

[25] R. Barriuso and A. Knies, *SHMEM User's Guide*, Cray Research Inc, 1994.

[26] C. Whitby-Strevens and al., "IEEE Draft Std P1355 – Standard for Heterogeneous Interconnect – Low Cost Low Latency Scalable Serial Interconnect for Parallel System Construction," 1993.

[27] A. Greiner, J. Desbarbieux, J. Lecler, F. Potter, F. Wajsburt, S. Penain, and C. Spasevski, *PCI-DDC Specifications*, UPMC / LIP6, Paris, France, December 1996, Revision 1.3.

[28] A. 26-1998, "Myrinet-on-vme protocol specification." 1998.

[29] "GM: A message-passing system for myrinet networks 2.0.12," 1995. [Online]. Available: http://www.myri.com/scs/GM-2/doc/html/

[30] *Connection Machine CM-5 Technical Summary*, Thinking Machine Corporation, Cambridge, Massachusetts, November 1992.

[31] A. Fenyo, "Conception et réalisation d'un noyau de communication bâti sur la primitive d'écriture distante, pour machines parallèles de type «grappe de pcs»," Thèse de doctorat, UPMC / LIP6, Paris, France, July 2001.

[32] P. Buonadonna, A. Geweke, and D. Culler, "An implementation and analysis of the virtual interface architecture," in *Supercomputing '98: Proceedings of the 1998 ACM/IEEE conference on Supercomputing (CDROM)*. IEEE Computer Society, 1998, pp. 1–15.

[33] Y. Chen, X. Wang, Z. Jiao, J. Xie, Z. Du, and S. Li, "Myvia: A design and implementation of the high performance virtual interface architecture," in *CLUSTER '02: Proceedings of the IEEE International Conference on Cluster Computing*. IEEE Computer Society, 2002, p. 160.

[34] N. Basil and C. Williamson, "Network traffic measurement of the x window system."

[35] J. Cao, W. S. Cleveland, D. Lin, and D. X. Sun, "On the nonstationarity of internet traffic," in *SIGMETRICS '01: Proceedings of the 2001 ACM SIGMETRICS international conference on Measurement and modeling of computer systems*. ACM Press, 2001, pp. 102–112.

[36] C. Williamson, "Internet traffic measurement," *IEEE Internet Computing*, vol. 5, no. 6, pp. 70–74, 2001.

[37] R. Gusell, "A measurement study of diskless workstation traffic on an ethernet," in *IEEE Transaction on Communications*, vol. 38, no. 9, September 1990, pp. 1557–1568

[38] H. Tezuka, A. Hori, and Y. Ishikawa, "Pm: a highperformance communication library for multi-user parallel environments," 1996. [Online]. Available: citeseer.ist.psu.edu/tezuka96pm.html

[39] R. Bhoedjang, T. Ruhl, R. Hofman, K. Langendoen, H. Bal, and F. Kaashoek, "Panda: A portable platform to support parallel programming languages," 1993, pp. 213–226. [Online]. Available: citeseer.ist.psu.edu/bhoedjang93panda.html

# Detailed Theoretical Considerations for a Suite of Metrics for Integration of Software Components

V. Lakshmi Narasimhan and B. Hendradjaya
*Faculty of Electrical Engineering and Computer Science*
*University of Newcastle, NSW 2308, Australia.*
*Email: {narasimhan, bayu}@cs.newcastle.edu.au, url: http://www.cs.newcastle.edu.au/~narasimhan*

**Abstract**.

*This paper defines two suites of metrics, which cater static and dynamic aspects of component assembly. The static metrics measure complexity and criticality of component assembly, wherein complexity is measured using Component Packing Density and Component Interaction Density metrics. Further, four criticality conditions namely, Link, Bridge, Inheritance and Size criticalities have been identified and quantified. The complexity and criticality metrics are combined into a Triangular Metric, which can be used to classify the type and nature of applications. Dynamic metrics are collected during the runtime of a complete application. Dynamic metrics are useful to identify super-component and to evaluate utilisation of components. In this paper both static and dynamic metrics are evaluated using Weyuker's set of properties. The result shows that the metrics provide a valid means to measure issues in component assembly.*

**Keywords:** Component Metrics; Component Assembly; CORBA Component Model

## 1. Introduction

The use of Component-Based Software Engineering (CBSE) has become very important for building large software systems, as it leads to shorter development time at a reduced cost [3, 24] by the easy process of assembling and wiring software component into a complete application. Although there is no IEEE/ISO standard definition that we know of, one of the leading exponents in this area, Szyperski [24], defines a software component as follows:

*"A software component is a unit of composition with contractually specified interfaces and explicit context dependencies only. A software component can be deployed independently and is subject to composition by third parties".*

Although the definition permits assembling several hundred of components to make a complete software application, a standard is needed to make the components work together. Some well-known standards are Sun's JavaBeans, Microsoft's .NET and Object Management Group's (OMG) Corba Component Model (CCM) standard. As with many standards, the systems designed using even the known standards still do not integrate well, primarily because of 1) the need to strictly conform to the given standard (which often may not be comprehensive) [1, 2, 5, 6] and 2) due to the differing behavioral characteristics of components themselves [21]. A more recent work to address these problems is by Oberleitner et. al. [16], who have developed the Vienna Component Framework to enable composition across different component standards.

Although many metrics have been defined in literature for software integration, these need to be amended or enhanced for CBSE based integration. For instance, Lorenz and Vlissides [15] identify that the CBSE development model is different to designing with object-oriented design. Sparling [22] describes how software component development life cycle has different activities compared to the traditional life cycle. Therefore their strong conclusion is that a series of new software metrics is required for each activity. Arsanjani [1] recognizes other issues in development and integration of software component. In a more recent work, Gill and Grover [13] identify clearer reasons why it is not suitable to use traditional software metrics.

This research is an attempt to build a suite of software metrics, with particular emphasis on software component assembly. The suite accommodates static and dynamic aspects of component assembly. The static metrics measure the complexity and criticality of component assembly, while the dynamic metrics characterize the dynamic behaviour of a software application by recognizing component activities during run-time.

The paper is organized as follows: section 2 presents the need for component integration metrics, while section 3 discusses the definition of the metrics and their description. Section 4 shows how to incorporate metrics using CCM with StockQuoter example. The metrics are evaluated using the nine Weyuker properties in section 5. The discussion and conclusions of the research are provided in sections 6 and 7 respectively.

## 2. Research Problem

Component based metrics have been proposed by several researches. For example, Dolado [9] validates the use of Lines Of Code (LOC) in counting the size of software, while Verner and Tate [26] estimate the number of LOC early in the software life-cycle. But since the use of LOC depends on the language of implementation, it is hard to predict the size of the software prior to implementing. This is more so for software components, which do not have information on their source code. Most other metrics on CBSE have aimed at the reusability of components [12, 20, 27, 28], while [10, 21] focus on the process and measurement framework in



*T. Sobh and K. Elleithy (eds.), Advances in Systems, Computing Sciences and Software Engineering*, 257–264.

developing software components. Ebert [11] suggests some classification techniques to identify critical items in software project, which can be applied for a CBSE project, but he does not tackle the criticality aspects of component integration.

Specific issues on integration are discussed by Sedigh-Ali et al. [21], where the complexity of interfaces and their integration is interpreted as quality metrics. Cho et. al. [8] define metrics for complexity, customisability and reusability. They calculate the complexity of metrics by using the combination of the number of classes, interfaces and relationship among classes. They combine the calculation of cyclometric complexity with the sum of classes and interfaces, which need information from the source code.

We propose static metrics and dynamic metrics for component assembly[1], which could be of substantial use in estimating the effectiveness of the overall integration process, during the specification and design stages. Static metrics are collected from static analysis of component assembly, while dynamic metrics are gathered during execution of complete application. We consolidate our work and validate the metrics suite through Weyuker's evaluation criteria [29]. Furthermore, the suite integrates existing metrics available in the literature as an integrated package.

## 3. Component Metrics

The proposed metrics use graph connectivity as a medium to represent a system of integrated components. Each node and link represents a component and their relationship with other components respectively. Interactions happen through interfaces and events arising or arriving in. If a component 'X' requires an interface that is provided by another component 'Y', then 'Y' will be the incoming interaction for 'X'. If a component 'X' publishes an event which is consumed by component 'Y', then 'X' is said to raise an outgoing interaction to 'Y'. OMG [30] defines a *provided interface* as a 'facet', a *required interface* as a 'receptacle', a *published event* as an 'event source' and a *consumed event* as an 'event sink'.

Fig. 1 shows a link between two components of system P and system Q where one could consider component B to be more complex than components A, D or X, because it has more links than the others. Therefore in one sense, B can be termed as a critical component. On the other hand, component X may not be complex, but it is critical for the correct operation of the integrated system. Here, component X functions as a bridge between two systems P and Q. The definition for criticality does not stop here and it has other dimensions (see more in section 3.2).

[1] For this paper, we differentiate the term Component Assembly and Component Integration as follows: A 'Component Assembly' refers to system in which components have been integrated, whereas 'Component Integration' refers to the process of integrating components.

Similarly the complexity of a system depends on packaging density of components. For example, vertically connected components can be easily integrated as compared to multiply connected components. In addition, facets of components also have roles in connectivity and quality metrics. Further, dynamic characteristics of components along with their constraints can aid the design of new type of metrics for quality, management and reliability measures.

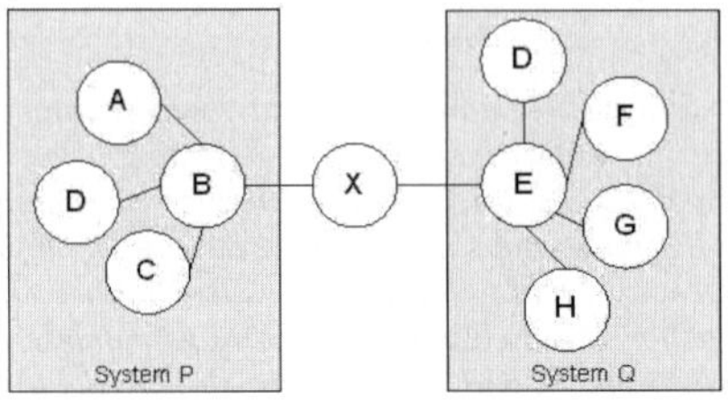


*Fig. 1 Interacting systems and components.*

In order to compute the value for criticality and complexity, we have identified several conceptual views of the components and systems, based on the analysis of the static and dynamic characteristics of integrated components. We give a formal definition of the metrics and their descriptions below. The detailed reasoning for each metrics is explained in [0].

### 3.1. Complexity Metrics

Complexity metrics belong to two categories: The first category deals with packing density of integrated component, while the second deals with interaction density among components in an integrated component. Using either CIDL [30], or ACOEL(A Component-Oriented Extension Language) [23] specifications, we can build a graphical representation of component assembly and derive *Component Packing Density* (CPD) metrics and *Component Interaction Density* (CID) metrics. The definitions of the metrics are given below:

1. Component Packing Density Metric

$\mathbf{CPD}_{constituent\_type}$ = **#constituent / #components**, where #<constituent> is one of the following: LOC, object/classes, operations, classes and/or modules in the related components.

2. Component Interaction Density Metric

**CID** = $\mathbf{\#I/\#I_{max}}$, where #I is the number of actual interactions and #Imax is the number of maximum available interactions.

3. Component Incoming Interaction Density Metric

**CIID** = $\mathbf{\#I_{in}/\#Imax_{in}}$, where $\#I_{in}$ is the number of incoming interactions used and $\#Imax_{in}$ is the number of incoming interactions available.

4. Component Outgoing Interaction Density Metric

**COID** = $\mathbf{\#I_{out}/\#Imax_{out}}$, where $\#I_{out}$ is the number of outgoing interactions used and $\#Imax_{OUT}$ is the number of outgoing interactions available.

5. Component Average Interaction Density Metric

**CAID = $\Sigma_n$CIDn/#components**, where $\Sigma_n CID_n$ is the sum of interactions density for component 1 to n and #components is the number of the existing component in the actual system.

### 3.2. Criticality Metrics

A critical component is typically a component that binds a system. In Fig. 1, the bridge is a critical component. For a software tester this component requires substantial testing effort. Every possible scenario for this critical component has to be tested, particularly if it is a base component, so that wrong operations are not inherited by the subcomponents. In this research, we propose four metrics to identify the critical components and, in addition, characterize the circumstances that make a component critical. The metrics are Link Criticality, Bridge Criticality, Inheritance Criticality and Size Criticality metrics.

1. Link Criticality Metrics

**$CRIT_{link}$= #link_component**, where #componentlinks is the number components, with their links more than a critical value.

2. Bridge Criticality Metric

**$CRIT_{bridge}$=#bridge_component**, where #bridgecomponent is the number of bridge components.

3. Inheritance Criticality Metrics

**$CRIT_{inheritance}$=#root_component**, where #root_component is the number of root components which has inheritance.

4. Size Criticality Metrics

**$CRIT_{size}$=#size_component**, where #componentx is the number of component which exceeds a given critical value.

5. #Criticality Metrics ($CRIT_{all}$)

$$CRIT_{all}= CRIT_{link}+CRIT_{bridge}+CRIT_{inheritance}+CRIT_{size}.$$

### 3.3. Triangular Metrics

The three metrics, namely, CPD, CAID and $CRIT_{all}$ have different points of view. While *Component Packing Density* (CPD) is calculated from the density in the integrated components, *Average Interaction Density* (CAID) is derived from density of interaction within an integrated component. Both metrics, however, represent the complexity of the system. The last metric ($CRIT_{all}$) is based on the criticality of the component.

In order to get a better view of the complexity metrics, we can combine CPD and CAID into two-axes diagram.

By examining their value, we deduce the characteristic of software as follows:

Case 1: A low value of CPD and a low value of CAID: This condition might happen within a system, having low data processing and low computation, such as a simple transaction processing system.

*Case 2*: A low value of CPD and a high value of CAID. This condition might happen within a system, having low data processing and high computation, such as a compute-intensive real time system.

*Case 3*: A high value of CPD and a low value of CAID. This condition might suggest a transaction processing system, which is characterised by high volume of data processing with many components, but has low interaction among components.

*Case 4*: A high value of CPD and a high value of CAID. This condition represents a very complex system, which might has many classes or constituents within its components and high interactions among components.

Fig. 2 illustrates these characteristics.

By combining information from the two axes diagram with new axis of criticality ($CRIT_{all}$), we can further characterize a system (see Fig. 3). A real-time system usually has higher criticality compared to a transaction-based system.

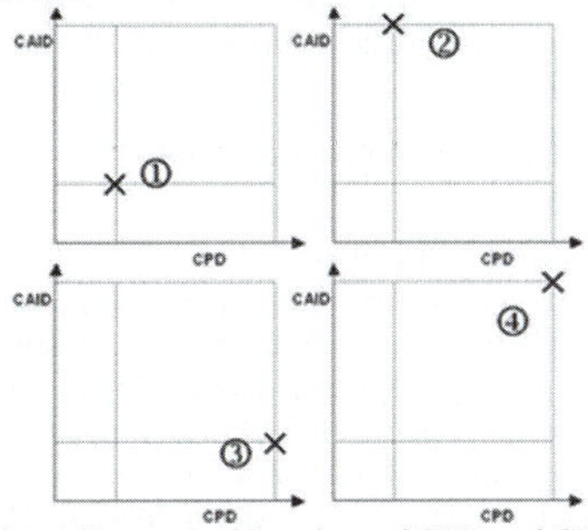


*Fig. 2 Four possible values of CPD and CID*

A business application tends to have more components to access data, than a real time application. A list of application types can be found in [17], while distinguishing characteristics between business type applications and real-time systems can be found in [4]. Our results concur with these observations for transaction and real-time system.

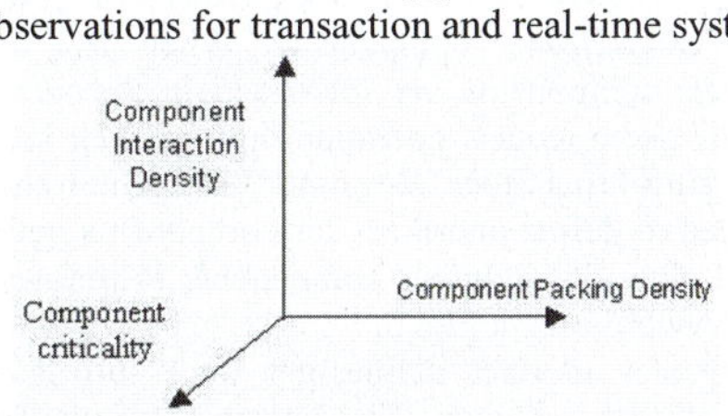


*Fig. 3 Three axes of component complexity and criticality*

### 3.4. Dynamic Metrics

Data on dynamic metric are collected during the execution of a complete application. Even though they cannot be used during the design phase, the results are still useful for testing and maintenance purposes. Gathering data for dynamic metrics is possible by instrumenting the source code, prior to compilation. When an application gets executed, components get called in various orders resulting in execution sequence cycles (ESC). When a component becomes active, its provided interface is used by other components.

1. Number of Cycle (NC) Metric

**NC = # cycles**, where #cycles is the number of cycles within the graph.

2. Average Number of Active Components

**ANAC = #activecomponents/$T_e$**, where #activecomponents is the number of active component and $T_e$ is time to execute the application (in seconds)

3. Active Component Density (ACD)

**ACD = #activecomponent / #components**, where #activecomponent is the number of active components and #component is the number of available components.

4. Average Active Component Density

**AACD = $\Sigma_n$ACDn/Te**, where $\Sigma_n ACD_n$ is the sum of ACD and $T_e$ is time to execute the application (in seconds). Execution time can be any of execution of a function, between functions or execution of the entire program.

5. Peak Number of Active Components

**$PNAC_{\Delta t}$ = max { $AC_1$,..,$AC_n$}**, where $\#AC_n$ is the number of active component at time n and $\Delta t$ is the time interval in seconds.

## 4. Incorporating the Metrics into CCM Application

Corba Component Model (CCM) extends the CORBA object model to support the implementation, packaging, assembling and deploying components. CCM adding new metatypes, tools and mechanism to Corba Object Model.

We illustrate the importance of the proposed metrics through the evaluation of a stock-broker system. This example has been shown at C/C++ Users Journal [18, 19].

A StockDistributor (SD) component monitors a stock database. If values in database change, this component generates an event via an event source (Notifier-out) to corresponding event sinks (Notifier-in). Several StockBroker (SB) components can respond with their event sinks. Fig. 4 below shows a StockDistributor component with two StockBroker components. When one or more StockBroker components are interested in the new value, they can invoke a request operation through their receptacle (GetQuoterInfo) to a StockDistributor facet (QuoterInfo).

We need to define interfaces for components before it is developed. Our StockQuoter components is defined using IDL 3.x keywords, as in Fig. 5.

StockQuoter interface defines get_stock_info to provide the value of changed stock. This interface returns the value of StockInfo that has the name and the value. The trigger interface handles the start and stop of stockdistributor, which runs as a daemon. To notify other components about new value in stock, we define a StockName eventtype data. This event is published via event mechanism. The StockBroker component has two ports. The first one consumes an event from StockDistributor component and the second one use information provided by StockDistributor component. The StockDistributor component also has two ports. The first port publishes StockName event to notify other components about a new value of stock and the second port defines a StockQuoter for other components. StockBrokerHome manages the creation and the destruction of StockBroker component, and so does StockDistributorHome to StockDistributor component.

Then we need to define the structure and state of the components. Component implementation definition of StockQuoter system is defined using CIDL as follows in Fig. 6. StockDistributor_impl and StockBroker_Impl show the name of the composition in the StockQuoter system. StockDistributorHome_Exec and StockBrokerHome_Exec are the home executors and they manage StockDistributor_Exec and StockBroker_Exec respectively.

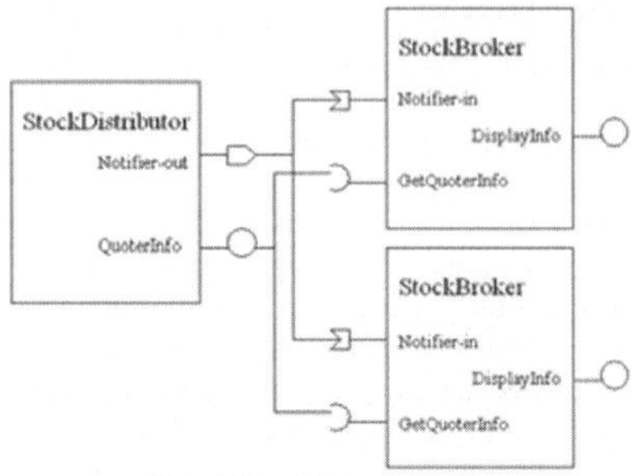


*Fig. 4 StockQuoter system*

```
module StockQuoterSystem {
interface StockQuoter {
    StockInfo get_stock_info
          (in string stock_name);  };
interface DisplayInfoStock {
    void Display(in StockInfo
               stock_detail);  };
struct StockInfo {
    string Name; ong   Val;  };
interface Trigger {
    void start(); void stop();  };
eventtype StockName {
    public string name;  };
component StockBroker {
    consumes StockName notifier_in;
    uses StockQuoter GetQuoterInfo;
    provides DisplayInfoStock
               DisplayInfo;  };
component StockDistributor supports
                        Trigger {
    publishes StockName notifier_out;
    provides  StockQuoter quoter_info;  };
home StockBrokerHome manages StockBroker {};
home StockDistributorHome manages
    StockDistibutor {}; }
```

*Fig. 5 IDL for StockQuoter System.*

The composition definitions in CIDL are compiled to generate skeleton code that associates the implementation of components and component homes. After this, component developer can implement the requirement stated for each particular components.

```
composition session StockDistributor_Impl {
    home executor
              StockDistributorHome_Exec
    {
      implements StockDistributorHome;
      manages StockDistributor_Exec;
    }
 }
composition session StockBroker_Impl
{
    home executor StockBrokerHome_Exec
    {
      implements StockBrokerHome;
      manages StockBroker_Exec;
    }
 }
```

*Fig. 6 CIDL for StockQuoter System*

The implementation of components can be packaged and deployed. The components can be connected to other components to form a component assembly. It is assembled in an assembly package. The component assembly package has a set of component packages and an assembly descriptor. The assembly descriptor is defined in XML. The component assembly descriptor for our examples is shown in Fig. 7.

With component assembly descriptor and implementation of component, we can generate our metrics as follows:

- Component Packing Density Metrics
  Our example only shows 'get_stock_info' and 'DisplayInfo' as the operations in the component assembly. DisplayInfo appears twice as two StockBrokers are included. With three components, we have:
  $CPD_{OPERATION} = 3/3 = 1$
- Component Interaction Density Metris
  Component SD has two available interactions (a facet and an event) and they are used for interaction. SB has three available interactions (a facet, a receptacle and an event) and only two are used for interactions.

$CID_{SD}=2/2=1$, $CID_{SB1}=2/3=0.667$, $CID_{SB2}=2/3=0.667$

SB has a receptacle and an event sink available for incoming interactions, and has a facet for outgoing interaction, but only the receptacle and the event sink are in used. SD has a facet and an event source for outgoing interactions, and both are in used.

- $CIID_{SD}=0$, $CIID_{SB1}=2/2 = 1$, $CIID_{SB2}=2/2=1$, $COID_{SD}=2/2 =1$, $COID_{SB1}= 0/1 = 0$, $COID_{SB2}= 0/1=0$, $CAID=(0+1+1)/3=2/3=0.667$

For the criticality of the StockQuoter system, we say that:

- For component SD, SB1 and SB2, we have link(SD) = 4, link(SB1)=2 and link(SB2) = 2 respectively. $CRIT_{LINK}$ is determined by the number of component that has link value exceeds a threshold value. Our examples has relatively small link values, which in this case $CRIT_{link}$ = 0.
- To examine bridge criticality, we have to identify a component, which functions as a bridge. In our example, CRITbridge = 0.
- Inheritance criticality is identified with a base component and its root. For this example, CRITinheritance = 0.
- For size criticality, we can use the number of operations in each component as a constituent. Each component in our example has one operation. CRITsize is determined from the number of components that has size value exceeding the threshold value. For this example, CRITsize = 0.

By summing up all the values above, we have CRITall = 0+0+0+0 = 0 which shows that the criticality of the system is 0. If it is stated that CRITall > 20 is a critical system, then our example is not a critical one.

To calculate triangular metrics, we use CPD=1, CAID = 0.667 and CRITall = 0. These values are relatively small and so it is concluded that the StockQuoter system is a simple application. For dynamic metrics, we simulate the activity of components for each given time (from t1 to t10 in seconds) as described in table 1. The table shows that at t1, SD and SB2 are active components and at t2, all the components are active and the rest exercise the same idea.

We cannot show a *number of cycle* metric in this example, since there is no actual cycle in the graph representation. But for active component metrics, we can see that at t7, t8 and t9, only one component is active, two components are active at t1, t3, t4, t5 and t10 and three components are active at t2 and t6.

The following are the metrics for active components.

- Average Number of Active Component
  ANAC = (2+3+2+2+2+3+1+1+1+2)/10 = 1.9
- Active Component Density
  $ACD_{t7} = ACD_{t8} = ACD_{t9} = 1/3$
  $ACD_{t1} = ACD_{t3} = ACD_{t4} = ACD_{t5} = ACD_{t10} = 2/3$
  $ACD_{t2} = ACD_{t6} = 3/3 = 1$
- Average Active Component Density
  AACD=(3*1/3+5*2/3+2*1)/10=0.633
  Peak number of Active Component
  The maximum number of active component are at t2 and t6 and thus: $PNAC_{\Delta t} = 3$

| | t1 | t2 | t3 | t4 | t5 | t6 | t7 | t8 | t9 | t10 |
|---|---|---|---|---|---|---|---|---|---|---|
| **SD** | 1 | 1 | 1 | 0 | 1 | 1 | 0 | 0 | 0 | 1 |
| **SB1** | 0 | 1 | 0 | 1 | 0 | 1 | 0 | 0 | 1 | 1 |
| **SB2** | 1 | 1 | 1 | 1 | 1 | 1 | 1 | 1 | 0 | 0 |

**Note: 1= component is active, 0= component is inactive**

*Table 1 Simulation of active component for 10secs.*

## 5. Metrics Evaluation Criteria

Weyuker has proposed an axiomatic framework for evaluating complexity measures [29]. The properties is not without critique as these have been discussed in [12] and [14]. The properties, however, have been used to validate the C-K metrics by Chidamber & Kemerer[7] and, as a consequence, we will employ the same framework for compatibility's sake. Table 2 shows the properties which are the modified definitions as provided by [12]; the original definitions are available at [29].

```xml
<componentassembly id = "StockQuoterSystem">
   ...
   ...
<connections>
<connectinterface>
 <usesport>
   <usesidentifier>GetQuoterInfo</usesidentifier>
   <componentinstantiationref idref="StockBroker1" />
 </usesport>

 <providesport>
   <providesidentifier>QuoterInfo</providesidentifier>
   <componentinstantiationref
               idref="StockDistributor" />
 </providesport>
 </connectinterface>

 <connectinterface>
   <usesport>
     <usesidentifier>GetQuoterInfo</usesidentifier>
     <componentinstantiationref idref="StockBroker2"/>
   </usesport>

   <providesport>
      <providesidentifier>QuoterInfo
      </providesidentifier>
      <componentinstantiationref
            idref="StockDistributor" />
    </providesport>
  </connectinterface>
  <connectevent>
    <consumesport>
          <consumesidentifier>
                Notifier-in</consumesidentifier >
          <componentinstantiationref
                 idref="StockBroker1" />
       </consumeport>
       <publishesport>
           <publishesidentifier>
               Notifier-out</publishesidentifier>
           <componentinstantiationref
               idref="StockDistributor" />
        </publishesport>
     </connectevent>

     <connectevent>
        <consumesport>
           <consumesidentifier>
               Notifier-in</consumesidentifier>
           <componentinstantiationref
               idref="StockBroker2" />
        </consumeport>

        <publishesport>
            <publishesidentifier>
               Notifier-out</publishesidentifier>
             <componentinstantiationref
               idref="StockDistributor" />
        </publishesport>
     </connectevent>
   </connections>

   </componentassembly>
```

*Fig. 7 XML description for StockQuoter System*

The properties are:

**Property 1**: There are programs P and Q for which M(P) ≠ M(Q)
**Property 2**: If c is non-negative number, then there are only finitely many programs P for which M(P)=c
**Property 3: There are distinct programs P and Q for which M(P)=M(Q)**
**Property 4**: There are functionality equivalent programs P and Q for which M(P) ≠ M(Q)
**Property 5**: For any program bodies P and Q, we have M(P) ≤ M(P;Q) and M(Q) ≤ M(P;Q)
**Property 6**: There exist program bodies P, Q and R such that M(P)=M(Q) and M(P;R) ≠ M(Q;R)
**Property 7**: There are program bodies P and Q such that Q is formed by permutting the order of statements of P and M(P) ≠ M(Q)
**Property 8**: If P is a renaming of Q, then M(P) = M(Q)
**Property 9**: There exist program bodies P and Q such that M(P)+M(Q) < M(P;Q)

**Property 1: There are programs P and Q for which M(P) ≠ M(Q)**

- An integrated component comprises various components having different constituents. As a consequence, the CPD metric satisfies property 1.
- For *Interaction Density Metrics (IDM)*, different configurations with integrated components yield different values, so they satisfies this property.
- For the *Criticality Metrics* (CM), different components can have different value of criticalities, so they satisfy property 1.
- The *triangular metric* (TM) uses *CPD*, *CAID* and critical metrics, which satisfy property 1 and the *triangular metric* value will be different for each component assembly and/or application type, so property 1 is satisfied.
- During run time of different applications, we can always find a different number of cycles, so the *Number of Cycle metric* (NC) satisfies property 1.
- For any executions of different component assemblies, we can always find a different number of active components at a time. Therefor0e all *Active Component* (AC) metrics satisfy property 1.

**Property 2: If c is non-negative number, then there are only finitely many programs P for which M(P)=c**

- For every application, there are a finite number of components with a finite number of constituents, so this property is met by the *CPD* metric.
- There are a finite number of interactions within a component, therefore *CID* satisfies property 2. The same logic goes for the *CIID*, *COID* and *CAID* metrics also.
- Every component has its own criticality. With a given criticality number, there is only a finite number of components, thus property 2 is satisfied.
- The TM satisfies property 2, as there are only a finite number of components in component assembly.
- The same NC value can be found from different executions of different component assembly. Therefore property 2 is satisfied.
- We can always find different applications with the same AC metrics value, thus property 2 is satisfied.

**Property 3: There are distinct programs P and Q for which M(P)=M(Q)**

- There is always possible to create a minimum of two combinations of components with their constituents and the metric value of *CPD* is the same. Therefore *CPD* metric satisfies this property.
- For IDM metrics, we can configure different interactions in more than one component that results in the same value, thus satisfying property 3.
- In integrated components, we can always find configuration of components, which have the same CM value and therefore property 3 is satisfied.
- Along the same logic as above, property 3 is satisfied by TM.
- Different component assembly can yield the same measurement for the NC and AC metrics. Therefore property 3 is satisfied by both metrics.

**Property 4: There are functionality equivalent programs P and Q for which M(P) ≠ M(Q)**

- If there are two integrated components, which perform the same functions, this does not imply that the CPD metric value will be the same. The same function can be built by different components and with different constituents. So the CPD metric satisfies this property.
- By the same logic, IDM also satisfies property 4.
- This property is satisfied by CM, since the same functionality with different implementation can have different criticality.
- The same function of component assembly can be built using different components. Thus the TM and all *dynamic metrics* (NC and AC) also satisfy property 4.

**Property 5: For any program bodies P and Q, we have M(P) ≤ M(P;Q) and M(Q) ≤ M(P;Q)**

- Let X is a component assembly and Y is a combination of X with other components.
- CPD value of X is no more complex than CPD value of Y. Therefore CPD metric satisfies property 5.
- For IDM, Y will present more interaction than X. So IDM metrics satisfy property 5.
- Combination of components yield higher criticality for Y than X. Thus property 5 is satisfied by CM.
- It is implied that TM satisfies property 5, since property 5 is satisfied by CPD, IDM and CM.
- Execution of Y yields more cycles than X and more active component involved. Therefore NC and AC satisfy property 5.

**Property 6: There exist program bodies P, Q and R such that M(P)=M(Q) and M(P;R) ≠ M(Q;R)**

- Let the component assemblies P,Q and R have respective CPD values of a/b, c/d and e/f, where a, c & e represent the number of constituents and b, d & f represent the number of components respectively. If the measurement on P is equal to Q, then ad = bc. Integration of (P;R) and (Q;R)yields (a+e)/(b+f) and (c+e)/(d+f), respectively. By working through the equation, we can conclude that (P;R) and (Q;R) will not have the same value, except when a = b = c = d = e = f = 1. This means that more than one component exists in an integration component and more than one constituent in a component. Therefore property 6 is satisfied.
- Using the above logic, one can note that the IDM satisfy property 6.
- Adding more components always gives more criticality values. Therefore property 6 is satisfied by CM.
- It is implied that TM satisfies property 5, since property 5 is satisfied by CPD, IDM and CM.
- Adding more components creates more cycles and more active component at run-time. Dynamic metrics (NC & AC) satisfy property 6.

**Property 7: There are program bodies P and Q such that Q is formed by permutting the order of statements of P and M(P) ≠ M(Q)**

Permutation on component assembly does not affect on the metric values statically and dynamically. Therefore all proposed metrics satisfy property 7.

**Property 8: If P is a renaming of Q, then M(P) = M(Q)**

Renaming the components does not affect on all the metrics proposed, since the measurement only concerns the number of components and their constituents and the number of interactions. So all proposed metrics satisfy property 8.

**Property 9: There exist program bodies P and Q such that M(P)+M(Q) < M(P;Q)**

- If we have two component assemblies having x1 constituents and y1 components and another configuration with x2 constituents and y1 components, then we can compute CPD1=x1/y1 and CPD2=x2/y2 and we can integrate both configurations. The resultant CPD3=(x1+x2)/(y1+y2). The value of *CPD1+CPD2* will always be lesser than *CPD3*. Therefore property 9 is not satisfied.
- The same logic can be used for *IDM* metrics, thus they do not satisfy property 9.
- For CM, combination of the metrics will always add more links, bridges, size and possibly inheritance also, so the resultant value will be always higher. Therefore CM satisfies property 9.
- For TM, let two different component assemblies P and Q, have triangular metrics of (x1,y1, z1) and (x2, y2, z2) respectively. Combining P and Q into one component assembly with (x3, y3, z3) as its metrics value, we cannot always have x1+x2<x3, y1+y2<y3 or z1+z2<z3. Therefore triangular metrics does not satisfy property 9.
- Let P and Q be two different component assemblies with certain number of circles. Combining P and Q into one component assembly does not always increase the number of circle possible. Therefore property 9 is not satisfied by NC metric.
- The above logic holds good for *active component metrics* also. Combination of component assemblies will increase the number of active components and therefore AC metrics satisfy property 9.

## 6. Discussion

We summarize the results of the paper through table 2. All proposed metrics satisfy property 1-6 and 8, but fail to satisfy property 7. Property 9 is only satisfied by *Criticality* and *Active Component* metrics.

Property 7 requires that permutation should affect the value of complexity. But in component integration, component ordering is not significant and therefore, this issue is not highly relevant.

| | Property | | | | | | | | |
|---|---|---|---|---|---|---|---|---|---|
| Metrics | 1 | 2 | 3 | 4 | 5 | 6 | 7 | 8 | 9 |
| CPD | Y | Y | Y | Y | Y | Y | N | Y | N |
| IDM | Y | Y | Y | Y | Y | Y | N | Y | N |
| CM | Y | Y | Y | Y | Y | Y | N | Y | Y |
| TM | Y | Y | Y | Y | Y | Y | N | Y | N |
| NC | Y | Y | Y | Y | Y | Y | N | Y | N |
| ANAC | Y | Y | Y | Y | Y | Y | N | Y | Y |
| ACD | Y | Y | Y | Y | Y | Y | N | Y | Y |
| AACD | Y | Y | Y | Y | Y | Y | N | Y | Y |
| PNAC | Y | Y | Y | Y | Y | Y | N | Y | Y |

*Table 2: Summary of metric properties*

Property 9 is not satisfied for CPD, *interaction density* and *triangular metrics*. Chidamber and Kermerer metrics [8] do not satisfy property 9 either and they suspect that this property is not suitable at the design level. We believe that for the same reason CPD, CID, CIID, COID and AID metrics do not satisfy property 9. In fact our metrics can be used both at the design level and at the implementation level. For the *number of cycle* metric, property 9 requires increase measurement value if we combine two component assemblies. But this metric relates on the behavior within each assemblies, which not always affected if combined. Finding a new super-component with NC metrics has close relation with design. So this conclusion is still coherent with [8].

## 7. Conclusions

This paper proposes a set of static and dynamic metrics. The static metrics should help developer in reasoning how complex a system is and locating critical areas in a component assembly. The dynamic metrics help identifying new super-components, and high extent of use of particular components.

The metrics suite can also be incorporated in a CASE (Computer Aided Software Engineering) tool. Object Constraint Language [30] can be embedded on the component model as an added constraint to system building. Adding the constraint in the proposed metrics could yield another method of measuring CBSE software development. Component relationship can also be visualised using information that is used to extract the metrics.

We believe that our metrics is based on measurement theory and have been validated using Weyuker's properties. Most metrics fulfill the Weyuker's property criteria, while a few do not. We intend to gather field data for validating the metrics empirically. A further study of complexity and criticality on software component metric would help provide a basis for significant future progress in this area.

## 8. References


[0] V. Lakshmi Narasimhan and B.Hendradjaya, "Some Considerations for Component Integration Metrics", Proc. of the 3rd International Information and Telecommunication Technologies Symposium (I2T2S), San Carlos-SP, Brazil, 6 - 9 December, 2004

[1] A. Arsanjani, Developing and Integrating Enterprise Components and Services, Communication of the ACM, Vol. 45, No. 10, October 2002, pp. 31-34.

[2] F. Berzal, I. Blanco, J. Cubero, N. Marin, Component-based Data Mining Frameworks, Communications of the ACM, Vol. 45, No. 12, December 2002, pp. 97-100.

[3] L. D. Blak, A. Kedia, PPT: A COTS Integration Case Study, Proceeding of 22[th] International Conference on Software Engineering (ICSE), Orlando, 2002, pp.41-48.

[4] B. Boehm, C. Abts, A. W. Brown, S. Chulani, B. Clark, E. Horowitz, R. Madachy, D. Reifer and B.Steece,

[5] A.W. Brown, Large-Scale, Component-Based Development, Prentice Hall PTR, 2000.

[6] L. Brownsword, T. Obendorf, C.A. Sledge, Developing New Processes for COTS-Based Systems, IEEE Software, July/August 2000, pp. 48-55.

Software Cost Estimation with COCOMO II, Prentice Hall, 2000.

[7] S. R. Chidamber and C. F. Kemerer. A Metrics Suite for Object-oriented Design, IEEE Transaction on Software Engineering, Vol. 20 No. 6, June 1994, pp. 476-493.

[8] E. S. Cho, M.S. Kim, S.D. Kim, Component Metrics to Measure Component Quality, The 8th Asia-Pacific Software Engineering Conference (APSEC), Macau, 2001, pp. 419-426.

[9] J. J. Dolado, A Validation of the Component-Based Method for Software Size Estimation, IEEE Transactions on Software Engineering, Vol. 26, No. 10, October 2000, pp. 1006-1021.

[10] R.R. Dumke, A.S. Winkler, Managing the Component-Based Software Engineering with Metrics, Proceeding of the 5[th] International Symposium on Assessment of Software Tools, Pittsburgh, June 1997, pp. 104-110.

[11] C. Ebert, Metrics for Identifying Critical Components in Software Projects, Handbook of Software Engineering and Knowledge Engineering, 2001.

[12] N. E. Fenton and S. L. Pfleeger, Software Metrics: A Rigorous & Practical Approach 2nd edition, PWS Publishing Company, 1997.

[13] N.S. Gill and P.S. Grover, Component-Based Measurement: Few Useful Guidelines, ACM SIGSOFT Software Engineering Notes, Vol. 23, Issue 6, November 2003, pp (4-4)1-4.

[14] B. Henderson-Sellers, Object-Oriented Metrics: Measures of Complexity, Prentice Hall, 1996.

*Ref. 15-30 can be obtained from the authors. They have been omitted due to space limitations.*

# Study on a Decision Model of IT Outsourcing Prioritization

XIE Xiang[1], GUAN Zhong-liang[2]

1 School of Economy and Management, Beijing Jiaotong University, P.R.China, 100044

2 School of Economy and Management, Beijing Jiaotong University, P.R.China, 100044

***Abstract*-In today's knowledge-based society, IT outsourcing has been used increasingly in many enterprises as a policy instrument for changing the way publicly funded services are provided. Full IT outsourcing strategy is an extreme one, there are many arguments for it, and selective IT outsourcing strategy often results in greater flexibility and better services. So, identifying IT object prioritization is very important to IT outsourcing. But, now most of researchers paid too much attention the decision of Application Service Provider (ASP) selection. Therefore, this paper proposes a decision model, which uses the fuzzy gray matter-element space theory (FHW) method, to help enterprises prioritizing the IT outsourcing objects. By this decision model, enterprises can make decision about selective IT outsourcing. At last, the paper gives an example as the illustration to show the practicability of the model.**

## I. INTRODUCTION

In today's knowledge-based society, the use of Information Technology (IT) has the potential to the major driver of enterprises' economic wealth. But, as IT applications increase in sophistication and complexity, enterprises need a lot of resources and effort for maintenance and repair of IT application. According to the research, in a number of enterprises, about 70% of IT investment has been spent on maintenance. Enterprises only can grow by creating values using their core competitive power and they want to concentrate their resources on core competitive power rather than IT maintenance and management, so IT outsourcing has been used increasingly in many enterprises as a policy instrument for changing the way publicly funded services are provided.

Generically outsourcing can be defined as "the transfer of previously in-house activities to a third party" [1]. Apply to IT, IT outsourcing can be defined as "the significant contribution by external vendors in the physical and/or human resources associated with the entire or specific components of the IT infrastructure in the user organization" [2]. IT outsourcing can be considered a significant administrative innovation where there is a significant shift in the mode of governance, significant change in the internal processes of user enterprises, and significant change in the corporate routines used to deal with the external environment.

The terminology of IT outsourcing was perhaps first used in 1989 when Eastman Kodak made the decision to make total outsourcing agreements with three large IS external providers [3]. Industrial analysts predicted that the global market for IT outsourcing would grow from $86 billion in 1996 to more than $137 billion in 2001 [4]. In addition, on conservative estimates, IT outsourcing may well represent, on average, 30 - 35% of IT budgets by 2002. Also, the outsourcing market has been predicted to grow to over $120 billion by the year 2002 and even $150 billion by 2004 [5].

The IT outsourcing framework consists of four major elements (see Fig.1): IT outsourcing subject, IT outsourcing object, IT outsourcing partner, and IT outsourcing decision. IT Outsourcing subject is the economic institution which plans to outsource (or not) its IT applications. The subject has to make the strategic outsourcing decision. IT Outsourcing objects are IT processes or IT process results which might be outsourced. From an industrial perspective, the IT outsourcing object is closely linked with the degree of enterprise information, including hardware, software, applications, networks, and business services etc. Outsourcing partners are all possible suppliers for the IT activities considered for outsourcing. This supplier not only could be an external vendor, but also be an in-house supplier, e.g. an independent business unit within a group of enterprises. Outsourcing decision is a process how to decide what objects to be outsource to what suppliers.

According to the scale of IT outsourcing objects, IT outsourcing has two distinguished forms, namely: full IT outsourcing and selective IT outsourcing.

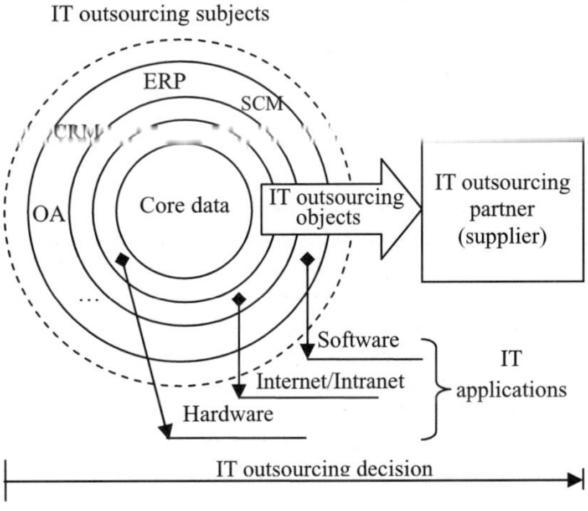


Fig.1 IT outsourcing Framework

Support by the National Natural science Foundation of China



*T. Sobh and K. Elleithy (eds.), Advances in Systems, Computing Sciences and Software Engineering*, 265–269.

In full IT outsourcing, all of the IT applications are outsourced to the Vendor. This is an extreme outsourcing strategy because the entire department information services duties are assigned to the outsourcing partner as in the case of Eastman Kodak. This, according to Pearlson, happens when an enterprise does not see "IT as a strategic advantage" that should be developed internally. Arguments for full outsourcing usually involve the allocation of organizational resources to areas that can add greater value to the enterprise's value chain or reduce cost per transaction due to economies of scale [6]. However, in selective outsourcing, only a range of services is selectively outsourced or contracted to a third party. It often results in greater flexibility and better services. So selective IT outsourcing should be considered as the first policy when one enterprise wants to outsource its IT applications.

Decision of selective IT outsourcing has two components: Application Service Provider (ASP) selection and IT objects selection. Up to now, most of literatures about IT outsourcing decision focus on the decision of ASP selection, such as analyzing the client–supplier relationship, its characteristics, its partnership quality, and the impact of these factors on outsourcing success. In contrast, few of literatures pay attention to the problem of IT objects selection.

So, our study focuses on the latter problem. This paper firstly analyzes those variables that will influence the benefits of IT outsourcing objects and then proposes a decision model of IT outsourcing prioritization, which is based on the fuzzy gray matter-element space theory (FHW) [7]. This decision model will help enterprise prioritize IT objects and decide what should be outsourced.

## II. THE INDEX SYSTEM OF IT OUTSOURCING OBJECT PRIORITIZATION

Before we construct the FHW model, it is important for us to look for the factors that will influence the benefits of IT outsourcing objects. In the evaluating IT outsourcing object consideration, it means to find out those factors that affect the benefits of the enterprise. Several factors were used before, such as transaction cost economics, and strategy or commodity etc. Different enterprises should have different considerations. Enterprises should include all factors which can affect enterprises benefit and potential risk as possible as they can. A careful examination of those factors mentioned above concludes that five dimensions or factors, operation, management, technology, economics and risk, should be employed [8]. According to the five factors, following comprehensiveness, dominant and feasibility principle, the index system of IT outsourcing object prioritization is presented, as shown in Fig.2.

For operation, enterprises need to optimize their business progress and improve the productivity. The enterprises can make strategic alliance with vendors to make up the shortage of resources or technology, develop new products, accelerate the time of product to market and widen the selling market. In addition other operational considerations also include: supplying the high reliability and excellent performance of IT and high quality customer service level.

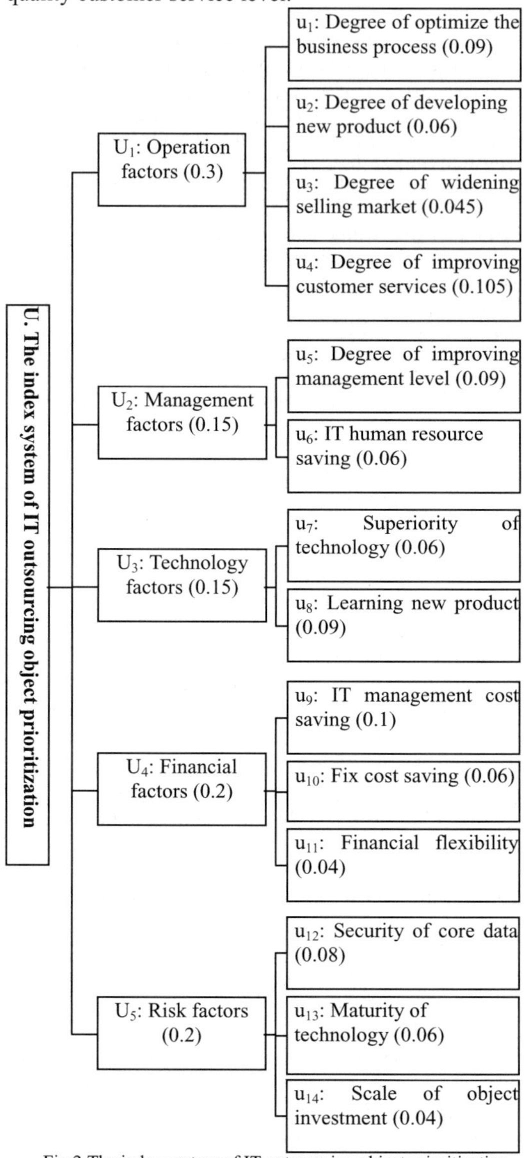

Fig.2 The index system of IT outsourcing object prioritization

For management, the problems that have to be dealt with include: insufficient performance of IT department, the floating and scarcity of employee, etc. Outsourcing can improve the performance of IT department, and enhance morale and reengineering the enterprise, so it has been regarded as an effective means of management by the high management level.

For technology, the fastest and most effective way to get the newest technologyof IT is to outsource. In-house workers can learn new technology of hardware management or software management and development from the vendor. But, enterprises have to take the superiority, practicability and security of IT objects into account.

For economics, the major consideration of a enterprise is to reduce the development and maintenance costs of IT applications. Another consideration of economics is financial flexibility. Because of outsourcing, the facilities and employee would be transferred to the vendor side, which transform fixed costs into variable costs, resulting in increasing financial flexibility.

For risk, when the investment of IT outsourcing object is more, the scale is greater and the relationship with the management is more closely, the risk of the object is much greater. From the outsourcing object angle, the major consideration of IT outsourcing risk is including: the security of core data, maturity of the technology and the scale of investment.

The weight of each index is computed by the analytic hierarchy process (AHP) method [9]. Because AHP is a well-known method and is used widely, this paper will leave out the detail of the compute process of AHP and present directly the weight of index.

## III. FHW DECISION MODEL

Because the decision of IT outsourcing prioritization involves five factors and each factor also has some different attributes, which consist of qualitative indices and quantitative ones, the decision model has to be a multicriteria analysis model. In this paper, the model is based on the fuzzy gray matter-element space theory (FHW), in that FHW is a theoretical method which is about science decision supporting to solve the problems of making decision and forecasting in the large complex systems. Based on Delphi method, it combines the BS and K.J methods together and synthesizes several theories including fuzzy mathematics, gray system and matter-element analysis. FHW adequately considers the subjective and objective evaluation and combines the qualitative and quantitative methods together.

The stepwise procedure of holding the FHW decision model as follows:

**Step1**: Establish the main index system and accessorial index system.

Enterprises can use the index system of IT outsourcing object prioritization, shown in Fig.2, as the main index system, and also can increase or decrease these indices that are suitable for themselves. The accessorial index system consists of these indices which are the most important indices in the main index system. The accessorial index is used to reflect on relationship of the advantage and disadvantage of objects between that at the present and that in future.

**Step2**: Select experts and assign weights to each expert.

Because FHW is expert decision method, the selection of experts has great affection on the veracity of IT outsourcing prioritization. So these experts should include economist, management experts, IT experts, and so on. And if the capability of experts was taken account, the weight should be assigned to each expert. AHP also can be used to assign expert's weight.

**Step3**: Evaluating each IT outsourcing object by questionnaires.

The questionnaire consisted of some categories with the main and accessorial index system and is designed by fuzzy and gray theory. Ten possible responses are provided and an absolute measurement with rank from 1 to 10 is used to score the degree of response. The survey questionnaires are distributed to each expert, and expert marks to each content in the questionnaire. So, the original expert estimation of each outsourcing object is gained:

$$(u_1, u_2, \ldots, u_n)、\{(p_1, a_1) \ldots, (p_m, a_m)\}、\{(q_1, b_1), \ldots, (q_m, b_m)\}. \qquad (1)$$

And,

$n$: the number of main indices;

$m$: the number of accessorial indices;

$u_i$: fuzzy evaluation of the $i$th main index ;

$(u_1, u_2, \ldots, u_n)$ : Fuzzy evaluation vectors;

$(p_i, a_i)$ is a gray matter-element. $p_i$: the present advantage of the $i$th accessorial index; and $a_i$: the future advantage of the $i$th accessorial index;

Same as $(q_i, b_i)$. $q_i$: the present disadvantage of the $i$th accessorial index; and $b_i$: the future disadvantage of the $i$th accessorial index;

Then, amend the original value by the weights of experts and get the final experts estimation:

$$(M_1, M_2, \ldots, M_n)、\{(P_1, A_1), \ldots, (P_m, A_m)\}、\{(Q_1, B_1), \ldots, (Q_m, B_m)\}. \qquad (2)$$

**Step4**: Analyze the final expert estimation, and compute four overall evaluation values.

$$M = \sum_{i=1}^{m} M_i w_i \qquad (3)$$

$M$: Total score; $w_i$: the weight of the $i$th main index.

$$C = \sum_{i=1}^{m} P_i v_i \Big/ \sum_{i=1}^{m} Q_i v_i \qquad (4)$$

$C$: White ratio of advantage to disadvantage. $C$ implies the radio of current advantage and current disadvantage of IT outsourcing objects, and if $C$ were less than 1, this white ratio value would not be considered.

$$D = \sum_{i=1}^{m} A_i v_i \Big/ \sum_{i=1}^{m} B_i v_i \qquad (5)$$

$D$: Gray ratio of advantage to disadvantage. $D$ implies the radio of future advantage and future disadvantage of IT outsourcing objects, and if $D$ were less than 1, this white ratio value would also not be considered.

$$N = (N' + N^2)/2 \qquad (6)$$

$$N' = 1 - \left| 0.5 + \frac{1}{2} \left( \sum_{i=1}^{m} P_i v_i - \sum_{i=1}^{m} A_i v_i \right) / \left( \sum_{i=1}^{m} P_i v_i + \sum_{i=1}^{m} A_i v_i \right) \right| \tag{7}$$

$$N^2 = 1 - \left| 0.5 + \frac{1}{2} \left( \sum_{i=1}^{m} Q_i v_i - \sum_{i=1}^{m} B_i v_i \right) / \left( \sum_{i=1}^{m} Q_i v_i + \sum_{i=1}^{m} B_i v_i \right) \right| \tag{8}$$

*N*: Total Gray Degree. *N* implies the uncertainty degree of the evaluation result and the incomplete degree of information. *N* value is greater, and the risk of outsourcing is greater.

**Step5**: Prioritizing IT outsourcing object and make IT outsourcing decision.

According to these overall evaluation values computed above, rank the IT outsourcing objects in descending. And, the object in the front of sequence has higher priority and which should be outsourced firstly.

## IV. CASE STUDY

In an enterprise, there are three IT objects, namely: A, B, C. Because of the insufficiency of funds, it is impossible to outsourcing all of them together. So, FHW decision model can be used to prioritize these three IT objects and confirm the best selective decision.

In this case, the main index system (Fig.3) is made up of all indices as shown in Fig.2, and the weights are same.

$$\begin{cases} \text{1 Degree of optimize the business progress} & w_1(0.09) \\ \text{2 Degree of developing new product} & w_2(0.06) \\ \text{3 Degree of widening selling market} & w_3(0.045) \\ \text{4 Degree of improve customer service} & w_4(0.105) \\ \text{5 Degree of improving management level} & w_5(0.09) \\ \text{6 IT human resource saving} & w_6(0.06) \\ \text{7 Superiority of technology} & w_7(0.06) \\ \text{8 Learning new technology} & w_8(0.18) \\ \text{9 IT management cost saving} & w_9(0.1) \\ \text{10 Fix cost saving} & w_{10}(0.06) \\ \text{11 Financial flexibility} & w_{11}(0.04) \\ \text{12 Security of core data} & w_{12}(0.08) \\ \text{13 Maturity of technology} & w_{13}(0.06) \\ \text{14 Scale of object investment} & w_{14}(0.04) \end{cases}$$

Fig.3 The main index system

The accessorial index system consists of six indices which weights are the greatest in the main index system. Amend the weights of accessorial indices based on their original weights. So, the accessorial index systems are shown in Fig.4.

$$\begin{cases} \text{1 Degree of optimize the business progress} & v_1(0.17) \\ \text{2 Degree of improve customer service} & v_2(0.19) \\ \text{3 Degree of improving management level} & v_3(0.17) \\ \text{4 Learning new technology} & v_4(0.18) \\ \text{5 IT management cost saving} & v_5(0.14) \\ \text{6 Security of core data} & v_6(0.15) \end{cases}$$

Fig.4 The accessorial index system

The survey questionnaires are distributed to 20 experts, which consists of economist, management experts, IT experts etc. In the case, we suppose the weight of each expert is same, so we can respectively obtain the expert estimation of main indices in Table I. The evaluation of accessorial indices includes: advantage evaluation and disadvantage evaluation, as shown in Table II and Table III, what consider the difference which result from the change between the benefits and loss of indices in current and those in future.

TABLE I
EXPERT ESTIMATION OF MAIN INDICES

| Main IT objects \ indices | A | B | C |
|---|---|---|---|
| 1 Degree of optimize the business progress (10) | 7.3 | 8.01 | 8.86 |
| 2 Degree of developing new products (10) | 6.01 | 7.76 | 6.79 |
| 3 Degree of widening selling market (10) | 7.37 | 6.64 | 6.07 |
| 4 Degree of improving customer service (10) | 6.8 | 7.2 | 7.45 |
| 5 Degree of improving management level (10) | 7.4 | 6.04 | 6.63 |
| 6 IT human resource saving (10) | 8.1 | 7.7 | 8 |
| 7 Superiority of technology (10) | 7.2 | 7.5 | 7.9 |
| 8 Leaning new technology (10) | 6.87 | 6.51 | 6.44 |
| 9 IT management cost saving (10) | 7.27 | 7.85 | 7.1 |
| 10 Fix cost saving (10) | 7.3 | 7.55 | 8.01 |
| 11 Financial flexibility (10) | 6.8 | 7.5 | 7.07 |
| 12 Security of core data (10) | 7.91 | 7.51 | 7.13 |
| 13 Maturity of technology (10) | 7.16 | 7.43 | 7.45 |
| 14 the scale of object investment (10) | 7.05 | 7.2 | 7.52 |

TABLE II
ADVANTAGE EVALUATION OF ACCESSORIAL INDICES

| Accessorial indices \ IT objects | A | | B | | C | |
|---|---|---|---|---|---|---|
| | current | future | current | future | current | future |
| 1 Degree of optimize business progress （10） | 7.19 | 7.12 | 7.9 | 8.35 | 8.8 | 9.01 |
| 2 Degree of improving customer service （10） | 6.6 | 6.9 | 7.2 | 7.25 | 7.4 | 7.5 |
| 3 Degree of improving management level （10） | 7.4 | 7.34 | 6.05 | 6.5 | 6.6 | 6.8 |
| 4 Leaning new technology （10） | 7.1 | 7.2 | 7.55 | 7.25 | 7.95 | 7.7 |
| 5 IT management cost saving （10） | 7.3 | 7 | 7.7 | 7.2 | 8.1 | 7.6 |
| 6 Security of core data （10） | 8.09 | 7.98 | 7.62 | 7.43 | 7.22 | 6.8 |

TABLE III
DISADVANTAGE EVALUATION OF ACCESSORIAL INDICES

| Accessorial indices \ IT objects | A | | B | | C | |
|---|---|---|---|---|---|---|
| | current | future | current | future | current | future |
| 1 Degree of optimize business progress （10） | 5.18 | 5.02 | 5.5 | 5.3 | 5.6 | 5.75 |
| 2 Degree of improving customer service （10） | 4.8 | 4.6 | 5.41 | 5.06 | 5.33 | 5.04 |
| 3 Degree of improving management level （10） | 5.8 | 5.5 | 5.77 | 5.03 | 4.92 | 4.71 |
| 4 Leaning new technology （10） | 3.32 | 3.51 | 4.2 | 4.88 | 4.32 | 4.55 |
| 5 IT management cost saving （10） | 6.41 | 6.19 | 5.24 | 5.83 | 4.8 | 4.2 |
| 6 Security of core data （10） | 4.81 | 4.31 | 5.26 | 5.02 | 5.11 | 4.88 |

Furthermore, analyze the expert estimations and obtain overall evaluation values in Table IV.

TABEL IV
OVERALL EVALUATION S OF MAIN AND ACCESSORIAL INDICES

| IT objects | Total score (M) | White Ratio (C) | Gray Ratio (D) | Total Gray Degree (N) |
|---|---|---|---|---|
| A | 7.1812 | 1.4510 | 1.5065 | 0.4951 |
| B | 7.3089 | 1.4016 | 1.4190 | 0.4988 |
| C | 7.3969 | 1.5291 | 1.5555 | 0.4949 |

From Total score $M$, White Ratio $C$ and Gray Ratio $D$, the values of object C is all biggest. These imply the whole benefit of object C is greatest.

From Total Gary Degree $N$, the value of C is smallest. And this implies that the uncertain factors do less harm to the benefit of object C.

So, these three objects are ranked: C>B>A. Therefore, enterprise can make decision that C object is going to be outsourcing firstly.

## V. CONCLUSION

Nowadays, with the prevalence of selective IT outsourcing, IT outsourcing prioritization has played more and more important role in enterprises information. The FHW decision model offers systematic steps and quantitative results to increase the precision of decision-making and help the enterprises make better IT outsourcing decisions.

## REFERENCES

[1] C. Lonsdale, Effectively managing vertical supply relationships: A risk management model for outsourcing supply chain management, An International Journal, vol.4, no.4, pp.176, 2001.
[2] L. Loh, N. Venkatraman, Determinants of information technology outsourcing, Journal of Management, Information Systems, vol.9 no.1, pp.24,1992.
[3] S. Slaughter, S. Ang, Employment Outsourcing in Information Systems, Communications of the ACM, vol.39, no.7, pp.47 - 54, July, 1996.
[4] A. Diromualdo, V. Gurbaxani, Strategic intent for IT outsourcing, Sloan Management Review, vol.39, no.4, pp.67 - 80, 1998.
[5] Abdulwahed, Mo. Khalfan, Information security considerations in IS/IT outsourcing projects: a descriptive case study of two sectors, International Journal of Information Management, vol.24, pp. 29 - 42, 2004.
[6] K. E. Pearlson, Managing and using information systems: a strategic approach, Chichester:Wiley, 2001.
[7] XIE Yan-qing, ZHANG Jiang, GUO Qiang, LIN Hua, Fuzzy gray Matter element space Theory and practical application and development-The policy decision supporting system in macro complex system (in Chinese), Engineering Science, vol.4, no.11, pp.56-66, 2002.
[8] Chyan Yang, Jen-Bor Huang, A decision model for IS outsourcing, International Journal of Information Management, vol.20, pp.239, 2000.
[9] T. L. Saaty, How to make a decision: The analytic hierarchy process, European Journal of Operational Research, vol.48, no.9, pp.26, 1990.

# AN EFFICIENT DATABASE SYSTEM FOR UTILIZING GIS QUERIES

Fawaz A. Masoud and Moh'd Belal Al- Zoubi

King Abdullah School for Information Technology

University of Jordan, Amman – Jordan

(Fawaz,mba)@ju.edu.jo

**Abstract-In this paper, we propose a new technique to answer GIS queries quickly. The technique uses phone zones as main indexes to narrow down the search processes. In this paper, we build modest GIS software system to handle different GIS queries efficiently. Using our new software, many benefits can be gained including search efficiency, simple interface, displaying spatial relationships, showing sales and service territories.**

KEYWORDS

GIS, spatial search, database directory, telephone zones.

## 1 INTRODUCTION

Geographic Information Systems (GIS) is a relatively new branch of information technology and the term GIS did not appear until the early 1960s [1]. GIS is a computer-based technology and methodology for collecting, managing, analyzing, modeling, and presenting geographic data for a wide range of applications [1]. GIS have often been used to identify suitable areas for land developments [2]. GIS can benefit many application areas such as real estate, health care, education, environment, petroleum industry, transportation public safety and public/private services.

GIS databases cover all conventional sources such as maps, census data, as well as recent high technology sources such as remote senses, Global Positions Systems (GPS) and Internet. These sources are housed in and managed via database management systems (DBMSs).

However, GIS has very limited capabilities for integrating geographical information with the decision maker's values and preferences and hence of limited use for decision support mechanisms [3]. In fact, the majority of today's DBMSs are either incapable of managing spatial data or are not user-friendly [4].

In this paper, we present a GIS software, equipped with efficient search techniques and simple interface. We use of phone zones (territories) as a main index to narrow down the search in order to efficiently answer different queries. Phone numbers themselves are served as sub-indexes.

## 2 RELATED WORK

There exist many on-line software systems. One of the earliest of these system is the yellow pages project [5] of Australia. The project searches Australian business directory with over 1.7 million business listings. Yellow pages can be searched to find out about businesses. So, phone and fax numbers, addresses, products and services can be shown. The system provides searches using three types of search; search by type of business or business name, or map based search and browse for a business.

Another work is the ATM finder of VISA corporation software system [6]. The system allows the user to enter the address or intersection anywhere in the United States, and it will then return a map showing the location of the three nearest ATM machines that accept VISA cards, plus information about those ATMs.

Yahoo maps [7] are also available for directing users within the United States and Canada territories

## 3 THE PROPOSED TECHNIQUE

Imagine searching for a house location to deliver certain service to it, Pizza, for example, and the only information you have to locate that house is its telephone number. This is exactly what happens when you order a pizza; they ask for your telephone number. In this article, we develop a GIS software that can help in locating places within the area of Amman, capital of Jordan, if the telephone number of a certain place is known.



*T. Sobh and K. Elleithy (eds.), Advances in Systems, Computing Sciences and Software Engineering*, 271–275.

Amman area is divided into seven sub areas; each has its own telephone zone territory. As shown in Figure 1, these sub areas are ABDALY, ASHRAFEYYEH, CENTRAL, NAZAL, SWAILEH, TLA EL-ALI and WADI ESSIER.

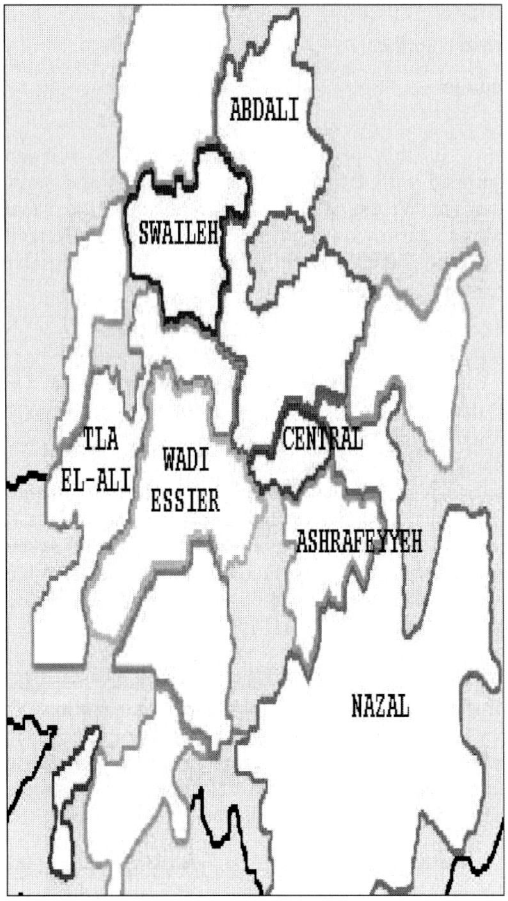


Figure 1 Amman area is divided into seven sub areas

Queries like: Where is the person calling for help located? What is the best route to reach her? are important queries in different service application areas such as emergency and Pizza-delivery serves. In this paper, we develop a software system to handle such and similar queries.

In addition, we are providing users with high-speed access to find locations. Once the telephone number is identified in the text-based directory (database), a map of the location is displayed. The display shows a red circle indicating the location plus an image for the surrounding streets, street names, and features in the immediate vicinity, just as if the user were looking in a street map directory.

We propose a tow-step algorithm to efficiently search for answers in the databases:

1. Filter step: In this step, we use phone zones (territories) as main index. The first two digits of a phone number represent a different zone buffer. By just entering the phone number for a location, the search starts at the series of phone numbers and this is done in step 2 below.
2. Refinement step: Here, the exact telephone number of each record in the database is examined. In this step, efficient indexing methods can be used to access data in a certain territory [8, 9]. A popular choice is the binary tree family of methods to index the data. Thus, for a small overhead, scanning the whole database can be avoided, and this is our first contribution in this paper which leads to more efficient analysis and faster decision making.

The results of combining the above two steps offer significant increase in efficiency over exhaustive methods when used to GIS data.

## 4 DATABASE MODELING REQUIREMENTS

Database systems are essential components of GIS. GISs are used to collect, analyzes, and presents information describing the physical and logical properties of the geographic real world problems. Geographic information system can store and analyze maps, weather data, and satellite images [10]. This force us to model the database carefully to get an efficient database system for geographic information system. There are four major functional units in the GIS:

- Database Model.
- Database Input.
- Database Manipulation.
- Result Presentation Facilities.

### *4.1 Database Model*

A conceptual database model is a type of data abstraction that hides the details of data storage [11]. It uses logical concepts, which may be easier for most users to understand. It supports data input, manipulation and result presentation. Many GISs are organized as a collection of themes. Each theme represents the values of a unique attribute of the geographic space. A theme may independently partition, decompose, and fragment the continuous space for a particular value (or value range) of the attribute. The partitions and fragments of space within each theme are often stored within the database and can be treated as entities or objects. The database entity relationship model of the GIS for finding location using telephone zones is shown in Figure 2.

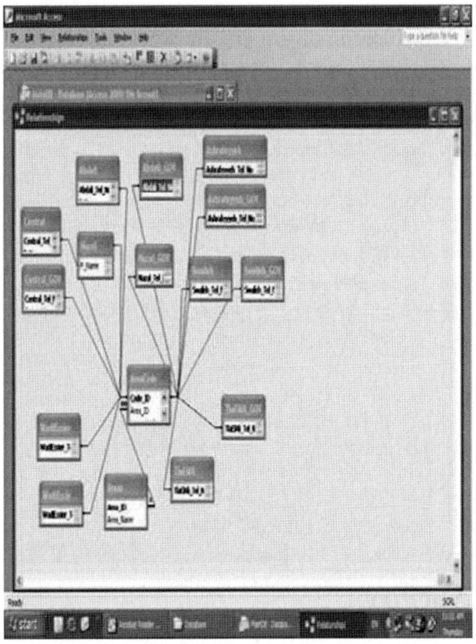

Figure 2 The ER Diagram of the GIS for finding location using telephone zones

### *4.2 Database Input*

The database entity relationship model of the GIS for finding location using telephone zones provides the relationships to support the database collection process and states that raw data is refined into processed data, which is refined into interpreted data.

### *4.3 Database Manipulation Capabilities*

Spatial relationships are added to the model to accommodate spatial operations in GIS. These relationships include topological relationships, direction relationships and metric relationships. These relationships serve as spatial predicates for queries in GIS. Directional relationships involve the location of the objects (examples are north of, south of, and north east of). Topological relationships involve the regions occupied by the objects (examples are adjacent to, inside, etc.).

### *4.4 Result Presentation Facilities*

The database entity relationship model of the GIS for finding location using telephone zones proposes that visual representation be specified in the database to declaratively specify the essential properties of the customer location. Visual representation consists of primitives such as text, icons, graphs and geometries like points and lines. These primitives are associated with a location and orientation inside a visual representation. In addition, visual constraints ensure the visual representation does not convey any information which is not present in the geographic data, and that the visual representation should convey all the information requested by the user. Visualization constraints on map road segments should remain connected in the visual representation and location that the distortion of the location of various objects should be maintained within acceptable ranges.

## 5 SOFTWARE DEMONSTRATION

As the software started, the user is asked to enter the telephone number and the area for which the telephone number belongs in highlighted (in a yellow color), as shown in Figure 3.

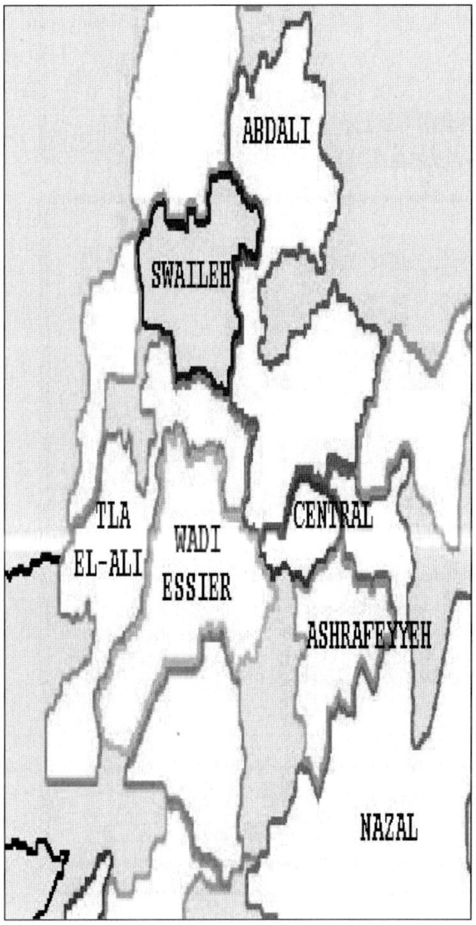


Figure 3 the telephone number and the area for which the telephone number belong is highlighted

Information about the location are then displayed, showing the name, address and other stored information, as shown in Figure 4.

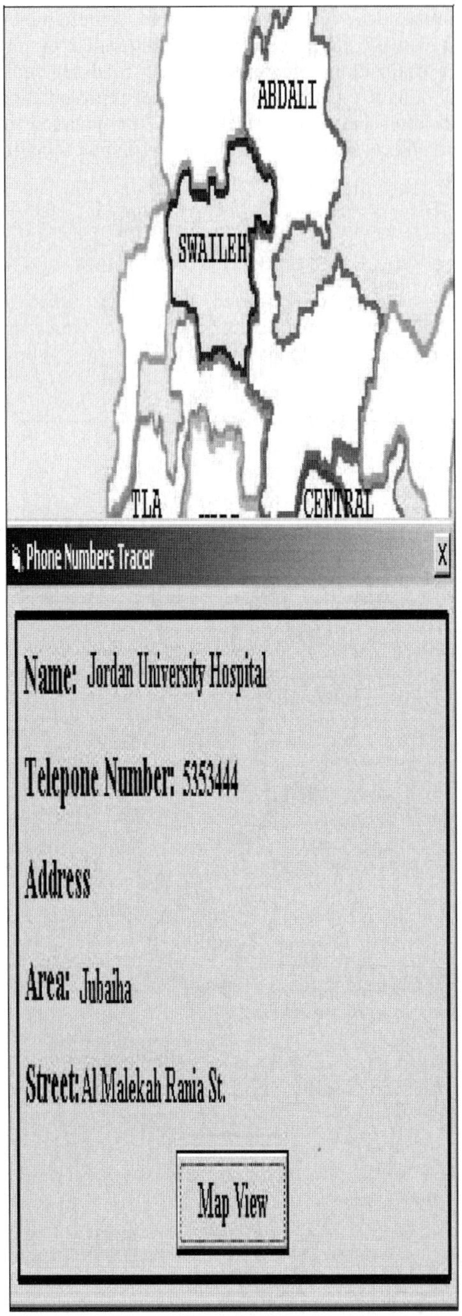


Figure 4 Information about the location

When, clicking on the Map View button, the map of Amman is displayed, showing the targeted locating in red color, as shown in Figure 5.

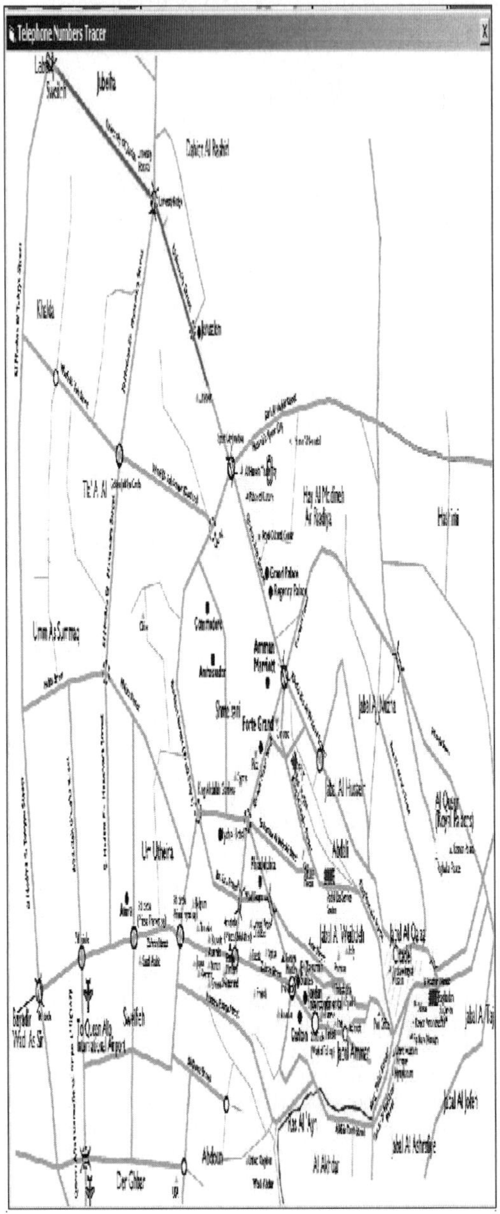


Figure 5 targeted locating in Amman

Figure 6 shows the targeted location enlarged, surrounded by a red circle, showing the location and the streets, and other features around it.

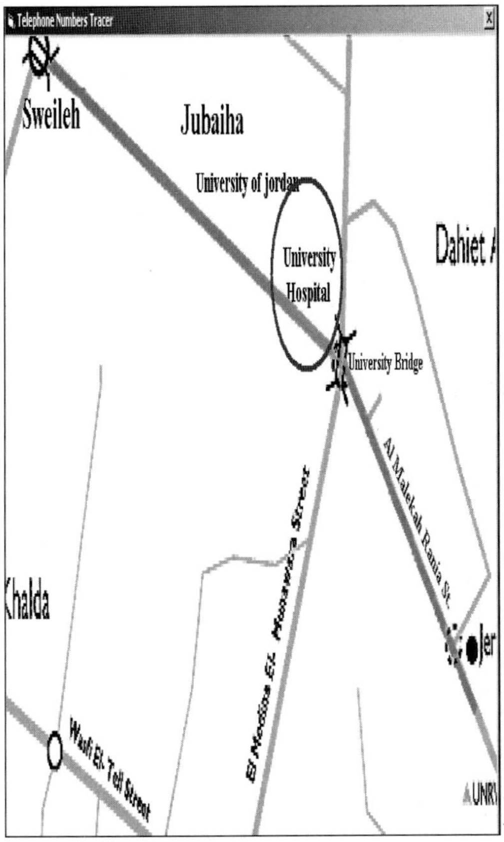


Figure 6 the targeted locating enlarged

The software is equipped with an efficient search technology and simple interface. Efficiency comes from the fact that Amman area is divided into seven sub areas. Each sub has its own series of phones.

## 6 CONCLUSION AND FUTURE WORK

In this paper, we present new GIS software. The software algorithm works in two steps. In the first step, called filter step we use phone zones (territories) as main index. In the second, called refinement step, we perform exact search. The results of combining these two steps offer significant increase in efficiency over exhaustive and other methods when used with GIS data. Thus, large CPU time savings are obtained.

Using our new software, many benefits can be gained including search efficiency, simple interface, displaying spatial relationships, showing sales and service territories.

For future work, we propose to add other features to the software such as providing distance functions, calculating least-cost distances and shortest paths.

## 7 REFERENCES

1. Davis, B., " GIS: A Visual Approach", OnWord Press, 2001.

2. Jones, R. and Barron. M. "Site Selection of Petroleum Pipelines", A GIS Approach to Minimize Environmental Lo, C.P. and Yeung, A., Concepts and Techniques of Geographic Information Systems, Prentice-Hall, New Jersey, USA, 2002.

3. Malczewski, J., "GIS and multicriteria decision analysis", John Wiley and Sons, 1999.

4. Shekhar, S and Chawla S., "Spatial databases: a tour", Prentice Hall, 2003.

5. Yahoo maps: www.yahoo.com.

6. VIZA ATM Finder: http://www.viza.com.

7. Yellow pages: www.yellowpages.com.au.

8. Hjaltason, R. and Samet, H., "Distance Browsing in Spatial Databases", ACM Transactions on Database Systems, Vol. 24, No.2, 1999, pp 26-42.

9. Ngu, A., Sheng, Q., Huynh, D. and Lei, R., "Combining multi-visual features for efficient indexing in a large image database", International Journal on Very Large Data Bases (VLDB), Vol. 9, No. 4, 2001, pp. 279-293.

10. Elmasri R., and Navathe S. B. "Fundamentals of database systems", Addison-Wesley, third edition, 2000.

11. Silberschatz A., Korth H. K., and Sudarshan S., "Database system concepts", McGraw-Hill, 5th edition, 2006.

# Effective Adaptive Plans

### A Hypothetical Search Process

**Kees P. Pieters**
de Batenburg 37
3761AX, Soest
the Netherlands
cees_pieters@wxs.nl

**Abstract- Many iterative search processes, or *adaptive plans*, that aim to find an optimal solution in a given problem domain, suggest that an optimal search process has an exponential character. Plans that consist of multiple strategies running in parallel, such as *bandit searches*, aim to demonstrate this pattern in probabilistic distributions of finding a best observed strategy amongst a number of alternatives.**
**This paper introduces a hypothetical adaptive plan that consists of three strategies. One strategy guarantees a better result with each iteration, one has comparable results, and one guarantees worse results. The idea behind this approach is the suspicion that every adaptive plan can basically be mapped to these three base strategies, and that the exponential character of an optimal plan is a trait of its recurrent character.**

## I. INTRODUCTION

An *adaptive plan*, active in a *problem domain P*, consisting of n+1 variables $\{x_0, \ldots, x_n\}$ usually develops a sequence of trials $\{B(1),\ldots,B(t)\}$, with $B(i) \subset P$, $1 \leq i \leq t$ [1].

This sequence aims to *rank* a subset of variables of P, in other words, to create pairs $\{x_i, r_i\}$, with $x_i \in P$ and $r_i$ a token that allows comparison between variables in P. In this paper a variable $x_i \in P$ will be considered 'better' than $x_j \in P$ if $r_i > r_j$, in other words if $x_i$ is *ranked higher* than $x_j$.

The goal of an adaptive plan is to find either the highest ranked variable(s) in P, or a sufficiently high-ranked one according to a certain external goal, such as a termination condition.

The sequence can be considered the result of a mix of strategies $\{\xi_0, \ldots, \xi_k\}$, some of which aim to randomly test variables in P (variation), while others may try to focus on certain 'interesting' areas based on the outcome of previous trials (optimisation).

Most literature on these parallel running strategies, such as those based on bandit searches [1,2], consider some of these strategies superior to others, and the goal of an optimal adaptive plan therefore is to allow the superior strategies to get more time to run than the poor ones, or to allow the portion of variables in B(t) provided by successful strategies to grow with respect to the remainder [1,2,3,5].

This paper introduces an hypothetical adaptive plan, consisting of three strategies of which one guarantees better results with each iteration step. This rule makes the adaptive plan inherently hypothetical, but it also provides an intuitive reason why the plan may be optimal; ideally this is what a search process should aim for.

The reason to take this approach is the idea that every adaptive plan can basically be *decomposed* into the three strategies, and that successful ones are those where the previous rule is predominant. This, of course depends on the characteristics of the problem domain and the implementation of the problem solver.

The interesting question is, whether the hypothetical plan described above will have the exponential character that is attributed to an optimal search process.

## II. AN HYPOTHETICAL ADAPTIVE PLAN

Suppose a hypothetical adaptive plan that develops a sequence B(t) variables and has the following characteristics:

1. A subset of b(t) variables in B(t), with $b(t) \geq 0$, is always better than the previous selection B(t-1)
2. A subset of c(t) variables in B(t), with $c(t) \geq 0$, is worse than the previous selection B(t-1)
3. The remaining variables in B(t) are equal to those in $\{B(0),\ldots,B(t-1)\}$

This set can be considered as belonging to three strategies $\{\xi^+,\xi^\sim,\xi^-\}$, one of which develops the subset of increasingly better variables, one that develops the increasingly poor variables ($\xi^-$) and a subset that remains equal ($\xi^\sim$).
Suppose now that an optimal adaptive plan would be one where the observed best strategy $\xi^+$ gets exponentially increasing trials with respect to the remainder [1]. One could also turn this criterion around: if the remaining strategies get an exponentially decreasing amount of trials with respect to the observed best, then the adaptive plan performs optimally, provided that the amount of variables in $B(t) < \infty$. This means that the remainder $\xi^\sim$ and $\xi^-$ get exponentially less trials, or



*T. Sobh and K. Elleithy (eds.), Advances in Systems, Computing Sciences and Software Engineering*, 277–282.

that their portion in B(t) decreases exponentially. But $\xi^+$ and $\xi^\sim$ can be combined to become one new strategy $\xi^*$, *which is still an observed best strategy.* So an adaptive plan would also be optimal if the amount of trials assigned to $\xi^-$ decreases exponentially with respect to $\xi^*$. An adaptive plan will therefore behave optimally if the strategy that "collects" the worst variables gets exponentially decreasing less trials with respect to "don't care" and "improving" strategies. This notion is captured in the following theorem.

*Theorem 1:*

Consider a problem domain P, consisting of n+1 variables $\{x_0,\ldots, x_n\}$ and take an adaptive plan consisting of three strategies $\{\xi^+,\xi^\sim,\xi^-\}$ that develops a sequence $\{B(1),\ldots, B(t)\}$ of trials $B(t) \subset P$, where:

- $\xi^+$ provides a subset of variables in B(t) that is better than those of $\{B(1),\ldots, B(t-1)\}$
- $\xi^\sim$ provides a subset of variables in B(t) that is equal to those of $\{B(1),\ldots, B(t-1)\}$
- $\xi^-$ provides a subset of variables in B(t) that is worse than those of $\{B(1),\ldots, B(t-1)\}$

Suppose now that an adaptive plan behaves optimally if the observed best strategy gets an exponentially increasing amount of trials with respect to the remainder. Then the adaptive plan will perform optimally if $\xi^-$ gets an exponentially decreasing amount of trials.

*Proof:*
An adaptive plan behaves optimally if the observed best strategy $\xi^*$ gets an exponentially increasing amount of trials, or if its portion $M_{\xi*}(t)$ of B(t) will increase exponentially with t, with respect to the remainder $M_{\xi-}(t)$ [1]. In other words:

$$\frac{dM_{\xi*}(t)}{dt} = c_1(t).M_{\xi*}(t) \qquad [1.1]$$

The change of $M_{\xi*}$ is inverse to that of the remainder $M_{\xi-}$, so:

$$\frac{dM_{\xi-}(t)}{dt} = -c_1(t).M_{\xi*}(t) \qquad [1.2]$$

If $M_{\xi*}(t)$ increases exponentially, then:

$$\frac{dM_{\xi-}(t)}{dt} \equiv -c_1(t).k_1.e^{\alpha t}, k_1 > 0 \qquad [1.3]$$

Therefore $M_{\xi-}(t)$ decreases exponentially.

{***}

Theorem 1 suggests that an adaptive plan will behave optimally if a rule can be found that discards an exponentially increasing number of poor variables with every trial. These variables will not affect $\xi^+$ and $\xi^\sim$ as these strategies 'collect' good variables, while $\xi^-$ will get less variables to choose from. In other words, if the amount of variables in the problem domain that are tested decreases exponentially:

$$n(t) = n.e^{-\alpha t} \qquad [1.4]$$

Theorem 1 presupposes that the observed best strategy gets an exponentially increasing amount of variables with respect to the remainder. It can be proven that this is inherent to the hypothetical adaptive plan. This proof is captured in the following three theorems.

*Theorem 2:*

Given an adaptive plan that develops a sequence $\{B(1),\ldots,B(t)\}$ characterised by the following rules:

1. One new variable is added to B(t) with each iteration step t
2. This variable is ranked higher than the variables in B(t-1)

The total amount of possible trials S(m) that can be made with the m variables (0<m) is:

$$S(m) = 1 + 2.S(m-1), \text{ with } S(1) = m \qquad [1.5]$$

*Proof:*
First note that the two rules can be considered as creating a subset P' from P, where a new and higher ranked variable is added to P' with each iteration step.

S(m-1) is the total number of possible trials that can be made with m-1 variables. One variable $x_m$ is added to P', making the total amount of variables m. This variable is ranked higher than the m-1 variables that are already contained in P' by rule 2. The total amount of possible trials that can be made with this new set equals:

- All the trials that can be made with m-1 variables S(m-1)
- The new sample $\{x_m\}$, that could be selected at iteration step t =1
- All the previous trials with the new sample added to the end because it is ranked higher

This means that the total amount of trials over all iterations is equal to [1.5]

{***}

Theorem 2 can be visualised by a tree with root $\phi$ (empty root) which connects to m variables with one variable. Every variable in turn is the root of a subtree that contains the same subtree as its parent which contains all its predecessors in the sequence. Take for instance a problem domain consiting of three variables P' = {A,B,C} and the ranking is according to the position in the alphabet (C is ranked highest).

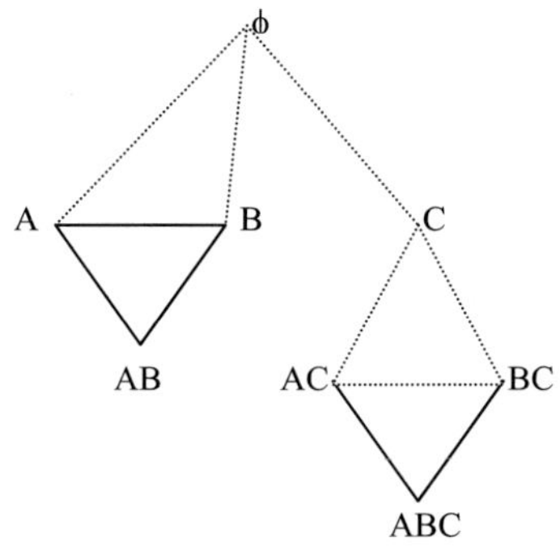


Figure 1: Expanding P' = {A,B} with a new variable {C}

Adding variable C to an initial set of {A,B} results in new possible trials {C} and the threesome {AC}, {BC} and {ABC}, which is a "skewed" copy of the original tree {A}, {B} and {AB}, with variable C added to the end (see figure 1).
The skewed character of the copied m-1 tree, results in the following theorem:

*Theorem 3:*

Given the adaptive plan described in theorem 2, then the amount of new trials $\psi_m(t)$ that are possible in P' at a given iteration step t for a total amount of m variables (m>1) equals:

$$\psi_m(t)=\psi_{m-1}(t)+\psi_{m-1}(t-1),\ with\ \psi_m(1)=m \qquad [1.6]$$

*Proof:*
The total amount of possible trials at t=1 is equal to m. According to theorem 2, adding one variable to the search space of m variables results in a copy of the m-1 tree being added to the newly added variable. Moving down the newly formed tree, starting from the root, every step t will consist of $\psi_{m-1}(t)$ possible trials of the original tree plus $\psi_{m-1}(t-1)$ trials of the skewed copy.

{***}

The tools that have been introduced can now be used to say something about *all* iterations over n variables in P.

*Theorem 4:*

Given the adaptive plan described earlier, with n variables, then the total amount of possible trials at a given iteration t is:

$$\psi_n(t)=\binom{n}{t}="n\ over\ t" \qquad [1.7]$$

*Proof:*

This can be proven with full induction. If [1.7] is correct for t=1, it should hold for t+1 if the theorem is supposed to be correct for t.

For t=1, $\psi_n(1)$=n:

$$\binom{n}{1}=n, so\ this\ is\ correct$$

Suppose the theorem is correct for n variables, then it is also correct for n-1 variables:

$$\psi_{n-1}(t)=\binom{n-1}{t}$$

According to [1.6], $\psi_n(t+1)$ should be:

$$\psi_n(t+1)=\binom{n-1}{t+1}+\binom{n-1}{t}=$$

$$=\frac{(n-1)!}{(t+1)!(n-t-2)!}+\frac{(n-1)!}{t!(n-1-t)!}=$$

$$=\frac{(n-1)!}{t!(n-t-2)!}\cdot\left[\frac{1}{t+1}+\frac{1}{(n-t-1)}\right]=$$

$$=\frac{(n-1)!}{t!(n-t-2)!}\cdot\left[\frac{n}{(t+1)(n-t-1)}\right]\Rightarrow$$

$$\psi_n(t+1)=\frac{n!}{(t+1)!(n-t-1)!}=\binom{n}{t+1}$$

{***}

This means that:

$$\Delta\psi_n(t)=\psi_n(t)-\psi_n(t-1)=\binom{n}{t}-\binom{n}{t-1}=$$

$$\Delta\psi_n(t)=\psi_n(t).(t^2-n.t+1) \qquad [1.8]$$

Or:

$$\frac{\Delta\psi_n(t)}{\psi_n(t)}=c(t)\approx -n.t,\ if\ n>>1,\ 1\le t<<n$$

Therefore, the possible trials decrease exponentially as a function of iteration steps t for this adaptive plans if n.t >> $t^2$+1. For most practical problem domains, n is very large so then this rule applies. It can easily be seen that this situation is similar to that of theorem 1, so an adaptive plan that ensures better solutions with each trial has an exponential character.
A practical plan therefore behaves optimally if the strategies that make up the plan, at *any* time, can be *mapped* to the three base strategies of the hypothetical plan. This notion is fundamentally different than the assumptions of most parallel

search processes that are based on multi-armed bandit problems. In the latter case, it is assumed that there are strategies that outperform others, and the search process aims to find these strategies. In this paper, *any* strategy of the plan may *at some time* find better solutions, and therefore contribute to the $\xi^+$ strategy. An optimal adaptive plan will therefore consist of a mix of strategies of which at least one is finding better variables at a certain point in the search process. This mapping process will be investigated next for a few well-known strategies.

## III. HILL-CLIMBING STRATEGIES

In the previous section, it was argued that *any* adaptive plan that ensures better variables with every iteration step complies with an optimal search. In practical settings, this situation can not be achieved continuously, but an adaptive plan can consist of *phases* when the search is optimal, for instance while hill-climbing. Hill-climbing can be compared with an adaptive plan that searches an optimum in a sub-domain of the actual problem domain[1]. This sub-domain has a number of variables with monotonously increasing or decreasing rank, and so this knowledge is used to get better variables by 'climbing up the slope' of the hill.

According to the previous argumentation, the search process will be optimal during the time the adaptive plan is hill-climbing, so if it develops a sequence with multiple variables, the plan will be most effective if at least one variable is in a hill-climbing phase during the entire search process. This will ensure the the adaptive plan as a whole behaves optimally. Seen from this angle, any adaptive plan will therefore perform three activities:

1. Finding slopes in the problem domain
2. Determining the direction of higher ranking variables on the slopes
3. Hill-climbing

The slope of a hill is important because this gives 'meaning' to the relative index positions of variables that are located on it. Therefore the concepts of 'near' and 'far' will be introduced next, in order to describe the relationship between two variables. A variable $x_i$ is defined as being 'near' another variable $x_j$ if both are situated on the same slope, with slope length $\sigma_h$, and:

$|i-j| < \sigma_h$

Two variables are 'far' if they are not on the same slope.
A strategy that utilises the fact that variables are near is called a *neighbourhood function.*

[1] Note that this fact that this makes the n in [1.8] smaller than the amount of variables in P. A very restrictive hill-climbing strategy would therefore not result in an optimum search process, but rather in a sequence of many suboptimal search processes carried out one at a time.

An optimal hill-climbing strategy would ideally span as much variables in the slope as possible, making the amount of variables in the subdomain of the optimising strategy as large as possible, but without adding 'meaningless' variables outside the slope that could limit the phase of optimisation of the strategy. Again, this optimum can at best be approximated.

## IV. GENETIC ALGORITHMS

Genetic Algorithms (GAs) are often associated with multi-armed bandit problems [1]. With GAs, the variables of the problem domain P are usually represented in a form that allows processing by genetic operators on the internal representation (chromosome). Although there are various forms of such a representation, one of the most used (and classic) forms is the familiar bit string. Alternative representations can be considered fundamentally comaparable [6], especially from the very pragmatic stance taken here.
Every bit of the internal representation can be manipulated individually, without relevance for the variable it represents. Genetic Operators therefore perform in a highly abstract domain.
There are three kinds of genetic operators:

- *Reproduction* re-selects (high-ranked) variables, and therefore behaves like the $\xi^-$ strategy that was defined earlier
- *Mutation* creates a random new variable based on a previous one (a *parent).* Usually this done by changing one token (bit) of the internal representation of the parent at random
- *Crossover (sexual recombination)* is usually considered the "power" operator of GAs, and is performed by taking two (successful) parents and swapping parts of their internal representation

### A. Mutation

Mutation is an operation where a new sample is created from one variable $x_p$ (parent). A random bit from the bit representation of the variable is inverted, thus creating a new variable, the child $x_c$. This is the "classic" form of mutation. There are many other alternative ways, but for this discussion they are fairly comparable. The following situations can be identified:

1. The child is far from its parent
2. The child is near its parent

The first situation will normally occur if the inversion is applied to the high order bits of the parent. No conclusion can be drawn from this situation, as the child can be situated anywhere in the problem domain. The operation can therefore be considered random.

The second situation will normally occur when a low order bit has been inverted. Suppose that both parent and child are situated on a slope, then there is a probability of 50% that the child is better ranked than its parent. The fact that higher-ranked variables are re-selected for further processing result in hill-climbing of successful mutants and therefore mutation can result in optimal phases when low-order bits are processed. Mutation therefore has two faces if the $\lambda^{th}$ bit is mutated:

- It has a 50% probability of being a hill-climbing function if $2^{\lambda} < \sigma_h$
- It is a random strategy otherwise

Mutation, even though it can have an optimal phase every now and then, will not often contribute to hill-climbing, and so its effect is mainly to provide random variables.

*B. Crossover*

A crossover operation takes two variables and swaps part of their internal representation. Although there are various ways that this can be done, normally a *crossover point* is determined based on the fitness (rank) of each variable, after which a set of bytes of the two parents are swapped. The crossover operation therefore splits the internal representation of a variable in a high order (h) portion and a low order (l) portion:

$$x_i = x_{i,h} + x_{i,l} \text{ with } x_{i,l} \leq x_{i,h}$$

Take for instance two samples with a chromosome length of ten, 1110001111 and 1011010101, and their fitness are ranked six and four respectively. Two new samples are made by the crossover operation (the bold face portions of the chromosomes are combined):

| | | |
|---|---|---|
| $x_1 =$ | **111000**\|1111 | 111000 \|**1111** |
| $x_2 =$ | 101101 \|**0101** | **101101**\|0101 |
| $x_3$ , $x_4 =$ | **111000\|0101** | **101101\|1111** |

As with mutation, $x_1$ and $x_2$ are called the parents of $x_3$ and $x_4$.

A closer look at the operator reveals that if $x_{1,l} > x_{2,l}$ then $x_4 > x_2$ and $x_3 < x_1$ The reverse applies if $x_{1,l} <$ $x_{2,l}$.

When looking at $x_{1,h}$ and $x_{2,h}$ two distinct situations can be identified:

1. $x_{1,h}$ and $x_{2,h}$ are not almost equal ($x_1$ and $x_2$ are "far" from each other)
2. $x_{1,h} > x_{2,h}$ are almost equal ($x_1$ and $x_2$ are "near" each other)

In the first case, one can not make any statement about the crossover operation. The children can be situated on a widespread area in the problem domain. Crossover therefore tends to create random variables in such a situation.

The second situation is more interesting. If the two variables are positioned on the same slope, the result of a crossover operation is that one of the resulting variables $x_3$ or $x_4$ is *always* better than one of its parents. This means that there is a 100% probability that one of the resulting variables is better than one of its parents, and 50% probability that one of the children is better than both. Under these circumstances, the adaptive plan is optimal! The only expense is that one child is definitely poorer than one of the parents, and that there is a 50% probability that it is worse than both. This reduces the effectiveness of the adaptive plan, *but it remains exponential*! The crossover operator therefore "sacrifices" one solution in order to create an exponential search on the slopes. If the selection of increasingly successful variables is taken into this equation, the following patterns emerge:

1. Crossover concentrates more and more on successful slopes
2. Selected parents start to flock more closely together, thereby narrowing down the distances between parent and child as hill-climbing progresses

Just like mutation, a crossover operation such as normally is carried out in a GA is mix of a random search strategy and an optimal search strategy, although the optimisation is much stronger and enduring than that of mutation.

In this view, it is fundamentally incorrect to associate a GA with bandit problems. A bandit problem consists of multiple strategies that do not influence each other, except for decisions made by the plan to select strategies at a certain stage. A GA on the other hand does more than this, as it actually swaps variables amongst the strategies. A good solution found through mutation is passed to crossover or selection strategies for further processing. In bandit terms, this is a bit like finding the lucky *coins* that causes the most wins (provided the coin is returned when the bandit cashes out). And even then, as was argued earlier, mutation and crossover strategies are mapped to hill-climbing and random search strategies.

## V. CONCLUSIONS

The hypothetical adaptive plan introduced in this paper offers a novel, although maybe disenchanting, perspective on the success of adaptive plans. A practical search process is considered to be a mix of random search (variation) and hill-climbing (optimisation), and its success as its ability to find better variables during the search process. This notion gives support to the many experimental comparisons between various search processes, that prove that there are no supreme universal problem solvers (no free lunch theorem [7]).

Specifically for genetic algorithms, the crossover operation, and to a (much) lesser extent mutation, is a very effective combination or random search and hill-climbing.

Understanding this, and especially further optimising the neighbourhood schemes in current GAs, may help to improve these implementations. The argumentation laid out in this paper can also be applied to other kind of search strategies that can not be defined in terms of genes or chromosomes.

## REFERENCES


[1] Holland, John H.,"*Adaptation in Natural and Artificial Systems*", MIT Press 1975, revised version 1992

[2] Rudolph, Günter, *"Reflections on Bandit Problems and Selection Methods in Uncertain Environments"*, University Dortmund, 1997

[3] Michalewicz, Zbigniew,*"Genetic Algorithms + Data Structures = Evolution Programs"*, Springer Verlag 1999

[4] Koza, John. R., *"Genetic Programming",* MIT Press 1992

[5] Bä ck, Thomas, *'Evolutionary Algorithms in Theory and Practice"*, Oxford University Press 1996

[6] Woodward, J. *"GA or GP. That is not the Question"*, CEC 2003

[7] Igel, C. & Toussaint, M., *"Recent Results on No-Free-Lunch Theorems for Optimization"*, Institute für NeuroInformatik, Bochum, 2003

# A Morphosyntactical Complementary Structure for Searching and Browsing

M. D. López De Luise - mlopez74@palermo.edu
Department of Informatics Engineering , Universidad de Palermo University
Av. Córdoba 3501, Capital Federal, C1188AAB, Argentina

***Abstract*—This paper is a proposal for the construction of a pseudo-net built with precisely defined tokens describing the content and structure of the original WWW. This construction is derived by morphosyntactical analysis and should be structured with a post-processing mechanism. It is provided also an in-depth analysis of requirements and hypothesis to be stated to accomplish with this goal.**

**An in-depth analysis of requirements and hypothesis to be stated to accomplish this goal is also provided. Such derived structure could act as an alternate network theme organization with a compacted version of the original web material. This paper does not describe nor study the post-processing approaches. Instead, it is posted here as a future work. A statistical analysis is presented here with the evaluation of the understanding degree of a hand-made structure built with some tokens derived under the hypothesis presented here. A comparison with the keyword approach is also provided.**

***Index Term*** — Web browsing, Web-Mining, morphosyntactical analysis.

## I Introduction

The special features of the WWW sometimes make it hard, if not almost impossible, to retrieve the exact information searched. Sometimes the results of a search activity include many syntactical matches and even semantic matches that are not related with the actual user searching. There are several well-known solutions for the retrieval activity when searching for specific information. Such solutions normally try to put in context a specific keyword (or a minimum set of them) in different ways: with sophisticated indexing methods, smart visualization alternatives, etc. In this paper provides another point of view: instead of studying the best way to locate and extract specific information, the way to filter and structure some *representative portion* of the content adding the associated site and web location is studied.

This paper presents a proposal (let us name it EC for *Estructura Complementaria* in Spanish, complementary structure in English), a morphosyntactical structure derived from the web that should provide support to certain searches. In this paper it will be shown that it is possible to construct an organized support structure to provide a suitable view of the same data without loosing meaning. This new organization of the data has components with good representation of the underneath information in the top level. Paradoxically, it will be shown here that this morphosyntactical approach can provide help in the construction of an alternate organization with good semantical representation. A set of components, parameters and minimal behavior (i.e. functionality) required to make an implementation of this proposal is presented. The management mechanism of this new construction will not be studied. The detailed structuring algorithm and browsing/querying are not described here.

The remainder of this paper is organized as follows: section II presents some background, related work, the constraints to be met by this proposal and its justification; section III describes the proposal; section IV describes a preliminary test, section V makes a critical analysis of the conceptual strengths and weaknesses of the structure and hypothesis and conclusions; and finally section **VI** states the main work to be done.

## II The proposal constraints

As stated earlier, EC is a proposal that depicts some representative portion of the web information and reorganizes it in any other way. It is hard to define the meaning of the word *representative* as it was used in the previous section, but it is the first step to provide a better understanding of any of the requirements stated here.

According to [2] one meaning for the word representative is serving to represent. Therefore, the requirements to be met in order to accomplish the previous definitions are:

I-provide an alternate way to present the same information.
II-provide a compact way to represent more complex and extensive information.
III-provide an alternate structure with the same information.
IV-be the basis for different visualizations of the same information.

In order to work with these requirements and some other derived from the special characteristics of the WWW[1], this proposal has been founded on the following hypothesis:

- The language reflects mental structures that depict a relationship between concepts in a precisely enough way.
- Most part of a language is composed of words or sets of words that represent any kind of objects.
- The contained information in any actual expression of the language can be represented with a number of EC components with a precisely defined alternative structure [14] (this hypothesis satisfy the requirements (II) and (III)).
- There are a finite number of objects in the real world to be represented.
- The searching and mining is user /language dependent (this hypothesis is related to requirement (IV)).
- Each document in the WWW is a kind of conceptual unit.
- There are unlimited resources to process (time and memory).

Furthermore, the following requirements have been stated:
- The EC components must be related to WWW.
- EC must represent implicit information (related with requirement (I))
- A self-adaptive structure.

[1] any of these special characteristics: heterogeneous, very large, dynamical, etc.



*T. Sobh and K. Elleithy (eds.), Advances in Systems, Computing Sciences and Software Engineering*, 283–290.

In the following subsections a more detailed justification for each one of these hypotheses is provided.

*A The language reflects mental templates that depict a relationship between concepts in a precisely enough way*

In this proposal the regularity is learned as a structure, which is afterwards clustered automatically without taking into account Chomsky's analysis for generality, precision and completeness. In the actual proposal the derived structures may lack of many, if not all these problems, but this is precisely the way the author handles the divergence between sentence grammaticality and its acceptability.

Noam Chomsky [12] defines the grammar as a product of a machine functioning in an intermediate point of precision between a Markow machine (strongest condition) and a Turing machine (weakest condition). Chomsky labels this as transformational grammar because it is founded in a set of components: set of morphophonemic rules (for phonological components), the phrase-structure and transformational rules (for syntactical components). Each of these components consisted of a set of rules operating upon a certain input to yield a certain output. In his proposal he was centered in syntactical problems and left out the semantic analysis.

Despite the fact that the natural language does not match exactly with the grammar, it has a deep correspondence with it. This hypothesis regards this correspondence to take its regularity as the basis to extract some structures represented by the actual expressions of the natural language.

*B Most part of a language is made of words or sets of words that represent any kind of objects*

This hypothesis takes the assumption that there is a limited set of objects to be modeled from the real world that is handled indirectly within the data.

Chomsky [13] also explains that the language can make infinite use of finite elements (for this infinite use he takes off the traditional concept that the natural order of the thinking is the same as the order in a sentence). These elements are concrete or abstract objects that constitute an internal representation learned from the outer reality.

*C An alternate structure of the same information can be generated*

This paper takes Chomsky's [12] concept of a subjacent regularity within any grammatical construction and makes a morphosyntactical analysis approach (morphological and syntactical because this proposal is based in words, symbols and the syntaxes related to them) to generate this alternate structure.

The need of alternate structuring and visualizing data can be thought as a consequence of the DB nature of the WWW [21] and makes it possible to help in the searching process [22].

Several proposals [17] have been made to present alternate structures for the web content some of them by metadata manipulation. Some alternatives for such a manipulation are hierarchically structuring of metadata clusters (SenseMaker [15]) for later querying; visually organization of metadata for browsing (Flamenco [16]), Semantic Web construction [18] followed by a Deep Web concept [19], etc.

Lawrence & Giles [1] make a good description of how hard and risky is the task of retrieving information from the Web and the organization and presentation of the web information to a user. This is also true for the semantics approaches. Alan D. [10], Gä rdenfors [11], Lesher [20] among others, remark and study some inherent difficulties with the concept of ambiguity, inconsistency and incompleteness that come up with sentences semantically manipulated. Several proposals have been made to manage semantics and its related problems from the Web.

*D There are a finite number of objects in the real world to be represented.*

This study takes part of this concept: each speaker manages a limited quantum of the objects from reality. This domain of objects is something represented through a limited and numerable set of precisely defined internal elements.

Carlos Peregrín Otero explains [35] that the sentences of a native speaker are related to his intelligence and history. The individual life is too small to be capable to learn each possible sentence exhaustively. Instead, a genetic program enables the learning of the subjacent regularity. These regularities are applied to a set of limited objects of the surrounding reality thanks to natural brain creativity. This creativity is based on certain apprehended rules: there isn't creativity without regularity.

Chomsky also defines a language as a set of statements each one with a finite length and built from a finite set of elements. This, of course, is applied to both artificial and natural languages.

*E The searching and mining is user/language dependent*

This proposal does not intend to define a visualization mechanism but it has to set the basis to minimize undesired restrictions to its construction.

There are many languages (Spanish, English, French, Japanese, etc.) to be processed by EC, but they must be translated to a common internal language. Browsing the Web could also be hard and basically depends upon the user. This is especially true if the client is not familiar to web activities. Furthermore, the heterogeneity, extent and complexity of the information saved is not helpful.

Many papers have detected and proposed alternatives to solve these problems. Just to mention a few: map visualization and browsing [7], an encapsulating layer that presents a personalized model of the data [23], a visual interface dynamically built with the navigation information [24], to filter customer preferences activating a software agent that learns from user navigation [25], adaptive design based on user profile filtering within a framework [26], to apply Markow chains on user navigation mining data for navigation prediction [27], to show brief information in an affordable visualization the extracted patterns by clustering over the remaining users' web mining [28], to visualize different SOM elaborations from data [29], [30], etc.

*F Each document in the WWW is a kind of conceptual unit.*

This hypothesis has been stated in order to reduce the complexity of the information structure in the Web, making it easier to model the information. It can be thought as a discard of the analysis for the relationship among sites, delimiting it to each site at a time.

*G There are unlimited resources to process*

To keep the focus on the main logical analysis, there will not be any hardware restriction considerations. They will be studied deeply in future works.

*H An alternate way to read the same information*

This proposal is supposed to support a mix of browsing and querying approaches to avoid the typical problems mentioned: it will browse over a set of possible short answers and to query the result of a statistically high probabilistic browsing activity.

From the traditional focus there are two main approaches: (1)browse within a hierarchical structure or (2)querying for specific information. When browsing is performed to search information over a structure, there is typically a set of representing keywords or visual tokens in the sense of (1) and (2) as in the Scent Trails [3], WebEyeMapper[4], GH_SOM[5], Narcissus [6], BibRelEx[8], etc.

But browsing a structure brings up the problem of how to make such structure and when to refresh the contained information. Some papers have studied the resulting impact on the repeatability and success of the navigation process by a user [7], [9]. Some other research have studied the relevance of the query formulation and interpretation ([1], [10], [11], etc.) in the precision and recall metrics.

*I The EC components must be related to WWW.*

As this proposal intends to be an alternate and compact formulation of the data saved in the WWW, there must be links that allow reaching the original data from the actual data in EC.

*J EC must represent implicit information*

According to requirement (I), the EC components must be extracted from the WWW and metadata, in such a way that makes them a real representation of the implicit information.

*K A self-adaptive structure*

The social evolution of the mankind can be easily shown [12], and therefore the consequent language evolution (as it is said to be dependent from the social and biological evolution).

Although many proposals have studied the adaptation to the web content ([6], [29], etc.) as it is a dynamical storage, they do not explicitly consider the language evolution.

## III The structure

The Fig. 1 shows the general disposition of the EC proposal as a whole and its relation with the WWW and the users.

The EC uses the data and metadata that are extracted from the actual WWW to feed the internal structure layer wich generate a virtual structure layer. The visual structure layer can be used to get information or browse it.

The three-layered structure of EC is a consequence of the three main activities to be performed:

*-Internal Structure*: gets data from the WWW and process it to derive certain Homogenized Basic Elements streams (HBE). It also solves the gathering, formatting and updating of the raw data and metadata from the Web These steams constitute some kind of translation from the actual language to an internal language. The internal language should be defined to overcome specific language problems. This layer should provide a uniform language vocabulary. It can be thought as an interface between the actual WWW and the Virtual Structure.

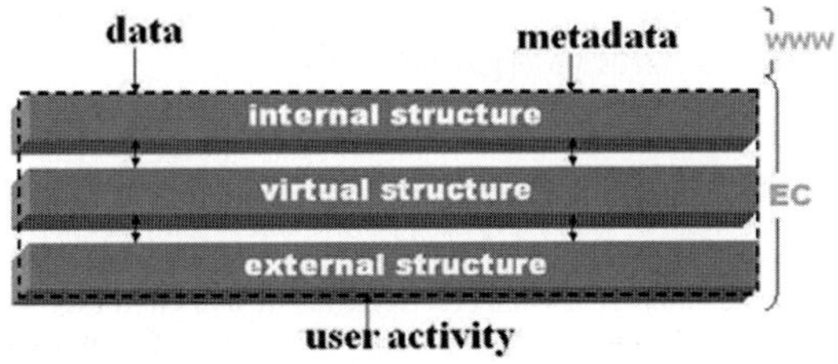


**Fig. 1. Three layered structure of the EC. Data and metadata are extracted from the WWW.**

*-Virtual Structure*: it analyzes the HBE, detects known regularities or learns new prominent regularities in the stream and organizes these regularities in compact structures (let us name it $E_{ci}$ for the name in Spanish *Estructura de Composición Externa*, External Composition Structure in English, a structure that reflects the closeness of the relationship between words within a phrase). These compact structures are also processed to set a hierarchical structure that point out its corresponding $E_{ci}$ and the original WWW location. This layer should provide a virtual view of the modeled data to the visual structure. This is performed by a set of organized elements labeled here as $E_{ce}$ (for the name in Spanish *Estructura de Composición Externa*, External Composition Structure in English, a supra-structure composed by a set of $E_{ci}$ structures all of them related to the same text. Such structure is intended to establish the way two o more $E_{ci}$s are related and reflect the main words they are built-up). It is the main part of this proposal.

*-Visual Structure:* to browse or query the virtual structure content as the user information requires it. It should provide alternate models for textual and/or graphical interaction with the user. It should be designed to handle elements like icons, flashing texts, hypertexts, links, etc., to make easier the Information browsing and searching activity. It can be considered as an interface between the Virtual Structure and any user.

This paper is intended to set the main Virtual Structure characteristics, as the other two layers can be thought as interfaces to and from this structure. Thus, it is important to define the Virtual Structure precisely.

*A Internal Structure*

Basically it processes any data from the WWW taking its specific language as a set or words, numbers and symbols. The specific activity will depend on the language and any language specific activity learned. For instance, one o more words may constitute a HBE, one o more symbols could also be a HBE, etc. (see Fig. 3) The specific behavior should be based on a set of learned rules to let the contents evolve with language. Despite the fact that this is not the central point of this work; a statistical analysis for a reduced set of words will be introduced in a next section as a background example.

*B Virtual Structure Description*

As stated before, this layer takes a number of HBE streams as input. Then it makes a special processing to generate a set of structures labeled here as $E_{ci}$.

Some components and their relation are shown in Fig. 2.
MC: (for the name in Spanish *Motor de Composition*, Composition Engine in English, is the $E_{ci}$ -factory) receives from an Internal Structure a set of HBE streams and makes a set of $E_{ci}$. This set of $E_{ci}$ can be thought as a sub layer within the Virtual Structure layer.

MA (Assimilation Engine) takes a set of $E_{ci}$ from MC and makes a set of $E_{ce}$ structures. Again, this set of $E_{ce}$ can be thought as another sub layer within the Virtual Structure.

Each $E_{ci}$ has derived from an individual statement. Each $E_{ce}$ is composed by one or more $E_{ci}$. The way $E_{ci}$ are generated depends on the content of the actual HBE stream. The $E_{ci}$ can have one of a number of predefined structures. When MC performs the $E_{ci}$ construction it determines the structure according the presence or not of some special HBEs.

The $E_{ce}$ is generated as a multi-layered structure where the lower layer is created by a sort of association of the $E_{ci}$-layer elements. The type of structure and the content of the $E_{ci}$ itself determines the nature of this association. The upper-layers are produced as a progressive factoring out of $E_{ci}$ elements.

The actual algorithm for MC and MA can change automatically. The way this is performed will be detailed in the next section.

*C Updating process*

To update information is not always an easy task in the web. For the metadata derived from this updated data, the consequence is the immediate obsolescence. Updating an EC data should be activated by any change in the original WWW.

In this proposal there is a kind of regulation mechanism for the Virtual Structure layer. This mechanism is implemented by the following main components (Fig. 4):

-A set of Thresholds:

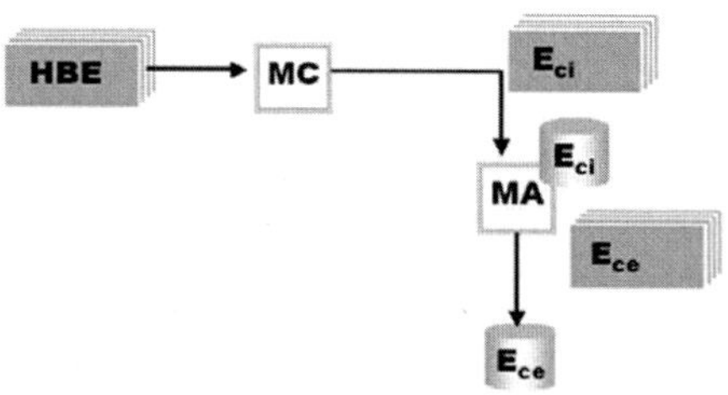


**Fig. 2. MC process HBE inputted and generates sets of $E_{ci}$. Afterwards MA makes a set of $E_{ce}$.**

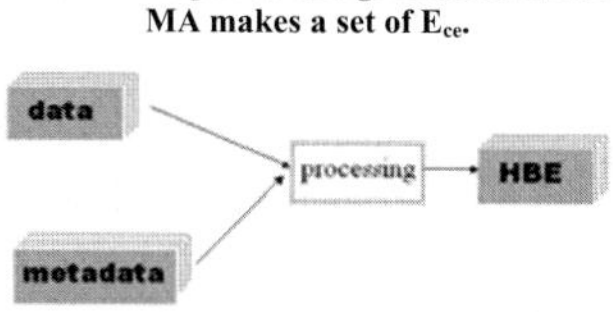


**Fig. 3. Translation from any language to an internal language**

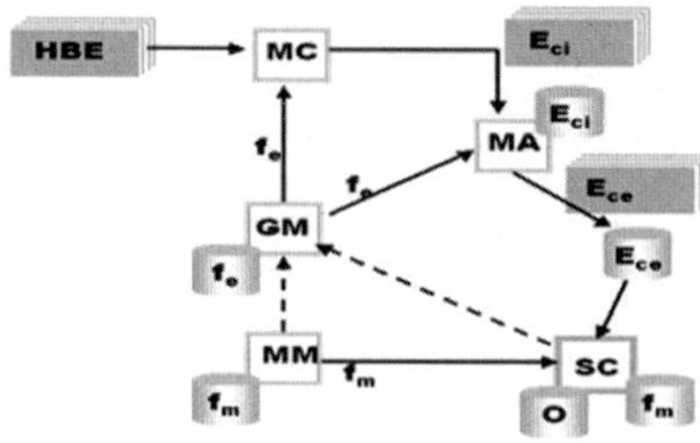


**Fig. 4. The updating mechanism.**

These thresholds are three and have correspondence with the three levels of processing within EC. The first one is labeled $O_{eci}$ and is related to the $E_{ci}$ sub layer. There is a $O_{ece}$ related to the $E_{ce}$ sub layer and finally there is an O related to the global Virtual Structure layer as a whole.

-A set of metric functions (fm):

The functions *fm* (for the name in Spanish *Función de Medición*, Metric Function in English) state a one-to-one relation with the thresholds O, $O_{eci}$ and $O_{ece}$. Therefore there are three of such functions: $f_m^S$, $f_m^{Eci}$, $f_m^{Ece}$. They return a numerical magnitude of the relevance of changing the portion of EC that is being revisited. This relevance is some kind of distance between the actual configuration and the corresponding threshold that estimates if EC is the best to reflect the data in WWW.

-A metric functions generator (MM):

As EC works, the actual fm evaluates a sort of distance between the $E_{ci}$ and the actual $O_{eci}$. The metric function generator can change the way it makes the evaluation. This change will take place when a controller system finds:

$$\mathbf{P(| f_m^{Eci}(O_{eci}, E_{ci})| > d)}\ , \qquad (1)$$

If this probability is high enough, then a new suitable $f_m^{Eci}$ function is generated.

-A set of effect-functions and its generator engine (GM):
This set composes the MA and MC functions. Part of the selected subset will constitute the set of temporary abilities for them. These sets of functions are supposed to perform the best suitable activity at that moment. Thus the $E_{ci}$ and $E_{ce}$ processing will change following the actual Threshold values. This set of functions is the core of the activity. The GM engine (for the name in Spanish *Generador de Métricas*, Metric generator Engine in English) makes any change in this set. Such a change could be a deletion, a factoring-out or a modification of any component of the actual set.

For MC: these functions could be learned HBE rules to consider the probability of certain type of structures and rules for reducing, classifying and sorting a stream. For instance, a set can state that if the HBE1 is present in the stream then the structure should be of type 12. But if the HBE500 is also present, then the structure should be of type 5. Let us say the stream is:

HBE2 HBE1 HBE67 HBE20

Let us say also the structure of type 12 requires each of the following HBE following it to be structured as shown in Fig. 5.

For MA: These functions could be a set of learned rules for $E_{ce}$ generations. For instance they could establish that two identical HBEs in different $E_{ci}$s can be linked as shown in Fig. 6.

-A Controller System (SC):

The SC (for the name in Spanish *Sistema Controlador*, Controller System in English) is a kind of controller that evaluates $P(|\ f_m^{Eci}(O_{eci}, E_{ci})| > d)$ and fires the GM if it is true. It has also the ability to change one o more thresholds, depending on the range of $f_m^{Eci}$.

When automatic updating is fired, the specific $f_m$ states if it makes sense to update any portion of EC (it could affect one o more $E_{ci}$). The updating of one or more $E_{ci}$ cascades up to $E_{ce}$ structures. The same could happen if one o more $E_{ce}$ should be updated.

### D *Internal Structure as interface*

To enable the construction of the $E_{ci}$ and $E_{ce}$, this layer should solve at least the following problems:

-Translation of text numbers and symbols.

-Avoid translation of less significant words, symbols, etc.

-Consider some kind of special internal word for reflecting the structure of the web as extra information. Cases of such data are: links, hyperlinks, sound, images, etc.

-Consider some kind of processing and later translation of main characteristic for images, sound and video.

-Consider a good translation for ambiguities, contradictions and missing information so as not to miss them as they are considered here as information too.

Finally, note that one o more HBE would represent one o more real world objects. There is no difference in the processing of abstract and concrete objects. The representation problems are solved with the traditional language artifacts.

### E *Visual Structure*

The visual structure should be able to search and/or browse the layered $E_{ce}$ structure for information. The preservation of the references to the actual Web should provide the facility to work on the Web directly if it is needed or desired.

The set of $f_m$ and $f_e$ should select prominent data to be promoted as HBE within upper $E_{ce}$ layers. Less important data should be part of the $E_{ci}$ level.

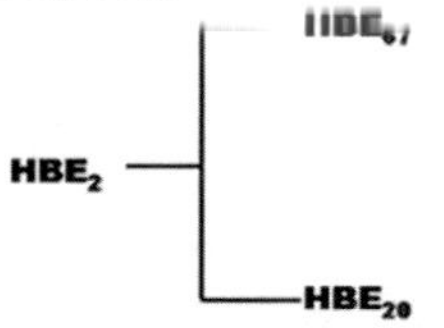


**Fig. 5. A visual structuring of a stream.**

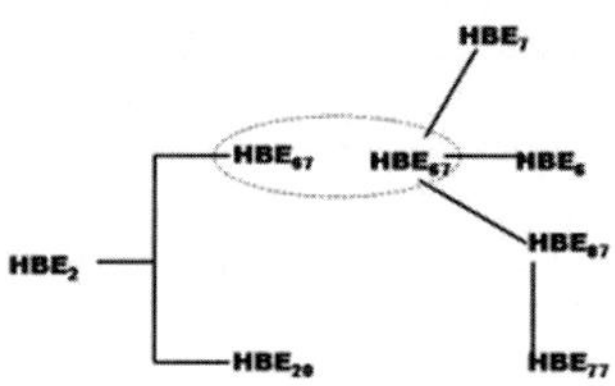


**Fig. 6. Two $E_{ci}$s with identical HBE**

As stated before, the visual structure should not have any further restriction if the resulting EC meets the expectations.

## IV Preliminary Testing

To find out the possibility of processing $E_{ci}$s as one of the complementary alternatives for keywording a poll has been made in Buenos Aires (Argentina), in the Universidad de Palermo, and in a private clinic. All the volunteers were native Argentine, Spanish speakers.

A questionnaire with six items had to be answered by 44 volunteers. The items were grouped in two subjects:

a) The four initial questions were related to a decease called *lymphedema*, which has the special characteristic of requiring a long medical treatment. Therefore, patients with this decease usually acquire a pretty good understanding of the lingo and learn some of its related medical information.

Afterwards, the polling was developed with three main groups of volunteers:

-Patients

-Health professionals with deep knowledge of the decease

-Other (mainly informatics students, for better evaluation of part of the test)

An $E_{ci}$ was written instead of a text titled *Normas de prevención de linfedemas* (*lymphedema* prevention). The Original text was never included in the form. Therefore the only extra information was the $E_{ci}$. It is important to note that the related processing for the $E_{ci}$ is not an optimal one, instead it is the result of a first analysis to approximate this analysis.

b) The two lasting questions were about a topic related to laws. As none of the volunteers had proved to have knowledge of this subject, it was selected to test how good could the inference mechanism be with a few words. These few words were the top level words (the *pointer-words*) from an $E_{ci}$ made with an original lawyer's text.

Following are some of the results obtained.

### A *Performance of the keyword selecting vs. noun selecting*

The volunteers were showed a drawing of one $E_{ci}$ and then asked to write three questions they would like to answer with the original text hidden. They were also asked to write a set of single words to replace each of the questions written before.

To make measurable the relation between writing question efficiency and the keywording, a special processing was done for the questions: the nouns were counted as keywords. In case of ambiguity, the word was also processed as a noun. Afterwards the number of matches between nouns and $E_{ci}$ words were counted. The results were compared with the number of keyword that matched the $E_{ci}$ words.

The Table 1 shows the general data analysis for each set of the noun matches (nm) and keyword matches (km). The Skewness is almost zero showing the typical Normal distribution symmetry around the Mean value. But the negative Kurtosis value denotes a tendency to flat the distribution.

It could be said that the Mean number of matches is higher for nm approach even considering its worse nm value (Mean – Confidence Level) against the best km value (Mean + Confidence Level) with a probability of 0.95.

This tendency is sustained by the Median and Mode values. Conversely, the Standard error, Standard Deviation (and of course the Sample Variance) although comparable, are higher for nm. It should be studied if this deviation could be lowered by improving the algorithmic behind the $E_{ci}$ construction, probably by studying deeply the best set of rules to be applied to words. In the following section some results will be shown, that could be thought as part of this analysis as well.

*B Performance of keyword and nouns by knowledge*

The same performance comparison as the previous section was performed for each of the three main subsets of the population (see Table 2, Table 3, Table 4).

The Mean value for all the sets is always high for nm. This indicates that the average number of matches is higher in all the cases. The same happens with the Median and Mode values.

As a counterpart, the standard error is higher for nm, but the difference narrows as the Count value increases (the number of individuals in the set). This could be the side effect of a small population and should be studied with larger sets to confirm or not the tendency. The Sample Variance (and of course its square root, the Standard Deviation) also presents a behavior similar to the Standard Error.

But in this case the corresponding values in the Table 4 are definitely lower for nm samples.

From the Confidence Level (set as 95.0 %) that for the first set worse value (Mean - CL) for nm is higher than the better value for the km (Mean + CL). For the other two sets these values have a small overlap.

TABLE 1
NM VS KM DATA ANALYSIS

| | nm | km |
|---|---|---|
| Mean | 0.619 | 0.404 |
| Standard Error | 0.043 | 0.036 |
| Median | 0.667 | 0.417 |
| Mode | 0.333 | 0.167 |
| Standard Deviation | 0.285 | 0.242 |
| Sample Variance | 0.081 | 0.059 |
| Kurtosis | -0.935 | -0.812 |
| Skewness | -0.021 | 0.081 |
| Range | 1.083 | 0.889 |
| Count | 44 | 44 |
| Confidence Level(95.0%) | 0.086 | 0.074 |

TABLE 2
DATA ANALYSIS FOR THE PATIENT SET

| Patients statistics | nm | km |
|---|---|---|
| Mean | 0.806 | 0.306 |
| Standard Error | 0.090 | 0.058 |
| Median | 0.833 | 0.333 |
| Mode | 1.00 | 0.333 |
| Standard Deviation | 0.270 | 0.174 |
| Sample Variance | 0.073 | 0.030 |
| Kurtosis | -0.535 | -0.768 |
| Skewness | -0.551 | -0.725 |
| Range | 0.833 | 0.500 |
| Count | 9 | 9 |
| Confidence Level(95.0%) | 0.208 | 0.134 |

TABLE 3
DATA ANALYSIS FOR THE PROFESSIONALS SET

| Patients statistics | nm | km |
|---|---|---|
| Mean | 0.442 | 0.338 |
| Standard Error | 0.088 | 0.064 |
| Median | 0.333 | 0.250 |
| Mode | 0.333 | 0.167 |
| Standard Deviation | 0.305 | 0.222 |
| Sample Variance | 0.093 | 0.050 |
| Kurtosis | 1.912 | 2.376 |
| Skewness | 1.337 | 1.613 |
| Range | 1.083 | 0.722 |
| Count | 12 | 12 |
| Confidence Level(95.0%) | 0.194 | 0.142 |

Finally, from the Kurtosis and Skewness the population behavior is almost a normal distribution. This is not true for the set in Table 3, perhaps due to the fewer number of samples.

Conclusion: for patients with some knowledge of the topic it has a notable better performance from nm than trying to select a keyword. This is true also for the people with no knowledge in the subject but with less difference with the km. The km sustains its performance in all three sets, but always below the nm alternative.

*C Performance of keyword and nouns by web searching skills*

It was studied from two perspectives: web skills and occupation. The web skill was measured by the quantity of web navigation hours per week. The Table 5 shows the average match for each nm and km alternatives.

Note that each peak in nm corresponds to a decrease in km. Conversely, peaks in km correspond to drops in nm. This behavior should be further studied to determine if it is a markable trend. According to this, there is no evident benefit in having more training in the web. The average value for people spending 31 to 50 hours in the web is almost the same as the value for people spending between 11 to 20 hours in the web. This trend is sustained in the km alternative.

Other external influence when trying to query the web could be the previous knowledge of the user in informatics. It could be thought as other kind of skill. The results apparently confirm this conjecture. Table 6 shows the frequency of nouns and keywords that match with the $E_{ci}$ content. Legend "S" denotes informatics experience.

TABLE 4
DATA ANALYSIS FOR "THE OTHER" SET

| Patients statistics | nm | km |
|---|---|---|
| Mean | 0.638 | 0.499 |
| Standard Error | 0.048 | 0.051 |
| Median | 0.750 | 0.542 |
| Mode | 0.833 | 0.667 |
| Standard Deviation | 0.231 | 0.242 |
| Sample Variance | 0.053 | 0.059 |
| Kurtosis | -0.924 | -0.110 |
| Skewness | -0.585 | -0.749 |
| Range | 0.833 | 0.889 |
| Count | 22 | 22 |
| Confidence Level(95.0%) | 0.099 | 0.107 |

TABLE 5
WEB SKILL INCIDENCE

| h./week | nm | km |
|---|---|---|
| 0-10 | 0.60 | 0.38 |
| 11-20 | 0.51 | 0.42 |
| 21-30 | 0.69 | 0.37 |
| 31-50 | 0.53 | 0.40 |
| >51 | 0.58 | 0.34 |

TABLE 6
MATCHING FREQUENCY FOR NM AND KM

| | nm | | km | |
|---|---|---|---|---|
| Avg. match | S | N | S | N |
| 0-0.2 | 1 | 3 | 5 | 7 |
| 0.2-0.4 | 5 | 5 | 1 | 1 |
| 0.4-0.6 | 2 | 4 | 2 | 4 |
| 0.6-0.8 | 6 | 3 | 6 | 3 |
| 0.8-1 | 8 | 7 | 8 | 7 |
| Count | 22 | 22 | 22 | 22 |
| Total | 44 | | 44 | |

As can be observed from both alternatives (volunteers with informatics knowledge are half of the total samples), there is a similar number of positive matching but a small improvement is obtained for nm in the low scoring range. These results should be confirmed with a larger sample size.

Conclusion: there is a complementary behavior in the matching performance for the two approaches. Navigation training has no high evident incidence in the matching score.

### D Representativeness of $E_{ci}$ information

To evaluate the $E_{ci}$ represents the original content, the 44 volunteers were asked to guess a title for a hypothesized text represented by the $E_{ci}$ structure. The Table 7 shows that almost all the guesses were correct. They were also asked to write down three questions that could be answered with the information in the hypothesized text.

In Table 8 the numbers show a good prediction for the three questions denoted as Q1, Q2, Q3. Note that these questions and title were correct even for those people who did not know about the subject.

Conclusion: the $E_{ci}$ is a good content representation of the original text.

### E Representativeness of pointer-words

As stated earlier some of the words in the $E_{ci}$ are denoted specially by the rules. It is expected that these words (named here as pointer-words) have a strong influence in the $E_{ci}$ representativeness. These special words had also been tested. A new set of $E_{ci}$s were developed from a completely new html page.

TABLE 7
TITLE PREDICTABLY

| % | correct Title? |
|---|---|
| Y | 90.7 |
| N | 9.3 |
| tot | 100 |

TABLE 8
TOPIC PREDICTABLY

| % | Q1 | Q2 | Q3 |
|---|---|---|---|
| Y | 97.73 | 100.00 | 100.00 |
| N | 2.27 | 0.00 | 0.00 |
| tot | 100.00 | 100.00 | 100.00 |

To assure null previous knowledge for all the volunteers, the subject was a new law regulation and its social consequences. Six pointer words were deduced. The polling form asked to guess three titles for this text using these words as the only knowledge about it.

Table 9 denotes that even with these few words, represented here as the column labeled with t1, the main title had a good chance of being correctly guessed. It is interesting to see that the first guessing was always the best one, since the other two titles have a lower probability to be correct.

To evaluate the convenience of adding location information, the same question was repeated after giving the domain of the page. Table 10 shows that it could serve as disambiguation information.

Conclusion: pointer words could be useful as text representation. Therefore they could be a kind of reduced index for the $E_{ci}$ . The $E_{ci}$ also could be a kind of index to the original text. These results should be studied for larger samples for a better verification.

## V CONCLUSIONS

A data treatment approach that relies on the concept that the data can be reduced losing and reordering data was presented. For this treatment a morphosyntactical structure was created. This structure is not designed to handle semantics directly but, the syntax and morphology of the language:

-Eci represents a good representation of keyworking.
-Eci performs well for people indepently of their knowledge on the specific subject
-Eci performs well for people indepently of their informatic knowkedge.
-Navigation training has no high evident incidence in the matching score.
-The Eci is a good content representation of the original text.
-Pointer words could be useful as textreduced index for the Eci.
-The Eci structure could be a kind of representation of the original text.

Remark: Virtual Structure is not a visual structure. Any figure or visual representation in this paper is for better illustration of the proposal.

## VI FUTURE WORK

There is a set of pending topics to be studied detailed:

-An implementation of an automatic adaptation to the dynamic structure and content of the data in the Web.
-The best selection of the words and the rules.
-The internal language, to overcome specific language problems (ambiguities, missing data, contradictions, etc.)
-A good design for the Visual Structure should be defined.
-A suitable treatment for multimedia data. This data should qualify for processing by EC defined components.
-Algorithmic and performance aspects.
-Algorithmic alternatives for generation and refinement of $f_m$ and $f_c$ functions.
-Extension of the HBE translation for special symbols.
-Extension to other languages.
-Performance and evolution of EC components with different languages.

TABLE 9
PERCENTAGE OF CORRECT GUESSES FOR THE TITLE

| % | t1 | t2 | t3 |
|---|---|---|---|
| Y | 88.37 | 76.74 | 69.44 |
| N | 11.63 | 23.26 | 30.56 |
| tot | 100.00 | 100.00 | 100.00 |

TABLE 10
PERCENTAGE OF CORRECT GUESSES FOR THE TITLE AFTER URL INFORMATION

| % | t1 | t2 | t3 |
|---|---|---|---|
| Y | 100 | 100 | 100 |
| N | 0 | 0 | 0 |
| tot | 100 | 100 | 100 |

-Consider avoiding the hypothesis that states: Each document in the WWW is a kind of conceptual unit.
-Consider avoiding the hypothesis: There are unlimited resources to process.

ACKNOWLEDGMENT

The author gratefully acknowledges the contributions of Dr. J. Ale and Prof. M. Bosch for their work on the original version of this document.

REFERENCES

[1] S. Lawrence, C. L. Giles, "Searching the World Wide Web", Science vol 280. Pp 98-100. April 1998.
[2] Merrian Webster Dictionary. http://www.m-w.com/
[3] C. Olston, Ed H. Chi. "Scent Trails: Integrating Browsing and Searching on the Web". ACM Trans. on Computer-Human Interaction, vol 10, No. 3, September 2003, pp 1-21
[4] R. Reeder, P. Pirolli, S. Card, "WebEyeMapper and WebLogger: Tools for Analyzing Eye Tracking Data Collected in Web-use Studies", UIR Technical report UIR-R-2000-06
[5] D. Merkl and A. Rauber, "Uncovering the Hierarchical Structure of Text Archives by Using an Unsupervised Neural Network with Adaptive Architecture", Pacific Asia Conference on Knowledge Discovery and Data Mining (PAKDD2000)
[6] A. Wood, R. Beale, N. Drew, B. Hendley, "HyperSpace: A World-Wide Web Visualiser and its Implications for Collaborative Browsing and Software Agents", HCI'95
[7] H. Chen, A. Houston, L. R.Sewell, B. Schatz," Internet Browsing and Searching:User Evaluations of Category Map and Concept Space Techniques", Journal of American Society for Information Science (JASIR).vol 49.nro 7.pp 582-603
[8] A. Brüggemann-Klein, R. Klein, B. Landgraf, "BibRelEx-Exploring Bibliographic Databases by Visualization of Annotated Content-based Relations", D-Lib Magazine.Vol 5.ISSN 1082-9873
[9] H. Chen, C. Schuffels, R. Orwing, "Internet Categorization and Search: A Self Organizing Approach", Journal of the visual communication and Image Representation.vol 7.pp 88-102
[10] D. Alan, "A Comparison of Techniques for the Specification of External System Behavior", Computing practices 1998.
[11] P. Gä rdenfors, "Meaning as Conceptual Structures", Tech Rep LUCS 40.Lund University Cognitive Studies.ISSN 1101-8453.
[12] N. Chomsky, "Syntactic Structures", Walter De Gruyter, Inc. 1974. ISBN: 9027933855.
[13] N. Chomsky, "Aspects of the Theory of SyntaxMIT Press. 1969. ISBN 0262530074 9.
[14] N. Chomsky et al., "Langue : Théorie Générative Étendue", Hermann. 1977. ISBN 2705658394.
[15] M. Baldonado, "An Interactive, Structure-Mediated Approach to Exploring Information in a Heterogenous, Distributed Environment", Ph.D. Dissertation, Stanford University, December, 1997.
[16] A. Elliott, "Flamenco Image Browser: Using Metadata to Improve Image Search During Architectural Design", Ame Elliott, Doctoral Consortium, in the Proceedings of the ACM CHI 2001.
[17] E. Amitay, "Web IR & IE", http://www.webir.org/
[18] T. Berners-Lee, J. Hendler, O. Lassila, "The Semantic Web", Scientific American. May 17, 2001.
[19] M. Bergman, "The Deep Web: Surfacing Hidden Value", Bright Planet. July 2001.
[20] W. G. Lesher , J. Moulton Bryan, D. J. Higginbotham, "Effects of Ngram Order and Training Text Size on Word Prediction", Dep. of Communication Disorders and Sciences. Univ. of New York at Buffalo.
[21] P.A. Bernstein, "Panel: Is Generic Metadata Management feasible?", Proc/ of the 26th Int. Conf. on Very Large Databases, Cairo, Egypt, 2000
[22] R. Davis, H. Srobe, P. Szolovits, "What Is a Knowledge Representation?", AAAI. SPRING 1993. Pp 17 – 18
[23] J. C. French, E. K. O'Neil, A. Grimshaw, C. L. Viles, "Personalized Information Environments", Poster. Darpa Contract N66001-97-C-8542
[24] A.M. Wood, N.S. Drew, R. Beale, R.J. Hendley, "HyperSpace: Web Browsing with Visualisation", Third International World-Wide Web Conference Poster Proceedings, Darmstadt, Germany, April, pp 21 – 25
[25] A.S. Pannu, K. Sycara, "Learning Text Filtering Preferences", Proc. of the AAAI Symposium on Machine Learning and Information Access (Stanford,CA,USA). 1996.
[26] M. Perkowitz, O. Etzioni, "Towards Adaptive Web Sites: Conceptual Framework and Case Study", Dep. of Computer Science and Engineering. Univ of Wachington. Seattle. USA. Artificial Intelligence 118 (2000) pp 245 – 275.
[27] B. Trousse, 'Evaluation of the Prediction Capability of a User Behaviour Mining Approach for Adaptive Web Sites", Inria - AID Research Group, B.P. - Sophia Antipolis Cedex. France.
[28] I. Cadez, D. Heckerman, C. Meek, P. Smyth, S. White, "Visualization of Navigation Patterns on a Web Site Using Model Based Clustering", Dep. of Information and Computer Science. Univ of California, Irvine. 2000.
[29] K. Langus, "Text Mining with the WEBSOM", Acta Polytechnica Scandinavica. Mathematics and Computer series No 110. 2000.
[30] K. Langus, T. Hokela, S. Kaski, T. Kohonen, "Self-Organizing Maps of Document Collections: A new Approach to Interactive Exploration", Simoudis E. Han J. Fayyad U. eds. Proc of the second Int Conf. On knowledge discovery and Data Mining, pp 238-243.AAAI Pres. Menlo Park. CA. 1996.
[31] L.G. Heings, D.R. Tauritz, "Adaptive Resonance Theory (ART): An Introduction", Technical Report 95-35, Leiden University, 1995.
[32] M. Jaczynski, B.Trousse, "Selective Markov Models for Predicting Web-Page Accesses", University of Minnesota, Department of Vcomputer Science/Army HPC Research Center Minneapolis. 2000.
[33] Britannica online.
[34] S. Kaski, "Data Exploration Using Self Organizing Maps", Acta Polytechnica Scandinavica,. No 82. Dr Tech. Thesis. Helsinki University of Tech. Finland. 1997.
[35] P. Otero, Introd. In "Estructuras Sintá cticas", Siglo XXI editores SA. 1974. Siglo XXI editores SA. 1974. ISSN: 1139-8736

# Remote Monitoring and Control System of Physical Variables of a Greenhouse through a 1-Wire Network†

N. Montoya[1], L. Giraldo [2], D. Aristizábal[3], A. Montoya [4]
Scientific and Industrial Instrumentation Group
Physics Department
Universidad Nacional de Colombia sede Medellín.
AA. 3840, Medellín, Colombia

**ABSTRACT - In this project a design and implementacion of a platform (hardware-software) for control and monitoring, using Internet, the physical variables of a greenhouse such as temperature and luminosity was made; for this a new type of microcontroller, TINI (Tiny InterNet Interfaces) was used, which can be used as a small server with additional advantages as being able to program it with JAVA language and to support a lot of communication protocols, like 1-wire protocol. Due to the form the platform was designed and implemented, this technology could be incorporated with very low cost in the PYMES (Small and Medium Companies) dedicated, for example to the production of flowers. An additional value of the platform is its adaptability in other applications like for example, laboratories, monitoring and control of manufactures, monitoring systems, among others.**

**KEY WORDS: 1-wire, TINI, Greenhouse, Java, Monitoring and Control.**

## I. INTRODUCTION

This article pretends to describe a hardware-software platform developed for remote monitoring and control of physical variables (e.g. temperature, relative humidity, luminosity, among others) of a greenhouse, to bring low cost and competitive solutions to the modern agricultural (or farming) methods.

Its development implements new technologies in electronics and programming languages, like 1-Wire communication protocol [1] and JAVA language.

The system in specific uses a special microcontroller called TINI (heart of the system), which has the ability of being programmable in JAVA language and it's oriented to innovating communication protocols, with big skills and low cost as two of it's principal characteristics [2].

The growth of Internet has motivated the use of systems plugged to the net. There are no longer important isolated systems, actually the connection to the big net it's necessary. New devices and its applications are bound to explode the opportunities given by global communication. The big net it's one of the new development pillars for the great amount of new technologies whose objective is making life easier for men. That's why a platform like the one designed and implemented it's going to help the PYMES in the integration of new applications based on Internet, letting increase their competitiveness in the globalize world.

## II. MATERIALS AND METHODS

The labors were divided in two different development areas (very specific but complementary one to the other); both have the same importance and complexity. These areas are:

- *Hardware development*: the electronics, sensors and controls, the equipment for the correct acquisition of data and the equipment for the delivery of data to Internet for a later interpretation of them in a distributed form.
- *Software development:* The JAVA programming, as well as for the management of the sensors and controls as for the correct delivery and reception of data and their sent to the final user too.

Both areas are related one to each other, so it was necessary a permanent retro alimentation between them as far as the project was being developed.

### *A. Hardware*

*TINI Board*

This is a small server developed by Maxim-Dallas Semiconductor (Fig.1). One of the most important features of this microcontroller it's the compatibility with a lot of communication protocols like TCP/IP (v4 – v6), RS232, 1-Wire, CAN (Controller Area Network), among others [3]. It has 1Mb of storage memory for applications and 512 Kb of RAM memory.

A very useful function of this device is that it's able to serve as a bridge for devices that doesn't have Internet connection in a way that they could be monitored and controlled by web from an anywhere host.

---

† Work made by Scientific and Industrial Instrumentation Group, Universidad Nacional de Colombia sede Medellín.

[1] e-mail: namontoy@unalmed.edu.co
[2] e-mail: lpgirald@unalmed.edu.co
[3] e-mail: daristiz@unalmed.edu.co
[4] e-mail: amontoya@unalmed.edu.co



*T. Sobh and K. Elleithy (eds.), Advances in Systems, Computing Sciences and Software Engineering*, 291–296.

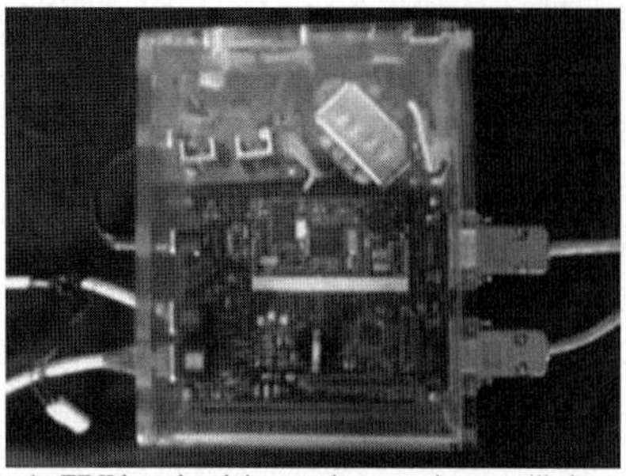

**Fig. 1.** TINI board and the supply power in an acrilic box.

This project uses in specific a special port of the TINI board to drive all tasks of 1-wire protocol, so this device could behave as a master of a 1-wire network.

*1-wire Devices network*

A 1-wire devices network it's an array (generally with bus architecture type) of very special devices that are capable to use 1-wire protocol. In this kind of array, there is a very particular element called master, which is the one that controls the entire network. The master controls all the elements plugged in the 1-wire network (slaves). The slaves are the ones who take the data and bring them to the master, besides, some of them have the capability of making specific tasks like switch devices connected to them and change the state of this devices (on/off) depending of the master orders.

In a 1-Wire network, the flown of data is half-duplex because only one element can transmit data at a time (no matter if it's the master or a slave); besides, only the master is capable of establish communication with the slaves (the communication between slaves it's not allowed).

This kind of network normally reaches a data rate of 14.4 Kbps (Kilo bits per second), could manage more than 100 1-wire elements plugged to the network and has a maximum length of 500 m. on the main line.

*Sensors*

The sensors chosen for this project belong to the family of 1-wire devices of Maxim-Dallas Semiconductor. They have the next specifications:

- They are compatibles with the TINI microcontroller.
- They receive instructions through programs implemented on the master.
- Economics.
- They have an unique MAC direction of 64 bits. This allows an easy identification of any element on the network no matter the quantity or type of devices connected to it.
- They are calibrated and have a very good resolution, this facilitates the implementation of the sensors and makes the measurement reliable.
- They required very few energy to work.
- They use CRC (Checksum Redundancy Cicle) to verify the correct transmission of data.
- They decrease the weight of the network because they only need two cables for its operation: Data and GND.

The 1-Wire sensors selected for this project are: DS18s20 (temperature), DS2450 (voltage, switch), DS2438 (temperature, luminosity, humidity).

*Controls*

There was designed and implemented a special circuit for the elements bound to do the control (switch on/off). This circuit has the possibility of switch voltage signals of any magnitude and form (Fig. 2), this means that it's possible to control a fan which needs 12 V DC to operate, but there it's also possible to control other devices, for example a 120 V AC lamp.

*Software*

The software was developed using JAVA language of Sun Microsystems [4].,[5]. and the latest technologies associated to it, like JSP (Java Server Pages) [6]., Servlets, JFreeChart [7]., TOMCAT; there was also use MySQL as database server [8]. The development tool used in the project was Netbeans 4.1 (of Sun MicroSystems). All of them are free.

The design of the software was made using the corresponding software engineering [9]. and there was used UML (Unified Modeling Language) as a visual support tool.

A special characteristic of the software design it's the carefully use of the powerful properties of Oriented Object Programming (OOP) such as serialization, polymorphism and inheritance.

*Network implementation details*

The developed network is based in a bus architecture with lateral extensions (like the branches of a tree). The entire network was made following all the recommendations of the experts in 1-wire protocol, except in the utilization of two cables for the power supply of some devices. These recommendations include all the schematics for each element of the network and the necessary protections. The exception consists in one modification that was made over two useless lines of the UTP cable (the recommended cable by Maxim-Dallas Semiconductors for a 1-wire network [1].). These two lines were used to send another GND line, and a 12V line that provides the necessary power for the DS2409 (branches of the network) and some others elements that require so (DS2450 and DS2438, for a better A/D conversion).

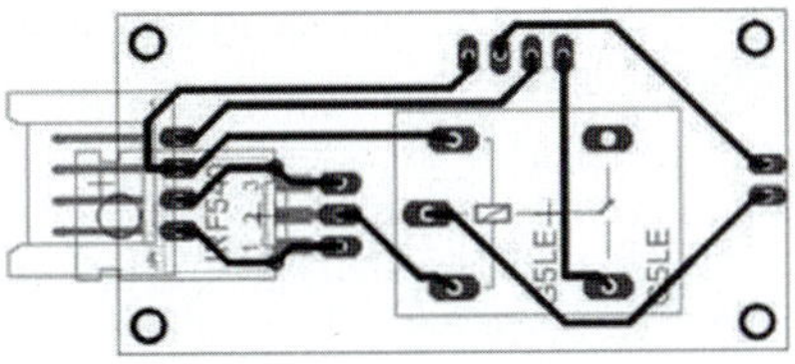

**Fig.2.** Schematic of the circuit implemented to control different devices.

## III. RESULTS

*Hardware*

The 1-wire network implemented in the monitoring and control system prototype has a length of 10 m. with three lateral extensions each one of 3 m. of length. Each lateral extension has three ports for 1-wire sensors (1 m. of separation between ports). Finally, the entire network has a total capacity for 17 elements: 6 DS18s20, 3 DS2409, 5 DS2438 and 3 DS2450. It has to be clear that each DS2450 works as one sensor and as three switches at a time, due to this, the entire network has 14 sensors (the DS2409 are not counted) and 9 control devices (switches. Fig. 3).

*Software*

The first software version for the TINI implements almost all the powerful JAVA tools like serialization, polymorphism and inheritance. The program has the capability of recognize each one of the elements plugged to the 1-wire network, register all sensors in the database (this process is completelly automatic), take data from sensors and delivery them to database. Also, the program implements a small server which is used to receive different kinds of petitions like the register of new elements, the actualization of the elements plugged to the network (renew an array which has the directions of the elements) and some other petitions like change the control devices state.

When a control device state is changed, the systems informs to database about the new state of the control device.

Moreover, the software was designed in a way that it's possible to change the time between data acquisition cycles.

The classes (of the software) responsible of the communication between TINI microcontroller and the server implement "sockets" and "serialization".

For each element plugged to the network, the program creates a serializable object (called "Paquete") which contains all the relevant information that the server needs (date and time of the data acquisition, element direction, acquired data or control device state, among others).

After every data acquisition cycle the data is sent to the server by a socket connection. Then, the information just received is saved on the database.

In general, the system spends about 3 s. to acquire data from a single sensor and to fill the corresponding serializable object Paquete. Therefore, the program takes about 45 s. between consecutive data acquisition (and delivery cycles) for the entire implemented 1-wire network. Figure 4 shows the class diagram of the definitive program implemented on TINI.

*The server*

There are two important parts of the system in the server: the server itself and the database. The server is the one who receives the data sent from TINI microcontroller, deserializes the objects, interprets all the information and saves it on database.

There is also a Servlets and JSP container (TOMCAT) in the server, which manages all the applications in charge of the remote system: the data reception and the graphic interface.

Figure 5 shows a general scheme of the entire system. There may be seen the 1-wire devices network, the 1-wire network master (TINI), the server (which contains the database and TOMCAT) and the final host. As it can be seen, there are three different kinds of network: The network conformed by 1-wire devices in which the network master is the TINI microcontroller, a network settle with TINI

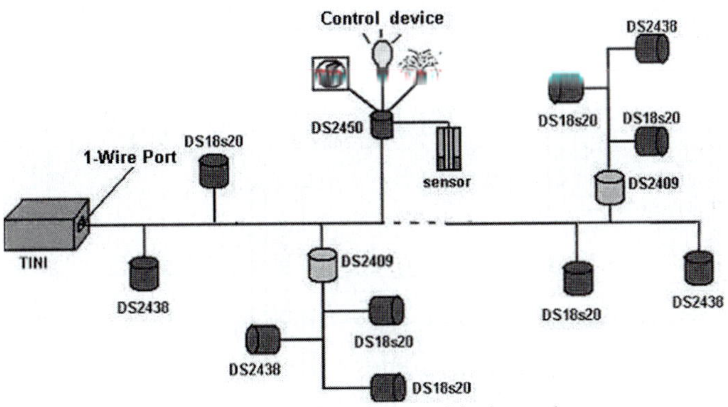


**Fig. 3.** Scheme of the 1-wire network implemented.

microcontrollers (each one of them represents a station, for example, for a flower greenhouse, the stations could be: roses, daisies, among others) in which each one of them communicates with the server and the network in which the final host is placed (Internet).

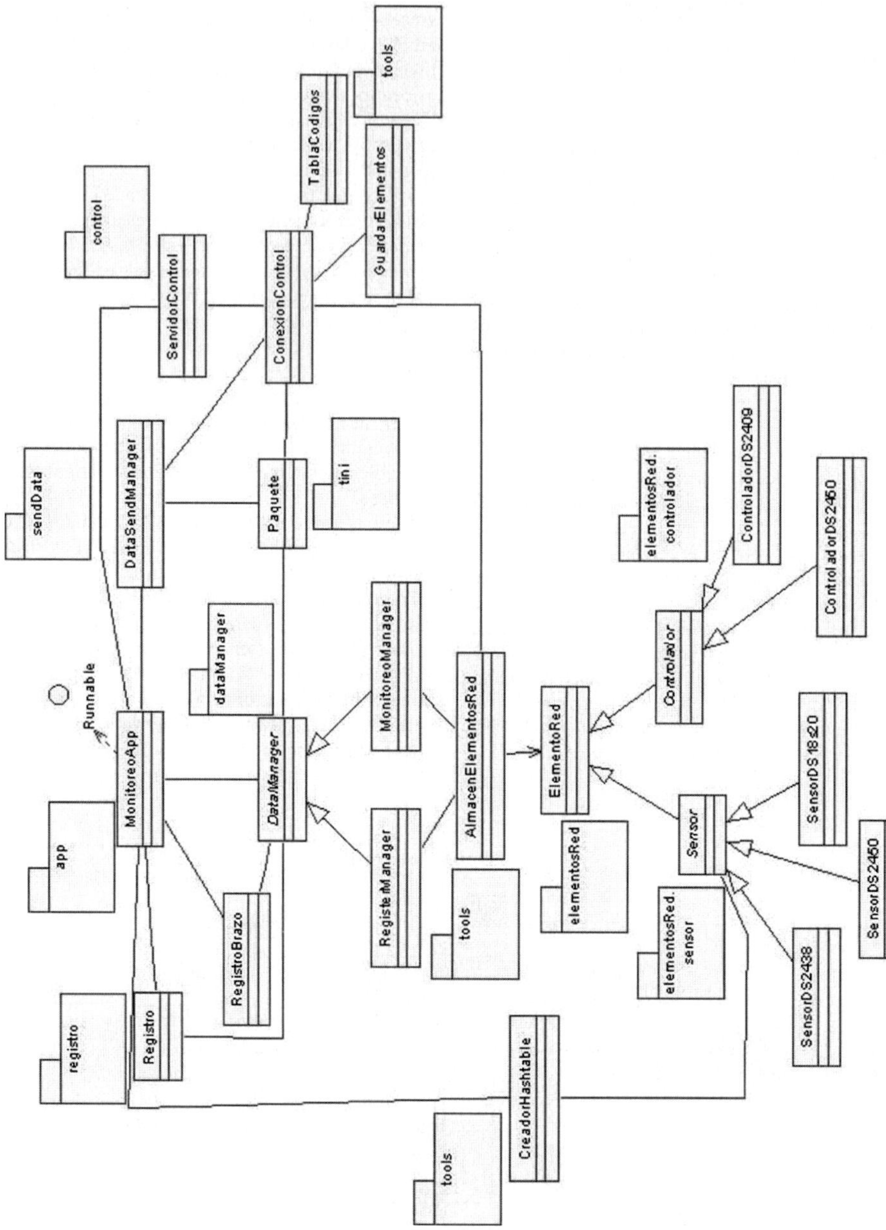


**Fig. 4.** Class diagram of the definitive program implemented on TINI.

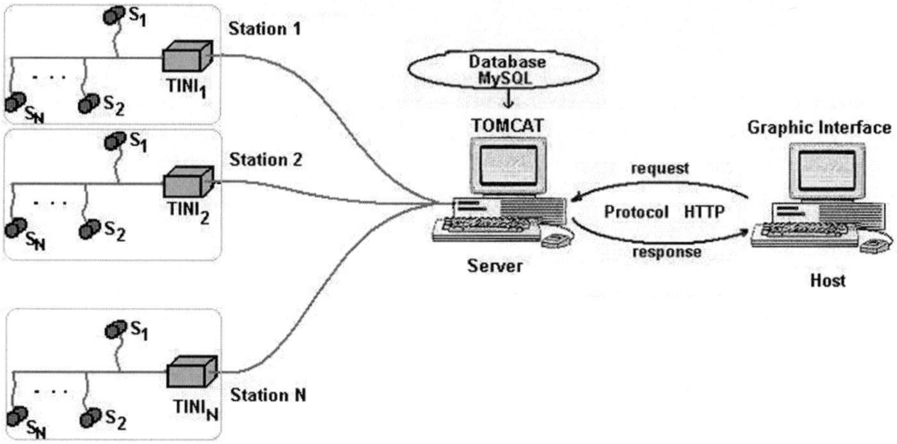


**Fig. 5.** System general scheme.

The final host graphic interface was developed using tools like JSP, servlets and Jfreechart.

The web application (Fig. 6) shows the variables and monitoring results and the corresponding state of the control devices. Then, the user may give instructions about the desired state of certain control device. When this instruction is made, the web application communicates with the server and this one sends the control instructions to TINI microcontroller. Then, when the control instruction is received by TINI, it sends a special command to the control device in which it orders to change the device's state and to immediately inform the new state of the element, so the new state could be registered in database and seen in the graphic interface.

Among the results of the multiple tests made in the completely integration of the system, there was found a very stable operation and performance: The program implemented on TINI registered all the sensors plugged on network satisfactory and the acquired data of each one of them were saved correctly on database. The sensors data were observed on the graphical interface from Internet and three different control devices were controlled (a fan, a lamp and a switch).

The system performance was satisfactory. It worked without interruption for 30 days, time in which 270000 data were saved on database, information which took only 24 MB on the server hard disc.

## IV. DISCUTION AND CONCLUTIONS

One of the most important profits is the management and acquisition of new technology like the TINI microcontroller, the 1-wire elements and 1-wire protocol, this one showed to be very useful in network applications and presented a lot of advantages because it helps the developer with the calibration of the sensors and the recognition of the elements plugged to the network is easier (thanks to the unique direction of each element). This last factor could be a very difficult matter for other kind of networks, for example, a network settle with serial devices.

The great system adaptability allows its incorporation in different kinds of environment like a house, a laboratory, a greenhouse or an industry. Besides, the developed system has the big advantage of being dynamic; this is because it allows the addition or extraction of an element on the network without any inconvenient while the program is still running. The unique system requirement is that the sensor that it's going to be added, has to be registered in a .txt file in which the TINI microcontroller finds out the variable in specific that each sensor measures. This register could be made from the graphical interface.

The incorporation of a database to the system brings an extra value because besides that the system is doing a monitoring of all environment variables, the information is being saved too, so it could be seen an analyzed in the future and allows a very detailed following of the environment variables for periods even of years.

The implementation of the serialization for the communication between the different system modules is highly convenient because it provides a better security; this is due to the fact that there can only be sent and received correspondent objects of the serializable class "Paquete".

The tests results showed that the system has good stability. This it's very important for applications that need an execution of 24 hours, all days of the year.

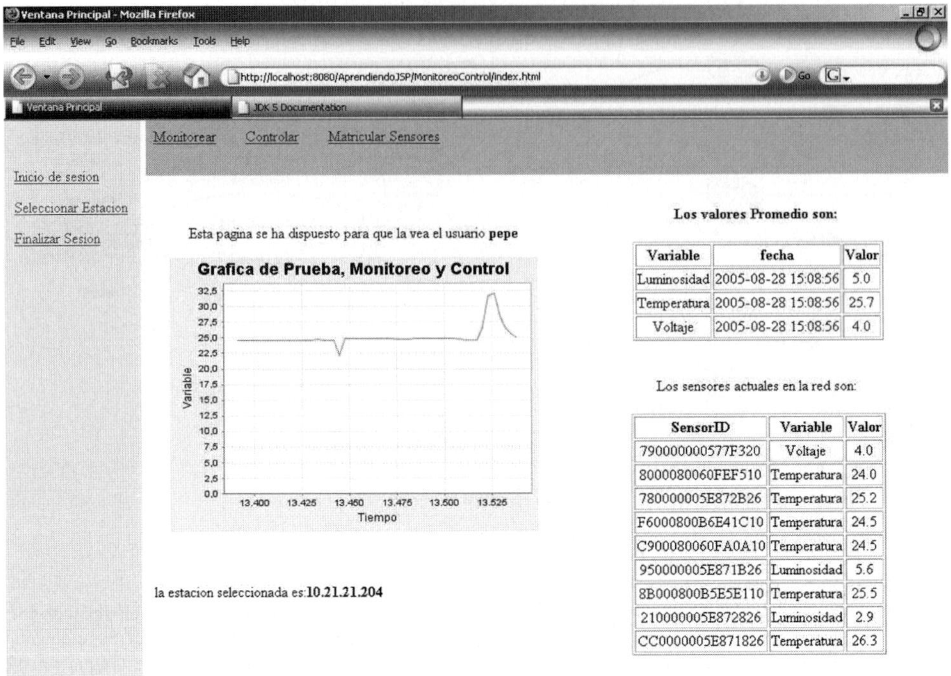


**Fig. 6.** Image of the monitoring and control system graphical interface.

The system was developed using free software tools, this lowed the development cost in a very substantial way, fact that facilitates its implementation in the small and medium companies.

Finally, the system was designed with enough robustness to be highly suitable. In other words, although the implemented network is small, the effort to grown it it's minimum; the reform of the database or software is not necessary, it is just necessary the addition of more stations (or sensors) and in this ones, new 1-wire networks are implemented. There must to be careful of not exceeding the maximum length (500 m.) for a 1-wire network.

## V. DEVELOPMENT FRONTS

The entire system is stable and responds to the initial established demands; however, some new demands rose while the system was developed. This new demands are very important for the future system robustness. These future development fronts are:

- Reduction of data acquisition time (actually 3s. per sensor).
- Improvement of the graphic interface.
- Reduction of the calls to garbage collector on TINI software.
- Reduction of unnecessary creations and destruction of objects.

## VI. REFERENCES


[1] MAXIM - Dallas Semiconductor, *Tech Brief 1: MicroLAN Design Guide*, [web] http://pdfserv.maxim-ic.com/en/an/tb1.pdf, [Last Access, 4 March 2005].

[2] Loomis, D., *The TINI$^{TM}$ Specification and Developer's Guide*, Addison-Wesley, 2001.

[3] MAXIM - Dallas Semiconductor, *TINI: Frequently Asked Questions*, [web] http://pdfserv.maxim-ic.com/en/an/AN1003.pdf, [Last Access, 20 January 2005].

[4] Froufe, A., *JAVA$^{TM}$ 2 Manual de usuario y tutorial*, Alfaomega, 2000.

[5] Franco, A., *Programación en el lenguaje JAVA*, [web] http://www.sc.ehu.es/sbweb/fisica/cursoJava/Intro.htm [Last Access, 20 April 2005].

[6] Tremblett, P., *Superutilidades para JavaServer Pages*, McGraw-Hill, 2002.

[7] Canales, R., *Adictos al trabajo Home*, [web] http://www.adictosaltrabajo.com, [Last Access, 10 Julio 2005].

[8] Van Gelder, W., *MySQL Tutorial*, [web] http://faculty.washington.edu/chungsa/research/technology/mysql/MySQLTutorial.pdf [Last Access, 15 July 2005].

[9] Larman, C., *UML y Patrones, Introducción al Análisis y Diseño Orientado a Objetos*, Pretice Hall, Pearson, 1999.

# Multimedia Content's Metadata Management for Pervasive Environment

Fitsum Meshesha
Addis Ababa University
Addis Ababa, Ethiopia
+251 091 1422944
fitsummk@yahoo.com

Dawit Bekele
Addis Ababa University
Addis Ababa, Ethiopia
+251 011 1222922
da8_bekele@yahoo.com

Jean-Marc Pierson
Lab. d'InfoRmatique en Images et Systèmes d'information (LIRIS)
INSA de LYON
20 Avenue Albert Einstein
69621 Villeurbanne, France
+33 4 72 43 88 97
Jean-Marc.Pierson@insa-lyon.fr

*Abstract*

**One of the major challenges of a pervasive environment is the need for adaptation of content to suit a client's specific needs and choices such as the client's preferences, the characteristics of the client device, the characteristics of the network to which the client is currently connected, as well as other related factors. In order for the adaptation to be efficient while satisfying the client's requirements and maintaining the semantics and quality of the content, the adaptation system needs to have adequate information regarding the content to be adapted, the client's profile, the network profile and others. The information regarding the content can be found from the content metadata. This work addresses the issue of content metadata management in a pervasive environment in relation to content adaptation. For this purpose, a distributed architecture for the management of metadata of multimedia content is proposed. The proposed architecture consists of components for storage, retrieval, update, and removal of metadata in the system. It also provides interfaces to external components through which metadata can be accessed. In addition, it proposes ways to specify, in the metadata, restrictions on the adaptations that may be applied on the content. This enables the content creator to impose constraints on adaptations that may potentially cause the loss of critical information or a decrease in the quality of the adapted content.**

## I. INTRODUCTION

With the advent of wireless communication and the development of mobile computing devices, it is now possible to have access to computing power being virtually anywhere. To make use of this computing power and provide access to information ubiquitously is the subject of pervasive computing [1]. However, as fascinating the idea of pervasive computing is, there are many challenges when it comes to the implementation. In a pervasive environment with many clients with different preferences and with various devices of different nature, it will be difficult to present the same content to all the devices in the same manner. For instance, a client may be using a PDA (Personal Digital Assistant) without graphics displaying capabilities. Another client may be using a powerful laptop computer. Hence, the client with the PDA can not view a document containing an image while the client with the laptop can. In this case, the system can replace all images with appropriate textual descriptions so that the client with the PDA can access the content of the document. In general, content adaptation is one of the major challenges in pervasive environments. And for content adaptation to be successful, the adaptation system needs to have adequate information regarding the content to be adapted and the user/device to which the content is being adapted. The information regarding the user/device can be found from the client profile and the information regarding the content can be found from the content metadata. Hence, both the client profile and the content metadata need to be properly managed so that the adaptation becomes efficient. This work deals with the management of metadata of multimedia content, particularly in relation to adaptation.

The rest of this paper is organized as follows: section II discusses metadata as well as its use and management in existing systems. Section III briefly discusses the adaptation/delivery infrastructure on which we base our metadata architecture. Our proposed architecture for the management of metadata is presented in section IV. Section V presents constraints on adaptation and how they can be modeled in the content metadata. Section VI discusses the implementation of the proposed architecture and section VII concludes this paper and outlines areas of future work.

## II. METADATA MANAGEMENT IN EXISTING SYSTEMS

Metadata can be formally defined as: structured information that describes, explains, locates, or otherwise makes it easier to retrieve, use, or manage an information resource [2]. Usually it is referred to as data about data.

Metadata is useful in providing important information about an information resource without the need to actually access the resource. Especially for multimedia content, metadata provides valuable information. Due to its nature, multimedia content is not



*T. Sobh and K. Elleithy (eds.), Advances in Systems, Computing Sciences and Software Engineering*, 297–304.

suitable for exact-match querying. In this case, metadata of the content can be used to perform queries against multimedia content.

In adaptation systems, metadata can provide the content adapter with information about the content to be adapted so that the appropriate decision can be made. For instance, if a client using a graphic capable PDA with display dimension of 320x240 pixels wants to access an image of dimension 500x500 pixels, the system needs to appropriately adjust the dimension of the image to fit within the display area of the client's PDA. In order to do this, the system needs to know the characteristics of the image such as its height, width, color depth, file format etc. These characteristics of the content are found in the metadata of the content. Hence, the content metadata provides a critical input to the process of content adaptation. Consequently, its proper and efficient management become inevitable.

For instance, in DCAF[1], a service-based adaptation driven framework developed by Girma Berhe at LIRIS[2] - INSA, Lyon, metadata is used by the decision engine (a core component that makes decision on the process of adaptation) to determine whether adaptation is necessary or not, if it necessary, what type of adaptation should be carried out, who should perform the adaptation, and so on. In addition, communication between the client and the system also takes place by making use of metadata. That is, the client's request for content is matched against the metadata of existing content and the system also responds by providing the client with the metadata of matching content [3].

We have used DCAF as a base framework to model our metadata management architecture; hence it will be discussed further in the next section.

Similarly, in the ADMITS[3] project, metadata is used to assist the process of adaptation of multimedia content [4]. In this project, metadata is used to describe variations of a video content, which are results of applying adaptation operations on the video content, and the relationship among the variations and the original content. The metadata also holds adaptation hints that help in the process of adaptation.

However, in both of the above works, the management of metadata is performed as part of the adaptation process. There is no clearly designated component that manages the metadata. In other words, the metadata management is attached to the adaptation process. This does not allow the exploitation of the full potential of metadata. In this work, we propose to detach the management of metadata from the rest of the system. We also propose a metadata management architecture specially for content adaptation purpose. In our architecture, the metadata will be separately managed by a designated set of components and it will be appropriately accessed by the other components by means of a standard interface provided by the metadata manager. In this case, it is possible to independently optimize or replace the metadata manager without requiring a change on the other components. Similarly, the other components can be improved or replaced without affecting the management of the metadata, by keeping the interface between the metadata manager and the other components the same.

## III. THE ADAPTATION/DELIVERY FRAMEWORK

Before introducing our architecture for the management of metadata, let us first introduce DCAF, the adaptation/delivery framework that we used to model the metadata management. DCAF is composed of clients, local proxies, content proxies, content servers, adaptation service registries, adaptation service proxies, and client profile repositories (see Fig. 1 below).

The functions of the various components of the architecture are briefly discussed below. The detailed description can be found from [3].

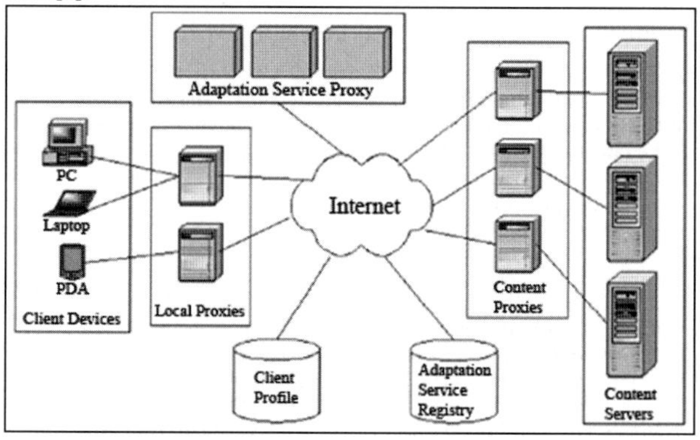


**Fig. 1. Service-based distributed content adaptation infrastructure**

- **Local Proxy:** responsible for retrieving client requests and profiles, analyzing client profile and content metadata, planning adaptation strategy, integrating and forwarding adaptation results to the user, caching of adapted content. One of the major components of the local proxy is the Content Negotiation and Adaptation Module (CNAM) which is responsible for creating an optimal adaptation plan according to the *delivery context* which is generated by analyzing client, content, and network profiles.
- **Content Proxy:** provides access to the content server, i.e. it provides access to multimedia content and content metadata. It also performs caching of content for efficient access.
- **Content Server:** stores multimedia content and the associated content profile or content metadata.
- **Client Profile:** stores profiles of clients. These include user preferences as well as device characteristics.
- **Adaptation Service Registry:** allows the lookup of adaptation services by storing service profile such as service type, location, supported media formats etc.

[1] Distributed Content Adaptation Framework
[2] Laboratoire d'Informatique en Images et Systèmes d'Information
[3] Adaptation in Distributed Multimedia IT Systems

- **Adaptation Service Proxy:** perform content adaptation on behalf of the user or content provider and are implemented as Web Services.

## IV. THE PROPOSED ARCHITECTURE

In this section, we present the proposed architecture for the management of metadata based on the DCAF architecture discussed in the previous section.

### *A. Requirements*

In order to develop an architecture for metadata management, it is first necessary to outline the issues required from the architecture. Accordingly, we have outlined the following requirements:

- Acquisition/removal of metadata to/from the system. The metadata manager should handle the insertion of metadata into the system when the content is first introduced into the system at the content servers. It should also handle removal of metadata from the system whenever the content is removed from the system in order to avoid the inconsistency of having a description (metadata) of a content that does not exist.
- The metadata management should provide appropriate access to the metadata for the other components of the system that make use of it, for example the CNAM – the component that makes adaptation decisions on a certain multimedia content. For this, the metadata manager should provide a well-defined interface through which users of the metadata can access the metadata in a transparent manner, i.e. the interface should hide the distribution and replication of the metadata as well as its underlying management.
- Updating of the metadata when there is a change on the content. If the content changes and this change affects the characteristics of the content recorded in the metadata, the metadata also have to be updated to reflect this change.
- Generation of metadata for newly created content. An adaptation of a given content produces a new content, which is a version of the original content. This new content should also have a metadata so that it can be used to serve similar requests in the future. The metadata manager should generate the metadata for this new content.
- Caching of metadata. In order to provide fast access to metadata, caching of metadata at the proxies is important. The metadata manager should handle the caching of metadata at the proxies to increase the efficiency of access to the metadata.
- Completing the metadata by extracting missing features. The metadata may not include all the necessary information required to serve client requests or to decide on the process of adaptation. In this case, the metadata manager should further enrich the metadata by employing the appropriate feature extraction module to extract the missing information.
- Metadata standard conversion. If the metadata of a newly inserted content is represented in a format that is different from the one that is used in the system, then the appropriate transformation (crosswalk) needs to be applied on the metadata in order to make it usable by the system. Hence, the metadata management should employ the appropriate crosswalk module to perform the transformation.
- Incorporating adaptation constraints in the content metadata so that the semantics of the content is preserved after adaptation.

### *B. Overview of the architecture*

To satisfy the requirements outlined above, we have developed an architecture for the management of metadata in a service-based adaptation driven framework. In the proposed system, the main work of management of metadata is performed by the *Metadata Manager (MDM)*.

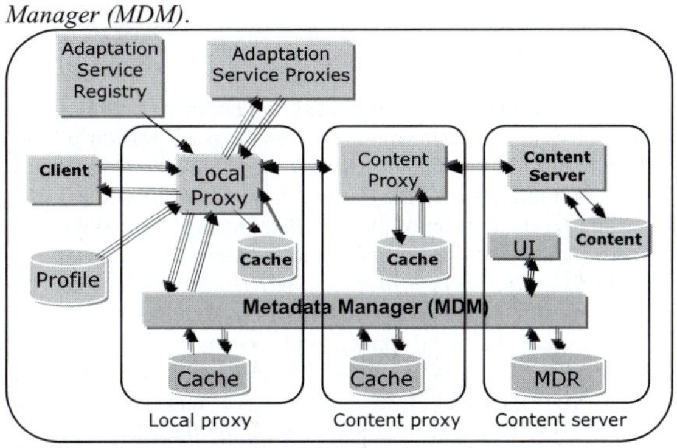


UI – User Interface, MDR – Metadata Repository

**Fig. 2. The proposed metadata management architecture**

The Metadata Manager is a distributed set of cooperating modules that execute operations in order to manage the metadata and to provide appropriate access for whoever needs it through well-defined interfaces (see Fig. 2 above).

The Metadata Manager is distributed at the various places where metadata is located and provides a transparent access to the metadata. These are the *content server* where content and metadata are originally inserted and stored permanently, the *content proxy* where content and metadata are cached to facilitate access to the content servers and to better serve similar requests efficiently, and the *local proxy* where adapted content and metadata may also be cached to increase the efficiency of similar future requests.

In this system, the metadata is stored in the Metadata Repository (MDR) located at the content server. In addition, for efficiency purposes, metadata is kept in caches at the content proxy and at the local proxy. On top of these storage components lies the Metadata Manager (MDM). The MDM is responsible for managing the metadata stored throughout the system (i.e. at the caches and at the MDR) and for providing an interface for accessing the metadata so that the other components of the system can access and use it. To achieve this, the MDM has different interfaces to external components depending on the location and nature of the components. Furthermore, the MDM has other internal modules that perform additional functions for the efficient management of metadata.

We have introduced a *User Interface (UI)* component in the architecture for inserting, deleting and updating metadata. The UI component is located at the content server and allows users (potentially metadata experts) to insert new metadata into the system, remove unwanted metadata from the system, or update existing metadata in the system. The UI communicates with the portion of the MDM at the content server to achieve the insertion, deletion, and update of metadata.

*C. Components of the architecture*

In this section, we outline the components of the architecture in further detail.

*i. The Metadata Manager (MDM)*

The Metadata Manager is composed of a number of modules each responsible for performing a certain logical portion of the overall metadata management task, as shown in Fig. 3 below. The modules are the *Local Proxy Interface (LPI)* responsible for the portion of the metadata management at the local proxy and providing access to the metadata for the local proxy, the *Content Proxy Interface (CPI)* responsible for management of metadata at the content proxy and serves as a bridge between the local proxy and content proxy, the *Content Server Interface (CSI)* responsible for the management of metadata at the content server and providing an interface for insertion, deletion, update of metadata, the *Feature Extraction Interface (FEI)* responsible for completing the metadata by extracting features from the content, and the *Crosswalk Interface (CWI)* responsible for transforming metadata from one standard to another so that it can be used in the system.

In this architecture, the communication between the MDM and the other components of the system will be conducted through the well-defined interfaces that are provided by the MDM itself. The communication among the three fragments of the MDM, located at the local proxy, content proxy and content server, is done by means of message passing using the Simple Object Access Protocol (SOAP) [5, 6]. SOAP is an XML-based lightweight protocol designed to work with existing Internet and XML open standards [7]. The fact that SOAP is XML based makes it an ideal choice in our case because the metadata that is going to be exchanged is also in XML. Hence, the metadata can be easily transferred among the three fragments of the MDM as a SOAP response message.

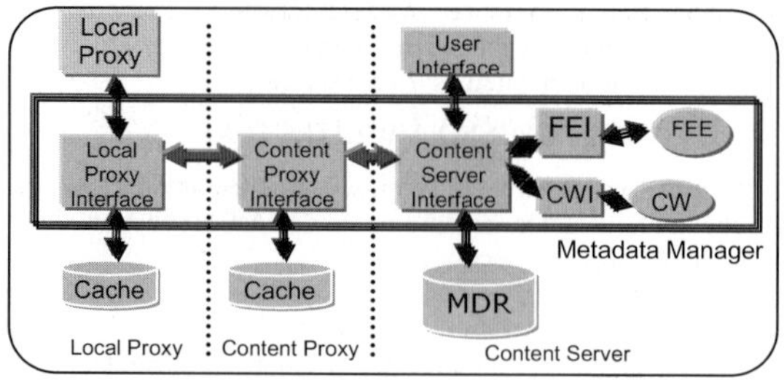


FEI – Feature Extraction Interface
FEE – Feature Extraction Engine
MDR – Metadata Repository

**Fig. 3. The Metadata Manager**

*The Local Proxy Interface (LPI)*

This component of the MDM is responsible for management aspects of the metadata at the local proxy. This involves providing an interface for the responsible component of the local proxy that wants to access the metadata, for instance the CNAM [3]. In addition, the LPI is responsible for caching metadata at the local proxy to increase the efficiency of metadata access. For requests of metadata that cannot be satisfied from the cache at the local proxy, the MDM is responsible for retrieving the required metadata from the neighboring cache, i.e. from the cache at the content proxy. This is achieved by having the LPI forward the request to the Content Proxy Interface (CPI) which is located at the content proxy.

Consequently, the LPI has to provide an interface to external components of the system for retrieval of metadata. This is basically presented in the form of a query against the metadata. The result of this query can be one or more metadata document(s) that satisfy the query or NULL if no metadata that matches the query is found. Hence the interface will have a general form of:
getMetadata (*query_parameters*) : metadata[ ]
where,
***query_parameters*** is a list of values that can be found in the metadata. For instance, a query can be made using *title* and *keywords* as query parameters.
**metadata[]** is a, possibly empty, set of metadata documents that are returned as a result of the query.

To reply to a ***getMetadata()*** query, the LPI performs a query against the cache at the local proxy. If this query returns NULL, the LPI has to forward the query to the CPI through the underlying network. This is achieved by composing a SOAP request message and sending it to the CPI. The request is simply the ***getMetadata()*** query, received from an external component of the system, wrapped in a SOAP envelope.

The component of the MDM that receives this message, in this case the CPI, extracts the parameter from the message and performs a local ***getMetadata()*** query against its cache. If this local query returns metadata, it wraps it with a SOAP envelope and sends it to the LPI as a SOAP response message. If, however, the CPI cannot find a matching metadata in its cache, it in turn forwards the request to the SCI as a SOAP request message. Then, it forwards the response it receives to the LPI as a SOAP response.

After receiving the SOAP response message, the LPI extracts the body of the message (i.e. the metadata) and returns it to the component that first initiated the ***getMetadata()*** call as a set of metadata documents.

*The Content Proxy Interface (CPI)*

This component is responsible for caching of metadata that are previously retrieved from the MDR so that requests for metadata from the LPI are satisfied without going to the MDR. The CPI accepts requests for metadata from the LPI. If it finds the requested metadata in its cache, then it returns it to the LPI. If the required metadata is not found in its cache, the request is forwarded to the MDR at the content server. When the CPI gets the metadata from the MDR, it in turn forwards it back to the LPI. The CPI communicates with the portion of MDM at the local proxy (LPI) and the portion of the MDM at the content server (CSI) by making use of SOAP over the underlying network. The request and response messages take the same form as the ones discussed for the LPI in the previous section and illustrated in Listing 1 and Listing 2, respectively.

*The Content Server Interface (CSI)*

This component is responsible for managing the metadata in the MDR located at the content sever, where the metadata is permanently stored. The CSI is responsible for providing an interface to the portion of the MDM at the content proxy, i.e. the CPI. In addition, it communicates with the User Interface (UI) component so that users (potentially metadata experts) can insert, delete, and/or update metadata. Moreover, the CSI is responsible for providing an interface to two other components of the MDM: the Feature Extraction Interface (FEI) and the Crosswalk Interface (CWI) which are responsible for invoking Feature Extraction Engines (FEE) and Crosswalks (CW), respectively. The FEEs extract features from multimedia content that are missing in the corresponding content metadata, while the CWs transform metadata from one standard to another. FEI and CWI are discussed further in the following sections.

The CSI provides different interfaces to different components of the MDM as well as external components of the architecture. First, it has to communicate with the CPI through a SOAP interface in order to accept SOAP request messages and to send metadata documents as SOAP reply messages.

Second, it has to provide interfaces for insertion, deletion, and update of metadata as per the request of the user through the UI. For this purpose the CSI provides the following interfaces:

- For insertion of metadata into the system:

insertMetadata (*metadata*, *error_detail*) : *status*
where,
***metadata*** is the metadata document, in XML, to be inserted to the system and ***status*** is a boolean flag returned to indicate the status of the operation. If successful, it will have a value of *True*; otherwise it will have a value of *False*. Whenever the **status** flag returns *False*, the caller can get details of the error that has occurred from the ***error_detail*** parameter, which gives a description of the error.

- For removal of metadata from the system:

removeMetadata(metadata_identifier, error_detail) : status
where,
***metadata_identifier*** is a unique identifier of the metadata document to be removed from the MDR and ***status*** and ***error_detail*** have the same meaning and function as for **insertMetadata()** discussed above.

- For update of metadata:

updateMetadata(metadata_identifier, attribute_list, value_list, error_detail) : status
where,
***metadata_identifier*** is a unique identifier of the metadata document to be updated, ***attribute_list*** is an array of attributes (such as *title*, *keyword*, etc) of the multimedia content that are to be updated and ***value_list*** holds the corresponding values of the attributes in the ***attribute_list***; and ***status*** and ***error_detail*** have the same meaning and function as for **insertMetadata()** discussed above.

Third, the CSI has to communicate with the Feature Extraction Interface (FEI) and Crosswalk Interface (CWI) of the MDM. Consequently, it needs to appropriately invoke the interfaces provided by these components. FEI and CWI are discussed in the following sections.

*The Feature Extraction Interface (FEI)*

This module is responsible for calling the appropriate *Feature Extraction Engine (FEE)* depending on the needs of the adaptation process. The feature extraction is required to enrich the metadata by including features that are not originally present but are required to fulfill some task. The actual implementation of the various feature extraction engines, however, is outside the scope of this work. In our architecture, we only provide an interface to invoke the appropriate extraction engine depending on the requirements of the system and consume the features that are extracted by incorporating them to the appropriate metadata.

The FEI has to provide an interface through which the CSI can request for feature extraction. For this purpose, the FEI internally maintains a list of FEEs, the corresponding features that are extracted by these FEEs, and the type of multimedia content they extract features from. The list takes the following form: (FEE, content_type, feature_list). The interface provided by the FEI for the CSI takes the following form:

extractFeature(content, content_type, feature_name_list, feature_value_list, error_detail) : status
where,
***content*** is the multimedia content from which features are to be extracted, ***content_type*** is the type of the multimedia content (text, image, audio or video), ***feature_name_list*** is the list of features to be extracted from the content, ***feature_value_list*** is the value of the features that are extracted from the content (it is empty when the call is made; and may be populated or empty after the call returns),

and ***status*** and ***error_detail*** have the same meaning and function as for **insertMetadata()** discussed above.

The FEI consults its internal list of FEE – feature_list pairs to determine which FEE(s) to invoke in order to reply to an ***extractFeature()*** call. There may not be an FEE that can extract all the requested features in the ***feaure_name_list***. Hence, the FEI may need to invoke several FEEs to carry out one ***extractFeature()*** call.

*The Crosswalk Interface (CWI)*

This component of the MDM is responsible for transforming metadata from one standard to another. This is important because the content adaptation/delivery system may employ a certain metadata standard but the metadata of a content can, possibly, be inserted into the system in a format different from that used by the system. For this purpose, the CWI invokes the appropriate *crosswalk (CW)*, that transforms the new metadata from its current standard to the standard used by the adaptation system, and inserts the resulting metadata into the Metadata Repository.

As there are several standards for metadata, there are (and will be) several crosswalks that implement the transformation of metadata from on standard to another. As in the case of the feature extraction engines, the implementation of the various crosswalks is outside the scope of this work. It should be performed by professionals who have a deep knowledge and experience in the various metadata standards [8]. Hence, the architecture is kept open in this regard by providing well-defined interfaces to invoke the appropriate crosswalk and make use of the result.

The CWI maintains a list of all the Crosswalks (CW) that it has under it. The list takes the form: (CW, from_standard, to_standard) and it keeps on growing as new CWs are introduced into the system. It is consulted when a request for a standard conversion is received from the CSI. For the CSI to be able to forward such requests, the CWI provides the following interface:

**convert_standard(*metadata, from_standard, to_standard, error_detail*) : *status***

where,

***metadata*** is the metadata document that is to be converted to a new standard, ***from_standard*** is the current metadata standard of ***metadata***, ***to_standard*** is the target metadata standard to which ***metadata*** is to be transformed, and ***status*** and ***error_detail*** have the same meaning and function as for **insertMetadata()** discussed above

*ii. The Metadata Repository (MDR)*

The Metadata Repository is where the metadata is stored. The metadata is going to be represented in XML. Hence, the MDR is responsible for storage of metadata encoded in XML.

In our architecture, the content of the metadata is categorized into three parts. The first part contains general information that can be used for querying. These include properties such as *title, content type, keywords, description, date of creation, creator,* etc. The second part contains technical information that is used in the process of adaptation. Such information is different for the different media types. Color information of images and videos, bit rates of audio, frame rates of videos are examples of content characteristics that fall in this second category. The third part contains constraints on adaptations that may be applied on the content. These are information that need to be taken into consideration when adapting the content (further discussed in the next section).

This classification of the content in the metadata facilitates the storage of metadata documents in databases tables. With this approach, each of the attributes in the first category can be considered as columns in the table while the attributes in the second category are stored under one column. The attributes in the third column too are stored under a single column. This simplifies the storage of the metadata in existing relational tables and facilitates querying for metadata documents using the SQL language through existing database programming interfaces. When a query for a metadata document is performed, the result of the query is a metadata document (in XML format) composed of the attributes from the three categories. Hence, the classification of the contents of the metadata content is hidden from the other components of the system. As far as the other components are concerned, the metadata is inserted and retrieved as a complete XML document.

When a new metadata is inserted into the system, the CSI parses the XML document and identifies the three categories of the content of the metadata discussed above and inserts them to the database.

When metadata is no more needed, the MDM will remove the metadata from the MDR and/or the caches. For instance, if a multimedia content is removed from the content server, then the Content Server Interface (CSI) will remove the associated metadata from the MDR and will inform the Content Proxy Interface (CPI) and Local Proxy Interface (LPI) to remove any cached metadata of the removed content from their caches. This avoids the presence of metadata that describe a content that does not exist in the system.

When content is updated at the content server, the CSI updates the corresponding metadata at the MDR and informs the CPI and LPI to remove cached metadata of the updated content from their caches. This avoids potential inconsistencies that may occur due to the presence of different metadata for a single content. Since multimedia content is not likely to be updated very frequently, this does not impose a serious drawback on the system performance. As another alternative, the updates may be transmitted to the caches so that any cached metadata is updated accordingly. This ensures that caches are still available and are up-to-date. However, additional processing of metadata is required at the proxies to update the document in the cache. Hence, the former approach is employed in the implementation of our architecture for experimental purposes.

## V. ADAPTATION CONSTRAINTS IN METADATA

Metadata for multimedia content contains a wide variety of information about the characteristics of the document. Such information is useful when making decision on the adaptation of the document based on a user's criteria. In this work, we consider the following common characteristics of each media type (text, image, audio, and video) to be included in the associated metadata, as shown in Table 1. But it should be understood that the metadata of multimedia content can contain far more characteristics depending on its applications [9]. The number and type of features also increases from time to time depending on various factors such as requirements of applications, availability of extraction tools, etc. Furthermore, depending on the domain of the application that is using the metadata, far more domain specific features of the content may be included in the metadata.

The main concern of this work is the usage of metadata for content adaptation. That is, the metadata should be managed in such a way that the appropriate decision, regarding the adaptation of the content, can be made. In this aspect, the metadata provides the necessary information about the content. In addition, it can also contain information regarding the types of adaptation that could and could not be applied on the content. Such information can be incorporated by the owner/creator of the content into the metadata so that the necessary information is not lost during adaptation, i.e. the semantics of the content is preserved after the adaptation.

**TABLE 1**

**PROPERTIES OF MULTIMEDIA CONTENT INCLUDED IN METADATA**

| Text | Image |
|---|---|
| Title | Title |
| Creator | Creator |
| Date | Date |
| Format (doc, pdf ...) | Format (gif, jpeg ...) |
| Size | Size |
| Language | Dimension (h X w) |
| URI | Color depth |
| Description | Description |
| Abstract | Keywords |
| Keywords | URI |
| Category | Category |
| Audio | Video |
| Title | Title |
| Creator | Creator |
| Date | Date |
| Format (mp3, wav ...) | Format (mpg ...) |
| Duration | Duration |
| Size | Size |
| Language | Language |
| Genre | Color |
| Bit rate | Frame rate |
| Category | Dimension (h X w) |
| Description | Category(movie, news ...) |
| Keywords | Description |
| URI | Keywords |
| | URI |

In particular, such information is critical in sensitive domains, such as the medical domain, since the multimedia content can be used for diagnosis purposes. For example, a description about a patient's medical history may be written in English. In order to be used by a physician who does not speak English, it needs to be translated to a language he/she can understand. But since the automatic language translation services currently available are not very reliable, the creator of the content may include a "don't translate" restriction in the metadata of the document. Hence, the physician who doesn't speak English needs to have the document translated by an appropriate human translator.

Less restrictive constraints can also be included in the metadata by the creator. For example let's consider a medical image with color depth of 24 bits. The image may contain information that is visible in high color depth values only. Hence, the creator may include a restriction such as "don't reduce color depth below 8 bits" in the image metadata. In this case, a color-reduction adaptation is possible but it should not reduce the color depth of the image below 8 bits. Another example of such a restriction can be seen regarding the dimension of the image. A "don't resize below 40% of the original size" constraint may be included in the metadata to prevent the reduction of the image dimension below a reasonable size which makes the image unusable due to loss of critical information.

Decrease in quality is also another reason for including constraints on adaptation (for example, reducing the bit-rate of an audio content)

In addition, the creator of the content may want the content to be adapted by a specific adaptation service. For instance, if the creator of a text content trusts only service X for language translation adaptation, he/she may include a constraint in the text metadata saying that the text should be translated by service X. The reason for selecting a specific service could also be due to security, financial, legal or other reasons. However, regardless of the reasons, the creator of the content should be able to include such type of constraints in the content metadata.

The modeling of adaptation constraints in the metadata that we propose can be precisely described using XML syntax so that it can be used by humans as well as machines. The constraint is incorporated into the metadata by humans who created the content and it is used by software that decides on the process of adaptation of the content. Hence, they need to be expressed in a format understandable by both, XML. Since the content metadata is also expressed in XML, incorporating the constraints into the metadata will not require much effort.

For this purpose, the following scheme is used. Listing 1 below shows a part of the metadata DTD for a multimedia content that concerns adaptation constraints. The DTD is defined based on the restrictions regarding a limited number of adaptation types on the four media. Namely, for text media: Text summary and Language translation; for image: Resizing and Color depth reduction; for audio: Bit rate reduction; for video: Resizing and Color reduction. This model does not exhaustively cover all the possible adaptation types, instead it considers a few, but common adaptation types on the four media types. However, the technique can be used to model constraints for a number of adaptation types that exist currently or that may come to existence in the future.

```
<!ELEMENT constraint (prefered_service*, ((reduce_bitrate?, stereo-to-mono?) |
(resize?, reduce_color?) | (translate?, summarize?)))>
<!ELEMENT prefered_service (service)>
<!ATTLIST prefered_service adaptation_type CDATA #REQUIRED>
<!ELEMENT service EMPTY>
```

```
<!ATTLIST service uri CDATA #REQUIRED>
<!ELEMENT resize EMPTY>
<!ATTLIST resize min_size CDATA #REQUIRED>
<!ELEMENT reduce_color EMPTY>
<!ATTLIST reduce_color min_color CDATA #REQUIRED>
<!ELEMENT reduce_bitrate EMPTY>
<!ATTLIST reduce_bitrate min_bitrate CDATA #REQUIRED>
<!ELEMENT stereo-to-mono EMPTY>
<!ATTLIST stereo-to-mono allow (yes|no) #REQUIRED>
<!ELEMENT translate EMPTY>
<!ATTLIST translate allow (yes|no) #REQUIRED>
<!ELEMENT summarize EMPTY>
<!ATTLIST summarize allow (yes|no) #REQUIRED>
```

**Listing 1. A DTD for modeling adaptation constraints in metadata**

Based on this DTD, we can formally express constraints on adaptations in the metadata of the content. For example, for a text document that should not be translated, we can have:

```
<constraint>
        <translate allow = "no"/>
</constraint>
```

## VI. IMPLEMENTATION AND RESULTS

The prototype developed in this work particularly aims at implementing the proposed metadata architecture in order to demonstrate its role in supporting the adaptation and delivery of content to clients. In addition, other components of the adaptation/delivery framework were also implemented to create an environment in which the metadata management can work.

Using the prototype, users can search for content based on various criteria. For this purpose, it provides an interface for the user through which he/she can login and select his/her device profile. After the user is logged in, he/she will be presented with a search interface that enables searching for content. The result of the search is a list of metadata that match the search criteria (Fig. 5). The user can then select and view (possibly after being adapted) one of the contents based on the information in the metadata.

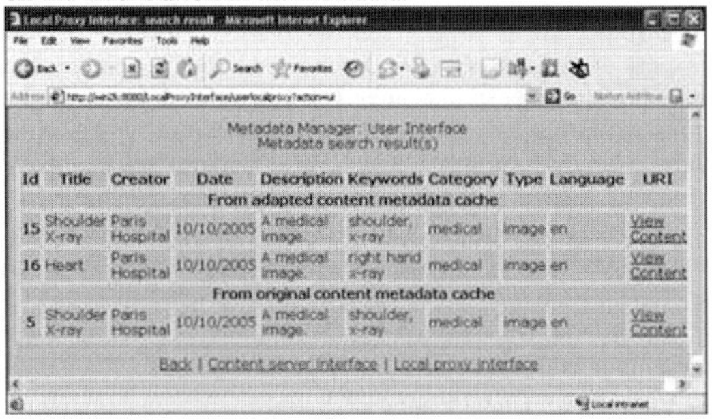


**Fig. 4. A metadata search result**

The implementation is carried out using Java Servlet technology with MySQL database.

## VII. CONCLUSION AND FUTURE WORK

In this work we have addressed the issue of metadata management in a pervasive environment by developing an architecture that facilitates the management of metadata. The architecture is composed of a distributed Metadata Manager, a Metadata Repository, metadata caches, and a User Interface. The Metadata Manager manages the metadata in such a way that as much information about the content as possible is supplied to the components of the adaptation/delivery framework that make use of the metadata. In addition, the metadata also provides information about the restrictions on the adaptation.

The modeling of constraints, however, could still be further studied to incorporate more constraints and to further generalize the modeling of constraints in metadata. In addition, with the popularity of web services, Feature Extraction Engines (FEE) and Crosswalks (CW) could be implemented as web services, in which case our architecture should also be appropriately improved to provide interfaces to invoke these services.

Regarding the implementation, we simulated most of the components of DCAF because we could not integrate our prototype with that of DCAF. Hence, the next step will be to incorporate our prototype into the prototype of DCAF and perform experiments so that the architecture can be further fine tuned.

## REFERENCE


[1] Mark Weiser, *The Computer for the 21st Century*, Scientific American, September 1991.

[2] NISO Press, *Understanding Metadata*, http://www.niso.org.

[3] Girma Berhe, Lionel Brunie, Jean-Marc Pierson, *Modeling Service-Based Multimedia Content Adaptation in Pervasive Computing*, Proceedings of the 1st conference on Computing frontiers, pages 60 – 69, Ischia, Italy, 2004.

[4] Lá szló Böszörményi, Hermann Hellwagner, Harald Kosch, Mulugeta Libsie, Stefan Podlipnig, *Metadata driven adaptation in the ADMITS project,* Signal Processing: Image Communication, Volume 18, Issue 8, pages 749-766, September 2003.

[5] Tom Clements, *Overview of SOAP,* http://java.sun.com/developer/technicalArticles/xml/webservices/, visited on March 3, 2005.

[6] The SOAP version 1.1 specification, http://www.w3.org/TR/SOAP.

[7] R. Allen Wyke, Sultan Rehman, Brad Leupen, *XML Programming*, Microsoft press, 2002, pages 277-279.

[8] Margaret St. Pierre, William P. LaPlant, Jr., *Issues in Crosswalking Content Metadata Standards,* http://www.niso.org/press/whitepapers/crosswalking.html , visited on May 16, 2005.

[9] Jane Hunter, Renato Iannella, *The Application of Metadata Standards to Video Indexing*, Proceedings of the Second European Conference on Research and Advanced Technology for Digital Libraries, pages 135 – 156, 1998.

# Data Loss Rate versus Mean Time To Failure in Memory Hierarchies

Novac Ovidiu*, Gordan Mircea** and Mihaela Novac**
* Department of Computers,
University of Oradea, Faculty of Electrotechnics and Informatics,
3700, 5, Armatei Române Str., Oradea, Romania, E-Mail: ovnovac@uoradea.ro

** Department of Electrotechnics, Measurements and using of Electrical Energy,
University of Oradea, Faculty of Electrotechnics and Informatics,
3700, 5, Armatei Române Str., Oradea, Romania, E-Mail: mgordan@uoradea.ro

***Abstract* – A common metric used for the assessment of overall reliability, in a memory hierarchy is the Mean Time To Failure (MTTF), but it doesn't take into account for the time of data storage in each level. We propose another metric, Data Loss Rate (DLR), for this reason. We will derive a recurrent formula for computing DLR, and we will validate it by computer simulation.**

## I. INTRODUCTION

Memory hierarchy is very important in a computer system. Practically if a computer system has a fast and efficient processor, if the memory hierarchy is slow and inefficient, memory access bottlenecks will arise and the overall system performance will be low. For this reason in any system design an adequate attention should be paid to the design of memory hierarchy subsystems, such as memory interface, cache, paging mechanism, TLB, CPU memory hierarchy support registers.

The memory devices of a computer system are organized in a hierarchy in order to achieve a good rate between performance (i.e. low access time) and cost per bit. A typical memory hierarchy of four levels is shown in figure 1. Proceeding down the memory hierarchy, its reliability improves owing to the storage technology in different levels. There is of course a tradeoff between high reliability and performance, which is influenced beside the construction, by the transfer policy used among levels.

Transfer policy is important because it directly affects the reliability of the overall hierarchy. A straightforward possibility is to write through to the most reliable, lowest level every time a data is transmitted from the CPU. This policy offers good reliability, but bad performance (high overhead for transferring data). On the other extreme, there is possible to write back data to lower level only when needed (for instance based on the amount of data in a level). This method yields a more reduced reliability (as data stays longer in a less reliable level), but better performance (less overhead for transferring data). At last, the third possibility is the delayed write, when data is written from level L to level L+1 after a time denoted $delay_L$. So $delay_L$ is the age of the data before it leaves level L and is written to level L+1. We can observe that $delay_L$, monotonically increase with L. This is an acceptable compromise between the previous two methods and we will consider only this policy from this point forward. [4].

Computer systems are composed of several types of storage, each with different degrees of reliability. We have obvious reasons for using a combination of storage devices with different reliability and performance. As example a backup tape is not fast enough, and a DRAM memory is not typically reliable enough to store all files.

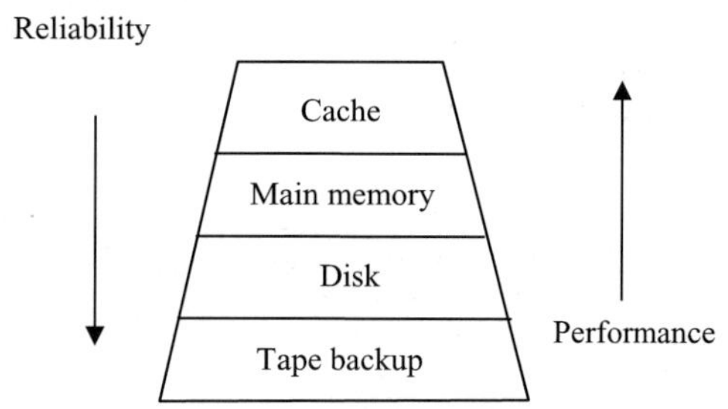


Figure 1.

## II. DLR AND MTTF METRICS

In order to evaluate the overall reliability of the hierarchy, some metrics are needed. One such a commonly used metric is MTTDL (Mean Time To Data Loss), which represents, in essence, the MTTF (Mean Time To Failure) of the whole hierarchy. Assuming a serial reliability block diagram of the system and for each level an exponentially distributed time to failure with a given MTTF ($MTTF_L$ for level L), MTTDL for a hierarchy of N levels can be computed with the equation (1).

However MTTDL is quite general and does not distinguish between the amount of data loss depending on the level where a failure occurs. For this reason a second metric, the DLR (Data Loss Rate) is proposed, which represents the fraction of data lost over time due to failures in the storage hierarchy.



*T. Sobh and K. Elleithy (eds.), Advances in Systems, Computing Sciences and Software Engineering*, 305–307.

$$MTTDL = \frac{1}{\sum_{L=1}^{N} \frac{1}{MTTF_L}} \quad (1)$$

Unlike in [1], for computing DLR we have used the following principle:

No. losses in L = No. losses due to L + No. losses due in (L+1) [ No. losses in L induced from a loss in (L+1) – No. losses already produced in L from those induced by (L+1)

On this basis we have obtained the following more accurate recurrent formula:

$$DLR_L = \frac{delay_L}{MTTF_L} + DLR_{L+1}\left(1 - \frac{delay_L}{MTTF_L}\right) \quad (2)$$

where, $DLR_L$ denotes the data loss rate in level L. (This formula does not allow to take in computation
a further error in a data item already corrupted in a previous level.) The last level is a special case, because it has no lower level where to transmit data. For this reason, we should consider $MTTF_N = \infty$, which yields $DLR_N=0$ (otherwise data loss rate of the whole system would be 100%, as all data will be lost in time).

## III. SIMULATION

In order to validate our model we have done a Monte Carlo simulation of the states of data in the memory hierarchy. The inputs of our program are: N (number of levels), $MTTF_L$, $delay_L$, n (number of runs). The output is DLR. For generating random failure times in each level we considered an exponential distribution of failure times according to the following cumulative distribution function

$$F_L = 1 - e^{-\lambda_L t}$$

where $\lambda_L=1/MTTF_L$ represents the failure rate of level L. In order to obtain a random failure time for level L ($t_L$), starting for uniformly selected random variable $u \in [0,1]$, we can write:

$$u = 1 - e^{-\lambda_L t_L}$$

$$t_L = \frac{1}{\lambda_L} \ln \frac{1}{1-u}$$

The partial results for some simulations are given in table 1a,b,c

| Run No. | 1 | 2612 |
|---|---|---|
| t1 | 20416098 | 28085836 |
| Success t1 > 2 | 1 | 1 |
| t2 | 67086060 | 200 |
| Success t2 >300 | 1 | 0 |
| t3 | 41379366 | 21943970 |
| Success t3 > 86400 | 1 | x |
| t4 | 2551845032950944200 | 1740184782032714500 |
| Success t4 > 1576800000 | 1 | x |
| DataLossRate[%] | 0.000000000 | 0.038284839 |

Table 1a. DLR calculated in run number 1 and 2612

| Run No. | 4733 | 4919 |
|---|---|---|
| t1 | 17793714 | 60183794 |
| Success t1 > 2 | 1 | 1 |
| t2 | 16958944 | 65691306 |
| Success t2 >300 | 1 | 1 |
| t3 | 78575 | 23102 |
| Success t3 > 86400 | 0 | 0 |
| t4 | 15732256004728312000 | 3165487150413436900 |
| Success t4 > 1576800000 | X | x |
| DataLossRate[%] | 0.105641242 | 0.121976011 |

Table 1b. DLR calculated in run number 4733 and 4919

| Run No. | 4999 | 5000 |
|---|---|---|
| t1 | 2845723 | 22188127 |
| Success t1 > 2 | 1 | 1 |
| t2 | 23453231 | 32012677 |
| Success t2 >300 | 1 | 1 |
| t3 | 143679403 | 31884861 |
| Success t3 > 86400 | 1 | 1 |
| t4 | 15015473684324770000 | 4565974819382920000 |
| Success t4 >1576800000 | 1 | 1 |
| DataLossRate[%] | 0.120024005 | 0.120000000 |

Table 1c. DLR calculated in run number 4999 and 5000

In tables 1a and 1b we cam observe that the DLR is 0 from run number 1 until run number 2611. When we get the first $t_n < delay_n$ we begin to calculate DLR. In our example in run number 2612, t2=200 , and the relation t2>300 is false, the DLR calculated is 0.038284839 %. In run number 4733, DLR calculated is 0.105641242 % and in run number 4919 we get the value 0.121976011 % for DLR.

The DLR in run number i is given by :

$$DLR_i = \frac{\overline{Z_1} + \overline{Z_2} + \ldots \overline{Z_i}}{i}$$

where i is the number of run and $Z_i$ is 1 if we get a success ($t_i > delay_i$ ) and 0 else.

We have considered a number of n=5000 simulation runs. The comparative results between mathematical model DLR and simulation DLR for different cases are centralized in table 2. It can be observed that the relative error (ε) between analytically DLR and simulated DLR is 3.38441401% in first case and 3.34568939% in the second case. Our recurrent formula for computing DLR is better that formula used in [1] because the difference between is analytically DLR and simulated DLR is under 3.4%.

| LEVELS | 4 | 4 |
|---|---|---|
| MTTF LEVEL[s] | 25920<br>504576<br>756864<br>92233720368547758 | 25920000<br>50457600<br>75686400<br>9223372036854775800 |
| DELAY LEVEL[s] | 2<br>300<br>86400<br>1576800000 | 2<br>300<br>86400<br>1576800000 |
| DLR model[%] | 11.46819529 | 0.11474915 |
| DLR simulated[%] | 11.08006408 | 0.11858830 |
| Epsilon[%] | 3.38441401 | 3.34568939 |

Table 2. Comparative results between mathematical DLR and simulated DLR

## IV. EXAMPLE

As an example, let us consider a memory hierarchy of four levels. For the data given in table 3, we have computed MTTDL and $DLR=DLR_1$ using equations (1) and (2). MTTDL of the whole hierarchy is not affected by the transfer policy between levels and is dominated by the cache boundary. For the assumed delayed-write policy, the resulting data loss rate is 0,1163%, that is about 10 hours of data loss per year.

This rate is dominated by the infrequent backing of data to tape. It is interesting to note, that increasing the reliability of a level, we can leave data to stay more time in that level, while maintaining the same overall data loss rate.

| Level | MTTF | Delay |
|---|---|---|
| 1. Cache | 10 month | 2 s |
| 2. Memory | 1.6 years | 5 min |
| 3. Disk | 2.4 years | 1 day |
| 4. Tape | 50 years | ∞ |
| MTTDL=5.3 month<br>DLR=0.001163 | | |

Table 3 Overall MTTDL and DLR of memory hierarchy.

## 5. CONCLUSIONS

In this paper we have used the metric Data Loss Rate instead of MTTF as a metric for evaluating the overall reliability of a memory hierarchy. It takes into account for the transfer policy between levels and is easy to compute based on our recurrent formula. Our model was validated by computer simulation. In an example we have demonstrated the usage of DLR, which allow us to consider the tradeoff between reliability and performance.

## REFERENCES

[1] Peter M. Chen, David E. Lowell - "Reliability Hierarchies", Workshop on Hot Topics in Operating Systems, 1999.

[2] Peter M. Chen, Wee Teck Ng, Subhachandra Chandra, Christopher M. Aycock, Gurushankar Rajamani, and David Lowell – „ The Rio File Cache: Surviving Operating System Crashes". In Proceedings of the 1996 International Conference on Architectural Support for Programming Languages and Operating Systems (ASPLOS), pages 74-83, October 1996.

[3] Yiming Hu, Tycho Nightingale, Qing Yang. – „ RAPID-Cache - A Reliable and Inexpensive Write Cache for High Performance Storage Systems", IEEE Transactions on Parallel and Distributed Systems, Vol. 13, No. 2, pages 1-18, February 2002.

[4] David A. Paterson, John L. Henessy – "Computer Architecture. A Quantitative Approach", Morgan Kaufmann Publishers, Inc. 1990-1996.

# Towards Living Cooperative Information Systems for Virtual Organizations Using Living Systems Theory

Shiping Yang, Martin Wirsing
Institut für Informatik, Ludwig-Maximilians-Universitä t München,
Oettingenstr. 67, München 80538, Germany
{yangs, mwirsing}@pst.informatik.uni-muenchen.de

***Abstract* - In order to conceive and design cooperative information systems to better adapt to the dynamics of modern organizations in the changing environment, this paper explores the requirements and approaches to build "living" cooperative information systems for virtual organizations -- both in terms of system flexibility and co-evolution of organization and information systems. The object of our case study is the Beijing Organizing Committee for the Games of the XXIX Olympiad. To meet the requirements of "living" cooperative information systems in the context of virtual organizations, we propose a unified peer-to-peer architecture based on the foundational concepts and principles of Miller's Living Systems Theory, which is a widely accepted theory about how all living systems "work". In our system framework, every peer belongs to one of the 6 organizational levels, e.g. peers, groups, organizations, communities, societies, supranational systems. Each level has the same types of components but different specializations. In addition, every peer has the same critical subsystems that process information. The case studies show how our architecture effectively supports the changing organization, dynamic businesses, and decentralized ownerships of resources.**

***Keywords* - virtual organization, cooperative information systems, living systems theory, peer-to-peer, olympic games.**

## I. INTRODUCTION

Virtual Organization (VO) generally refers to a new organizational form characterized by a temporary or permanent collection of autonomous individuals, groups, or organizations that share a common goal and work across space, time, and organizational boundaries made possible by the use of information and communication technologies to achieve the common goals. It is less rigidly structured and much more flexible than classical organizations. The business rationale for creating VOs in almost all cases is to address rapid change in a turbulent business environment. However, in all cases the turbulent business environment is a business environment where business change occurs faster than the involved organizations can change themselves. A successful business is constantly adapting to change, and therefore so should the information systems, which must support rather than hinder this changeability. Therefore, what is required for "living" businesses in the changing world are "living" information systems, which should be more flexible to respond quickly to the needs of business as they evolve.

However, how information systems can be conceived and designed to better adapt to the dynamics of modern organizations -- both in terms of system flexibility and co-evolution of organization and information systems, and thereby improve their performance, effectiveness, and satisfaction -- are significant challenges. This paper was motivated by the author's own experiences in designing cooperative information systems (CIS) to support distributed cooperative work in the Beijing Organizing Committee for the Games of the XXIX Olympiad (BOCOG), which employs hundreds of people to start its work with a period of planning followed by a period of organization that culminates in the implementation or operational phase. The objective of the work presented in this paper is to explore the requirements and approaches to build "living" cooperative information systems (LCIS) for VOs. For this purpose, BOCOG becomes the object of our case studies.

The remaining of this paper is organized as follows. Section II introduces the example of VOs, BOCOG. Then we develop detailed requirements of LCIS for VOs in section III. In section IV, Miller's Living Systems Theory is introduced to describe the "living" nature of the LCIS. In section V, a novel unified peer-to-peer conceptual architecture is proposed to meet the derived requirements, and the case studies in section VI briefly show how our architecture effectively supports the changing organization, dynamic businesses and decentralized ownerships of resources. Section VII describes related work. Finally, section VIII concludes this paper and outlines our topics for further research.

## II. BOCOG: AN EXAMPLE OF VIRTUAL ORGANIZATIONS

Olympic Games require a tremendous investment of human, financial, and physical resources from the communities which stage them. The dynamic characteristics of VOs (e.g. flexible boundaries, complementary core competencies / the pooling of resources, spatial dispersion, dynamic memberships, variable longevity, electronic communication) [1] in our opinion, are inherent to BOCOG, too.

### *A. Background*

BOCOG was established on December 13, 2001, five months after Beijing won the right to host the 2008 Games [2]. From the time of its constitution to the end of its liquidation, BOCOG must comply with the Olympic Charter, the contract entered into between the International Olympic Committee (IOC), the National Olympic Committee (NOC) and Beijing



*T. Sobh and K. Elleithy (eds.), Advances in Systems, Computing Sciences and Software Engineering*, 309–316.

(the Host City Contract) and the instructions of the IOC Executive Board. According to the definition of VO presented in section I, BOCOG is a virtual organization, where the "virtuality" consists of the fact that BOCOG is a network of many organizations (e.g. IOC, International Sports Federations (ISFs), International Paralympics Committee (IPC), Government, China State Sports Administration Bureau (CSSAB), Coordination Committees (CCs), media, partners, sub-committees (Shanghai, Shengyang, Tianjing, Qingdao, Qinhuangdao)), who make full use of social resources to build small or large teams to work on specific projects (see Fig. 1). There are some teams working close within one office or one building at the same physical location. Others spatially span over organizations, cities, and even countries.

Internally, BOCOG currently consists of 18 departments looking after everything from venue planning to environmental management [2]. Every department usually sets up several divisions to manage specific projects separately. The hierarchical internal structure of BOCOG is showed in Fig. 2.

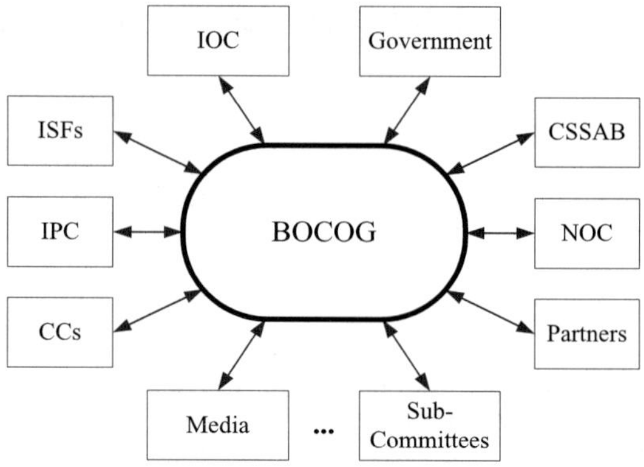


Fig. 1: The external structure of BOCOG

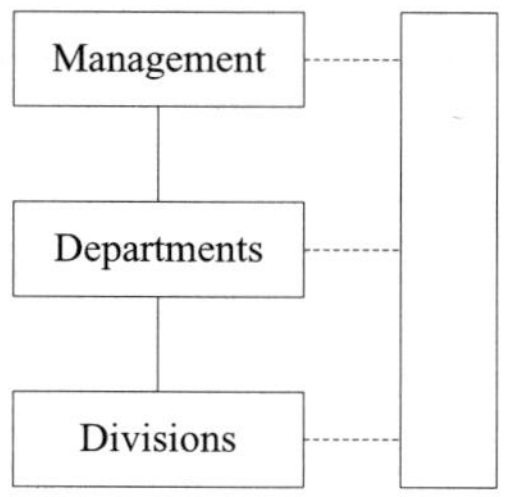


Fig. 2: The internal structure of BOCOG

### B. *Characteristics*

Every Olympic Games can be seen as a learning experience, for host cities and nations, for the Olympic Movement and for sport at large. Especially, this is the first time for China to hold Olympic Games. In this section, we describe the dynamic characteristics of BOCOG that arises from three areas of concern -- organization, businesses, and resources.

- *Organization is changeful*

BOCOG start its work with a period of planning followed by a period of organization that culminates in the implementation or operational phase. During this process, BOCOG will adjust organizational structures step by step to satisfy the increasing requirements of organizing activities. The organization has the following characteristics:

*Organizations are purpose-driven.* BOCOG are formed for specific reasons and with clear overall common goals and objectives in mind. All members have agreed on these goals and objectives. Each participant in this collaborative effort plays different roles, contributes to the realization of the end goal, and forms a link in the functional process of Olympic Games. Besides, the lifetime of BOCOG is explicitly restricted; it will be dissolved when its overall goal has been achieved in 2008.

*Organizational structure is dynamic.* From the outside, the external structure of BOCOG (see Fig. 1) may change continuously. For example, it may extend if additional tasks are identified which cannot be covered by existing members or if an external organization seems to be a useful supplement for the Olympic Games. From the inside (see Fig. 2), it has to deal with a permanent demand driven restructuring process with respect to functional units like divisions or departments. By the year 2008, there will be more than 30 departments under the BOCOG umbrella [2]. In addition, with the completion of sport venues and several new management centers that are built out to meet the functional requirements of Olympic Games, BOCOG may change its current co-located management mode into venue-centered distributed operation mode.

*Olympic Games bring people with often diverse backgrounds.* It involves a large, varying, and indeterminate number of participants. The total number of BOCOG staff is expected to reach around 4,000 by 2008 when the Olympic Games are held. Besides, about 60,000 volunteers will be employed to provide kinds of public services to Olympic family. Issues such as the attitudes, beliefs, customs, desires, and expectations of people, can have a great impact on organizations and the task performed. The members change quite often, and more importantly all members are not known from the beginning.

- *Businesses are dynamic and uncertain*

Work is a situated activity, and therefore its organizational context is the context of the processes within which it is performed. The organizational context of BOCOG is subject to changes and undergoes changes with different speed and different visibility, which cause various influences on work performed in BOCOG. For example:

*Most business processes are uncertain.* The uncertainty of a business process relates to the difficulty of predicting the individual circumstances of the business process, the tasks, and resources that the business process requires [3]. As a result of a high level of uncertainty, the specific activities required to perform a business process cannot easily be predicted up front, and even at the time of process occurrence, a significant amount of individual judgment might be required regarding the most appropriate measures to be taken. Such consensus is learned from experience by members of BOCOG every day. Moreover, the policies of government can affect organizations, too.

*There is a strong need of a decentralized control of business activities.* This is because organizational culture or work practices are seldom uniform throughout the organization. Collaboration in work is often changing and can take many forms. Some elements of the process may be consistent and well understood. Others may be ad-hoc, highly dependent on information associated with that work instance or subject to the surrounding context of the activity. Well understood elements of a work process may be automated or nearly automated, while others, such as planning and problem solving, may require significant human involvement to complete.

*Processes involve a number of organizations, resources, and applications.* In BOCOG, most business processes are composed of tasks involving a number of organizations, databases, and specialized applications. The activities within BOCOG often interact across organizational boundaries, e.g. interaction with customers, relationships with partners, and work performed by outside contractors. Within the overall process, the tasks are related and share various control, data, and temporal dependencies. Moreover, the role of a member of cooperative work often changes during the development of business processes.

- *Resources are dispersed*

Resources, which are produced and used by participants during their activities, play an important role in coordination and communication.

*Resources are variable.* In BOCOG, a large amount of data, represented in various formats and in different languages, are generated or received every day and stored in different types of data sources, e.g. pre-relational or relational DBMS or even flat files, which are dispersed via a heterogeneous network. In addition, different departments and organizations have their own information systems, with their own idiosyncratic representations, for managing their resources.

*There is a decentralized ownership of resources.* Organizations join the Olympic Games intentionally. Each organization attempts to maximize its own profit within the Olympic Games. This implies that they are intended to and will remain autonomous and independent. They control how their resources are consumed, by whom, at what cost, and in what time frame. Within an organization, there is also a decentralized ownership of the tasks, information, and resources managed by different organizational units involved in the business process.

*Different stakeholders require different views of resources.* Stakeholders may be technical or non-technical, both inside and outside an organization. Likewise, tools, techniques, management styles, work ethics, and environments vary across organizations. Each of these may require presentation to an end-user in different ways or highlighting different aspects or importances.

- *Summary*

Across our general observation regarding above three areas of concern, it seems that continuous *change* is the one constant theme in all of them and affects each other. We are aware that all of these problems described above are not problems pertaining to BOCOG. Actually, most of the problems appear in classical organizations, too.

## III. Requirements

CIS are described as "*next generation information systems*" in the literature [4], and the paradigm of CIS has been growing and gaining substance in technological infrastructure (e.g., middleware and Web technologies) and application areas (e.g., Business Process Management, e-Commerce, e-Government, and VOs) [5]. Enterprise CIS is a dynamic mix of organisation (people), businesses, and resources. However, current research communities focus on one or, at best, two of the three aspects as described in section II. During our investigation of BOCOG, there is an urgent need to develop information systems capable of supporting and managing change in all its manifestations and ramifications. This section describes the requirements of LCIS.

### *A. Why Living?*

Information systems have become the backbone of modern organizations. While information systems and the means of their development have improved over the last 30 years, it may be argued that many systems, if not all, are disappointed. High profile system and project failures, alongside increasing development and maintenance costs, bear witness to this [6]. This disappointment is mostly with the customer, be that the user or the owner, but also with the analysts and developers.

One reason is the project-based nature of development [7]. A project-based approach is inherently finite time horizon driven whereas the environment is infinite horizon. This generates a number of differences between the desire to build something, and the need to make it work. Another important reason is the constantly changing nature of the environment, where "*every feature of social organization -- culture, meaning, social relationships, decision processes and so on -- are continually emergent, following no predefined pattern*" [8]. Our problem the information systems specialist has often to face is that, when a specific information system is delivered, the organizational model at that time is frequently different from the initial model due to the dynamic nature of organizations and changes in the environment [9].

Today's business environment is becoming more complex, organisations are facing increased competition, global challenges, and market shifts together with rapid technological developments. A successful business is constantly adapting to change, and therefore so should the information systems, which must support rather than hinder this changeability. The problem then is not how to build information systems which share goals with their organizational environment, human users, and other existing systems at some time point. Rather, the problem is how to build information systems which continue to share goals with their organizational environment, human users, and other existing systems as they all evolve. Therefore, what is required for "living" businesses in a changing world are LCIS, which should be more flexible to respond quickly to the needs of business as they evolve.

### B. The Requirements

Enterprise information systems are, by their very nature, large and complex. A LCIS is not a collection of databases, applications, and interfaces. In order to be able to continually support organizations, the design, implementation and use of LCIS must reflect and support the continuously ongoing changes as described in section II. In our opinion, a LCIS must have the following characteristics to be effective:

- *Extensible.* Not everything needs to be, and often cannot be, known beforehand, especially in the changing environment. As mentioned in section II, user requirements are inherently uncertain and rarely static. User requirements evolve not only during the software development but also during the testing and use of the system [10]. In addition, the number of possible exceptions is very large; their scope and the great variety of possible contexts make it practically impossible to specify all exceptions statically and in advance. Furthermore, when information system is introduced, it changes users' work [11]. Therefore, if the modules of LCIS are "plug-and-play" compatible, the systems can thus be constructed from parts and survive over time, and survive reuse in multiple environments.

- *Autonomous.* People are demanding more autonomy, and more sense of ownership, shared power and participation [12]. As described in section II, there is a decentralized ownership, control, and perspective of different resources. The differences present between group members -- their varying roles, needs, skills -- and the differences between groups as a whole are a serious obstacle to achieving uniform acceptance and use of the systems, especially when the systems treat all people and groups identically. Each of these may require the systems to support different levels of freedom and current working customs in cooperative network, as well as necessary constraint.

- *Accessible.* To accomplish their work, people need to not only access data distributed in the various sources from both inside and outside the organization, but also use various tools/systems developed by different technologies to manipulate data and contact people. That is, end-users require to access resources independent of time and space distances. To enable "the right information to the right people at the right time", it requires seamless connections among people, software agents, and various kinds of IT systems.

- *Dynamic.* A key feature of the LCIS is that the systems (and the models) accurately represent the organization at all times. As business processes and organization models change, such changes need to be reflected in the system behavior. However, as time passes and an organization changes, the more likely it is that the system does not accurately represent the current organizational environment. The system must maintain currency to provide value to the organizations and members. In this way, the systems drive the organizations and the organizations drive the systems. This ensures system realism and "believability" [13].

## IV. The Living Systems Theory

A description of the "living" nature of our LCIS can be applied from James Grier Miller's Living Systems Theory (LST) [14], which is a widely accepted theory about how all living systems "work", about how they maintain themselves, and how they develop and change.

### A. Basic Concepts and Principles

By definition, living systems are open and thus constantly interact with their environment by means of information and matter/energy exchanges. The central concept of livings systems is that of a *system*, which is defined by Miller as follows:

*"A system is a set of interacting units with relationships among them. The word 'set' implies that the units have some common properties. These common properties are essential if the units are to interact or have relationships. The state of each unit is constrained by, conditioned by, or dependent on the state of other units. The units are coupled."* [14]

LST identifies basic principles that underlie the structure and processes of all living things. According to the way we perceive natural systems, Miller's basic model of living systems introduces two principle views of a living system:

- *A "structure" view.* A system is made of components, which in turn are systems. According to Miller *"...the universe contains a hierarchy of systems, each more advanced or 'higher' level made of systems of the lower levels"* [14]. He identifies 8 distinct levels for living systems: cells, organs, organisms, groups, organizations, communities, societies, supranational systems. Every living system belongs to one of the 8 defined levels and is composed of components that themselves are systems on the next lower level. The higher-level system, of which a system is a component, is called supra-system. For example, an organ is the supra-system of several cells. These organizational levels are characterized by the fact that each level has the same types of components but different specializations. Systems at higher levels are more complex than systems at lower levels. This distinction is tightly linked to our experience in perceiving and studying the world of living systems. It holds that these levels are practical (but not necessarily optimal) to describe the reality with enough simplicity. It is perfectly conceivable to define other levels. Miller chose these levels as they allowed him to do an extensive literature review related to these individual levels, resulting in the suggestion on how systems on all these levels could be modeled using a common meta-model.

- *A "process" view.* A system has subsystems, which carry out the essential processes of the system. Subsystems interact to achieve a desired overall matter/energy or information processing. Subsystems of a system are always realized by the collaboration of one or more components. The notion of an echelon is used to refer to a hierarchical organization of components within a subsystem taking over certain types of activities. According to LST, every living system has the same

# Research and Application of A Integrated System

## —Textile Smart ProcessUsing ANN combined with CBR

Xianggang YIN and Weidong YU*

(College of Textiles, Donghua Universituy, Shanghai, People's Republic of China 200051)

**Abstract: In this paper, the disadvantages and advantages of artificial neural networks (ANNs) and Case-base Reasoning (CBR) have been briefly introduced respectively. The capacity of network can be improved through the mechanism of CBR in the dynamic processing environment. And the limitation of CBR, that could not complete their reasoning process and propose a solution to a given task without intervention of experts, can be supplied by the strong self-learning ability of ANN. The combination of these two artificial intelligent techniques not only benefits to control the quality and enhance the efficiency, but also to shorten the design cycle and save the cost, which play an important role in promoting the intelligentized level of the textile industry. At the same time, utilizing ANN predicting model, the sensitive process variables that affect the processing performances and quality of yarn and fabric can be decided, which are often adjusted during solving the new problems to form the desired techniques.**

* Corresponding address e-mail: wdyu@dhu.edu.cn



*T. Sobh and K. Elleithy (eds.), Advances in Systems, Computing Sciences and Software Engineering*, 317–324.

## 1 Introduction

Owing to the inherent non-linear relationships that exist between the process parameters, material variables, and the resulting textile properties, the recently main predictive models such as first-principles modeling and empirical modeling [1] are inadequate for accurately modeling and optimizing complex non-linear processes like textiles processing courses [2]. From the mid-1990s, as an evolving branch of AI called *connectionist learning* (Artificial Neural Networks, ANNs) presents an attractive alternative to doing predictive modeling, neural networks have been used in various textile-related applications [3], such as the fiber's recognition [4,5], the prediction of yarn quality [6], spinning property prediction [7,8], dyeing defects [9], classifying [10], fabric properties and handle or style [11~13] and so on. Most of these researches, however, only theoretically prove the appealing characteristics of handling non-linear relationships of ANNs, it is seldom studied how to realize the optimization of process route and the quality control of textiles in practice according to the predicted results. On the other hand, there exist some disadvantages of ANNs, such as the contradiction of memorizing and generalizing, the lack of explanatory capabilities for rule extraction and the lack of the theoretical background of selection of the Network topology and its parameters with a "trial and error" matter *et al* [2, 14, 15], all which are impossible to be entirely solved by artificial neural networks itself.

Recent advance in artificial intelligent research have shown that both symbolic Artificial Intelligence (AI) and ANNs approaches to solving real-world problems have both their strengths and limitations [16]. Current beliefs suggests that these approaches are complementary, rather than competitive, and that a methodology drawing on the strengths of both these approaches can result in a more powerful problem solving capability than could be accomplished by either technology on its own[17~19]. Very complex problems may not be solved by the application of just one technique. Research in the area of hybrid system forms a relatively new field. In particular, it is beneficial to use a hybrid system when there is not enough (symbolic) knowledge about a system to use a symbolic algorithm or not enough data to train a neural network [20]. Case Based Reasoning and Artificial Neural Networks are complementary problem solving techniques [21]: CBR systems are able to reuse information from past experience; ANNs can generate adaptive structures using large data sets. The integration of ANNs with CBR has been investigated by a relatively small number of researchers [22~25].

In this paper, the advantages and disadvantages of CBR and ANN are discussed respectively based on the previous researches. In order to improve the capability of ANN used in the optimizing and controlling, the mechanism of CBR is combined with the training, predicting and decision-making of ANN, which forms a integrated intelligent model for textile manufacturing in the computer.

## 2 Case-Based Reasoning System

As a significant branch of AI, CBR has received more and more research attention. Much work has been dedicated to this topic, including its basic principle [26], methodologies [27] and application [28]. This approach focuses on how to exploit human experience, instead of rules, in problem-solving, and thus improving the performance of decision support systems. A typical CBR system is composed of four sequential steps which are called into action each time that a new problem is to be solved [26, 27]:

**(1)** ***Retrieve*** the most relevant case(s),

**(2)** ***Reuse*** the case(s) to attempt to solve the problem,

**(3)** ***Revise*** the proposed solution if necessary, and

**(4)** ***Retain*** the new solution as a part of a new case.

The basic idea behind CBR is that reasoning and problem-solving are based on the most specific experiences available instead of general abstract knowledge or rules. In this way, case-based reasoning provides a new method for building intelligent systems.

For a new problem or situation, a set of the most similar cases is retrieved by the CBR, based on similarity assessment criteria, and the solution, corresponding to these previous cases, is adapted to propose a solution for the given problem based on the adaptation criteria. It is obvious that the criteria for both case similarity assessment and case adaptation are inextricably linked, which means that the solution suggested is intimately connected to the similarities and differences between new and old cases. More precisely, such a statement may be described as

$$\text{Suggested Solution}=A[S(\{C^{Old}\}, C^{New})],$$

where $\{C^{Old}\}$ and $C^{New}$ denote a set of old cases and a new case, respectively. ***S*** ( , ) is the similarity assessment criterion, and ***A***, an operator, is defined by the adaptation criteria. Therefore, in essence, the CBR reasoning process is a pattern matching and classification process.

However, although CBR is simple in principle, and has been successfully used in engineering problem-solving, it still lacks a generally theoretically sound framework. A systematically mathematical model has not been established for CBR system design, analysis, comparison and test. Consequently, most CBR systems could not complete their reasoning process, and propose a solution to a given task without intervention of domain experts or system managers [28]. On the other hand, for a quantitative description model-based CBR system, the system design may become more flexible compared with symbolic description model-based systems. This is because a lot of mathematical approaches and optimization techniques are available for defining, synthesizing and analyzing the case similarity assessment and case adaptation criteria [25].

As is well known, ANN approaches have many appealing characteristics, such as parallelism, robustness, adaptability and generalization. Neural networks have the potential to provide some human characteristics of problem-solving that are difficult to describe and analyze using the logical approach of expert system. More importantly, learning and problem-solving can be incorporated naturally and effectively by using a network [25].

## 3 Artificial Neural Networks

Artificial Neural Networks, otherwise known as *connectionist* models, are information-processing mechanisms that consist of a large number of densely interconnected simple computational elements. ANNs represent a set of very powerful mathematical techniques for modeling, controlling, and optimizing that 'learn' directly from historical data. The main feature that makes neural nets the ideal technology for modeling the production process is that they are non-linear-regression algorithms that can model high-dimensional systems and have a very simple, uniform user interface [2]. The back-propagation learning algorithm was proposed by Rumelhar et al [29] in 1986 to modify connection weights in multilayered feed-forward networks, which is popularly used in practice. The back-propagation algorithm is an iterative gradient descent algorithm that minimizes the sum of the squared error between the desired output and actual output of a set of patterns through adjusting the connection weights. For a more detailed mathematical approaches and algorithm, see references [30, 31]. Once training is completed, the weights are set and the network can be used to find outputs for new inputs.

In textile industry, samples are constantly extracted from the production line and tested to ensure that the products being produced meet specifications and quality standards. Processes that have large quantity of historical data, but little knowledge about the complicated interactions between the inputs and outputs of system, are ideal candidates for building predictive model using artificial neural networks. The neural net can be trained with a large subset of the process data. When the net has finished learning on the data set, it can be tested with the remaining subset of process data to determine how well it can perform on data it has not seen. If the net predicts accurately on data it has never seen, the network can be said to generalize. If the generalization level is high, then the confidence is high, and the net has captured the process dynamics [2].

However, although the trained neural networks can predict accurately the output for the never seen input in the future, a net does not provide any understanding of why an input set of material and process parameters results in the predicted level of the forecasted product indices [15]. This analysis becomes the responsibility of the researcher. Obviously, it is impossible for a new technician, even seasoned technologist, to obtain a perfect suitable explanation from the complicated textile manufacturing process. Fortunately, CBR systems are particularly appropriate when rules from the knowledge domain are difficult to discern or the number and complexity of the rules is too large for the normal knowledge acquisition process; they have the potential to provide some of the human characteristics of problem solving that are difficult to simulate using the logical, analytical techniques of knowledge based systems and standard software technologies.

## 4 A Hybrid System

In this paper, a hybrid architecture integrating a ANN system with a CBR component has been developed to use general and specific knowledge in the optimizing the process parameters and controlling the textile's quality, shown as Fig. 1.

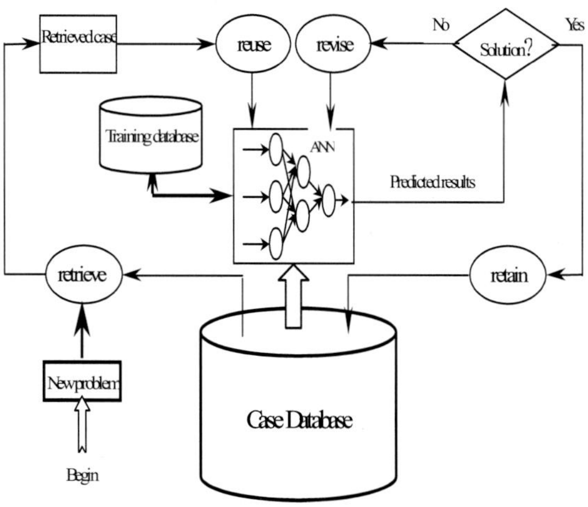


Fig 1. Flowchart of the hybrid system of ANN integrated with CBR

In such a hybrid system, the neural network itself is not involved in case matching, case retrieval and the reasoning process. Instead, the network is trained to learn general domain knowledge and, as a result, the trained network can be used as a source of general knowledge and decision-making. Furthermore, knowledge stored in the network is interpreted as a symbolic representation, which is combined with specific cases to support the reasoning. Because that one of the main advantages of neural networks is that the trained neural networks can be used in sensitivity analysis [1], which can be used to reveal the measured and control variables to which the process is sensitive.

On the other hand, the ability of the ANN to generalize may produce better results than CBR when it has been trained with an appropriate data set. In this situation it would be possible to rely on the ANN and its generalizing abilities. However, if the system is inserted in a dynamic environment, whose characteristics change in an un-predictable manner, an agent will require an adaptive mechanism capable of reacting to such changes in the environment. It is possible to endow a CBR mechanism with the ability to detect large changes in the environment. Normally, however, large changes will decrease the performance of the ANN, which is more effective when it has been trained with a significant amount of data representative of the whole possible data set. In such a situation CBR may be more effective than an ANN. As a result, the ANN will be required to be retrained using the most recently retrieved cased, using CBR mechanism [20].

Retrieval, reuse, revise and retain are the four key technology links of CBR reasoning. As shown in Fig. 1, combined with ANN, the optimization of textile process case is executed as follows. At first, according to the query information of the new problem, the similar cases are retrieved to form the retrieved cases, and the relevant weight is also provided in advance in terms of the practical requirement. Secondly, the most similar case selected from the retrieved cases is input into the trained neural networks. If the predicted results are agreed with the demanded quality, the extracted case can be directly applied in the light of the former process technics, parameters and materials. Otherwise, it must be revised to form a new case based on the sensitive variables provided by the trained net. Then the next predicting will be executed until the desired quality is obtained. Lastly, the modified successful case is validated and retained in the case database. At the same time, when the processing environment has been changed, that is to say, most of the past products have been replaced with the news, ANN model will be retrained using the new data to be combined again with CBR.

## 5 Application and Results

### 5.1 The Establish of the Entire Woolen Process Models

The single-hidden-layer networks with the back-propagation algorithm are structured using the data of 77 lots extracted using CBR mechanism from the mill historical database, in which 69 lots of them are considered as the training samples, and the other 8 lots as the test set. In order to expediently control the quality and optimize the variables, the entire woolen manufacturing process is constituted of spinning model, weaving model, and finishing model. These three models can not only be solely executed to predict and deduct, but also be integrated as one synthetic virtual processing system for decision-making. That is, what the quality of yarn or fabric can be understood in advance according to the material and process variables and what kind of material should be required before the yarn or fabric is manufactured. As one integrated system, the results of the former model can be regarded as the input neurons of the latter.

The capability of the trained models, that predicts the required accuracy and stability of the product quality and processing performance indices, is assessed by mean accuracy (**MA**) and relative coefficient (**R**) between the predicted values and the desired results of the test set, as shown in table 1.

**Table 1** The results of property analysis of ANN models

| Model | Properties | MA | R |
|---|---|---|---|
| Spinning model | Yarn unevenness(%) | 0.9890 | 0.9912 |
| | Thick places per kilometer (+50%) | 0.8494 | 0.9950 |
| | Thin places per kilometer (-50%) | 0.9546 | 0.9906 |
| | Yarn neps (+200%) | 0.9173 | 0.9366 |
| | Hairiness (m/m) | 0.9082 | 0.6420 |
| | Strength(cN) | 0.9675 | 0.9543 |
| | Extension at break(%) | 0.9184 | 0.9565 |
| | Ends-down 1000 spindle hours | 0.9393 | 0.9460 |
| Weaving model | Weaving efficiency(%) | 0.9719 | 0.8267 |
| | Woven defects(cm) | 0.9506 | 0.9811 |
| Finishing model | Steam shrinkage of warp(%) | 0.9189 | 0.8514 |
| | Steam shrinkage of weft(%) | 0.7075 | 0.5895 |
| | Seam slippage of warp(mm) | 0.8764 | 0.7922 |
| | Seam slippage of weft(mm) | 0.9061 | 0.5719 |
| | Crease recovery angle(°) | 0.9882 | 0.5548 |
| | washed strain rank | 0.8607 | 0.9475 |
| | Water shrinkage of warp(%) | 0.7982 | 0.9180 |
| | Water shrinkage of weft(%) | 0.8167 | 0.9142 |

### 5.2 Characteristic Variables and Similar Cases Retrieval

As mentioned foregoing, the main variables that can characterize products should be firstly considered for CBR system. In this paper, the characteristic parameters designed the worsted wool fabric include fiber mean diameter ($x_1$), warp/weft count ($x_2/x_3$)), compactness ($x_4$), unit weight ($x_5$), and wool percentage ($x_6$), stitch structure ($x_7$), which are considered as the retrieved rules in the case database. Sometimes, the cloth physical properties under the corresponding process technics should be retrieved and compared, such as fuzzing and pilling ($x_8$), steam shrinkage of warp and weft (%) ($x_9/x_{10}$), seam slippage of warp and weft (cm) ($x_{11}/x_{12}$), crease recovery angle of warp and weft (°) ($x_{13}$), washed strain rank ($x_{14}$). Some of them can be retrieved according to demanded fabric quality by the clients. For their respective importance when solving the new problem, i.e. the corresponding weights of each variable, which are varied with the practical demands, is experientially decided in this research using the professional knowledge and experts experiences when a new problem will be carried out. The detailed values are listed in table 2. In other words, not all the characteristic variables must be considered when a similar case is retrieved.

For example, when an order form maybe be accepted none but warp count (68s/2), weft count (46s/1), and stitch structure (2) are entirely satisfied, these three design characteristic variables should be firstly considered. Therefore, the weights of these three factors are evaluated as 1.00. On the other hand, the steam shrinkage of warp and weft is not exceeded 0.5%, and the seam slippage of warp and weft not exceeded 4.00 mm respectively. Because the requirement of fabric physical property is not very rigorous, the weights are evaluated as 0.50. According to the above mentioned and CBR mechanism, the retrieved results are seen in table 2. SIM denotes the similarity degree between the new problem and the former cases in the historical database in terms of the design characters and the cloth physical properties.

### 5.3 Analysis and Optimization of Most Similar Cases

For worsted fabric, because the stitch structure is firstly taken into account, the stitch structure of case 4# is not accorded with the others, which should be obviously excluded. Therefore, the adjusting and optimizing of process route and variables ought to aim at the cases 1#, 2#, or 3#, among which the case 1# is highest. Then the data of case 1# is extracted from the historical database in order to be prepared for analyzing and utilizing. In addition, the steam shrinkage of weft direction ($x_{10}$) for 2# and 4# fabric is negative value, which means the width of these fabrics to be increased after steamed. Maybe this is related with the finishing process, and not to be discussed detailedly in this paper.

**Table 2** The retrieved results of the similar cases

| Var. | $w_i$ | $f_i$ | 1# | 2# | 3# | 4# |
|---|---|---|---|---|---|---|
| $x_2$ | 1.00 | 68s/2 | 68s/2 | 68s/2 | 68s/2 | 84s/2 |
| $x_3$ | 1.00 | 46s/1 | 46s/1 | 44s/1 | 52s/1 | 44s/1 |
| $x_7$ | 1.00 | 2 | 2 | 2 | 2 | 1.5 |
| $x_9$ | 0.50 | 0.50 | 0.50 | 0.70 | 0.50 | 0.40 |
| $x_{10}$ | 0.50 | 0.40 | 0.40 | -0.90 | 0.40 | -0.80 |
| $x_{11}$ | 0.50 | 3.00 | 3.24 | 4.60 | 3.24 | 5.60 |
| $x_{12}$ | 0.50 | 3.50 | 3.52 | 4.00 | 5.30 | 4.20 |
| **SIM** | — | — | 1.000 | 0.9858 | 0.9592 | 0.8422 |

Under the unaltered process and variables of case 1#, the properties of the worsted product are predicted by ANNs models. The values that are listed in table 3 (Predicted results before adjusting, **PRBA**) mean that the results of the properties predicted are almost accorded with the actual quality shown in table 3 (Predicted results after adjusting, **PRAA**). However, some properties of the product are undesirable like thick places per kilometer (+50%), thin places per kilometer (-50%), and ends-down per 1000 spindle hours. Because these will affect the weaving performance and even the handle of the end-used products in the future, the techniques adjusting must be adopted for improving the processing and the quality.

Using ANNs advantages mentioned in *section* 4 before a new processing begins in order to improve the end-used worsted quality, the sensitive variables consist of fiber length discrete coefficient (*Ldc*), short fiber content (*Sfc*), top weight (*Tw*), top weight unevenness (*Twu*), fore-spinning total doubling ratio (*Ftdr*), sizing percentage (*Sp*), loom speed (*Ls*) through the large number of experiments. The predicted values are listed in table 3 (Measured Results, **MR**) after the sensitive parameters of the old case are adjusted, which indicates that the quality has been sharply improved. And thick places per kilometer (+50%), thin places per kilometer (-50%), and ends-down per 1000 spindle hours all declined from 112.00, 660.00, and 100.00 to 36.13, 208.11, and 35.54 respectively. In addition, yarn neps has also been obviously improved.

**Table 3** The predicted results before and after adjusting variables

| parameters | **MR** | **PRBA** | **PRAA** |
|---|---|---|---|
| Yarn unevenness(%) | 21.30 | 20.96 | **19.19** |
| Thick places per kilometer(+50%) | 112.00 | 112.00 | **36.13** |

| | | | |
|---|---|---|---|
| Thin places per kilometer (-50%) | 660.00 | 659.84 | **208.11** |
| Yarn neps (+200%) | 24.00 | 24.00 | **3.00** |
| Hairiness (m/m) | 2.30 | 3.80 | 3.80 |
| Strength(cN) | 201.00 | 200.37 | 184.56 |
| Extension at break(%) | 19.20 | 18.93 | 17.37 |
| Ends-down 1000 spindle hours | 100.00 | 95.31 | **35.54** |
| Weaving efficiency(%) | 85.00 | 85.00 | 86.36 |
| Woven defects(cm) | 200.00 | 203.25 | **102.70** |
| Steam shrinkage of warp(%) | 0.50 | 0.44 | 0.55 |
| Steam shrinkage of weft(%) | 0.40 | 0.60 | 0.44 |
| Seam slippage of warp(mm) | 3.24 | 4.13 | 4.07 |
| Seam slippage of weft(mm) | 3.52 | 3.71 | 3.66 |
| Crease recovery angle(° ) | 318.00 | 315.17 | 313.58 |
| washed strain rank | 3.00 | 2.85 | 2.64 |
| Water shrinkage of warp(%) | 0.20 | 0.46 | 0.52 |
| Water shrinkage of weft(%) | 0.60 | 0.90 | 0.94 |

Because all these could be finished rapidly in the computer, avoiding practical experimental time that the trial sample is designed and processed in the machines using the certain quantity of fibers. Therefore, the manufacturing cycle of the production sharply shortened. At the same time, the quality and progressing performances of the worsted can be understood in advance, which benefits to control the quality and economize cost.

## 6 Conclusion

As a set of very powerful mathematical techniques for modeling, controlling, and optimizing that 'learn' directly from historical data, ANNs is suitable and valid for solving non-linear relationship during textile processing. CBR is an analogical reasoning method, and the similar cases can be rapidly and exactly retrieved through the similarity.

Utilizing ANN predicting model, the sensitive process variables that affect the performances and quality of yarn and fabric have been extracted in this research. After these variables are adjusted, the predicted results are more satisfied and the new process line can be formed to be put into practice.

The combination of ANN and CBR not only sharply shorten the design cycle and save the cost, but benefit to control the quality and enhance the efficiency, which play an important role in promoting the intelligentized level of the textile industry.

**Acknowledgement**

Support for this project was provided by the State Economic and Trade Commission and Shangdong RuYi Group. The efforts of the following people of Machine college of Donghua University in software programme are gratefully acknowledge: Doctor Xiang Qian and Doctor Lv Zhijun. And thanks Mr. Zhao Hui in RuYi Group for sampling and testing.

**Reference**

[1] J. Keeler. Vision of neural networks and fuzzy logic for prediction and optimization of manufacturing processes. In "Applications of Artificial Neural Networks" III (SPIE Vol. 1709) ,1992, 447-456

[2] M.C. Ramesh, R. Rajamanickam, and S. Jayaraman. The prediction of Yarn Tentile Properties by Using Artifical Neural Networks. *J. Text Inst.*,1995, 86, 459~469

[3] A. Guha, R. Chattopadhyay, and Jayadeva. Predicting yarn tenacity: A comparison of mechanistic, statistical, and neural networks models. Journal of the Textile Institute, 2001, 92(2), 139~145

[4] F.H. She, L.X. Kong, S. Nahavandi, and A.Z. Kouzani. Intelligent Animal Fiber Classification with Artificial Neural Networks. *Textile Res. J.*, 2002, 72(7), 594~600

[5] C. Luo and C. Hossein. Color Grading of Cotton Part Ⅱ: Color Grading with an Expert System and Neural Networks. *Textile Res. J.*,1999, 69(12), 893~903

[6] C. Luo and D. L. Adams. Yarn Strength Prediction Using Neural Networks Part Ⅰ: Fiber Properties and Yarn Strength Relationship. *Textile Res. J.*,1995, 65(9), 495~500

[7] F. Pynckels, P. Kiekens, S. Sette, L. Van Langenhove, and K.Impe. Use of Neural Nets for Determining the Spinnability of Fibers. *J. Textile Inst.*, 1995, 86(3), 425~435

[8] S. Sette, L. Boullart, L. Van Langenhove, and P. Kiekens. Optimizing the Fiber-to-Yarn Production Process with a Combined Neural Network/Genetic Algorithm Approach. *Textile Res. J.*, 1997, 67(2) , 84~92

[9] Huang Chang chiun and Yu Wenhong. Fuzzy Neural Network Approach to Classifying Dyeing Defects. *Textile Res. J.*, 2001, 71(1), 100~104

[10] Chen Peiwen, Liang Tsair-chun, Yau Hon-fai, Sun Wan-li, Wang Nai-chueh, Lin Horng-chyi, and Lien Rong-cherng. Classifying Textile Faults with a Back-Propagation Neural Network Using Power Spectra. *Textile Res. J.*, 1998, 68(2), 121~126

[11] J. Fan and L. Hunter. A worsted Fabric Expert System Part Ⅱ: An Artificial Neural Network Model for Predicting the Properties of Worsted Fabrics. *Textile Res. J.*, 1998, 68(10) , 763~771

[12] C.L. Hui, T.W. Lau, and S.F. Ng. Neural Network Prediction of Human Psychological Perceptions of Fabric Hand. *Textile Res. J.*, 2004, 74(5) , 375~383

[13] A.S.W. Wong, Y. Li, and P.K.W. Yeung. Predicting Clothing Sensory Comfort with Artificial Intelligence Hybrid Models. *Textile Res. J.*, 2004, 74(1), 13~19

[14] A. Vellido, P.J.G. Lisboa, and J. Vaughan.Neural networks in business: a survey of applications (1992 – 1998). Expert Systems with Applications, 1995, 17, 51~70

[15] R. Rajamanickam, S. M. Hansen, and S. Jayaraman. Analysis of the Modeling Methodologies for Predicting the Strength of Air-Jet Spun Yarns. *Textile Research Journal*, 1997, 67(1), 39~44

[16] M. Colilla, C. J. Fernandez, and E. Ruiz-Hitzky. Case-based reasoning (CBR) for multicomponent analysis using sensor arrays: Application to water quality evaluation. Analyst, 2002,127:1580~1582

[17] B. Lees and C. Irgens. Knowledge based support for quality in engineering design. *Procs. Eleventh International Conference on Expert Systems and Their Applicatio*ns, Avignon, May 1991,257~266

[18] B. Lees, N. Rees, and J. Aiken. Knowledge-based oceanographic data analysis. Procs. Expersys-92, October 1992,561~565

[19] R. Sun and L. Bookman. How do symbols and networks fit together? AI Magazine, 1993, 14(2):20~23

[20] B. Lees and J. Corchado. Integrated case-based neural network approach to problem solving. Proceedings of the 5th Biannual German Conference on Knowledge-Based Systems: Knowledge-Based Systems - Survey and Future Directions. March 1999, 157~166

[21] J. M. Corchado, B. Lees, and N. Rees. A multi-agent system "test bed" for evaluating autonomous agents. Proceedings of the first international conference on Autonomous agents, California, United States, 1997, 386~393

[22] S. Krovvidy and W. C. Wee. Wastewater treatment system from case-based reasoning. Machine Learning, 1993, 10:341~363

[23] G. Kock. The neural network description language CONNECT, and its C++ implementation. *Technical report, GMD FIRST Berlin*, Universitat Politechnica de Catalunya, August 1996

[24] J. Corchado and B. Lees. A hybrid case-based model for forcasting. *Applied Artificial Intelligence*, 2001, 15(2/1):105~127

[25] Daqing Chen and P. Burrell. Case-based reasoning system and artificial neural networks: Areview. Neural Computing & Application, 2001, 10:264~276

[26] J. Kolodner. Case-Based Reasoning, San Mateo, CA: Morgan Kaufmann, 1993

[27] A. Aamodt and E. Plaza. Case-based reasoning: foundational issues, methodological variations, and system approaches. AI Communication, 1994, 7(1): 39~59

[28] I. Waston. Applying Case-Based Reasoning: Techniques for Enterprise Systems. Morgan Kaufmann, 1997

[29] D. E. Rumelhart, G.E. Hinton, and R.J. Williams. Learning internal representations by error propagation. In: Parallel Distributed Processing: Explorations in the Microstructures of Cognition, ,D.E. Rumelhart an J.L. McClelland, Eds., MIT Press, Cambridge, MA, 1986, vol. 1, 318~362

[30] E. Rich and K. Knight/ artificial Intelligence. McGraw-Hill, New York, NY, USA, 1991, 487~524

[31] J. Hertz, A. Krogh, and R.G. Palmer. Introduction to the theory of neural computation. Addison-Wesley, Reading, MA, USA, 1991

# Proposed Life--Cycle Model For Web-Based Hypermedia Applications Development Methodologies.

Shazia Arshad ,M. Shoaib and A. Shah
Department of Computer Science & Engineering
University of Engineering and Technology
Lahore, Pakistan

***Abstract***-Developing WBHAs is moving fast due to an explosive increase of Internet/WWW use. Furthermore, the ability to use the internet/WWW tecnologies in closed environements (e.g., intranet and extranet) have also helped in the spread of WBHAs development. However, the classical life-cyle model reveal several limitations in term of describing the process of developing WBHAs. Recently, several design methods such as RMM [15], HDM [1], OOHDM [25], EORM [3] have been proposed for the design requirements of WBHAs. However, none of them have addressed the life-cyle of WBHAs development. Hence, in this paper, we first study different WBHA architectures, then we identify a list of requirements for WBHAs development. After that, we show how the waterfall model lacks the capability to satisfy these requirements. Finally, we propose a new life-cycle model for WBHAs development.

## 1. Introduction

During the last years, a large number of commercial WBHAs emerged on the Internet. Many companies are using this medium as a new means of communication with their customers. Their extent is growing with the increasing number of tasks they fulfill. The WWW provides a unified view of large companies, their product catalogue, their divisions, their various commericial, social and cultural activities. By means of gateway-scripts to inhouse information systems the WWW becomes a platform-independent, easy-to-use interface to the company's information resources [22, 23, 24].

Developing WBHAs is not a trivial task. WBHAs consisting of a large number of interlinked pages and several gateway scripts have to be developed in a systematic way. Consequently, the need for the design of WBHAs has been realized and Web design has become one of the major topics of important conferences and meetings in this area. There are several approaches for the design of traditional hypermedia applications, a field very closely related to WBHAs design. HDM, EORM or OOHDM, adopt modeling constructs of database design methodologies and add hypermedia related features in order to integrate the aspect of navigational design. RMDM, the design technique of the Relationship Management Methodology (RMM) starts with an Entity Relationship diagram and extends the notation by access structures, leading to a navigational model of the resulting hypermedia application. Therefore, RMM offers an easy way for system analysts to extend an existing data model towards a hypermedia application.

Most of the proposed WBHAs development methods are only concerned with the design aspects of WBHAs. However, WBHAs does not only need to be designed, but, they need to be analyzed, designed, then implemented. Moreover, a life-cyle describing WBHAs development is also desirable. In this paper, we discuss the requirements needed to be supported by WBHAs development methods. We also describe the need for a new life-cycle model for WBHAs development. This leads us to propose a new life-cycle model for WBHAs development.

This paper is organized as follows. Section one describes different WBHA architectures. For each architecture, both components that constitute the architecture and relationships between these components are described. Section two describes the classical life-cycle model. Section three describes the requirements of WBHAs development. Section four describes why the classical life-cycle is not appropriate for WBHAs development. Finally the proposed life-cycle model for WBHAs development is described in section five.

## 2.0 BackGround

In this section we describe the background knowledege needed to understand the goal of this paper.

### 2.1 WBHA architectures:

Developing WBHAs is moving fast due to an explosive increase of Internet/WWW use. Now, there are many languages/technologies that can be used in developing WBHAs such as Hypertext Markup Language (HTML), Common Gateway Interface (CGI), Java applets, gateway to relational databases. These languages/technologies have enabled the creation of different architecture types for WBHAs. These architecture types are as follows:

- Static-content model.



*T. Sobh and K. Elleithy (eds.), Advances in Systems, Computing Sciences and Software Engineering*, 325–332.

- Server-side operation model.
- Server-side database interface model.
- Client-side dynamic-content model.
- Distributed computation model.
- Mixed environment model.

**2.2 The Waterfall Model**

The software life-cycle is simply the entire existence of a software product. Another way of looking at the life cycle is to consider it as the process model; i.e. a model for the development and use of software. The waterfall life-cycle model views the development process as including a series of discrete phases. In its simplest form each phase is completed and 'signed-off' before commencing the next stage. The stages that are typically used are to analyze and specify the system, then to design the system, then to implement the software, test the final system, and finally to operate and maintain the software [5,14,15,16].

Although this form is useful from the point of project management, in practice the various stages typically overlap, and feedback is usually provided from each stage to the previous. However, there are a number of criticisms of the waterfall model. These include: First, it freezes the specification at too early a stage of the development, and that it makes iteration difficult. Second, A working version of the program will not be available until late in the project timespan. Finally, it suits a specific class of software applications [5, 17,18,19,20].

Consequently, a number of other models attempt to resolve these problems such as prototype model and spiral model. The prototype model as shown in Figure 4.1has extended the water fall model to include theprototype phase. In the prototype model, the prototype phase is the first phase of software development followed by development phases of the waterfall model. The prototype phase involves developing a quick program that helps both customers and developers in understanding the requirements. Then, This program is used as a starting point from which analysis, design, and implementation phases are performed [1,2,5,21].

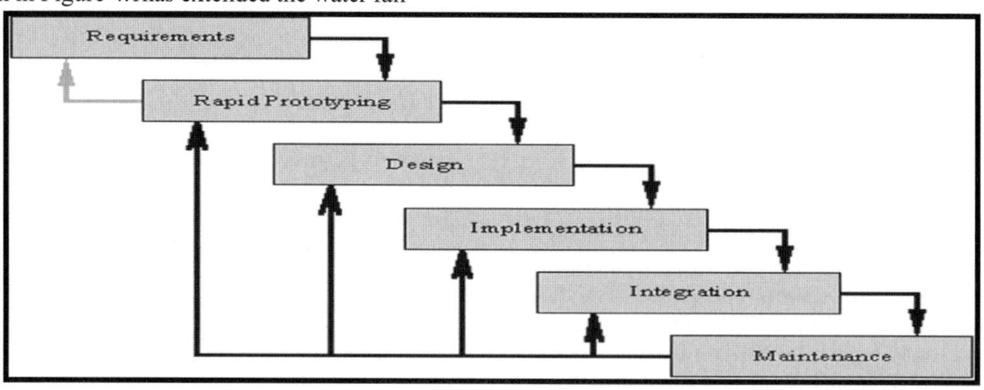


**Figure 4.1** Prototype model

The spiral model has been developed to encompass the best features of both the waterfall model and the prototype model with the addition of *risk analysis*. Typical Cycle of the Spiral model (see Figure 4.2) are as follows:

**1.** Identification of :

- objectives (of the portion of the product being elaborated)
- alternative means of implementing this portion of the product (design A, design B, buy)
- constraints imposed on the application of the alternatives (cost, schedule, copyright)

**2.** Evaluate the alternatives relative to the objectives and constraints

- Identify areas of uncertainty that are significant sources of risk.
- Evaluate the risks (e.g., by prototyping, simulation, benchmarking).

**3.** The next step is determined by the relative remaining risks. Each level of software specification and development is followed by a validation step.

**4.** Each cycle is completed by the preparation of plans for the next cycle and a review involving the primary people or organizations concerned with the product. The review's major objective is to ensure that all concerned parties are mutually committed to the approach for the next phase [3,4,5].

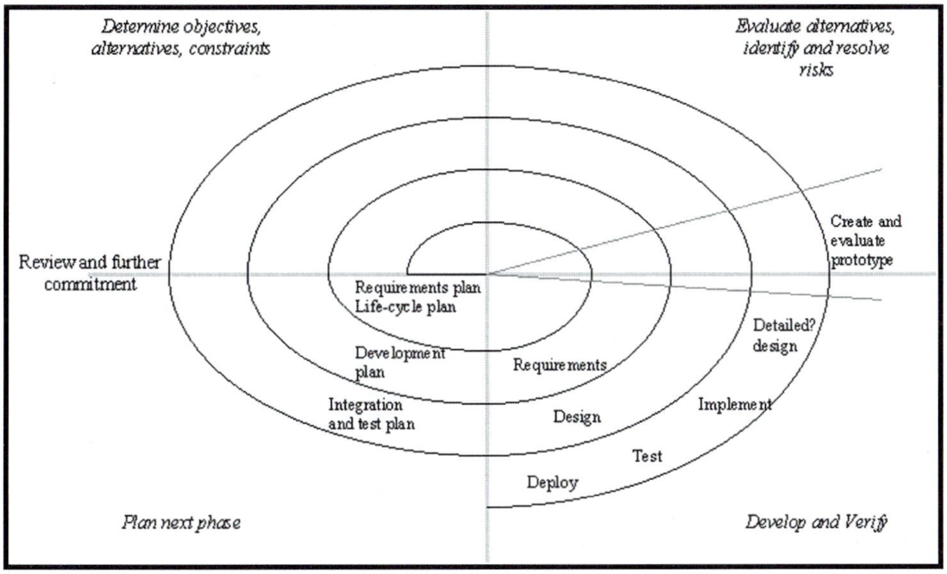


**Figure 4.2:** Spiral Model

## 3.0 Requirements of WBHAs development.

In this section, we describe the essential requirements of WBHAs development. These requirements are desirable to be supported by both the life-cycle model and development methods proposed for WBHAs development.

- **Migration issue:** Many organizations are aiming at migrating from their traditional software application to WBHAs. However, following an ad-hoc approach may result in the increase of both cost and time needed for the migration process. Furthermore, the final WBHA may suffer from poor performance, hard maintenance, and low usability.

Migration issue is a big challenge, and need to be handled and specified carefully in the life-cycle model of WBHAs development. Traditional life-cycle model can not handle the migration issue becuase it assumes that software applications are created from scratch. What is needed, is a new life-cycle model that is desirable to be capable of both taking advantage of existing software components and reducing cost and time of migration by specifying where the migration process should start.

- **WBHAs architectures issue:** When the WWW was initially developed, its primary goal was to be used as a simple GUI retrieval system of information stored in the Internet. Now, The WWW supports many techniques/languages that can be used to build different application architectures (see section one). The development method is desirable to be capable of supporting the development process needed to create these different application architectures.

- **Integration Issue:** WBHAs development is a very complex task, it involves collecting a huge number of multimedia information and linking them together. For example, *Microsoft WWW site* contains now over 300,000 pages [11,12,13]. To be able to control this huge number of interlinked Web pages, the development process should be performed in a decentralized way, which means that each division and group in an organization will be responsible to develop a sub-WBHA and then integrate these sub-WBHAs to represent the global view of the final WBHA.

Integration issue is not only important and useful because it simplifies the development of large-scale WBHAs. But, because actually, many WBHAs are developed in this way. Typical examples include:

- The management often only wants to determine a very global structure of the WBHA. The details of the model are designed by the functional departments of the company. Thus e.g. the marketing department is responsible for the presentation of the firm's products and the personnel department informs about new job offers or trainee programs.
- Many multinational organizations maintain local Web-servers at their national subsidiaries. It is necessary to integrate the various models to a unified one, not only to document the

WBHA but to detect redundancies and lead to a consistent, readable WBHA.

- A company's Web-site might, for example, actually consist of two or more completely different server systems: One of them is maintained by the marketing department and describes the company's products while the other one, a secure server operated by the company's internet provider, is used to perform the actual transactions. To generate locally distributed WBHA (relative and absolute URLs) out of the specified models it is necessary to consider this aspect in the design phase.
- Many organizations maintining their intranet. Then they decided to integare these intranets together to create an extranet. Therefore, a global view is needed to let users access information residing on these intranets easily.

The need for the collaborative development of WBHAs is obvious. But, the question that need to be answered is: *does the implelemntation environement support it?* The answer is yes. Mainely because: First, the WWW is a distributed system. This means that both data and processing can be stored and performed respectively in any computer connected to the internet. Second, the transparent use of physically seperated WBHAs. Third, the technology that allows different intranets to be connected toghether (i.e., extranet). Therfore, there is a strong need to handle a distributed development process.

- **Incremental development issue:** Most of the WBHAs running on the Web are almost developed incrementally. For example, when *Microsoft Web site* was created in 1994, it was containing a few thousands of Web pages. Now, *Microsoft Web site* contains more than 300,000 Web pages [9,10]. Actually, WBHAs are usually created in small sizes. But, as more multimdia information becomes available and more users are requesting more functionality, the WBHA gets larger incrementally. Incremental development need to be handled and specified in WBHAs development methods. Otherwise, as the WBHA gets larger as the maintenance gets difficult, and you may end up with unusable WBHA.

. **Multimedia information issue:** Traditional development methods are usually concerned only with building the metadata of a software application. however, in WBHAs this is not the case since the WBHAs development involves creating both the metadata and the data itself. Furthermore, WBHAs usually contain a huge amount of multimedia information that require special skills, equipment, and people to create. It is also a time consuming and costly task. Therefore, the process of gathering, creating and linking multimedia information need to be handled and specified in a development method of WBHAs.

. **Hyperlink issue:** The degree of success of a WBHA is directly related to the developer's ability to capture and organize the structure of complex software problems in such a way as to make it clear and accessible to a wide range of users. To control the potential explosion of the number of hyperlinks, a WBHA developer does not really use every word as a hyperlink, but rather tries to directly interconnect only the most important and meaningful parts of the information to convey the overall meaning of that part of information in a more natural way [1,6,7,8]. Furthermore, linking related multimedia information requires determining the location of associated information. For large-scale WBHAs, the need to remember the location of related information becomes a major cognitive burden on WBHAs developers affecting the overall quality of work and productivity.

Therefore, The development method is desirable to support the process of linking multimedia information. it is also desirable to support ways of determining related hypermedia information. Finally, it is desirable to support developers with an insight of the WBHA pages and how they are linked together without going into much of the details of their implementation and the multimedia information they contain.

## 4.0. Proposed WBHAs Development Life-Cycle Model (WDLCM).

In this section, a new life-cycle model for

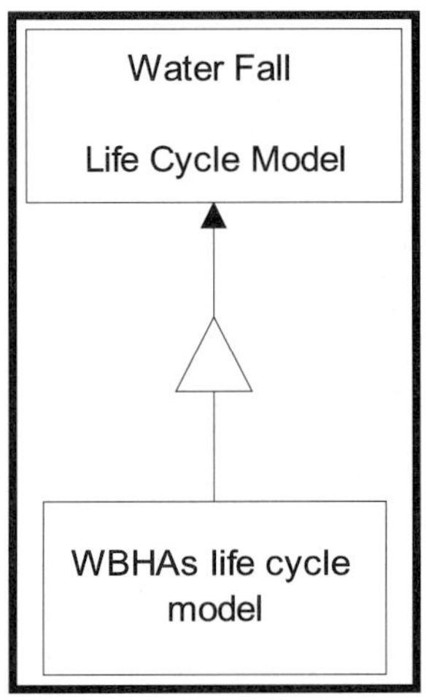


**Figure 4.3:** The relationship between waterfall model and WDLCM

WBHAs development is proposed. WDLCM extends the waterfall model to satisfy the requirements of WBHAs development. Figure 4.3 shows the relationship between the WDLCM and the waterfall model. The relationship is an inheritance relationship where the waterfall model represents the supper class and the WDLCM is its subclass. The WDLCM subclass inherits all the features of the waterfall model class and extends it to incorporate the special requirements of WBHAs development.

### 4.1 Overview of the WDLCM

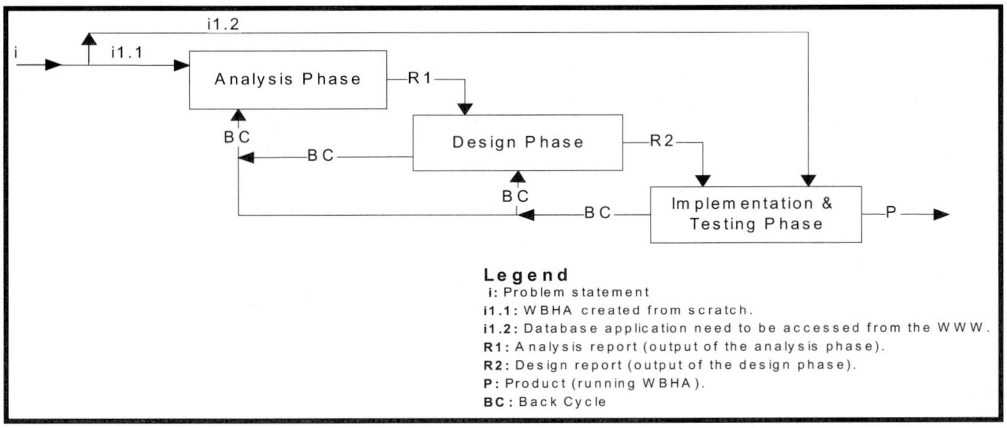


**Figure 4.4:** The proposed life-cycle model for WBHAs development

Figure 4.4 shows the proposed WDLCM. It includes the same development phases of the waterfall model. However, it extends the waterfall model to meet requirements of WBHAs development: migration issue, integration issue, incremental development, metadata and data creations, and multimedia information gathering process.

In Figure 4.4, The input to the analysis phase is represented with an arrow labeled with **i**. The input could be of two types. First, a problem statement describing the need for creating a WBHA (labeled with **i1**). Second, a problem statement describing the need for either migrating from existing application to WBHA or integrating existing WBHAs into one single WBHA (labeled with **i2**). The output of the analysis phase is the analysis report shown in Figure 4.4 with an arrow labeled **R1**. The input to the design phase could be of two types. First, the analysis report labeled with **R1.2**. Second, the need for either migrating to distributed computing model labeled or integrating existing WBHAs into single WBHA (labeled with **i2.1**). The output of the design phase is the design report. Finally, the input to the implementation phase could be of four types. First, The design report labeled with **R2**. Second, The analysis report labeled with **R1.2**. Third, the multimedia information gathered form different media sources. Fourth, the need for migrating to one of three architecture types: static-content model, server-side operation model, or server-side database interface model. The output of the implementation phase is a working version of the WBHA.

Figure 4.5 shows a typical senario followed while developing WBHAs. The senario starts with the requirements for creating a new WBHA. In this case, the paths **i** then **i1** is followed.

After that, new requirements come specifing the need for incremental development. In this case, path **u** is followed, then developers have the choice to follow **u1**, **u2**, or **u3** paths. Moreover, New requirements come specifing the need for migrating from existing database application to **server-side database interface model**. In this case, path **i**, **i2**, then **i2.2** is followed. Finally, new requirements come specifing the need to migrate two WBHAs. In this case, the path **i**, **i2**, then **i2.1** is followed.

| **Requirements** | **Path** |
|---|---|
| 1- Create WBHAs (**WBHA_1**) | **(i) . ( i1)** |
| 2- Incremental development | **(u) . (u1\| u2\|u3)** |
| 3- Migrate Database App. to Server-side database-interface model (**WBHA_2**) | **(i) . (i2) . (i2,2)** |
| 4- Integrate **WBHA_1** and **WBHA_2** into **WBHA_3** | **(i) . (i2) . ( i2.1)** |
| 5- Incremental development | **(u) . (u1\| u2\|u3)** |

**Figure 4.5:** A typical senario of WBHAs development

In the analysis phase, the WBHA to be developed is analyzed and specified. This involves defining: the information structure, user types, user tasks, user access paths, classifying objects into static and dynamic objects, and defining operations performed by the WBHA. The analysis report is used to design the WBHA. The design phase involves specifing: Web pages structure, access primitives, partitioning dynamic objects into client and server objects. Finally, the implementation phase goes through three steps: multimedia information gathering, Web page templates creation, and multimedia information linking. In the implementation phase, the analysis report is used to perform the multimedia information gathering, and The design report is used to create HTML templates. The last step of the implementation phase involves the linking of multimedia information and filling HTML templates.

### 4.2 How the WDLCM is meeting the requirements of WBHAs development:

In this section, we describe how the proposed WDLCM meets requirements of WBHAs development.

**4.2.1 Migration issues:** The migration process is defined as the process of converting a traditional software application to a WBHAs. Migration can be of different types:

- Migration to static-content model.
- Migration to server-side operation model.
- Migration to server-side database interface model.
- Migration to distributed computation environment.
- Migration to mixed environment.

**Migration to static-content model:** This type of migration is useful only if the existing application is a data intensive application. Typical examples of data intensive applications include:

1. Traditional hypermedia applications such as MS Encarta and MS windows help.
2. Relational database application.
3. Object-oriented database application.

In this type of migration, developers can go directly to the implementation phase assuming that both analysis and design phases are performed. For example, converting **MS Encarta** to static-content model requires converting each screen of **MS Encarta** to a web Page. Then linking those Web pages in the same way **MS Encarta** links the screens. Therefore developers can follow the line labeled **i2.2** in Figure 4.4.

**Migration to server-side operation model.** Figure 4.6 shows the work process of server-side operation model. In the migration process, developers need to write two things:

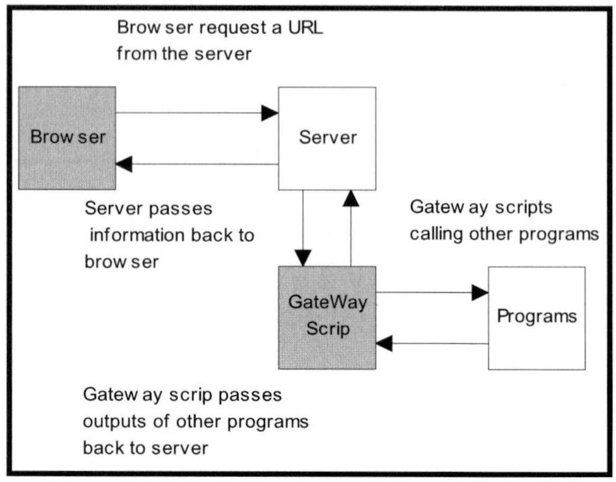


**Figure 4.6:** Working process of server-side operation model

the gateway scripts that communicate with CGI programs and the HTML forms that are used for both entering user data and displaying information resulted from invoking those gateway scripts. To perform these two activities, developers can go directly to the implementation phase. Therefore developers can follow the line labeled **i2.2** in Figure 4.4.

**Migration to server-side database interface model.** The working process of server-side database interface model is the same as the server-side operation model except that the server CGI program is an RDBMS. For example, to enable a client-server database application running under ***oracle*** RDBMS as a server and ***MS access*** as a client to be accessed from the WWW, we need in the implementation phase to map the ***Ms Access*** forms into HTML forms and to write simple CGI scripts that communicate with ***oracle*** RDBMS to perform the required SQL quires. Therefore, the migration process can start from the implementation phase following the line labeled with **i2.2**.

- **Migration to distributed computation model:** In this type of migration, we need to partition the objects/classes into client and server. An optimal object partitioning will greatly improves the overall performance of WBHAs. The existing application shows the objects that constitute the application. But it does not show the partitioning of these objects into clients and servers. In the proposed life-cycle model, the object partitioning is considered a design concern. Therefore, the migration process can start from the design phase following the line in Figure 4.4 labeled with **i2.1**.

- **Migration to mixed environment:** In this type of migration, we need to specify which objects/classes will be static and dynamic. Then for dynamic objects we need to partition the objects into client and server. In the proposed life-cycle model, deciding which object will be static and which object will be dynamic is considered an analysis concern. Therefore, the migration process can start from the analysis phase following the line labeled with **i1** in Figure 4.4.

**4.2.2 Integration issue:**

The development of WBHAs is sometimes a complex task that can not be perfomed in a centralized way. Therefore, a more reasonable approach is to first allow the different departements of an organization to build their own WBHA and then to integrate them to represent the global view of the complete WBHA. This type of collaboratrive development is will supported by the WWW since it is a distributed system.

The problem with this approach is that it is not will supported by the waterfall model. The waterfall model assumes that the software to be built is created from scratch but not as existing components that need to be integrated toghether. In the proposed WDLCM, the integration of the different components of a WBHA is a design concern. Therefore, when existing different WBHAs need to be integrated together, the integration process can start from the design phase directly. In the design phase, designers have to check for redundancies and inconsistencies and assigns a uniform layout to all WBHAs. Moreover, in the implementation phase the developer can determine, whether sub-WBHAs will be implemented locally or on a remote server.

**4.3 Incremental Development.**

Incremental development suggests that new requirements to existing WBHAs are categorized into three different types:

- New requirements that affect analysis components such as adding a new object or functionality to the WBHAs.
- New requirements that affect design components such as access primitives, objects partitioning, or Web page components.
- New requirements that affect implementation components such as the behavior or appearance of multimedia objects, or storage directory structure.

Figure 4.4 shows three different paths for incremental development: 1) when new requirements affect analysis, the development process is desirable to start from analysis phase. 2) If new requirements affect design only, the development process is desirable to start from design phase. 3) Finally, if the new requirements affect only implementation phase, then the development process is desirable to start from the implementation phase.

**4.4 Metadata and data creation:**

The classical waterfall model use the problem statement as input to create the metadata of intended software application. The creation of data is assumed to be after the system is installed. However, for WBHAs, developers usually submits to users both the working version of the program and the data the program operate on. In the proposed life-cycle model for WBHAs development, the implementation phase accepts as input both the design report resulted from the design phase and the multimedia information collected from external sources. The implementers will use the design report to build Web page templates, then they use the collected multimedia information to fill in the page templates. After that, a working version of the WBHA will be ready to be installed on the WWW.

**4.2.5 Multimedia information gathering:**

The multimedia information gathering process is an implementation concern. However, it has to be started as early as possible. In the proposed WDLCM, the multimedia information gathering process is started after the analysis phase is finished. This way, the development process will be less time consuming and less costly.

**5.0 Conclusions and Future Works:**

In this paper, we presented the requirements need to be supported by analysis and design methodlogies for WBHAs. Then, we described how the waterfall model lacks the capabilities to support those requirements. Finally, we presented our poposed lifecycle model for WBHAs development. The proposed life-cycle is augemented to meet the requirements of WBHAS development. In the fuature, we plan to complete this work by proposing a full detailed WBHAs development method that covers analysis, design, and implementation phases.

**References**

[1 ] F. Garzotto, L. Mainetti, and P. Paolini, "HDM- A Model Based Approach to Hypermedia Application Design," *ACM Transactions on Information Systems*, Vol. 11, No. 1, January 1993, pp. 1-25.

[2 ] P. Balasubramaniam, T. Isakowitz, and E. Stohr, "Designing Hypermedia Applications,"

*Conf. Proc. of the 27th Annual Hawaii International Conference on System Sciences, IEEE Computer Society Press*, January 1994, pp. 354-364.

[3 ] D. Lange, "An Object-Oriented Design Method for Hypermedia Information Systems," *Conf. Proc. of the 27th Annual Hawaii International Conference on Systems Sciences, IEEE Computer Society Press*, January 1994, pp. 366-375.

[4 ] J. Rumbaugh, M. Blaha, W. Premerlani, F. Eddy, and W. Lorensen, " Object Oriented Modeling and Design," Prentice Hall Inc. 1991.

[5] J. Sommerville,"Software Engineering," Addision-Wesley, 1992.

[6] R. Pressman," Software Engineering," McGraw-Hill International Editions, 1992.

[7] F. Garzotto, P. Paolini, and D. Schwabe, "Authoring-in-the-Large: Software Engineering Techniques for Hypermedia Application Design," Conf. *Proc. 6th IEEE Int. Workshop on SW Specification and Design*, October 1991, pp. 193-201.

[8 ] J. Tomek, S. Khan, T. Muldner, M. Nassar, G. Novak, and P. Proszynski "Hypermedia-Introduction and Survey," *Journal of Microcomputer Applications*, 1991, pp. 63-103.

[9] F. Garzotto, L. Mainetti, P. Paolini, and P. Milano, "Navigation Patterns in Hypermedia DataBases," *Conf. Proc. of the 26th Annual Hawaii International Conference on System Sciences, IEEE Computer Society Press*, 1993, pp. 269-379.

[10] A. Ginige, D. Lowe, and J. Robertson, "Hypermedia Authoring," *Conf. Proc. On Multimedia*, IEEE, 1995, pp. 24-35.

[11] D. Schwabe and G. Rossi, "Building Hypermedia Applications as Navigational Views of Information Models," *Conf. Proc. of the 28th Annual Hawaii International Conference on System Sciences, IEEE Computer Society Press*, 1995, pp. 231-240.

[12] V. Balasubramanian and M. Turoff, "A Systematic Approach to User Interface Design for Hypertext Systems," *Conf. Proc. of the 28th Annual Hawaii International Conference on System Sciences, IEEE Computer Society Press*, 1995, pp. 241-250.

[13 ] S. Guo, W. Sun, Y. Deng, W. Li, Q. Liu, and W. Zhang, "Panther: An Inexpensive and Integrated Multimedia Environment," *Conf. Proc. of the 27th Annual Hawaii International Conference on Systems Sciences, IEEE Computer Society Press*, January 1994, pp. 382-391.

[14] Herman and G. Reynolds, "MADE: A Multimedia Application Development Environment," *Conf. Proc. of the 27th Annual Hawaii International Conference on Systems Sciences, IEEE Computer Society Press,* January 1994, pp. 184-194.

[15] D. Isakowitz, E. Stohr, and P. Balasubramanian, "RMM: A Methodology for Structured Hypermedia Design," *Communication of the ACM*, Vol. 38, No. 8, August 1995, pp. 34-44.

[16] J. Walker, "Requirements of an Object-Oriented Design Method," Software Engineering Journal, March 1992, pp. 102-113.

[17] J. Nerson, "Applying Object-Oriented Analysis and Design," *Communication of The ACM*, September 1992, pp. 63-74.

[18] M. Thuring, J. Hannemann, and J. Haake, "Hypermedia and Cognition: Designing for Comprehension," *Communication of the ACM*, Vol. 38, No. 8, August 1995, pp. 57-66.

[19] H. Fernandes, "Online and Hypermedia Information Design," *Conf. Proc. On Expanding Technologies for Technical Communication,* IEEE, 1991, pp. 28-32.

[20]C. Meghini, F. Rabitti, and C. Thanos, "Conceptual Modeling of Multimedia Documents," IEEE Computer, October 1991, PP. 23-30.

[21] E. Seidewitz, "General Object-Oriented Software Development: Background and Experience," *Conf. Proc. of the 21st Annual Hawaii International Conference on System Sciences, IEEE Computer Society Press*, January 1988, pp. 262-270.

[22] E. Cho, S. Kim, S. Rhew, S. Lee, and C. Kim," Object-Orinted Web Application Architectures and Development Stratigies," IEEE computer, 1997, pp. 322-331.

[23] E. Evana and D. Rogers," Using JAVA Applets and CORBA for Multi-User Distributed Applications," IEEE Computer, June 1997, PP. 43-58.

[24] E. Yourdon," JAVA, the Web, and Software Development," IEEE Internet, Augus 1996, pp. 25-32.

# Reverse Engineering Analyze for Microcontrollers' Assembly Language Projects

M. Popa, M. Macrea, L. Mihu
Computer and Software Engineering Department
Faculty of Automation and Computers
University "Politehnica" Timisoara
2 Blv. V. Parvan, 300223 Timisoara, ROMANIA

***Abstract*-The problem of reverse engineering assembly language projects for microcontrollers from embedded systems is approached in this paper. A tool for analyzing projects is described which starts from the source files of the project to be analyzed, grouped in a Project Folder and from a Configuration files and generates diagrams for describing the program's functionality. The tool is useful for the programmer and for the project manager in different phases of a project: code review, design review and development. It is also useful for understanding and documenting older projects.**

## I. INTRODUCTION

There are many situations in which older software projects which must be updated or used in new bigger projects are very difficult to understand because of their poor documentation. To complete the documentation there is the handmade solution, difficult, time consuming and error prone or an automated solution, but it needs specific software tools and the operation is called reverse engineering.

According to [1], reverse engineering, in software, "aims at obtaining high level representations of programs. Reverse engineering typically starts with a low level representation of a system (such as binaries, plain source code or execution traces) and try to distill more abstract representations from these (such as, for example… source code, architectural views or use cases, respectively). Reverse engineering methods and technologies play an important role in many software engineering tasks, such as program comprehension, system migrations and software evolution". Reverse engineering can be done at different levels: from binaries to assembly language, from assembly language to high level language, from one high level language to another high level language or between a language to a more abstract visual representation.

The problem of reverse engineering assembly language projects for microcontrollers from embedded systems is approached in this paper. A tool for analyzing microcontrollers' assembly language projects in embedded systems is described. It receives assembly language project and a configuration file and offers a modular and symbolic representation, using diagrams, of the program. Several types of diagrams are obtained giving a complete visual representation of the program functionality, at high level.

The next section presents other similar works, the third section describes the tool which analyses the assembly language projects, in the fourth section the used technologies are detailed, the fifth section describes a case study and the last section outlines the conclusions.

## II. RELATED WORK

The reverse engineering process was treated in other papers too. Many tools for analyzing high level language projects were developed but fewer for assembly language projects.

Reference [2] describes a tool for transforming a high level language project into an assembly language one. A project written in ZIL language, which is a high level language close to C, is transformed in a project written in the Z86 microcontroller's assembly language. The decompilation process is approached in [3]. The application generates high level C or C++ source code having as inputs the compiled executables. The migration software process is addressed in [4]. A set of tools are implemented for adding the migration of C code into Java programs. Reference [5] deals with the disassembly process. Two commonly used disassembly algorithms are examined and their disadvantages are outlined. Another solution is described which make a better disassembly, detecting the situations where the disassembly may be incorrect and limiting the extend of errors. The work from [6] is closer to the embedded systems. It presents a reverse engineering tool for assembler language to ANSI-C for DSPs. A similar level transforming, from IBM 370 assembler language to C is described in [7]. The achievement from [8] deals with real-time assembly code. It transforms Z86 assembly language programs for several microcontrollers into ZIL Language.

Unlike the above mentioned achievements, in which the transformation is made from a language to another one, the tool described in this paper analyses an assembly language program for transforming it in a modular and visual representation using diagrams. It is a more abstract level showing the relationships between the modules and the functionality of the program.

## III. ASSEMBLY LANGUAGE PROJECTS ANALYZE

A tool was developed for reverse engineering analyze for microcontrollers' assembly language projects. The projects' files are grouped into modules, based on their names. All files with the same name, but different extensions will constitute a module and each of them will be parsed independently.



*T. Sobh and K. Elleithy (eds.), Advances in Systems, Computing Sciences and Software Engineering*, 333–338.

Variables, constants, macros, functions and function like macros are module members identified for each module.

Starting from the input project files (with .stc and .inc extensions) and from the Configuration file (with .xml extension), which contains the application's configuration, the tool will generate: Control and Data Flow diagrams between modules and Flowcharts and Call Trees diagrams for each function and function like macro. Microsoft Visio was used for drawing the diagrams.

The Control Flow diagram shows all function calls and function like macro substitutions between two or more modules. An example of such a diagram is presented in fig. 1.

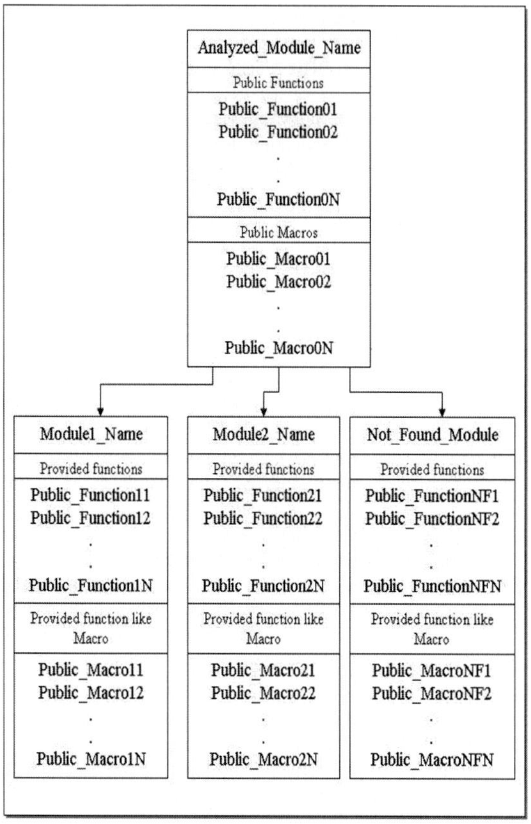


Fig. 1 A Control Flow diagram

The Control Flow diagram represent all the call made from the analyzed module, placed in the top of the diagram, and other modules, further called interfering modules.

The analyzed module is represented in a box which contains the module name and the public functions and public macros exported from that module. All the interfering modules are represented in the diagram only if a relation exists with the analyzed module. The control relationship between the analyzed module and an interfering module is indicated by an arrow. An interfering module is represented in a box that contains the module name and two sections: one for the functions required by the analyzed module from that interfering module and one for the function like macros imported by the analyzed module from that interfering module.

When parsing the source code of the analyzed module, if there are functions or function like macros required but not provided by any module belonging to the project, they will be assigned to a virtual module called "Not_found_module". It means that the project folder loaded does not contain all the modules that interact with the analyzed module.

The Data Flow diagram should represent all data exchanged between the analyzed module and the other modules of the project. It must show the origin of the data, the data flow and the access type (read, write or read/write). The Data Flow diagram will indicate for each module, all provided and required variables, constants, macros and access type. Fig. 2 presents a Data Flow diagram.

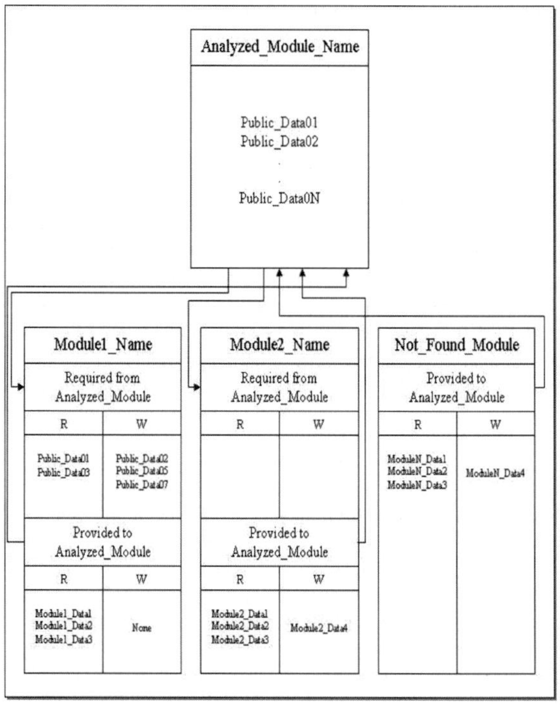


Fig. 2 A Data Flow diagram

The analyzed module is represented in a box, placed in the top of the diagram, which contains the module name and all the public data exported from the module. The interfering modules are represented in the diagram only if a relation exists between them and the analyzed module. The box which contains an interfering module has in its top the module name and two more sections, called: Required from Analyzed Module and Provided to Analyzed Module. The first section contains all the data exported from the analyzed module and used in the interfering module and the second section contains all the data exported from the interfering module and used in the analyzed module. Each section is further divided into two subsections: R (Read) and W (Write). In the Read subsection the data that is just being read is listed and in the Write subsection the data that is being altered is listed.

The data relationships between the analyzed module and the interfering modules are indicated by arrows. An arrow pointing to the analyzed module shows data imported from the interfering module. An arrow pointing to an interfering module shows that the interfering module uses data exported from the analyzed module. All the interactions shown concern the relationship between the analyzed module and the interfering modules and that's why the interactions between the interfering modules are not represented.

When parsing the source code of the analyzed module, if a data (variable, constant or macro) is found as required but not provided by any other module belonging to the project that data will be assigned to a virtual module called "Not_found_module". This module will contain only the Imported in Analyzed Module section because there is no way of finding if exported data from the analyzed module is used in other modules that are not loaded as part of the project.

The Flowchart diagram will be generated for each function or function like macro from the entire project. By analyzing the function or function like macro, the tool will distinguish each instruction type, will create adequate diagram blocks, and will connect the blocks with each other for obtaining a flowchart. Fig. 3 presents a Flowchart diagrams.

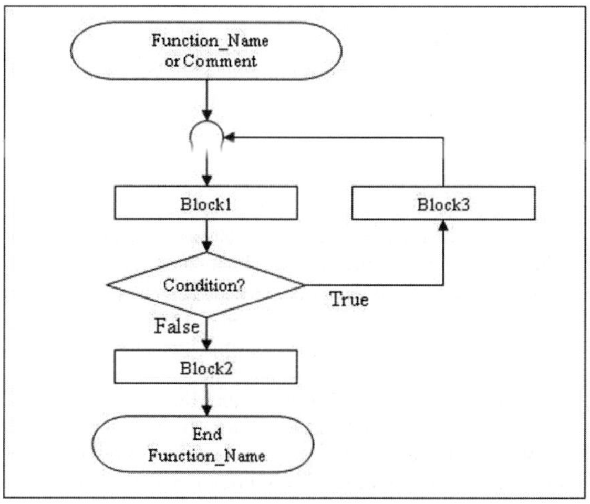


Fig. 3 A Flowchart diagram

The blocks will contain either the lines extracted from the code or the comments associated based on the rules defined further. A block is a sequence of code that will be associated with a single rectangle on the flowchart.

The Call Tree diagram will result from analyzing all the function or function like macro calls made from the analyzed method. The tree will be expanded until the leafs have empty call lists or the leafs are not found in the analyzed project, meaning the loaded project is not complete. Also if recursive calls are made, the tree will contain the call that produces the recursivity but will not be expanded anymore. Both the function and the function like macro calls must appear in the diagram. An example of a Call Tree is shown in the fig. 4.

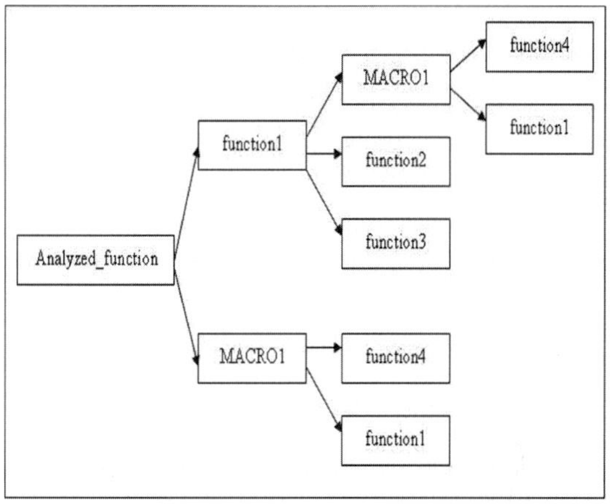


Fig. 4 A Call Tree

## IV. USED TECHNOLOGIES

XML was used for the configuration file and for storing the module to file structure. The configuration file is the file that offers to the tool the flexibility regarding the assembly language used. The format of the configuration file is presented below:

```
<configuration>
  <source_file>source_extension</source_file>
  <header_file>header_extension</header_file>
  <regex_public_names>Reg_exp</regex_includes>
  <regex_functions>Reg_exp</regex_functions>
  <regex_function_declaration>Reg_exp</regex_function_declaration>
  <regex_f_l_macros>Reg_exp</regex_f_l_macros>
  <regex_constants>Reg_exp</regex_constants>
  <regex_variables>Reg_exp</regex_variables>
  <regex_macros>Reg_exp</regex_macros>
  <instructions>
    <branches>
```

```
        <instruction mnemonic="instr_name" no_params="n" separator="c" type="XYZ" label="l" to_instr="+n/-n"/>
      </branches>
      <return>
        <instruction mnemonic="instr_name"/>
      </return>
      <other>
        <instruction mnemonic="instr_name"/>
      </other>
      <keywords>
        <instruction mnemonic="instr_name"/>
      </keywords>
    </instructions>
  </platform>
</configuration>
```

Next, the tags will be described:

- <platform> tag: a <platform> tag must be defined for each platform that will be supported; the *"name"* attribute contains the name of the platform;
- <source_file> tag: tag for the source file extension for the specified platform; only one source file extension per platform is allowed;
- <header_file> tag: tag for the header file extension for the specified platform; only one header file extension per platform is allowed;
- <regex_public_names> tag: contains regular expression for public names;
- <regex_includes> tag: contains regular expression for included files;
- <regex_functions> tag: contains regular expression for functions;
- <regex_function_declaration> tag: contains regular expression for function definition;
- <regex_f_l_macros> tag: contains regular expression for function like macros;
- regex_constants> tag: contains regular expression for constants;
- <regex_variables> tag: contains regular expression for variables;
- <regex_macros> tag: contains regular expression for macros;
- <instructions> tag: contains all the instruction classes that are taken into consideration during source code parsing.
  - <branches> tag: contains all the branch instructions of the specified platform;
    - <instruction> tag: an <instruction > tag must be presented for every branch instruction; it must have the following attributes:
      - *mnemonic* – the mnemonic of the instruction;
      - *no_params* – the number of parameters of the instruction;
      - *separator* – the separator between the parameters;
      - *type* – a three letter code XYZ meaning: X = J/C (Jump/Call), Y = N/C (Unconditioned/Conditioned), A/R/X (Absolute/Relative/Indexed);
      - *label* – the number of the parameter which contains the label to jump to (for absolute jumps and for calls);
      - *to_instr* – the number of instructions to jump over from the current position (+n/-n) (for relative jumps);
  - <other> tag: contains all other instructions of the specified platform;
    - <instruction> tag: an <instruction> tag must be presented inside <other> tag for every instruction of the specified platform, except for the ones described in the other categories; the only attribute contained is the instruction mnemonic;
  - <return> tag: contains all the return instructions of the specified platform;
    - <instruction> tag: an <instruction> tag must be presented inside <other> tag for every return instruction of the specified platform; the only attribute contained is the instruction mnemonic;
  - <keyword> tag: contains all keywords in order not to be detected as function like macro calls; the keyword list is necessary in order not to be taken as function like macros substitution; when an instruction is analyzed this configuration file is being read and a list of all instructions and keywords found in this file is being created; if the analyzed line contains an instruction that is not found in this list then it is considered a function like macro substitution.

In order to change the assembly language this configuration file is the first that must be changed.

XML is also used for the File to Module structure. The format of the described XML file is shown below:

```
<ORGANIZATION>
<MODULE module_name\>
  <FILE file_name\>
...
<\MODULE>
...
<\ORGANIZATION>
```

The chosed structure is simple and intuitive. The file will contain a list of modules and each module will contain a list of files.

Regular expressions were used in developing the reverse engineering tool. They were used several times:

- in the parsing process: to identify the module members, to identify the coding templates and when parsing the function and function like macro headers to identify needed information from the headers;
- in the drawing process: to identify code lines, to identify line containing instructions and to identify line containing labels.

Regular expressions were used because they offer a powerful and flexible way of searching. In order to be as independent as possible from the assembly language, regular expressions offer a flexible way of defining the information that needs to be extracted from the analyzed files. The choice of searching with regular expressions was made because of the string matching character of the regular expression.

Between all those regular expressions mentioned above only the expressions used for identifying the module members are present in the configuration file and contribute the most to the language flexibility of the tool. Only those expressions will be further mentioned. They are: public names, functions, function declaration, function like macros, constants, variables and macros. They all start with the ^(?: |\t)* character sequence and end with the (\n|$) sequence. The first character has the role to match only the expressions that begin at a new line and not within a line and the character sequence from the end has the role to match an expression that ends with a new line. Only two descriptions will be given:

Public Names:
^(?: |\t)*(public|Public)(?: |\t)*((?:[A-Z]|[a-z]|_|[0-9])+)(?:[^:\n])*(\n|$)

This regular expression is used to identify all public names. This expression has one storable sub expression: "((?:[A-Z]|[a-z]|_|[0-9])+)" where the public name will be stored. As seen in this sub expression the public name can be built out of the following character set: [A-Z], [a-z], [0-9] and _. When a match occurs the tool will read the public name from the first sub expression.

Variables:
^(?: |\t)*(public|Public)?(?: |\t)*((?:[A-Z]|]a-z]|_|[0-9])+)(?: |\t)*:(?: |\t)*(?:DS|ds)(?:|\t)*(?:(?:[A-Z]|[a-z]|[0-9]|_)+)(?: |\t)*(?:(?:;(?: |\t)((?:[^\n]*?)))?(?:\n|$)

This regular expression is used to identify all variable definitions. The presented expression is used to identify variable definitions under the STARC platform. It is divided into five main parts:

1. "(public|Public)?": the first part detects if the variable definition is a public one or not; the "public" or "Public" keyword can appear one or zero times in the matched expression; this part is the first sub expression, from here the tool will read either one of the keywords mentioned above or the empty string;
2. "((?:[A-Z]|[a-z]|_|[0-9])+)": the second part identifies the name of the variable, the character set accepted for a variable name definition is the same as accepted for the public name definition;
3. "(?:DS|ds)": the third part is a non storable sub expression; it means it won't count as a sub expression, its purpose is to identify the variable definition keyword "ds";
4. "(?:(?:[A-Z]|[a-z]|[0-9]|_)+)": the forth part is a non storable sub expression that will match the values attributed to the variable; it defines the same set of character as that for the name of the variable;
5. "(?:(?:;(?: |\t)*)((?:[^\n]*?)))?": the last part of this expression has the role to identify any comments attached to the variable; this part contains also the third sub expression from where the tool will read the attached comments or the empty string; from the sub expression the tool will read only the text of the comment.

All other regular expressions are hard coded and much simpler defined as the ones shown above. The main target of the used regular expressions is to make the finding of text more flexible, quicker and easier. The choice of enabling the user to modify the regular expression of the module members lead to an assembly language independent tool.

Microsoft Visio is a tool which was used for drawing all the generated flowcharts and diagrams. It offered a collection of predrawn shapes from a template and it offered also the possibility to create new templates.

## V. A CASE STUDY

The tool was verified for the assembly language of STARC and NEC microcontrollers but can be easily adapted to other assembly languages too.

Next, an analyze for an assembly language project of a NEC microcontroller will be shown. The analyzed code is:

```
;------------------------------------------------------------
;FUNCTION  :"arrcvtgb_captureIT"
;DOES   :Nec function
; COMMENT   :<compulso ry if complicated.Interactions,
;PARAMETERS :
;RETURN VALUES  :
;VARIABLES AFFECTED :<rub1>(r);<rub2>(w); used for df
;RESOURCES USED   :
;BROKEN REGISTERS   :
;------------------------------------------------------------
arrcvtgb_captureIT
    ;check lower limit of T15
    cmpw  ax,#T_REC_T15_LEFT
        ;maybe T1?
    bc   s_ri_check_T1
    ;T15 detected: insert bit(s)
        ;check for lower limit of T2
    cmpw  ax,#T_REC_T2_LEFT
        ;above T15 and below T2: wrong edge, reini
        bc   s_ri_good_T1
    ;T2 detected: insert bits

s_check_ITload


    mov a,S:rub_RecBytes  ;move current byte to a
    set   cy   ;1 to be inserted
    rolc a,1  ;insert the carry flag to LSB
    mov S:rub_RecBytes,a  ;store new value of current byte
    set1 rbi_prev_bit  ;update previous bit
    dec  S:rub_SyncEdgeCntTG     ;decrement bit counter
        ;byte complete; further steps
    br   leave_IT

    ;above T1 and below T15: wrong edge, reinit
s_ri_check_T1:
```

```
    br    leave_IT
s_ri_good_T1:
    SET_DEB_PORT_P50
    CLR_DEB_PORT_P50
    ;increment counter
    inc   S:rub_SyncEdgeCntTG
leave_IT:
    movx  ax,hl
    movw  rpuw_TxBufferpnt,ax
    pop   ax
    pop   hl
    ret
;ENDF
END
```

Based on the option flags set by the Configuration file, several diagrams will be generated. Fig. 5 shows a Flowchart diagram when all option flags are not checked.

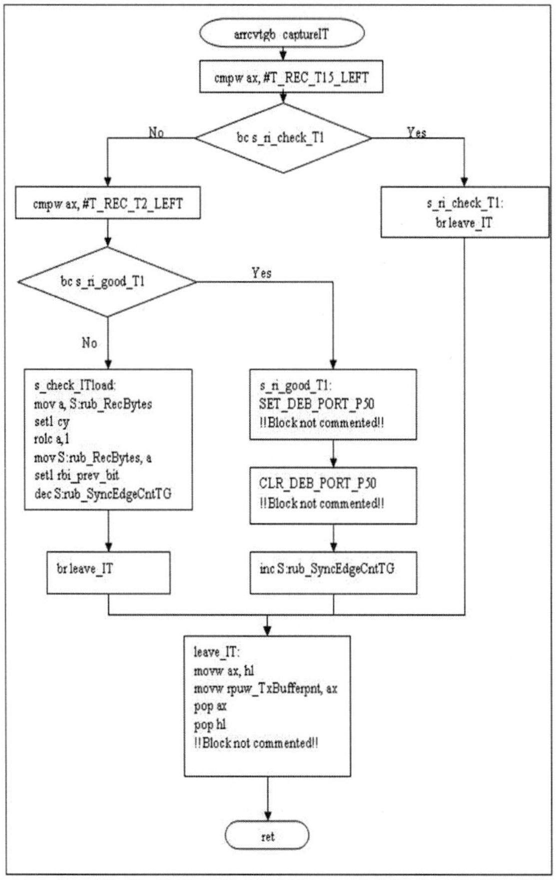


Fig. 5 Flowchart diagram (all option flags not checked)

## V. CONCLUSIONS

The tool was developed for the programmer and the project manager. The tool helps the programmer to review code, to understand an already written project and to create the documentation of a newly created project. The tool helps the project manager to obtain a better view over the entire project and to better enable planning and tracking over the growth of the project.

Further improvements are possible, by adding several modules and features, such as:

- code simulation module: it will be useful for replacing the affected variables header section and to detect all variable changes by analyzing the code; it must take into account the microcontrollers architecture;
- code generation module: it will be used for generating empty module with the proper template sections when a new module is added to the tool structure and for generating empty module members with all header and comment requirements when those members are being added to the target module;
- artificial intelligence module for flowchart generation: it will increase the analyzing possibilities;
- module for regular expressions generation from a visual interface;
- adding new outputs and improve the existing outputs: e. g. an inclusion tree which will help seeing the include dependencies between modules;
- visual relation between data and the functions or function like macros where it is modified: this feature will help identifying the locations were a data is being modified and the way it is modified;
- own Layout module to compensate Visio's layout for a more exact representation.

## REFERENCES

[1] M. Naik, J. Palsberg, "Compiling with Code – Size Constraints", *ACM Transactions on Embedded Computing Systems,* vol. 3, no. 1, February 2004, pp. 163 - 181

[2] E. S. Imsand, A. C. Sachitano, J. A. Hamilton, "Reverse Engineering Vulnerabilities in Simulation Software", in *Proc. of Advanced Simulation Technologies Conf.,* Orlando, USA, 2003, pp. 105 – 110

[3] J. Martin, H. Mulller, "C to Java Migration Experiences", in *Proc. of the 6th European Conference for Software Maintenance and Reengineering,,* Budapest, Hungary, March 2002

[4] B. Schwarz, S. Debray, G. Andrews: "Disassembly of Executable Code Revisited", in *Proc. of the Ninth Working Conference on Reverse Engineering (WCRE'02), IEEE Computer Society,* Richmond, Virginia, USA, October – November 2002

[5] A. Johnstone, E. Scott, T. Womak, "Reverse compilation of Digital Signal Processor: a working example", in *Proc. of the 33th Annual Hawaii International Conference on System Sciences (HICSS – 33), IEEE Computer Society,* New Jersey, January 2000

[6] M. P. Ward, "Assembler to C migration using FermaT transformation system", in *Proc. of IEEE International Conference on Software Maintenance, ICSM'99,* Oxford, UK, August – September 1999, pp. 67 – 76

[7] J. Palsberg, M. Wallace, "Reverse Engineering of Real – Time Assembly Code", in *Proc. of POPL'98, 25th Annual ACM SIGPLAN – SIGACT Symposium on Programming Languages,* San Diego, USA, 1998

# COMPOSITION LINEAR CONTROL IN STIRRED TANK CHEMICAL REACTORS

A. Regalado Méndez[1] and J. Álvarez-Ramírez[2]
[1]Universidad del Mar. Área de Ingeniería Ambiental.
Ciudad Universitaria, S/N, Puerto Ángel,
San. Pedro Pochutla., Oaxaca, 70902, México.
[2]Universidad Autónoma Metropolitana-Iztapalapa. DIPH.
San Rafael Atlixco 186.
Col Vicentina. México, D.F., 09340, México.

**Abstract**

This work studies loop control composition in continuous chemical reactors with simple structures, due to its large acceptance in chemical industry. A linear cascade composition control (master/slave) is proposed, designed with basic control structures based on *Laplace* tools. Two configurations are designed, which were evaluated in a dynamic model of continuous stirred tank. From a stability analysis it is noted that, for such configurations, system assent time is 7 to 8 times reduced if compared to the assent time without loop control. Besides, the system shows a good performance when coming to the asked reference. Implementation of such control configurations can solve the problem of loop control composition.

*Keywords*: composition controls, cascade controls; gain, feedback, loop control.

## I. INTRODUCTION

The chemical reactors are the heart of the chemical process. The continuous reactor of jacked tanks is the most used in chemical processes industries. The most common example for the production of paints and plastics is polymerization. These systems usually have up to three equilibrium points, and if the desired equilibrium point is unstable, this corresponds to the optimal production in the process [7]. If the disturbances disappear slowly, or if the settling time is too long, we can ask ourselves it's possible to make it stable?, improve it's stability, or diminish the settling time?. The answer is *yes*. For example, to improve the stability we increase the slope of the hot elimination line.

A control application can be essential to operate some chemical processes, as the techniques of control used will have a prime importance on quantity and quality of products. The use of a control technique also implies an increase of the security level. From a contaminant diminution point of view, a good process is achieved if the chemical process limits unnecessary losses or gaseous emissions to the atmosphere. Such control systems could be implemented in water treatment plants to comply with the minimum quality required by authorities, discharged or reused water.

The design of a particular control technique will depend on operational necessities and the kind of parameters that can be measured or estimated.

Aris and Admunson started to design and investigation integral proportional controls (**PI**) to maintain and determine the stability of Continuous Stirred Tank Reactors (**CSTRs**) [2].

The integral proportional control (**PI**) is a popular control method since the proportional control cannot by itself compensate for steady state deviations when the process is submitted to disturbances. A *Laplace* analysis in a closed loop system presents sufficient integral action to compensate for steady state deviation which reaches the desired **setpoint**.

The cascade control is one of the most of use tools for the design of advanced controls. It has a special capacity to eliminate disturbances and reduce time constants, which improves the performance controls. This type of strategy is used in chemical processes in which the measurement of some parameters is delayed or in great capacities process such as distillation columns. In such systems, the conventional control types (proportional and integral controls) lead to a delay in control action and poor performance as the disturbances and the measurement noises will not be detected in enough time to allow a corrective action [1].

A cascade control is proposed to solve the problem of disturbances and measurement delays which effects are more quickly detected in the outlet of secondary control than in a controlled variable.

The use of a cascade control is recommended for those processes in which the dynamic of a secondary control loop is quicker than the one of a primary control [9].

In these conditions, due to the difference between them the interaction between the two control schemes is practically inexistent. Therefore, these controls do not present significant synchronization problems.

## II. SYSTEM DESCRIPTION

The dynamics of **CSTR** in which $m$ reactions take place involving n (n > m) chemicals species can be described by:

$$\begin{aligned} \dot{C} &= \theta\left(C_{in} - C\right) + Er\left(C,T\right) \\ \dot{T} &= \theta\left(T_{in} - T\right) + Hr\left(C,T\right) + \gamma\left(u - T\right) \end{aligned} \tag{1}$$



*T. Sobh and K. Elleithy (eds.), Advances in Systems, Computing Sciences and Software Engineering*, 339–344.

where:

- $C \in \mathbb{R}^n$ is the vector of concentrations of chemical species.
- $C_{in} \in \mathbb{R}^n$ is the vector non-negative and constant feed concentrations.
- $T \in \mathbb{R}$ is the vector temperature.
- $T_{in} \in \mathbb{R}$ is the vector feed temperature.
- $r(C,T) \in \mathbb{R}^m$ is the smooth, non-negative, bounded vector of reaction kinetics, with $r(C,T) = 0 \;\; \forall \;\; t \leq 0$.
- $E \in \mathbb{R}^n \times \mathbb{R}^m$ is the stoichiometric matrix.
- $H(C,T) \in \mathbb{R}^m$ is the smooth, bounded row vector of reaction enthalpies, with $H(C,T) = 0 \;\; \forall \;\; t \leq 0$.
- $\theta$ is the reactor dilution rate (i.e. flow rate/volume).
- $\gamma$ is the heat transfer parameter.
- $u$ is the jacket or wall temperature, which is taken as the as the *control input*.

Non-linearity in models of equations 1 are introduced by the reaction kinetics $r(C,T)$. Commonly, $r(C,T)$ has a polynomial or rational dependency on C and has an Arrhenius dependency on T. Due to this kinetics, **CSTRs** can display a great variety of dynamic behaviors from multiplicity of steady states to sustained oscillations, including odd attractors [3].

## III. SLAVE CONTROL DESIGN

The expression of proportional control (**P**), is represented as follows [6, 8]:

$$u = \bar{u} + kp\left(T^r - T\right) \tag{2}$$

where:

$\bar{u}$ is the jacket cool temperature nominal not dimensional

$T^r$ is the reference temperature

$kp$ is the process gain

Applying the *Laplace* transformation to the equation (2), one get:

$$\frac{u}{T} = \frac{\Delta u(s)}{\Delta T(s)} \approx -kp \tag{3}$$

The Ordinary Differential linear Equation (**ODE**), in terms of deviation variables, is given by:

$$\dot{T} = \left(k_{pr}/\tau_o\right)u - \tau_o^{-1}T \tag{4}$$

where:

$k_{pr}$ is the process gain, temperature unities

$\tau_o$ is the process time

The linear **ODE** is presented next without being submitted to any perturbations, as:

$$\dot{T} = -\tau_c^{-1}T \tag{5}$$

where:

$\dot{T} = -\tau_c^{-1}T$ is the time characteristic

Equaling equations 4 and 5 and order to search for the controller parameters in terms of process parameters, with can be easily calculated with the **IMC** method, we get:

$$\left(k_{pr}/\tau_o\right)u - \tau_o^{-1}T = -\tau_c^{-1}T \tag{6}$$

Reordering $u/T$:

$$\frac{u}{T} = k_{pr}^{-1}\left[1.0 - \tau_o/\tau_c\right] \tag{7}$$

To get the gain of slave control, equations 7 and 3 give:

$$kp = -k_{pr}^{-1}\left[1.0 - \tau_o/\tau_c\right] \tag{8}$$

As shown in the following equation, the process time without disturbances is directly proportional to the real process time:

$$\tau_c = \alpha\tau_o \tag{9}$$

Therefore:

$$\frac{\tau_o}{\tau_c} = \frac{1}{\alpha} \tag{10}$$

Substituting equation (10) into equation (8):

$$kp = k_{pr}^{-1}\left[\frac{1}{\alpha} - 1\right] \tag{11}$$

The $\alpha$ term can't be zero, as this leads to $1/\alpha \approx \infty$, and this is not acceptable. The other boundary extreme would be $1/\alpha = 1$. This only happens if $\alpha = 1$, which means that the control would no longer exist. Thus $\alpha$ indicates the rate of control response, with $0 < \alpha \leq 1.0$.

Finally, substituting equation 11 in equation 1 we have the **P** controller:

$$u = \bar{u} + k_{pr}^{-1}\left[\frac{1}{\alpha} - 1\right]\left(T^r - T\right) \tag{12}$$

where $k_{pr}$ is the process gain which calculated is by the **IMC** method.

Equation 12 is the S*lave Control* law, which is directly proportional to the temperature where the proportional constant is the process gain.

## IV. MASTER CONTROL DESIGN

The master control law will be used next it's worth noting that the principal reason that motivates the action of the integral control is related to the fact that the proportional control cannot compensate for the come up to out of steady state when the process is submitted to fluctuations. Thus, the integral action is able to compensate from deviations to steady state which means that the integral action is used to increase the set-point convergence.

Combining the action of the integral concentration control and the proportional temperature control, we get a cascade control. (**P$_T$I$_C$**) which is the fundamental part of the present work, expressed as:

$$T^r = \bar{T} + K_{I,C} \int \left(C^r - C\right) dt \tag{13}$$

$$K_{I,C} = 1.0/k_{pri}\tau_I \tag{14}$$

where:

$\tau_I$ is the proportional control convergence time

$K_{I,C}$ is the integral control gain, concentration$^{-1}$ unities

$k_{pri}$ is the concentration process gain, concentration$^{-1}$ for time$^{-1}$ unities

The equation 13 is the Master Control Law which is directly proportional to integrate of the concentration, from time zero to time t, where the proportional constant is the gain of the integral control.

The integral control process gain can be obtained from the slope of the tangent, which is calculated by the relationship of the concentration to the reactor temperature. Mathematically this is represented as:

$$k_{pri} = \frac{dC}{dT} \cong \frac{\Delta C}{\Delta T} \tag{15}$$

Different control configurations will be computed. These will be tested and evaluated for a sample case.

Since the main problem of this work consists in designing and evaluating the cascade control strategies of composition for **CSTRs**, control linear techniques were employed. To reach them, a secondary control was first designed followed by a primary control. It is important to mention that two configurations of cascade control were designed which are proven using numerical simulation. The linear cascade type control design of composition in it's configuration **P$_T$I$_C$** consists of two parts: a master control that corresponds to equation 15 and a slave control, which corresponds to equation 16.

*Cascade control of concentration configuration* **P$_T$I$_C$**

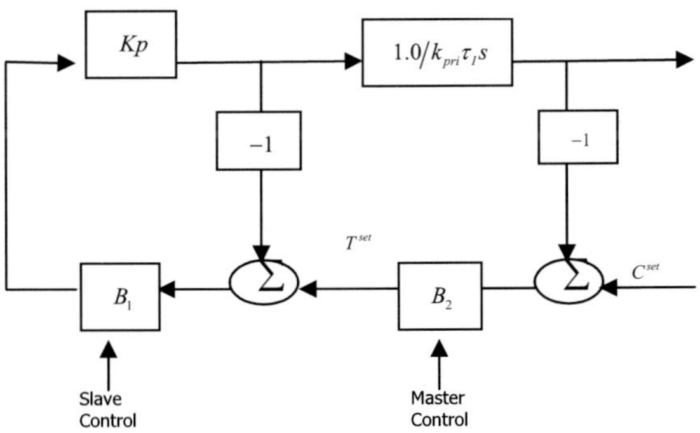


Fig.1. Blocks of diagram **P$_T$I$_C$** control configuration.

$$u = \bar{u} + k_{pr}^{-1}\left[\frac{1}{\alpha} - 1\right]\left(T^r - T\right) \tag{15}$$

$$T^r = \bar{T} + K_{I,C} \int \left(C^r - C\right) dt \tag{16}$$

$$K_{I,C} = 1/k_{pri}\tau_I \tag{17}$$

*Cascade control of concentration configuration* **PI$_T$I$_C$**

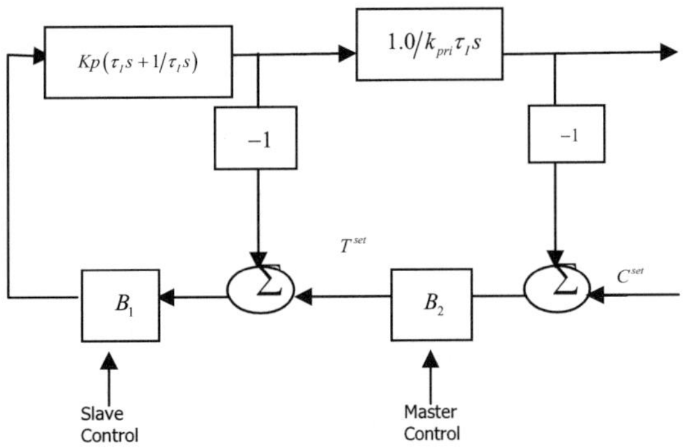


Fig.2. Blocks of diagram **PI$_T$I$_C$** control configuration.

$$u = \bar{u} + k_{pr}^{-1}\left[\frac{1}{\alpha} - 1\right]\left(T^r - T\right) + K_{I,T} \int \left(T^r - T\right) dt \tag{18}$$

$$T^r = \bar{T} + K_{I,C} \int \left(C^r - C\right) dt \tag{19}$$

$$K_{I,C} = 1/k_{pri}\tau_I \;;\; K_{I,T} = Kp/\tau_O \tag{20}$$

where:

$k_{pri}$ is the process of concentration gain, concentration$^{-1}$temperature$^{-1}$ unites
$\tau_0$ is the time of the process
$\tau_I$ is the time of convergence **P** control

## V. APLICATION EXAMPLE

The system consists of a catalytic reaction of benzene to form maleic anhydride. The equations that describe the dynamics of this study case, in which the control structures previously designed will be implemented, are presented next [5]:

$$\xi_i = \exp\left(\beta_i T / \left(1 + T/\beta_m\right)\right) \tag{21}$$

$$k_i = A_{i0} \exp\left(\frac{-E_i}{RT_f}\right) \tag{22}$$

$$rxn_1 = a_5\left(k_{1f}\xi_1 + k_{3f}\xi_3\right)C_{bz} \tag{23}$$

$$rxn_2 = a_5\left(k_{1f}\xi_1 C_{bz} - k_{2f}\xi_2 C_{ma}\right) \tag{24}$$

$$rxn_3 = a_3\left(\begin{array}{l}\left(\Delta H_1 k_{1f}\xi_1 + \Delta H_3 k_{3f}\xi_3\right)C_{bz} \\ +\Delta H_2 k_{2f}\xi_2 C_{ma}\end{array}\right) \tag{25}$$

$$\dot{C}_{bz} = \theta_c\left(C_{bzin} - C_{bz}\right) - rxn_1 \tag{26}$$

$$\dot{C}_{ma} = \theta_c\left(C_{main} - C_{ma}\right) + rxn_2 \tag{27}$$

$$\dot{T} = \theta_T\left(T_{in} - T\right) - rxn_3 + \gamma\left(U - T\right) \tag{28}$$

In Figure 3 it's observed that the settling time is nearly 39 minutes. This time will serve to justify the use of the control system.

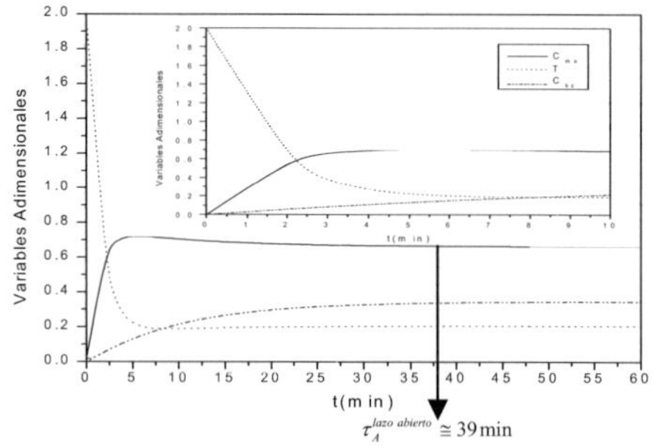


Fig.3. Numerical simulation of open loop dynamic system.

The stability of the reactor is presented in the next analysis. To obtain the equilibrium points of the case of study, finding that after a extended work of simulation only one equilibrium point exists and thus this point is stable, it remains unnecessary to realize the stability probes. However, Figure 3 shows clearly the equilibrium point in a way that you can read the not-dimensional equilibrium point. Nevertheless, it's necessary to find the operation point of the dynamic system studied, as only one equilibrium point is not the optimum point of production. Thus, we need to perform an analysis of the reaction speed in order to find the highest speed of the anhydride maleic reaction. This will allow us to find the vector of operation utilizing: $\vec{\psi} = [Cbz \quad Cma \quad Tr]$ (which is like unstable equilibrium points which are more difficult to control). It is therefore necessary to generate the sections of not-dimensional reaction speed ($V_{rxn}$) for each of the species presented in the case study.

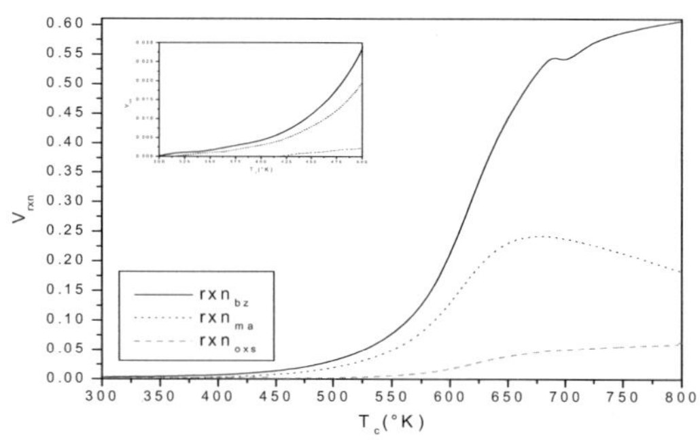


Fig.4. Speed reaction profile of the species in the dynamic system of the case study in the temperature cooling jacket function.

In the line marked with dots in Figure 4, we obtain the maximum with the type $F\left(T_C(K), V_{rxn}\right) = (670, 0.245)$ to realize the simulation of the system dynamic and find the vector optimum point of operation with: $\vec{\psi} = [0.41 \ \ \%mol \ \ 0.774 \ \ \%mol \ \ 710.144 \ \ K]$. Besides, we simulated the system dynamics changing the temperature of the cooling jacket in a way that we obtain Figures 5 and 6, in which the optimum vector of operation fits with the maximum point in the concentration-temperature map of the reactor and of the cooling jacket temperature.

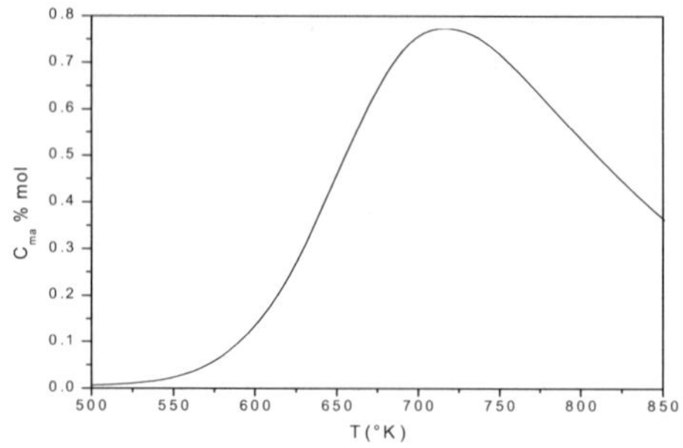


Fig.5. Steady-state map concentration $C_{ma}$ in reactor temperature T(K) function.

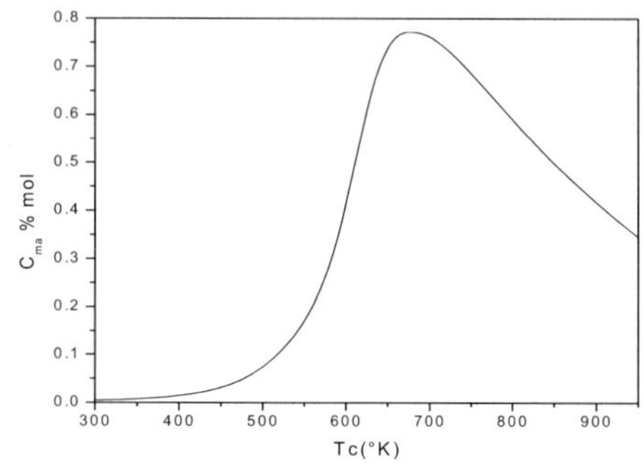


Fig.6. Steady-state map concentration $C_{ma}$ in cooling jacket temperature $T_c$(K) function.

Figures 5 and 6 are the stationary state maps which are used to show the dynamic behavior of the simulated system without a control law. These maps give an idea of the maximum reference point that could be fed in both control configurations. In Figures 5 and 6, we observe that the maximum composition is 0.774% mole. It's important to obtain the maximum concentration value because in the simulation, if a higher value is inputted the conversion couldn't be reached. Also, it is not interesting to work with the maximum point, as problems with the synchronization are generated. Thus it was set in a point close to the optimum point, which is designed by Tc = 640 K.

We performed a step test with temperature functions of the cooling jacket using:

$$T_c = 640 + u \quad where \quad u = \pm 5\% \tag{29}$$

These functions are plotted in Figures 7 and 8, according to a positive or negative step response of temperature, respectively. Based on these plots, **P** control parameters for the temperature are calculated. In Figure 7, we observe that there is an approximate delay of 6 minutes between the initial and final stationary state whereas in Figure 8 it delays approximately 8 minutes. Comparing this with theoretical time, it can be calculated as 4 times the inverse of the dilution rate and mathematically as: $\tau = 4(1/\theta_c)$, $\tau = 6.38\,\text{min}$ where $\theta_c = 0.6265$ and $1/\theta_c = 1.59\,\text{min}^{-1}$. For that reason $\tau_+ < \tau < \tau_-$, where $\tau_+ = 6\,\text{min}$ and $\tau_- = 8\,\text{min}$.

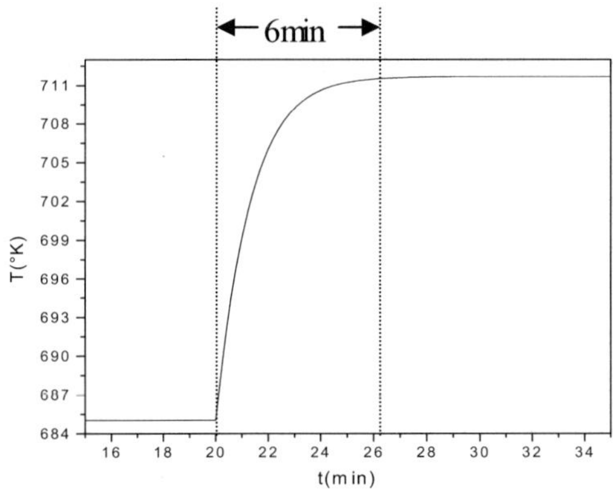


Fig.7. Step positive response of temperature in time function, with step change from Tc = 640 K to 672 K.

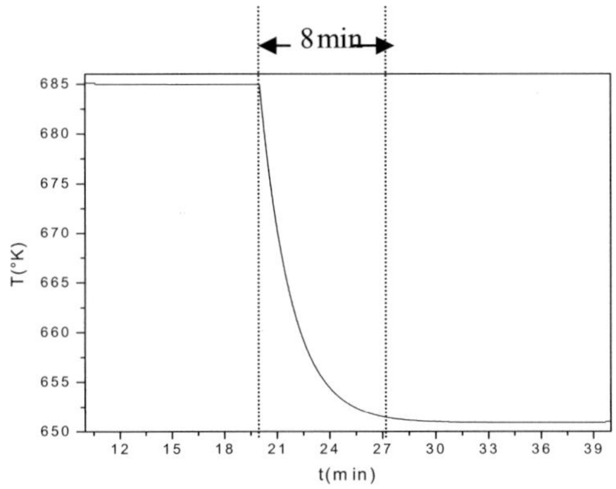


Fig.8. Step negative response of temperature in time function, with step change from Tc = 640 K to 608 K.

With base in the central theory, we obtain the parameters of the control configurations $\mathbf{P_TI_C}$ and $\mathbf{PI_TI_C}$ which are: $k_{pr}$ = 0.833 $K^{-1}$, $k_{pri}$ = 0.055% mol/K, $\tau_0$ = 8.313 min, $\tau_1$ = 1.35 min $\overline{C}_{ma} = 0.733\%mol$, $\overline{T} = 692.244$ K, $\overline{T}_c = 645.0$ K.

Figures 9 and 10 show the performance of the two control configurations described previously, which clearly show that the settling time ( $\tau_A$ ) decrease as:

$$\left(\tau_A^{control\,PI_TI_C} = 5.3\,\text{min}\right) << \left(\tau_A^{control\,P_TI_C} = 6.5\,\text{min}\right) << \left(\tau_A^{openedknott} = 39\,\text{min}\right)$$

, in a value of α = 0.1.

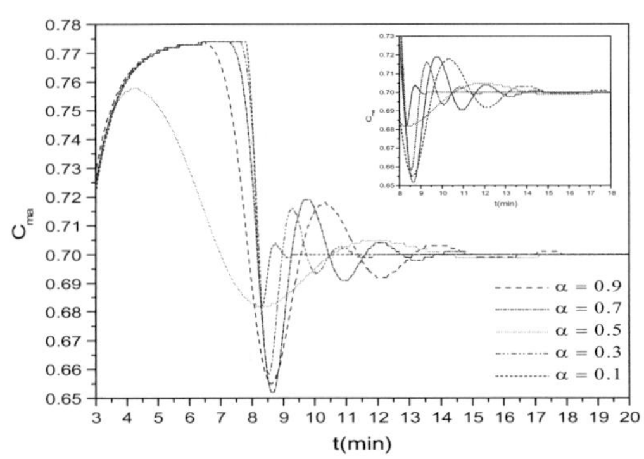


Fig.9. Performance of composition in the configuration $\mathbf{P_TI_C}$.

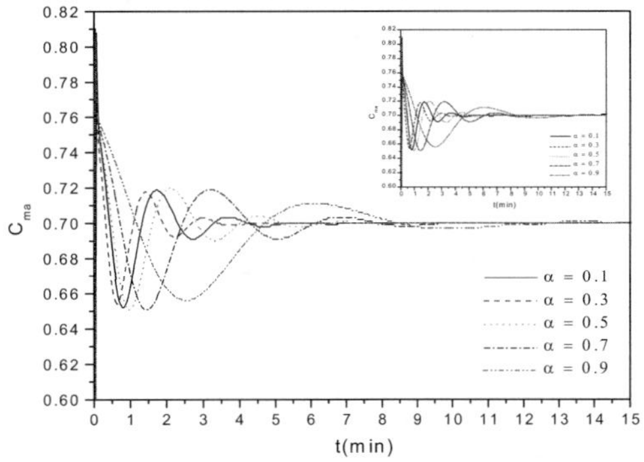


Fig.10. Performance of composition control in the configuration $\mathbf{PI_TI_C}$.

Figures 11 and 12 shows the description face of the proposed configurations control in which we see how the curves converge to the set point, which corresponds to the optimum operation $\left(710.144K \quad ,0.774 \quad \%mol\right)$, found after performing the dynamic system analysis.

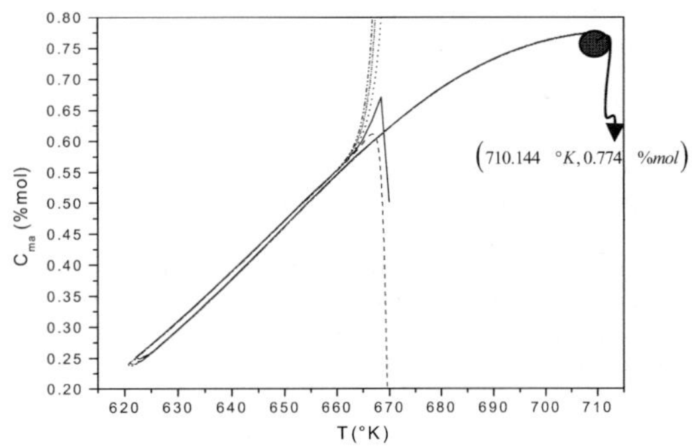


Fig.11. Phase portrait of configuration $\mathbf{P_TI_C}$ with $C_{ma}^{set} = 0.774 \quad \%mol$ of set-point.

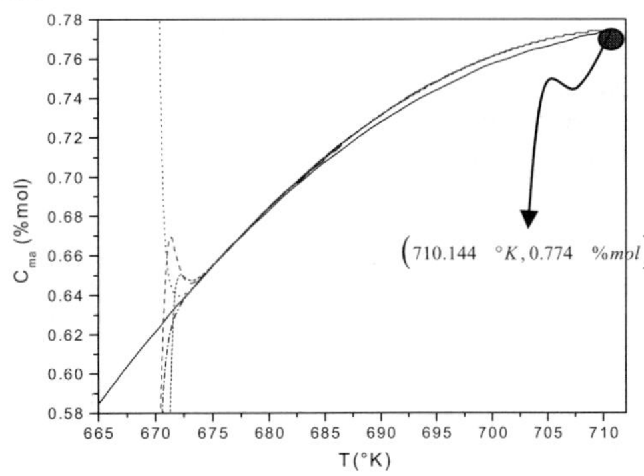


Fig.12. Phase portrait of configuration $\mathbf{PI_TI_C}$ with $C_{ma}^{set} = 0.774 \quad \%mol$ of set-point.

Figures 9 and 10 present similar structures, as is obtained when one uses unstable equilibrium points. This means that the previously calculated optimum production point calculated is unstable.

Note that the control design was performed under the conditions that all the variable states are available in the feed-forward. However, in the practice not all the variables state is available to the feed-forward. So, we need to calculate the variables state that is not disposal. That is the case of composition. Nevertheless, some methods exist to estimate the state variables which are not measured by a differentiation process. These methods are based on a mechanism that estimates and observes the variables state.

## VI. CONCLUSIONS

Simple sketches of the cascade type control were designed, based on simple in-out models in-out of low rank obtained by a first order lineal differential equation. The type cascade control sketches don't present the disadvantage of classic lineal controls (i.e. not considering dynamics model contempt and the external disturbances) designed by a first order lineal theoretical model. This is due to the fact that non-modeled dynamic effect and non-modeled external disturbances effects are included in the control derivations by one term which groups the systems errors.

The cascade control $\mathbf{P_TI_C}$ and $\mathbf{PI_TI_C}$, show favorable performance because they maintain the requested reference. The scheme of control in the $\mathbf{P_TI_C}$ configuration has the structure of a classic **PI** control, with two adjustable parameters: the proportional temperature constant, and the integral constant of concentration.

For the $\mathbf{PI_TI_C}$ configuration, the structure corresponds to an advanced $\mathbf{PI^2}$ control with three adjustable parameters: the proportional constant of temperature, the time constant of temperature, and the integral constant of concentration.

For dynamic systems, which have one only equilibrium point, an optimal production point of the desired specie can be obtained through an analysis of the reaction speed. Using this operation, the optimum point will be similar to an instable operation point.

As proposed this work demonstrates that **CSTRs** present satisfactory performances with simple structures of cascade type linear control.

## ACKNOWLEDGMENT:

The authors wish to thank the Consejo Nacional de Ciencia y Tecnología (CONACYT) for the grant, and their work colleagues Dr. Aitor Aizpuru and Benny Gueck.

## REFERENCES

1. Alvarez-Ramírez, J. and Morales A., "**PI** control continuously stirred tank reactors: stability and performance", *Chemical Engineering Science*, 55, Pp. 5497-5507, 2000.
2. Alvarez-Ramírez, J. and Puebla H., "On Classical **PI** Control of Chemical Reactors", *Chemical Engineering Science*, 56, Pp. 2111-2121, 2001.
3. Álvarez-Ramírez *et al.*, "Composition Cascade Control for Chemical Reactors", *International Journal of Robust and Nonlinear Control*, 12, Pp. 1145–1171, 2002.
4. Andersen H. *et al.*, "Tuning of Dual-Composition Distillation Column Control", *Chemical Engineering Science*, 44, Pp. 619-630, 1998.
5. Aoufoussi H. *et al.*, "Feedback Linearizing Control of Fluidized Bed reactor", *The Canadian Journal Of Chemical Engineering*, 70, Pp. 356–367, 1992.
6. Luyben, W. L., *Process Modeling, Simulation and Control for Chemicals Engineers*, Pp. 205–300 and 404-414, McGraw-Hill, U. S. A., 1995.
7. Perlmutter D. D., (1972), *Stability of Chemical Reactors*, Pp. 5-6 and 51-55, Prentice-Hall, U. S. A.
8. Smith, C. A. and Corripio, A. B., *Control Automático de procesos*, Pp. 177-223, Limusa, México, 1994.
9. Stephanopoulos G., *Control Systems with Multiple Loop In Chemical Process Control*, Capítulo 20, Prentice Hall, U. S. A., 1984.
10. Vieira M., Sayer C., Lima L. E., y Pinto J. C., "Closed-Loop Composition and Molecular Weight Control of a Copolymer Latex Using Near-Infrared Spectroscopy", *Industrial Engineering Chemicals Research*, 35, Pp. 475-484, 2002.
11. Vicente M. *et al.*, "Maximizing Production and Polymer Quality (MWD and Composition) in emulsion polymerization Reactors with Limited Capacity of Heat Removal", *Chemicals Engineering Research*, 58, Pp. 215-222, 2003.
12. Wolff E. and Skogestad S., "Temperature Cascade Control of Distillation Columns", *Industrial Engineering Chemicals Research*, 41, Pp. 2915-2930, 1996.
13. Zaldo F. and Alvarez J., A "Composition – Temperature Control Strategy for Semibatch Emulsion Copolymer Reactors", *IFAC (International Federation of Automatic Control) International Symposium on Dynamics and Control of Process Systems (DYCOPS 5)*, Pp. 217 – 222, 1998.

# Intelligent analysis to the Contamination in the city of Santiago from Chile

**Delgado C. Miguel, Sanchez F. Daniel**
University of Granada
DECSAI-Spain
mdelgado@ugr.es , daniel@decsai.ug.es

**Zapata C. Santiago, Escobar R. Luis, Godoy M. Jimmy**
University of Technological Metropolitan
Computer Science Department - Chile
szapata@utem.cl , laescoba@utem.cl, jimy.godoy@ciberutem.cl

## Abstract

Air contamination is one of the biggest problems that affects to the countries in almost any part of the world. The increase in the quantities of gases and particles that are potentially harmful to health and the environment has been verified to world scale, and each day it becomes more obvious that the answer to these problems should concentrate in the search of intelligent solutions.

In Chile, by law, the obligation settles down of developing decontamination plans in areas where the levels of pollutants systematically exceed the environmental norms, and plans of prevention where these norms are in danger of being overcome.

During the autumn-winter season the population of Santiago's city centre is affected by a sudden increase in the levels of air contamination. This situation is known as a critical episode of atmospheric contamination, and it takes place when there are register high levels of concentration of pollutants during a short period of time. These episodes originate from the convergence of a series of meteorological factors that impede the ventilation of Santiago's basin due to an increase in the previous emissions to the episode.

According to the existing by-law, the criteria to decrease a critical episode of contamination is referred to in the index ICAP, that is generated by the data of the Net of Mensuration of gases and particles that are entered into a prognostic method developed by the physicist J. Cassmassi. [44], [47], [55].

This way, the authority makes the decisions with regard to the critical episodes that can affect Santiago's city centre, depending on the predictions that gives a pattern.

Our investigational work is framed in the line of looking for intelligent methodologies for prediction and control of environmental problems in Santiago's city centre and its Metropolitan Area.

**Words Key**: Intelligent Analysis, Decision Support Systems, Knowledge Data Discovery, Contamination, Pollution.

## 1. INTRODUCTION

### 1.1. The Problem

The contamination of the air in Santiago's city centre originates mainly during autumn-winter season, in these months the population of the Metropolitan Region is affected by a sudden increase in the levels of air contamination.

These situations, known generically as critical episodes of atmospheric contamination, take place when they register high levels of certain polluting agents' concentrating during short periods. These episodes originate from the convergence of a series of meteorological factors that impede the ventilation of Santiago's basin, and also due to an increase in the emissions of these polluting agents.

The Quality of the Air for Material Particle Index, ICAP [1], [2], [3] is the indicator that through the data of Cassmassi's diagnostic model serves as an antecedent so that the government's authority can determine that Santiago's city centre in the presence of a critical episode of contamination. Cassmassi's model predicts the maximum value of concentration in an average of 24 hours of breathable particle material is PM10, for the period of 00 to 24 hours of the following day.

- For this forecast of episodes, a team of expert meteorologists prepares a daily forecast of meteorological conditions associated with episodes of atmospheric contamination for the Metropolitan Region, based on modernized meteorological information, national and international, and from the data of quality of air of the automatic Net of Mensuration of the air quality and meteorology Net MACAM [2], [45].
- It is important to highlight that because the pattern predicts the air quality for the following day, the declaration of an episode on the part of the authority does not imply that the air has worsened, but rather that it could end up worsening. That is to say, the episodes are decreased in preventive manner to avoid reaching the predicted indexes, and this way to protect the population's health.



*T. Sobh and K. Elleithy (eds.), Advances in Systems, Computing Sciences and Software Engineering*, 345–352.

This way of approaching the problem and making decisions presents some undesirable results, such as;

- The pattern of Cassmassi considers eight observation stations in isolated form, and in many occasions it has been observed that during critical incident episodes, the stations interact with each other.
- The maintenance a reductionist point of view (in midst of the systemic focus) has induced the occurrence of countless prediction errors.
- In a study carried out by the CONAMA [3], [48], [49] it showed that this model is ineffective to predict critical levels of contamination. In period of March 20, 2003 to August 10, 2003 they have correctly predicted a critical incident only 8 times, with 30 false alarms and 12 underestimated incidences.
- Due to the global climatic change, each year is more difficult to carry out reliable predictions, but this model has not adapted to these new situations (that is to say, it maintains its stable parameters, although the conditions in the study change).
- Another element that stands out is that Cassmassi's model has not been modernized since it was designed, and six years have past since its development. It has also been ignored that the technicians of the Cenma, the organization of the University of Chile in charge of carrying out the predictions to make decisions, have been requesting its modernization for some time.

- With regard to the primary norm of air quality for thick particle material (PM10) it is said that the current pattern of prediction only centers the attention in the values' means and not in the variations per hour, which means that the maximum contamination of each day are hidden in the average of 24 hours.

### 1.2. Administration of the Problem

At the present time and to world level, the study and control of problems of environmental contamination, has been approached with the support of the calls by DSS (Decision Support Systems) that are a specific class of systems of information that support the processes of making decisions in the organizations. These DSS is interactive systems that help to the making of decisions by facilitating the handling of the data, documents, knowledge model that are used to solve problems inside the organizations.

One of the areas that the DSS is very useful is in the support in the making of decisions for analysis and control of environmental problems, in particular the support that it can offer to the Data Mining, when allowing to extract patterns, models, relationships, tendencies, etc. that finally allow to find "rules " or " patterns " (" knowledge ") from the data and to then communicate them to the user through the DSS.

### 1.3. Objectives of the Investigation

The general objective of our investigation is to propose a methodology currently based on the DSS, like an alternative to the method used to measure and make decisions with regard to the environmental emergencies in Santiago's city centre.

## 2. CURRENT ADMINISTRATION OF THE CONTAMINATION IN THE CITY DE SANTIAGO

### 2.1. Introduction

One of the biggest problems that the Metropolitan Region possesses, especially in the municipality of Santiago, Chile is the concentration in the atmosphere of polluting particles that damage the health of people that live and commute through the city. For this reason it is important to have a good support system in the decisions that allow for the analysis of the data to obtain results can help to make better decisions of prevention for the citizens.

A responsible entity that takes charge of looking after the care of the environment is the SESMA [44] who through a net of mensuration called Net MACAM [45] register the levels of contamination in the environment on a daily basis.

### 2.2. MACAM Net

The official net of Automatic Monitoring of Air Quality and Meteorology of the city of Santiago, Chile (Net MACAM) began in 1988 with 5 monitoring stations, mostly located in Santiago's central sector. In the winter of 1997, this net was renovated and enlarged to 8 stations of automatic monitoring stations, in what today is known as the Net MACAM-2, clerk of the Metropolitan Service of Health of the Atmosphere (SESMA). The website to obtain the data is as follows:

**http://www.minsal.cl/sesma/pvcasesma/default.htm**

### 2.3. Chilean laws of Quality of the Air

#### 2.3.1. Laws

The Resolution N° 369, from the Ministry of Health on April 12, 1988, published in the official newspaper on April 26, 1998, establishes the indexes of air quality to determine the level of atmospheric contamination in the Metropolitan Region.

### 2.3.2 ICA E ICAP Index

As much the ICA as the ICAP, they give origin to the following level according to the obtained value:

| INDEX | LEVEL |
|---|---|
| 0 – 100 | GOOD |
| 101 – 200 | REGULATE |
| 201 – 300 | BAD |
| 301 – 400 | CRITICIZE |
| 401 – 500 | DANGEROUS |

## 2.4. Predictive Model of Joseph Cassmassi

### 2.4.1. Introduction

This model is used to predict the episodes of Alert, Pre-emergency or Emergency whose measures that take the authority have been described in the introduction of this present investigation, the information surrendered by the pattern implies being ahead of the excessive increase in the Indexes of Air Quality considering the meteorological variables and the levels of contamination for particle material in suspension.

The Pattern of Prediction uses eigh monitoring stations distributed geographically in the city of Santiago, Chile and for its estimate it considers: the historical behavior of the atmospheric contamination and their association with meteorological parameters, levels of atmospheric contamination and their daily variation. All these factors are entered to a mathematical Polynomial that was developed by the American physicist Joseph Cassmassi in 1999, based on the classic statistical techniques of multiple lineal regressions [47].

### 2.4.2. I model of presage for episodes

The Plan of Prevention and Atmospheric Decontamination for the Metropolitan Region (PPDA) has established the necessity to create a model of predicting episodes of contamination, due to the importance of anticipating the measures of control and of the receipt of the population in front of the occurrence of critical situations.

On the other hand, the Supreme Ordinance N° 59 of 1998 of the Ministry of the Secretary General of the Presidency that establishes the norm of primary quality for breathable particle material (MP10), it included the requirements that it should satisfy a methodology of predicting of air quality.

In accordance with these requirements, by means of the resolution N° 12.612of 1998 of the Metropolitan Service of Health of the Atmosphere of the Metropolitan Region (SESMA), the first official application of a model of predicting of air quality in Santiago's city, starting in July 1998.

As a form of optimizing this tool, in 1999 CONAMA took charge a study to improve the methodology of predicting of air quality in the Metropolitan Region. The study's results gave a new predicting model, denominated model Cassmassi.

### 2.4.3. Used variables

The methodology of prediction of concentrations of PM10 is based on calculation algorithms developed by means of application of statistical techniques of multiple regression, focused to find relationships between possible predictive variable and the preceding. The predictors include observed meteorological variables, indexes of observed and predicted meteorological conditions, observed concentrations of pollutants, and indexes of prospective variations of emissions and others.

### 2.4.4. Meteorological potential of atmospheric contamination (used Algorithms)

The operational application of this methodology considers two prediction algorithms for each monitoring station:

The first algorithm includes the "Index of Meteorological Potential of Atmospheric Contamination" (PMCA) predicted for the following day. Figure 2.1 shows that the PMCA has a good correspondence with the averages of 24 hours of the observed concentrations of PM10.

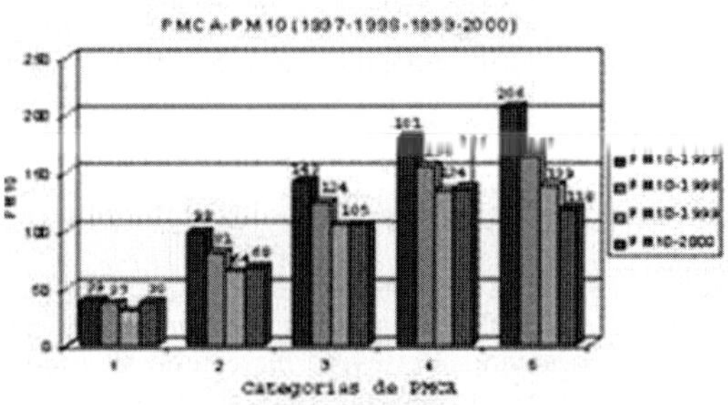


**Figures 2.1. - Relationship PMCA and contamination for PM10.**

The second algorithm is only based on observations (of the same day and of the previous day). This way, if the first algorithm cannot be applied due to an inadequacy of information, the second algorithm is used.

### 2.4.5. Validation of the pattern

In 1999/2000, an evaluation of the pattern Cassmassi was carried out on the part of the Department of Geophysics of the University of Chile. For this evaluation the information from the period of April 1, 1999 to September 17, 1999 was used. This way, the independence of the data was guaranteed and validated the pattern of those used in its construction. The validation shows that the pattern Cassmassi fulfills the requirements of the Supreme Ordinance N° 59 of 1998.

The detailed description of the Cassmassi model is in the study "Improvement of the Forecast of Air Quality and of the Knowledge of Local the Meteorological Conditions in the Metropolitan Region" (Cassmassi 1999), [48], [49].

### 2.5. Flaws of Cassmassi's Predictive Pattern

The predictive pattern is inefficient to predict critical levels of contamination, for example, in the year 2003 alone it was right only 8 times on predicting critical situations, with 30 false alarms and 12 underestimated episodes. According to the studies (Journal That it Happens, May 16, 2003), [50] it indicates that one of the main causes of the flaws is that it has not been modernized in 5 years, and that besides using Cassamassi's official predictive pattern, it depends of 5 other unofficial models, to estimate the level of certainty of the first one. For example, one day any prediction indicated that there was 80% of probabilities of environmental alert, but with an uncertainty of 35% that is transformed in environmental pre-emergency.

This model considers 8 monitoring stations in isolated form and in many occasions it has been observed that during the occurrence of critical episodes, the stations interact with other. Adopting the reductionist focus instead of the systemic focus has generated the occurrence of countless errors.

### 2.6. Decisions starting from the results of the pattern

The noxious effects to a person's health brough on by the contamination of the air in Santiago can differ from person to person. However, these effects become significantly more dangerous in cases of exposure to high levels in short periods of duration. This is precisely what occurs during critical episodes. Therefore, it is fundamental to anticipate high risk situations to be able to prevent the negative consequences of an environmental warning.

Regional authorities use the results from diagnostic readings as a preventative tool. Based on these diagnostic readings, should the level surpass any one of the eight stations:

The **level ICAP 200** according to this model should be declared as an **Environmental Alert**
The **level ICAP 300** according to this model should be declared as an **Environmental Pre-emergency**
The **level ICAP 500** according to this model should be declared as an **Environmental Emergency**

### 2.7. Cost of the errors of the Pattern Predictivo

According to economic studies, the costs associated to the errors of the predictive pattern exceeded 2 billion pesos [2.7 million Euros] in the year 2001, 5 billion pesos [6.9 million Euros] in 2002, and it was estimated to double for the year 2003 reaching 12 billion pesos. [50], [52].

As a corrective measure, these studies recommended the analytical review of the effectiveness of the adopted measures used during critical episodes; evaluate the costs of paralyzing the industrial sector; and thirdly, to revise and improve the quality of the preventive models used.

## 3. PROPOSAL OF WORK

### 3.1. Introduction

Based on the problems previously demonstrated when administering to the environmental contamination of the city of Santiago, we provide a solution that is detailed below.

The **current treatment** of the presented problem is demonstrated in figure 3.1., in which the Cassmassi model (utilizing classic statistical tools) has not been modified from its original form created almost six years ago. It does not justify clearly the critical episodes and bases it prediction on the average behavior of the involved variables from the previous day. It is not systemic as it does not relate to the (synergetic) effects from the generated data from the different registry stations which are distributed geographically throughout the city of Santiago.

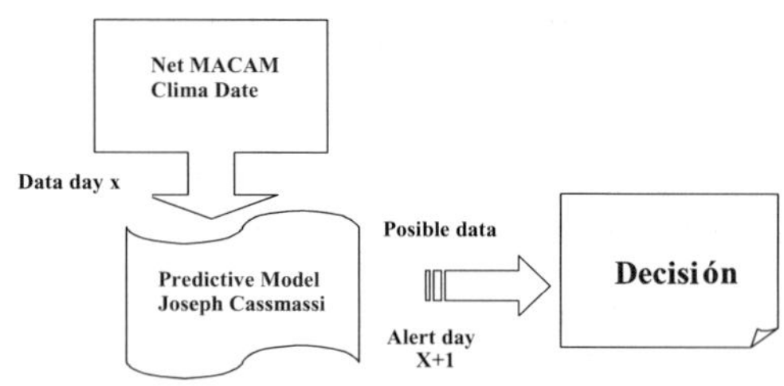


**Fig. 3.1. Current treatment of administration of environmental decisions**

The **alternative proposal,** intends to prepare the basic data surrendered by the monitoring stations of the Net MACAM and the meteorological information provided by the climate predictors from the meteorological services of the city into a DataWarehouse based on a Starchema model.

The DataWarehouse should be designed to respond to queries (not forseen) relative to the activities of the organization and analyzed from different points of view.

For example:

- Number of pre emergencies declared in the winter.
- Alerts declared to be correct.
- Municipality with the most pre emergency alerts in the year, grouped by each monitoring station.

### 3.2. Design of the DataWarehouse SESMA

The intial step is to design a Star or Snowflake [Starchema or Snowflake] for each activity analyzed in the organization. In this case, the activity to analyze is the **SITUATION**, this will be our facts chart which will contain the important information of our organization.

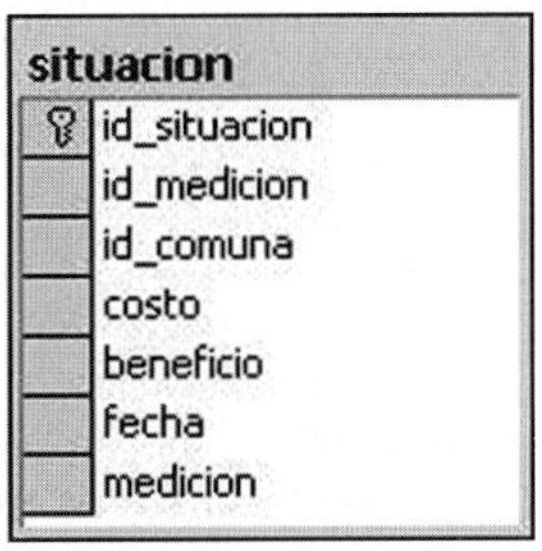


**Fig. 3.2. Table based on a Star Model**

This table contains the columns *id_situation, id_measurement and id_municipality,* it also contains the column *cost* which indicates to us the cost associated to a certain situation. The column *benefit* indicates the benefit associated to a certain situation. The columns *date and measurement* indicates the measurement made in a certain day (this measurement was calculated previously from the original database OLTP)

Each row from the facts table contains the observable data (cost, benefit, measurement) of the activity and the references to the dimensions that characterize them (*id_measurement, id:municipality, dates*).

Generally, the facts table is a many to many relationship with their respective dimensions.

The second step is to define our dimensions table and its properties, they are the relevant dimensions in our observations.

The dimensions that we define for our DataWarehouse are: **measurement, municipality** and a dimension of time **dates.**

With this and after normalizing to avoid the concatenation of charts to the moment to carry out the consultations, the outline shatters it is like it is shown in the figure 3.3

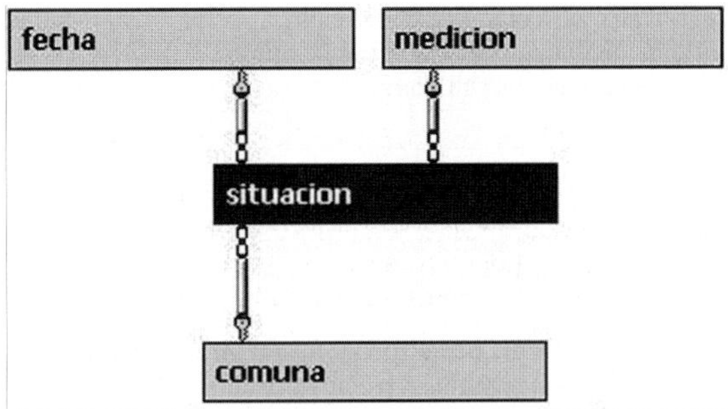


**Fig. 3.3. Star Model DataWarehouse SESMA**

### 3.3. QUERYING THE DATAWAREHOUSE

We can now in the position to use the datawarehouse, for example, to show a date in that the "situation" that was a "pre-emergency" in the "municipality" of "Pudahuel" in the "period" of "winter ", this would be made in the following way:

select medicion.nombre, fecha.fecha from situacion, medicion, comuna, fecha
where situacion.id_medicion=medicion.id_medicion and
medicion.nombre='preemergencia' and
situacion.id_comuna=comuna.id_comuna and
comuna.nombre='pudahuel' and
fecha.id_fecha=situacion.fecha and
fecha.estacion='tinvierno'

| | nombre | fecha |
|---|---|---|
| 1 | preemergencia | 2002-07-13 00:00 |
| 2 | preemergencia | 2002-07-14 00:00 |

**Fig. 3.4. Result when querying date and situation in period of winter in the municipality of Pudahuel**

In the previous case, we introduced the dimension of time for the analysis and included a new additional level of data over the type of date, this is known as drill-down

Another query that would be of interest to us would be to generate reports that would allow us to make critical decisions: number of pre-emergencies declared in winter, the code for this query would be written the following way:

```
select COUNT (*)
From situacion, medicion, comuna, fecha
where situacion.id_medicion=medicion.id_medicion
and situacion.id_comuna=comuna.id_comuna
and fecha.id_fecha=situacion.fecha
and medicion.nombre='preemergencia'
and fecha.estacion='tinvierno'
```

| | (Sin nombre de c( |
|---|---|
| 1 | 3 |

**Fig. 3.5. Result when querying for number of pre-emergencies declared in winter.**

In relation to the previous query we could show the municipalities and the date when the pre-emergency was declared during the winter. The query would be written following way:

```
select fecha.fecha, comuna.nombre
From situacion, medicion, comuna, fecha
where situacion.id_medicion=medicion.id_medicion
and situacion.id_comuna=comuna.id_comuna
and fecha.id_fecha=situacion.fecha
and medicion.nombre='preemergencia'
and fecha.estacion='tinvierno'
```

| | fecha | nombre |
|---|---|---|
| 1 | 2002-07-13 00:00:00 | PUDAHUEL |
| 2 | 2002-07-14 00:00:00 | PUDAHUEL |
| 3 | 2002-07-13 00:00:00 | CERRO NAVIA |

**Fig. 3.6. Results by row when querying dates and municipality with pre-emergencies declared in winter**

This design of DataWarehouse allows us to know important data when making the final decision on a course of action regarding the quality of the air in Santiago from Chile, the final decision is the basis of reports, graphics, etc.

Just as important as the design of the DataWarehouse is the design of a powerful system to negotiate the data and that can provide useful and reliable information for making decisions.

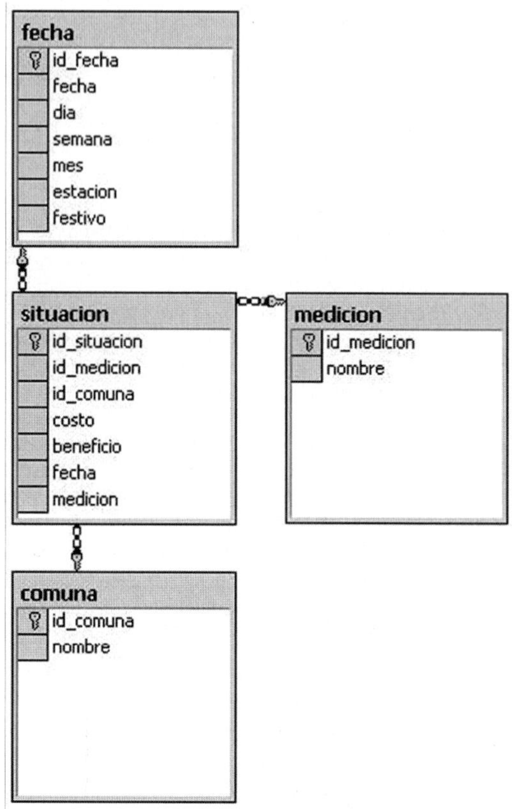


**Fig. 3.7. Star Outline of DataWarehouse**

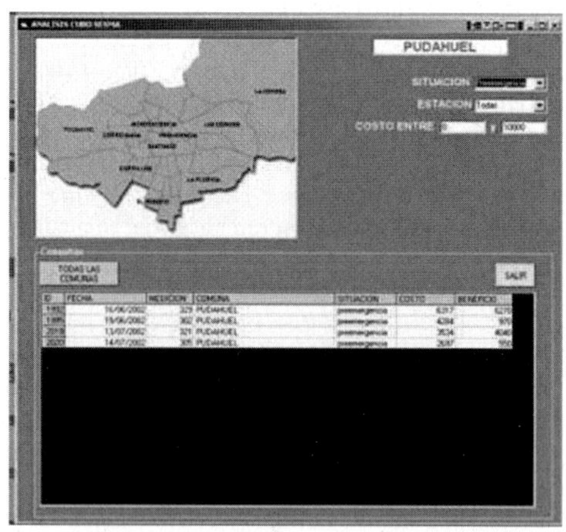


**Fig. 3.8. Typical search results screen**

### 3.4. REQUIREMENTS OF THE SYSTEM

The used database motor is SQL Server 2000, on an operating system Windows XP Home Edition, with a processor Intel Pentium IV of 2000 GHz with 256 Mb of RAM, the software used to design the graphic face for the consultations SQL is Visual Microsoft Basic 6.0, [56], [57].

### 3.5. CONCLUSIONS OF THE CONSULTATIONS

The queries on a DataWarehouse are more useful than those in a relational database, given that a lot of information is filtered before filling in the facts table and their respective dimensions. Also it provides great help in the normalizing to obtain simpler code and provides data of relevant information for making decisions.

The design adopted for the Data Warehouse allows to know important data for the taking of Decisions Preposition the quality of the air in the city of Santiago from Chile, decisions based on reports, graphics, etc.

It leaves important of the investigation, besides the design of the Data Warehouse, this in design of a powerful system dates to negotiate the data and that can give useful and reliable information for the taking of decisions

Future work is promising, finding alternative solutions to the problem of contamination in the city of Santiago, Chile has become a humanitarian crusade. The problems for the population's health and the economy of the country are reaching levels that surpass what is acceptable for any community.

## 5. BIBLIOGRAPHY


[1] www.sesma.cl/sitio/pag/aire/Indexjs3aire005.asp

[2]www.sesma.cl/sitio/pag/aire/Indexjs3airesist_monit.asp

[3] http://www.conama.cl/portal/1255/channel.html

[4] R. L. Ackoff "Science and the Systems Age: Beyond I.E., O.R. and M.S.". Operations Research, May-June 1973.

[5] J. G. Saxe (1816 – 1887)

[6] W. Churchman "The Systems Approach" Dell Pub. (1968)

[7] http://w3.mor.itesm.mx/~albreyes/sistemas/

[8] Power, D. J. A Brief History of Decision Support Systems. DSSResources.COM, World Wide Web, http://DSSResources.COM/history/dsshistory.html, version 2.6, November 12, 2002 .

[9] Ferguson and Jones. Management Decision System: Computer-Based Support for Decision Making, 1969.

[10] Gorry, G.A. y Scott-Morton, M.S. (1971): "A Framework for Management Information Systems", Sloan Management Review, vol 13, nº 1, fall 1971., Scott Morton, M.S. (Eds.), The Corporation of the 1990s, Oxford University Press, Oxford, 1991.

[11] Sprague, R., Carlson E. Building Effective Decision Support Systems, Englewood Cliffs, NJ, Prentice Hall, May 1982

[12] J. P. Shim, Merril Warkentin, James F. Courtney, Daniel J. Power, Ramesh Sharda, Christer Carlsson Past, Present, and Future of Decision Support Technology, Decision Support System 32 (2002) 111-126, www.elsevier.com/locate/dsw

[13] Martha Patricia Bobadilla, El impacto en los negocios del DSS, 07 /05 / 2001 http://www.claveempresarial.com/soluciones/notas/nota010507b.shtml

[14] http://www.bitam.com.mx/AcercaDeBI.htm

[15] Diana Cristina Romero Sánchez, La Importancia De Los GDSS En El Trabajo Colaborativo, dromero@campus.zac.itesm.mx http://www.netmedia.info/netmedia/articulos.php , Netmedia S.A.

[16] Los PDAs, la última frontera para los MSS http://www.pdaexpertos.com/Articulos/Experiencias_de_Usuarios/7.shtml , Everardo Alpuche M., Ingeniero en Sistemas Computacionales, estudiante de Maestría en Administración de Tecnologías de Información, Universidad Virtual del ITESM, México.

[17] Turban, Efraim & Aroson, Jack E. (1998). Decision Support Systems and Intelligent Systems. EEUU: Prentice-Hall. Quinta Edición.

[18] Hernández A., Federico (1997). "Revista Soluciones Avanzadas", Sistema Gerencial Administrativo para el Soporte de Decisiones, Infolatina, 14 de Enero de 2001.

[19] Porter, Michael E., Competitive Strategy: Techniques for Analyzing Industries and Competitors, Free Press, New York, 1980.

[20] Power, D. J. Decision Support Systems Web Tour. World Wide Web, http://dssresources.com/, version 4.2, June 15, 2002.

[21] Scott Morton, M.S. (Eds.), The Corporation of the 1990s, Oxford University Press, Oxford, 1991.

[22] BARRY DE VILLE, Data Mining en SQL Server 2000, http://www.w2000mag.com/sqlmag/atrasados/04_mar01/articulos/portada_1.htm

[23] Tuya Javier, Adenso Díaz; "Los Decisión Support Systems: Arquitectura y Aplicaciones Empresariales", ETS Ingenieros Industriales e Informaticos, Universidad de Oviedo, http://www.di.uniovi.es/~tuya/pub/ati-98-dss-resumen.html

[24] Sprague, R. H., "A Framework For the Development of Decision Support Systems". Management Information Systems Quarterly, No. 4. pp 1-26, 1980. Sprague, R.H., y H.J. Watson, Decision Support for Management, Prentice Hall, New Jersey, 1996.

[25] Marín Ruiz Nicolás, Introducción al Data Warehousing, Decsai, Universidad de Granada, Granada, España.

[26] W.J. Frawley, G. Piatetsky-Shapiro y C.J Matheus. "Knowledge Discovery in Datbase: an Overview". Knowledge Discovery in Datbase. AAAI-MIT Pres, Menlo Park, California 1991, páginas 1-27.
[27] U.M. Fayyad "Data Mining to Knowledge Discovery: making sence out of data". En IEEE Expert Vol. 5, octubre 1996, páginas 20-25.
[28] Artículo de Internet con el título "Data Mining: torturando los datos hasta que confiesen", por Luis Carlos Molina Felix, coordinador del programa de data Mining (UOC), 2001
[29] IBSNAT. 1989. Decision Support System for Agrotechnology Transfer (DSSAT) Version 2.1, Dept. of Agronomy and Soil Science, College of Tropical Agriculture and Human Resources. University of Hawaii, Honolulu. www.mic.hawaii.edu/dev_tech/software/dssat.html
[30]Copyright 1995-2002 by: ESS Environmental Software and Services GmbH AUSTRIA http://www.ess.co.at/AIRWARE/
[31]http://www.ucar.edu/communications/newsreleases/2003/mdss.htmlhttp://www.amazings.com/ciencia/noticias/270103a.html
[32] http://www.baronams.com/EDSS/
[33]http://strategis.ic.gc.ca/Ces_Web/_providers_info_.cfm?CES_ESTBLMT_NO=2641&target=English
[34] http://www.riks.nl/PROJECTS/KRIM http://www.krim.uni-bremen.de/englisch/indexenglisch.html
[35]http://www.aaas.org/international/lac/plata/baethgeninfo.shtml http://www.inia.org.uy/disciplinas/agroclima/index.html
[36] Biss, A. "Dynasty Triage Advisor Enables Medical Decision-Support", 2002, at URL DSSResources.COM. www.dynasty.com. http://dssresources.com/cases/dynasty.html
[37]http://www.mega-supply.com/websites/megasupply/ap1-2.htm http://www.ap1soft.com/
[38]http://www.iiasa.ac.at/Research/WAT/docs/desert.html.
[39]http://www.netmeye.net/enespañol/experiencias/Soporte_de_decisiones.htm http://www.inta.gov.ar/bariloche/ssd/rn
[40] http://www.water.ncsu.edu/watershedss/index3.html
[41] http://www.riks.nl/projects/Xplorah. http://www.riks.nl/RiksGeo/projects/xplorah/XplorahBrochurePrint_es.pdf
[42] http://www.geogra.uah.es/Proyectos/sedis.htm
[43] http://www2.ing.puc.cl/gescopp/investigacion.html
[44] SESMA (Servicio de Salud Metropolitano del Ambiente), www.sesma.cl
[45] RED MACAM (Red de Monitoreo Automático de Calidad del Aire y Meteorología), www.conama.cl/rm/568/article-1114.html
[46] Universidad de Santiago de Chile, "Normativa Ambiental en Aire. (Fuentes Fijas)", http://lauca.usach.cl/ima/sesma-1.htm
[47] http://www.cleanairnet.org/lac/1471/article-40847.html.
[48] http://www.conama.cl/rm/568/article-1183.html.
[49] http://www.conama.cl/rm/568/article-2581.html.
[50] http://www.quepasa.cl/revista/2003/05/16/t-16.05.QP.NAC.CONTAMINACION.html
[51] http://www.geofisica.cl/papers/pherman.htm.
[52]http://www.sustentable.cl/portada/Descontaminacion/2453.asp
[53]http://www.sesma.cl/sitio/pag/aire/Indexjs3aire005.asp
[54] WEKA (Waikato Environment for Knowledge Analysis) http://www.cs.waikato.ac.nz/~ml/
[55] http://www.geofisica.cl/English/pics3/FUM6.htm http://www.geofisica.cl/Meteorol.htm
[56] http://www.microsoft.com/spain/sql/
[57] http://www.microsoft.com/latam/sql/

# Designing Human-Centered User –Interfaces For Technical Systems

Salaheddin Odeh, PhD
Department of Computer Engineering, Faculty of Engineering, Al-Quds University,
P.O. Box 20002, Abu Dies, Jerusalem, Palestine
Email: sodeh@eng.alquds.edu

**Abstract** – This contribution emphasizes the appropriateness of using human experience for designing rule-based user displays for controlling technical systems. To transform the human (operator) experience into a computer presentation, suitable means are needed. The proposed method and technique for designing user-machine interfaces for system control uses fuzzy logic techniques to convert human experience into a computer representation, and to compute values to animate graphical objects on the user display. This modest contribution investigates and clarifies the reasons for considering the appropriateness of fuzzy logic in designing rule-based human-machine interfaces for technical system control.

## I. INTRODUCTION

A novel design technique for the construction of human-machine interfaces will be presented. This technique seeks to increase the users' orientation adapting human-machine interfaces to the cognitive structures of human users. Different models of the human experience that are available in analogous or conceptual form have to be considered, in the design of both a graphical user display and in the design of the underlying information management system of the user-interface. Fuzzy logic is used for translating natural language procedures acquired from operators into rule-base objects. Thus, technical systems can be considered and controlled from different viewpoints. It is noted that technical systems are mostly presented by conventional user displays based on topological representations and flow diagrams [1]. For large and complex systems, views in these kinds of user displays are split up according to system/sub-system hierarchies. Such views become intricate if complex plants have to be presented. The design method and technology investigated in this study is based on different models of human operators and the technical system. These models that are acquired from experienced operators through task and system analysis must be used in an integrated manner to achieve a powerful work environment for human operators.

## II. ANIMATION OF GRAPHICAL OBJECTS ON THE USER DISPLAY

The user-interface and its information management system should include different models of the technical system to provide the super visioning and controlling system with a comprehensive database ([2], [3]). To convert these models into a computer representation suitable means are needed. However, there are several methods and techniques of which fuzzy logic is chosen available for this purpose. The research presents and shows the main reasons that underlie the appropriateness of fuzzy logic for modeling the rule-base of the user-interface and for calculating values for the user-display.

In the first place, to increase the interaction between human and machine (technical system), it has been necessary to develop and to expand the fuzzy formalisms and inference mechanisms for storing and for processing of the technical system. This fulfills the requests of system operators in their control and supervision activities with regard



*T. Sobh and K. Elleithy (eds.), Advances in Systems, Computing Sciences and Software Engineering*, 353–357.

to ergonomic system presentation and sufficient adaptation of their work context.

Secondly, an easy conversion of the experience acquired by the system operators through inquiries into a formal representation demands a suitable means. With the help of the operators' experience, it should be possible to present system states as well as the necessity for doing actions by the operators.

It is possible to apply the programming language Prolog [4], which offers the possibility to deal easily with the predicate logic. Prolog represents a computer-reference language for formal relationships. However, it should be noted that our goals, especially the fuzzy system presentation, cannot be realized through the usage of the programming language Prolog. The system presentation using fuzzy logic allows the graphical representation of system values with different degrees of truth. This third reason highlights additional problems that are likely to appear when different software systems developed with various programming languages have to be integrated within a human-machine system.

In the following section, we will go through the reasons which clarify why fuzzy logic techniques are suitable for use in fuzzy rule-based user displays.

The experience acquired by people exists from the very beginning in a colloquial form. Therefore, the usage of means that allows the direct conversion of this experience into a computer representation has several advantages; firstly, the time needed to develop the human-machine interface can be shortened, and secondly, it will be possible to achieve representations of technical systems that correspond to the ways of how people think. These representations consist of graphical objects animated by values that are calculated by the fuzzy-inference machine. After Zadeh ([5], [6]), fuzzy logic offers methods and techniques for formalizing the human behavior and storing it into a computer system.

At higher levels of the system control, operators do not think in absolute numbers or values, rather they predominantly do it in linguistic concepts. At such levels of system control, an operator strives to build a rough overview of a system situation. This means the designing of human-machine interfaces that give an overview picture of a technical system which must offer qualitative system information. The representation of concrete system values that could be coded in different forms have to be represented at the lower levels of the human-machine interface.

It is agreed that display elements with symbolic information contents play a central role within contemporary human-computer interfaces [7]. Modern user displays lack graphical elements for the presentation of system states, symbolic alarms, goal-fulfillment grades and strengths of urgency for doing actions. Most of today's user displays for system control include graphical elements for visualizing system alerts. The computation of the alarm values in such conventional user displays is based on duel-logic principles; its representation on the graphical user display consists of different visual-encoding forms.

It should be noted that the human infers in approximate manner. This means that he often uses rules, whose condition parts and conclusions are more or less true, e.g. "if the pressure is somewhat high, then the danger caused through this state is still minimal". This conclusion is designated as an approximate conclusion. This fact can be now used in developing and in designing of user displays for system presentation, and therefore we can designate it as „ fuzzy system presentation“.

On user displays, it will be possible to code graphical display elements with symbolic information contents not only with binary values, but moreover with different magnitudes from the interval „ not true“ and „ true“ or „ zero“ and „ one“.

A further problem that occurs during the presentation of system states and goals lies in the characterization of the corresponding variables and intervals. Fig. 1 illustrates several possibilities to characterize a system variable. In this example, the goal of the operator is to keep the magnitude of a system-state variable within the represented interval.

If the selected intervals are too thin (see Fig. 1, first possibility) then system-state changes on the graphical user-display described with different system variables characterized by means of such intervals behave too sensitive. This means that this kind of characterization causes that negligible system changes are unnecessarily prematurely mediated. If several system violations occur concurrently, then the large offer of information by the user-display confuse system operators during their task of system control. Therefore, system operators are forced to observe and to identify malfunctions at different places of the user display.

If we now enlarge the width of the interval (second possibility of Fig. 1) in order to avoid the problem mentioned above, the presentation of system changes will be now delayed. This method, however, does not provide a satisfying solution. The combination of both possibilities (see Fig. 1) is conceivable and desirable. The internal interval [LBi, UBi] is assigned a truth degree of one. Whereas the external intervals [LBo, LBi] and [UBi, UBo] will be assigned truth degrees described by a mathematical function whose values are located between 0 and 1 (see Fig. 1, combination of possibility 1 and 2). In fuzzy logic, these functions are known as membership functions.

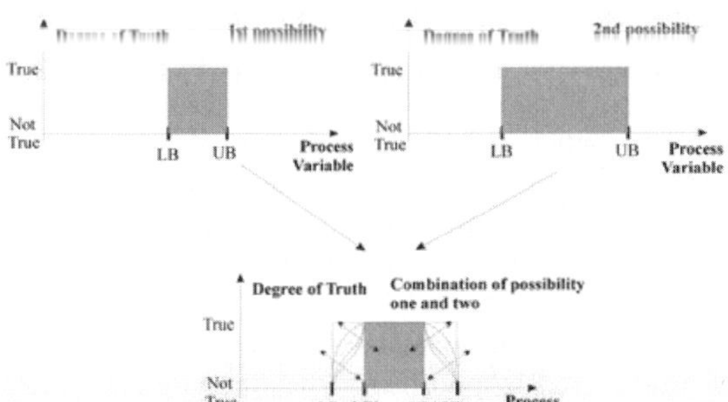


Fig. 1: Possibilities for characterization of variable attributes.

Fuzzy-logic modeling has numerous advantages regarding the innovation of new graphical display elements. One of these lists these advantages is that the possibility to present symbolic information with different strengths: For instance, instead of presenting alarm values with two states "danger exists" or "danger does not exist", new kind of display elements representing dangers with different strengths are achieved. Another advantage is the reduction of graphical elements on the user display: Usually, several graphic objects are necessary to represent the different alarm stages. This technique allows us to integrate various graphical elements to a single one. A third advantage is Implication of tendencies: Because through the use of fuzzy logic strengths of system violations or danger situations increase or decrease progressively, system operators will be gradually warned or unwarned. This type of information can be considered as tendencies.

The realization of a rule-based human-machine interface including methods of fuzzy logic as a fundamental means for representing and for processing of the experience has large advantages regarding the integration of the different part systems within the total human-machine system. In our case, all subsystems have to be implemented by means of the object-oriented programming language C# [8].

## III. A USER -DISPLAY BASED ON FUZZY LOGIC

In order to achieve effective interaction with the human-machine interface a complete computerized system containing both the software management system and a graphical user display should be available. This tool allows, technical systems to be considered and controlled through different representations such as topology, causal coherence, system goals, states, necessity for doing tasks and security classes. In this section, some presentation aspects of the qualitative rule-based user display, the HCUI[1] user display, will be presented. Fig. 2

HCUI: Human-Centered User Interface [1]

shows the architecture of the HCUI for presentation of qualitative states and goal fulfillment - grades of a technical system. The upper portion of the screen presents different information about the system such as operational situations and security classes. An overview includes system information about states and system goals of the critical subsystems of the technical system. This overview window cannot be covered by other displays and, thus, the operators are always informed about the current system state.

The goal icons that can be presented by pie-chart symbols mediate whether the goals of the influence-function units in the different critical subsystems are fulfilled or not. As illustrated above, the calculation of every pie-chart value is based on fuzzy rules and is executed by a local fuzzy inference-machine.

Details of a critical subsystem can be enlarged to get more information about the macro-states (sub-states) and necessities for doing tasks. Moreover, detailed information about the fulfillment-grades of component functionalities within a function unit can also be shown. These icons can be used to activate the corresponding topological views of the system. The operator can select any given macro state represented by a gauge icon to gather information about the rules underlying this specific macro-state. These rules are presented in the window shown on the lower right side of Fig. 2.

Another possible representation form supported by the HCUI is the task-dependent state presentation managed by a rule-based module called security classes. Within this object-oriented module, system violations describing illegal or undesirable combinations of system values are modeled. These combinations of system values are not related to one specific state. This means that these system violations can occur during all operational situations. System violations with different strengths can be presented on the left-middle side of the desktop. Qualitative values for terms of security classes represented in the overview of the technical system are calculated by means of the MAX operator [9].

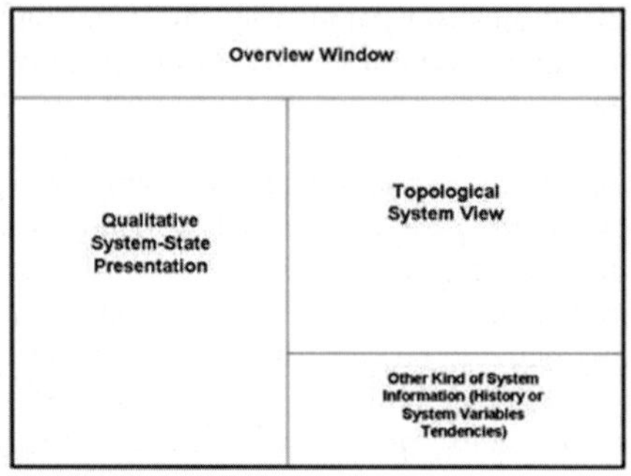


Fig. 2: The architecture of the proposed human-centered user-display HCUI during presentation of qualitative system states.

The MAX operator combines all results of fuzzy rules belonging to a class to a single value.

Furthermore, locations of system violations can be achieved by sub-premises of rules represented by icons. These sub-premises containing system variables and values margins can be selected to activate topological views in order to go to the location of the components that must be controlled and handled. This state presentation, as mentioned above, is task-dependent. This means that after completion of correct actions, a system violation will disappear although the state variables are still have undesirable values.

## IV. CONCLUSION

It has been shown that numerous advantages can be achieved by using fuzzy logic for storing human (operator) experience as well as for computing system-control values for life-critical systems. In addition, it is noted that the symbolic fuzzy presentation results through combination of physical system values and experiences acquired by the system operators. The new technique strives towards the creation of a powerful human-machine interface that support system operators in their control and supervision activities with regard to ergonomic system presentation and sufficient adaptation of their work context. By means of this method, a graphical user display called HCUI was developed and compared, for evaluation purposes, with other two displays. Through these displays that based on different design philosophies, the same technical system can be controlled and monitored. The purposes for developing these user displays were to analyze supervision problems in system control as well as to use them as reference systems in the evaluation mentioned above. The presentation technique achieved through this method is called roughly 'rule-based system presentation'. The main goal of this presentation is the optimum integration of the operator in the whole control loop of the human-machine system. This will, hopefully, lead to better work satisfaction of the human as well as more security and economic efficiency.

## REFERENCES


[1] VDI/VDE Guideline: Process control using display screens, The Association of German Engineers. Duesseldorf: VDI-Verlag, 2005.

[2] E.A. Averbukh and G. Johannsen, Knowledge-based interface systems for human supervisory control. In: Proc. Japan-CIS Symposium on Knowledge Based Software Engineering '94, Pereslavl-Zalesski, 1994, pp. 229 - 232.

[3] B. Shneiderman, C. Plaisant, Designing the User Interface: Strategies for Effective Human-Computer Interaction (4th Edition). Addison Wesley Longman, 2004.

[4] I. Bratko, Prolog Programming for Artificial Intelligence. Wokingham, England: Addison-Wesley, 1986.

[5] L. A. Zadeh, Calculus of fuzzy restrictions. In: Zadeh, L. A. et al. (ed): Fuzzy sets and their applications to cognitive and decision processes, 1-39. New York: Academic Press, 1975.

[6] L. A. Zadeh, The concept of a linguistic variable and its application to approximate reasoning. Information Sciences, 8, 199-249(I), 8, 301-357(II), 9, 43-80(III), 1975.

[7] G. Johannsen, S. Ali, J. Heuer, Human-Machine Interface Design Based on User Participation and Advanced Display Concepts. Post HCI 95 Conf., Seminar on Human-Machine Interface in Process Control, Kyoto, July 1995.

[8] H. M. Deitel, P. J. Deitel, J. A. Listfield, R. T. Nieto, C. H. Yaeger, M. Zlatkina, C# for experienced programmers . Prentice Hall, 2002.

[9] H.-J. Zimmermann, Fuzzy Set Theory and its Applications. Boston, MA.: Kluwer Academic Publishers, 1991.

# Service Selection Should be Trust- and Reputation-Based

Vojislav B. Mišić
Department of Computer Science
University of Manitoba
Winnipeg, Manitoba, Canada
vmisic@cs.umanitoba.ca

***Abstract*—While existing Web Services standards provide the basic functionality needed to implement Web Service-based applications, wider acceptance of the Web Service paradigm requires improvements in several areas. A natural extension of the existing centralized approach to service discovery and selection is to rely on the community of peers—clients that can share their experiences obtained through previous interactions with service providers—as the source for both service discovery and service selection. We show that distributed algorithms for trust maintenance and reputation acquisition can effectively supplant the current centralized approach, even in cases where not all possible information is available at any given time. Some preliminary results demonstrate the feasibility of trust and reputation-based service selection approach.**

***Keywords–Web Services, service selection, trust and reputation, P2P systems***

## I. WEB SERVICES: WHAT'S MISSING

Many approaches have been proposed to combat the ever-increasing complexity of modern software systems and the so-called service-based computing paradigm has recently attracted a lot of attention. In particular, the importance of a family of service-based applications known as Web Services has rapidly grown [5]. Web Services may be succinctly defined as a family of service-based applications where (1) clients and service providers are permanently hosted on different computers, and (2) the interactions take place by exchanging messages written in an XML compliant language over the Internet (i.e., using the TCP/IP family of protocols).

However, despite their perceived importance for e-business and other applications, Web Services are still lacking important functionality which impedes their further adoption. A number of open standards pertaining to different aspects of Web Services has been introduced, most importantly Web Services Description Language (WSDL) for WS description, Universal Description, Discovery, and Integration (UDDI) for WS discovery, and Simple Object Access Protocol (SOAP) for WS invocation, binding, and message exchange. All these standards use XML-compliant languages, and all of them rely on the ubiquitous HTTP protocols for communications. A service provider publishes WSDL description(s) of its service to the registry; a client requests information from the registry via UDDI; having obtained it, the client contacts the provider directly to negotiate service binding and actual interaction, as shown in Figure 1. . The registry, in fact, acts as a simple directory providing basic information about services and service providers to any client that requests it.

The research presented here was in part supported by the NSERC Discovery Grant.

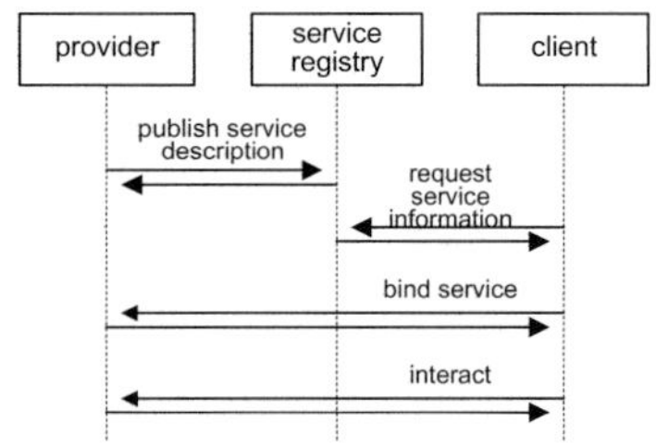


Figure 1. Operation of Web Services according to the current paradigm.

The idealized picture we have just described suffers from a number of major problems. The first problem is the quality of individual services: when a search for a web service produces a list of services that fulfill the specification used, there is little to help the client in the process of selecting the right service to be bound and subsequently invoked. This decision could be made on the basis of functional and nonfunctional qualities and guarantees which are commonly referred to as Quality of Service (QoS), yet the existing standards do not address QoS-related issues at all. While a number of proposals aim to remedy this situation – Web Service Level Agreement language (WSLA) [3], Web Service Management language (WSML) [9], DARPA Agent Markup Language for Web Services (DAML-S) [1], and Web Services Offerings Language (WSOL) [11], among others – none of them has achieved widespread acceptance as yet.

While the knowledge of operational parameters (described through the appropriate WSDL description) and QoS guarantees (described using any of the proposed WSDL extensions) is crucial, it is not nearly sufficient to make an informed choice. Adherence to well defined standards and specifications may be expected within a single organization (well, almost), but in case of Web Service-based applications, the client and the service provider belong to different



*T. Sobh and K. Elleithy (eds.), Advances in Systems, Computing Sciences and Software Engineering*, 359–364.

organizations. Consequently, the description of QoS guarantees is just a list of promises, and the client application has no way of knowing whether they will actually be met when the service is invoked. Interaction must, then, be based on QoS guarantees as well as the belief about the truthfulness of those guarantees [6], as is common in other situations where two independent parties interact. This belief is commonly referred to as trust [10].

## II. TRUST – YES, BUT HOW TO GET IT?

Trustworthiness of individual service providers might be assessed with the aid of a dedicated certification entity, which effectively functions as a parallel registry, as proposed in [8]. The operation in this setup is shown in Figure 2. In this case, the certification registry would have to be able to store data about both operational parameters of its services and QoS-related parameters. Changes would have to be made to both WSDL and UDDI standards to accommodate QoS-related information. Furthermore, the protocols for service publishing and discovery would become slightly more complicated since either one of the operations, or perhaps both of them, involves consultation with the certification registry, in addition to the interaction with the current directory registry.

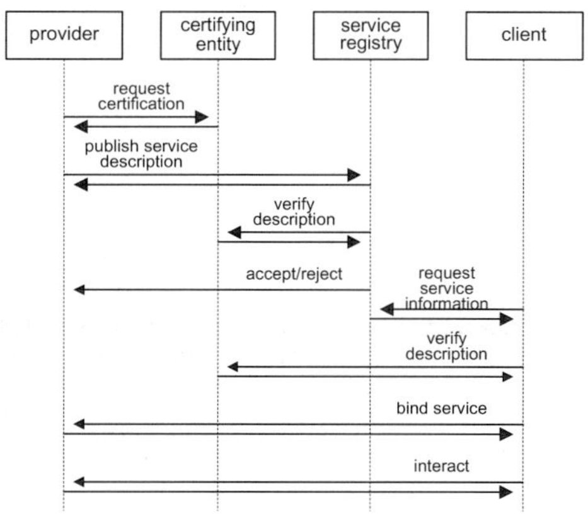


Figure 2. Reputation maintenance with the aid of dedicated certification entity or registry.

Unfortunately, the implementation of a single such entity poses two problems that seem rather hard to solve. First, who would operate this authority, and through what mechanism would the service providers be forced to get their services certified? Second, the certification means that the services being certified are actually exercised through suitable applications. This is simply infeasible both conceptually and computationally (one should remember that similar proposals regarding certification of software components have been virtually abandoned).

Both of the above mentioned shortcomings could be simply overcome by relying on the community of clients, rather than on any individual client or independent entity. The clients that have already interacted with the service (or service provider) in question should simply record their experience(s) and share them with others upon request. In that manner, any client could rely on collective experience of other clients to aid in the selection of the most reliable provider. (Note that the word `reliable' is used in its everyday sense, i.e., as being dependable, able to yield the same or compatible results repeatedly.)

Shared information may be collected and kept in a centralized manner, by the directory registry. When a provider publishes its services, the registry may request in broadcast that the community of clients send in their experiences regarding the QoS parameters of the service. Furthermore, when a client finishes its interaction with a service, it may send a digest of its experiences to the registry and/or broadcast them to the community. In this manner, the community of clients can assist the registry in maintaining the reputation of services and their providers, as shown in Figure 3.

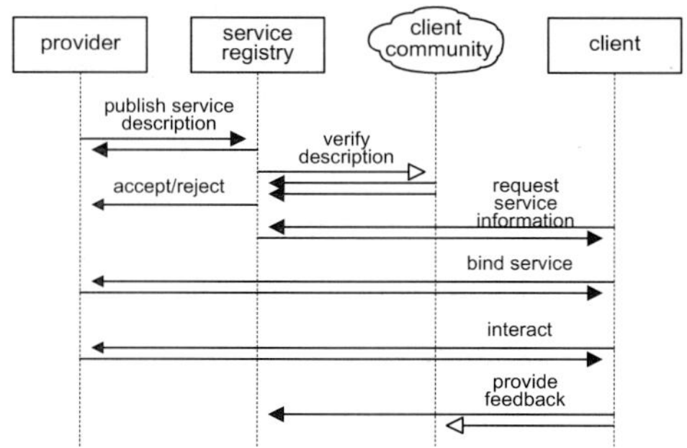


Figure 3. Community-assisted reputation maintenance (arrow with a hollow head indicates a multicast/broadcast message).

The notion of trust comes into play in another likely scenario. Namely, there is nothing in the current Web Services-related standard that prohibits the existence of several service registries. The client that wants to find a particular service may query and receive appropriate responses from several registries; the question is which one of the responses should be used? The availability of QoS-related information is obviously important in this case; but the degree of trust in the results returned by a particular registry, and by extension, the trust in the registry itself, should not be ignored [6]. One should not ignore the fact that multiple registries would certainly improve the overall reliability of Web Services applications, since a failure of one registry would not bring the entire system down.

If service selection is to rely on the community of clients and if multiple service registries may exist, it seems that the client community could be used not only as a secondary source of trust-related information, but as the main source for both service discovery and service selection, as shown in Figure 4.

The notions of trust and reputation have been investigated for some time within the autonomous agents community, with the initial impetus coming from the development of electronic markets and market agents. A number of schemes to develop and maintain trust indicators have been described [10], and similar notions have recently been applied to Web Services [7]. Much remains to be done, however, before the Web Services can be transformed to make use of the community-based discovery and selection mechanism. Fortunately, the flexibility of XML content models allows for easy extension of existing languages, and it also facilitates backward compatibility since hosts which do not support extended features can simply disregard them in communication.

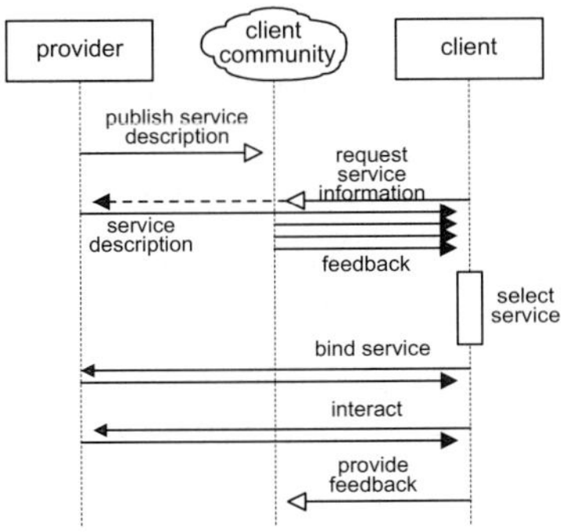


Figure 4. Community-only reputation maintenance.

As a small step in this direction, we present some preliminary results that demonstrate the feasibility of incorporating the concepts of trust and reputation in the process of Web Service discovery and selection.

## III. Recording and Evaluating Trust and Reputation

Our inspiration comes from the business area where the selection of business partners is often made on the basis of trust, i.e., our own belief that the other party will act according to its promises, and reputation, i.e., the shared belief of the community about the party in question [10]. This belief is a summary of experiences obtained through previous interactions with the service and its provider, both those of our own (i.e., trust) and those by others (i.e., reputation). The use of such collective experience helps the client community to select the most reliable (or, rather, dependable) partners - in this case, service providers.

Trust information might be collected and kept in a centralized manner, e.g., by extending the functionality of the UDDI registry or registries. While such changes are indeed possible, we focus on a more promising approach that relies on the client community as the main source for both service discovery and service selection, as shown in Figure 4. It may be worth noting that this approach is similar in spirit to the increasingly popular P2P schemes for file sharing [2].

Furthermore, a number of schemes to develop and maintain trust and reputation indicators have been described in relationship with electronic markets of various flavors; as mentioned above, similar attempts have been made in the area of Web Services, but the results are still inconclusive [7]. As a small step in this direction, we present some preliminary results that demonstrate the feasibility of incorporating the concepts of trust and reputation in the process of Web Service discovery and selection.

For simplicity, let us assume that each service $S_i$ has an associated value promise $V_p(S_i)$, which may be obtained as a suitable combination of price and QoS guarantees. (While individual properties can be introduced, the ultimate decision to select or skip a particular service will likely be based on such an aggregate value.) The customer satisfaction after invoking and using a given service will be proportional to the difference between the delivered value $V_d$ and the promised one $V_p$. The trust held by client $C_j$ in the provider $P_k$ can, then, be obtained as the sum of all satisfaction values for the services provided by $P_k$.

The time dimension of trust should also be taken into account, by assigning more weight to more recent information; we propose to use a simple exponential averaging of the form $T' = \alpha t + (1-\alpha)T$ to model the aging, with the number of interactions used as a proxy measure of elapsed time.

The reputation of a particular provider will be the sum of trust values for that same provider, but held by other clients. If the trust information is recorded in a centralized manner, each client will have access to the recorded trust of all other clients – i.e., perfect reputation accuracy can be obtained. In a distributed, P2P-like approach not all clients will respond to the reputation query, either because they don't want to respond, or because they are too slow and their response arrives after the client has already made the decision. Consequently, the reputation accuracy will be less than perfect.

## IV. Simple Selection – No Trust, No Reputation

In order to validate the basic concepts of community-assisted service provider selection based on trust and reputation, we have built a simulator with 200 clients and 25 service providers, offering a total of 201 distinct services. Each provider offers exactly 134 services, i.e., two thirds of the total number; providers and services are randomly paired to give offers. Each service offer is described with a random price and performance attributes, jointly referred to as the service promise. Each provider, on the other hand, is equipped with a reliability attribute; reliability is uniformly distributed in the range of zero to one. The actual satisfaction of a service call is calculated as a function of the provider reliability independently of the underlying service promise; satisfaction values range from -1 (indicating totally unsatisfactory call) to 1 (indicating total satisfaction, i.e., the service has been performed exactly as promised).

Then, we ran several experiments in which the simulator runs for 25 cycles. In each cycle, each client requests a randomly chosen service and selects the provider according to the designated criterion, as discussed below. Therefore, each cycle corresponds to 200 service calls, and each experiment includes for a total of 5000 service calls.

After each call, the caller client determines the satisfaction with the service it obtained; the satisfaction values are aggregated by provider and recorded. When trust is used, this information is used to select the provider; reputation involves using the satisfaction information obtained from other clients. The following diagrams show the results of some sample runs.

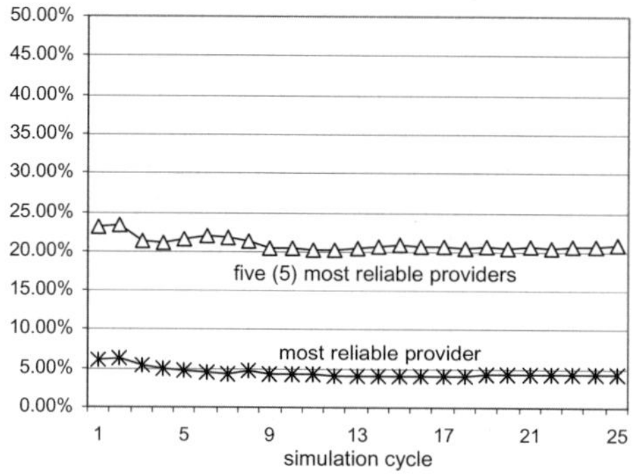


Figure 5. Percentage of service calls to the most reliable provider (asterisks) and five most reliable providers (triangles). Service selection is made at random.)

In our first experiment, the clients have chosen services at random. The proportion of calls directed towards the most reliable provider and five most reliable providers is shown in Figure 5. As can be seen, after some initial fluctuations which are due to the random choice of services, the percentage of calls serviced by the most reliable providers quickly stabilizes to give approximately uniform distribution of calls per provider. The actual percentages of calls serviced by the most reliable provider and five most reliable providers are close to 4 and 20 percent, respectively. This should come as no surprise, since the system contains a total of 25 service providers, and each provides gets an approximately equal share of service calls.

Then, the clients were made to choose the services (and their respective providers) according to the advertised service promise. Two values of discovery probability were used: 1.0, which corresponds to the ideal case where one centralized registry holds the information about all services, and 0.5, which may be interpreted as the case in which the discovery process is distributed and only about 50% of available offers are found in each call. This latter case models the behavior of a P2P system in which not every client may be expected to respond to every query: some clients may be offline at the moment the query is issued, while others may not want to share their information with the community. As can be seen from the diagrams in Figure 6. , there is little difference between these two cases, apart from fluctuations due to the random selection of services.

An important conclusion that can be made on the basis of these findings is that the community-only reputation maintenance is able to function as expected even when not all members of the community partake in the process. In other words, even a small community will be able to effectively share reputation information and thus facilitate the service selection process.

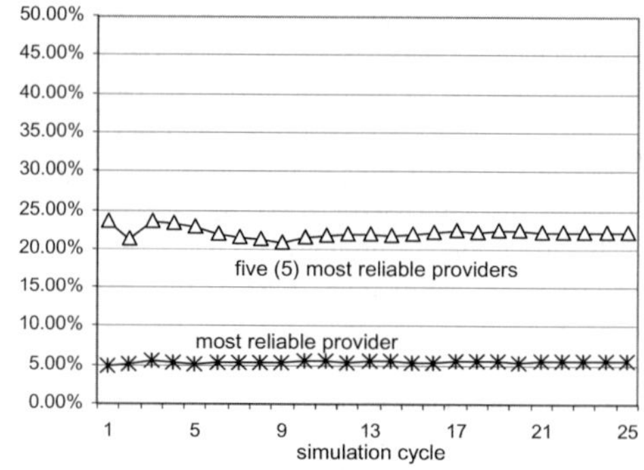


(a) discovery probability 1.0

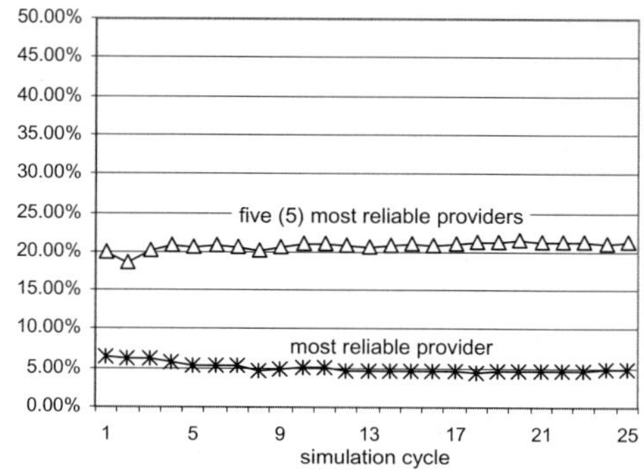


(b) discovery probability 0.5

Figure 6. Performance of centralized vs. distributed service discovery. Percentage of service calls to the most reliable provider (asterisks) and five most reliable providers (triangles). Service selection is made according to the service promise. Discovery probability as noted above.

## V. SELECTION USING BOTH TRUST AND REPUTATION

The next step was to include trust information in the selection process; the corresponding results are shown in Figure 7. As can be seen, the call distribution favors more reliable providers, with limiting values being more than 30% for the single most reliable providers, and well over 70% for the five most reliable providers. In other words, clients clearly favor more reliable service providers over those less reliable ones. Furthermore, the convergence toward the most reliable provider, or providers, is reasonably fast: in the experimental setup, it takes 17 to 18 cycles to reach 90% of the final value.

When both trust and reputation are used (i.e., when trust information is shared with other peers), the call distribution may be expected to give even more preference to more reliable providers. The convergence to the asymptotic value may be expected to be faster too. This has been confirmed by the call distribution diagrams shown in Figure 8. (a). The experimental data were obtained under the assumption of ideal reputation sharing (which means that each client has access to all other clients' data).

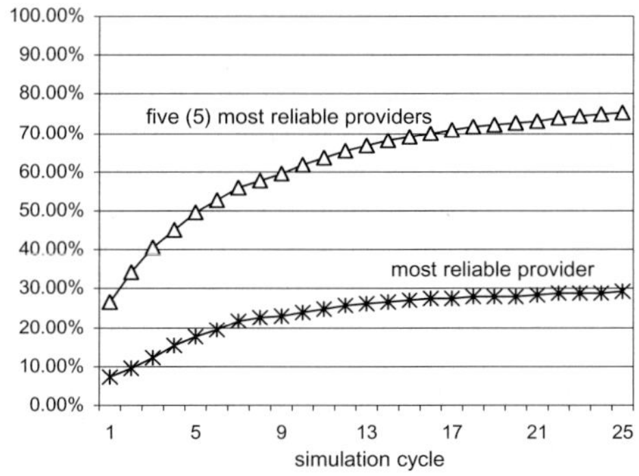


Figure 7. System performance when the selection of the service provider is made according to the service promise (40\%) and trust in provider performance (60\%). Asterisks denote the percentage of services provided by the most reliable provider, while triangles denote the percentage of the calls provided by five most reliable providers.

As can be seen from the diagrams, the five most reliable providers quickly take up more than 90% of all calls. To further strengthen out conclusion, we have also plotted the number of service providers that have never been selected, shown in Figure 8. (b). As can be seen, this number is initially high and then slowly drops to a value somewhere below ten, which means that nearly forty percent of least reliable providers are never selected at all.

When reputation sharing is less than perfect, the bias towards most reliable providers is somewhat milder, but still quite noticeable. The diagrams in Figure 9. show the service call distribution and the number of idle providers when only 75\% of other clients' information is available to each client. As can be seen, the portion of all calls serviced by most reliable provider(s) is somewhat smaller, and the convergence toward asymptotic values is slower. The number of providers that have received no calls at all is also smaller, which means the selection process is less biased toward the most reliable providers.

This trend is continued when the sharing is further decreased. When the sharing is at the 50\% level (for brevity, the corresponding diagrams are not shown), the portion of calls serviced by most reliable providers is even smaller, as could be expected; the number of providers that receive no calls starts with three in the first cycle and drops to zero in subsequent cycles, which is why it's not shown. In fact, even the least reliable providers were sometimes selected despite their unfavorable promise-to-performance ratio.

Overall, the system behavior confirms our intuitive understanding of the concepts of trust and reputation.

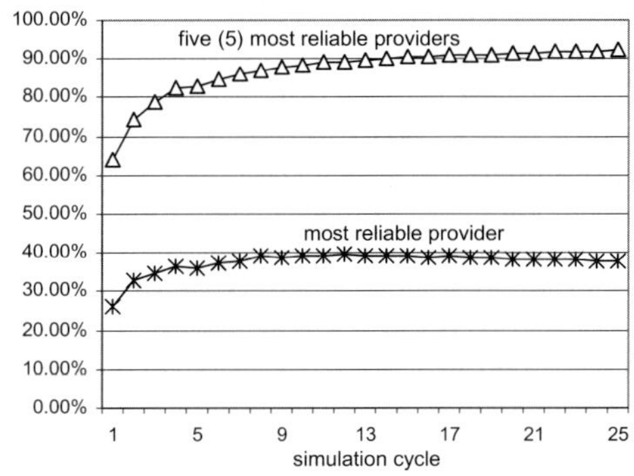


(c) Service call distribution.

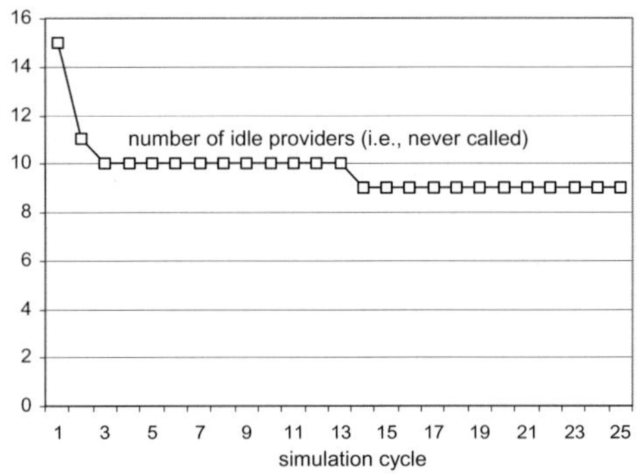


(d) Number of service providers that have never received a call.

Figure 8. Service call distributiond when the selection of the service provider is made according to service promise (40\%), trust in the provider performance (40\%), and reputation of the provider (20\%). Perfect sharing of reputation is assumed..

## VI. WHAT NEXT?

Issues to be addressed in future research include:

- Investigate extensions to existing languages and protocols, WSDL and UDDI in particular, that will facilitate community-based service discovery and selection.
- Authentication and security policies and algorithms to be utilized in the service discovery and selection process.
- Algorithms and policies for trust and reputation formation and maintenance.
- Design and development methodologies for Web Service-enabled applications that will make use of extended facilities available through the community-based service discovery and selection.

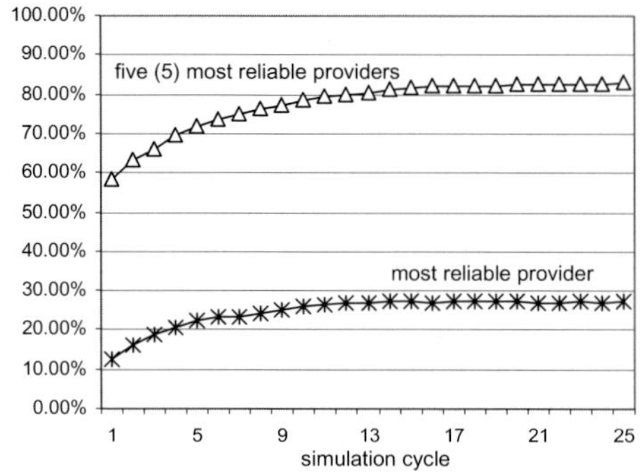


(e) Service call distribution.

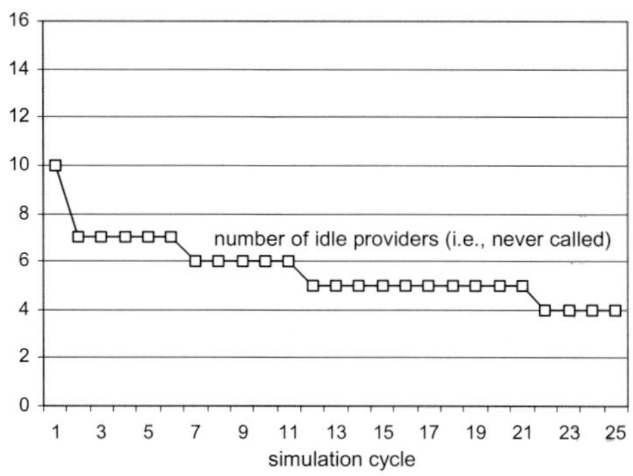


(f) Number of service providers that have never received a call.

Figure 9. Service call distributiond when the selection of the service provider is made according to service promise (40\%), trust in the provider performance (40\%), and reputation of the provider (20\%). Perfect sharing of reputation is assumed..

An investigation into different strategies that malicious providers may employ in order to undermine the reputation system and trick the clients is also important. An in-depth understanding of those strategies can facilitate the development of techniques that will enable the clients to detect such attempts and thwart them

## ACKNOWLEDGMENT

The author wishes to thank Messrs. Duraid Ibrahim and Michael W. Rennie for their useful suggestions and comments on an earlier version of this manuscript.

## REFERENCES

[1] A. Ankolekar, M. Burstein, J. R. Hobbs, O. Lassila, D. L. Martin, D. McDermott, S. A. McIlraith, S. Narayanan, M. Paolucci, T. R. Payne, and K. Sycara. DAML-S: Web service description for the semantic web. *Proceedings of the First International Semantic Web Conference (ISWC)*, Sardinia, Italy, June 2002

[2] E. Damiani, S. De Capitani di Vimercati, S. Paraboschi, and P. Samarati. Managing and sharing servents' reputations in P2P systems. *IEEE Transactions on Knowledge and Data Engineering*, **15**(4):840-854, July/August 2003

[3] A. Dan, R. Franck, A. Keller, R. King, and H. Ludwig. Web service level agreement (WSLA) language specification, August 2002

[4] P. Fletcher and M. Waterhouse, editors. *Web Services Business Strategies and Architectures*. Chicago, IL: Expert Press, 2002

[5] K. Gottschalk, S. Graham, H. Kreger, and J. Snell. Introduction to web services architecture. *IBM Systems Journal*, **41**(2):170-177, 2002

[6] M. N. Huhns and M. P. Singh. Service-oriented computing: Key concepts and principles. *IEEE Internet Computing*, **6**(1):75-81, January/February 2005

[7] E. M. Maximilien and M. P. Singh. Conceptual model of Web Service reputation. ACM SIGMOD Rec., **31**(4):36-41, 2002

[8] S. Ran. A model for Web Services discovery with QoS. *SIGecom Exch*., **4**(1):1-10, 2003

[9] A. Sahai, A. Durante, and V. Machiraju. Towards automated SLA management for web services. Research Report HPL-2001-310 (R.1), Hewlett-Packard Laboratories, Palo Alto, CA, July 2002

[10] M. P. Singh. Trustworthy service composition: Challenges and research questions. *Proceedings of the Autonomous Agents and Multi-Agent Systems Workshop on Deception, Fraud and Trust in Agent Societies*, Bologna, Italy, July 2002

[11] V. Tosic, B. Pagurek, B. Esfandiari, and K. Patel. Management of compositions of e- and m-business Web Services with multiple classes of service. *Proceedings IEEE/IFIP Network Operations and Management Symposium NOMS 2002*, pp. 935-937, Florence, Italy, April 2002

# A short discussion about the comparison between two software quality approaches Mafteah/MethodA[i] method and the software Capability Maturity Model[ii] Integration.

G.R. Kornblum
Academic College of Judea and Samaria
Department of Industrial Engineering and Management
Software Engineering Chair
44837Ariel, Israel

Conference Paper for International Joint Conferences
on Computer, Information, and Systems Sciences, and Engineering
(CIS$^2$E 05)

*Abstract*- **Two different methods for obtaining software programs with predictable quality are compared by positioning these models in a product/process and confirmation/improvement framework. The Mafteah/MethodA method can be placed in the confirmation segment, whereas the software Capability Maturity Model can be positioned in the improvement/process quadrant.**

## I. INTRODUCTION

At the end of the eighties and the beginning of the nineties of the former century it became clear that one universal software development method could not be the answer on the increasing demands for software programs with a higher quality but also with lower maintenance costs. The answers were formulated [1] in the direction of a modular and consistent framework for the development of software. A toolkit should be made that could "tailor" the specific software development method for the organization at hand and the situation. It was in that direction that in Israel a governmental and private cooperation resulted in the so-called Mafteah/MethodA [2] framework. In the USA the development of a capability maturity model (CMM) [3] was started, which evaluated into the Capability Maturity Model Integration (CMMI). These two practical answers were formulated in about the same time and will be placed here alongside each other. The ensuing discussion will focus on the main differences between these "methods" along the lines of a description of the methods [4] concerning amongst others the basic philosophy, the scope and a generalized framework [11] in the form of a Boston Matrix. The positioning in the matrix is according to a confirmation/improvement characterization pair on the horizontal axis and a focus on process or product quality on the vertical axis.

## II. THE MAFTEAH/METHODA SOFTWARE PROCESS MODEL

In 1989 Mafteah/MethodA was jointly developed by the Israeli government [5] and MethodA Computers Ltd. It is continuously developed and version [illegible] is operative at the moment (2005). It consists of a knowledge base, along with a procedural framework that breaks down every project into phases and standard components. The standard components are ushered through the project life cycle to adapt to the specific system demands.

The so-called procedural framework is mostly represented as a table (see Fig.1) with the system life cycle (the phases) on the horizontal axis and the (standard) components on the vertical axis. In other words organizational issues vertically and horizontally project management subjects. The meaning of the letters in Fig.1 is for the standard phases on the horizontal axis. These phases are part of system development Lifecycle:



*T. Sobh and K. Elleithy (eds.), Advances in Systems, Computing Sciences and Software Engineering*, 365–370.

I= initiation,

C=Characterization,

A= Analysis,

D&B= Design and Build,

T&E= Test and Evaluation,

R&T= Repair and Test-runs,

O&M= Operation and Maintenance.

Mafteah/MethodA is used as a comprehensive procedural framework for handling Information Technology within an organization, both for each individual project and for the organization as a whole.

- At the project level (the individual system), that is depicted as the system development lifecycle on the horizontal axis, Mafteah/MethodA defines how to manage, develop, and maintain a computerized system.
- At the organizational level, indicated as the system component tree -SCT- (standard components), the vertical axis of the table, Mafteah/MethodA contains a variety of procedures and tools for managers, ranging from annual work planning through project follow-up, management techniques, and overall Information Technology supervision, and extending to outsourcing strategies.
- At both organizational level and project level, the Mafteah/MethodA approach combines its procedural framework with an open methodology.

This framework resembles thus very much the "toolkit" approach mentioned in the introduction [1]. The table can also be viewed as the 2D projection of the essentially 3D framework. The third dimension is formed by the knowledge base. The knowledge base is made up of the different toolkits (TK's).

Interesting to know is the fact that Mafteah/MethodA was raised a compulsory standard by the Israeli government, for all governmental information processing systems, including both those developed by the government ministries themselves and those built in cooperation with outside (hardware/software) contractors. The fact is interesting because the Mafteah method is relatively unknown outside Israel.

The result is that it has also been adopted by

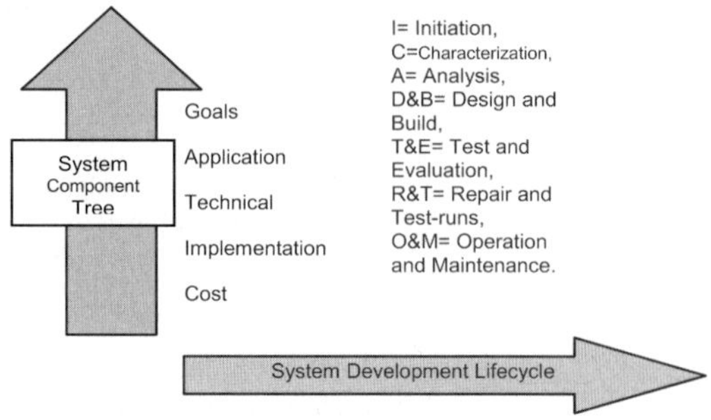


Figure 1: The Mafteah/MethodA table

numerous non-governmental organizations and companies in the public and private sectors in Israel. Many computer companies and software houses follow it, particularly when working with the government and with other Mafteah/MethodA-compliant organizations.

The model presented in Fig. 1 is schematic, and it does not necessarily reflect the way a particular project is developed or managed in practice. It could be thought that only systems could be developed in a linear way (waterfall). For actual work, Mafteah/MethodA includes a variety of practical and dynamic models that cover a number of ways of managing a project, such as:

- Sequential projects
- Iterative projects (delivery units)
- Spiral development, etc

In order to get an impression how the matrix of Fig.1 can be used, consider an example of a project that is well underway and the management wants to know what the project status is. The project status is given in Fig.2. It gives the same matrix of Fig.1 but with red and blue arrows. The red arrows indicate how far the standard components are and the blue arrows emphasize how the system's progress intersects various control points and milestones, at which the content and other aspects of the system may be assessed. The red arrows are unequal, showing that - as is certainly possible – not all the components have progressed to the same degree. For example, it could be that the user interface is fully specified, designed, and even prototyped as a working model while the system's training and integration program remains somewhat unformed.

Though, as noted, a situation like this is possible, still it is obvious that large discrepancies in progress are unsatisfactory in real life.

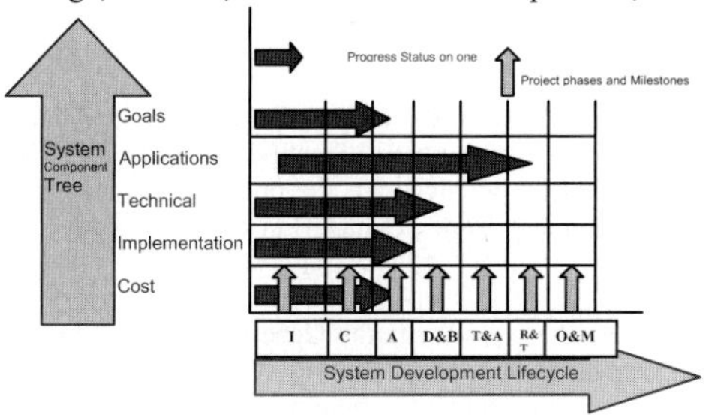


Figure 2: Project Status indication with the Mafteah matrix

*IMPRESSION OF PARTIALLY FILLED IN MAFTEAH/METHODA TABLE*

Now it will be interesting to know what actually will be the content of the cells at the intersections of the standard components (SCT) and the system development life cycle. In the matrix of Fig.3 some of the cells are filled in. This is only an example and it will be clear that several layers (the kits) behind this model are needed to reach this outcome.

| **System Component Tree** | | | | | | | |
|---|---|---|---|---|---|---|---|
| **Goals** | partly defined | finally designed | | no changes | standing check | | Small changes |
| **Application** | general idea | largely defined | | finished | system test | | manage the changes |
| **Technology** | not known/constraint | architecture | | acquisition | check of the equipment | | manage the changes |
| **Implementation** | | largely defined | | improvement | project evaluation | | Configuration management |
| **Cost** | obstructions/first estimates | assessment | | final assessment | budget control | | Maintenance |
| | Initiation | Characterization | Analysis | Design & Build | Test & Evaluation | Repair & test-run | Operation & maintenance |

**System Development Lifecycle**

Figure 3: Example of a partially filled in Mafteah table.

## IV. THE SOFTWARE CMM INTEGRATION APPROACH

Humphrey [6] realized that the software development process can be controlled, measured and improved. He developed a software maturity framework which has evolved into the capability maturity model (CMM). The capability maturity model approach is based upon the premise that the quality of a system not only is determined by the quality of the resulting product but also by the processes leading to it. These premises are established facts in manufacturing and owe tribute to the principles of statistical quality control of Walter Shewart in the 1930s [7]. These principles were further developed by W. Edwards Deming [8] and Joseph Duran [9] in the second half of the last century.

At this moment (2005) the capability maturity model integration (CMMI) consists of Models, Modules and Appraisal Methods[1]). The models are separated into four disciplines and two representations. The models comprise the representations which mean the staged representation and the continuous representation. The addition of the term "integration" to the former CMM so that it became CMMI means that four disciplines are added within the model structure as will become clear when looking at Table 1.

At the core of the model however the SE/SW discipline remains, whereas the other disciplines are "extended" disciplines.

| Abbr. | Full Name |
|---|---|
| SE | Systems Engineering |
| SW | Software Engineering |
| IPP | Integrated Product and Process Development, with focus on early continuing stakeholder involvement |
| SS | Supplier Sourcing, |

Table 1: The four CMMI disciplines, abbrevations (Abbr.) and full names of these disciplines.

The focus here will be on the continuous representation and we limit ourselves to the Software Engineering (SW) discipline.

In this continuous representation we will find so called four process areas categories –PAC's (Table 2). In this table is also added the Process Categories of ISO 15504[10] so that a comparison with ISO is possible.

| CMMI Process Area Categories, Continuous Representation | ISO 15504 Process Categories |
|---|---|
| 1- Process Management | Organization |
| 2- Project Management | Management |
| 3- Engineering | Engineering |
| 4- Support | Support |
| | Customer-Supplier |

Table 2: CMMI Process Area Categories and ISO 15504 Process Categories.

The PAC's are made up of 25 process area's (PA's) in the following way (Table 3):

[1] ) Not mentioned here are the training modules, because there are not considered relevant in the current comparison.

| Number of Process Areas (PA's) | CMMI Process Area Categories (PAC's), Continuous Representation |
|---|---|
| 5 | Process Management |
| 8 | Project Management |
| 6 | Engineering |
| 6 | Support |

Table 3: The number of Process Areas (PA's) within each PAC

Each PA contains up to three Specific Goals (SG). Each of those goals comes with several Specific Practices (SP). The specific goals and practices are the operational trajectories and lead to typical work products (an example is given in Table 4).

Additional to these goals are the Generic Goals (5) and the accompanying 18 Generic Practices (GP), which are most easily seen as the core of a generic quality improvement cycle [10]. Now the total hierarchy of the CMMI can be overseen and is presented in left hand column of Table 4.

The flexibility of the model lies in the fact that each Process Area can be individually assigned a capability level in the continuous representation which leads to a so called capability level profile for an organization. The capability level contains all the Process Area's with its respective mark (assignment level).

| Model Element Name | Actual Name / Short Description |
|---|---|
| Representation (2) | Continuous |
| Process Area Category (4) | Project Management |
| Process Area (22) | Risk Management |
| Generic Goal (5) | Achieve specific goals |
| Generic Practice (18) | Identify work scope |
| Specific Goal (48) | Identify and analyze risks |
| Specific Practice (168) | Evaluate, categorize, and prioritize risk. |

Table 4: Hierarchical Overview of CMMI

## IV. COMPARISON

In order to perform the comparison we shall look at the Process Areas Categories of CMMI listed in Table 2. The Project Management (PAC2) is comparable to the Development Life Cycle Phases of Mafteah (the horizontal axis in the framework- Fig.1 and Fig.2). The process management PAC1 of CMMI compares with the System

Component Tree –vertical axis in Fig.1- of the Mafteah/MethodA framework.
The remaining PAC's of CMMI i.e. Engineering (PAC3) and Support (PAC4) must be detailed further to make a reasonable comparison possible. The same applies for the Mafteah/MethodA framework and knowledge base.

| **Engineering (PAC3)** | **Mafteah/ MethodA** | **Component** |
|---|---|---|
| REQM: Requirements Management | Goals | SCT |
| RD: Requirements Development | Application | SCT |
| PI: Product Integration | Implementation | SCT |
| TS: Technical Solution | Technical | SCT |
| VAL: Validation | Verification & Validation | TK |
| VER: Verification | Verification & Validation | TK |

Table 5: CMMI Engineering and Mafteah/MethodA;
SCT and TK are components of Mafteah/MethodA; SCT=system component tree; TK=(tool) kit

The knowledge base is implemented as a series of (tool) kits (TK's).
For the comparison we pick the relevant components from the Mafteah table and knowledge base and confront these with the PAC's of CMMI (Table 5 and Table 6).
It should be noted that in this process the components for the Mafteah Method can not be extracted from one component only e.g. only the System Component Tree (SCT). They have to be retrieved also from the extensive (tool) kit collection (TK) inherent in the Mafteah framework knowledge base.

From the tables 5 & 6 it appears that already on the uppermost level of the PAC's for CMMI and the framework for Mafteah a rather good mutual coverage exists. Only for two PA's the Decision Analysis and Resolution (DAR) and

| **Support (PAC4)** | **Mafteah/MethodA** | Component |
|---|---|---|
| CM: Configuration Management | Configuration Management | TK |
| PPQA: Process and Product Quality Assurance | Quality Assurance (QA) | TK |
| MA: Measurement and Analysis | Metrics | TK |
| DAR: Decision Analysis and Resolution | -- | |
| OEI: Organizational Environment for Integration | Goals | TK |
| CAR: Causal Analysis and Resolution | -- | |

Table 6: CMMI Support and Mafteah/MehodA; legend see table 5

Causal Analysis and Resolution (CAR) an equivalent can not simply be found at this level in the Mafteah Method. Within the current limitation of this paper a more extensive investigation including other levels is not considered as meaningful. Later investigations might involve comparison on more detailed levels.

## V. CONCLUSION

Both methods strive to increase the quality of the delivered software components. The comparison can be rounded off with the process/product dichotomy on the one hand and the conformance/ improvement at the other hand [11] as positioning matrix. Quality criteria can be imposed on product or process. Techniques, procedures or tools may be designed to ensure that the product or process conforms to these quality criteria. At the other hand, schemes can be contrived that will improve the quality of product or process.
From the comparison made here and the respective philosophies of both methods, it will become clear that Mafteah, which is based upon the meta modeling school or Methodology Engineering[1] will be positioned more at the conformance part and CMMI coming from TQM[6, 9], more at the improvement part (see Fig.4). The Mafteah method places more emphasis on the final products, whereas CMMI has more focus on processes. However both methods take firmly process approaches into account.

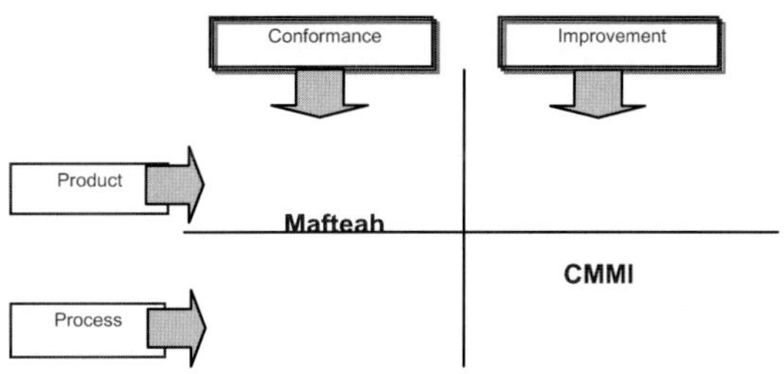


Figure 4: The relative positioning of both methods in the conformance/improvement and product/process matrix

## REFERENCES

[1] R.J. Welke, K. Kumar and H. G. van Dissel, Methodology Engineering, Informatie, nr05(1991).
[2] A. Yuval; MethodA Executive summary, MethodA Computers (2005)
www.methoda7.com/global_interface/main_fs.htm
www.hakirya.ac.il/MethodA/MethodA_60_Academic/H_MAmodel/H_MAmodel.htm
[3] Software Engineering Institute; Carnegie Mellon University; www.sei.cmu.edu/cmmi;
[4] G.R. Kornblum, Methods for (knowledge based) system development, in: Knowledge in Organizations, Applications and Theory of knowledge based systems, R.J. Jorna, J.L. Simons (Eds.) Couthino, Muiderberg, The Netherlands (1992).
[5] The Israeli State Comptroller's Office and the Budget Division of the Ministry of Finance of Israel.
[6] Humphrey W. S. Managing the Software Process, SEI series in Software Engineering, Addison Wesley, (1989).
[7] Shewhart W.A., Economic Control of Quality of Manufactured Product, New York, USA: D. Van Nostrand Co.; (1931)
[8] Deming W.E. Out of the Crisis. Cambridge: Massachusetts Institute of Technology(1986).
[9] Juran, J.M., Juran On Quality By Design: The New Steps For Planning Quality Into Goods And Services. The Free Press, a division of McMillan Inc. New York, N.Y. (1992).
[10] Keefer G., Lubecka, H. The CMMI in 45 minutes, Avoca, Stuttgart (2002) .
[11] van Vliet, H. Software Engineering, Principles and Practice, page 103, 2nd Ed, John Wiley and Sons (2002).

---

[i]) Mafteah and MethodA are registered trademarks.
[ii ii]) Capability Maturity Model and CMM are registered trademarks in the U.S. Patent and Trademark Office.

# Using Association Rules and Markov Model for Predict Next Access on Web Usage Mining

**Siriporn Chimphlee** [1] **Naomie Salim** [2] **Mohd Salihin Bin Ngadiman** [3] **Witcha Chimphlee** [4]

[1,4] *Faculty of Science and Technology*
*Suan Dusit Rajabhat University, 295 Rajasrima Rd, Dusit, Bangkok, Thailand*
*Tel: (+66)-2445675, Fax: (+66) 2445675, Email:* [1]*siriporn_chi@dusit.ac.th,* [4]*witcha_chi@dusit.ac.th*

[2,3] *Faculty of Computer Science and Information Systems,*
*University Technology of Malaysia, 81310 Skudai, Johor, Malaysia*
*Tel: (607) - 5532070, Fax: (607) 5565044, Email:* [2]*naomie@fsksm.utm.my,* [3]*msn@fsksm.utm.my*

**Abstract- Predicting the next request of a user as visits Web pages has gained importance as Web-based activity increases. A large amount of research has been done on trying to predict correctly the pages a user will request. This task requires the development of models that can predicts a user's next request to a web server. In this paper, we propose a method for constructing first-order and second-order Markov models of Web site access prediction based on past visitor behavior and compare it association rules technique. In these approaches, sequences of user requests are collected by the session identification technique, which distinguishes the requests for the same web page in different browses. We report experimental studies using real server log for comparison between methods and show that degree of precision.**

**Index Terms**—Markov Model, Association rules, Prediction

## 1 INTRODUCTION

World Wide Web is storing huge useful information. One important data source for this study is the web-server log data that trace the user's web browsing actions. The web log data consist of sequences of URLs requested by different clients bearing different IP Addresses. Association rules can be used to decide the next likely web page requests based on significant statistical correlations. The result of accurate prediction can be used for recommending products to the customers, suggesting useful links, as well as pre-sending, pre-fetching and caching of web pages for reducing access latency [1]. The work by Liu et al. [2] and Wang et al. [3] considered using association rules for prediction by selecting rules based on confidence measures, but they did not consider the sequential classifiers [1]. It has been observed that user tend to repeat the trails they have followed once [4]. So, better prediction of a user's next request could be made on the data pertaining to that particular user, not all the users. However, this would require reliable user identification and tracking users among sessions. This is usually achieved by sending cookies to a client browser, or by registering users. Both require user cooperation and might discourage some of potential site visitors. As a result, many web sites choose not to use these means of user tracking. Also, building prediction models on individual data would require that users have accessed enough pages to make a prediction, which is not usually the case for a university website that has many casual users [5].

In the network system area, Markov chain models have been proposed for capturing browsing paths that occur frequently [4, 6]. However, researchers in this area did not study the prediction models in the context of association rules, and they did not perform any comparison with other potential prediction models in a systematic way. As a result, it remains an open question how to construct the best association rule based prediction models for web log data [1].

This paper is organized as follows. In section 2, we discuss the background of study and review the past works in related research. In section 3, we present the experimental design. In section 4, we discuss the experimental result. We conclude our work in section 5.

## 2 BACKGROUND OF STUDY

### 2.1 Web Mining

Web mining is the term of applying data mining techniques to discover automatically and extract useful information from the World Wide Web documents and services [7]. The web mining system uses information determined from the history of the investigated web system. When valuable hidden knowledge about the system of interest has been discovered, this information can be incorporated into a decision support system to improve the performance of the system. These rules, patterns typically reflecting real world phenomena, are mined from the web server logs, proxy server logs, user' profiles, registration data etc. Three major web mining methods are web content mining, web structure mining and web usage mining. Web content mining is the process of extracting knowledge from the content of documents or their descriptions. Web structure mining aims to generate structural summaries about web sites and web pages [8]. Web usage mining is to discover usage patterns from web data, in order to understand and better serve the needs of web-based application. It is an essential step to understand the users' navigation preferences in the study of the quality of an electronic commerce site. In fact, understanding the most likely access patterns of users allows the service provider to personalize and adapt the site's interface for the individual user, and to improve the site's structure.



*T. Sobh and K. Elleithy (eds.), Advances in Systems, Computing Sciences and Software Engineering,* 371–376.

## 2.2 Association rules

Agrawal and Srikant [9] were proposed to capture the co-occurrence of buying different items in a supermarket shopping. Given a set of transactions, where each transaction is a set of items, an association rule is an expression X=>Y, where X and Y are sets of items. The intuitive meaning of such a rule is that transactions in the databases which contain the items in X tend to contain also the items in Y [10]. For instance, 98% of customers who purchase tires and auto accessories also buy some automotive services; 98% is called the confidence of the rule. The support of the rule X=>Y is the percentages of transactions that contain both X and Y. Association rule generation can be used to relate pages that are most often referenced together in a single server session [11]. In the context of Web usage mining, association rules refer to set of pages that are accessed together with a support value exceeding some specified threshold. The association rules may also serves as a heuristic for prefetching documents in order to reduce user-perceived latency when loading a page from a remote site [11]. These rules are used in order to reveal correlations between pages accessed together during a server session. Such rules indicate the possible relationship between pages that are often viewed together even if they are not directly connected, and can reveal associations between groups of users with specific interests. A transaction is a projection of a portion of the access log. In their work Liu et al. [2] and Wang et al. [3] considered using association rules for prediction without considering sequential classifiers. In contrast, Lan et al. [12] developed an association rules mining technique for the pre-fetching of web document from the server's disk into the server's cache. They constructed rules of the form $D_i \rightarrow D_j$, where $D_i$ and $D_j$ are documents (URLs). The intuitive interpretation of such rules are that document C is likely to be requested by the same user sometimes after document $D_i$ has been requested and there is no other request between the requests for $D_i$ and $D_j$, since it is usually the case according to the log. The counting of support is done differently than in Agrawal and Srikant [9] since the ordering of documents is considered [13]. They defined confidence is the ratio support $(D_iD_j)$ /support($D_i$), support ($D_i$)is the total number of occurrences of document $D_i$ in the transactions over the total number of the transactions and support $(D_iD_j)$ is the total number of occurrences of a sequence $D_iD_j$ in the transactions over total number of transactions. Only consecutive subsequences inside a user transaction are supported. For instance, the user transaction ABCD supports the subsequences: AB, BC, and CD. Yang et al. [1] studied different association-rule based methods for web request prediction. Their analysis is based on a two dimensional structure and real web logs are used as training and testing data. First dimension is named antecedent of rules. They consist of five rules; called subset rule, subsequence rule, latest-subsequence rule, substring rule and latest-substring rules. These representations build the left-hand-side of association rules using non-empty subsets of URLs, non-empty sequences of URLs, non-empty sequences of URLs which end in the current time, non-empty adjacent sequences of URLs, and non-empty adjacent sequences of URLs which end in the current time. Second dimension is named criterion for selecting prediction rules. It represents as three prediction methods called longest match selection, most confident selection and pessimistic selection. The longest-match method selects the longest left-hand-side of the rule and matches an observed sequence from all rules with the minimum support rules. Most confident selection always chooses a rule with highest confidence and minimum support rules. Pessimistic selection combines the confidence and support for a rule to form an unified selection measure. The result showed that the latest substring rule coupled with the pessimistic-selection method gives the most precision prediction performance. In this work, we also used the latest-substring to represent the prediction rules.

## 2.3 Markov model

Markov models [14] have been used for studying and understanding stochastic processes, and were shown to be well-suited for modeling and predicting a user's browsing behavior on a web-site. Markov models have been widely used to model user navigation on the web and predicting the action a user will take next given the sequence of actions he or she has already performed. For this type of problems, Markov models are represented by three parameters < A, S, T >, where A is the set of all possible actions that can be performed by the user; S is the set of all possible states for which the Markov model is built; and T is a $|S| \times |A|$ Transition Probability Matrix (TPM), where each entry $t_{ij}$ corresponds to the probability of performing the action j when the process is in state i [15]. In general, the input for these problems are the sequence of web-pages that were accessed by a user and the goal are to build Markov models that can be used to model and predict the web-page that the user will most likely access next. Padbanabham and Mogul [16] use N-hop Markov models predicted the next web page users will most likely access by matching the user's current access sequence with the user's historical web access sequences for improving pre-fetching strategies for web caches. Pirolli and Pitkow [4] predict the next web page by discovering the longest repeating subsequences in the web sessions, and then using a weighted scheme to match it against the test web sessions. Sarukkai [17] used first-order Markov models to model the sequence of pages requested by a user for predicting the next page accessed. Cadez et al. [18] clustered user behaviors by learning a mixture of first-order Markov models using the Expectation-Maximization algorithm. They then display the behavior of a random sample of users in each cluster along with the size of each cluster. They also applied to the visualization of web traffic on the msnbc.com site.

## 3 EXPERIMENTAL DESIGN

In this paper, we study association rules and Markov models for predicting a user's next web requests. The prediction models that we build are based on web log data that correspond with users' behavior. They are used to make prediction for the general user and are not based on the data for a particular client. This prediction requires the discovery of a web users' sequential access patterns and using these patterns to make predictions of users' future access.

The experiment on used web data, collected from www.dusit.ac.th web server (see example in Fig. 1) during 1st December 2004 - 31st December 2004. The total number of web pages with unique URLs is equal to 314 URLs, and there are 13062. These records are used to construct the user access sequences (Figure 2). The user sessions are split into training dataset and testing dataset. The training dataset is mined in order to extract rules, while the testing dataset is considered to evaluate the predictions made based on these rules. We experimentally evaluated the performance of the proposed approach: first-order markov model, second-order markov model, and association rule mining and construct the predictive model.

```
1102801060.863 1897600 172.16.1.98 TCP_IMS_HIT/304 203
GET http://asclub.net/images/main_r4_c11.jpg - NONE/- image/jpeg
1102801060.863 1933449 172.16.1.183 TCP_MISS/404 526
 GET http://apl1.sci.kmitl.ac.th/robots.txt DIRECT/161.246.13.86
text/html
1102801060.863 1933449 172.16.1.183
TCP_REFRESH_HIT/200 3565
GET http://apl1.sci.kmitl.ac.th/wichitweb/spibigled/spibigled.html -
 DIRECT/161.246.13.86 text/html
```

Fig.1. Web log data

### 3.1 Web log preprocessing

Web log files contain a large amount of erroneous, misleading, and incomplete information. This step is to filter out irrelevant data and noisy log entries. Elimination of the items deemed irrelevant by checking the suffix of the URL name such as gif, jpeg, GIF, JPEG, jpg, JPG. Since every time a Web browser downloads a HTML document on the Internet, several log entries such as graphics and script are downloaded too. In general, a user does not explicitly request all the graphics that are in the web page, they are automatically down-loaded due to the HTML tags. Since web usage mining is interested in studying the user's behavior, it does not make sense to include file requests that a user does not explicitly request. The HTTP status code returned in unsuccessful requests because there may be bad links, missing or temporality inaccessible pages, or unauthorized request etc: 3xx, 4xx, and 5xx. Executions of CGI script, Applet, and other script codes are also eliminated. This is due to the fact that there is not enough Meta data to map these requests into semantically meaningful actions, as these records are often too dynamic and contain insufficient information that makes sense to decision makers.

### 3.2 Session identification

After the preprocessing, the log data are partitioned into user sessions based on IP and duration. Most users visit the web site more than once. The goal of session identification is to divide the page accesses of each user into individual sessions. The individual pages are grouped into semantically similar groups. A user session is defined as a relatively independent sequence of web requests accessed by the same user [19]. Fu et al. [20] identify a session by using a threshold idles time. If a user stays inactive for a period longer than the max_idle_time, subsequent page requests are considered to be in another episode, thus identificated as another session. Most researchers use heuristic methods to identify the Web access sessions [21] based on IP address and time-out does not exceeding 30 minutes for the same IP Address. A new session is created when a new IP address is encountered after a timeout. Catledge and Pitkow [22] established a timeout of 25.5 minutes based on empirical data. In this paper, we use time-out of 30 minutes to generate a new user (see example in Fig. 2).

```
Session 1 : 900, 586, 594, 618
Session 2 : 900, 868, 586
Session 3 : 868, 586, 594, 618
Session 4 : 594, 618, 619
Session 5 : 868, 586, 618, 900
```

Fig.2. User session from data set

We assume the access pattern of a certain type of user can be characterized by a certain a minimum length of a user's transaction, and that the corresponding future access path is not only related to the last accessed URL. Therefore, users with relatively short transactions (e.g. 2-3 accesses per transaction) should be handled in a different way from users with long transactions (e.g. 10-15 accesses per transaction) [23]. In this study, we proposed a case definition design based on the transaction length. User transactions with length of less than three are removed because it is too short to provide sufficient information for access path prediction [23].

### 3.3 Prediction using Association rules

To capture the sequential and time-limited nature of prediction, we define two windows. The first one is called antecedent Window (W1), which holds all visited pages within a given amount of user requests and up to a current instant in time. A second window, called the consequent window (W2), holds all future visited pages within amount of user requests from the current time instant.

The web log data are a sequence of entries recording which documents was requested by a user. We extracted a prediction model based on the occurrence frequency and find the last-substring [1] of the W1. The last-substrings are in fact the suffix of string in W1 window. We will refer Left-Hand-Side as LHS and

Right-Hand-Side as RHS. These rules not only take into account the order and adjacency information, but also the newness information about the LHS string. We used only the substring ending in the current time (which corresponds to the end of window W1) qualifies to be the LHS of a rule [1]. We give a simple example to illustrate the prediction scheme of association rules clearly (see Table 1). The input data for training the model consists of web sessions, where each session consists of the sequence of the pages accessed by a user during his/her visit to the site. The training data of our example, shown in Figure 2, is from five user sessions.

TABLE 1
THE LATEST-SUBSTRING RULES

| W1 | W2 | The latest-substring rule |
|---|---|---|
| 900, 586, 594 | 618 | {594} -> 618 |
| 868, 586, 594 | 618 | |

From these rules, we extract sequential association rules of the form LHS->RHS from the session [1]. The support and confidence are defined as follows:

$$supp = \frac{count(LHS\text{-} > RHS)}{number\ of\ sessions}$$

$$conf = \frac{count(LHS\text{-} > RHS)}{count(LHS)}$$

Our goal is to output the best prediction on a class based on a given training set. Therefore, we need a way to select among all rules that apply. In a certain way, the rule-selection method compresses the rule set. If a rule is never applied, then it is removed from the rule set. In association rule mining, a major method to construct a classifier from a collection of association rules is the most-confident selection method [2]. The most confident selection method always chooses a rule with the highest confidence among all the applicable association rules, where support values are above the minimum support threshold. For example, suppose a testing set has a previous sequence of (A, B, C). Using the most-confident rule selection method, we can find three rules which can be applied to this example,

Rule 1: (A, B, C) →D with confidence 35%
Rule 2: (B, C) →E with confidence 60%
Rule 3: (C) →F with confidence 50%

In this case, the confidence values of rule 1, rule 2 and rule 3 are 35%, 60% and 50%, respectively. Since Rule 2 has the highest confidence, the most-confident selection method will choose Rule 2, and predict E as the next page to be accessed.

### 3.4 Prediction using Markov models

The Markov model has achieved considerable success in the web prefetching field [4, 24, 25]. However the limit of this approach in web prefetching is that only requested pages are considered. The state-space of the Markov model depends on the number of previous actions used in predicting the next action. The simplest Markov model predicts the next action by only looking at the last action performed by the user. In this model, also known as the first-order Markov model, each action that can be performed by a user corresponds to a state in the model (Table 2). A somewhat more complicated model computes the predictions by looking at the last two actions performed by the user. It is called the second-order Markov model, and its states correspond to all possible pairs of actions that can be performed in sequence (Table 3). This approach is generalized to the Kth-order Markov model, which computes the predictions by looking at the last K actions performed by the user, leading to a state-space that contains all possible sequences of K actions [15]. We also used training user sessions, shown in Figure 2. In this example, we find first-order and second-order Markov Model and set the support threshold as 2. Based on this training set, the supports of different order sequences are counted. Prediction rules and their predictions confidence are shown in Table 4.

TABLE 2
SAMPLE FIRST-ORDER MARKOV

| 1 st order | Support count | | | | | |
|---|---|---|---|---|---|---|
| Second item in sequence / First item in sequence | 586 | 594 | 618 | 619 | 868 | 900 |
| 586 | 0 | 2 | 1 | 0 | 0 | 0 |
| 594 | 0 | 0 | 3 | 0 | 0 | 0 |
| 618 | 0 | 0 | 0 | 0 | 1 | 1 |
| 619 | 0 | 0 | 0 | 0 | 0 | 0 |
| 868 | 3 | 0 | 0 | 0 | 0 | 0 |
| 900 | 1 | 0 | 0 | 0 | 0 | 1 |

TABLE 3
SAMPLE SECOND-ORDER MARKOV

| 2 nd order | Support count | | | | | |
|---|---|---|---|---|---|---|
| Second item in sequence / First item in sequence | 586 | 594 | 618 | 619 | 868 | 900 |
| 586->594 | 0 | 0 | 2 | 0 | 0 | 1 |
| 594->619 | 0 | 0 | 0 | 1 | 0 | 0 |
| 868->586 | 0 | 1 | 1 | 0 | 0 | 0 |
| 900->868* | 1 | 1 | 0 | 0 | 0 | 0 |

* as url web page

## 4 EXPERIMENTAL RESULTS & DISCUSSIONS

The most commonly used evaluation metrics are accuracy, precision, recall and F-Score. Deshpande and

Karypis [24] used several measures to compare different Markov model-based techniques for solving the next-symbol prediction problem: accuracy, number of states, coverage and model-accuracy. Haruechaiyasak [26] and Zhu et al. [27] used precision and recall to evaluate the performance of method.

TABLE 5

CONFUSION MATRIX

| Actual | Predicted | |
|---|---|---|
| | positive | negative |
| Positive | TP | FN |
| Negative | FP | TN |

The precision measure the accuracy of the predictive rule set when applied to the testing data set. The recall measures the coverage or the number of rules from the predictive rule set that matches the incoming request [26]. To evaluate classifiers used in this work, we apply precision and recall, which are calculated to understand the performance of the classification algorithms. Based on the confusion matrix computed from the test results, several common performance metrics can be as Table 5, where TN is the number of true negative samples; FP is false positive samples; FN is false negative samples; FP is true positive samples. Precision and recall can be as Table 5.

$$recall = \frac{TP}{TP+FN} \qquad precision = \frac{TP}{TP+FP}$$

### 4.1 Results

The results plotted in Figure 3 and show comparison of the different algorithms. We divided the web log as training data and testing data. As can be seen from the Figure 3, the first-order Markov model consistently gives the best prediction performance. The second-order worst when the recall less than 50% but the association rule is worst after the precision less than 50%.

### 4.2 Discussions

Web usage mining is the application of data mining techniques to usages logs of large Web data repositories to produce results that can be used in the design tasks. In this experiment, the three algorithms are not successful in correctly predicting the next request to be generated. The first-order Markov Model is best than other because it can extracted the sequence rules and chose the best rule for prediction and at the same time second-order decrease the coverage too. This is due to the fact that these models do not look far into the past to discriminate correctly the difference modes of the generative process.

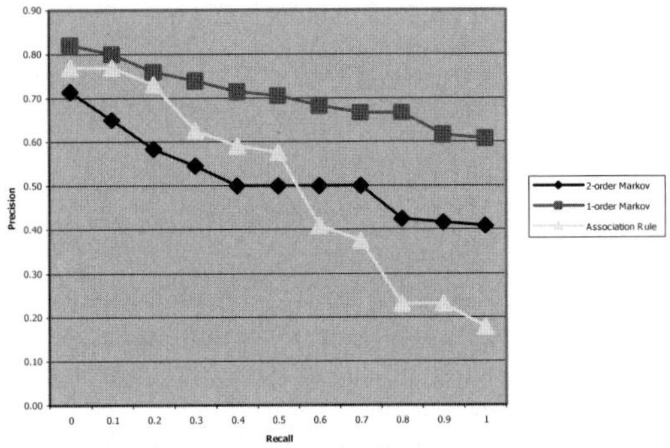


Fig.3. Result compare among three techniques

## 5 CONCLUSION AND FUTURE WORK

Web servers keep track of web users' browsing behavior in web logs. From log file, one can builds statistical models that predict the users' next requests based on their current behavior. In this paper we studied different algorithm for web request prediction. Our analysis based on three algorithm and using real web logs as training and testing data and show that the first-order Markov model is the best prediction after compared to use. In the future, we plan to use rough sets for prefetching to extract sequence rules.

## ACKNOWLWDGMENT

This research was supported in part by grants from Suan Dusit Rajabhat University at Bangkok, Thailand.

## REFERENCES

[1] Q.Yang, T.Li, K.Wang, "Building Association-Rules Based Sequential Classifiers for Web-Document Prediction", *Journal of Data Mining and Knowledge Discovery*, Netherland: Kluwer Academic Publisher, vol. 8, 253-273, pp. 2004.

[2] B.Liu, W.Hsu, and Y.Ma, "Integrating Classifi- cation and Association Mining", *Proc. of the Fourth International Conference on Knowledge Discovery and Data Mining* (KDD-98), 1998.

[3] K.Wang, S.Q.Zhou, and Y.He, "Growing Decision Trees on Association Rules", *Proc. of the International Conference of Knowledge Discovery in Databases*, 2000.

[4] J.Pitkow and P.Pirolli, "Mining Longest Repeating Subsequences to Predict World Wide Web Surfing", *Proc. USENIX Symp. On Internet Technologies and Systems*, 1999.

[5] O.Chakoula, "Predicting Users' Next Access to the Web Server" Master diss., Dept. of Computer Science, University of British Columbia, 2000.

[6] Z.Su, Q.Yang, Y.Lu, and H.Zhang, "What next: A Prediction System for Web Requests using N-gram Sequence Models", *Proc. of the First International*

*Conference on Web Information Systems and Engineering Conference*, Hong Kong, 2000.

[7] O.Etzioni, "The World Wide Web: Quagmire or Gold Mine", *Communications of the ACM*, vol.39 (11), 1996, pp.65-68.

[8] S.K.Madria, S.S.Bhowmick, W.K.Ng, and E.Lim, "Research Issues in Web Data Mining", *Proc. First International Conference on Data Warehousing and Knowledge Discovery*, Italy, Florence, 1999, pp.303-312.

[9] R.Agrawal and R.Srikant, "Fast algorithms for mining association rules", *Proc. of the 20th VLDB Conference ages*, Santiago, Chile, 1994.

[10] R.Srikant, R.Agrawal, "Mining Generalized Association Rules", *Proc. of the 21st Int'l Conference on Very Large Databases*, Zurich, Switzerland, Sep., 1995. Expanded version available as IBM Research Report RJ 9963, June 1995.

[11] J.Srivastava, R.Cooley, M.Deshpande and P.Tan "Web Usage Mining: Discovery and Applications of Usage Patterns from Web Data", *SIGKDD Explorations*. vol.1 (2), 2000, pp.12-23.

[12] B.Lan, S.Bressan, B.C.Ooi, and Y. Tay, "Making Web Servers Pushier", *Proc. Workshop Web Usage Analysis and User Profiling*, 1999.

[13] A.Nanolopoulos, D.Katsaros, and Y.Manolopoulos, "A Data Mining Algorithm for Generalized Web Prefetching", *IEEE Transactions on Knowledge and Data Engineering*, vol. 15 (5), 2003, pp.1155-1169.

[14] A.Papoulis, "*Probability, Random Variables, and Stochastic Processes*", NY: McGraw Hill, 1991.

[15] M.Deshpande, "Prediction/Classification Technique for Sequence and Graphs", Ph.D. dissertation, University of Minnesota, January 2004.

[16] N.Padmanabhan and J.C.Mogul, "Using predictive prefetching to improve World Wide Web latency", *ACM SIGCOMM Computer Communication Review*. vol. 26 (3), 1996, pp. 22-36.

[17] R.R.Sarukkai, "Link prediction and path analysis using Markov Chain", *Proc. of the 9th international World Wide Web conference on Computer networks: The international journal of computer and telecommunications networking*, Amsterdam, Netherlands, 2000.

[18] I.Cadez, D.Heckerman, C.Meek, P.Smyth, and S.White, "Visualization of Navigation Patterns on a Web Site Using Model Based Clustering", *Technical Report MSR-TR-00-18, Microsoft Research*, 2000.

[19] R.Cooley, P-N Tan, J.Srivastava, "Discovery of Interesting Usage Patterns from Web Data", *Springer-Verlag LNCS/LNAI series*, 2000.

[20] Y.Fu, K. Sandhu, and M.Y.Shih, "Clustering of Web Users Based on Access Patterns", *Proc. of the 5th ACM SIGKDD International Conference on Knowledge Discovery & Data Mining*, San Diego: Springer, 1999.

[21] G.Pallis, L.Angelis, and A.Vakali, "Model-based cluster analysis for web users sessions", *Springer-Verlag Berlin Heideberg*, 2005, 219-227.

[22] L.Catledge and J.E. Pitkow, "Characterizing Browsing Behaviors on The World Wide Web", *Computer networks and ISDN Systems*, vol.27 (6), 1995.

[23] C.Wong, S.Shiu and S. Pal, "Mining Fuzzy Association Rules for Web Access Case Adaptation", *Proc. of the workshop Programme at the fourth International Conference on Case-Based Reasoning*, Harbor Center in Vancouver, British Columbia, Canada, 2001.

[24] M. Deshpande and G. Karypis, "Selective Markov Models for Predicting Web Page Accesses", *In Workshop on Web Mining at the First SIAM International Conference on Data Mining*, 2001.

[25] B. Mobasher, H. Dai, T. Luo, M. Nakagawa, "Using Sequential and Non-Sequential Patterns in Predictive Web Usage Mining Tasks", *(ICDM 2002)*, 2002.

[26] C. Haruechaiyasak, "A Data Mining and semantic Web Framework for Building a Web-based Recommender System", Ph.D. diss., University of Miami, 2003.

[27] J.Zhu, J.Hong, and J.G.Hughes, "Using Markov Chains for Link Prediction in Adaptive Web Site", *Proc. of Soft-Ware 2002: First International Conference on Computing in an Imperfect World*, Lecture Notes in Computer Science, Springer, Belfast, 2002.

# New Protocol Layer for Guaranteeing Trustworthy Smart Card Transaction

Zheng Jianwu, Liu Mingsheng, and Liu Hui
Department of Information Engineering,
Shijiazhuang Railway Institute, Hebei 050043, China,
{zhengjw, liums, liuhui}@sjzri.edu.cn

*Abstract*—**We first point out that trustworthy information exchange between the card application and the smart card should be guaranteed, given the application system wants to leverage the enhanced security and privacy functionality of the smart card to make itself an attack-proof system, then suggest to introduce a new protocol layer for ensuring trustworthy transaction into current smart card protocol stack, and determine the position where the new protocol layer should be in the stack.**

**Furthermore, protocol detail of the new layer, i.e. cryptographic policy in essence for establishing trustworthy channel between the card application and the smart card is proposed, specifically, how measures for ensuring transaction information privacy & integrity, and fighting fraudulent transaction are taken and integrated. Moreover, STS (Station to Station) protocol for attestation and authenticated key negotiation, which is the key to the success of the new protocol layer, is introduced.**

## I. Introduction

Because the smart card is an intrinsically secure computing platform, it is often incorporated in the application system to enhance the security of the system or realize some special security purposes. Fig. 1 depicts two scenarios of the networked application system.

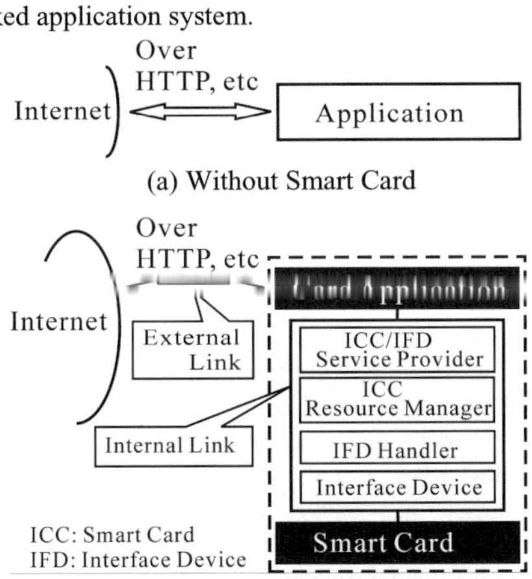


(a) Without Smart Card

(b) With Smart Card

Fig. 1. Two Scenarios of the Networked Application System

After incorporating the smart card, on the one hand, it is possible for the system to exploit the enhanced security and privacy functionality of the smart card, and on the other hand, the system is complicated, namely, besides the card application and the smart card, so many intermediary parties, including ICC/IFD Service Provider, ICC Resource Manager and so on are involved in the application terminal.

We always pay more attention to the external link (Internet link), such as implementing SSL/TLS, IPSec and so on for securing the traffic between application terminals through Internet, than that to the internal link connecting the card application and the smart card, precisely, it is always neglected to take measures to guarantee trustworthy information exchange between the card application and the smart card. Consequently, attacks initiated by spyware, virus software and so forth find favourite entrances here and bring disastrous consequences to the system.

In reality, two links mentioned above are equally important in terms of the security of the system, and therefore should be thoroughly safeguarded with proper policies and measures. This paper aims to propose measures to secure message exchange between the card application and the smart card in the application terminal.

### *A. Transmission Model and Protocol Stack*

Fig. 2 (taken from PC/SC Specification Version 2.0 [1] with modifications) and Fig. 3 depicts transmission model and protocol stack of smart card transaction executed in the application terminal depicted in the bottom scenario of the Fig. 1 respectively.

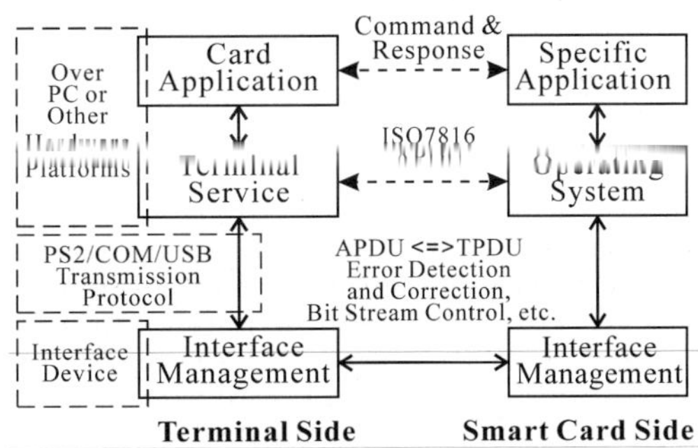


(APDU: Application Protocol Data Unit
TPDU: Transmission Protocol Data Unit )

Fig. 2. Transmission Model of Smart Card Transaction

From Fig. 1 and Fig. 2, it is clear that the card application is the entrance to the Internet, connecting the external world over



*T. Sobh and K. Elleithy (eds.), Advances in Systems, Computing Sciences and Software Engineering*, 377–381.

specific network protocols. Moreover, it is instructive to view the smart card as three separate components depicted at the right side of the Fig. 2, especially for intuitively understanding peer-to-peer protocol stack shown in Fig. 3.

| | |
|---|---|
| 5 | Card Application Layer (CAL)<br>(Smart Card Aware Applications) |
| 4 | Terminal Application Layer (TAL)<br>(ISO7816 APDU) |
| 3 | Terminal Transport Layer (TTL)<br>(APDU<=>TPDU) |
| 2 | Data Link Layer<br>(Data Frame, Timing,<br>Error Correction and Detection, etc.) |
| 1 | Physical Layer<br>(Bit Stream Control, etc.) |

Fig. 3. Protocol Stack of Smart Card Transaction

### B. *Defining the Problem*

Information exchange between the card application and the smart card of the application terminal is completed in two steps.

1) Card application initiates the transaction and sends command down through the stack of terminal side, then the command is moved back up the stack to specific application of smart card side.
2) After completing the command, smart card returns response down the stack of smart card side layer by layer, and the response is moved back up the stack of terminal side until card application gets the response.

Therefore, the task of guaranteeing trustworthy [2][3] transaction can be essentially divided into the task of mutually authenticating each other and the task of establishing trustworthy transaction channel between the card application and the smart card; we suggest to introduce a new protocol layer into the protocol stack shown in Fig. 3 for accomplishing these two tasks.

## II. Cryptographic Measures for Trustworthy Transaction

It is clear that data privacy, data integrity, and identity legitimacy are three fundamental requirements should be satisfied in terms of guaranteeing trustworthy smart card transaction. Furthermore, frustrating transaction commands or responses originated by adversaries is very important for achieving the goal, we therefore should incorporate proper security services for fighting these fraud transaction attacks aiming to destroy normal transaction processes. Namely, fraud transaction prevention should be specified for trustworthy transaction, as one of requirements besides data confidentiality and integrity.

Although a wide variety of measures can be implemented to satisfy security requirements specified above, the most commonly adopted mechanisms are based on the use of cryptographies.

### A. *Measures for Data Confidentiality and Integrity*

Both secret key cryptography [4][5][6] and public key cryptography [7][8] can be used to secure essential message exchanges between the card application and the smart card.

Message digest algorithms and private or public cryptographies can be combined to effectively detect intentional or unintentional data modifications (however, cryptography does not protect message from being altered). When secret key cryptography is used, FIPS-113, Computer Data Authentication [9], specifies a standard technique for calculating a MAC (message authentication code) for integrity verification. Public key cryptography verifies integrity by using of secure hashes (FIPS-180, Secure Hash Standard [10]) and public key signatures (FIPS-186, Digital Signature Standard [11]).

This paper selects MD5 algorithm [12] (certainly, other message digest algorithms can also be OK, such as algorithms of SHA serie [10].) and secret key cryptography for guaranteeing data confidentiality and integrity. When securing data transmission with secret key cryptography, a secret session key should be used; how to securely and dynamically negotiate the session key before any essential transaction exchange is discussed in subsection IV-B.

### B. *Measures for Identity Authentication*

Authentication relates to source authentication (i.e. confirmation of the identity of communication source) or peer entity authentication, which assures one entity of the purported identity of the other correspondent. Either secret key or public key cryptography can be implemented for identity authentication.

When utilising secret key cryptography, authentication can be implemented by demonstrating the knowledge of the cryptographic key. In this paper, Public key cryptography is selected for peer identity authentication. Specifically, the card application and the smart card authenticate each other with the certificates.

### C. *Measures for Fraud Transaction Prevention*

In order to resist malicious attacks, the tough work is to introduce reliable and feasible mechanisms for distinguishing authentic (or legitimate) transaction information from fraudulent information, e.g. attacks launched by adversaries.

A special byte is introduced into every transaction message, whose value obeys following two rules. The first rule is that value of the byte in every legitimate command-response transaction pair is unique, i.e. one specific value is only effective in one command-response pair. The second one is that value of special bytes in different command-response pairs should vary according to initial value and principle agreed on by two legitimate communicating parties; consequently bytes introduced into transaction messages from the first to the last establish a logical relationship among them during one session time or life time of a secret session key. This special byte is named as Transaction Serial Number Byte SN. Therefore, the card application and the smart card can conclude that the transaction information received is a fraud when value of SN byte in the message is inconsistent with two rules specified above.

The mathematical mechanism adopted for agreeing on the initial value of SN (in the first command-response pair) and principle, according to which value of SN byte varies, is identical to the mechanism adopted for negotiating the secret session key.

## III. Position of the New Protocol Layer

Before advancing further, we should answer the question -- where the new protocol layer should be? Because the new layer is introduced into the stack for implementing cryptographic measures mentioned above, it is therefore preferable that the transaction message (Command and Response) from the neighbour source layer to the new layer should be as simple as possible.

As command (resp. response) moves down the protocol stack of terminal side (resp. smart card side), layers of the stack not only adjust (specifically, complicate) the command layer by layer, such as pre-appending or/and post-appending some mandatory information to the message for labelling, they but also take measures for some mandatory requirements, e.g. for perfect communication, which further complicate the transaction message; therefore, the upper the new layer is in the stack, the simpler it is to accomplish the task of implementing cryptographic measures for guaranteeing trustworthy smart card transaction.

However, the new layer cannot be set upon the stack, otherwise it is independent to the stack. Clearly, position under CAL and above TAL is preferable to any other possible places in the current protocol stack.

Fig. 4 depicts the modified protocol stack after introducing a new layer into the stack depicted in Fig. 3. The new layer and the new stack are named as Terminal Security Layer (TSL) and trustworthy protocol stack respectively.

| | |
|---|---|
| 6 | Card Application Layer (CAL)<br>(Smart Card Aware Applications) |
| 5 | Terminal Security Layer (TSL)<br>(Encryption, Decryption, Authentication, etc.) |
| 4 | Terminal Application Layer (TAL)<br>(ISO7816 APDU) |
| 3 | Terminal Transport Layer (TTL)<br>(APDU<=>TPDU) |
| 2 | Data Link Layer<br>(Data Frame, Timing,<br>Error Correction and Detection, etc.) |
| 1 | Physical Layer<br>(Bit Stream Control, etc.) |

Fig. 4. Trustworthy Protocol Stack of Smart Card Application System

## IV. How the TSL Accomplishes the Task

### A. Establishing Trustworthy Transaction Channel

Task of establishing trustworthy transaction channel is in essence to propose strategy for trustworthily securing APDU transmission between two legitimate parties, i.e. the card application and the smart card.

#### 1) Structure of APDU Message

As shown in the trustworthy protocol stack, card application initiates and sends a command to smart card through TSL, TAL, etc., layer by layer; and smart card returns a response via TSL after processing the command to the card application. A specific response corresponds to a specific command, referred to as a command-response pair. An application protocol data unit (APDU) contains either a command data unit (C-APDU) or a response data unit (R-APDU).

Fig. 5 depicts the C-APDU message output by CAL of terminal side, consisting of a mandatory header of four bytes, i.e. CLA, INS, P1 and P2, and a conditional body ([...]) of variable length, i.e. possibly including Lc, IDATA and Le.

[CLA][INS][P1][P2] [Lc] [IDATA] [Le].

Fig. 5. Structure of C-APDU Message

Fig. 6 depicts the R-APDU message output by CAL of smart card side, consisting of a conditional body ODATA of variable length and a mandatory trailer of two bytes. Two mandatory bytes are status bytes SW1 and SW2; two status bytes code the status of the receiving entity after processing command APDU.

[ODATA] [SW1] [SW2].

Fig. 6. Structure of R-APDU Message

#### 2) Defining the Symmetric Cryptographic Functions

Two symmetric cryptographic functions are defined without specifying any particular symmetric algorithm, one for encryption, and the other for decryption.

The encryption function is defined as C = GE(M, K), and the decryption function is defined as M = GD(C, K). M and C are plaintext and cipher respectively, and K is a secret session key. The session key used for cryptographic manipulations should be negotiated after mutually authenticating identities of two intending communication parties.

#### 3) Strategy for Securing APDU Transmission

The strategy consists of strategy for C-APDU transmission and strategy for R-APDU transmission, which will be implemented in the TSL layer. As strategy for C-APDU and strategy for R-APDU are built on the same idea, only strategy for C-APDU transmission is detailed.

Assuming that card application wants to send C-APDU P to smart card, a new C-APDU P' is constructed from original P in the following steps, which will be sent from card application to smart card instead of original P.

The original C-APDU P is expressed as P=[CLA][INS][P1][P2] [Lc] [IDATA] [Le].

1) If Lc field is absent in P, insert a zero byte to P at Lc position (directly after P2). Original message P is updated as
   P=[CLA][INS][P1][P2][Lc] [IDATA] [Le].
2) Call MD5 to hash the message P,
   H = MD5(P).
   H is the output message digest, whose length is exactly 16 bytes.
3) Concatenate output message digest H with IDATA in

original message P. If IDATA is absent in P, IDATA' is H after concatenation. Concatenating operation is illustrated as
IDATA' = H || *[IDATA]*.

4) Concatenate a transaction serial number byte SN, and IDATA' is modified as
IDATA' = SN || IDATA' .

5) Call message encryption function GE() to encrypt IDATA' above with secret session key K. Encryption is expressed as
C = GE(IDATA', K) .

6) Construct a new command APDU P'. P' is constructed as
P' =[CLA] [INS] [P1] [P2] [Lc'] [C] *[Le]*.
Four mandatory bytes are taken directly from original message P. Value of Lc' in P' is different from value of Lc in original message P, and its value is Lc plus 17 (if the message digest algorithm is not MD5, the value of Lc' is Lc plus the byte length of message digest calculated by the hash algorithm selected, and further plus one for accommodating the SN byte.). Le field in new command message P' is the same as Le field in original P; if Le is absent in P, Le will not be present in P', otherwise, Le will be present in P'.

7) Send P' to smart card instead of original command message P down through the protocol layers of the trustworthy protocol stack under the TSL layer until it attains the smart card.

*4) Discussion to the Strategy*

1) Adversaries cannot read or understand the kernel transaction information by intercepting or tampering messages exchanged between the card application and the smart card without authority of accessing the proper session key. Legitimate parties who possess right session key can recover any encrypted information to its intelligible representation. Therefore, the card application and the smart card can confidentially exchange transaction messages with extra encryption and decryption operations.

2) To verify data integrity of transaction message received, the recipient possessing the proper secret session key can recalculate a hash code and compare it with the hash code calculated and encapsulated in the message by the sender. If even a single binary bit in the message is altered, the recipient will generate a different hash code. With data integrity, both legitimate parties involving in the transaction know that what they are seeing is exactly what the other party sent.

3) Because the initial SN byte and the principle are all trustworthily negotiated by legitimate parties before any essential transaction information exchange, the card application and the smart card can therefore distinguish authentic transaction information from fraudulent information by verifying the value of SN byte in the message received. Namely, the value privacy of initial transaction serial number byte and the privacy of principle deciding the value variation of SN lie behind the possibility.

## *B. Attestation and Negotiation*

Authenticating trustworthiness of the card application and the smart card is to convince them that the card application (resp. smart card) is really the party it claims to be by evidencing notarised document, such as public certificate signed by a proper (publicly recognised) certificate authority.

Furthermore, as mentioned in subsection IV-A, in order to establish trustworthy transaction channel between the card application and the smart card, secret session key K, initial value of SN and so forth, should be trustworthily negotiated .

Namely, identity authentication and authenticated information negotiation are work left for guaranteeing trustworthy smart card transaction. W. Diffie's STS (Station to Station) protocol [13] supplies a complete and detail solution to the task. This paper will not repeat the protocol steps of STS. Some further discussions about generating K, SN and so on, are as follows.

1) The confidential message agreed on by the card application and the smart card with STS protocol is of variable bit length (its value range is [1, P-1], P is a big odd prime, whose bit length should be more than 512.), therefore it can not be directly taken as cryptographic key needed by symmetric cryptographic functions. Hash algorithms supply good solutions to transforming message of variable length to message of a fixed length.

2) More than one confidential messages are needed for establishing trustworthy transaction channel, including secret session key K (may contain many sub-keys), initial value of SN, and initialisation vector $IV_0$ (needed in some particular modes of cryptographic operation), then it is needed to cycle STS protocol many times.

3) In order to increase execution efficiency, identical {P, q} (Diffie-Hellman Parameters) can be used for cycling STS protocol during effective time of a secret session key negotiated with the identical parameters.

# V. Conclusion

It has been demonstrated that the new protocol layer (TSL) introduced in this paper can be easily incorporated into the current smart card protocol stack to guarantee trustworthy transaction between the card application and the smart card, consequently a trustworthy client for the networked application system is constructed. Cooperating with measures or securing the network traffic through the Internet, the security of the networked application system will be actually enhanced, and special security purposes will be eventually achieved.

# References

[1] PC/SC Workgroup, PC/SC Specification Version 2.0, Part 1: Introduction and Architecture Overview, August 2004.

[2] Trusted Computing Group. https://www.trustedcomputinggroup.org/
[3] Microsoft Corporation, NGSCB. http://www.microsoft.com/resources/ngscb/default.mspx
[4] FIPS Publication 46-3, Data Encryption Standard. October 25, 1997.
[5] FIPS Publication 81, DES Modes of Operation. May 31, 1996.
[6] FIPS Publication 197, Advanced Encryption Standard. November 26,2001.
[7] R. Rivest, A. Shamir, and L. Adleman, A method for obtaining digital signatures and public-key cryptosystems, Communications of the ACM, 1978, 21(2): 120-126.
[8] RSA Laboratories, PKCS \#1 V2.1, RSA Cryptography Standard . June 14, 2002.
[9] FIPS-113, Computer Data Authentication. May 30, 1985.
[10] FIPS-180-2, Secure Hash Standard (SHS). August 1, 2002.
[11] FIPS-186-2, Digital Signature Standard (DSS). January 27, 2000.
[12] IETF RFC 1321, The MD5 Message-Digest Algorithm. April 1992.
[13] W. Diffie, P. C. van Oorschot, and M. J. Wiener, Authentication and authenticated key exchanges, Designs, Codes and Cryptography, 1992, (2): 107-125.

# Metadata Guided Statistical Information Processing

Wilfried Grossmann and Markus Moschner
Work Group Data Analysis and Computing,
Department of Computer Science,
University of Vienna, A-1010 Vienna, Austria

**Abstract - Metadata may be used for handling of statistical information. Some metadata standards have already emerged as guiding lines for information processing within statistical information systems. Conceptual models for metadata representation have to address beside the dataset itself additional data objects occurring in connection with a dataset. A unified description framework for such data objects is discussed with respect to metadata handling. Basic ideas on integration and translation of metadata standards are given with a focus on the framework. Hereby principles of ontology engineering play a key role as starting point.**

## I. Introduction

Economic and business decisions rely many times on empirical information contained in data. Such data get frequently assembled as information base in data warehouses. Hereby statistical methods are used. Building and using such empirical information needs not only data itself but also additional information about the collected data, which is usually summarized under the topic meta-information. The term meta-information means here that we consider more than the data itself, especially descriptions of the data, so called metadata, which are necessary for obtaining and handling the information. Hence, it is not surprising that metadata play a key role in statistical information systems for a long time. The earliest reference is Sundgren [11] who introduced the concept of metadata in statistics for a so called infological approach towards statistical information. This approach has been developed further in many ways by a number of researchers as well as statistical agencies and has lead to a number of metadata standards. One line of development focused on metadata standards for highly aggregated data in data warehouses with special emphasis on data exchange, for example the SDDS-standard [9]. Another approach stressed more the data archive aspect, considered metadata requirements of such archives and lead to tools like NESSTAR [7].

Due to the different starting points of these approaches it is rather cumbersome to integrate data in cases where the definition of the data schemes is based on such different documentation schemes. Following the developments of intelligent information processing in recent years the field of statistical information processing has seen a number of efforts to develop the idea of metadata further into the direction of ontology (see for example the *MetaNet* project [6]). In fact, statistical metadata practice includes to far extent information needed in ontology-engineering.

In this paper we present a foundational view on metadata usage for data representation, for data retrieval, for statistical data processing and for mapping between different metadata standards. Section II outlines the basic elements of statistical data models together with a rather flexible structure for a unified representation of these elements, so called *composites*. The design is oriented towards all kinds of statistical data processing. Section III considers metadata formalization and in section IV we present first ideas on formal processing.

## II. Composites

An important contribution in matters of metadata standards was made by the METANET project [3, 6], a thematic network within the fifth EU-research framework. This project aimed at a coherent framework for data collection and processing. In focus have been the requirements of the ontology definition of Gruber [5] ("ontology" is a specification of a conceptualization) by formalizing the statistical data paradigm, taking into account the representational, operational and functional semantics of statistical information. First steps towards an ontology for data description were set by outlining a taxonomy [3].

*Elementary Notions*

Starting points are basic information objects like dataset, statistical population, or statistical variable, which constitute the categories for the ontology. These categories constitute the basic statistical objects. Datasets include elements with some information. Such elements belong to a universe of basic objects. The information units *Population* and *Unit* represent the universe from which the dataset is a special model. A representation of a dataset always uses *Variables*, which represent measures of interesting properties for the units. *Variables* give some "explanation" for the dataset as descriptions of values for attributes taken from a proper defined *Value Domain*. *Additional Attributes* ease handling of computational results for a dataset.

The development of an ontology (for metadata) starts from the dataset with its accompanying information units. Hereby an arbitrary number of information units of specific types may occur. Basically the accompanying populations resemble more or less the structure of the dataset; units can be considered as elements of the set representing the population and the basic structure for the value domains are also sets. For all these objects one can define basic operations, which are the building



T. Sobh and K. Elleithy (eds.), *Advances in Systems, Computing Sciences and Software Engineering*, 383–386.

blocks for all kinds of processing activities, rather independent from given datasets.

*Datasets, Population, Unit, Variable* and *Value Domains* are the basic objects, the so called categories of a unifying description framework. For documentation of these categories the idea of facet classifications (as known to the librarians) is used. Descriptions get organized with respect to four different categories, which give rise to the corresponding view facets of statistical categories.

The following four views were distinguished:

- The conceptual category view represents the subject-matter definition of any category instance and builds the bridge to reality. Relation concepts for the considered objects are described here (conceptual data modeling may be done here). Validity of the definition of the subject matter concept gets usually restricted by temporal and geospatial constraints.
- The statistical (or: methodological) category view describes the statistical properties of the category instance by using a number of (type and role) parameters, which have to be defined in such way that specific properties of the different categories are taken into account. Data structures and relations to other objects are described here. A distinction between type and role parameters is given, what resembles roughly a distinction between context independence and dependence.
- The data management category view is geared towards machine supported manipulation storage and retrieval of the category instance data.
- The administrative category view addresses management and book-keeping of the structures.

Based on these view facets a representation scheme can be defined, which takes account of typical needs for statistics though not restricted to. Proper usage of datasets depends sometimes on more than object descriptions, but also derived from these objects. Traditionally such data are called metadata. Thus a uniform framework for representation of objects and the corresponding description seems to be a desirable aim.

A first sketch of such a model was presented in [2]. All the necessary information gets packed into so called *composites*. Composites are acyclic tree-like graphs where the subtrees represent so called data buckets for which corresponding *bucket schemes* are presumed. In the sense of uniformity each object type has bucket types analogous to the description views: conceptual, statistical, data management and administration buckets.

*Technical Issues on Composites*

Processing of *composites* can be composed from a sequence of transformations *T* of *composites* producing new ones out of already existing:

$$(C_i)_{1 \leq i \leq p} \rightarrow (T_j((C_i)_{1 \leq i \leq p}))_{1 \leq j \leq k}$$

Informally, a transformation *T* is describable by definition of the origin *composites*, the output *composites* and the operators for computation with specifications of operator parameters.

Transformations are assumed to start with request for necessary operations onto the data set. Within a *composite* one can check the semantic correctness of operations what should depend on the information inside of a *composite*. One operation may require a bunch of additional operations for other parts of a *composite*. A transformation plan depends on the specific entries of the various *bucket schemes* and *data buckets*, yet nevertheless three main steps will be encompassed:

1. Metadata processing:
   An admissibility check of a transformation is done by using only those data in the *composites* necessary for the description. The implied side effects of an operation are listed in the case of approval.
2. Data processing:
   Operations for various *data buckets* of *composites* get executed according to the specification of metadata processing in the above step.
3. Output Generation:
   A new result *composite* is generated. In case of unary operations it may be sufficient to add buckets for additional attributes, though new composites have to be defined for binary operations.

The last steps follow only if the first one (on metadata processing) does not result in an interrupt.

An example of such processes may be found in [4].

## III. METADATA

The outline of processing in the previous section puts metadata into a central role for all kinds of activities. Metadata play also a role of something like a Kerberos – not only guiding but permitting process steps, too. A *composite* is not fixed with respect to data models or metadata definitions. It is assumed that one sticks to a former decision of choice. Yet not everybody will choose equally. That reason and interest into automated processing gives a motivation for metadata interoperability. Today's ontologies [10] within knowledge representation can act as a starting point for formal metadata definitions (as basis for conceptual models for metadata) – and henceforth metadata standards. In fact there are different ontologies and henceforth metadata standards, and it is to expect that such numbers grow.

*Formal Metadata Descriptions*

Here we concentrate only on such standards which basic principles and formulations can be systematically treated by humans from a bird's eye view. Only from such a unifying treatment formalizations are tackled, as there are knowledge representations, data types (order sorted algebras), mathematical approaches and statistical notions (units, population and statistical variable).

*A) Basic Concept:*

Metadata standards can get formulated by a common basis of fundamental concepts (foundational basis). That means that there are atomic notions within a partial order where such an order means something like sub- or super-concept. Set-like operations (join, meet and complement) get induced by such an "order". The used vocabulary comprises attribute-like properties and restricted quantifications like in description logic [8]. Herewith the basis for the intended fundamental concepts is formulated such that differences and relevant properties get included. It will not be realistic to aim at world knowledge nor will all technical issues be settled. For special cases even different ontological approaches might be chosen (e.g. [1] ).

The aforementioned basic statistical objects (population, units and variables) have to fit into the chosen conceptualization.

A set-like formulation for a fundamental structure $K$ lies at hand. Formally it is defined by

$$K = < BNotion, BRel, Subsump, Meet, Join, Compl, Null >,$$

a pool of Variables (there might be more than one sort), and properties of elements of K analogous to predicates (roles) and quantifiers in description logics ([8]).

*B) Data Types:*

Until now only abstract concepts have been considered. Handling of values as numbers is one part of statistical information processing. Analogous to programming languages specification of data types with concrete value domains is mandatory. There are data types like numbers and strings which do not share much or even nothing (from a conceptual viewpoint). Furthermore some data types form (nontrivial) order sorted algebras. One example are natural, integer, rational and real numbers where operations are also extended. Intervals will play a prominent role with respect to numbers. The possibility of (domain) restrictions needs the concept of attributes - monadic predicates which may be used as generators for new sorts. One may distinct between sub- and super sorts with respect to the partial order of the conceptual basis and sorts associated with subset properties.

A data type $D = < DDom, DPred >$ represents a domain with predicates defined on it (operations have to be formulated as some sort of equations - what seems rather naturally for most cases). For each data type we have an instantiation (or: intended interpretation) which maps $D$ into some grammatical structure.

Usual database models do not reflect such data type properties. Since (partial) orders are represented by (binary) relations that is not a problem for matters of storage. Data type properties are needed first for guiding the processing (and decision making). The data management view tackles matters of archivation.

*C) Formalization of Mathematical Concepts:*

There is a strong relation between sorts of formal mathematical content and data types. Formalization brings a large body of ordered sorts. It is not always necessary to have a correspondence between formal sorts and ontological concepts. Formalization gives (semi-)automatic processing of mathematical statements, thus only basic or important mathematical notions need a correspondence to ontological concepts.

*D) Statistical Variables:*

A statistical variable $SV$ is a (partial) mapping from $K$ into instantiations of data type domains.

There is not only ontological meaning behind. Statistical variables give also concrete values for abstract notions. In that sense they play a central role in being a bridge between abstract notions and value domains: Thus it is legitimate to see them as interpretations of metadata standards into the available order sorted algebras. The statistical notions of unit and population get hereby a determination with concrete values. One needs abilities for comparing and processing mathematical notions, since they play a role in understanding statistical variables.

*E) Specification for Data Repositories:*

This topic covers merely technical specifications like data base or storage organization. Detailed inspection of data management views is beyond the scope of this paper.

*F) Metadata Formalization:*

Concrete metadata standards get reformulated pertaining to the fundamental concepts. The original formulation of metadata standards gets simply mapped onto ontological notions whereas totality is not compulsory. What is likely to appear is that especially for technical notions such a standard might be finer grained than the used ontology. Either start again with an enhanced foundational basis or construct an according coarse mapping. Such a mapping does not mean to forget such more fine grained standard notions – yet only its aggregation with certain ontological notions. That is not only a matter of lowering the work for ontology construction but sometimes one does not need or has no justification for differentiation between some technical terms. In matters of the ontology these are too near related.

It is to expect that such metadata standards are like taxonomies. At least there has to be a set $TC$ of taxonomical concepts which is partially ordered.

*G) Morphisms between Metadata Formalizations:*

Translations between metadata standards may be tackled when the foundations are established. Since ontological foundations shall be taken into account such morphisms are not simply between taxonomical standards itself. Yet these are between the relations of the fundamental concepts and metadata standards. These morphisms need to be isotone in the sense that the partial order of concepts with respect to subsumption in the foundational basis has to be kept for mapped pairs.

Such a morphism *M* can be seen technically as subrelation of the support of *K* and *TC*. Hereby isotony applies as a basic constraint, yet further may be useful.

The morphism between the standards is effectively constructed by use of the fundamental concepts. If done by hand humans play the role of the fundamental concepts. Thus some explication of this activity is demanded here – a task that is too often underestimated.

*Relation to View Facets*

The points above correspond in a certain way to the view facets of statistical categories: Topic A) corresponds to the conceptual category view where matters of knowledge structures are addressed in B) – D). F) and G) comprise statistical approaches as well as concrete object properties. That makes them counterparts of statistical categories. Topic E) resembles the data management and administrative category view.

## IV. On Formal Processing

At least there are two possible applications of metadata guided processing:

- Automated checking for incompatibilities between data;
- Interpretation or classification of incompatibilities for the given context.

Automatic processing demands efficiently treatable knowledge representations with semantic capabilities. Some exclusive conditions arise of these requirements. Often such conditions are informally labeled as ``good practice''. Reflection on such issues should not be underestimated, especially with respect to avoidance and exclusion criteria.

Different *composites* may involve information which shall be merged somehow. First one has to check if the presupposition of feasibility exists, what involves a (metadata) check for *bucket schemes*. This has to begin with examinations on the used metadata standards. In case of (partial) translatability from one to another, the generation of new *composites* for one metadata standard may follow. The resulting *composites* have again to be checked by usage of metadata but now involving the *data buckets*. This is necessary because of the translatability with respect to taxonomies does not resolve all questions or problems with respect to data operations. Resulting datasets have to be checked for compatibility with respect to the intended operation(s). Here we are back at the issues of section II.

The detection of differences between different descriptions is a goal itself. Requirements of translatability act as additional demand for accurate conceptions of metadata descriptions and methodologies. Full conceptions are not used sometimes. Thus nonequivalence need not be a failure, but then subsumption properties need some handling. Information about data quality may be available from such comparisons. Here reliability and efficiency (or lucidity) are two antagonistic yet fundamental issues.

## V. Conclusion and Further Remarks

Processing of statistical information is discussed with regard to composites([2]) and metadata. Basic problems and principles are debated and specified. Especially the main focus lies on the use of metadata for control of composite processing. The importance of mathematical structures for statistical life is indicated through remarks on the formalization of mathematical knowledge. Statisticians or users of statistics will not always feel confident with tools neglecting mathematical knowledge. Thus acquaintance with such tools and representations of mathematics is a subject of future research. Herewith interrelationship between datatypes and mathematical representation determines some choices. Such choices cannot be made without consideration of the ontological specifications. This paper intends to be some starting steps into this field of problems.

## References


[1] T. Andreasen, H. Bulskov, R. Knappe, " On Automatic Modeling and Use of Domain – Specific Ontologies," in *Proc. 15th Int. Symp. ISMIS* 2005, M-S Hacid, Z.W. Ras, S. Tsumoto Eds. LNAI 3488, Springer Verlag, 2005.

[2] M. Denk, K.A. Froeschl, W. Grossmann, "Statistical Composites: A Transformation Bound Representation of Statistical Datasets," in *Proc. 14th Int. Conf. on Scientific and Statistical Database Management*, J. Kennedy, Ed. IEEE Los Alamitos, California/USA (2002), pp. 217 - 226.

[3] K.A. Froeschl, W. Grossmann, V. delVecchio, "The Concept of Statistical Metadata," *MetaNet* (IST-1999-29093) Work Group 2, Deliverable 5, 2003.

[4] W. Grossmann, M. Moschner, "Towards an Ontology for Data in Business Decisions," in *Proc. Int. Conf. PAKM 2004,* D. Karagiannis, U. Reimer, Eds. LNAI 3336, Springer—Verlag, 2004.

[5] T.R. Gruber, "A Translation Approach to Portable Ontologies," *Knowledge Acquisition,* vol. 5, pp. 199-220, 1993.

[6] The *MetaNet* project, http://www.epros.ed.ac.uk/metanet.

[7] NESSTAR – Networked Social Science Tools and Resources,http://www.nesstar.org

[8] J.Z. Pan, I. Horrocks, "Web Ontology Reasoning with Datatype Groups;" in *Proc. of ISWC 2003,* D. Fensel, et al. Eds. LNCS 2870, Springer Verlag, Berlin 2003.

[9] SDDS – Special Data Dissemination Standard, Dissemination Standards Bullentin Bord, http://dsbb.imf.org/Applications/web/sddshome/\#metadata.

[10] J.F. Sowa, "Ontology, Metadata, and Semiotics," in *Conceptual Structures: Logical, Linguistics, and Computational Issues,* in: B. Ganter, G.W. Mineau, Eds. LNAI 1867, Springer Verlag, Berlin 2000.

[11] B. Sundgren, "*An Infological Approach to Data Bases*" Urval, Nr. 7, National Central Bureau of Statistics, Stockholm, Sweden.

# Combinatorial Multi-attribute Auction (CMOA) Framework for e-auction

*A paper for the CIS²E 05 Conference, December 10 – 20, 2005.*

*Aloysius Edoh City University London and University of East London*

**Abstract: The traditional auction protocols (aka Dutch, English auctions) that Ebay, Amazon and Yahoo use; although considered success stories [9], have limited negotiation space: The combinatorial auction (CA) and multi-attribute auction (MAA) [10], [17] have been developed to address these shortages but even these do not allow users to negotiate more than one attribute at a time. As an answer to these limitations a new e-auction protocol has been created to enable agents negotiate on many attributes and combinations of goods simultaneously. This paper therefore shows how the automated hybrid auction was created to reduce computational and bid evaluation complexity based on CMOA bidding language and Social Construction of Technology (SCOT) principles. SCOT states that (i) the 'relevant social group'; (ii) their 'interpretative flexibility"; and "workability"/ functionality of the technology must be considered for the development of system such as e-auction as an alternative to E-Bay. SCOT is of the view that technologies emerge out of the process of choice and negotiations between 'relevant social groups' [15], [8], in this case the bidders, auctioneers, sellers and auction house.**

**This paper represents a collaboration of studies in progress - The Combinatorial Multi-attribute Auction as an online auction as compared to existing e-auction protocols such as Amazon, eBay, and application of intelligent software and Agent UML in e-auction.**

## 1.0 Introduction

Automated auction is the most commonly used method for buying and selling goods and service. Some examples of automated auctions (aka e-auction) are eBay, Amazon and Yahoo. Currently many organisations use e-auction for selling airport landing and departure slots, sales of railway lines with freight and passenger services and sales of frequency spectrum. The automated auction uses the same set of rules for negotiation and evaluation as the physical auction markets. Unlike the physical auction market, in automated auction the users are connected to the market through the Internet or other electronic means and they need not be physically present therefore we have a virtual market. Again the offers from the bidders are sent to the auctioneer by electronic means and the bids are evaluated using automated algorithms. Auctions are designed such that they have a set of strict lay down rules, which govern the procurement, negotiation and allocation procedures, this is called auction protocol. Some of the traditional auctions protocol used in e-auction are English, Dutch and sealed bid auctions [6]. According to [1] E-bay, uses the Dutch and sealed bid protocol whiles Amazon uses English auction. Below are the rules that govern some of the traditional auctions:

1. Sealed bid auction is a process that allows bidders to submit only one bid which is kept secret from other bidders. The bids are evaluated and the bidder with the highest price is declared the winner;
2. In Dutch auctions, bidders are allowed to submit secret bids as many times as possible within a specified time. The bidding continues within the duration until the anticipated value is obtained;
3. In the English auctions, a base (aka private) value or attribute (usually price) is announced and the bidders are allowed to give an offer. The various offers are announced and bidders who wish to give a higher offer can do so until the highest bid is achieved. The various bids and offers are made open and the highest bidders get the offer.

The main elements in any auction are the protocols, the users which are the bidders (buyer of the goods or services), auctioneers (the middleman) and the sellers (providers of the services or goods). In physical auctions, the auctioneers are given the evaluation function but in automated system the evaluation algorithms are incorporated into the programmes on the bidding server or auction house server. Again in automated auction, almost all the key elements can be converted into software or intelligent agents therefore there is no need for the physical market. The advantages of having automated auction with in-built bid evaluation mechanism are that it reduces travelling costs, creates a global virtual market, attracts bigger audiences, saves travelling time, improves the bid allocation and distribution. Thus with the increased



*T. Sobh and K. Elleithy (eds.), Advances in Systems, Computing Sciences and Software Engineering*, 387–395.

computer power and improved Internet security, many auction houses are using automated auction like E-bay.

All this seems successful but have a limited negotiation space and not flexible [9]. This means they can be used to bid for only one item at a time and negotiation carried out on only one attribute price. To address these limitations, advanced auctions namely combinatorial auctions (CA) and multi-attribute auction (MAA) were introduced [10] [17] [18][20]. Advanced auctions are market mechanisms that allow auction users to bid and negotiate on combinatorial goods or on multiple attributes (such as price or quantity), that is, on multidimensional parameters. Unlike the traditional auctions, advanced auctions are designed to handle complex negotiations and have the ability to incorporate one or more of the traditional auction protocols. However, even these advanced auctions do not allow users to negotiate on combination of items and more than one attribute at a time. As an answer to these limitations we have introduced a new e-auction protocol that has increased negotiation space and is more flexible compared to CA and MAA. The new auction called Multi-attribute Combinatorial Auction or CMOA in short is a hybrid of the two advanced auctions with an in-built evaluation mechanism. This paper shows how the automated hybrid was created to reduce computational and bid evaluation complexity using CMOA bidding language with CNF and Social construction of Technology (SCOT) principles. The paper is structured as follows; it reviews the limitations of advanced auctions and presents the CMOA framework as the solution. This is followed by a discussion on CMOA and CA protocol such as E-bay, a conclusion is then presented.

## 2.0 Background:

### 2.1 Combinatorial auctions (CA)

In combinatorial auctions (CA) the bidding agents negotiate on a combination of goods. The advantage of this auction is that it allows bidders to negotiate on combinations of items or services, however some of its drawbacks are: The bidders have limited negotiation space and flexibility because they can only negotiate on one attribute, namely, price; secondly, when bidders are allowed to bid for arbitrary combinations of goods, this would lead to computational complexity (i.e. exponential expansion), which is impossible to solve. This is called the Combinatorial Auction Problem (CAP); thirdly, is the problem of how to allocate the bid or determine the winner in order to maximise the auctioneer's revenue or utility function and at the same time satisfy the individual participants [10]. This is called winner determination or Bid Allocation Problem. Fourthly is the issue of how to express the bid and communicate it to all participants in order to avoid ambiguity;

A lot of research work has been conducted on the above complexity issues, particularly on bid representation, bid allocation and communication complexity [6]. The two main research directions are led by [10][6]. Sandhold [10] proposed an algorithm for determining the optimal winner which is based on Integer Programming (IP). However attractive such approach maybe, it is complex, difficult and computationally inflexible. [6], on the other hand, suggested a winner determination model based on Linear Programming (LP), which in turn has limitations, particularly when prices (i.e. attributes) are not attached to individual combinatorial bids. Again another proposed solution is the introduction of bidding languages to address the problem of avoiding bid ambiguity and representation. Compared with MAA, the bidding languages for CA are fairly well developed and there are three different types of languages: $L_B$, $L_G$ and $L_{GB}$. Recently another new CA bidding language has been proposed by [18][19]. Although these bidding languages enable user to express their preferences they have not been incorporated into the bid evaluation mechanism. Thus there is no seamless transition from bid representation to bid evaluation and allocation.

### 2.2. Multi-attribute auction (MAA)

Multi-attribute auction (MAA) is another type of advanced auction protocol aka multi-objective auction. In this auction the bidders negotiate on multiple attributes such as price and quality simultaneously, which gives the bidding agents more negotiation flexibility. However MAA has the some limitations: The main problem is bid representation, which involves preparing a bid format that accurately represents and matches the seller's requirements. It must also express the bidder's preferences so as to maximise the agent's utility [16]. Another limitation is how to develop a system for evaluating the various different attributes and determine the winner [17]; lastly, unlike in CA, bidders can negotiate only on one item at a time; this limits the bidder's choice and negotiation space. Again, a new bidding language has also been proposed for MAA by [17] but this is yet to be fully developed.

The CMOA framework addresses all the issues discussed above except the communication complexity. Communication complexity means during an auction the amount of information that must be transferred between the participants can grow exponentially and therefore it would be difficult to send it over a communication channel [6][19]. This is an auction and distributed computing problem that involves the handling and transfer of the large volume of information generated during an auction. The transfer of large information across the computer networks between the bidders and auctioneers during the negotiation stage must be addressed.

## 2.3 CMOA

In the real physical market, Combinatorial Multi-Objective Auction (CMOA) is the negotiation process that most people would apply when buying or selling items in the market. In an automated configuration, this will attract more participants (i.e. audience) who are located globally and they do not need to be physically present for the auction (i.e. creating a virtual market). It is less expressive than the physical room-based auctions because the participants only have to connect to the virtual market through the Internet. Again we can use software agents to enhance the negotiation process. Considering these advantages and the limitations of advanced auction, we proposed a new type of automated CMOA auction protocol that has the following benefits:

1. This new auction will have a larger negotiation space than CA or MAA. The negotiation space of CMOA is such that bidders can hold detailed discussions on combinations of goods as well as on all the conditions and attributes which are related to the goods in the bid. Therefore the negotiations will not be based only on price as in CA but also on non- price attributes (i.e. warranty or quantities) as well as on combinations of goods;
2. The CMOA framework will reduce the computational complexity of bid evaluation and bid allocation. This was achieved by developing a new bidding language that reduces the number of steps required to evaluate the accepted offer. It reduces the number of bid allocation steps by converting the different attributes in each bid into a common utility value that forms the bases for bid evaluation. This algorithm is different from the existing multi-attribute bid evaluation algorithms, where the evaluator must go through the scoring function for each attribute before allocating the bid. This improves the efficiency of bid evaluation and bid allocation because there would be fewer steps. It also leads to improved negotiation time, number of iterations, execution time, and negotiation space;
3. This auction will assist both sellers and bidders to freely and fully express their preferences without limitations. In the proposed framework, the seller and auctioneer can express all their preferences before the auction starts, in a Request To Bid document (RTB). This is not possible in CA or MAA auctions;
4. In the CMOA automated auction, the participating agents can negotiate as in a real-world market. This means that the seller and bidding agents can negotiate on a combination of items and on different attributes at the same time. This is not possible in CA and MAA auctions;

The CMOA framework can be used to determine the combination of components or items that will maximise the utility function of the system and improve its output. Secondly, it can be used to determine the component that has the optimal utility value in any combinatorial system. Again, CMOA will assist developers to design and implement multipurpose auction platforms, which can handle the traditional auctions ranging from Dutch and English auction protocols to sealed-bid and at the same time be used for advanced auction protocols. In addition to the above the most appealing characteristic is that the CMOA framework can be used in other fields.

## 2.4 CMOA framework

The operation of the CMOA framework

In the CMOA auction protocol, the Request To Bid (RTB) documents are sent to the bidding agents. The format of the RTB is based on the CMOA bidding language, in which the bidding agents fill in the value of each attribute of the item in the combinatorial auction and use the correct logical connectives for the bids in the auction. The logical connectives refer to the type of combination of goods that can be used in the auction and these are complementary goods (aka *AND*), substitutable goods (aka *OR*) and high value goods (aka *XOR*). The quotation sent by each bidder must have attached to each item in the combination a specific set of attribute values. For example some of the attribute values can be price (*P*), delivery days ($D$), speed($S$) or quantities ($Q$). On receiving the quotations (i.e. bids), the CMOA framework performs the following operations:

1. Express the RTB in logical format using logical connectives (i.e. use the proposed bidding language)
2. Converts the various attributes into a common utility unit ($U$) using weights and multi-attribute utility theory (MUAT);
3. The combinatorial goods in the logical statements must be attached to the overall utility value $U$ using the equivalent logical connector;
4. The framework then converts the logical expressions into Conjunctive Normal Form (CNF);
5. The resulting CNF expression is transformed into a set of inequalities, aka constraints matrix table;
6. When determining the item that has the optimal value, the sum of the utility unit of each item becomes the coefficient of the utility function of the good;
7. The optimal item is determined by using objective function and constraints matrix table;
8. To determine the winner, all accepted bids are given 1 and the bids which are not accepted are given 0 (zero) then the bid allocation is calculated using a bid evaluation mechanism based on Linear Programming (LP) techniques.

The framework was developed using C programming language running on the MATLAB optimisation toolbox; however any language and LP system can be used.

### 3. 1 Discussion

We compared the CMOA with the e-bay auction protocol using the following assumptions: Firstly, that the traditional protocols namely Dutch, sealed-bid or English auction used by E-bay can be incorporated and applied **in** advanced auction. This means we can bid for a combination of goods on E-bay using CA protocol which incorporates either Dutch or English protocol. Secondly the evaluation algorithm used for both CA and CMOA would have a utility units that is summed in a linear manner because the goods complement each other.

Based on these assumptions we analysed a situation where a medical informatics company wants to upgrade their computing system by incorporating wireless and Internet facilities. They have decided to buy computers with Pentium 4 Intel processors operating at 4GHz and have hard drives of 40GB. The computers must be supplied with matching all in one printer, copier and scanner: The printing speed for (Black/White) must be 25 ppm and speed of colour is 10ppm. These combinations of equipment must be supplied with corresponding 4Mb/s broadband facilities that connects to the Internet and use wireless devices. The bidders are required to bid the three complementary equipments as a turnkey package. The combinatorial items in each offer namely $G_1$, $G_2$, $G_3$ are evaluated using an evaluation function that is based on price $P_i$, and speed $S_i$. For evaluation, the weight on price is 30 and speed is 20.

The CA and CMOA bidding processes were based on sealed – bid protocol and three quotations we received from three bidders. The framework works as follows: For step 1, the RTB are sent to the bidders in the CMOA bidding language format. CMOA language (aka $L_{CMOA}$) is a mathematical format that uses logical connectives, brackets and comas to describe the relationship between goods or services and their attributes in a bid. Logical connectives such as *AND, OR* and *XOR* are used to formalise logical reasoning however in CMOA each of these logical connectives has its syntax and semantics In $L_{CMOA}$, the bid expression uses the logical connective *(AND)* $\wedge$, to indicate complementary goods, *(OR)* $\vee$ for substitutable goods, and *XOR* for highly valued goods. This paper will focus on complementary goods and below is a CMOA expression for the three complementary equipments required in the above scenario.

$$B_1: [(G_1 \wedge G_2 \wedge G_3), P_1, S_1, P_2, S_2, P_3, S_3] \quad (3.1)$$

A CA auction has one attribute ($P$ *or* $S$), therefore it can be expressed as follows:

$$[(G_1 \wedge G_2 \wedge G_3), P_1, P_2, P_3] \quad (3.2)$$

$$[(G_1 \wedge G_2 \wedge G_3), S_1, S_2, S_3] \quad (3.3)$$

For MAA auctions where only one item $G_1$ is to be auctioned but has many attributes, the expression will be written as

$$[(G_1), P_1, S_1] \quad (3.4)$$

The CMOA expression has multiple attributes therefore it must be converted into common virtual currency called utility unit. To obtain the common virtual currency, we applied the multiple attribute utility theory (MAUT) to convert all the attributes in the formulae into utility units. As explained in the assumption in complementary goods auction each item matches the other therefore the utility unit of the items are added in linear manner.

To map the attributes in the formulae to utility units, the MAUT principle states that each attribute must be given a weight that corresponds to the auctioneer's preference. In this analysis the weight on each bidder $x_i$ attribute is represented as $w_i$ and relationship between the value of the attribute quoted and the estimated value (auctioneer's value) as $f(s))$. Thus the weight $w_i$ on the $x_i$ attribute would be subsumed as linear function $s$ (i.e. $f(s))$, which is the relation between the estimated (private) valve and value quoted. This principle of converting the attributes to utility unit is summarised as $U$ = sum $[w_i\, f(s_i)]$. When applied to the scenario, the utility unit for item 1 from bidder 1 becomes:

$$Ug_1 = \sum (w_p\, f(sp_1) + w_d\, f(sd_1) + w_q\, f(sq_1)) \quad (3.5)$$

Therefore the concept can be summarised in the following equation.

$$B_1\text{: } [(G_1 \wedge G_2 \wedge G_3)\ Ug_1, Ug_2, Ug_3] \quad (3.6)$$

The sum of the utility unit of each item then becomes the total utility unit offered by each bidder; therefore for bidder 1 we have $Ug_1 + Ug_2 + Ug_3$ equal to $U_1$. To determine the winner in the auction we used Linear programming techniques as indicated in steps 7 and 8 of the framework. In step 7 the objective function is formulated using the sum of the individual unit of each bidder to optimise the bid. The objective function to be optimized will be $U_1 X_1 + U_2 X_2 + U_3 X_3$ from the generic linear program expression

$$Z = \sum_{i}^{n} (U_1 X_1 + U_2 X_2 + \ldots + U_n X_n) \quad (3.7)$$

In step 8, we define the constraint matrix to be used to evaluate the bids. In this approach, each bidder is represented as $X_i$ and the quotations from each bidder is converted into the utility unit and represented as $U_i$. To formulate the constraints for solving the bid allocation problem, a matrix table Table 3.1 is formed. The constraint matrix is made up of rows representing the items the bidder offered. All accepted items are marked 1 while the items the bidder did not quote for are marked zero. Finally the objective function and constraints matrix obtained are used to determine the bidders whose offer maximises the auctioneer's utility function. We used the LP in the MATLAB optimization toolbox to determine the winner.

Applying the framework to the scenario, we first evaluated the offers from each bidder's quotation to determine the overall utility unit for the bidder's bid. The attributes used in the experiments were price and speed of the computers and Internet facilities. The estimated price ($P_{01}$) for item1, $G_1$, is £30,000 and the speed for item1 $G_1$, ($S_{01}$) is 10GHz.. The estimated price $P_{02}$ for item 2 ($G_2$) is £ 40,000 and the speed $S_{02}$ is10 Mb/s. The weight on price is 30 and that of the speed is 20. For easy of calculation the values £10, 000 and £8,000 were made £10 and £8 respectively.

Quotation from Bidder 1 is as follows:

$U_{11}$ is utility unit for good 1: the quotation is $Pg_1 = £10$ and speed $Sg_1$ of 4 GHz

$U_{12}$ is utility unit for good 2: the quotation is $Pg_2 = £8$ and speed $Sg_2$ of 2 Mb/s

Calculation of the overall utility unit, $U_1$, for Bidder 1 is as follows:

The utility unit $U_{11}$ for good1 is $\sum( w_p \text{ x } P_1/ P_0 + w_s \text{ x } S_1/S_0)$ = (30 x 10/30 + 20 . 4/10) = 18; Then for good 2 the price, $Pg_2$ = £ 8 and the speed, $Sg_2$ is 2 Mb/s. The utility unit for good 2 $U_{12}$ is $( w_p \text{ x } P_2/ P_0 + w_d \text{ x } S_2/S_0)$ = (30 x 8/40 + 20 x 2/10) = 10. Therefore the overall utility unit for bidder 1 is $U_1 = U_{11} + U_{12} = 18 + 10 = 28$.

Similarly, the utility units of all the three bidders were calculated and the results are $X_1$ = 28, $X_2$ = 18 and $X_3$ = 20. Again applying step 8, the accepted items were marked 1 and none accepted item given 0. The result is depicted in Table 3.1.

Table 3.1: Showing that not all the items were quoted for by the bidders

| Bidder / Items | Bidder1 $X_1$ | Bidder2 $X_2$ | Bidder3 $X_3$ |
|---|---|---|---|
| $G_1$ | 0 | 1 | 1 |
| $G_2$ | 1 | 0 | 0 |
| $G_3$ | 0 | 1 | 1 |

The winner determination problem is then rewritten using the calculated utility units, LP expression (3.7) and Table 3.1 as follows: Optimize the objective function $28\ X_1 + 18\ X_2 + 20\ X_3$ subject to the constraint matrix in Table 3.1. The bid allocation problem was solved using MATLAB to find the optimal valuation and the maximum utility value in the auction. The optimal integer results obtained were $X_1 = 1$ and $X_3 = 1$ whiles $X_2 = 0$. Therefore the winners were $X_1$ with utility unit of 28 and $X_3$ was 20. This is because both $X_2$ and $X_3$ quoted the same items whereas $X_1$ quoted for only one.

### 3.2 Evaluation

In this paper we first analysed the computational space complexity, which is the amount of space required to evaluate the bids and determine the winner. For this analysis we used mathematical operations that involve converting the logical bid expression into CNF and then using Kaman's transformation table [12] calculate the set of constraints. We used the expression 3.1 for CMOA and 3.2 for CA. Our investigation revealed that the number of CNF calculated is equal to the set of inequalities. The number of inequalities calculated from the mathematical operations was plotted against the number of goods in the combinatorial bid expressions. The figure 3.1 depicts that the space required for CMOA is two times lower than the space required for CA. In complementary CMOA auctions, it was observed that as the number of goods in the auction increases, the number of constraints generated also increases in a polynomial function. The function is $y = x + 1$, which means the constraints $(y)$ increases by one more than the number of goods$(x)$ see figure 3.1.

In CA complementary auction, the constraints $(y)$ generated increases 2 times faster than the number of goods in the combinational bid with function $y = 2x$ see figure 3.2. In practice, this means the number of constraints required to solve the CA optimization problem increases sharply and thus affect the required computational space. Thus the increase in the number of constraints will create more space complexity problems in CA than in CMOA.

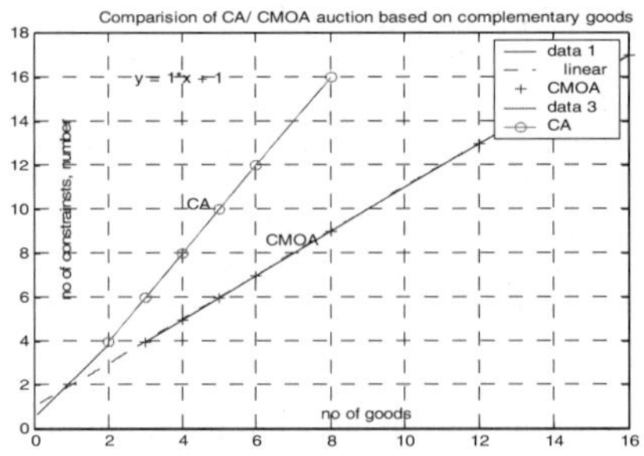


Figure 3. 1: shows the comparison between the constraints generated for CA and CMOA auctions using complementary goods (*AND*).

Secondly, we used the LP optimization package in MATLAB to analyse the stability of the results from the winner determination mechanism in section 3.1. The stability analysis was conducted using sensitivity test, which involved changing the utility unit of the next lowest bidder till it is equal to the winner's value. A series of experiments were conducted and the result of the sensitivity analysis is depicted in Figure 3.2. It was observed that the upper boundary at point 4 moved downwards till it reached the same value as the highest utility unit, which remained at 0 throughout the experiment. This means the original winner remained the highest bidder throughout the incremental raise till the values of the two bidders became the same. It was observed that the upper boundary

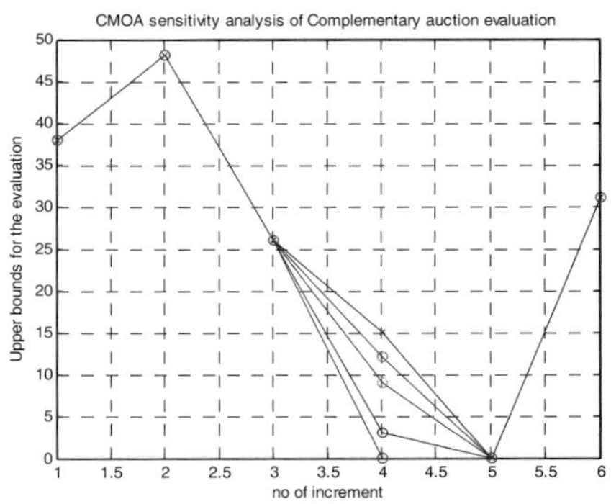


Figure 3.2: Winner determination sensitivity analysis using complementary goods auction

moved downward till the utility units of both bidders were the same. The original winner determined by the framework remained stable till the next bidder had the same utility value as the first thus the protocol is robust and stable.

### 3.3 Technological changes and social perspective

We were motivated by Richard Susskind analysis of how group of people who resist change act towards ICT. He used legal industry as the case study and indicated that there must be a link between the legal practitioners, consumers and technological change [14]. In his paper "Transforming the Law", he systemises his hypothesis by introducing "the Grid" which is made of four quadrants. The Grid focuses on how the legal ICT designer should look at and do, what the clients want how the clients and consumers are using the technology, how it affects the traditional role of lawyer (medics) and how competitors are using the technology [14].

Using the Internet as example he suggested "Disintermediation" and "re – intermediation". This is a process where due to technological changes customers have little choice but to use the state of the art such as WWW. Therefore all professions must be proactive and "work under the same virtual roof" as client. Thus all customers are been complied by E-bay, Amazon etc to do business over the Internet. Figure 3.3 depicts how the application of modern computer technologies such as intelligent agents, Agile and Agent UML has influenced software development and lead to the rapid growth from the first to the fourth quadrant. .

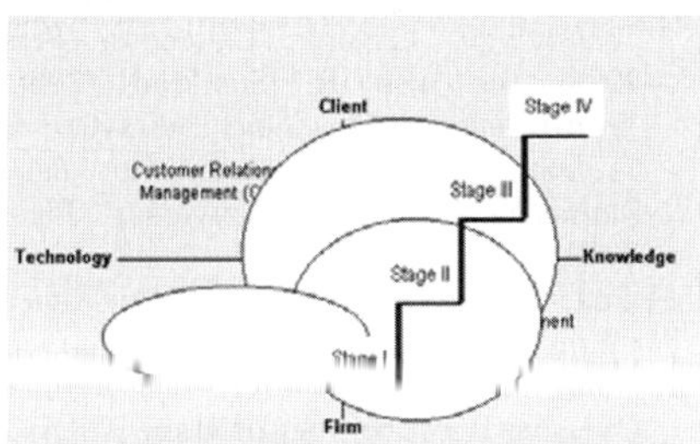


Figure 3.3: Stages of Growth Linking Knowledge Management and Legal Web Advice [20].

Again, other works on technology change and society theories are Technological Determinism (TD) and the Social Shaping of technology (SST) approach. Under SST are theories like the Social construction of technology (SCOT), actor-network theory and systems approach among other. In this paper, our focus is on how SCOT can be used as an analytical framework to look at how automated auction systems can be adopted to users' actual needs.

### 3.4 SCOT as an analysis framework for automated auction systems

The SCOT principles include: (i) the 'relevant social group', (ii) their 'interpretative flexibility", (iii) functionality of the technology and (iv) closure of technology. **'Relevant social group':** The SCOT analysis is based on the SCOT theory, which is of the view that technological changes emerge out of the process of choice and negotiations between 'relevant social groups' [15],[8], in this case the bidders, auctioneers, sellers and auction house. These social groups then interpret the technology is different ways thus attributing it different functionalities.

The **'Interpretative flexibility'** of a technology, is the way in which different social groups involved with a technology understand the technology. The e-auction systems may have been originally developed to benefit the auction house more and so the number of attributes that a good had did not matter. As more users come on board they desired a system that had multiple attributes and allowed several functionalities at the same time. Based on this user driven demand for change therefore, the CMOA has been able to move the e-auction protocol two steps further; (i) the protocol enables the determination of combinations of components and items that maximise the utility function of the system and improves output and (ii) can be used to determine the component that has the optimal utility value in any combinatorial system. This however enables CMOA automated auction protocols reach 'closure'.

'**Closure**' deals with the stage in the life of an artefact when it is considered to be 'working'. SCOT considers an artefact as '**working**'/functional only if it meets the needs for which it is required by a specific social group[2].

A critical aspect that was ignored in the development of e-auction system (i.e. eBay) is the fact that technology is a social construct and as such the same technology would mean different things to different social groups. In this analysis an automated auction system has been looked at as a social product, object or processes, which took on several meanings when experienced in everyday life by different users. A number of users would like to be able to bid for more than one item at a time and also negotiate on price simultaneously but this is restricted. A re-evaluation of this system has resulted into an alternative that according to its architect (Aloysius Edoh) is a protocol that not only increases negotiation space but also allows for flexibility in terms of price and quantity.

### 4.0 Conclusion

We have explained that E-bay like other e-auction systems use the traditional Dutch and English auctions protocol for their operations. It has been established that these can be upgraded to advanced auctions particularly CA, however these will still have limited negotiation space and flexibility. Our mathematical analysis revealed that for evaluating complementary auction, the computational space required for CA is two times higher than CMOA. This means when e-bay enhanced their auction protocol with CA instead of CMOA they will required more computational space for evaluation.

Our investigation also shows that CMOA is stable protocol as indicated in the graph. It is therefore subsumed that CMOA can be used for E-bay without any complication, however, further work is required if the number of items in the combinatorial is increased arbitrary. SCOT, as a social shaping theory, although criticised for attacking technology and trying to pull society back into to a 'mythical natural state' [13], actually offers a useful analytical framework for studying the impact of implementation of technology on society and its environment.

### References

[1] David. M. Goldschlag, Michael G. R., Paul F. Syverson (2000). The Cocaine Auction Protocol. 13. 8. eBay. http://www.ebay.com/. 9. Minneapolis, Minnesota, USA.

[2] Finnegan, R., Salaman, G. and Thompson, K. (Eds.) (1987) *Information technology: Social issue, A Reader;* London: Hodders & Stoughton

[3] Fox, R. (ed.) (1996). *Technological Change: Methods and Themes in the History of Technology*, Amsterdam: Overseas Publishers Association

[4] Friedman, B. (Ed.) (1997): *Human values and the design of computer technology*. Cambridge: Cambridge University Press

[5] Heap, N., Thomas, R., Einon, G., Mason, R. and Mackay, H. (1995) *Information technology and society*; London: Sage

[6] [6] Nisan, N. Bidding and allocation in combinatorial auctions; Electronic Commerce Proceedings of the 2nd ACM conference on Electronic commerce

[7] Shigeo Matsuba (2001). Accelerating information revelation in ascending-bid auctions: avoiding last minute bidding; Electronic Commerce; Proceedings of the 3rd ACM conference on Electronic Commerce; Tampa, Florida, USA Pages: 29 – 37

[8] Webster, F. (1995): *Theories of the information society*. London: Routledge

[9] Bunnel. D and luecke R 2000 The Ebay Phenomenon: Business Secrets Behind the World's Hottest Internet Company John Wiley & Sons Inc, September 1, 2000),

[10] [Sandholm2002] Sandholm, T. An algorithm for optimal winner determination in combinatorial auctions, *Artificial Intelligence*, vol 135, pp1-54, (2002) *and as In: Proceedings of IJCAI-99,*

[11] [David 2002] David E and Azulay-Scwartz. R An English Auction Protocol for Multi-Attribute Items, *In: AMEC – IV LNCS No 2531* (2002).

[12] [Raman1994] Raman, R and Grossman, I. E. Modelling and computational techniques for logic based integer programming, *Computers and Chemical Engineering*, vol 18, pp 563. (1994).

[13] Mackenzie, D. and Wajcman, J. (Eds.) (1999). *The Social Shaping of Technology,* 2nd edn. Buckingham: Open University Press

[14] R. Susskind (1998), *The Future of Law. Facing the challenges of Information Technology*. Oxford: Clarendon Press 1998, see also R. Susskind (2000), *Transforming the Law: Essays on Technology, Justice and the Legal Marketplace*, Oxford University Press, 2000.

[15] Fox, R. (ed.) (1996). *Technological Change: Methods and Themes in the History of Technology*, Amsterdam: Overseas Publishers Association

[16] Jeffrey Teich , Hannele Wallenius , Jyrki Wallenius "An extension of *negotiauction* to multi-attribute, multi-unit combinatorial bundle auctions" In MCDM 2004, Whistler, B. C., Canada, August 6-11, 2004

[17] Bichler, M and Kalagnanam, J.: Bidding Languages and Winner Determination in Multi-Attribute Auction, In: *IBM Research Report* 2247 (2001)

[18] Boultier C and Hoos H Solving combinatorial auctions using stochastic local search, In: *Proceedings of AAAI-00*, pp 22-29. (2000)

[19] Jian CHEN, He HUANG , "On-line Multi-attributes Procurement Combinatorial

Auctions Bidding Strategies" In proceedings ICCS 2005; 5th International Conference May 22-25 2005 Atlanta GA, USA

[20] Y Narahari Pankaj Dayama, Combinatorial auctions for electronic business *Sadhana* vol 30, part 2 & 3 April /June 2005 pp 179 -211. Printed in India

**Aloysius Edoh** of the School of Computing and Technology (SCOT), University of East London and a MPhil/PhD student at City University. He has a BSc Electrical/Electronic Engineering and MSc Information technology,. He has worked in the areas of Software Agents, Distributed Systems, Engineering and Medical Informatics. He is a Member of the British Computer Society and MIEE.

**Contact:**
School of Computing and Technology (SCOT), University of East London, Longbridge RM8 2AS
Email: ek717@soi.city.ac.uk and edoh@uel.ac.uk

# Measuring of Spectral Fractal Dimension

J. Berke
Department of Statistics and Information Technology
University of Veszprém, Georgikon Faculty of Agronomy
H-8360 Keszthely, Deák Ferenc street. 57. HUNGARY

***Abstract-There were great expectations in the 1980s in connection with the practical applications of mathematical processes which were built mainly upon fractal dimension mathematical basis. Results were achieved in the first times in several fields: examination of material structure, simulation of chaotic phenomena (earthquake, tornado), modelling real processes with the help of information technology equipment, the definition of length of rivers or riverbanks. Significant results were achieved in practical applications later in the fields of information technology, certain image processing areas, data compression, and computer classification. In the present publication the so far well known algorithms calculating fractal dimension in a simple way will be introduced as well as the new mathematical concept named by the author 'spectral fractal dimension' [8], the algorithm derived from this concept and the possibilities of their practical usage.***

## I. Introduction

In the IT-aimed research-developments of present days there are more and more processes that derive from fractals, programs containing fractal based algorithms as well as their practical results. Our topic is the introduction of ways of application of fractal dimension, together with the mathematical extension of fractal dimension, the description of a new algorithm based on the mathematical concept, and the introduction of its practical applications.

## II. The fractal dimension

Fractal dimension is a mathematical concept which belongs to fractional dimensions. Among the first mathematical descriptions of self similar formations can be found von Koch's descriptions of snowflake curves (around 1904) [18]. With the help of fractal dimension it can be defined how irregular a fractal curve is. In general, lines are one dimensioned, surfaces are two dimensioned and bodies are three dimensioned. Let us take a very irregular curve however which wanders to and from on a surface (e.g. a sheet of paper) or in the three dimension space. In practice [1-4], [8-19] we know several curves like this: the roots of plants, the branches of trees, the branching network of blood vessels in the human body, the lymphatic system, a network of roads etc. Thus, irregularity can also be considered as the extension of the concept of dimension. The dimension of an irregular curve is between 1 and 2, that of an irregular surface is between 2 and 3. The dimension of a fractal curve is a number that characterises how the distance grows between two given points of the curve while increasing resolution. That is, while the topological dimension of lines and surfaces is always 1 or 2, fractal dimension can also be in between. Real life curves and surfaces are not real fractals, they derive from processes that can form configurations only in a given measure. Thus dimension can change together with resolution. This change can help us characterize the processes that created them.

The definition of a fractal, according to Mandelbrot is as follows: A fractal is by definition a set for which the Hausdorff-Besicovitch dimension strictly exceeds the topological dimension [16].

The theoretical determination of the fractal dimension [1]: Let $(X,d)$ be a complete metric space. Let $A \in H(X)$. Let



*T. Sobh and K. Elleithy (eds.), Advances in Systems, Computing Sciences and Software Engineering*, 397–402.

$N(\varepsilon)$ denote the minimum number of balls of radius $\varepsilon$ needed to cover $A$. If

$$FD = \lim_{\varepsilon \to 0} \left\{ Sup \left\{ \frac{LnN(\bar{\varepsilon})}{Ln(1/\bar{\varepsilon})} : \bar{\varepsilon} \in (0, \varepsilon) \right\} \right\} \quad (1)$$

exists, then $FD$ is called the fractal dimension of $A$.

The general measurable definition of fractal dimension (FD) is as follows:

$$FD = \frac{\log \frac{L_2}{L_1}}{\log \frac{S_1}{S_2}} \quad (2)$$

where $L_1$ and $L_2$ are the measured length on the curve, $S_1$ and $S_2$ are the size of the used scales (that is, resolution). There have been several methods developed that are suitable for computing fractal dimension as well. [8], [18], (see Table 1).

TABLE 1

METHODS OF COMPUTING FRACTAL DIMENSIONS

| Methods | Main facts |
|---|---|
| Least Squares Approximation | Theoretical |
| Walking-Divider | practical to length |
| Box Counting | most popular |
| Prism Counting | for a one dimensional signals |
| Epsilon-Blanket | To curve |
| Perimeter-Area relationship | To classify different types images |
| Fractional Brownian Motion | similar box counting |
| Power Spectrum | digital fractal signals |
| Hybrid Methods | calculate the fractal dimension of 2D using 1D methods |

## III. Measuring fractal dimension

Fractal dimension, which can be the characteristic measurement of mainly the structure of an object in a digital image [19], [16], [4], [1], can be computed applying the Box counting as follows:

1. Segmentation of image
2. Halving the image along vertical and horizontal symmetry axis
3. Examination of valuable pixels in the box
4. Saving the number of boxes with valuable pixels
5. Repeat 2-4 until shorter side is only 1 pixel

To compute dimension, the general definition can be applied to the measured data like a function (number of valuable pixels in proportion to the total number of boxes).

## IV. Spectral fractal dimension

Nearly all of the methods in Table 1 measure structure. Neither the methods in Table 1 nor the definition and process described above gives (enough) information on the (fractal) characteristics of colours, or shades of colours. Fig. 1 bellow gives an example.

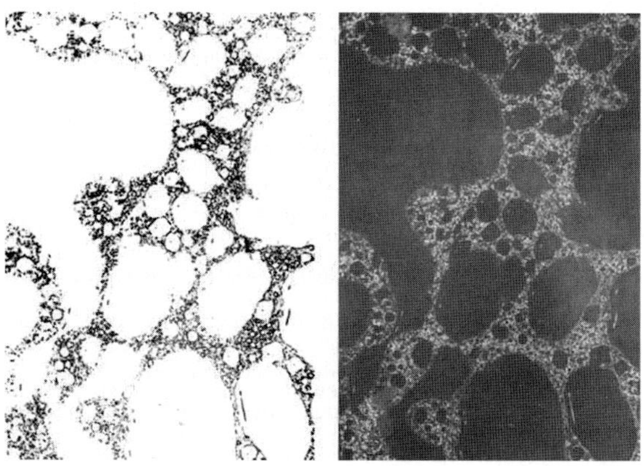

**Fig. 1** The fractal dimension measured with the help of the Box counting of the two images is the same (FD=1.99), although the two images are different in shades of colour.

Measuring with the help of the Box counting the image on the right and the one on the left have the same fractal dimension (FD=1.99) although the one on the left is a black and white (8 bit) image whereas the other one on the right is a 24-bit coloured image containing shades as well - the original images can be found at www.georgikon.hu/digkep/sfd/index.htm, [20] -, so there is obviously a significant difference in the information they contain. How could the difference between the two images be proven using measurement on the digital images?

Let spectral fractal dimension (SFD) be:

$$SFD = \frac{\log \frac{L_{S2}}{L_{S1}}}{\log \frac{S_{S1}}{S_{S2}}} \quad (3)$$

where $L_{S1}$ and $L_{S2}$ are measured spectral length on N-dimension colour space, $S_{S1}$ and $S_{S2}$ are spectral metrics (spectral resolution of the image).

In practice, N={1, 3, 4, 6, 32, 60, 79, 126}, where

- N=1 black and white or greyscale image,
- N=3 RGB, YCC, HSB, IHS colour space image,
- N=4 traditional colour printer CMYK space image
- N=6 photo printer $CC_pMM_pYK$ space image or Landsat TM, Landsat ETM satellite images
- N=32 DAIS7915 VIS_NIR or DAIS7915 SWIP-2 sensors
- N=60 COIS VNIR sensor
- N=79 DAIS7915 all
- N=126 HyMap sensor

In practice the measure of spectral resolution can be equalled with the information theory concept of {$S_i$=1, …, $S_i$=16, where i=1 or i=2} bits.

Typical spectral resolution: (4)

- Threshold image -1 bit
- Greyscale image - 2-16 bits
- Colour image - 8-16 bits/bands

On this basis, spectral computing is as follows:

1. Identify which colour space the digital image is
2. Establish spectral histogram in the above space
3. Half the image as spectral axis
4. Examine valuable pixels in the given N-dimension space part (N-dimension spectral box)
5. Save the number of the spectral boxes that contain valuable pixels
6. Repeat steps 3-5 until one (the shortest) spectral side is only one (bit).

In order to compute dimension (more than two image layers or bands and equal to spectral resolution), the definition of spectral fractal dimension can be applied to the measured data like a function (number of valuable spectral boxes in proportion to the whole number of boxes), computing with simple mathematical average as follows:

$$SFD_{measured} = \frac{n \times \sum_{j=1}^{S-1} \frac{\log(BM_j)}{\log(BT_j)}}{S-1} \quad (5)$$

where

- n – number of image layers or bands
- S - spectral resolution of the layer, in bits – see Eq. 4
- $BM_j$ - number of spectral boxes containing valuable pixels in case of j-bits
- $BT_j$ - total number of possible spectral boxes in case of j-bits

The number of possible spectral boxes ($BT_j$) in case of j-bits as follows:

$$BT_j = (2^S)^n \quad (6)$$

With Eqs. (5) and (6) the general measurable definition of spectral fractal dimension is as follows, if the spectral resolution is equal to all bands (SFD $_{\text{Equal Spectral Resolution}}$ – $SFD_{ESR}$):

$$SFD_{ESR} = \frac{n \times \sum_{j=1}^{S-1} \frac{\log(BM_j)}{\log((2^S)^n)}}{S-1} \quad (7)$$

If the spectral resolution is different in bands/layers, the general measurable definition of spectral fractal dimension (SFD $_{\text{Different Spectral Resolution}}$ – $SFD_{DSR}$) is as follows:

$$SFD_{DSR} = \frac{n \times \sum_{j=1}^{(\min(S_i))-1} \frac{\log(BM_j)}{\log(2^{(\sum_{k=1}^{n} S_k)})}}{(\min(S_i))-1} \quad (8)$$

where,

- $S_i$ - spectral resolution of the layer i, in bits

During computing:

1. Establish the logarithm of the ratio of BM/BT to each spectral halving
2. Multiply the gained values with *n* (number of image layers or bands)
3. Find the mathematical average of the previously gained values

On this basis, the measured SFD of the two images introduced in figure 1 show an unambiguous difference ($SFD_{\text{left side image}}$=1.21, $SFD_{\text{right side image}}$=2.49).

### V Practical Application of the Algorithm

A computer program that measures SFD parameter has been developed in order to apply the algorithm above. The measuring program built on this method has been developed in two environments (MS .NET, C++). According to the measurements so far it can be stated that there is no significant difference between the computing times of the algorithm running in either environment (in case of P4, 2,6GHz, 512 MB RAM it is nearly around half a minute in both cases a 3 Mpixels image). This means that the algorithm can be applied well in object oriented programming environments as well.

Practical testing has been completed using images where the expected (theoretical) measurement results are unambiguous – see Table 2.

TABLE 2

SFD MEASUREMENT RESULTS OF TEST IMAGES

| Type of images | Theoretical SFD | Measured SFD |
|---|---|---|
| 3 image bands, intensity of every pixel is zero - black coloured 3-bands image | 0 | 0 |
| 3 image bands, intensity of every pixel is 255 – white-coloured 3 bands image | 0 | 0 |
| 3 image bands, intensity of every pixel is the same, different from zero and 255 - one colour 3-bands image) | 0 | 0 |

The SFD results measured by the program are invariant for identical scale pixels with different geometric positions in case the number of certain scales is the same and shade of colour is constant.

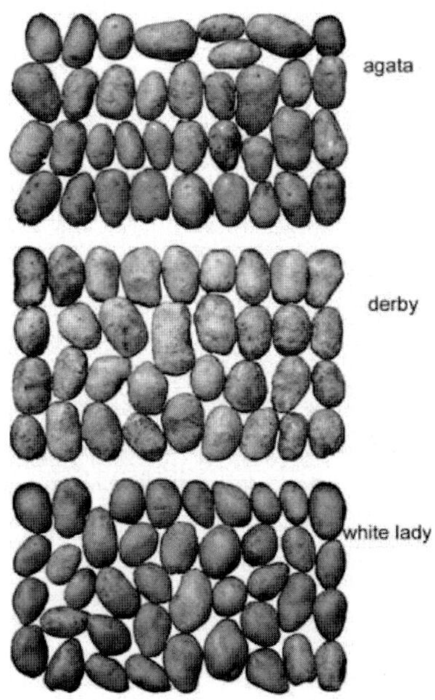


**Fig. 2** Qualification of potato by SFD

Successful practical application of SFD at present [4], [5], [6], [7], [8], [9]:

- Measurement of spectral characteristics of satellite images
- Psychovisual examination of image compressing methods
- Qualification of potato seeds /see Fig. 2/ and chips[1]
- Temporal examination of damage of plant parts
- Classification of natural objects /see Fig. 3/
- Resistance of potato by multispectral and multitemporal images [10], [14], /see Fig. 4/
- Virtual Reality based 3D terrain simulation

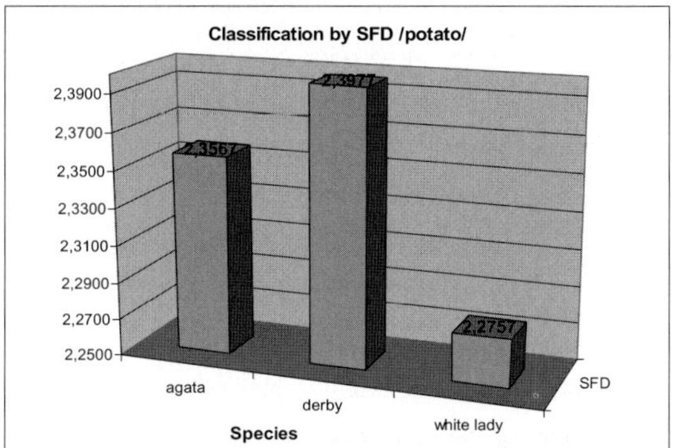


| | agata | derby | white lady |
|---|---|---|---|
| SFD | 2,3567 | 2,3977 | 2,2757 |

**Fig. 3** SFD based classification of potato seeds by visual images

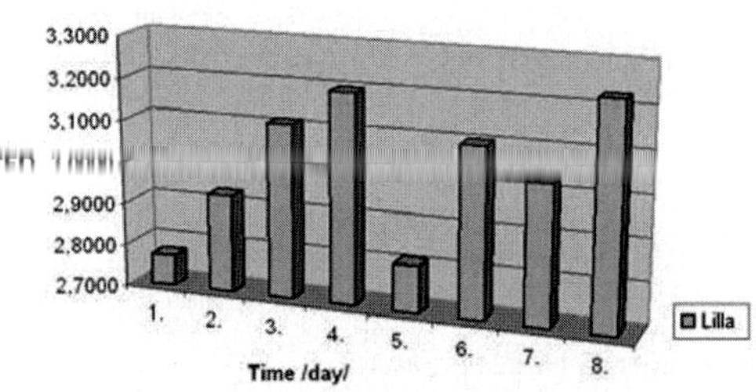


**Fig. 4** Resistance of potato by multispectral (visual - 3 layers, infra – 1 layer and thermal – 1 layer) and multitemporal (1-8 days) images

[1] This work has been supported by the National Office for Research and Technology, IKTA ref. No.: 00101/2003

## VI Conclusion

When examining digital images where scales can be of great importance (image compressing, psychovisual examinations, printing, chromatic examinations etc.) SFD is suggested to be taken among the so far usual (eg. sign/noise, intensity, size, resolution) types of parameters (eg. compression, general characterization of images). Useful information on structure as well as shades can be obtained applying the two parameters together [8], /see Fig. 5/.

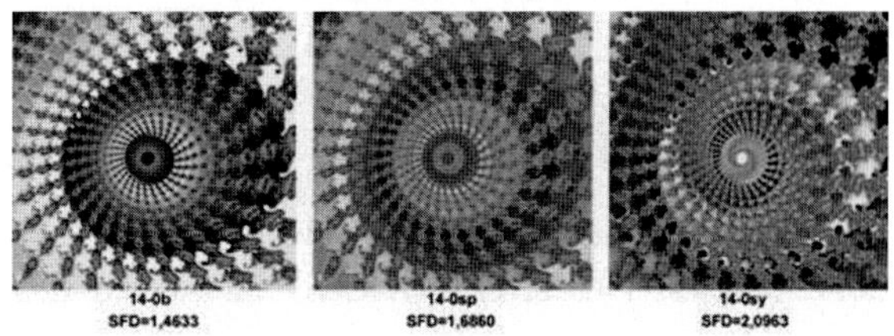


**Fig. 5** SFD of different colourized fractal

/Mandelbrot set: Re: -0,74745 … -0,74637, Im: 0,10671 … 0,10779/

Several basic image data (aerial and space photographs) consisting of more than three bands are being used in practice. There are hardly any accepted parameters to characterize them together. I think SFD can perfectly be used to characterize (multi-, hyper spectral) images that consist of more than three bands /see Fig. 6/.

On the basis of present and previous measurements it can be stated that SFD and FD are significant parameters in the classification of digital images as well.

SFD can be an important and digitally easily measurable parameter of natural processes and spatial structures besides the structural parameters used so far. These measurements are being carried out at present already - using the above method applicable in practice - and are to be accessed by anyone [20].

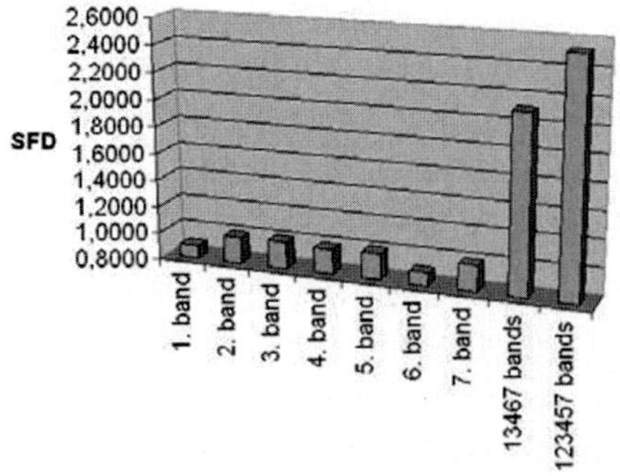


| 1. band | 2. band | 3. band | 4. band | 5. band | 6. band | 7. band | 13467 bands | 123457 bands |
|---|---|---|---|---|---|---|---|---|
| 0,8943 | 0,9934 | 0,9976 | 0,9847 | 0,9951 | 0,8943 | 0,9934 | 2,1099 | 2,5288 |

**Fig. 6** SFD of Landsat TM images

The applied method has proven that with certain generalization of the Box method fractal dimension based measurements – choosing appropriate measures- give practically applicable results in case of optional number of dimension.

### VII References

[1] Barnsley, M. F., *Fractals everywhere*, Academic Press, 1998.

[2] Barnsley, M. F. and Hurd, L. P., *Fractal image compression*, AK Peters, Ltd., Wellesley, Massachusetts, 1993.

[3] Batty, M. and Longley, P. *Fractal cities*, Academic Press, 1994.

[4] Berke, J., Fractal dimension on image processing, *4th KEPAF Conference on Image Analysis and Pattern Recognition*, Vol.4, 2004, pp.20.

[5] Berke, J., Real 3D terrain simulation in agriculture, *1st Central European International Multimedia and Virtual Reality Conference*, Vol.1, 2004, pp.195-201.

[6] Berke, J., The Structure of dimensions: A revolution of dimensions (classical and fractal) in education and science, *5th International Conference for History of Science in Science Education*, July 12 – 16, 2004.

[7] Berke, J. and Busznyák, J., Psychovisual Comparison of Image Compressing Methods for Multifunctional Development under Laboratory Circumstances, *WSEAS Transactions on Communications*, Vol.3, 2004, pp.161-166.

[8] Berke, J., Spectral fractal dimension, *Proceedings of the 7th WSEAS Telecommunications and Informatics (TELE-INFO '05)*, Prague, 2005, pp.23-26, ISBN 960 8457 11 4.

[9] Berke, J. – Wolf, I. – Polgar, Zs., Development of an image processing method for the evaluation of detached leaf tests, *Eucablight Annual General Meeting*, 24-28 October, 2004.

[10] Berke, J. – Fischl, G. – Polgár, Zs. – Dongó, A., Developing exact quality and classification system to plant improvement and phytopathology, *Proceedings of the Information Technology in Higher Education*, Debrecen, August 24-26, 2005.

[11] Burrough, P.A., Fractal dimensions of landscapes and other environmental data, *Nature*, Vol.294, 1981, pp. 240-242.

[12] Buttenfield, B., Treatment of the cartographic line, *Cartographica*, Vol.22, 1985, pp.1-26.

[13] Encarnacao, J. L. – Peitgen, H.-O. – Sakas, G. – Englert, G. *eds. Fractal geometry and computer graphics*, Springer-Verlag, Berlin Heidelberg 1992.

[14] Horváth, Z. – Hegedűs, G. – Nagy, S. – Berke, J. - Csák, M., *Fajtaspecifikus kutatási integrált informatikai rendszer*, Conference on MAGISZ, Debrecen, August 23, 2005.

[15] Lovejoy, S., Area-perimeter relation for rain and cloud areas, *Science*, Vol.216, 1982, pp.185-187.

[16] Mandelbrot, B. B., *The fractal geometry of nature*, W.H. Freeman and Company, New York, 1983.

[17] Peitgen, H-O. and Saupe, D. *eds. The Science of fractal images*, Springer-Verlag, New York, 1988.

[18] Turner, M. T., - Blackledge, J. M. – Andrews, P. R., *Fractal Geometry in Digital Imaging*, Academic Press, 1998.

[19] Voss, R., *Random fractals: Characterisation and measurement*, Plenum, New York, 1985.

[20] Authors Internet site of parameter SFD - www.georgikon.hu/digkep/sfd/index.htm.

# Design and Implementation of an Adaptive LMS-based Parallel System for Noise Cancellation

Kevin S. Biswas
Department of Electrical and Computer Engineering
University of Windsor
Windsor, Ontario, Canada
biswas4@uwindsor.ca

Jason G. Tong
Department of Electrical and Computer Engineering
University of Windsor
Windsor, Ontario, Canada
tong4@uwindsor.ca

***Abstract*— When a desired signal is encompassed by a noisy environment, active noise cancellation may be implemented to remove the background noise. The presented algorithm is based on the standard Least Mean Squares (LMS) algorithm developed by Bernard Widrow. Modifications to the LMS algorithm were made in order to optimize its performance in extracting a desired speech signal from a noisy environment. The system consists of two adaptive systems running in parallel, with one having a much higher convergence rate to provide rapid adaptation in a non-stationary environment. However, the output of the higher converging system results in distorted speech. Therefore, the second system, which runs at a lower convergence rate but regularly has its coefficients updated by the first system, provides the actual output of the desired signal. All of the algorithm development and simulation were initially performed in Matlab, and were then implemented on TMS320C6416 Digital Signal Processor (DSP) evaluation board to produce a real-time, noise-reduced speech signal.**

***Keywords*— *active noise cancellation, adaptive, signal processing, LMS, least-mean-square, interference cancelling, DSP***

## I. Introduction

Bernard Widrow has defined an adaptive system as a system whose structure is alterable or adjustable in such a way that its behaviour or performance (according to some desired criterion) improves through contact with its environment [1].

The most suitable system for our application is the interference canceling adaptive concept. The system can be represented in block diagram in figure 1.

Here the desired signal, *s*, is corrupted by additive noise, *n*, and a distorted but correlated version of the noise, *n'*, is also available.

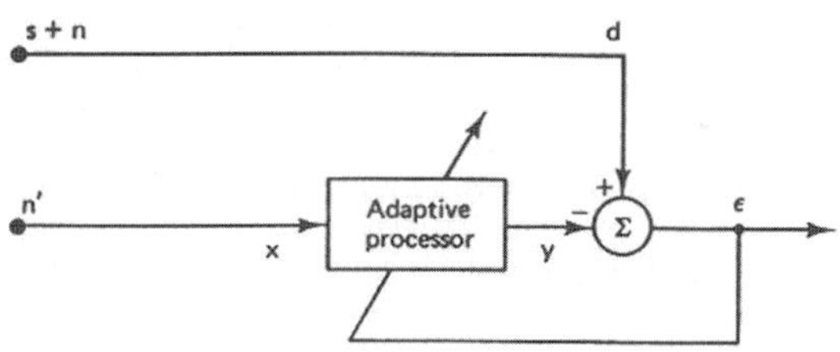


Figure 1: General Interference Cancelling Adaptive System

The goal of the adaptive system is to produce an output *y*, that resembles *n* as closely as possible, in order for the overall output of the system, *e*, to resemble *s*. It can be mathematically shown that the optimal filter settings occur when the mean square value, or signal power, of the system output is minimized.

The adaptive filter is simply an adaptive linear combiner (also known as non-recursive adaptive filter), whose weights are time-varying. Much of the design process involves determining a method for the adaptive filter to reach the optimum set of weights accurately and as fast as possible. It will be shown that there is a tradeoff between speed and accuracy in terms of the quality of the output signal. The algorithm used for descending towards the optimum weight vector, is Bernard Widrow's LMS method [1], where:

$$\mathbf{W}_{k+1} = \mathbf{W}_k + 2\mu e_k \mathbf{X}_k. \qquad (1)$$

$\mathbf{W}_{k+1}$ is the next weight vector, $\mathbf{W}_k$ is the current weight vector, $\mu$ is the convergence factor, $e_k$, is the output error of the system, and $\mathbf{X}_k$ is the correlated reference noise input to the adaptive filter. This formula allows for easy computation and efficient gradient estimation [1].

## II. Effect of Convergence Factor in a Basic Adaptive LMS System

Results from a basic adaptive noise cancellation system, which was first implemented in Matlab, was analyzed. In preliminary tests, the effect of changing the convergence factor $\mu$ was investigated.

Five second WAVE files with sampling rate 44.1 kHz (220501 samples) were used. A suitable periodic noise was used for the background noise. A voice signal was used for the desired signal. Waveforms of the input signal plus noise and the overall system output were compared. The order of the system was kept constant for each run. A relatively small value of $\mu$ was first used ($\mu$ = 0.01). The following results were generated using this value. The large peaks represent the desired signal, while smaller peaks are created by the background noise. Figures 2 displays the desired signal mixed with noise and figure 3 shows the output of the simple LMS algorithm. These graphs are generated using MATLAB.



*T. Sobh and K. Elleithy (eds.), Advances in Systems, Computing Sciences and Software Engineering*, 403–409.

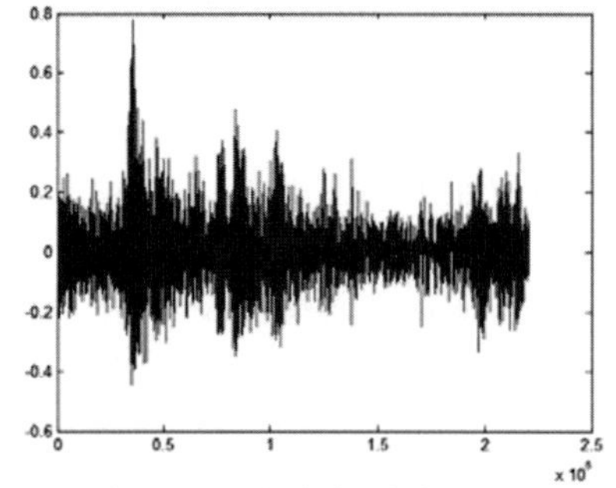

Figure 2: Desired Signal Plus Noise

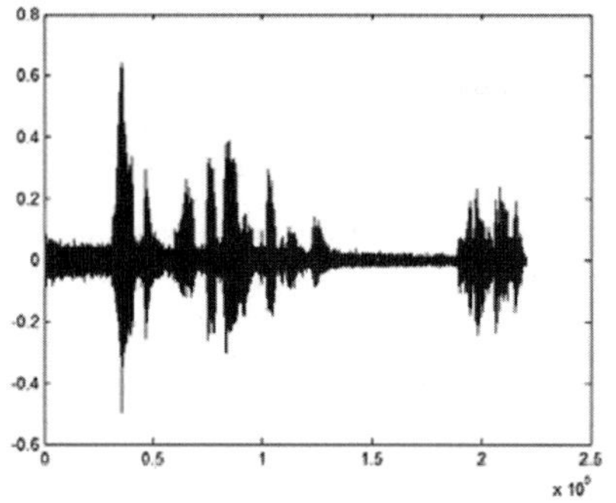

Figure 3: System Output (After Processing) $\mu$ =0.01

The above waveforms show that noise was indeed attenuated. It took quite some time for the adaptive filter to reach its optimum weights, due to the low value of $\mu$. However, the quality of the desired signal was essentially retained, indicating that the optimum weight vector was accurately reached (i.e. $n_0$ very close to $y$). Because of the small value of $\mu$, the optimum weights were reached in a relatively non-oscillatory fashion, which resulted in a good sounding desired signal. But once again, the process was slow, and much of the background noise was heard at the output at the beginning of the run.

Significant noise attenuation was apparent almost instantly, when using $\mu = 0.5$ (figure 4). However, the desired signal was obviously distorted. The speed at which the filter approaches the optimum weight vector is much faster than the previous example. Once this happens, however, the large value of $\mu$ causes the system weights to fluctuate dramatically about the optimal point of the system's performance surface. Unlike the previous example, the optimal weights are not accurately reached, resulting in an adaptive filter output which causes distortion to the overall system output (i.e. $y$ does not stay very close to $n_0$). The large fluctuation in weights actually causes the desired signal to exhibit a time-varying characteristic.

It is evident from these preliminary simulation results, that the basic LMS system has difficulty in accomplishing noise cancellation effectively. Noise cancellation can be done quickly with a high convergence factor, but at the expense of

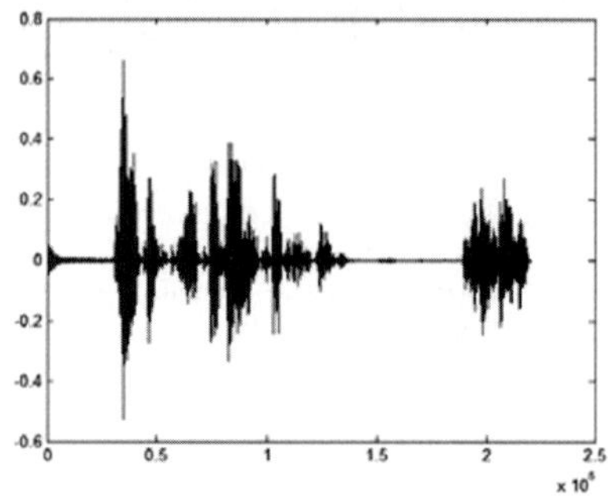

Figure 4: System Output (After Processing) $\mu = 0.5$

poor quality of the desired signal. Reducing the convergence factor produces an almost perfect desired signal at the output, but the system requires a relatively longer period of time to achieve this. This suggests that a system with a variable convergence factor may work better in achieving design requirements.

## III. Development of Parallel System Algorithm

As previously discussed, initial simulation results suggested that it was difficult to achieve a high quality output signal using the standard LMS algorithm in a noise cancelling system. The two most important aspects of the output signal, which are the ability to achieve a high degree of noise cancellation and the minimization of the distortion to the speaker's voice, were two divergent requirements. Even after optimizing the performance of the standard LMS algorithm in a noise cancelling environment, it is impossible to fully satisfy both of the requirements. The above results lead to the development of a system which relies on some of the fundamental aspects of the LMS algorithm, but whose performance is not limited by the chosen value of the convergence parameter. In other words, our motivations were to simultaneously achieve a fast convergence to the optimal weights, while minimizing distortion to the output speech signal. Ideally, a high convergence value should be used in the absence of the desired signal (i.e. pauses between words), and a low convergence factor when the desired signal is present (i.e. talking). However, detection for the presence of speech proved to be an inefficient and difficult task.

A significant observation was made over the course of the study. It was observed that when the system maintained a consistently high value of the convergence parameter, the pauses between words were most obvious at the output of the system. This can easily be explained by the fact that the system was constantly attempting to adapt to the noise signal, and therefore achieved a very high level of cancellation during the pauses between words. Even when the noise was not stationary, the system quickly adapted to the noise signal and produced a minimal output when the speaker was not talking. This can be observed in the figure below, which shows the output of the system with a large convergence parameter. Notice that between the waveforms of each word, there is a significant number of samples where the output is almost zero.

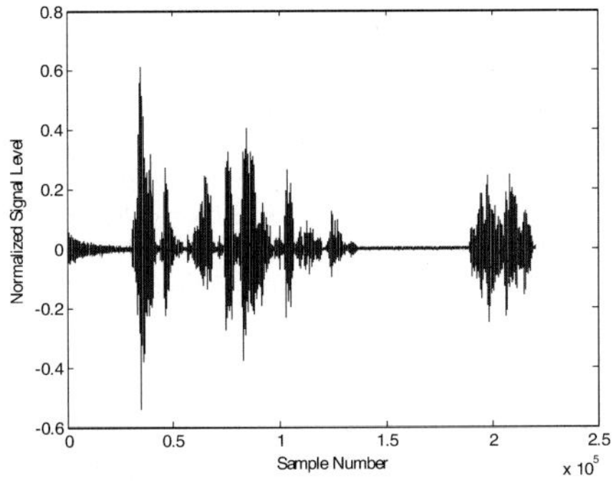


Figure 5: Fast Convergence System Output

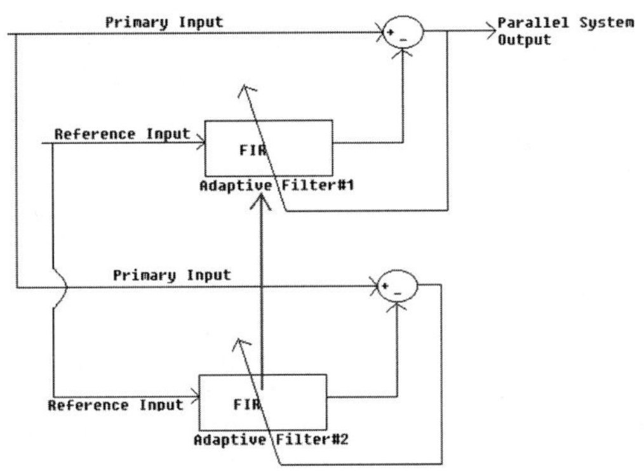


Figure 6: Block Diagram of the Parallel System

Due to the continuous adaptation of a system that maintains a large value of the convergence parameter, the concept of the parallel system was developed (figure 6). The parallel system is based on the premise that two noise cancelling systems which implement optimized LMS algorithms could operate in parallel on the input signals. One of the systems would be used exclusively to provide the processed speech signal, while the other system would be entirely in charge of adapting to any changes in the input noise statistics. The first adaptive system, which provides the desired output, would periodically receive updated weights from the second adaptive system. In addition, the first adaptive system operates with a very small convergence parameter, in order to prevent it from introducing distortion to the speech signal. The second adaptive system operates at a very high value of the convergence parameter, in order to quickly calculate the optimal weights, which are then passed back to the first system. Figure shows the block diagram of the parallel system. Note that the red arrow represents the periodic transfer of the weights from the second adaptive system to the first adaptive system.

*A. Determination of Optimal Weight Transfer Period*

As previously discussed, the second adaptive system (which operates at a high convergence rate), periodically provides a copy of its weights to the first adaptive system. It is important that this occurs rapidly enough so that when the input noise statistics change, the noise is not present at the output of the first adaptive system for a significant period of time. Otherwise, the quality of the processed speech signal would quickly degrade. However, there is also a bound on the minimum amount of time that can be used as the period of the weight transfers. This minimum bound is due to the statistical nature of the LMS algorithm, which requires a minimum amount of data before a significant result can be obtained. This is due to the noisy estimate of the gradient vector that is used to estimate the direction of the optimal weights. In addition, it was found that if the weights of the second adaptive system are transferred to the first system too frequently, a time-varying distortion is introduced to the speech signal. The characteristics of the distortion are very similar to those of the distortion produced at the output of the previously described single adaptive systems, when they are operated with a constantly high value of the convergence parameter.

In order to determine the optimal time period between weights transfers, simulations were carried out in Matlab. These simulations were conducted with recorded signals containing various types of noise sources, and at various sampling frequencies. The outputs of the parallel system were compared in terms of the quality and speed of the noise cancellation process, as well as the level of distortion introduced to the speech signal. In order to detail the results of the Matlab simulation, the following terminology must be introduced. A 'frame' refers to a block of audio data of a specific size, which will be used as the fundamental unit of processing in the algorithms. In other words, whenever a new frame of data is available, the algorithm processes the frame. Obviously there is a direct relationship between frame size and the period between weight transfers, and therefore the results of the simulation will be described in terms of the optimal frame size. Table 1 presents the optimal frame sizes at various frequencies.

The selected frame sizes were produced by averaging the results from many simulations involving different noise sources. There is an obvious relationship between the optimal frame size and the sampling frequency. In general, in order to provide rapid adaptation to changing noise characteristics while minimizing distortion, the frame size should be such that the weights of the first adaptive system are updated approximately 8 times per second.

*B. Determination of Convergence Coefficients*

There exists a bound on the maximum value of the convergence parameter for an adaptive system. This is due to the feedback of the error signal to the adaptive filter. Even though the filter itself is a necessarily stable FIR filter, the feedback causes the entire system to be recursive in nature, thus the potential for instability exists.

TABLE 1
OPTIMAL FRAME SIZES FOR VARIOUS SAMPLING FREQUENCIES

| | Sampling Frequency (kHz) | | | |
|---|---|---|---|---|
| | 8 | 16 | 32 | 44.1 |
| Optimal Frame Size (in Samples) | 1000 | 2000 | 4000 | 5000 |

However, it is desirable that our second adaptive system have as high a value of the convergence parameter as possible. Therefore, it is necessary to dynamically calculate the upper bound of the convergence parameter. The upper bound is dependent on the order of the adaptive filter, which does not vary with time, but also the potentially time-varying input signal power. The value of the convergence coefficient for the second adaptive system is calculated each time a new frame of data is available. This coefficient then remains constant during the processing of the current frame of data. In order to quickly adapt during a limited period of time, the coefficient is chosen such that it is maximized, while still guaranteeing stability. The power of the input signal will be upper-bounded by selecting the sample with the largest normalized magnitude in the reference frame's current set of data. This value, which has an absolute value between zero and one, is squared in order to generate the maximum input signal power. While this approach may seem to grossly overestimate the power of the input signal, it was observed that a very small number of samples whose power is above the calculated signal power can quickly lead to instability. If 'P' represents the estimate of the signal power, and 'L' represents the number of filter weights, the value of the convergence coefficient is calculated as follows [2]:

$$\textit{convergence coefficient}\ (\mu) = 1/(L{\cdot}P) \qquad (2)$$

With regards to the first adaptive system, a less mathematical approach was used to determine the optimal convergence coefficient. Based on simulations conducted in Matlab, it was found that a convergence coefficient of 0.005 or less did not produce a noticeable distortion in the output speech signal. As well, it allowed adaptation to occur, albeit slowly. It could easily be argued that the speed of adaptation at this low value is insignificant, due to the frequent set of new weights provided by the second adaptive system. In fact, it is possible to achieve good performance by assigning a value of zero to the convergence coefficient of the first system, which decreases the number of operations of the system. The sequence of operations of the parallel system algorithm is shown in figure 7.

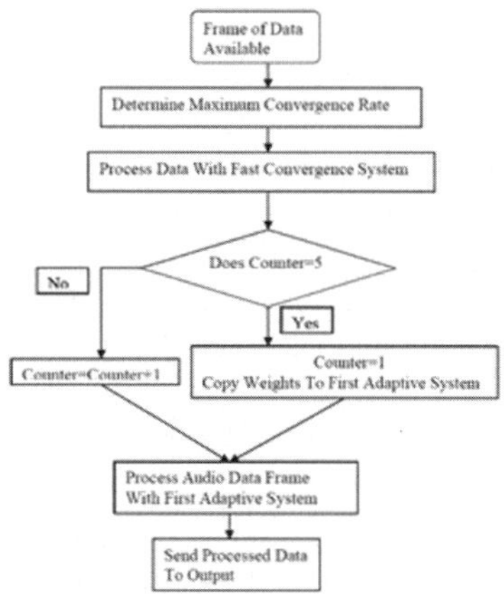


Figure 7: Sequence of Operations in Parallel System Algorithm

## IV. EXPERIMENTAL RESULTS

The performance of the system can be readily evaluated by comparing the original speech signal that was contaminated with noise, to the cleaned speech signal that is output from the adaptive noise cancellation system. Simulation was done in Matlab. Comparison was most naturally done by simply listening to the audio signals. However, by examining the waveforms of these two signals, as well as their power spectrums, a more mathematical description of the results was also obtained.

The first significant result is that our designed parallel system algorithm provides faster adaptation than a simple implementation of an LMS adaptive system. In order to fairly compare the two systems, the convergence coefficient of the simple LMS adaptive system was chosen to match the convergence coefficient of the slowly converging filter in the parallel system algorithm. This ensured that both systems contributed similar distortion to the output speech signal, since this distortion effect is a very subtle time-domain distortion of the waves, but that significantly affects output quality. Unfortunately, this distortion is easily detected by human hearing, but it is not easy to detect graphically.

For the following results, a value of 0.005 was assumed for the convergence coefficient of the simple LMS system as well as the slowly converging system in the parallel algorithm. As well, a filter order of 20 was assumed for each adaptive filter, and a sampling rate of 8 kHz for the audio signals.

The following graphs (figures 8-10) represent the results of the parallel algorithm adaptive system and a simple LMS adaptive system, for the case of an interfering sinusoid whose frequency undergoes changes at different intervals in time. While a single sinusoid may seem to be a trivial implementation of noise cancellation, it provides a very clear comparison of the convergence rate of different adaptive systems.

Note that adaptation is required whenever the frequency of the sinusoid undergoes an instantaneous change. The above graphs clearly show the increase in speed of adaptation due to the parallel system algorithm. Notice that the first second consists entirely of noise, and therefore a comparison of the output signal levels from both algorithms provides a measure of adaptation speed. A detailed examination of the graphs provides the following results:

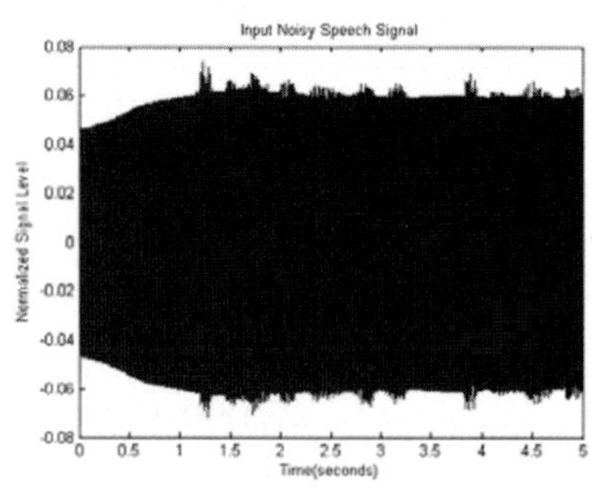


Figure 8: Input Noisy Speech Signal – Interfering Tones

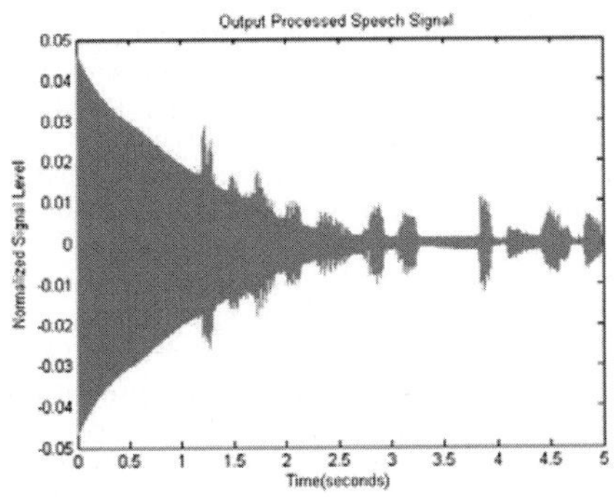


Figure 9: Simple LMS Adaptive System

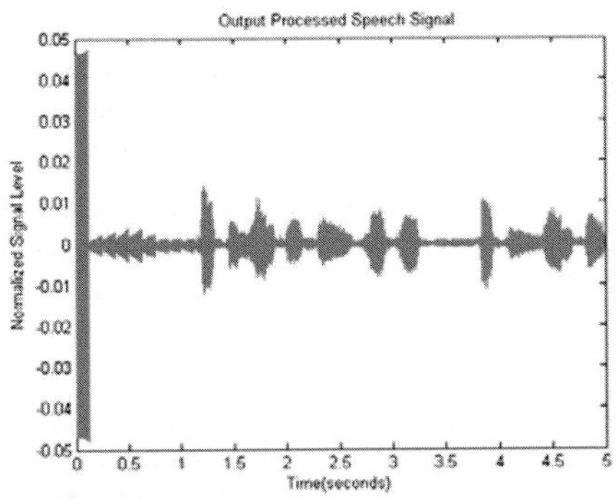


Figure 10: Parallel Algorithm Adaptive System

In addition to the speed of adaptation, a second important performance measure is a comparison of the signal-to-noise ratio(SNR) in the original signal, to the SNR in the output signal. The SNR is difficult to measure during adaptation, since the noise level decreases rapidly, and is only measurable when a speech signal is present. In order to provide a fair comparison between the original noisy signal and the output signal, the peak signal power and the peak noise level will be evaluated at several distinct instances of time, after adaptation is complete

TABLE 2
NORMALIZED PEAK OUTPUT SIGNAL LEVEL

| | Normalized Peak Output Signal Level | | | | | |
|---|---|---|---|---|---|---|
| | t=0.125s | t=0.250s | t=0.375s | t=0.5s | t=0.625s | t=0.75s |
| **Simple LMS** | 0.04 | 0.035 | 0.032 | 0.0295 | 0.02725 | 0.0242 |
| **Parallel** | 0.00065 | 0.0009 | 0.0012 | 0.0017 | 0.00096 | 0.00058 |

TABLE 3
SIGNAL-TO-NOISE RATIO

| | Normalized Signal Level | | |
|---|---|---|---|
| **Normalized Signal Level** | **t=1.25s** | **t=2.85s** | **t=3.9s** |
| Speech Signal | 0.012 | 0.006 | 0.009 |
| Input Noise | 0.06 | 0.06 | 0.06 |
| Output Noise | 0.001 | 0.0006 | 0.0007 |
| Output SNR (dB) | 21.58 | 20 | 22.18 |
| Input SNR (dB) | -13.98 | -20 | -16.48 |

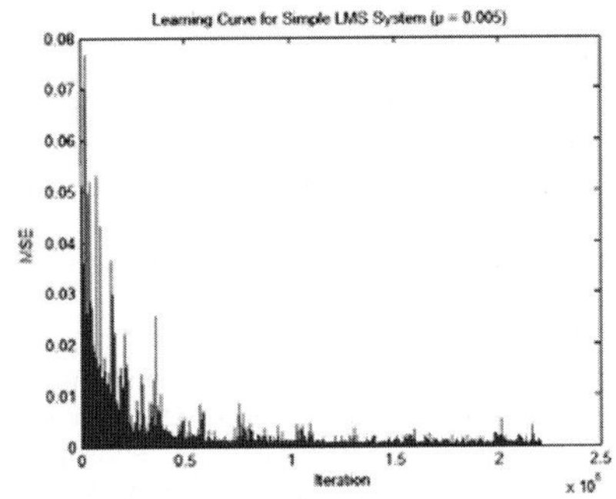


Figure 11: Learning Curve for Simple LMS System

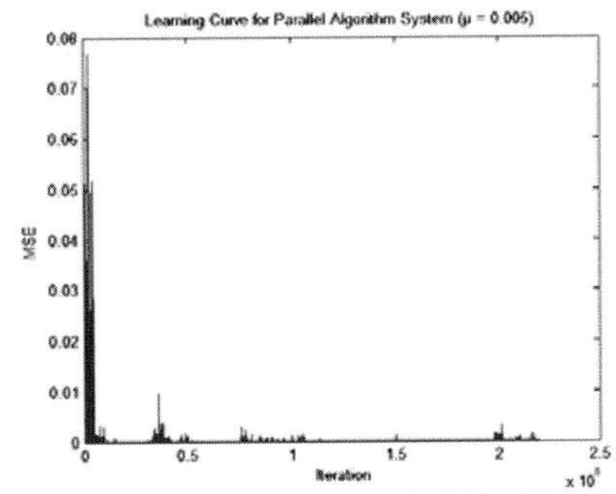


Figure 12: Learning Curve for Parallel Algorithm System

As can be seen from the tables 2 and 3, the SNR is increased by approximately 30-40dB, which allows an originally unintelligible signal to be made intelligible.

Mean square error of the system output can also express system performance. Minimization of this value results in a minimized difference between adaptive filter output and the corrupting noise. The system learning curves shown in figures 11 and 12 clearly show that speed and quality of adaptation of the parallel system is superior to a simple LMS noise cancellation system.

## V. IMPLEMENTATION WITH TMS320C6416 DSP CONTROLLER

The designed noise cancellation algorithm was implemented with Texas Instruments TMS320C6416 DSP controller. It contains the DSP microprocessor operating at a frequency of 600MHz, AIC23 Stereo CODEC, 16MB of synchronous RAM, 512 KB of non-volatile Flash memory and the Code Composer Studio (CCS) Software Development Kit (SDK) that acts as an interface with the board. The algorithm was programmed into the CCS using C programming language and was downloaded into the DSP board. In the SDK package contained the "Reference Framework 3" (RF3) which was used as the main driving force of the project. It was significantly modified to suit the needs of the parallel system. Since the RF3 framework is beyond the scope of explanation for this paper, only the major components and software functions will be shown.

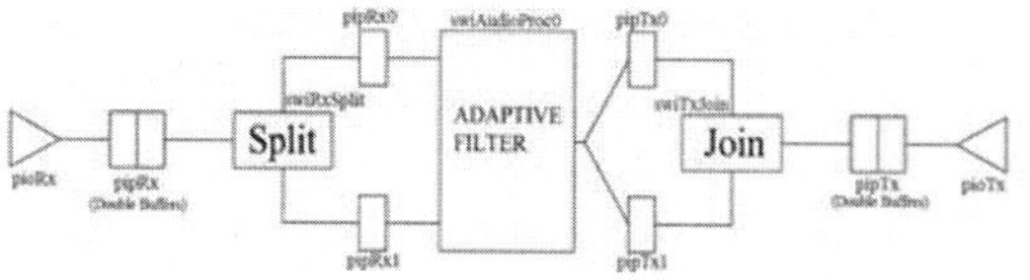


Figure 13: Modified Reference Framework

### A. *Reference Framework 3*

The original configuration of the RF3 architecture supplied by Texas Instruments had to be modified in order to be suitable for our designed algorithm. In addition to inserting the parallel algorithm into the processing functions, the overall structure of the application had to be modified to meet the requirements of an adaptive noise cancellation system. The adaptive system must accept 2 input audio signals, and provide a single mono audio output signal. This is accomplished with the line-in jack of the DSP's CODEC, which allows stereo input, and the mono audio output is paralleled with the stereo output jack of the CODEC. Figure 13 shows the modified reference framework that was used to implement the designed algorithm.

### B. *Operation of Modified Reference Framework*

The correct operation of the system in figure 13 relies on a system of software interrupts, mailboxes, and pipe sizes, which are specific to our application. From the diagram above, it can be seen that pioRx is the first element in the processing sequence. It represents a pipe that maintains a handle to a device external to the DSP, and in this case it is the stereo CODEC. The pioRx stores the left and right audio channel sample values, and transfers a block of data to the pipRx pipe when a full frame of data is available.

The selected frame size for operation at 8 kHz was 200 sampled audio values. The sampled audio data from the CODEC is quantized into a 16-bit integer value, which is half the size of the DSP's word size (32-bit). Therefore, in order to store 200 sample values, the frame size must be 100 words. However, the pipRx and pipTx pipes must simultaneously store both the left and right channel values, and therefore their frame sizes must be twice the frame size of pipRx0, pipRx1, pipTx0, pipTx1, which store the sample values for a single channel. In addition, the pipRx and pipTx are ping-pong buffers, since they consist of two separate pipes in series. A ping-pong buffer allows data to be simultaneously read from one pipe while the other pipe receives data from its writer [3]. This is necessary, since the audio data is constantly being sampled, and any loss of audio data would be fatal to the application.

When a full frame of audio data is available in the pipRx pipe, and both pipRx1 and pipRx0 are empty, the swiRxSplit software interrupt is called and the Split function is called. This function separates the left and right channel data, and stores the results in pipRx0 and pipRx1 respectively. The samples in pipRx0 represent the noisy speech signal, and the samples in pipRx1 represent the reference noise signal. When both pipRx1 and pipRx0 contain a full frame of data, and both pipTx0 and pipTx1 are empty, the adaptive filter function is called by the swiAudioproc0 software interrupt. This is the core of the application, since it contains the parallel system algorithm that was originally designed and simulated in Matlab. The designed algorithm was easily converted to C code from the original Matlab files, and is identical in functionality to the Matlab algorithms. When the algorithm completes its processing of the current frames of audio data, the output of the system is written to the pipTx0 and pipTx1 pipes.

When both pipTx1 and pipTx0 contain a full frame of data, and the pipTx output pipe contains an empty frame of data, the swiTxJoin software interrupt is called and the Join function is called. This combines the identical contents of pipTx1 and pipTx0 into a stereo output signal, and is stored in the available pipTx pipe. These audio samples represent the output of the adaptive system, which is the processed audio signal.

In order to complete the process, the pipTx pipe is read by the pioTx pipe, which has a handle to the output of the CODEC. When a frame of data is available, the pioTx pipe passes it directly to the CODEC with intervention from the DMA controller. This completes the process and allows the output processed speech signal to be heard via a speaker connected to the line-out jack of the evaluation board.

Results from the DSP board essentially matched the simulations carried out in Matlab, and further allowed for real-time signal processing.

## VI. CONCLUSIONS

The Matlab-simulated and DSP-implemented results show that the presented parallel algorithm adaptive system is capable of extracting a voice signal from a noisy signal. Its performance is most impressive when the original noise completely masks the speaker's signal, since the output of the system provides a clearly intelligible speech signal with a much greater SNR. This could potentially be implemented in order to allow communications in highly noisy environments, such as inside the cockpit of a fighter jet.

The parallel algorithm provides significantly better convergence performance over a standard LMS adaptive filter system. As can be seen in the results section, the convergence rate is approximately 5 times faster than the equivalent single adaptive filter system. In general, significant convergence is accomplished in 0.125 seconds. The design of the parallel algorithm was the most significant aspect of study, allowing fast convergence, low distortion, and real-time DSP-based noise cancellation of noisy speech signals.

In highly noisy environments where the original speech signal is completely overpowered by noise, a combination of the parallel system algorithm and a spectral subtraction method may provide even higher quality results. This is due to the fact that the two methods complement each other very well. The parallel system algorithm excels at increasing the SNR of the speech signal, while spectral subtraction techniques use frequency domain techniques to further improve signal quality. However, spectral subtraction techniques are not effective if the original speech signal has a very low SNR, and therefore the parallel system algorithm could be used to increase the SNR prior to spectral subtraction techniques. The collaboration of these two techniques in a single system is a potential topic for future work.

## References


[1] B. Widrow and S. D. Stearns, *Adaptive Signal Processing*, New Jersey: Prentice-Hall, Inc., 1985.

[2] S. Haykin, *Adaptive Filter Theory*, New Jersey: Prentice-Hall, Inc., 1996.

[3] D. Magdic, A. Campbell and Y. DeGraw, *Reference Frameworks for eXpressDSP Software; RF3, A Flexible, Multi-Channel, Multi-Algorithm, Static System*, Santa Barbara: Texas Instruments Application Report, April 2003.

[4] K. C. Ho and P. C. Ching, "Noise Cancellation Using a Split-Path Adaptive Filter," *IEEE China 1991 International Conference on Circuits and Systems*, pp. 149-152, June 1991.

[5] Ahmed I. Sulyman and Azzedine Zerguine, "Convergence and Steady-State Analysis of a Variable Step-Size Normalized LMS Algorithm," *Proceedings of the Signal Processing and its Applications*, July 2003.

[6] M. Akiho, M. Tamura, M. Haseyama and H. Kitajima, "Performace Improvements on MEFX-LMS Based Noise Cancellation System in a Vehicle Cabin," *Proceedings of the 2000 IEEE International Symposium on Circuits and* Systems, pp. 353-356, May 2000.

# Requirements Analysis: A Review

Joseph T. Catanio
LaSalle University
Mathematics and Computer Science Department
Philadelphia, PA 19141
1.215.951.1142
catanio@lasalle.edu

*Abstract* - **Many software organizations often bypass the requirements analysis phase of the software development life cycle process and skip directly to the implementation phase in an effort to save time and money. The results of such an approach often leads to projects not meeting the expected deadline, exceeding budget, and not meeting user needs or expectations. One of the primary benefits of requirements analysis is to catch problems early and minimize their impact with respect to time and money.**

**This paper is a literature review of the requirements analysis phase and the multitude of techniques available to perform the analysis. It is hoped that by compiling the information into a single document, readers will be more in a position to understand the requirements engineering process and provide analysts a compelling argument as to why it should be employed in modern day software development.**

## I. INTRODUCTION

Requirements analysis, as it relates to software projects, is the process of studying, determining and documenting user needs and expectations of the software system to be designed that solves a particular problem. The process is referred to as requirements engineering and entails feasibility studies, elicitation, specification, and validation process steps. The process generates software requirement documents that capture what is to be implemented by fully describing the software system's functionality, performance, design constraints, and quality attributes. Precisely documenting what to build helps to reduce uncertainty and equivocality [1]. Determining and documenting the requirements of an information system, is arguably the key to developing successful information systems [2]. Not getting the correct final software system requirements at the project onset is largely responsible for the cost and schedule overruns plaguing the information system development process [3][4][2]. Table 1 shows that the earlier in the development process an error occurs and the later the error is detected, the more expensive it is to correct [5].

Table 1 Relative Cost to Repair a Software Error in Different Stages

| Stage | Relative Repair Cost |
|---|---|
| Requirements | 1-2 |
| Design | 5 |
| Coding | 10 |
| Unit Test | 20 |
| System Test | 50 |
| Maintenance | 200 |

Thus, performing software requirements analysis at the project onset helps to identify and correct problems early. Consequently, the relative repair cost is low and reduces the chances of project cost overruns.

Much of the literature describes the requirements analysis process as three sub-processes and compares it to an engineering methodology [6][7][8][2][9]:

- Elicitation
- Specification
- Validation & Verification

Each sub-process addresses the problem definition aspects from different angles during the requirements creation process to collectively describe the software system to be built. The software specification documents generated, as a result of the analysis, captures the user needs and describes the operation of the proposed software system. The software system description is generally comprised of three types of documents [6][9][10].

- Functional Requirements
- Non-functional Requirements
- Design Constraints

Functional requirements describe what the software system should provide and how it should behave. In contrast, non-functional requirements are not concerned with specific functionality delivered by the system. Instead non-functional requirements relate to system properties such as



*T. Sobh and K. Elleithy (eds.), Advances in Systems, Computing Sciences and Software Engineering*, 411–418.

reliability, response time, memory space, portability, maintainability, ease-of-use, robustness, security, and re-usability [10][9]. Design constraints describe the requirements that are specific to characteristics of the problem domain that do not easily categorize into the other two types of documents.

The hardest single part of building a software system is determining and documenting precisely what to build [12][10]. The difficulties of documenting and specifying software requirements are primarily due to human problem-solving limitations [13]. Davis also points out that these limitations are because human beings have a limited ability to process information. Much of the time humans leave or filter out information to prevent information overload. To help include all the available information, methodologies have been developed to provide a systematic repeatable approach to the description and development of software requirements and systems [14][15][16][17][10]. Methodologies help to structure the problem and solution domain into a collection of smaller sub-problems. These sub-problems are then individually described and eventually implemented. The aggregate of all the components, describe the entire problem domain. This approach helps to divide large complex problems into smaller, more manageable components. In addition, complexity is reduced since the amount of information that a sub-component must consider is also reduced. Therefore, the documentation of sub-problems help achieve a goal of requirements analysis, namely to understand and capture what is to be solved in a component-oriented manner using all available information.

## II. ELICITING REQUIREMENTS

Requirements elicitation is the process of identifying the application domain by determining the desired software system functionality. This activity should involve many different kinds of people that have a stake in the system being built. These stakeholders work together to define and scope out the application domain. Each participant is a stakeholder and represents a different interest in the project. These interests are dependent upon the role the individual performs. Therefore, the elicitation process should include all people that are either directly involved with the project or indirectly affected. Once the stakeholders are identified, the process enters the problem domain description phase. The description is realized either in an independent or a team-oriented collaborative manner. Gause points out that many organizations reinforce a negative image of cooperative work, encouraging instead competition among employees by such devices as individual achievement awards [18]. However, software development projects should not be realized alone and need the diversity and ability that a collaborative team approach can provide. Much of the literature supports the concept that groups generate more and better solutions to identifying, describing, and solving a problem [19][20][21][22][10]. The general consensus is that problem definition is likely to be more complete when realized by participation in a collaborative team environment.

The literature identifies many different techniques that are possible to elicit requirements of a computer or information system in both a group team and single person team approach [23][24][25][9]. Some examples of a non-group approach are:

- Introspection
- Questionnaires
- Interviews
- Protocol Analysis
- Ethnography

The introspection technique attempts to elicit requirements of the desired computer or information system by having the development team members individually imagine the system they want. Thus, many perspectives and interpretations will result from introspection. Many viewpoints help to identify all aspects of the problem domain, but these do not necessarily reflect the needs of the end-users of the system. The literature does not consider this technique a practical way to elicit requirements due to its apparent lack of end-user involvement.

Similar to introspection, questionnaires and interviews attempt to elicit requirements by asking questions in a non-group oriented fashion. Questions are presented to individuals in either a written or verbal format, and the answers recorded. Although this is a systematic process Suchman argues that these approaches lead to multiple interpretations in both the questions and the answers [26]. To reduce misinterpretations, the interview technique can be extended to permit dialog between the interviewer and interviewee. However, the literature indicates that the interview process usually involves assumptions concerning the interaction among participants. Goguen strongly argues that assumptions and misinterpretations that can result from questionnaires and interviews make this technique impractical to elicit computer and information systems requirements [23].

Protocol analysis is a process in which a person performing a task does so aloud while his or her thought processes are observed and recorded. This represents direct verbalization of specific cognitive processes [27]. This technique helps to understand an individual's approach to problem solving. Therein lies the problem; it is not a team-oriented approach. The project team consists of many different kinds of people operating in different roles. Some of these individuals have knowledge about the business and organizational needs, while others have technical knowledge. The process of eliciting requirements from these different types of people possessing different types of knowledge is a social endeavor requiring group communications. Protocol analysis is an individual process, not a social interaction method. Although protocol analysis has greatly influenced cognitive psychology it is inappropriate for the requirements analysis and elicitation process [23].

Ethnography is an observational technique to develop an understanding of work processes through observing as opposed to interviews in which people act differently than they say. These observations help to understand the social and organizational requirements. This is a time consuming process but contains much depth by helping to identify implicit system requirements. Implicit requirements are more easily identified by a third party observer because workers often perform tasks out of habit and rarely consider these tasks as part of their work process. The literature describes many ethnographic studies that showed a worker's actual work practices were much more detailed and complex than these individuals were able to describe [28][29]. Thus, ethnography has been shown to be very effective at discovering the way in which people actually work rather than the way in which process definitions say they should work [9]. In software system project development, ethnography may be best suited to determine how to modify an existing system to make it more effective as opposed to at the onset of a completely new project.

The aforementioned techniques are non-collaborative and tend to be inaccurate, inflexible, costly, time consuming, and do not represent natural interactions among people. The literature does indicate that more effective techniques to elicit requirements of a computer or information system are collaborative by design. Three such techniques are:

- Rapid Application Development (RAD) Focus Groups
- Viewpoint-oriented Requirements Definition (VORD)
- Use-case Scenarios

Rapid application development focus groups are a type of group interview that permits interactions among people to discuss requirements of the desired system. These interactions are both formal and informal depending on the organization performing the RAD and can reduce the cycle-time by 50% in the definition and development of the system [30]. RAD is an iterative process that employs interactive sessions of developer, customer and end-user to identify and define the requirements of the desired system. This collaborative effort affords the project team the ability to openly discuss the system that is to be built. The successes of RAD have been attributed to the inclusion of the end-users during the system definition process [31].

In addition to end-users, viewpoint-oriented elicitation also includes other stakeholders in the viewpoint-oriented requirements definition (VORD) approach to identify requirements. A viewpoint represents a certain perspective of the system. The VORD process recognizes the different viewpoints provided by the different stakeholders and incorporates their perspectives into the requirements specification process [9][32]. The VORD method includes four steps [25][9]:

- Viewpoint Identification
- Viewpoint Structuring
- Viewpoint Determination
- Viewpoint-System Mapping

These four steps utilize a highly structured text–based method to document a viewpoint's attributes, events, and scenarios during brainstorming sessions. These viewpoints are documented using viewpoint and service templates that help to identify both the functional and non-functional requirements. Nuseibeh uses a similar approach but augments the process by allowing incomplete scenarios to exist during elicitation [33]. As the process of determining the desired system to be built continues, the incomplete scenarios are more rigorously defined, specified, and resolved by final specification.

Use-case analysis is a scenario-based technique for requirements elicitation. Use-cases capture requirements from the user's point of view [34] and helps describe what functionality is contained within the system. The literature indicates that use-case analysis is the most widely accepted method to eliciting requirements for software systems.

These three techniques are more successful than other techniques because they adopt a team-oriented user-inclusive strategy. It is important to include the eventual end-users of the system since they make or break the product. Therefore, the team should consist of the customers, developers, and end-users. However, it should be noted that the inclusion of customers, developers and end-user stakeholders during the elicitation process does have six primary difficulties [9].

- Stakeholders often do not know what they want from the computer system except in the most general terms, finding it difficult to articulate what they want from the system.
- Stakeholders make unrealistic demands because they are unaware of the cost of their requests.

- Stakeholders in a system naturally express requirements in their own terms and with implicit knowledge of their own work. Others must try to understand these terms and relate them to the application domain.
- Different stakeholders have different requirements and may express them in different ways. Requirement engineers have to discover all potential sources of requirements and discover commonalities and conflict.
- Political factors may influence the requirements of the system. These may come from managers who demand specific system requirements because these allow them to increase their influence in the organization.
- The economic and business environment is dynamic, which can lead to inevitable changes during the process. The importance of particular requirements may change. New requirements may emerge from new stakeholders who were not originally consulted.

Regardless of these difficulties, the literature suggests that the most effective way to elicit requirements is utilizing a team-oriented user-inclusive strategy.

## III. DOCUMENTING REQUIREMENTS

During the software system elicitation process, software requirements must be captured in a written format to help stakeholders clarify the operational needs of the software system. A document, known as the concept of operations document (ConOps), provides a well-defined operational concept definition of system goals, missions, functions, and components. Thayer lists the essential elements to be included in a ConOps document as [6]:

- Description of current system or situation
- Description of needs that motivate development of a new system or the modification of an existing system
- Modes of operation
- User classes and characteristics
- Operational features
- Priorities among operational features
- Operational scenarios for each operational mode and class of user
- Limitations
- Impact analysis

The ConOps document represents a bridge between the description of the user needs and the technical specifications of the software system specification process [6]. Both the U.S. Department of Defense and the IEEE support the creation of the ConOps document when developing or modifying a software system. The document serves as a framework to guide the analysis and provides the foundation document for all subsequent system development activities.

The literature indicates that the ConOps document can be developed anytime during the system life-cycle but is most beneficial at the beginning of the software development process [32][5][6]. Developing the document at the onset of the development process affords all parties involved the opportunity to repeatedly review and revise the document until all stakeholders agree on the content. This iterative process helps bring to the surface many viewpoints, needs, wants, and scenarios that might otherwise be overlooked. In addition, the creation of formal specifications forces a detailed system analysis that could reveal errors and inconsistencies. This error detection is perhaps the most powerful argument for developing a formal system specification [35].

Developers utilize the ConOps document to create a Software Requirements Specification (SRS). The SRS describes exactly what is to be built by capturing the software solution. This represents a transition from the problem analysis to the technical analysis. Faulk outlines the roles of the SRS document [5].

- Customers: Provide a contractual basis for the software project
- Managers: Provide a basis for scheduling and measuring progress
- Designers: Provide a basis for what to design to
- Coders: Provide tangible outputs that must be produced
- Quality Assurance: Provide a basis for test planning and verification
- Marketing: Provide a basis for marketing plans and advertising

A properly written SRS satisfies both the semantic and packaging properties [5]. An SRS satisfies the semantic properties if it is complete, implementation independent, unambiguous, precise, and verifiable. To satisfy the packaging properties, the SRS must be readable, modifiable, and organized for reference and review. There is not a specific industry-wide system specification standard. The literature indicates that the IEEE Std. 830-1998 is widely accepted at documenting the SRS [5][6][9].

## IV. VERIFYING AND VALIDATING REQUIREMENTS

Verification and validation (V&V) is the process that determines if the software conforms to the specifications and meets the needs of the customer. Verification involves checking that the software conforms to the specifications. Validation checks that the software meets the expectations of the users [9]. Boehm expresses this subtle difference as follows [3]:

- Verification: Are we building the product right?
- Validation: Are we building the right product?

The V&V process of system checking is realized using two techniques, namely software documentation inspection and software system testing. The major software V&V activities are outlined in Table 2 [36].

Table 2 Major Software Verification and Validation Activities

| Activity | Tasks |
|---|---|
| Software V&V Management | Planning<br>Monitoring<br>Evaluating Results<br>Reporting |
| Software Requirements V&V | Review of Concept Documentation<br>Traceability Analysis<br>Software Requirements Evaluation<br>Interface Analysis<br>Initial Planning for Software Integration Test<br>Reporting |
| Software Design V&V | Traceability Analysis<br>Software Design Evaluation<br>Interface Analysis<br>Initial Planning for Unit Test<br>Initial Planning for Software Integration Test<br>Reporting |
| Code V&V | Traceability Analysis<br>Code Evaluation<br>Interface Analysis<br>Completion of Unit Test Preparation<br>Reporting |
| Unit Test | Unit Test Execution<br>Reporting |
| Software Integration Test | Completion of Software Integration Test Preparation<br>Execution of Software Integration Tests<br>Reporting |
| Software System Test | Completion of Software System Test Preparation<br>Execution of Software System Tests<br>Reporting |
| Software Installation Test | Installation Configuration Audit<br>Reporting |
| Software Operation and Maintenance V&V | Impact of Change Analysis<br>Repeat Management V&V<br>Repeat Technical V&V Activities |

Static, dynamic, and formal verification techniques can be performed to fulfill the requirements of the V&V process activities [18][36][9].

- Static analysis reviews the structure of the software product prior to its execution. The analysis is performed on the requirements documents, design documents, and source code utilizing review and inspection oriented techniques.
- Dynamic analysis detects errors by executing the software and testing the actual outputs against the expected outputs as outlined in the SRS documents.
- Formal analysis is the use of rigorous mathematical techniques to analyze the algorithms of a software solution [9].

The software V&V process should begin at the onset of the project and continue throughout the entire software development life-cycle process. Prior to software construction, the V&V process should examine preliminary documents such as the ConOps document to help determine if the system to be built is feasible. Subsequently, the V&V process examines the SRS to ensure that the requirements are complete, consistent, accurate, readable, and testable. The approach helps to find errors in the software requirements before the software implementation phase. Also, the software requirements analysis conducted is necessary in order to develop relevant test plans. This early error detection helps to reduce cost overruns, late deliveries, poor reliability, and user dissatisfaction [4]. However, the predominant V&V technique is software testing. Software testing involves executing the software product and examining its operational behavior. Testing is used to find discrepancies between the software program and the corresponding specifications and is referred to as defect testing. In contrast, statistical testing determines the software's performance and reliability by noting the number of system failures [37].

Test management organizes the testing process and utilizes test plans to determine the objectives for unit, integration, and system testing. Table 3 outlines the structure of a software test plan [38][9]:

Table 3 Software Test Plan Structure

| | |
|---|---|
| The Testing Process | A description of the major phases of the testing process. |
| Requirements Traceability | Testing should be planned so that all end-users requirements and expectations are individually tested. |
| Tested Items | The products of the software process to be tested should be specified. |
| Testing Schedule | The overall testing schedule should indicate resources needed to perform this task in the time frame allotted. |
| Test Recording Procedures | The test results must be systematically recorded and repeatable. |
| Hardware and Software Requirements | This section should outline the software tools required and hardware requirements. |
| Constraints | A description of constraints affecting the testing process. |

The test plan document is dynamic and may change to support the dynamic nature of the software life-cycle process. As changes are made to the software requirements, management of all documents, including the test plan document, becomes crucial. The V&V process must determine how software requirement changes affect the overall testing plans, which may also affect the deliverables schedule. Thus, managing the documents is crucial to incorporating change into the overall software development life-cycle process.

## V. REQUIREMENTS MANAGEMENT

Requirements management is a life-cycle process that attempts to understand and control change to software system requirements. Change to requirements is constant due to the inability to fully define the problem upfront, therefore creating incomplete software requirements and specifications. As the software development process progresses the understanding of the problem also changes. These changes cause modifications to the original desired software system [39][40]. These changes must be incorporated into the requirements and specifications in a systematic and traceable manner. The requirements change management process coordinates change requests and must ensure that the modifications are performed at all levels. Changes affect requirements, specifications, design documentation, implementation, verification & validation test plans, and operational procedures.

The change management process assesses the cost of changes and consists of three stages [41][9].

- Problem Analysis and Change Specification: The change proposal is analyzed to determine if the request is necessary, attainable, and verifiable.
- Change Analysis and Costing: The cost of making the change is estimated in terms of modifications to the requirements documents, specification documents, design documents, test plans and implementation.
- Change Implementation: Modifications to the requirements documents, specification documents, design documents, test plans, and implementation are scheduled and performed.

The ability to effectively trace changes in the baseline documents help to link requirements to stakeholders resulting in decision-making accountability [42]. Traceability provides an audit trail as to what changes were requested by whom and for what purpose. The traceability process helps to ensure accountability, which generally results in modification requests that are necessary, attainable, and verifiable.

The current state of the practice of traceability management is realized using Computer Aided Software Engineering (CASE) tools that support a change and version control system. Cadre TeamWork for Real-Time Structured Analysis (CADRE), Requirements Traceability Manager (RTM), SLATE, DOORS, Requirements Driven Design (RDD), Foresight, Software Requirements Methodology (SREM), Problem Statement Language / Problem Statement Analyzer (PSL/PSA), and Requirement Networks (R-Nets) are some of the more widely used software industry CASE tools [43][44][45][46]. These automated tools help to maintain a history of all the changes by capturing the following requirement attributes:

- Document Name
- Version Number
- Creation Date
- Modification Date
- Author
- Responsible Manager
- Approval
- Sign-Off
- Change Description
- Priority

These attributes help requirements management establish version control, change control, and traceability process procedures. These activities help to create a well-defined requirement definition process. These processes should be incorporated into the software life-cycle to help ensure that the documents describing the software product contain all features and operational behaviors of the released software product.

Version and release management are the processes of identifying and tracking different versions and releases of the software system [9]. CASE tools provide version numbering management and ensure that each version is uniquely identifiable. This type of configuration control generally uses some type of check-in/check-out procedures to help aid both developers and coordinators manage the process. Whereas version management identifies each component version, release management handles all the steps that are necessary to package the software solution to the customer. This packaging includes executable code, configuration files, data files, installation program, and documentation. Configuration management (CM) is the term used to identify configuration control processes. CM is separate from developmental procedures and helps to coordinate the release process[40]. Thus, version control, change control, and traceability are intertwined and are essential components of the requirements management process.

## VI. SUMMARY

Requirements analysis is the process of studying, determining and documenting user needs and expectations of the software system to be designed that solves a particular problem. The process generates software requirements documents that capture what is to be implemented by fully describing the software system's functionality, performance, design constraints, and quality attributes. The software system attributes are fully described in the functional requirements, non-functional requirements, and design constraints documents. In addition to defining the static and dynamic features of the system, these documents describe end-user system needs. Consequently, understanding end-user needs is essential to provide them a system they expect and will use.

ACKNOWLEDGMENTS

Many thanks to Teri, Chason, John, and Shads for their continual support and encouragement.

## REFERENCES

[1] Daft, R., & Lengel, R. (1986). "Organizational Information Requirements, Media Richness, and Structural Design," *Management Science*, Vol. 32, No. 5, pp. 554-571.

[2] Vessey, I. & Conger, S. (1994). "Requirements Specification: Learning Object, Process, and Data Methodologies," *Communications of the ACM*, Vol. 37, No. 5, pp. 102-113.

[3] Boehm, B. (1981). *Software Engineering Economics*, Prentice-Hall, Englewood Cliffs, New Jersey.

[4] Abdel-Hamid, T., & Madnick, S. (1991). *Software Project Dynamics,* Prentice Hall, Englewood Cliffs, New Jersey.

[5] Faulk, S. (2000). "Software Requirements: A Tutorial," *Software Requirements Engineering,* Second Edition, IEEE, Los Alamitos, California, pp. 158-179.

[6] Thayer, R., & Dorfman, M. (2000). "Software System Engineering Process Models," *Software Requirements Engineering,* Second Edition, IEEE, Los Alamitos, California, pp. 453-455.

[7] Ram, S., & Khatri, V. (2005). "A Comprehensive Framework for Modeling Set-based Business Rules During Conceptual Database Design", *Information Systems*, Vol. 30, No. 2, pp.89-118.

[8] Camille, B., Achour, S.,Opdahl, A., & Matti, R. (2002). "Requirements Engineering : Foundations for Software Quality," *ACM SIGSOFT Software Engineering Notes*, Vol. 27, Issue 2, pp. 35-49.

[9] Sommerville, I. (2001). *Software Engineering*, Sixth Edition, Addison-Wesley Publishers, Massachusetts.

[10] Letier, E., and Lamsweerde, A. (2002). "Deriving Operational Software Specifications from System Goals," *Proceedings of the 10th ACM SIGSOFT Symposium on Foundations of Software Engineering*, pp. 119-128.

[11] Mylopoulos, J., Chung, L., & Yu, E. (1999). "From Object-Oriented To Goal-Oriented Requirements Analysis," *Communications of the ACM*, Vol. 42, No. 1, pp. 31-37.

[12] Brooks, F. (1987). "No Silver Bullet: Essence and Accidents of Software Engineering," *Computer*, pp. 10-19.

[13] Davis, G. (1982). "Strategies for Information Requirements Determinations," *IBM Systems Journal*, Vol. 21, No. 1, pp. 4-30.

[14] DeMarco, T. (1978). *Structured Analysis and System Specification,* Yourdon Press, New York.

[15] Gane, C., & Sarson, T. (1979). *Structured Systems Analysis: Tools and Techniques*, Prentice-Hall, Englewood Cliffs, New Jersey.

[16] Rumbaugh, J. (1991). *Object-Oriented Modeling and Design*, Prentice-Hall, New Jersey.

[17] Booch, G. (1991). *Object Oriented Analysis and Design*, Second Edition, Benjamin/Cummings Publishing Company, California.

[18] Gause, D., & Weinberg, G. (1989). *Exploring Requirements: Quality Before Design*, Dorset House Publishing, New York.

[19] Baroudi, J., Olson, M., & Ives, B. (1986). "An Empirical Study of the Impact of User Involvement on System Usage and Information Satisfaction," *Communications of the ACM*, pp. 232-238.

[20] DeSanctis, G., & Gallupe, B. (1987). "A Foundation for the Study of Group Decision Support Systems," *Management Science*, Vol. 33, No. 5, pp. 589-609.

[21] Gallupe, B., DeSanctis, G., & Dickson, G. (1988). Computer Based Support for Group Problem-Finding: An Experimental Investigation," *MIS Quarterly*, Vol.12, No. 2, pp. 277-296.

[22] Connolly, T., Jessup, L., & Valacich, J. (1990). "Effects of Anonymity and Evaluation Tone on Idea Generation in Computer-Mediated Groups," *Management Science*, Vol. 36, No. 6, pp. 305-319.

[23] Goguen, J., & Linde, C. (1993). "Techniques for Requirements Elicitation," *Proceedings from the International Symposium on Requirements Engineering*, pp. 152-164.

[24] Rumbaugh, J. (1994). "Getting Started: Using Use Cases to Capture Requirements," *Object-Oriented Programming*, pp. 8-12.

[25] Kotonya, G., & Sommerville, I. (1996). "Requirements Engineering with Viewpoints," *Software Engineering Journal*, Vol. 11, No. 1, pp. 5-18.

[26] Suchman, L., & Jordan, B. (1990). "Interactional Troubles in Face-To-Face Survey Interviews, " *Journal of the American Statistical Association*, Vol. 89, No. 409, pp. 232-241.

[27] Simon, H.A., and Ericson, K. (1984). "Protocol Analysis: Verbal Reports as Data," *MIT.*

[28] Suchman, L. (1983). "Office Procedures as Practical Action," *ACM Transaction on Office Information Systems*, Vol. 1, No. 3, pp. 320-328.

[29] Myers, M. (1999). "Investigating Information Systems with Ethnographic Research," *Communications of the Associations for Information Systems.*

[30] Engler, N. (1996). "Bringing in the Users," *Computerworld.*

[31] Gonzales, R., & Wolf, A. (1996). "A Facilitator Method for Upstream Design Activities with Diverse Stakeholders," *Proceedings of the International Conference on Requirements Engineering*, IEEE Computer Society, pp. 190-197.

[32] Silva, A. (2002). "Requirements, Domain and Specifications: A Viewpoint-based Approach to Requirements Engineering," *Internal Conference on Software Engineering*, pp. 94-104.

[33] Nuseibeh, B., Kramer, J., & Finkelstein, A. (1994). "A Framework for Expressing the Relationships Between Multiple Views in Requirements Specification," *IEEE Transactions on Software Engineering*, Vol. 20, No. 10, pp. 760-773.

[34] Zhang, L., Xie, D., and Zou, W. (2001). "Viewing Use Cases as Active Objects," *ACM SIGSOFT Software Engineering* Notes, Vol. 26, Issue 2, pp. 44-48

[35] Hall, A. (1990). "Seven Myths of Formal Methods," *IEEE Software*, Vol. 7, No. 5, pp. 11-20.

[36] Wallace, D., & Ippolito, L. (2000). "Verifying and Validating Software Requirements Specifications," *Software Requirements Engineering*, Second Edition, IEEE, Los Alamitos, California pp. 437-452.

[37] Musa, J., & Ackerman , A. (1989). "Quantifying Software Validation: When to Stop Testing?," *IEEE Software*, Vol. 6, No. 3, pp. 19-27.

[38] Frewin, G., & Hatton, B. (1986). "Quality Management: Procedures and Practices," *Journal of Software Engineering*, Vol. 1, No. 1, pp. 29-38.

[39] Bennett, K., & Vá clav T. Rajlich, V. (2000). "Software Maintenance and Evolution: a Roadmap," *Proceedings of the Conference on The Future of Software Engineering*, pp.73-87.

[40] Estublier, J. (2000). "Software Configuration Management: A Roadmap," *Proceedings of the Conference on The Future of Software Engineering*, pp.279-289.

[41] Hooks, I. (2001). "Managing Requirements," *NJIT Requirements Engineering Handout*, pp. 1-8.

[42] Ng, P. & Yeh, R. (1990). "Software Requirements: A Management Perspective," *System and Software Requirements Engineering*, IEEE, Los Alamitos, California, pp. 450-461.

[43] Nallon, J. (1994). "Implementation of NSWC Requirements Traceability Models," *IEEE Transaction on Software Engineering*, pp. 15-22.

[44] Rundley, N., & Miller, W. (1994). "DOORS to the Digitized Battlefield: Managing Requirements Discovery and Traceability," *CSESAW*, pp. 23-28.

[45] Vertal, M. (1994). "Extending IDEF: Improving Complex Systems with Executable Modeling," *Proc. Ann. Conf. for Business Re-engineering.*

[46] Palmer, J. (2000). "Traceability," *Software Requirements Engineering*, Second Edition, IEEE, Los Alamitos, California, pp. 412-422.

# Nuclear Matter Critical Temperature and Charge Balance

A. Barrañón-Cedillo
Dept. of Basic Sciences, Universidad Autónoma Metropolitana
Av. San Pablo esq. Eje 5 Nte
México, D.F. 02200 México.

J.A. López-Gallardo
Dept. of Physics, The University of Texas at El Paso
University Avenue
El Paso, TX 79969 USA

F. de L. Castillo-Alvarado
Av. Instituto Politécnico Nacional esq. Ticomá n.
Dept. of Physics, Instituto Politécnico Nacional
México, D.F. 07730 México.

***Abstract*- An iterative algorithm was developed to fit Fisher Law for Heavy Ion Collisions with distinct charge balance, obtaining different critical temperatures in agreement with recent theoretical and experimental results. This way is confirmed the influence of charge balance on the caloric curve of nuclear matter.**

**Keywords: Iterative Algorithm, Polynomial Fitting, Critical Temperature, Nuclear Caloric Curve**

## I. Introduction

Weiszä cker nuclear energy contains a term related to charge balance, namely the difference between the number of neutrons and protons, which should lead to a change in the limit temperature of the nuclear caloric curve[1]. Weiszä cker's expresion for nuclear energy is given by [2]:

$$E=-a_V A+a_S A^{2/3}+a_a\frac{(N-Z)^2}{A}+a_C\frac{Z(Z-1)}{A^{1/3}} \qquad (1)$$

where the first term represents bulk symmetric nuclear matter energy excluding coulombian interactions. The rest of the terms compensate nuclear energy disagreement with the asymmetric nuclear matter bulk limit. These terms are related to surface energy, to the asymmetry between neutrons N and protons Z and to coulombian repulsion, respectively.

Nuclear matter critical temperature is related to other variables such as system finite size and entropy [3], therefore the results hereby reported may help to reach a better understanding about the factors influencing the characteristics of the nuclear caloric curve. Other authors [4] have reported nuclear limit temperatures taking on account the balance between charge Z and the number of neutrons N, via the correlation between the factor N/Z for the fragment projectile and the isobaric ratio Y(3H)/Y(3He), obtaining a limit temperature in the range of 6 to 7 MeV. Souza et al. [5] have applied an improved statistical multifragmentation model (ISMM) to proton and neutron rich sources, obtaining a temperature plateau close to 6 MeV, with a minimum difference between the limit temperatures of the neutron and proton rich sources.

Fisher Law for nuclear matter can be written as [6]:

$$\ln\langle n_A\rangle=\ln q_0-\tau\ln A+A\frac{\Delta\mu}{T}-\frac{c_0\varepsilon A^{\sigma}}{T} \qquad (2)$$

using Fisher Liquid Droplet Model where $q_0$ is a normalization constant that depends only on the value of $\tau$, $\tau$ is the critical topological exponent related to system dimensionality that can be computed through a tridimensional random walk in a closed surface, $c_0\varepsilon A^{\sigma}$ is the free surface energy of a droplet of size A,



*T. Sobh and K. Elleithy (eds.), Advances in Systems, Computing Sciences and Software Engineering*, 419–423.

$c_0$ is surface energy coefficient, $\sigma$ is the critical exponent related to the ratio of the surface dimensionality and volume dimensionality; and $\varepsilon = \left(1 - \frac{T}{T_C}\right)$ is the control parameter measuring the distance to the critical point whose temperature is given by $T_C$ [7]. For a system with only a hundred of particles:

$$q_0 = 1/\sum_{A=1}^{A_{\max}} A^{1-\tau} \tag{3}$$

where A is the fragment size.

An iterative algorithm was developed to fit Fisher Law using Heavy Ions Collisions with different charge balance (N-Z), obtaining critical temperature estimates whose difference is in the range of 1.2 MeV. This way is confirmed the minimal influence of the term (N-Z) on nuclear matter critical temperature.

## II. METHODOLOGY

Forty thousand Heavy Ion collisions were simulated using Latino Model [8], where system evolves following a Newtonian dynamics via a Verlet algorithm. Internucleonic forces are computed with a Pandharipande potential. Fisher Law parameters were approximated by a fourth order polynomial in the excitation. The iterative algorithm floats the Fisher Law critical exponents, performing a Least Squares Fit for the polynomial coefficients as explained in the following. Fisher Law is written as:

$$\ln\frac{\langle n_A\rangle A^{\tau}}{q_0} = A\frac{\Delta\mu}{T} - \frac{c_0\varepsilon A^{\sigma}}{T} \tag{4}$$

the difference of chemical potentials $\Delta\mu$ as well as the surface energy adimensional coefficient $C_0$ are approximated by polynomials on the excitation $e^*$ :

$$\mu = P_{l.m}^{(1)}\left(e^*\right) \tag{5}$$

$$C_0 = P_{l.m}^{(2)}\left(e^*\right) \tag{6}$$

where:

$$P_{l.m}(x) = \sum_{n=0}^{m} a_n x^n \tag{7}$$

leading to a least squares problem whose solution is given by:

$$\mathrm{M}^T Y = \mathrm{M}^T \mathrm{M} A \tag{8}$$

where $A$ is a vector containing the polynomial coefficients of $P_{l.m}$, M is a matrix and $Y$ is a vector obtained from the system of equations:

$$\left[\ln\frac{\langle n_A\rangle A^{\tau}}{q_0}\right]_i = A\left[\frac{P_{l,m}^{(1)}}{T}\right]_i - \left[\frac{P_{l,m}^{(2)}\varepsilon A^{\sigma}}{T}\right]_i \tag{9}$$

where i is an index related to each collision replica.

Random variables $\eta_x$ are generated to float Fisher Law critical exponents as well as the critical excitation:

$$\tau^{(t+1)} = \tau^{(t)} + \eta_\tau \tag{10}$$

$$\sigma^{(t+1)} = \sigma^{(t)} + \eta_\sigma \tag{11}$$

$$e_C^{*\,(t+1)} = e_C^{*\,(t)} + \eta_{e_C^*} \tag{12}$$

During the iteration the values of these critical exponents that optimize the statistical $\chi^2$ are kept [9] until the best fit is attained. In this fashion were simulated twenty thousand replicas of $Zn^{76}$+$Ca^{40}$ and $Zn^{76}$+$Ar^{40}$ Heavy Ion Collisions. This way an estimator for the critical excitation for these collisions was obtained.

Once the best estimator was obtained, it was used to estimate the critical temperature via the following relation :

$$e_C^* = E_{x,c}/A = T_C^2/13 \tag{13}$$

This relation assumes a Fermi degenerate gas behavior and is in better agreement with experiment results than the empirical thermometers based on isotope ratios, as shown by Moretto et al. in a study about the evaporation in compound nucleus decay [10]. Nevertheless, other studies have criticized the hypothesis of a Fermi degenerate gas behavior close to the critical point, as long as the relation E/A=a$T^2$ holds for fragments produced at moderate low temperatures and in the case of intermediate energy collisions there are fast particles emitted in the final state from the region where the projectile and fragment overlap.

Heavy Ion collisions were simulated using LATINO semiclassical model where binary interaction is reproduced with a Pandharipande potential given by:

$$V_{nn} = V_{pp} = V_0 \left( \frac{e^{-\mu_0 r}}{r} - \frac{e^{-\mu_0 r_C}}{r_C} \right) \quad (14)$$

and:

$$V_{np} = V_r \left( \frac{e^{-\mu_r r}}{r} - \frac{e^{-\mu_r r_C}}{r_C} \right) - V_a \left( \frac{e^{-\mu_a r}}{r} - \frac{e^{-\mu_a r_a}}{r_a} \right) \quad (15)$$

made up from linear combinations of Yukawa potentials whose coefficients are designed to reproduce nuclear matter properties and to fulfill Pauli Exclusion Principle [11].

Clusters are detected using an Early Cluster Recognition Algorithm that optimizes the configurations in energy space. Most Bound Partition is obtained minimizing the sum of cluster energies for each partition:

$$\{C_i\} = \arg\min \left[ E_{\{C_i\}} = \sum_i E_{\text{int}}^{C_i} \right] \quad (16)$$

where the cluster energy is given by:

$$E_{\text{int}}^{C_i} = \sum_i \left[ \sum_{ij \in C_i} K_j^{CM} + \sum_{j,k \in C_{iu}, j \le k} V_{jk} \right] \quad (17)$$

in this expression the first sum is on the partition clusters, $K_j^{CM}$ is the kinetic energy of particle j measured in the cluster mass center, and $V_{ij}$ is the internucleonic potential. The algorithm uses the technique of "simmulated annealing" to find the most bound partition in energy space.

Ground states of neutron or proton rich sources, were built up starting from a random configuration with a given kinetic energy and confined in a parabolic potential. Nucleon speed was gradually reduced until the system was bound, afterwards the parabolic potential was supressed and a frictional method was applied until the system reached its theoretical binding energy (Fig. 1).

Projectile is boosted on target with a given kinetic energy for distinct impact parameters. System evolution was simulated using a Verlet algorithm [12], where two Taylor expansions are substracted, one of them forwards and the other backwards on time:

$$\vec{r}(t+\Delta t) = 2\vec{r}(t) - \vec{r}(t-\Delta t) + \vec{a}(t)h^2 \quad (18)$$

$$\vec{v}(t+\Delta t) = \vec{v}(t) + 0.5 * [\vec{a}(t+\Delta t) + \vec{a}(t)]h \quad (19)$$

$$\vec{a}(t+\Delta t) = -(1/m)\nabla V(\vec{r}(t+\Delta t)) \quad (20)$$

Fig 1.- Shows the ground state of the Heavy Ion $Zn^{76}$ obtained starting from a random configuration, subsequently confined in a parabolic potential and finally cooled by a frictional method until it attains its theoretical binding energy.

Excitation is computed by the temperature attained by the projectile-target compound, when the maximal compression is reached. This temperature is estimated using Kinetic Theory for the n nucleons in the compound:

$$\frac{3}{2} nT = K^{CM} \quad (21)$$

Projectile energy is varied in the range going from 600 up to 2000 MeV and system evolves until its microscopic composition remains frozen (Fig. 2), although some monomers might be ejected. This time can be determined using the Microscopic Persistence coefficient defined as the probability of having two particles linked in a cluster of partition X still bound in a cluster of partition Y:

$$P[X,Y] = \frac{1}{\sum_{cluster} n_i} \sum_{cluster} n_i a_i / b_i \quad (22)$$

where $b_i$ is equal to the number of particles that belong to cluster $C_i$ of partition $X \equiv \{C_i\}$ and $a_i$ is equal to the number of particle pairs belonging to cluster $C_i$ of partition

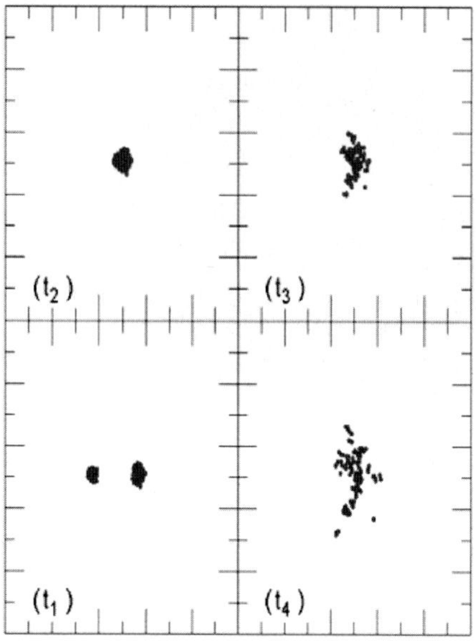


Fig 2.- Shows the evolution of central collision $Zn^{76}$+$Ar^{40}$ for a projectile energy equal to 1600MeV, simulated with model LATINO.

$X \equiv \{C_i\}$ that also belong to a given cluster $C_i'$ of partition $Y \equiv \{C_i'\}$. $n_i$ is the number of particles in cluster $C_i$. Fig. 3 shows that persistence attains an asymptotic limit value once the biggest fragment size (FM), as well as the logarithmic derivative of the kinetic energy transported by light fragments and the logarithmic derivative of the number of intermediate fragments are altogether stable.

The ratio of isotope yields:

$$Y_2(N,Z)/Y_1(N,Z) \propto \exp(\alpha N + \beta Z) \qquad (23)$$

has been commonly used as a signature of thermodynamic equilibrium. Nevertheless Latino Model has been used elsewhere to prove that this signature holds at early stages of the Heavy Ion Collision, when the biggest fragment temperature is out of equilibrium [13]. Fig. 4 shows a $\chi^2$ fit of the ratio of isotope yields for one hundred thousand replicas of $^{76}Zn + ^{40}Ar$ and $^{76}Zn + ^{40}Ca$ Heavy Ion Collisions using Latino Model, obtaining Isoscaling Parameters,

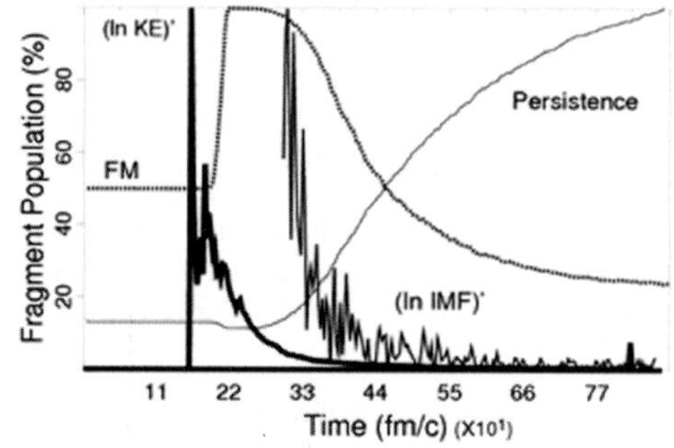


Fig 3.- Persistence attains an asymptotic limit value once the biggest fragment size (FM), as well as the logarithmic derivative of the kinetic energy transported by light fragments and the logarithmic derivative of the number of intermediate fragments are altogether stable. Central Collision $Zn^{76}$+$Ar^{40}$ with a projectile energy equal to 1600MeV.

$\alpha = 0.35$ and $\beta = -0.40$, in agreement with those experimentally reported by Liu et al. in [14].

## III. Results

Applying an iterative algorithm that floats Fisher Law critical exponents, an estimator is obtained for the critical excitation that optimizes Fisher Law fitting using data from Heavy Ion Collisions simulations for distinct charge balances (N-Z).

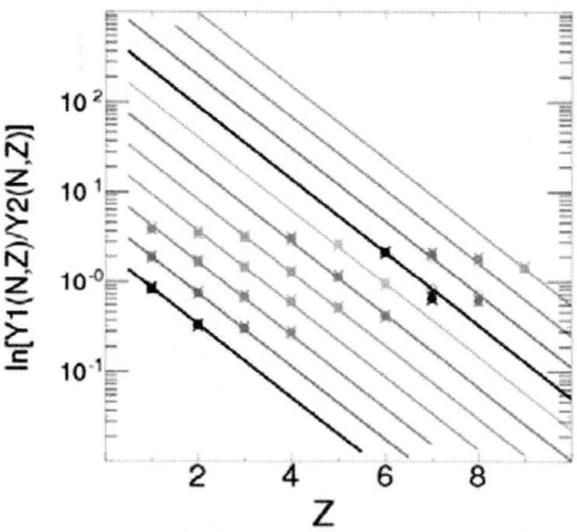


Fig. 4.- Best $\chi^2$ fit of the ratio of isotope yields for one hundred thousand replicas of $^{76}Zn + ^{40}Ar$ and $^{76}Zn + ^{40}Ca$ Heavy Ion Collisions using Latino Model

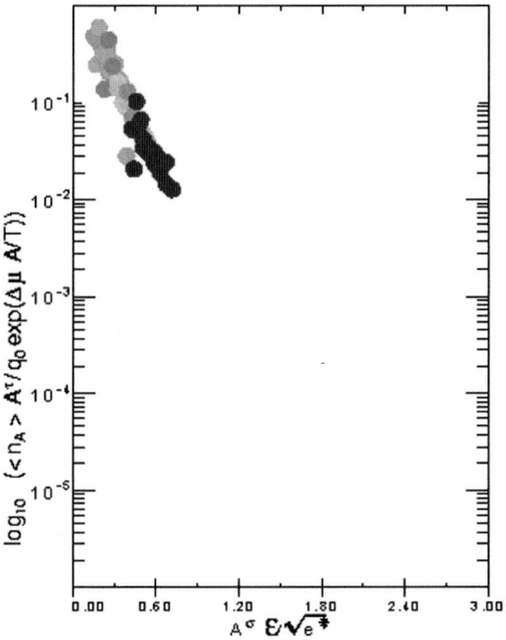


Fig 4.- Best $\chi^2$ fit of Fisher Law, when twenty thousand replicas were performed for $Zn^{76}$+$Ca^{40}$ collision using LATINO model.

This way, critical temperature estimates equal to 7.5MeV for collision $Zn^{76}$+$Ca^{40}$ (Fig. 5) and 8.7 MeV for collision $Zn^{76}$+$Ar^{40}$, were obtained. These values are close to those obtained by Tapas Silk et al. [15], using as criticality signature the maximum of the constant volume heat capacity. And the difference between the critical temperatures hereby estimated comes out to be minimal, in agreement with the results reported by Souza et al. [5].

## IV. Conclusions

An iterative algorithm was developed to obtain computational evidence about the influence of the term (N-Z) on nuclear matter critical temperature, comparing estimates for collisions with distinct values of (N-Z). The difference between the critical temperatures estimated was of about 1MeV for collisions $Zn^{76}$+$Ca^{40}$ and $Zn^{76}$+$Ar^{40}$, in agreement with experimental and theoretical results recently reported about the minimal influence of isospin on nuclear matter critical temperature.

## Acknowledgment

Authors acknowledge ready access to the computational resources of UAM-A and The University of Texas at El Paso.

## References

[1] A. Barrañón, J. Escamilla-Roa and J. A. López, "The Transition Temperature of the Nuclear Caloric Curve", *Brazilian Journal of Physics*, vol. 34, pp. 904-906, 2004.

[2] P. Danielewicz. "Surface Symmetry Energy", *Nucl. Phys.* A727, pp. 233-268, 2003.

[3] A. Barrañón, J. Escamilla-Roa and J. A. López, "Entropy in the nuclear caloric curve", *Physical Review C*, vol. 69, Number 1, pp. 1-6 , 2004.

[4] M. Veselsky, R.W. Ibbotson, R. Laforest, E. Ramakrishnan, D.J. Rowland, A. Ruangma, E.M. Winchester, E. Martín and S.J. Yennello, "Isospin dependence of isobaric ratio Y(3H)/Y(3He) and its relation to temperature", *Phys. Lett. B*, vol. 497, pp. 1-7, 2001.

[5] S.R. Souza, R. Donangelo, W.G. Lynch, W.P. Tan and M.B. Tsang, "Isoscaling bearing information on the nuclear caloric curve," *Phys. Rev. C,* vol. 69, 031607, 2004.

[6] M.E. Fisher. "The theory of equilibrium critical phenomena", *Rep. Progr. Phys.* 30, pp. 615-730, 1967.

[7] J.B. Elliott, L.G. Moretto, L. Phair and G.J. Wozniak. " Nuclear multifragmentation, percolation and the Fisher Droplet Model: common features of reducibility and thermal scaling", *Phys. Rev. Lett.* 85, 1194, 2000.

[8] A. Barrañón, A.Chernomoretz, C.O.Dorso, J.Lopez, and J.Morales, "LATINO: A Semiclassical Model to Study Nuclear Fragmentation.", *Rev. Mex. de Fís.*, vol. 45 Supl. 2, Oct., 1999.

[9] P. Hoel. *Introduction to Mathematical Statistics*, New York: J. Wiley, 4 ed., 1971, p. 254.

[10] L. G. Moretto, J. B. Elliott, L. Phair and G. J. Wozniak, "Liquid-vapor phase transition in nuclei or compound nucleus decay?", *Preprint arXiv: nucl-ex*/0209009, 2002. Reporte LBNL-51306.

[11] R. J. Lenk, T. J. Schlagel and V. R. Pandharipande, "Accuracy of the Vlasov/Nordheim approximation in the classical limit". *Phys. Rev. C* 42, pp. 372-385, 1990.

[12] L. Verlet, "Computer Experiments on Classical Fluids. I. Thermodynamical Properties of Lennard-Jones Molecules "*Phys. Rev.* 159, 98, 1967. L. Verlet, *Phys. Rev.* 165, 201, 1967.

[13] C. O. Dorso, C. R. Escudero, M. Ison and J. A. López, "Dynamical aspects of isoscaling" *ArXiv: nucl-th/0504036*, Apr. 2005.

[14] T.X. Liu, X.D. Liu, M.J. van Goethem, W.G. Lynch, R. Shomin, W.P. Tan, et al.. "Isotope yields from central 112,124Sn+112,124Sn collisions, dynamical emission?". *Phys.Rev. C* 69 014603, 2004.

[15] T. Sil, S.K. Samaddar, J.N. De and S. Shlomo, "Liquid-Gas phase transition and its order in finite nuclei," *ArXiv:nucl-th/0210021*, Oct. 2002.

# Activation-Adjusted Scheduling Algorithms for Real-Time Systems

Alex A. Aravind, *Member, IEEE*, and Jeyaprakash Chelladurai
*(csalex, chellad)@unbc.ca*
University of Northern British Columbia,
Prince George, BC, CANADA V2N 4Z9.

***Abstract*** **- Scheduling in real-time is an important problem due to its role in practical applications. Among the scheduling algorithms proposed in the literature, static priority scheduling algorithms have less run-time scheduling overhead due to their logical simplicity. Rate monotonic scheduling is the first static priority algorithm proposed for real-time scheduling[1]. It has been extensively analyzed and heavily used in practice for its simplicity. One of the limitations of rate monotonic scheduling, as shown recently in [26], is that it incurs significant run-time overhead due to high preemptions. The main objective of this paper is to propose static priority scheduling algorithms with reduced preemptions.**

**We present two frameworks, called** *off-line activation-adjusted scheduling (OAA) and adaptive activation-adjusted scheduling (AAA),* **from which many static priority scheduling algorithms can be derived by appropriately implementing the abstract components. The proposed algorithms reduce the number of unnecessary preemptions and hence: (i) increase processor utilization in real-time systems; (ii) reduce energy consumption when used in embedded systems; and (iii) increase tasks schedulability. We conducted a simulation study for selected algorithms derived from the frameworks and the results indicate that the algorithms reduce preemptions significantly. The appeal of our algorithms is that they generally achieve significant reduction in preemptions while retaining the simplicity of static priority algorithms intact.**

***Index Terms*** **- Scheduling, rate monotonic, earliest deadline first, real-time systems, context switch, preemptions.**

## I. INTRODUCTION

### *A. Background*

Real-time scheduling is one of the active research areas for a long time since the seminal work of Liu and Layland[1], due to its practical importance. The field is getting renewed interest in recent times due to pervasiveness of embedded systems and advancement of technological innovations.

Real-time scheduling algorithms are generally preemptive. In operating systems' context, preemption incurs a cost and also has an effect on the design of the kernel[13]. Preemption involves activities such as processing interrupts, manipulating task queues, and performing context switch. This cost is significantly high if the system uses caches in single or multi-levels and cache memory is used in almost all systems today[3], [9], [17]. However, in real-time systems' context, the cost of preemption has been considered negligible and therefore assumed to be zero for a long time. As the availability of advanced architectures with multi-level caches and multi-level context switch (MLC)[10] is becoming increasingly common, the continued use of the popular scheduling algorithms like rate monotonic (RM) will likely to experience cascading effect on preemptions. Such undesirable preemption related overhead may cause higher processor overhead in real-time systems, high energy consumption in embedded systems, and may make the task set infeasible[23]. This paper deals with the issue of preemptions in static priority scheduling for real-time systems.

### *B. Motivation*

In real-time systems' context, Rate Monotonic (RM) and Earliest Deadline First (EDF), introduced in [1], are widely studied and extensively analyzed[2], [26]. RM is a static priority based scheduling and EDF is dynamic priority based scheduling, and they are proved optimal in their respective classes[1].

Though EDF increases schedulability, RM is used for most practical applications. The reasons for favoring RM over EDF are based on the beliefs that RM is easier to implement, introduces less run-time overhead, easier to analyze, more predictable in overloaded conditions, and has less jitter in task execution. Recently, in [26], some of these claimed attractive properties of RM have been questioned for their validity. In addition, the author observes that most of these advantages of RM over EDF are either very slim or incorrect when the algorithms are compared with respect to their development from scratch rather than developing on the top of a generic priority based operating system kernels. Some recent operating systems provide such support for the developing user level schedulers[16].

One of the unattractive properties of RM observed in [26] is that, it experiences a large number of preemptions compared to EDF and therefore introduces high overhead. The preemption cost in a system is significant, as indicated earlier, if the system uses cache memories[3], [9], [12], [17]. As a matter of fact, most computer systems today use cache memory. This brought us to a basic question that, is it possible to reduce the preemptions in static priority scheduling algorithms in the real-time systems while retaining their simplicity intact. This paper attempts to answer this question.

### *C. Contribution*

In this paper, we present two frameworks, called *off-line activation-adjusted scheduling (OAA) and adaptive activation-adjusted scheduling (AAA)*, from which many static priority scheduling algorithms can be derived by appropriately



*T. Sobh and K. Elleithy (eds.), Advances in Systems, Computing Sciences and Software Engineering*, 425–432.

implementing the abstract components. Most of the algorithms derived from our frameworks reduce the number of unnecessary preemptions, and hence they:

- increase processor utilization in real-time systems;
- reduce energy consumption when used in embedded systems; and
- increase tasks schedulability.

We have listed possible implementations for the abstract components of the frameworks. There are two components, one is executed off-line and the other is to be invoked during run-time for the effective utilization of the CPU. These components are simple and add a little complexity to the traditional static priority scheduling, while reducing the preemptions significantly.

We conducted a simulation study for selected algorithms derived from the frameworks and the results indicate that some of the algorithms experience significantly less preemption compared to EDF and RM. The appeal of our algorithms is that they generally achieve significant reduction in preemptions, while retaining the simplicity of static priority algorithms intact.

### D. Related Works

The idea of delaying the activations of the tasks, from their default activation points - the beginning of the periods, have been explored in [7], [20] for specific objectives. In [7], it has been used to reduce the mean response time of soft tasks. The algorithm, referred as dual priority scheduling, uses three level priority queues - middle level priority queue for soft tasks and high and low priority queues for real-time tasks. In this algorithm, each real-time task is delayed in the low priority queue for a precomputed time called promotion delay. In [20], the delay time is used to reduce preemptions for restricted task set.

The approaches to reduce the number of preemptions in fixed priority scheduling have been presented in [14], [18], [19], [23]. In the approaches presented in [14], [18], [19], the tasks are assigned a threshold value in addition to their priorities such that they can be preempted only when other tasks have priorities higher than the threshold. This is similar in essence to dual priority system and requires simulating preemption threshold using mutexes - generally not desirable or not possible in all systems. In [23], an approach is presented based on an involved off-line analysis of preemption dependencies based on fixed execution times and can be effective if the actual execution times are same as the assumed execution times.

Static priority scheduling has been widely studied[1], [2], [6], [21], [22], [26], and preemptions related issues has been analyzed for real-time scheduling in [3], [10], [12], [14], [15], [18], [19], [24], [25], [26], and preemptions in static priority scheduling has been investigated in [8], [9], [14], [17], [23], [26].

### E. Organization

The rest of the paper is organized as follows: Section II presents the system model and problem statement. Section III gives an overview of static priority scheduling algorithms. A framework for Off-Line Activation-Adjusted Scheduling Algorithms (*OAA*) is presented first and then many *OAA* algorithms have been derived and discussed in Section IV. Section V introduces a framework for Adaptive Activation-Adjusted Scheduling Algorithms and then shows the derivation and analysis of *AAA* algorithms. A simulation study and the experimental results comparing RM and EDF with our algorithms is presented in Section VI. The paper is concluded in Section VII.

## II. System Model And Problem Statement

This section introduces the task model and the notations used throughout the paper. Consider a single processor system with a set of $n$ periodic tasks. We introduce the following terminology.

- A periodic task $\tau_i$ is a ***triple*** $< T_i, C_i, P_i >$, where
  - $T_i$ is the length of the period,
  - $C_i$ is the worst case execution time (WCET), and
  - $P_i$ is the priority.
- A set of $n$ periodic tasks is called a task set and is denoted by $\Gamma = (\tau_1, \tau_2, ..., \tau_n)$. Without loss of generality, we assume that $\tau_1, \tau_2, ..., \tau_n$ are ordered by decreasing priority, so that $\tau_1$ is the highest priority task.
- The absolute periods for $\tau_i$ are: $[0, T_i]$, $[T_i, 2T_i]$, $[2T_i, 3T_i]$, ....The end of the periods $T_i$, $2T_i$, ..., are defined as absolute deadline for $\tau_i$ in the respective periods.
- We denote the absolute activation time for $\tau_i$ in the $k^{th}$ interval as $a_{i,k}$.
- $1/T_i$ is defined as the *request rate* of the task $\tau_i$.
- The ratio $u_i = C_i / T_i$ is called the *utilization factor* of the task $\tau_i$ and represents the fraction of processor time to be used by that task.

We adopt the following assumptions from [1].

- All tasks are independent and preemptive.
- The priority of each task is fixed.

The problem is to design a scheduling algorithm that determines the task to be executed at a particular moment so that each task $\tau_i$ in the system complete its $k^{th}$ execution within its $k^{th}$ period $[(k-1)T_i, kT_i]$, $\forall\, k = 1, 2, 3, ....$

## III. Static Priority Scheduling Algorithms

The basic idea behind static priority scheduling algorithms is simple, that:

- the priority of the tasks are assumed to be fixed through-out the execution,
- at any time, the scheduler selects the highest priority runnable-task for execution, and
- the selected task runs until either it completes its execution in that period or another task with priority higher than it is ready for run.

An implementation scheme for fixed priority schedulers is described in [5] as follows. The scheduler maintains essentially two queues: *ready queue* and *wait queue*. The ready queue contains the tasks which are ready to run and the wait queue contains the tasks that have already run and are waiting for their next period to start again. The ready queue is ordered by priority and the wait queue is ordered by earliest start time.

When the scheduler is invoked, it examines the tasks in the wait queue to see if any task should be moved to the ready queue. Then it compares the head of the ready queue to current running task. If the priority of the task in the head of the ready queue is higher than the priority of current running task, then the scheduler invokes a context switch. The scheduler is invoked by an interrupt from either an external event or a timer.

## IV. OFF-LINE ACTIVATION-ADJUSTED SCHEDULING ALGORITHMS

### *A. Introduction*

The motivation for our algorithm results mainly from a recent observation that the representative static priority algorithm RM incurs high preemptions compared to the popular dynamic priority algorithm EDF[26]. The objective of our algorithms is to reduce the number of preemptions, while retaining the run-time overhead low - an attractive property of static priority algorithms.

Preemption occurs when a higher priority task is activated during the execution of a lower priority task. A lower priority task would experience more preemptions as it stays longer in the ready queue. Therefore, to reduce the chance of the system experiencing high preemptions, it is necessary to reduce the life time of lower priority tasks in the ready queue. One way to reduce the life time of lower priority tasks is to delay the activation of higher priority tasks if possible to increase the chance for the lower priority tasks to utilize the CPU as much as they can. This is the basic idea behind our first class of algorithms. Here, the delay is computed off-line and incorporated in the periods to get adjusted-activations. We illustrate the idea using the following simple example.

*Example 4.1*: Consider a task set consisting of three tasks $\tau_1$, $\tau_2$, $\tau_3$ with $C_1 = 1$, $T_1 = 3$, $C_2 = 3$, $T_2 = 9$, $C_3 = 2$, $T_3 = 12$. For this task set, the schedule generated by RM has been shown in Fig. 1.

From Fig. 1, we observe four preemptions for the task $\tau_2$, and two preemptions for the task $\tau_3$, as they are preempted by $\tau_1$. In Fig. 1, the preemption points are indicated by P. The task $\tau_1$ will never experience preemption, because it has the highest priority and therefore can get the CPU without any interruption from other tasks.

Fig. 2 illustrates how the number of preemptions for $\tau_3$, the lowest priority task, can be reduced by delaying the activation of the tasks $\tau_1$ and $\tau_2$.

The delay times for $\tau_1$ and $\tau_2$ are computed using the equation 3 and they are 2 and 4 respectively. The tasks $\tau_1$ and $\tau_2$ are being delayed by their delay times and $\tau_3$ is activated immediately. From Fig. 2, we can observe that preemptions for $\tau_3$ have been reduced by one.

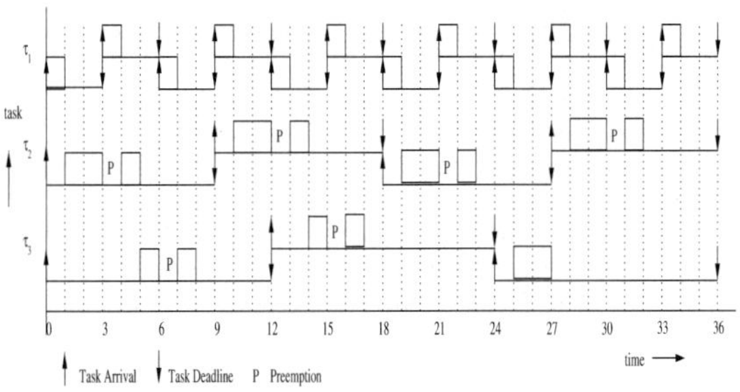


Fig. 1. Execution by RM

If the activation-adjustments are done only for a subset of tasks, then by varying the subset, many algorithms can be derived. Next we present the general framework for these algorithms.

### *B. Framework*

We consider the **Off-line Activation-Adjusted Scheduling** (*OAA*) as a *quadruple* $< \mathrm{r}, f_{AT}, AT, S' >$, where

r : a task set of size *n*.

$f_{AT}$: a function defined as follows. $f_{AT}(\mathrm{r}) = \Pi$, where $\Pi$ is a subset of r for which the activation times are to be adjusted.

*AT*: a set of pairs $(N_{a,i}, a_{i,1})$, where $N_{a,i}$ is the next activation time and $a_{i,1}$ is the offset of activation adjustment of $\tau_i$. For every task $\tau_i$ in $\Pi$, the absolute activation time $a_{i,k}$ is computed as follows.

$$\begin{cases} a_{i,1} = T_i - R_i \\ a_{i,k} = a_{i,k-1} + T_i, \quad \forall k > 1 \end{cases} \tag{1}$$

where $R_i$ is the worst case response time of $\tau_i$. $R_i$ is calculated iteratively using the equation 3.

For every task $\tau_j$ not in $\Pi$, the absolute activation time $a_{j,k}$ is

$$\begin{cases} a_{j,1} = 0 \\ a_{j,k} = a_{j,k-1} + T_j, \quad \forall j > 1 \end{cases} \tag{2}$$

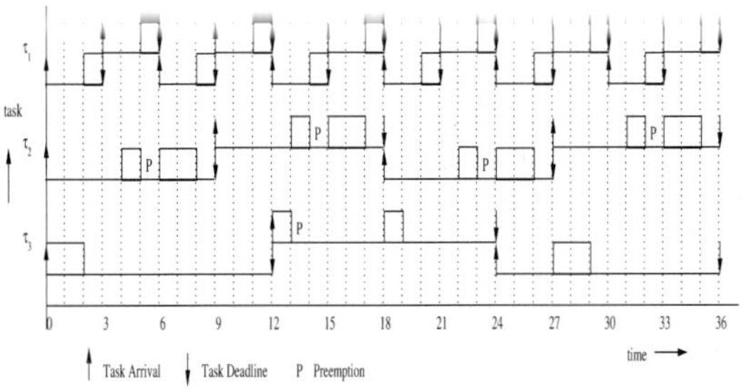


Fig. 2. Altered execution by delaying the activations of $\tau_1$ and $\tau_2$

$S'$: *the scheduler*. The scheduler component is a *triple* $< W_q, R_q, S_p' >$, where

$W_q$: a queue of tasks waiting to be activated, ordered by increasing absolute activation time.

$R_q$: a queue of ready tasks, ordered by decreasing priority.

$S_p'$: the scheduling policy. The scheduler $S'$ can be invoked either by the completion of a task or by a timer expiry. When the scheduler $S'$ is invoked, then

1) If the invocation of $S'$ was by the completion of a task, then
   * $S'$ places the completed task in $W_q$ with next activation time set.
2) Else, if the invocation of $S'$ was by timer interrupt, then
   * If a running task is interrupted by the timer then $S'$ places the interrupted task in $R_q$.
3) $S'$ checks $W_q$ to see if any tasks are to be transferred from $W_q$ to $R_q$ and then transfers such tasks to $R_q$.
4) If $R_q$ is not empty, then
   * Let $\tau_i$ be the task in the head of $R_q$, with priority $p$. $S'$ scans $W_q$ starting from the head and identifies the first task, say $\tau_k$, with priority greater than $p$.
   * $S'$ sets the timer to $\tau_k$'s next activation time.
   * $S'$ schedules $\tau_i$ for execution.
5) $S'$ waits for invocation.

**Note:** $W_q$ and $R_q$ may be implemented more efficiently, as mentioned in [26], by splitting them into several queues, one for each priority.

*C. Computing $R_i$*

The *worst-case response time* $R_i$ of the task $\tau_i$ can be computed iteratively using the following formula[26]:

$$\begin{cases} R_i(0) = C_i \\ R_i(k) = C_i + \sum_{j \in hp(i)} \left\lceil \dfrac{R_i(k-1)}{T_j} \right\rceil C_j \end{cases} \qquad (3)$$

where $hp(i)$ is the set of higher priority tasks than $\tau_i$ which causes interference for $\tau_i$ and hence preempting it [6]. The worst case response time of $\tau_i$ is given by the smallest value of $R_i(k) = R_i(k\text{-}1)$.

*D. OAA Scheduling Algorithms*

The crux of *OAA* algorithms is in the implementations of $f_{AT}$ in the framework. From simple set theory, $f_{AT}$ can have $2^n$ possible implementations. We list only a few meaningful implementations below:

For a given task set r,

1) $f_{AT}(\mathrm{r}) = \{\}$
2) $f_{AT}(\mathrm{r}) = \mathrm{r}$
3) $f_{AT}(\mathrm{r}) = \{ \tau_1, \tau_2, \ldots, \tau_m \}$, where $1 \leq m < n$

Next we present *OAA* scheduling algorithms for RM assigned priorities.

*E. OAA-RM Scheduling Algorithms*

RM is the most used scheduling algorithm for real-time applications because it is supported in most OS kernels[11]. The key component of RM scheduling is its priority assignment scheme. In RM, high frequency tasks are assumed to be of higher priority than low frequency tasks (that is, tasks with high activation rate get higher priorities and hence the name rate monotonic).

With RM assigned priority, many *OAA* algorithms can be obtained by suitably choosing $f_{AT}$. We refer these algorithms as *OAA-RM*. The representative *OAA-RM* algorithms are:

*OAA-RM1*: $f_{AT}(\mathrm{r}) = \{\}$. This is same as RM.

*OAA-RM2*: $f_{AT}(\mathrm{r}) = \{\tau_1\}$. Only the highest priority task is delayed activation.

*OAA-RM3*: $f_{AT}(\mathrm{r}) = \{\tau_1, \tau_2, \ldots, \tau_{n/2}\}$. The lower half of the task set is delayed activation.

*OAA-RM4*: $f_{AT}(\mathrm{r}) = \{\tau_1, \tau_2, \ldots, \tau_{n-1}\}$. Expect the lowest priority task, all other tasks are delayed activation.

*OAA-RM5*: $f_{AT}(\mathrm{r}) = \mathrm{r}$. All the tasks are delayed activation.

We simulate *OAA-RM3* and compare with RM and EDF in the simulation study section.

*F. Analysis*

Compared to traditional static priority algorithms, *OAA* algorithms have additional off-line computation costs: computing $f_{AT}$ of the task set and generating the values of *AT*. This one-time cost may be justified by the reduction of run-time costs. *OAA* algorithms generally performs better than RM (as you can see in the simulation study section later) in terms of reducing preemptions, when the CPU utilization is high. We observed that the delayed activation often creates CPU idle time clusters. Allowing the potential tasks to utilize these idle time clusters might reduce the chance of task preemptions. We illustrate this using the task set considered in the Example 4.1.

*Example 4.2*: The task set consisting of three tasks $\tau_1$, $\tau_2$, $\tau_3$ with $C_1 = 1$, $T_1 = 3$, $C_2 = 3$, $T_2 = 9$, $C_3 = 2$, $T_3 = 12$, as in Example 4.1. Delaying the tasks $\tau_1$ and $\tau_2$, as shown in Fig. 2, reduces just one preemption. The key observation that we can make from Fig. 2 is that there are free CPU times from time instance ($t$) 3 to 4, 9 to 11, etc., even though the tasks $\tau_1$ and $\tau_2$ are ready for the execution at time instance $t$ = 3 and 9. Allowing the tasks to utilize such free times by adaptively relaxing the delayed activation might reduce the contention for CPU and hence reduce preemptions. This is shown in Fig. 3.

By comparing Fig. 1 and Fig. 3, we can see that the number of preemptions has been reduced for the task $\tau_2$ from 4 to 2, the task $\tau_3$ from 2 to 0, and the overall preemptions from 6 to 2. This

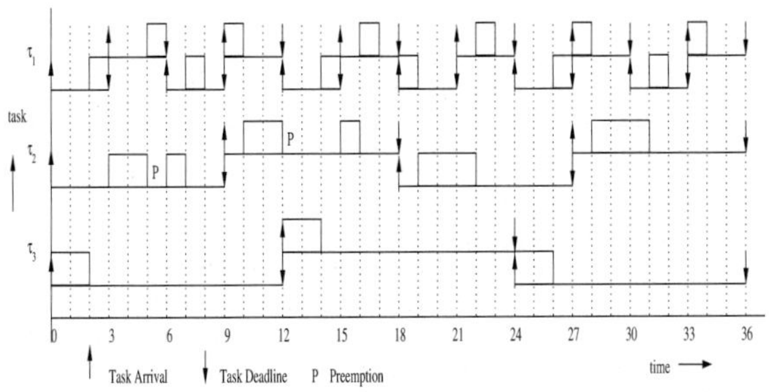


Fig. 3. Execution by Adaptive Delay

is the motivation for our second class of algorithms called adaptive activation-delayed scheduling algorithms.

## V. Adaptive Activation-Adjusted Scheduling Algorithms

### A. Introduction

The basic idea behind adaptive activation-adjusted scheduling algorithms is that the activation of the tasks are delayed only when needed. For the sake of simplicity in implementation, the algorithm delays the activations of all tasks to their adjusted-activation times and then wisely revokes the delays of some tasks to utilize the free CPU. The algorithm is same as *OAA* if the CPU is always busy. When the CPU becomes free, that is when $R_q$ is empty, then the scheduler looks at $W_q$ to look for an eligible task to schedule.

*Definition 5.1:* Assume that a task $\tau_i$ has completed its $k^{th}$ execution and it is waiting in $W_q$ for its next execution. The task $\tau_i$ is **eligible for its next execution** at time *t*, if $t \geq kT_i$.

Next we present the framework incorporating this idea.

### B. Framework

We consider the **Adaptive Activation-Adjusted Scheduling** (*AAA*) as a *quadruple* $< \Gamma, f_{AT}, AT, S'' >$, where

$\Gamma$ : a task set of size *n*.

$f_{AT}$: a function defined as follows. $f_{AT}(\Gamma) = \Pi$, where $\Pi$ is a subset of $\Gamma$ for which the activation times are to be adjusted.

$AT$: a set of pairs $(N_{i,1}, a_{i,1})$, where $N_{i,1}$ is the next activation time and $a_{i,1}$ is the offset of activation adjustment of $\tau_i$. For every task $\tau_i$ in $\Pi$, the absolute activation time $a_{i,k}$ is computed as stated in *OAA*.

$S''$: *the scheduler*. The scheduler component is a *quadruple* $< W_q, A_p, R_q, S_p'' >$, where

- $W_q$: a queue of tasks waiting to be activated, ordered by increasing absolute activation time.
- $R_q$: a queue of ready tasks, ordered by decreasing priority.
- $A_p$: a policy to select an eligible task from $W_q$ to transfer to $R_q$. It returns either the id of first eligible task or the id of a task which will become eligible in the nearest future.
- $S_p''$: the scheduling policy. The scheduler $S''$ can be invoked either by the completion of a task or by a timer expiry. When the scheduler $S''$ is invoked, then

1) If the invocation of $S''$ was by the completion of a task, then
   * $S''$ places the completed task in $W_q$ with next activation time set.
2) Else, If the invocation of $S''$ was by timer interrupt, then
   * If a running task is interrupted by the timer then $S''$ places the interrupted task in $R_q$.
3) $S''$ checks $W_q$ to see if any tasks are to be transferred from $W_q$ to $R_q$ and then transfers such tasks to $R_q$.
4) If $R_q$ is not empty, then
   * Let $\tau_i$ be the task in the head of $R_q$, with priority *p*. $S''$ scans $W_q$ starting from the head and identifies the first task, say $\tau_k$, with priority greater than *p*.
   * $S''$ sets the timer to $\tau_k$'s next activation time.
   * $S''$ schedules $\tau_i$ for execution.
5) Else[1],
   * **$S''$ calls $A_p$, and let $A_p$ returns $\tau_k$ and *t* be the current time.**
   * **If $N_{a,k}$ - $a_{k,1} \geq t$, then**
     - **$S''$ transfers $\tau_k$ from $W_q$ to $R_q$.**
     - **go to step 4.**
   * **Else,**
     - **$S''$ sets the timer to min(timer, $N_{a,k}$- $a_{k,1}$).**
6) $S''$ waits for invocation.

### C. AAA Scheduling Algorithms

We can derive many *AAA* algorithms by suitably implementing $A_p$ and $f_{AT}$ from the framework. We have listed the selection choices for $f_{AT}$ in section IV-D. Here we list some choices for $A_p$.

We assume that the task search for $A_p$ starts from the head of the $W_q$ and returns a task which will be eligible in the nearest future[2] satisfying the following criteria:

*AP1*: The first task in $W_q$.

*AP2*: The lowest priority task in $W_q$.

*AP3*: The highest priority task in $W_q$.

*AP4*: The first lowest priority task in $W_q$.

*AP5*: The first highest priority task in $W_q$.

*AP6*: The task with minimum $C_i$ in $W_q$.

*AP7*: The task with maximum $C_i$ in $W_q$.

---

[1] This is extra component over traditional static priority algorithms and therefore highlighted in bold face.

[2] A task which is eligible now is also eligible in the nearest future.

*AP8*: The task with best-fit[3] $C_i$ in $W_q$.

Next we present *AAA-RM* algorithms.

*D. AAA-RM Scheduling Algorithms*

With RM assigned priority, many *AAA* algorithms can be obtained by suitably choosing $f_{AT}$ and $A_p$. Some of the algorithms are as follows.

*AAA-RM1:* $f_{AT}(\Gamma) = \{\tau_1, \tau_2, \ldots, \tau_{n/2}\}$ and $A_p = AP1$.

*AAA-RM2:* $f_{AT}(\Gamma) = \{\tau_1, \tau_2, \ldots, \tau_{n-1}\}$ and $A_p = AP1$.

*AAA-RM3:* $f_{AT}(\Gamma) = \{\tau_1, \tau_2, \ldots, \tau_{n/2}\}$ and $A_p = AP2$.

*AAA-RM4:* $f_{AT}(\Gamma) = \Gamma$ and $A_p = AP1$.

*AAA-RM5:* $f_{AT}(\Gamma) = \Gamma$ and $A_p = AP2$.

*AAA-RM6:* $f_{AT}(\Gamma) = \{\}$ and $A_p = AP1$. This behaves the same way as RM.

We simulate *AAA-RM4* to compare with RM and EDF.

*E. Analysis*

When compared with static priority algorithms, our *AAA* algorithm has an extra run-time step (step 5 in the framework) in addition to the offline-computation of $f_{AT}$. Note that the step 5 in the framework (and hence in the algorithm) will be executed only when $R_q$ is empty. That is, step 5 consumes only the free CPU which otherwise would have been wasted. But the benefit gained in preemption reduction due to step 5 is significant, as witnessed in the simulation study.

## VI. Simulation Study

For our simulation, we built and used a Java based discrete event simulator to simulate the algorithms. We are interested in studying the cost involved in context switches, success ratio and average number of deadline misses.

Context Switch is an activity of switching the CPU from one task to another task. This activity generally involves a nonzero cost and varies from system to system, based on so many factors such as cache usage, scheduler complexity, context size, etc. However, for most analysis in the real-time systems, it is assumed as either zero or fixed. Also, the cost varies depending upon the reason for the occurrence of context switch: completion of the current task or request from a higher priority task.

*A.. Terminology*

*Definition 6.1*: If the context switch occurs due to task completion then the cost is loading/restoring the context of the new task. We call this cost as **task-switching cost**.

*Definition 6.2*: If the context switch occurs due to an interrupt from a higher priority task then the cost is saving the context of the current task and loading/restoring the context of the new task. We call this cost as **preemption cost**.

We assume that this cost is constant for a task set and varies from 0% to 25% of the mean worst case computation times of the task set.

*Definition 6.3*: The **average context switch cost** is the average of task-switching cost and preemption cost.

*Definition 6.4*: The ratio of the number of feasible task sets to the total number of task sets is called **success ratio**.

*Definition 6.5*: **Hyperperiod** of a task set is defined as the smallest interval of time after which the schedule repeats itself and is equal to the least common multiple of the task periods [26].

Since the schedule repeat itself after the hyperperiod, our simulations are run for the first hyperperiod. We denote the utilization of the task set by *U* and the number of task sets to be generated by *N*.

*B. Experimental Setup*

The period and the worst case computation time for each task are generated using uniform distribution. Task sets with larger LCM are rejected.

We conducted four experiments. The parameters for the experiments are as follows.

*Experiment 1* uses the following setup similar to the one used in [23].

- The LCM for each task set is randomized between 10 to 20 times of $n$ where $n$ is the number of tasks in the task set.
- Task periods $T_i$ for each $\tau_i$ are randomized in the interval $\left[\frac{LCM}{n}, LCM\right]$.
- Computation time $C_i$ for each task $\tau_i$ is generated for a given utilization $U$.
- The values are averaged for 50 task sets.

*Experiment 2, 3, and 4* uses the following setup similar to one used in [24].

- Task periods $T_i$ for each task $\tau_i$ are generated uniformly in the range [10ms, 100ms].
- An initial utilization $u_i$ for each task $\tau_i$ is uniformly assigned in the range [0.05, 0.5].
- Computation times $C_i$ for each task $\tau_i$ is set to $u_i.T_i$.
- The values are averages for 100 task sets.

*C. Experiments and Result Analysis*

In this section we present our simulation results and observations.

*Experiment 6.1:* In this experiment, we compare the behavior of *OAA-RM3* and *AAA-RM4* with RM, and EDF for the total number of preemptions as a function of utilization *U*.

*Observation 6.1:* We observe from Fig. 4 that RM, EDF and *OAA-RM3* almost have same number of preemptions at lower utilization. The number of preemptions starts to diverge as the utilization increases, because the lower priority tasks are frequently preempted by higher priority tasks. Preemptions in

[3] The maximum $C_i$ less than the remaining timer value.

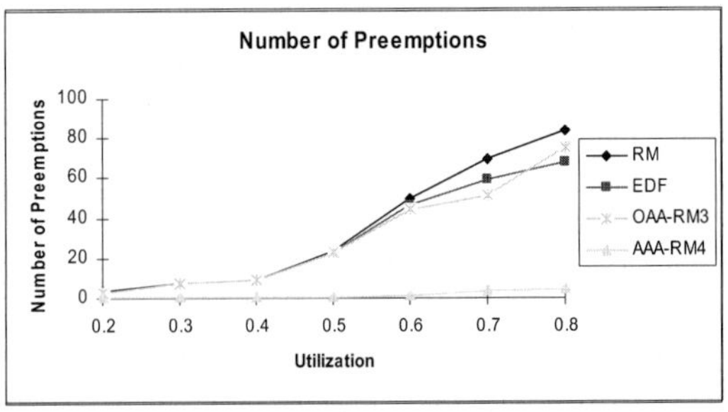


Fig. 4. Number of Preemptions vs. Utilization

RM is the highest and the preemptions in *AAA-RM4* is the lowest. In fact, *AAA-RM4* experiences almost no preemptions until 0.7 utilization and a very few preemptions after 0.7 utilization. EDF outperforms RM and *OAA-RM3* performs generally better than both RM and EDF.

In *OAA-RM3* the preemptions are reduced because the activation times for higher priority tasks are delayed and the lower priority tasks are activated immediately. This allows lower priority tasks to complete their executions with less interference. Further reduction in preemptions in *AAA-RM4* is due to the effective utilization of free CPU.

*Experiment 6.2*: In this experiment, we compare the behavior of *AAA-RM4* with RM and EDF for the number of preemptions as a function of number of tasks *n*, by fixing the utilization to *U* = 80%.

*Observation 6.2*: From Fig. 5 we see that for smaller number of tasks, the number of preemption increases. Then the preemption decreases for the larger number of tasks for RM, EDF, and *AAA-RM4*. This can be explained as follows. For smaller number of tasks, the chances for a task to be preempted increase with an increase in the number of tasks in the system. As the number of tasks gets higher, the task computation times get smaller on average, to keep the processor utilization constant. Hence chances for a lower priority task to be preempted have been reduced.

*Experiment 6.3*: In this experiment, we study the behavior of RM, EDF, and *AAA-RM4* for success ratio as function of total average context switch cost.

*Observation 6.3*: From Fig. 6 we observe that as the total average context switch cost increases from 5% to 25%, the success ratio drops for RM, EDF, and *AAA-RM4*. This is due to the fact that an increase in the total average context switch cost becomes significant and accounts for undesired higher processor utilization, making the task set unschedulable. For *AAA-RM4* success ratio drops gradually and is always higher than RM and EDF because of less preemptions. This reduction in preemptions allows more task sets to be schedulable.

*Experiment 6.4*: In this experiment, we compare the average number of deadline misses for RM, EDF, and *AAA-RM4* as a function of total average context switch cost.

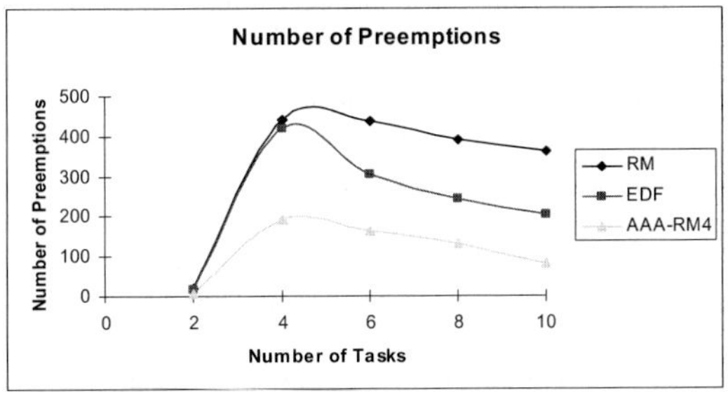


Fig. 5. Number of Preemptions vs. Number of Tasks

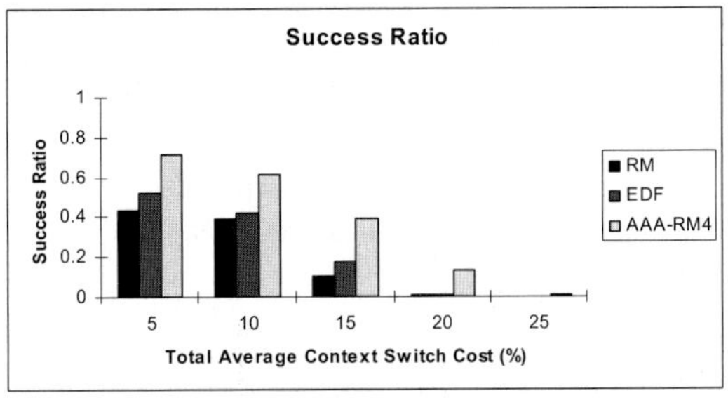


Fig. 6. Success Ratio vs. Total Average Context Switch Cost

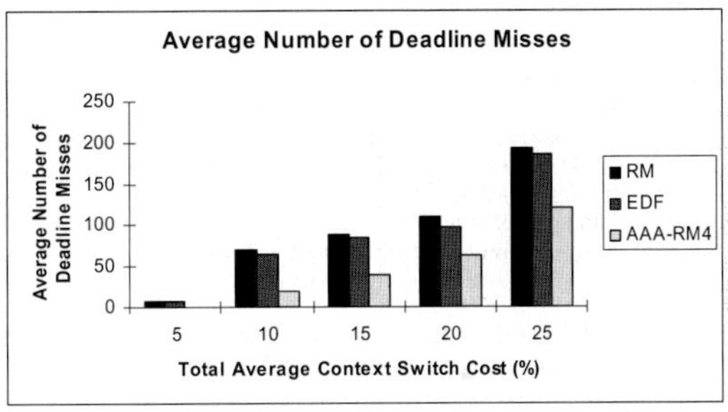


Fig. 7. Average Number of Deadline Misses vs. Total Average Context Switch Cost

*Observation 6.4*: We see that average number of deadline misses for RM, EDF, *AAA-RM4* from Fig. 7 increases with the increase in percentage of total average context switch cost. In case of *AAA-RM4*, the average number of deadline misses is less compared to RM and EDF. The achieved reduction in deadline misses is due to reduced preemptions.

## VII. CONCLUSION

In this paper we introduced two frameworks for static priority scheduling algorithms. Many algorithms, including variations to rate monotonic, can be derived from the frameworks with varying characteristics. Although the static priority scheduling algorithm rate monotonic has been widely used in practice due to its many attractive properties, its runtime overhead has been observed as a limitation[26]. Many algorithms derived from our frameworks alleviate this limitation while retaining the simplicity of the original algorithm intact. We conducted a simulation study on the variations of RM derived from our frameworks and the results indicate that our algorithms reduce preemptions significantly compared to RM and EDF.

There are many directions for further study. Some of them are:

- Schedulability analysis and other theoretical analysis for the algorithms derived in the frameworks can be explored.
- In our simulations, the algorithms assume that all tasks use their worst case computation time (WCET). The algorithms can be studied by relaxing this assumption. That is, study of the algorithms when actual computation time is less than WCET.
- Extending the algorithms for multiprocessor systems and embedded systems.

## ACKNOWLEDGMENT

The authors would like to thank Mahi, Srinivas, Bharath, and Pruthvi for their constructive comments on this paper.

## REFERENCES

[1] C. L. Liu and J. W. Layland, Scheduling Algorithms for Multiprogramming in a Hard Real-Time Environment, *Journal of the ACM, 20*(1):46-61, 1973.

[2] J. Lehoczky, L. Sha, and Y. Ding, The Rate Monotonic Scheduling Algorithm: Exact Characterization and Average Case Behavior, *Proc. Of the IEEE Real-Time Systems Symposium*, 166-171, 1989.

[3] J. C. Mogul and A. Borg, The effect of context switches on cache performance, *Proc. of the fourth international conference on Architectural support for programming languages and operating systems*, 75-84, 1991.

[4] K. Jeffay and D. L. Stone, Accounting for Interrupt Handling Costs in Dynamic Priority Task Systems, *Proc. of the 14th IEEE-Real Time Systems Symposium*, 212-221, 1993.

[5] D. I. Katcher, H. Arakawa, and J. K. Strosnider, Engineering and Analysis of Fixed Priority Schedulers, *IEEE Transactions on Software Engineering*, 19(9):920-934, 1993.

[6] N. Audsley, A. Burns, M. Richardson, K. Tindell, and A. J. Wellings, Applying New Scheduling Theory to Static Priority Pre-emptive scheduling, *Software Engineering Journal*, 284-292, 1993.

[7] R. Davis and A. Wellings, Dual Priority Scheduling, *Proc. of IEEE Real-Time Systems Symposium*, 100-109, 1995.

[8] A. Burns, K. Tindell, and A. Wellings, Effective Analysis for Engineering Real-Time Fixed Priority Schedulers, *IEEE Transactions on Software Engineering*, 21(5):475-480, 1995.

[9] C.-G. Lee, J. Hahn, Y.-M. Seo, S. L. Min, R. Ha, S. Hong, C. Y. Park, M. Lee, and C. S. Kim, Analysis of Cache-Related Preemption Delay in Fixed-Priority Preemptive Scheduling, *IEEE Transactions on Computers*, 47(6):700-713, 1998.

[10] J. Jonsson, H. Lonn, and K. G. Shin, Non-Preemptive Scheduling of Real-Time Threads on Multi-Level-Context Architectures, *Proc. of the IEEE Workshop on Parallel and Distributed Real-Time Systems*, LNCS, 1586:363-374, 1998.

[11] Y. Shin and K. Choi, Power Conscious Fixed Priority Scheduling for Hard Real-Time Systems, *Proc. of the Design Automation Conference*, 134-139, 1999.

[12] S. Lee, S.L. Min, C.S. Kim, C.G. Lee, and M. Lee, Cache-Conscious Limited Preemptive Scheduling, *Real-Time Systems*, 17(2/3):257-282, 1999.

[13] A. Silberschatz and P. B. Galvin, *Operating System Concepts*, John Wiley & Sons, Inc., 1999.

[14] M. Saksena and Y. Wang, Scheduling Fixed-Priority Tasks with Preemption Threshold, *Proc. of the IEEE Real-Time Computing Systems and Applications*, 328-335, 1999.

[15] M. Saksena and Y. Wang, Scalable Real-Time System Design Using Preemption Thresholds, *Proc. of the IEEE Real-Time Systems Symposium*, 25-26, 2000.

[16] M. A. Rivas and M. G. Harbour, POSIX-compatible application defined scheduling in MaRTE OS, *Proc. of the Euromicro Conference on Real-Time Systems*, 2001.

[17] C.-G. Lee, J. Hahn, Y.-M. Seo, S. L. Min, R. Ha, S. Hong, C. Y. Park, M. Lee, and C. S. Kim, Bounding Cache-Related Preemption Delay for Real-Time Systems, *IEEE Transactions on Computers*, 27(9):805-826, 2001.

[18] S. Kim, S. Hong, and T.-H. Kim, Integrating Real-Time Synchronization Schemes into Preemption Threshold Scheduling, *Proc. of the 5th IEEE Intl. Symp. on Object-Oriented Real-Time Distributed Computing*, USA, 2002.

[19] S. Kim, S. Hong, and T.-H. Kim, Perfecting Preemption Threshold Scheduling for Object-Oriented Real-Time System Design: From the Perspective of Real-Time Synchronization, *Proc. of the Languages, Compilers, and Tools for Embedded Systems*, Germany, 2002.

[20] M. Naghibzadeh, K. H. Kim, A Modified Version of Rate-Monotonic Scheduling Algorithm and its Efficiency Assessment, *Proc. of the Seventh IEEE International Workshop on Object-Oriented Real-Time Dependent Systems*, 289-294, 2002.

[21] E. Bini, G. C. Buttazo, G. M. Buttazzo, Rate Monotonic Analysis: The Hyperbolic Bound, *IEEE Transactions on Computers*, 52(7):933-942, 2003.

[22] E. Bini and G.C Buttazo, Schedulability Analysis of Periodic Fixed Priority Systems, *IEEE Transactions on Computers*, 53(11):1462-1473, 2004.

[23] R. Dobrin and G. Fohler, Reducing the Number of Preemptions in Fixed Priority Scheduling, *Proc. of the Euromicro Conference on Real-Time Systems*, 144-152, 2004.

[24] R. Jejurikar and R. Gupta, Integrating Preemption Threshold Scheduling and Dynamic Voltage Scaling for Energy Efficient Real-Time Systems, *Proc. of the Intl. Conference on Real-Time and Embedded Computing Systems and Applications*, August 2004.

[25] W. Kim, J. Kim, and S. L. Min, Preemption-Aware Dynamic Voltage Scaling in Hard Real-Time Systems, *Proc. of the ACM/IEEE Symposium on Low Power Electronics and* Design, 393-398, 2004.

[26] G. C. Buttazzo, Rate Monotonic vs. EDF: Judgment Day, *Real-Time Systems*, 29:5-26, 2005.

# Index